Lecture Notes in Computer Science 650

Edited by G. Goos and J. Hartmanis

Advisory Board: W. Brauer   D. Gries   J. Stoer

Series Editors

Gerhard Goos
Universität Karlsruhe
Postfach 69 80
Vincenz-Priessnitz-Straße 1
W-7500 Karlsruhe, FRG

Juris Hartmanis
Cornell University
Department of Computer Science
4130 Upson Hall
Ithaca, NY 14853, USA

Volume Editors

Toshihide Ibaraki
Kyoto University, Kyoto 606, Japan

Yasuyoshi Inagaki
Nagoya University, Nagoya 464, Japan

Kazuo Iwama
Kyushu University, Fukuoka 812, Japan

Takao Nishizeki
Tohoku University, Sendai 980, Japan

Masafumi Yamashita
Hiroshima University, Higashi-Hiroshima 724, Japan

CR Subject Classification (1991): F.1-2, G.2-3, H.3, I.3.5

ISBN 3-540-56279-6 Springer-Verlag Berlin Heidelberg New York
ISBN 0-387-56279-6 Springer-Verlag New York Berlin Heidelberg

© Springer-Verlag Berlin Heidelberg 1992
Printed in Germany

Typesetting: Camera ready by author/editor
Printing and binding: Druckhaus Beltz, Hemsbach/Bergstr.
45/3140-543210 - Printed on acid-free paper

T. Ibaraki   Y. Inagaki   K. Iwama
T. Nishizeki   M. Yamashita (Eds.)

Algorithms and Computation

Third International Symposium, ISAAC '92
Nagoya, Japan, December 16-18, 1992
Proceedings

Springer-Verlag

Berlin Heidelberg NewYork
London Paris Tokyo
Hong Kong Barcelona
Budapest

Preface

The papers in this volume were presented at the third International Symposium on Algorithms and Computation (ISAAC'92). The symposium took place December 16-18, 1992, at Nagoya Congress Center, Nagoya, Japan. It was organized by Nagoya University in cooperation with the Technical Group on Theoretical Foundation of Computing of the Institute of Electronics, Information and Communication Engineers (IEICE) and the Special Interest Group on Algorithms (SIGAL) of the Information Processing Society of Japan (IPSJ).

Its purpose is to provide a forum in the Asian and Pacific regions as well as other parts of the world to exchange ideas on algorithms and computation theory. The first such symposium was held 1990 in Tokyo, Japan as "SIGAL International Symposium on Algorithms," and the second one was held 1991 in Taipei, Republic of China as "Second Annual International Symposium on Algorithms."

In response to the program committee's call for papers, 98 papers were submitted; 62 from Asia, 16 from North America, 18 from Europe and 2 from Oceania. From these submissions, the committee selected 45 for presentation at the symposium. In addition to these contributed papers, the symposium included six invited presentations. The Best Paper Award was given to the paper "Parallel and On-Line Graph Coloring Algorithms" by M. Halldórsson.

Papers authored or coauthored by the committee members were allowed to be submitted to ISAAC'92, but were reviewed in a conventional manner after excluding the relevant members. We hope that contributions by the committee members will become unnecessary in the future.

The papers in these proceedings generally represent preliminary reports on continuing research, and it is anticipated that most of them will later appear in more complete and polished form in archival journals.

December 1992

T. Ibaraki
Y. Inagaki
K. Iwama
T. Nishizeki
M. Yamashita

Symposium Chairs
> Y. Inagaki (Nagoya U., Japan)
> T. Ibaraki (Kyoto U., Japan)

Symposium Co-Chairs
> Y. Igarashi (Gunma U., Japan)
> A. Maruoka (Tohoku U., Japan)

Program Committee
> D. Avis (McGill U., Canada)
> F. Chin (U. Hong Kong, Hong Kong)
> B. Courcelle (U. Bordeaux I, France)
> P. D. Eades (U. of Newcastle, Australia)
> O. Ibarra (U.C. Santa Barbara, U.S.A.)
> K. Iwama (Co-Chair; Kyushu U., Japan)
> R. C. T. Lee (Tsing Hua U., R.O.C.)
> T. Nishizeki (Chair; Tohoku U., Japan)
> M. Oyamaguchi (Mie U., Japan)
> K. Sugihara (U. Tokyo, Japan)
> S. Toda (U. Electro-Comm., Japan)
> T. Tokuyama (IBM TRL, Japan)
> M. Yamashita (Hiroshima U., Japan)

Finance
> T. Hirata (Nagoya U., Japan)

Local Arrangement
> T. Sakabe (Nagoya U., Japan)

Publicity
> O. Watanabe (TITech, Japan)

ISAAC Advisory Committee
> F. Chin (U. Hong Kong, Hong Kong)
> W. L. Hsu (Academia Sinica, R.O.C.)
> T. Ibaraki (Kyoto U., Japan)
> D. T. Lee (Northwestern U., U.S.A.)
> R. C. T. Lee (Tsing Hua U., R.O.C.)
> T. Nishizeki (Chair, Tohoku U., Japan)

Supported by
> Ministry of Education, Science and Culture
> International Communication Foundation
> International Information Science Foundation
> Shimazu Science Foundation

Contents

Session 3A: Graph Algorithms II

Session 3B: Parallel Algorithms

Session 4: Invited Papers

Session 6B: Computation Theory

Session 7: Invited Papers

Session 8A: Geometry and Layout

Session 8B: Complexity Theory II

Session 9A: Combinatorial Algorithms

Session 9B: Data Structures

Methods in Parallel Algorithmics and Who May Need to Know Them?

Uzi Vishkin*

University of Maryland &
Tel Aviv University

EXTENDED ABSTRACT

The first part of the talk is a general introduction which emphasizes the central role that the PRAM model of parallel computation plays in algorithmic studies for parallel computers. In particular, the PRAM model is useful as a virtual model of computation to be efficiently emulated on parallel machines. This is much in the same way that the RAM model is a virtual model of serial computation, which is efficiently emulated on almost every available computer. This part will begin by mentioning the following observation, which is due to the companion papers [MV-84] and [V-84]: *The PRAM model can be formally emulated, in an efficient manner, by a synchronous distributed computer, where each processor is connected to a small number of others in a fixed pattern.* This observation played a major role in convincing the academic computer science community (theoretical and other) that PRAM algorithmics is important. Elegant refinements of this observation were given in [KU-86], [R-87], [U-89] and [ALM-90]. For more on this topic see [L-92a] and [L-92b].

An alternative approach which is based on identifying the theory of parallel algorithms as an emerging main stream paradigm for computer science was suggested recently in [V-91b]. For background, we note that,

- The new generation of "serial machines" is far from being literally serial, as explained next. Each possible state of a machine can be described by listing the contents of all its *data cells*, where data cells include memory cells and registers. For instance, pipelining with, say s, single cycle stages, may be described by associating a data cell with each stage; all s cells may change in a single cycle of the machine. More generally, a transition function may describe all possible changes of data cells that can occur in a single cycles; the set of data cells that change in a cycle define the *machine parallelism* of the cycle; a machine is *literally serial* if the size of this set cannot be more than one. We claim that literally serial machines hardly exist and that considerable increase in machine parallelism is to be expected. Machine parallelism comes in such forms as pipelining, or in connection with the Very-Long Instruction Word (VLIW) technology, to mention just a few.

*Partially supported by NSF grants CCR-8906949 and 9111348.

- In contrast to this, textbooks on algorithms and data-structures still educate students, who are about to join the job market for at least forty years, to conceive computers as essentially literally serial; such textbooks are not to blame, of course, since software for currently widely available machines does not enable the typical user to express parallelism.

- Parallel algorithms can be exploited to make more effective uses of machine parallelism; the paper [V-91b] demonstrates how to use parallel algorithms for prefetching from slow memories (using pipelining); other means include using the parallelism offered by VLIW; there, a CPU can process simultaneously several operands, and essentially provide the same effect as having separate processors working in parallel on these operands.

The alternative approach suggests a multi-stage strategy.

- The agenda for the first stage (or stages) consists of the following.

  - *Agenda for the education system:* Teach students in undergraduate courses, as well as professional programmers, parallel algorithms and data-structures and educate them to do their programming in parallel. Explain to them that *parallelism is a resource that can be traded in several ways for taking advantage of machine parallelism.* In other words, the answer to the question in the title of this abstract is: everybody in computer science.

  - *Agenda for computer manufacturers:* Update the design of (serial) machines, including their software, so that parallel algorithms can be implemented on them; a particular emphasis should be given to allow parallel algorithms to take advantage of machine parallelism. The intention is that the first few generations of such computers will still provide competitive support for serial algorithms (including existing code). We note that Fortran 90, the new international standard described in [MR-90], requires that manufacturers will continue to provide support for serial algorithms, and at the same time will also do so for new array instructions that essentially enable representing PRAM algorithms. Needless to say that this support should be effective.

- Later stages will advance from aiming at moderate machine parallelism to massive parallelism coinciding with the previous approach.

The main advantages of the alternative approach are:

- A gradual, rather than abrupt, transition of general purpose computing from serial to parallel algorithms. Using the terminology of [HP-90], this gives the advantage of being more evolutionary in its demands from the user (and less revolutionary).

- The first generations of machines that will be built this way will cost much less than massively parallel machines. (The alternative approach can be nicknamed a "poor-person" parallel computer, as opposed to a massively parallel approach that can be called a "rich-person" one.)

- A much wider access to implementation of parallel algorithms. This will enable students and professional programmers to be trained, over a period of several years, towards the emerging age of parallel computing. As was explained above, the alternative approach does not mean to replace the effort for building massively parallel machines, but rather augment it, since both approaches share the same ultimate goal.

Some of the collective knowledge-base on non-numerical parallel algorithms can be characterized in a structural way. Each structure relates a few problems and technique to one another from the basic to the more involved. The second part of the talk will:

- overview several of these structures; and

- zoom in on some methods.

These structures are described in [V-91a]. An update for randomized fast algorithms can be found in [GMV-91].

We finish this abstract by mentioning two recent books on parallel algorithms: [J-92] and [R-92].

Acknowledgement. Helpful comments by Clyde Kruskal and Yossi Matias are gratefully acknowledged.

References

[ALM-90] S. Arora, T. Leighton and B. Maggs. On-line algorithms for path selection in a nonblocking network. In *Proc. of the 22nd Ann. ACM Symp. on Theory of Computing*, pages 149-158, 1990.

[GMV-91] Y. Gil, Y. Matias, and U. Vishkin. Towards a theory of nearly constant time parallel algorithms. In *Proc. of the 32nd IEEE Annual Symp. on Foundation of Computer Science*, pages 698–710, 1991.

[HP-90] J.L. Hennessy and D.A. Patterson. *Computer Architecture a Quantitative Approach*. Morgan Kaufmann, San Mateo, California, 1990.

[J-92] J. JáJá. *Introduction to Parallel Algorithms*. Addison-Wesley, Reading, MA, 1992.

[KU-86] A. Karlin and E. Upfal. Parallel hashing – an efficient implementation of shared memory. In *Proc. of the 18th Ann. ACM Symp. on Theory of Computing*, pages 160–168, 1986.

[L-92a] F.T. Leighton. *Introduction to Parallel Algorithms and Architectures: Arrays, Trees, Hypercubes*. Morgan Kaufmann, San Mateo, California, 1992.

[L-92b] F.T. Leighton. Methods for message routing in parallel machines. In *Proc. of the 24th Ann. ACM Symp. on Theory of Computing*, pages 77–96, 1992.

[MR-90]   M. Metcalf and J. Reid. *Fortran 90 Explained.* Oxford University Press, New York, 1990.

[MV-84]   K. Mehlhorn and U. Vishkin. Randomized and deterministic simulations of PRAMs by parallel machines with restricted granularity of parallel memories. *Acta Informatica,* 21:339–374, 1984.

[R-87]   A.G. Ranade. How to emulate shared memory. In *Proc. of the 28th IEEE Annual Symp. on Foundation of Computer Science,* pages 185–194, 1987.

[R-92]   J.H. Reif, editor. *Synthesis of Parallel Algorithms.* Morgan Kaufmann, San Mateo, California, 1992.

[U-89]   E. Upfal. An $O(\log N)$ deterministic packet routing scheme. In *Proc. of the 21st Ann. ACM Symp. on Theory of Computing,* pages 241–250, 1989.

[V-84]   U. Vishkin. A parallel-design distributed-implementation (PDDI) general purpose computer. *Theoretical Computer Science,* 32:157–172, 1984.

[V-91a]   U. Vishkin. Structural parallel algorithmics. In *Proc. of the 18th Int. Colloquium on Automata, Languages and Programming,* pages 363–380, 1991, Lecture Notes in Computer Science 510, Springer-Verlag.

[V-91b]   U. Vishkin. Can parallel algorithms enhance serial implementation? TR-91-145, University of Maryland Institute for Advanced Computer Studies (UMIACS), College Park, Maryland 20742-3251, 1991.

Rectilinear Paths among Rectilinear Obstacles*

D. T. Lee **

Department of Electrical Engineering and Computer Science
Northwestern University
Evanston, IL 60208
dtlee@eecs.nwu.edu

Abstract. Given a set of obstacles and two distinguished points in the plane the problem of finding a collision-free path subject to a certain optimization function is a fundamental problem that arises in many fields, such as motion planning in robotics, wire routing in VLSI and logistics in operations research. In this survey we emphasize its applications to VLSI design and limit ourselves to the rectilinear domain in which the goal path to be computed and the underlying obstacles are all rectilinearly oriented, i.e., the segments are either horizontal or vertical. We consider different routing environments, and various optimization criteria pertaining to VLSI design, and provide a survey of results that have been developed in the past, present current results and give open problems for future research.

1 Introduction

Given a set of obstacles and two distinguished points in the plane, the problem of finding a collision-free path subject to a certain optimization function is a fundamental problem that arises in many fields, such as motion planning in robotics, wire routing in VLSI and logistics in operations research. In this survey we emphasize its applications to VLSI design and limit ourselves to the rectilinear domain in which the goal path to be computed and the underlying obstacles are all rectilinearly oriented, i.e., the segments are either horizontal or vertical. We give results pertaining to routing in VLSI according to routing environment, and optimization criterion.

In VLSI layout, a basic function unit or circuit module, is represented by a rectilinear polygon, whose shape may or may not be (rectilinearly) convex. The *pins* or *terminals* of the same *net* that lie on different modules need to be interconnected by wires that mostly run either horizontal or vertical. In a single-layer model, the total *length* of wires in the interconnection is often used as an objective function, as it affects cost of wiring and total delay. On the other hand, the *number of bends* on a path affects the resistance and hence the accuracy of expected timing and voltage in chips. In a two-layer model in which horizontal

---

** Supported in part by the National Science Foundation under Grants CCR-8901815 and INT-9207212.

* This is an abridged version of [25] jointly with C. D. Yang and C. K. Wong.

wires are restricted on a *horizontal* layer, and vertical wires are restricted on a *vertical* layer, the number of *bends* used in the interconnection becomes an important measure as well, as each bend corresponds to a *via*. Having more *vias* not only introduces unexpected resistance and higher fabrication costs, but also decreases the reliability of the interconnection. It is desirable to have these routing measures, i.e., length and the number of bends, minimized. Unfortunately they cannot be both optimized *simultaneously* in general. Therefore a "best path" can be categorized by either minimizing each of these measures individually, giving them different optimizing priorities, or giving one a certain bound and minimizing the other. In addition to these optimization factors of rectilinear paths, the routing models and the *types* of obstacles also affect the complexity of the problems.

Most of the previous results [1, 7, 10, 14, 19, 21, 22, 28, 32, 33, 34, 45, 46, 47] deal with problems under the assumption that paths do not cross any obstacles, i.e., the path is *collision-free* (Figure 1a). Among these results, many aim at finding a *shortest* path among a set of *non-intersecting* obstacles in the plane[7, 10, 19, 32, 33, 34, 39, 46, 47]. In the hierarchical representation of VLSI layout, some pre-routed modules or lines may be re-designed or re-routed if it is subsequently found to be more beneficail to have some paths cross over the area previously occupied by them. Under these circumstances, such modules can be considered *penetrable* and assigned *weights* properly representing the cost to re-construct them (Figure 1b). In this setting, the total routing cost is measured by adding the extra cost (due to penetration) to the original measure.

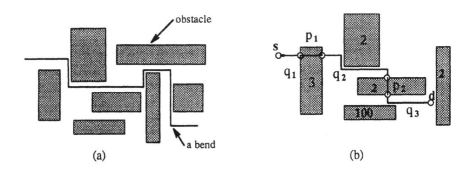

(a)   (b)

Fig. 1. Finding paths among obstacles

As to the models where routing is performed, researchers have considered two or more layers with restrictions reflecting the needs or limitations of VLSI design. In a *two-layer model*, obstacles are scattered on two layers separately and the routing orientation in each of the layers is fixed, e.g., allowing horizontal routing in one layer and vertical routing in the other. Paths can switch to the other layer when they need to change directions. Assorted paths with respect to the criteria mentioned above, i.e., *length* and *number of bends*, are also sought in this model. In addition, finding paths by minimizing the total length of their *subpaths* in one of the layers has also been studied. This is motivated from practical considerations that different layers are fabricated using different materials, e.g., silicon and metal, and that electric current runs at a different speed in each separate layer. Since electric current runs much faster in metal layer than in silicon layer, minimizing only the path length in silicon layer may be sufficient to give an approximation to the actual delay.

In the next section, we formally define the problems. In section 3, we briefly discuss previous and current results, and summarize them in two tables. In section 4, we give some interesting open problems.

2 Preliminaries

A *rectilinear path* is a path consisting of only vertical or horizontal segments. The *length* of a rectilinear path is the sum of the lengths of all its segments. An obstacle is said to be *rectilinear* if all its boundary edges are either horizontal or vertical. The point adjacent to a horizontal segment and a vertical segment is called a *bend*. Let $d(\pi)$ and $b(\pi)$ denote, respectively, the length and the number of bends of a path, π. ¿From here on, 'path' refers to rectilinear path and all the obstacles are rectilinear unless otherwise specified.

Definition 1. Given a set of disjoint (non-penetrable) obstacles, a source point s and a destination point d, we define the following problems:

Shortest Path (SP) Problem: Find a shortest path from s to d, denoted as sp.

Minimum-Bend Path (MBP) Problem: Find a path from s to d with a minimum number of bends, denoted as mbp.

Smallest Path (SMALLP) Problem: Find a path from s to d such that it is shortest and has a minimum number of bends *simultaneously*. Such a path is called a *smallest path* and denoted as *smallp*.

Minimum-Bend Shortest Path (MBSP) Problem: Find a path with the minimum number of bends among all the shortest paths from s to d. Such a path is called the minimum-bend shortest path and denoted as $mbsp$.

Shortest Minimum-Bend Path (SMBP) Problem: Find a shortest path among all the minimum-bend paths from s to d. Such a path is called the shortest minimum-bend path and denoted as $smbp$.

Minimum-Cost Path (MCP) Problem: Find a minimum-cost path from s to d, where the cost is a non-decreasing function of the length and the number of bends of a path, denoted as mcp.

Bounded-Bend Shortest Path (BBSP) Problem: Find a sp among all the paths from s to d with the number of bends no greater than a given bound. Such a path is called the bounded-bend shortest path and denoted as $bbsp$.

Bounded-Length Minimum-Bend Path (BLMBP) Problem: Find a mbp among all the paths from s to d with length no greater than a given bound. Such a path is called the bounded-length minimum-bend path and denoted as $blmbp$.

As mentioned before, obstacles may be penetrated at an $extra$ cost. We represent the extra cost of penetrating an obstacle as the $weight$ of the obstacle, and define the $weighted$ path length as follows.

Definition 2. A path Π_{st} from point s to t in the presence of $weighted$ obstacles can be decomposed into subpaths $q_1, p_1, q_2, p_2, \ldots, q_k, p_k$, where q_i is a subpath without intersecting the interior of any obstacle and p_i is a subpath completely within some obstacle R_{j_i} (Figure 1b). Subpaths q_1 and p_k may be empty. The $weighted\ length$ of Π_{st}, denoted $dw(\Pi_{st})$, is defined as $dw(\Pi_{st}) = \sum_{i=1}^{k}(d(q_i) + d(p_i)) + \sum_{i=1}^{k} d(p_i) * R_{j_i}.w$, where $R_{j_i}.w$ is the weight of obstacle R_{j_i}.

Note that the first term contributing to the weighted distance is the (unweighted) path length $d(\Pi_{st})$, and the second term reflects the extra cost of going through the interior of the obstacles. We have the following problems with respect to different types of obstacles. Let $X \in \{sp, mbp, smallp, mbsp, smbp, mcp, bbsp, blmbp\}$.

Definition 3. Given a problem instance as in Definition 1, we define the following problems.

Weighted X Problem: Find the X path when the obstacles are weighted. The distance measure is the weighted distance of a path.

(Weighted) $X_\square$ Problem: Solve the (weighted) X problems when the given obstacles are all rectangular.

The $two\text{-}layer\ model$ is defined as follows. We are given two layers as the routing environment. On the $horizontal\ routing\ layer$ (denoted as H_{layer}), only horizontal segments are allowed, and on the $vertical\ routing\ layer$ (denoted as V_{layer}) only vertical segments are allowed. Obstacles are given on each layer and they are disjoint on the same layer. A path in this model consists of horizontal segments on H_{layer} and vertical segments on V_{layer}. When a path switches layers, a bend or via is introduced (Figure 2a.) The total length of the segments on H_{layer} is called H-distance or x-distance of the path. V-distance or y-distance of a path is similarly defined.

Definition 4. Given a problem instance as in Definition 1, we define the following problems.

2-Layer$_{r|ur}$ X Problem: Find the X path in a two-layer model where each obstacle is specified to be in one of the two layers. There are two variants: the one with subscript r restricts the routings on the two layers to be horizontal and vertical respectively; the other with subscript ur allows both vertical and horizontal routing in each layer. Subscript r is assumed in the two-layer model we consider.

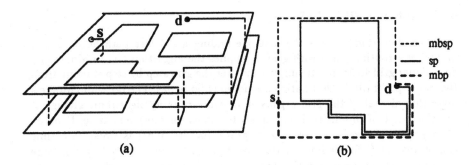

Fig. 2. Finding paths on multi-layer and single layer.

2-Layer$_{r|ur}$ 1D-X Problem: Find the X path in a two-layer model measuring only the distance of path in *one* specified layer. As before, routings in each layer may be restricted to be vertical and horizontal respectively (subscript r) or may be unrestricted (subscript ur).

1D-X Problem: Find the X path in the single-layer model measuring only the x- or y-distance of the path. It can be viewed as solving a 2-Layer 1D-X problem when the obstacles are considered present on both layers.

Multi-Layer (mD-)X Problem: Find the X path in a multi-layer model where routing in each layer may or may not be restricted to some directions. The distance measure may refer to the total length of subpaths on some specified layers.

Z^{pol} Problem: Solve Z problem within a given simple rectilinear polygon, where Z is any problem name defined above.

In the query versions, obstacles are fixed and given in advance. We have two cases depending on whether the source is given a priori or not.

Definition 5. Query$_{1|2}$ Z Problem: Find paths designated by Z from a fixed source to any query destination (subscript 1) or between two query points (subscript 2) among a fixed set of obstacles, with preprocessing allowed.

3   Previous and Current Results

We give a survey of previous and current results based on the problem classification defined above. Let n denote the number of obstacles, e the number of obstacle edges and V the set of obstacle vertices including the source and destination.

3.1 Problems SP and $SP_\square$

The Lee algorithm [20] was the first one finding a shortest path between two points on a routing plane with grid. It applies the algorithm due to Moore [37] on a planar grid structure. It simply runs over the grid step by step starting from the source point till the destination is reached (Figure 3a). The Lee's and Moore's algorithm are called the *grid expansion* or *maze-running* algorithm. Though it has the merit of simplicity, it consumes too much memory and time. It needs $O(m^2)$ space to represent an $m \times m$ grid structure and has worst case time complexity $O(l^2)$ for finding a path crossing l grid points. To reduce the large amount of memory required for maze-running algorithms, a *gridless model* was introduced, where obstacles are each represented by a set of line segments on the plane. Larson and Li [19] studied the problem of finding all shortest paths among

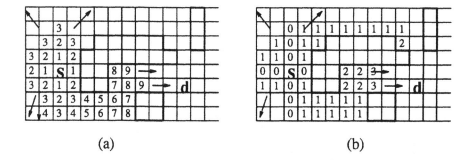

(a)   (b)

Fig. 3. Maze-Running approach

a set of m source-destination pairs with polygonal obstacles. By transforming the problem into a network problem, they solved it in $O((m + e)^2)$ time. deRezende *et al.* [10] proposed a $\theta(n \log n)$ time algorithm for $SP_\square$ along with a proof of its optimality. They utilized the monotonicity of the paths and used a so-called *plane-sweep* approach[41] to find the path. Mitchell [32] proposed a *45-degree wave-front* approach and presented an $O(e \log^2 e)$ algorithm for SP.[3] He used 45-degree line segments, called *wave-fronts*, to carry distance information as they travel away from the source. A wave-front can make turns around obstacles, can split to two wave-fronts if it hits an obstacle and can be deleted if it is subsumed by other wave-fronts. Wave-fronts are processed in ascending order of the distances they carry. The distance of a shortest path can be reported when a wave-front first hits the destination and the actural path can be constructed

---

[3] The time complexity is represented as $O(E \log e)$, where E is the number of events while dragging wave-fronts on the plane. E is proved to be $O(e \log e)$ whereas it is conjectured to be $O(e)$.

by tracing backwards. For SP, two of the best results were due to Clarkson, Kapoor and Vaidya [7] and Wu, Widmayer, Schlag and Wong [47]. One runs in $O(e \log^{3/2} e)$ time and space[7] or alternatively in $O(e \log^2 e)$ time using $O(e \log e)$ space[6], and the other in $O(e \log t + t^2 \log t)$ time[47], where t is the number of extreme edges among those given obstacles. An obstacle edge is *extreme* if its two adjacent edges lie on the same side of the line containing the edge. The algorithm by Wu, *et al.*[47] may have better performance than the one by Clarkson, *et al*[7], if t is much less than e. Both of their algorithms use *graph-theoretic* approach where a so-called *path-preserving* graph is constructed which provides sufficient distance information for one to apply graph-searching algorithms to find the goal path. Yang, *et al.* [51] recently proposed a method combining the features from these two path-preserving graphs and obtained a better algorithm which runs in $O(e + t \log^{3/2} t)$ time and space.

3.2   Problems MBP and SMALLP^{pol}

The problem of finding a minimum-bend path has been studied and gained more attention recently. In Euclidean case, several minimum-link path algorithms have been proposed (e.g., [18, 36, 43]). These algorithms may provide approximations for rectilinear case. Although the Lee algorithm [20] can be modified to find a minimum-bend path, (Figure 3b) it has the drawback that it is neither time nor space efficient. The Mikami-Tabuchi line-searching algorithm [31] finds a *mbp* in a plane or in the two-layer model. Hightower's algorithm [16] is similar to the Mikami-Tabuchi algorithm, but it does not always find a *mbp*. Combining both maze-running approach and line-searching approach, Heyns *et al.* [15] proposed a *line-expansion* algorithm. It continues searching along all possible escape lines leaving the zone reachable from the current line. It [15] has the advantages of assuring a connection and preventing duplicated searching, but has a drawback that it does not always find a *mbp* (cf. [40]). Ohtsuki [40] provided suitable data structures and analyzed their algorithms [15, 16, 31] that all have quadratic time complexity. Ohtsuki [39] proposed an algorithm for finding a *mbp*, which is modified from line-searching algorithms. The algorithm runs in $O(e \log^2 e)$ time using $O(e)$ space. Alternatively, one can use the data structure given by Imai and Asano [17] and apply Ohtsuki's algorithm to solve MBP in $\theta(e \log e)$ time and space. Sato *et al.* [42] independently presented a line-searching algorithm which finds a *mbp* in $O(e \log e)$ time. Their algorithm has an advantage that iterative path-searching can be done efficiently. A tile-expansion router obtained by Margarino, *et al.* [29] and derived from line-expansion approach, finds a *mbp* but no analysis of its time complexity is given.

Other variants of SP and MBP are also studied. de Berg [9] showed that there always exists a smallest path within a simple rectilinear polygon and presented an algorithm to solve SMALLP^{pol} (or alternatively SP^{pol} or MCP^{pol}) in $O(e \log e)$ time and space. His algorithm was basically designed for query version of the problem (see section 3.6). McDonald and Peters [30] later presented a linear time algorithm for SMALLP^{pol}. Lipski proposed a *Manhattan Path* problem asking for a rectilinear path (not necessarily a *sp*) connecting two specified segments such that the path is covered (constructed) by a given set of n horizontal or vertical line segments. He presented an $O(n \log n \log \log n)$ time algorithm [26] in which a *mbp* tree from the source segment to every other reachable segment is constructed, which is later improved to run in $\theta(n \log n)$ time and space[27]. For a generalized Manhattan Path problem where the segment intersection of the path is required to lie in a given rectangular area, Asano [2] presented an algorithm that runs in $O(n \log n)$ time and space or alternatively $O(n \log^2 n)$ time using $O(n)$ space.

3.3   Problems MBSP$_\square$, MBSP, SMBP, MCP, BBSP and BLMBP

Algorithms which find the shortest paths do not necessarily produce a path with a minimum number of bends, and vice versa. For example, the algorithm due to Wu, *et al.* [47] finds a *sp* but leaves the number of bends in the solution path uncertain. The *sp* obtained by Clarkson, *et al.* [7] tends to have more bends than necessary, as shown by the solid lines in Figure 2b. Similarly, the *mbp* obtained by Ohtsuki[39] is by no means close to being the shortest.

Using *graph-theoretic approach*, Yang, Lee and Wong [49] constructed a *path-preserving* graph which is assured to retain desired goal paths, *mbsp, smbp, mcp, bbsp* and *blmbp*. By applying the shortest path searching algorithm of Fredman and Tarjan [12] to such a graph, an $O(K + e \log e)$ algorithm is derived for MBSP, SMBP and MCP, where K, bounded by $O(et)$, is the number of intersections between *tracks* derived from extreme points and convex vertices. (Recall that t is the number of extreme edges of the obstacles.) The time complexity can be improved to $O(K + \frac{e \log e}{\log \log e})$, if one applies the *trans-dichotomous* algorithm of Fredman and Willard [13].[4] For BBSP an $O(B(K + e \log e))$ time algorithm was given in [49] based on the same graph, where $B = \min\{b^*, b\}$, and b^* and b are, respectively, the bend number of the *mbsp* and the given bound for the number of bends. As for BLMBP, a similar algorithm, which runs in $O(b^*(K + e \log e))$ time was given [49]. When obstacles are rectangular, the *mbsp* can be found in optimal $\theta(e \log e)$ time using plane-sweep approach (see Yang, *et al.* [52].) A new *dynamic-searching* approach is devised by Yang, Lee and Wong [51] by which MBSP and MCP can be solved in $O(e \log^{3/2} e)$ time and space. Instead of pursuing a smaller path-preserving graph, the authors [51] construct a graph just for guidance purpose. The so-called *guidance graph* is assured to contain homotopy information of the goal path and allows one to perform searching on it and compute the goal path on the fly. It is a combination of both graph-theoretic and continuous searching (e.g., wave-front approach) approaches. For SMBP, an optimal $\theta(e \log e)$ time algorithm was also obtained by Yang, *et al.* [50] using a *horizontal wave-front* approach.

---

[4] The trans-dichotomous algorithm utilizes addressing of random access machine and speeds up computation in a way falling outside the framework of comparison based algorithms.

3.4   Weighted X and Weighted X$_\square$ Problems

In [35], Mitchell and Papadimitriou first introduced the Euclidean *weighted region problem* with applications to motion planning in robotics. They proposed an algorithm to find the shortest *weighted distance* path among weighted regions that may be adjacent. Their algorithm runs in $O(n^8 L)$ time, where L is the precision of the problem instance (including the number of bits required to specify the largest integer among the weights and the coordinates of vertices) and n is the number of the weighted regions.

The version where paths are rectilinear and obstacles are non-adjacent was studied by Lee, Yang and Chen [23], who extended the $O(e \log^{3/2} e)$ result of Clarkson, Kapoor and Vaidya [7] for SP to Weighted SP. For Weighted SP$_\square$ problem Yang, Chen and Lee [48] proposed an optimal $\theta(n \log n)$ time algorithm using plane-sweep approach and a weighted segment tree structure.[5]

3.5   2-Layer$_{r|ur}$ X Problems and 1D-X Problems

Ohtsuki and Sato [38] studied the 2-Layer 1D-SP problem and presented an $O(e \log e)$ algorithm. They used graph-theoretic approach to construct a linear-sized graph providing suitable interconnections in the routable region and then applied the Dijkstra's algorithm [11] to solve the 1D-SP problem. They also studied 2-Layer$_{ur}$ MBP problem. For this problem, a bipartite graph is constructed to represent the interconnections on both layers, and then searching is applied to the graph. Their algorithm runs in $O(e \log^2 e)$ time using $O(e)$ space or alternatively, in $O(e \log e)$ time and space if one adopts the data structure of Imai and Asano [17] (see also [3]). A simpler version of the problem is to consider the x-distance or the y-distance of the paths in a plane. Yang, Lee and Wong [50] presented optimal $\theta(e \log e)$ time and $O(e \log e)$ space algorithms for 1D-MBSP, 1D-SMBP and 1D-MCP by using the horizontal wave-front approach. Their results can be modified to solve 2-Layer 1D-MBSP, 2-Layer SMBP and 2-Layer 1D-MCP by adopting a two-layer-to-one-layer problem transformation [24].

The problem transformation proposed by Lee, Yang and Wong [24] provides a general solution for most of the problems in the two-layer model. They defined a *skeleton* which is a set of segments that can be placed in a single goal-layer and which represents sufficiently the blocking behavior of all obstacles on two layers. After making the skeleton the new obstacle set, most of the previous path-finding algorithms in one-layer can be applied and the resultant path can be easily transformed to the goal path in the two-layer model. The transformation can be done in $O(e \log e)$ time which is dominated by the time complexities of most path-finding algorithms. The correctness of the path transformation depends on a property that is required of the applied path-finding algorithms. The property, which is called the *alignment property*, requires each segment of the goal path to be aligned with obstacle edges. As mentioned in [24], most algorithms have the property or can be modified to have it by post-processing.

---

[5] Weighted MBSP$_\square$ is also solved in $\theta(n \log n)$ by Yang, Lee and Wong [52].

3.6 Query$_{1|2}$ Z Problems

Along with solving SP$_\square$, deRezende, Lee and Wu [10] also solved Query$_1$ SP$_\square$ in $O(\log n)$ time with $O(n \log n)$ preprocessing. Recently, Atallah and Chen [4] presented an algorithm for solving Query$_2$ SP$_\square$ in $O(\log e)$ time after $O(e^2)$ preprocessing. Variations where query points are restricted on the obstacle boundary or on the boundary of a polygon which encloses all obstacles are also discussed in [4].

In [32], Mitchell constructed a *shortest path map* structure (originally defined in [22]) of size $O(e \log e)$, with which the shortest distance from the source to a query point can be found in $O(\log e)$ time.

de Berg, Kreveld, Nilsson and Overmars [8] considered the case where the obstacles are all horizontal or vertical segments and a *mcp* from a fixed source to a given destination is queried. The input segments are allowed to intersect each other. As preprocessing their algorithm takes $O(n^2)$ time and $O(n \log n)$ space to compute one-to-all *mcp*'s from the source to every endpoint, where n is the number of segments. They used a *slab tree* data structure and reported a *mcp* from the fixed source to the query destination in $O(\log n)$ time. The slab tree structure can also be used as a pre-constructed structure for query version of other path-finding problems. Once all the goal paths from the fixed source to all obstacle vertices are found, one can construct a slab tree that partitions the plane and supports iterative queries. For example, Query$_1$ SP can be solved in $O(\log e)$ time after $O((t^2 + e) \log t)$ preprocessing time if the SP algorithm by Wu, *et al.* [47] is used to find the one-to-all shortest paths, or $O(e \log^{3/2} e)$ preprocessing time if the SP algorithm by Clarkson, *et al.* [7] is used.

de Berg [9] also presented algorithms that solves Query$_2$ SP^{pol} and Query$_2$ MBP^{pol} in $O(\log e)$ time after $O(e \log e)$ time and space preprocessing.[6] He also considered the variant when the query points are vertices of the polygon.

3.7 Tables of Problems

In this section, we summarized in two tables all the *recent* results by the problem factors under consideration. For algorithms with alternative time complexities due to time-space trade-offs, only the best time complexity is listed in the table. Blank entries indicate that no result is known to the author.

Numbers correspond to entries in Tables I and II.

{1} SP$_\square$. $\theta(n \log n)$ time and $O(n)$ space algorithm by deRezende, Lee and Wu [10] using plane-sweep approach due to the monotonicity of the goal path. Query$_1$ SP$_\square$ is also solved in $O(\log n)$ time with $O(n \log n)$ preprocessing time.

---

[6] The query problem that finds the shortest geodesic (respectively minimum link distance) path in a simple polygon between two query points is solved in $O(\log e + k)$ time by Guibas and Hershberger [14] (respectively Suri [44]).

Table I (objectives × problem models)

$\{x\}_f^{(special\ case)}$: for special case, if any, with complexity f. N/A: meaningless.

	rectilinear obstacles	weighted obstacles	on two layers 2-lyr distance	on two layers 1-lyr distance
sp	$\{1\}_{\theta(n \log n)}$ $\{2\}_{O(e \log^{3/2} e)}$ $\{3\}_{O((e+t^2)\log t)}$ $\{4\}_{O(e \log^2 e)}$ $\{5\}_{O(e+t \log^{3/2} t)}$	$\{11\}_{\theta(n \log n)}$ $\{12\}_{O(e \log^{3/2} e)}$	$\{13\}_{O(e \log^{3/2} e)}$	$\{8\}_{\theta(e \log e)}$ $\{14\}_{\theta(e \log e)}$
mbp	$\{6\}_{\theta(e \log e)}$	N/A	$\{6,13\}_{\theta(e \log e)}$	$\{6,13\}_{\theta(e \log e)}$
$mbsp$	$\{7\}^{\square}_{\theta(n \log n)}$ $\{8\}_{O(et+e \log e)}$ $\{9\}_{O(e \log^{3/2} e)}$	$\{7\}^{\square}_{\theta(n \log n)}$	$\{9,13\}_{O(e \log^{3/2} e)}$	$\{10,13\}_{\theta(e \log e)}$
$smbp$	$\{8\}_{O(et+e \log e)}$ $\{10\}_{\theta(e \log e)}$	N/A	$\{10,13\}_{\theta(e \log e)}$	$\{10,13\}_{\theta(e \log e)}$
mcp	$\{8\}_{O(et+e \log e)}$		$\{8,13\}_{O(et+e \log e)}$	
$bbsp$	$\{8\}_{O(B(et+e \log e))}$		$\{8,13\}_{O(Be^2)}$	
$blmbp$	$\{8\}_{O(b^*(et+e \log e))}$		$\{8,13\}_{O(b^* e^2)}$	

{2} SP. $O(e \log^{3/2} e)$ time and space (alternatively $O(e \log^2 e)$ time and $O(e \log e)$ space) algorithm by Clarkson, Kapoor and Vaidya [7] using graph-theoretic approach. Query$_1$ SP can be solved in $O(\log e)$ time with $O(e \log^{3/2} e)$ preprocessing time.

{3} SP. $O((e+t^2)\log t)$ time and $O(e+t^2)$ space algorithm by Wu, Widmayer, Schlag and Wong [47] using graph-theoretic approach, where t is the number of extreme edges of obstacle.

{4} SP. $O(e \log^2 e)$ time and $O(e \log e)$ space algorithm by Mitchell [32] using 45-degree wave-front approach.

{5} SP. $O(e \log t + t \log^{3/2} t)$ time and $O(e + t \log^{3/2} t)$ space (alternatively $O(e \log t + t \log^2 t)$ time and $O(e + t \log t)$ space algorithm by Yang, Lee and Wong [51] using graph-theoretic approach, where t is the number of extreme edges.

{6} MBP. $O(e \log^2 e)$ time and $O(e)$ space algorithm by Ohtsuki [39] using line-searching approach. Adopting the data structure from Imai and Asano [17], the algorithm runs in $\theta(e \log e)$ time and space.

{7} MBSP$_\square$ and Weighted MBSP$_\square$. $\theta(n \log n)$ time and $O(n)$ space algorithm by Yang, Lee and Wong [52] using plane-sweep approach and a monotonicity property of the goal paths.

{8} MBSP, SMBP, MCP, BBSP and BLMBP. $O(et + e \log e)$ time and $O(et)$ space algorithm by Yang, Lee and Wong [49] for finding $mbsp$, $smbp$ and mcp using graph-theoretic approach, where t is the number of extreme edges on obstacles. $O(B(et+e \log e))$ time algorithm for BBSP, where $B = min\{b^*, b\}$, b is the given bound and b^* is the bend number of a $mbsp$. $O(b^*(et + e \log e))$ time algorithm for BLMBP. Adopting the slab tree structure from de Berg,

Table II (objectives × problem models)

Superscripts *pre*: preprocessing time, $\square$: rectangular obstacles, $Q_{1|2}$: query types.

	Query version given obstacles	within a simple rectilinear polygon	paths on given n segments (Manhattan Path)
sp	$\{1,15\}_{O(\log n)}^{Q_1,\square,pre:O(n\log n)}$ $\{2,15\}_{O(\log e)}^{Q_1,pre:O(e\log^{3/2} e)}$ $\{19\}_{O(\log n)}^{Q_2,\square,pre:O(n^2)}$	$\{16\}_{O(e)}$ $\{18\}_{O(\log e)}^{Q_2,pre:O(e\log e)}$	
mbp		$\{16\}_{O(e)}$ $\{18\}_{O(\log e)}^{Q_2,pre:O(e\log e)}$	$\{17\}_{O(n\log n)}$
smallest p		$\{16\}_{O(e)}$ $\{18\}_{O(\log e)}^{Q_2,pre:O(e\log e)}$	
mcp	$\{15\}_{O(\log e)}^{Q_1,pre:O(e^2)}$ and as below		
$mbsp$ or $smbp$	$\{8,15\}_{O(\log e)}^{Q_1,pre:O(e^2)}$ $\{9,15\}_{O(\log e)}^{Q_1,pre:O(e\log^{3/2} e)}$		

et al.[8], Query$_1$ (MBSP|SMBP|MCP) can also be solved in $O(\log e)$ query time with $O(e^2)$ preprocessing.

{9} MBSP, SMBP, MCP. $O(e\log^{3/2} e)$ time and space (alternatively $O(e\log^2 e)$ time and $O(e\log e)$ space) algorithm by Yang, Lee and Wong [51] using dynamic-searching approach. Adopting the slab tree structure from de Berg, *et al.*[8], Query$_1$ (MBSP|SMBP|MCP) can also be solved in $O(\log e)$ query time with $O(e\log^{3/2} e)$ preprocessing.

{10} MBP, SMBP, 1D-SMBP, 1D-SP, and 1D-MBSP. $\theta(e\log e)$ time and $O(e\log e)$ space algorithm by Yang, Lee and Wong [50] using horizontal wave-front approach.

{11} Weighted SP$_\square$. $\theta(n\log n)$ time and $O(n)$ space algorithm by Yang, Chen and Lee [48] using plane-sweep approach and the monotonicity of the path.

{12} Weighted SP. $O(e\log^{3/2} e)$ time and space (alternatively $O(e\log^2 e)$ time and $O(e\log e)$ space) algorithm by Lee, Yang and Chen [23] using graph-theoretic approach.

{13} 2-Layer$_r$ X Problems. $O(e\log e)$ time algorithm by Lee, *et al.* [24] transforming problem instances in the two-layer model into ones in one-layer model. $O(e\log^2 e)$ time and $O(e\log e)$ space algorithm[24] for solving 2-Layer SP using graph-theoretic approach without problem transformation.

{14} 2-Layer$_r$ 1D-SP. $\theta(n\log n)$ time and $O(e)$ space algorithm by Ohtsuki and Sato [38] using graph-theoretic approach.

{15} Query$_1$ MCP. $O(\log e)$ query time by de Berg, Kreveld, Nilsson and Overmars [8]. Obstacles are segments that may have intersection. It needs $O(e^2)$ preprocessing time and uses $O(e\log e)$ space for the slab tree data structure, which can be used to solve other query problems.

{16} SMALLP^{pol}. Linear time and space algorithm by McDonald and Peters [30]. It relies on the linear time polygon triangulation algorithm of Chazelle [5].

{17} Manhattan Path. $\theta(n \log e)$ time algorithm by Lipski [27].

{18} SMALLP^{pol}. $O(e \log e)$ time and space algorithm by de Berg [9], in which Query$_2$-SMALLP^{pol} is solved in $O(\log e)$ time.

{19} Query$_2$ SP$_\square$. $O(\log e)$ query time with $O(e^2)$ preprocessing time by Atallah and Chen [4].

4 Research Directions and Open Problems

The upper bound for solving SP has long been conjectured to be $O(e \log e)$. $\theta(e \log e)$ time algorithm for finding sp, $mbsp$ and mcp remain to be seen. Algorithms with less preprocessing time for Query$_1$ SP or Query$_1$ MCP are worth studying. Solving Query$_2$ SP in linear or sublinear query time is very useful and practical. Algorithms solving these problems dynamically where obstacles can be deleted or inserted are crucial in iterative routing and are of great importance. This type of problems, called the *dynamic version* of path-finding problem has not been addressed.

Weighted rectilinear obstacles in previous studies are non-adjacent. That is, two obstacles can be as close to each other as possible and yet they always admit a path going in between without paying any costs. It would be nice to eliminate this assumption to allow adjacent weighted obstacles.

Finding paths in multi-layers is of interest since multiple layers have been used also as a routing model. The paths on each layer have different resistance and can be modeled by assigning different weights to the layers. (Recall that we have studied the case where paths on one of the two layers in the two-layer model have *zero* weight.) In this model, some of the layers may have fixed-orientation routing and others may allow free routing (e.g., for ground net). A general scheme for solving problems on multi-layers will be very useful.

We list some of these interesting open problems as follows.

Optimal Algorithms for SP, MBSP and MCP Find $\theta(e \log e)$ algorithms for SP, MBSP and MCP.

Adjacent Obstacles in Weighted Case Find assorted paths among weighted obstacles where obstacles can be adjacent such that a path going in between two adjacent obstacles pays a certain amount of cost.

Linear or Sublinear Algorithms for Query$_2$ Versions Preprocess given obstacles and report the assorted paths between two query points in linear or sublinear time.

Improved Algorithms for BBSP and BLMBP Find more efficient algorithms for BBSP or BLMBP.

Multi-layer nD-X Problem Formulate and solve path-finding problems in multi-layer model where the routing orientation may or may not be fixed on different layers and the paths on different layers may have different weights.

References

1. T. Asano, T. Asano, L. Guibas, J. Hershberger and H. Imai, "Visibility of disjoint polygons," *Algorithmica*, 1, 1986, 49-63.
2. T. Asano, "Generalized Manhattan path algorithm with applications," *IEEE Trans. CAD*, 7, 1988, 797-804.
3. T. Asano, M. Sato and T. Ohtsuki, "Computational geometry algorithms," in *Layout Design and Verification*, (T. Ohtsuki Ed.), North-Holland, 1986, 295-347.
4. M. J. Atallah and D. Z. Chen, "Parallel rectilinear shortest paths with rectangular obstacles," *Computational Geometry: Theory and Applications*, 1,2, 1991, 79-113.
5. B. Chazelle, "Triangulating a simple polygon in linear time," *Proc. 31st FOCS*, Oct. 1990, 220-230.
6. K. L. Clarkson, S. Kapoor and P. M. Vaidya, "Rectilinear shortest paths through polygonal obstacle in $O(n \log^2 n)$ time" *Proc. 3rd ACM Symp. on Computational Geometry*, 1987, 251-257.
7. K. L. Clarkson, S. Kapoor and P. M. Vaidya, "Rectilinear shortest paths through polygonal obstacles in $O(n \log^{3/2} n)$ time," submitted for publication.
8. M. de Berg, M. van Kreveld, B. J. Nilsson, M. H. Overmars, "Finding shortest paths in the presence of orthogonal obstacles using a combined L_1 and link metric," *Proc. SWAT '90, Lect. Notes in Computer Science*, 447, Springer-Verlag, 1990, 213-224.
9. M. de Berg, "On rectilinear link distance," *Computational Geometry: Theory and Applications*," 1,1, 1991, 13-34.
10. P. J. deRezende, D. T. Lee and Y. F. Wu, "Rectilinear shortest paths with rectangular barriers," *Discrete and Computational Geometry*, 4, 1989, 41-53.
11. E. W. Dijkstra, "A note on two problems in connection with graphs," *Numerische Mathematik*, 1, p.269.
12. M. L. Fredman and R. E. Tarjan, "Fibonacci heaps and their uses in improved network optimization algorithms," *J. ACM*, 34,3, July 1987, 596-615.
13. M. L. Fredman and D. E. Willard, "Trans-dichotomous Algorithms for Minimum Spanning Trees and Shortest Paths," *Proc. 31st IEEE FOCS*, 1990, 719-725.
14. L. J. Guibas and J. Hershberger, "Optimal shortest path queries in a simple polygon," *Proc. 3rd ACM Symp. on Computational Geometry*, 1987, 50-63.
15. W. Heyns, W. Sansen, and H. Beke, "A line-expansion algorithm for the general routing problem with a guaranteed solution," *Proc. 17th Design Automation Conf.*, 1980, 243-249.
16. D. W. Hightower, "A solution to line-routing problem on the continuous plane," *Proc. 6th Design Automation Workshop,*, 1969, 1-24.
17. H. Imai and T. Asano, "Dynamic segment intersection search with applications," *Proc. 25th IEEE FOCS*, 1984, 393-402.
18. Y. Ke, "An efficient algorithm for link-distance problems," *Proc. 5th ACM Sympo. on Computational Geometry*, 1989, 69-78.
19. R. C. Larson and V. O. Li, "Finding minimum rectilinear distance paths in the presence of barriers," *Networks*, 11, 1981, 285-304.
20. C. Y. Lee, "An algorithm for path connections and its application," *IRE Trans. on Electronic Computers*, V.EC-10, 1961, 346-365.
21. D. T. Lee, "Proximity and reachability in the plane," PhD. thesis, Univ. of Illinois, 1978.
22. D. T. Lee and F. P. Preparata, "Euclidean shortest paths in the presence of rectilinear barriers," *Networks*, 14, 1984, 393-410.

23. D. T. Lee, C. D. Yang and T. H. Chen, "Shortest rectilinear paths among weighted obstacles," *Int'l J. Computational Geometry & Applications*, 1,2, 1991, 109-124.
24. D. T. Lee, C. D. Yang and C. K. Wong, "Problem transformation for finding rectilinear paths among obstacles in two-layer interconnection model", submitted for publication.
25. D. T. Lee, C. D. Yang and C. K. Wong, "Rectilinear paths among rectilinear obstacles," Tech. Rep., Dept. EE/CS, Northwestern University, Aug. 1992.
26. W. Lipski, "Finding a Manhattan path and related problems," *Networks*, 13, 1983, 399-409.
27. W. Lipski, "An $O(n \log n)$ Manhattan path algorithm," *Inform. Process. Lett.*, 19, 1984, 99-102.
28. T. Lozano-Perez and M. A. Wesley, "An algorithm for planning collision-free paths among polyhedral obstacles," *Comm. ACM*, 22, 1979, 560-570.
29. A. Margarino, A. Romano, A. De Gloria, F. Curatelli and P. Antognetti, "A tile-expansion router," *IEEE Trans. CAD*, 6,4, 1987, 507-517.
30. K. M. McDonald and J. G. Peters, "Smallest paths in simple rectilinear polygons," *IEEE Trans. CAD*, 11,7 July 1992, 864-875.
31. k. Mikami and K. Tabuchi, "A computer program for optimal routing of printed circuit connectors," *IFIPS Proc.*, Vol.H47, 1968, 1475-1478.
32. J. S. B. Mitchell, "Shortest rectilinear paths among obstacles," Technical report NO. 739, School of OR/IE, Cornell University, April 1987. A revised version titled "L_1 Shortest paths Among Polygon Obstacles in the Plane," *Algorithmica*, to appear.
33. J. S. B. Mitchell, "A new algorithm for shortest paths among obstacles in the plane," Technical report NO. 832, School of OR/IE, Cornell University, Oct. 1988.
34. J. S. B. Mitchell, "An optimal algorithm for shortest rectilinear paths among obstacles," *Abstracts of 1st Canadian Conference on Computational Geometry*, 1989, p.22.
35. J. S. B. Mitchell and C. H. Papadimitriou, "The weighted region problem: finding shortest paths through a weighted planar subdivision", *Journal of the ACM*, 38,1, January 1991, 18-73.
36. J. S. B. Mitchell, G. Rote and G. Wöginger, "Minimum link path among a set of obstacles in the planes," *Proc. 6th ACM Symp. on Computational Geometry*, 1990, 63-72.
37. E. F. Moore, "The shortest path through a maze," *Annals of the Harvard Computation Laboratory*, Vol.30, Pt.II, 1959, 185-292.
38. T. Ohtsuki and M. Sato, "Gridless routers for two layer interconnection," *IEEE Int'l Conference on CAD*, 1984, 76-78.
39. T. Ohtsuki, "Gridless routers — new wire routing algorithm based on computational geometry," *Int'l Conference on Circuits and Systems*, China, 1985.
40. T. Ohtsuki, "Maze running and line-search algorithms," in *Layout Design and Verification*, (T. Ohtsuki Ed.), North-Holland, 1986, pp.99-131.
41. F. P. Preparata and M. I. Shamos, Computational Geometry, *Springer-Verlag*, NY, 1985.
42. M. Sato, J. Sakanaka and T. Ohtsuki, "A fast line-search method based on a tile plane," in *Proc. IEEE ISCAS*, 1987, 588-591.
43. S. Suri, "A Linear Time algorithm for minimum link paths inside a simple polygon," *Computer Vision, Graphics and Image Processing*, 35, 1986, 99-110.
44. S. Suri, "On some link distance problems in a simple polygon," *IEEE Trans. on Robotics and Automation*, 6, 1990, 108-113.

45. E. Welzl, "Constructing the visibility graph for n line segments in $O(n^2)$ time," *Info. Proc. Lett.*, 1985, 167-171.

46. P. Widmayer, Y. F. Wu, C. K. Wong, "On some distance problems in fixed orientations," *SIAM J. Computing*, 16, 1987, 728-746.

47. Y. F. Wu, P. Widmayer, M. D. F. Schlag, and C. K. Wong, "Rectilinear shortest paths and minimum spanning trees in the presence of rectilinear obstacles," *IEEE Trans. Comput.*, 1987, 321-331.

48. C. D. Yang, T. H. Chen and D. T. Lee, "Shortest rectilinear paths among weighted rectangles," *J. of Information Processing*, 13,4, 1990, 456-462.

49. C. D. Yang, D. T. Lee and C. K. Wong, "On Bends and Lengths of Rectilinear Paths: A Graph-Theoretic Approach," *Int'l Journal of Computational Geometry & Application*, 2,1, March 1992, 61-74.

50. C. D. Yang, D. T. Lee and C. K. Wong, "On Bends and Distances of Paths Among Obstacles in Two-Layer Interconnection Model," manuscript, 1991.

51. C. D. Yang, D. T. Lee and C. K. Wong, "Rectilinear path problems among rectilinear obstacles revisited," manuscript, 1992.

52. C. D. Yang, D. T. Lee and C. K. Wong, "On minimum-bend shortest path among weighted rectangles," manuscript, 1992.

Linear Time Algorithms for k-cutwidth Problem *

Maw-Hwa Chen and Sing-Ling Lee

Institute of Computer Science and Information Engineering
National Chung Cheng University
Chaiyi, Taiwan 62107, ROC

Abstract

The Min Cut Linear Arrangement problem is to find a linear arrangement for a given graph such that the cutwidth is minimized. This problem has important applications in VLSI layout systems. It is known that this problem is NP-complete when the input is a general graph with maximum vertex degree at most 3. In this paper, we will first present a linear time algorithm to recognize the small cutwidth trees. The approach we used in this algorithm can then be easily extended to recognize the general graphs with cutwidth 3 in $O(n)$ time.

1. Introduction

A linear arrangement (also called numbering, labeling, or layout) of an undirected graph $G = (V, E)$, $|V| = n$, is a 1-1 function $f : V \longrightarrow \{1, 2, \ldots, n\}$. The cutwidth of G under a linear arrangement f, denoted by $cw(G, f)$, is

$$\max_{1 \le i < n} |\{(u, v) \in E | f(u) \le i < f(v)\}|.$$

The cutwidth of a graph G, denoted by $cw(G)$, is the minimum of $cw(G, f)$ taken over all possible linear arrangement f. The k-cutwidth problem is to determine whether or not there exists a linear arrangement of the vertices such that the graph has cutwidth at most k. Clearly, the k-cutwidth problem is the decision version of the Min Cut Linear Arrangement Problem. Example of the Min Cut Linear Arrangement problem is shown in Fig. 1.1.

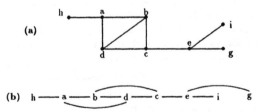

Fig. 1.1. (a) Source graph. (b) $f(h) = 1, f(a) = 2, f(b) = 3, f(d) = 4, f(c) = 5, f(e) = 6,$ $f(i) = 7, f(g) = 8,$ and $cw(G, f) = 3.$

* This work was supported by the National Science Council, Taiwan, R.O.C. Grant NSC 80-0404-E194-03

The Min Cut Linear Arrangement problem has many applications in VLSI CAD design. We can use the graph to abstract the representation of the circuit. The vertices of the graph are the circuit elements and the edges form the needed interconnection between them. In some approaches to VLSI design, the circuit elements must be first placed in rows or in a single line such that the cutwidth is minimized [6,12,14,15]. It is known that the MIN CUT problem is NP-complete even when restricted to graphs of degree at most 3 [12]. However, deciding whether a given graph with n vertices has cutwidth at most k or not can be done in time $O(n^k)$ [8]. If the graph is a tree T, then Yannakakis has found an $O(n \log n)$ to construct $cw(T)$ [17]. In this paper, we will first present a linear time algorithm to recognize the small cutwidth trees. The approach we used in this algorithm can then be easily extended to recognize the general graphs with cutwidth 3 in $O(n)$ time.

In the next section, we introduce some properties of the Min Cut Linear Arrangement problem on trees. Section 3 outlines the main ideas of our approach and the description of the algorithm to solve the 3-cutwidth problem on trees. The computational complexity of the algorithm is also analyzed in this section. In section 4, the extension of our method given in section 3 will be discussed. In section 5, we use similar approach to solve the 3-cutwidth problem on general graphs. Finally, we give the conclusion in section 6.

2. Preliminaries

In this section we will introduce several properties of the Min Cut Linear Arrangement problem on trees given in [4]. The approaches of our algorithm are based on these properties.

Let $T = (V, E)$ be a tree, $|V| = n$. Define P_f as the path connecting the two vertices u,v with $f(u) = 1$ and $f(v) = n$ under the Min Cut Linear Arrangement f in T and $T - P_f$ as the forest formed by removing the edges of P_f from T (but let the vertices of P_f stay). Then there always exists an optimal linear arrangement f^* with $cw(T, f^*) = cw(T)$ satisfying the following properties :

(1) The leaf property : The two vertices u,v with $f^*(u) = 1$ and $f^*(v) = n$ are leaves. (See Fig. 2.1(b).)

(2) The monotone property : Suppose P_{f^*} has vertices $v_0, v_1, \ldots, v_t$ with v_i adjacent to v_{i+1}. Then
$f^*(v_i) < f^*(v_{i+1})$ for $i = 0, 1, 2, \ldots, t - 1$ or
$f^*(v_i) > f^*(v_{i+1})$ for $i = 0, 1, 2, \ldots, t - 1$. (See Fig. 2.1(b).)

(3) The block property : Suppose $T - P_{f^*}$ has trees $T_1, T_2, \cdots, T_m$ ($m \geq 1$). Then for each vertex $u \in T_i$ ($1 \leq i \leq m$), there exists no two vertices $v, w \in T_j$ ($j \neq i$) such that $f^*(v) < f^*(u) < f^*(w)$, i.e. all vertices in T_i ($1 \leq i \leq m$) are labeled by a set of consecutive integers. (See Fig. 2.1(b) and (d).)

(4) The hereditary property : The induced labeling for each subtree $T_i = (V_i, E_i)$ of $T - P_{f^*}$ is optimal with respect to the Min Cut Linear Arrangement. (The induced labeling f_i of f^* on T_i is the one-to-one mapping from vertex set V_i to the set $\{1, 2, \cdots, |V_i|\}$ such that for any edge $\{u, v\}$ in E_i, $f_i(u) < f_i(v)$ if $f^*(u) < f^*(v)$.) (See Fig. 2.1 (b) and (d).)

Directly from the above properties, we now have the following lemma.

Lemma 2.1. Suppose f^* is optimal with respect to the Min Cut Linear Arrangement for T. Then $cw(T) = 1 + \max_{1 \leq i \leq m} cw(T_i)$ where T_i is a subtree in $T - P_{f^*}$.

At the end of this section, we will give an example that will show these properties and lemmas in this section. The detail is shown in Fig. 2.1.

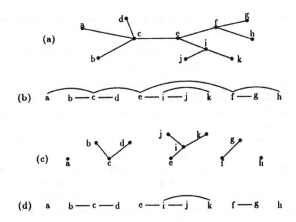

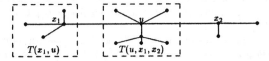

Fig. 2.1. (a) The tree T. (b) The optimal layout of this tree T. ($f^*(a) = 1, f^*(b) = 2, f^*(c) = 3, f^*(d) = 4$, $f^*(e) = 5, f^*(i) = 6, f^*(j) = 7, f^*(k) = 8, f^*(f) = 9$, $f^*(g) = 10, f^*(h) = 11$, $P_{f^*} = \{a, c, e, f, h\}$, and $cw(T) = 3$.) (c) The forest $T - P_{f^*}$. (d) The induced layout of the forest $T - P_{f^*}$.

3. An Algorithm for 3-cutwidth problem on trees.

In this section, we will give a linear-time algorithm for the 3-cutwidth problem on trees. At first, we will show the general characterization of trees with cutwidth 3. Let T be a tree and let $u, x_1, \ldots, x_r$ be arbitrary vertices in T. Define $T(u, x_1, \ldots, x_r)$ as the largest subtree of T that contains u but does not contain any of $x_1, \ldots, x_r$. This definition is illustrated in Fig. 3.1.

Fig. 3.1.

Lemma 3.1. [5] Let T be a tree of n vertices. $cw(T) \leq 3$ if and only if every vertex u of degree more than two has neighbors x_1, x_2 such that $cw(T(u, x_1, x_2)) \leq 2$.

It follows from Lemma 3.1 that the tree in Fig. 3.2 has cutwidth more than two since vertex u does not have neighbors x_1, x_2 such that $cw(T(u, x_1, x_2)) \leq 1$.

Fig. 3.2. Example of the lemma 3.1.

Clearly a graph has cutwidth one if and only if it is a simple chain. Now we will discuss the 2-cutwidth problem on trees. The lemma 3.2 will give the property of trees which have cutwidth 2.

Lemma 3.2. A tree $T = (V, E)$ has cutwidth 2 if and only if it has maximum vertex degree at most four and there is a path P in T which contains all degree 3 and 4 vertices.

From the above lemma, it is easily obtained that the 2-cutwidth problem can be solved in linear time.

Lemma 2.1 suggests that if we can find a path P in T such that the cutwidth of the forest $T - P$ is at most 2, then the cutwidth of this tree T is at most 3. The path P described from above is defined as a *basic path* for tree T and denoted by $BP(T)$. The forest is denoted by $T - BP(T)$. Unfortunately, the basic path for a tree T can not be obtained easily. Our method to find the basic path is illustrated as follows.

At first we select an arbitrary path P and some subtrees in $T - P$ have cutwidth greater than 2. It is obvious that the path P is not a basic path and we must modify the path P so that the new path will cross over these subtrees in $T - P$ which have cutwidth greater than two. The basic path in T can be constructed by iterating the above steps. The following lemma is obtained from the lemma 3.1 which will show that there are at most two subtrees in $T - P$ which have cutwidth greater than two for any path P in T.

Lemma 3.3. Let $T = (V, E)$ be a tree which has cutwidth at most 3. Then there are at most two subtrees in $T - P$ which has cutwidth greater than 2 for any path P in T.

The preceding discussion suggests that the divide-and-conquer strategy can be used in finding the basic path for a tree with cutwidth at most 3. Suppose we treat the recursive calls as a sequence of stages. The original problem is in stage 0. The works we do at stage i ($i \geq 0$) is to select an arbitrary path $P^{(i)}$ in $T^{(i)}$ and mark the connected components in $T^{(i)} - P^{(i)}$ which have cutwidth greater than 2. From the result of lemma 3.3., there are at most two connected subtrees in $T^{(i)} - P^{(i)}$ which have cutwidth greater than 2. We consider the three cases.

Case 1 : If all the connected subtrees in $T^{(i)} - P^{(i)}$ have cutwidth at most 2, then the basic path in $T^{(i)}$ is $P^{(i)}$.

Case 2 : If there is only one subtree (denoted by $T^{(i+1)}$) in $T^{(i)} - P^{(i)}$ which has cutwidth greater than 2, then the basic path in $T^{(i)}$ can be constructed by joining the path $P^{(i)}$ with the basic path in $T^{(i+1)}$.

Case 3 : If there are two subtrees (denoted by $T_1^{(i+1)}$ and $T_2^{(i+1)}$ respectively) in $T^{(i)} - P^{(i)}$ which have cutwidth greater than 2, then the basic path in $T^{(i)}$ can be constructed by joining the path $P^{(i)}$, the basic path in $T_1^{(i+1)}$, and the basic path in $T_2^{(i+1)}$.

The following lemma will show that the case 3 is occurred at most once among all the stages for the recursive calls.

Lemma 3.4. Let T be a tree with cutwidth at most 3. Suppose stage k is the first stage that there are two connected subtrees $T_1^{(k+1)}$, $T_2^{(k+1)}$ in $T^{(k)} - P^{(k)}$ such that $cw(T_1^{(k+1)}) > 2$ and $cw(T_2^{(k+1)}) > 2$. Then there is at most one connected subtree $T_j^{(i+1)}$ in $T_j^{(i)} - P_j^{(i)}$ such that $cw(T_j^{(i+1)}) > 2$ ($j = 1, 2, i \geq k + 1$).

Before describing the algorithm, we need to introduce some definitions and remarks which will be used in our algorithm. A *centroid* of a tree T of n vertices is a vertex such that when this vertex is deleted, the resulting forest has no connected component with more than $\lfloor n/2 \rfloor$ vertices. It is well known that a tree either has exactly one centroid or has exactly two adjacent centroids [3]. A "from-leaves-toward-center" method can be employed to get the centroids of a tree. This algorithm is linear.

From the lemma 3.3, we know that there are at most two subproblems will be considered at each stage in our algorithm. We then use the concept of a centroid in T to successively divide the tree into small pieces for the recursive calls, i.e. to reduce the size of each subproblem as small as possible. The detail will be illustrated in lemma 3.5.

Lemma 3.5. Let T be a tree with n vertices which has cutwidth at most 3. P is a path in T that contains a centroid z and crosses over the subtree with largest sizes when z is removed. Then the size of any tree in $T - P$ is at most $\lfloor 5n/6 \rfloor$.

We now describe an algorithm to determine whether the given tree T has cutwidth at most 3 or not. The main approach in our algorithm is to try to find the basic path P such that the cutwidth of the forest $T - P$ has cutwidth at most 2. The procedure to find the basic path will be shown in Fig. 3.3, and our main algorithm is shown in Fig. 3.4.

```
procedure Find_Basic_Path (T,v)
input: A tree T = (V, E) of n vertices.
output: find a basic path BP(T) for T, if no such path, return {} and set failure.
1.    Using v as an endpoint, construct a path P as described in lemma 3.5. Suppose the
2.       sequence of path P is v₀, v₁, ..., vₘ and T − P contains T₀, T₁, ..., Tₘ.
3.    Determine the cutwidth for each subtree T₀, T₁, ..., Tₘ.
4.    if (cw(Tᵢ) ≤ 2 for each i, 0 ≤ i ≤ m)
5.       then return BP(T) = P.
6.    if (only one subtree Tₖ such that cw(Tₖ) > 2) then
7.       call Find_Basic_Path(Tₖ,vₖ).
8.       if cw(T − {BP(Tₖ) + {v₀, ···, vₖ}}) ≤ 2
9.          then return BP(T) = {v₀, ···, vₖ} + BP(Tₖ).
10.         else return BP(T) = BP(Tₖ) + {vₖ, ···, vₘ}.
11.   if (only two subtrees Tᵢ,Tⱼ such that cw(Tᵢ) > 2 and cw(Tⱼ) > 2
12      and not occurred) then
13.         call Find_Basic_Path(Tᵢ,vᵢ).
14.         call Find_Basic_Path(Tⱼ,vⱼ).
15.         set occurred = true.
16.         return BP(T) = BP(Tᵢ) + {vᵢ, ···, vⱼ} + BP(Tⱼ).
17.   set failure = true.
18.   return BP(T) = {}.
end Find_Basic_Path
```

Fig. 3.3. The procedure Find_Basic_Path.

```
Algorithm 3-MINCUT_T
input: A tree T = (V, E) of n vertices.
output: determine whether there exists a Min Cut Linear Arrangement
        for T such that its cutwidth is at most 3.
1.    begin
2.       set failure = false.
3.       set occurred = false.
4.       call Find_Basic_Path(T,v).
5.       if (not failure and cw(T − BP(T)) ≤ 2) then
               print("There exists a Min Cut Linear Arrangement in T
                     which its cutwidth is at most 3").
6.       else
               print("The cutwidth of this tree T is greater than 3").
7.    end 3-MINCUT_T
```

Fig. 3.4. The 3-MINCUT_T algorithm

At the end of this section, we will analyze the time complexity of our algorithm. We know that there are at most two subtrees at each stage to be considered in our algorithm from the lemma 3.3. But from the lemma 3.4, we know that the condition of dividing the original problem into two small subproblems is occurred at most once among all the stages in our algorithm. Suppose $T(n)$ denote the time complexity for procedure Find_Basic_Path and case 3 is not occurred. We know that there is an $O(n)$ time algorithm to find the centroid and to determine the given tree which has cutwidth at most 2 or not. Furthermore, the size of each subproblem is bounded by $\lfloor 5n/6 \rfloor$ from the result of lemma 3.5. We have the following.

$$T(n) = T(\lfloor 5n/6 \rfloor) + O(n)$$
$$T(n) = O(n)$$

If case 3 is occurred, the total time complexity to solve the 3-cutwidth problem on tree T will be bounded by $2 \times T(n)$ from the result of lemma 3.4, which its value is also $O(n)$.

Theorem 3.1. The 3-cutwidth problem for a tree T can be solved by algorithm 3-MINCUT$_T$ in $O(n)$ steps.

4. Extension

In this section, we will give an extension to the k-cutwidth problem on trees, where k is a constant. We use similar approach as in the previous section to solve this problem. The following lemmas are very similar to the lemmas in section 3.

Lemma 4.1. Let T be a tree. $cw(T) \leq k$ if and only if every vertex u of degree more than two has neighbors x_1, x_2 such that $cw(T(u, x_1, x_2)) \leq k - 1$.

Lemma 4.2. Let T be a tree which has cutwidth at most k. Then there are at most two subtrees in F_{T-P} which have cutwidth greater than $k - 1$ for any path P in T.

Lemma 4.3. Let T be a tree with cutwidth at most k. Suppose stage l is the first stage that there are two connected subtrees $T_1^{(l+1)}, T_2^{(l+1)}$ in $T^{(l)} - P^{(l)}$ such that $cw(T_1^{(l+1)}) > k - 1$ and $cw(T_2^{(l+1)}) > k - 1$. Then there is at most one connected subtree $T_j^{(i+1)}$ in $T_j^{(i)} - P_j^{(i)}$ such that $cw(T_j^{(i+1)}) > k - 1$ $(j = 1, 2, i \geq l + 1)$.

Lemma 4.4. Let T be a tree of n vertices which has cutwidth at most k. P is a path in T that contains a centroid z of T and crosses over the subtree with largest sizes when z is removed. Then the size of any subtree in $T - P$ is at most $\lfloor (2k - 1/2k) \cdot n \rfloor$.

Theorem 4.1. If k is a fixed constant then the k-cutwidth problem for a tree T can be solved by an algorithm in $O(n)$ steps.

Proof: The basic strategy to solve the k-cutwidth problem on trees is the same as we used in solving 3-cutwidth problem (find the basic path). But the definition of the basic path shall be made a little modification. The *basic path* $BP(T)$ for the k-cutwidth problem is a path in tree T such that the cutwidth of the forest $T - BP(T)$ is at most $k - 1$. Let $T_k(n)$ denote the time complexity to find the basic path in k-cutwidth problem when the case 3 does not occur. From the result of lemma 4.2, we have the following.

$$T_k(n) \leq T_k(\lfloor (2k - 1)/2k \cdot n \rfloor) + 2 \cdot T_{k-1}(n)$$
$$T_k(n) \leq 4 \cdot k \cdot T_{k-1}(n)$$

$$T_k(n) \leq 4k \cdot 4k - 1 \cdots 4 \cdot 2 \cdot T_1(n)$$
$$\leq 2^{2k-3} \cdot k! \cdot O(n)$$
$$\leq O(2^{2k} \cdot k! \cdot n)$$

The total time complexity to solve the k-cutwidth problem will be bounded by $2 \times T_k(n)$ from the result of lemma 4.3. It is obvious that there is a linear time algorithm to solve the k-cutwidth problem, when k is a constant; moreover, it also explains why our method can not be extended to solve the general case. ∎

5. Min Cut Linear Arrangement of General Graph

We have discussed the k-cutwidth problem on trees in the previous sections. In this section, we will discuss this problem on general graph. At first, we will discuss a simple characterization of graph with cutwidth 2. We need some definitions first. An *outerplanar* graph is a graph which can be embedded in the plane so that no edges cross and all it's vertices share one face. The edges of the face are called *sides*; the remaining edges are called *chords*. A special kind of outerplanar graph, called a *cactus* [2], is a graph such that no two cycles share a common edge. In other words, a cactus is an outerplanar graph without any chords in its biconnected components. The following lemma will show the characterization of graphs with cutwidth 2.

Lemma 5.1. A graph G has cutwidth 2 if and only if :
  (1) all nodes in G have degree at most 4;
  (2) it is a cactus;
  (3) there is a chain of biconnected components $C = C_1, C_2, \cdots, C_m$ ($m \geq 1$) in G such that for each i ($1 \leq i \leq m-1$), C_i and C_{i+1} share an articulation point a_i, and the chain C contains all the simple cycles, degree 3 and degree 4 vertices;
  (4) for each simple cycle C_i ($1 \leq i \leq m$), the articulation points a_{i-1} and a_i are the only two points which can be attached to C_i by other subgraphs of G.

Since every graph G can be decomposed into its biconnected components, and it is easily obtained that if H is a subgraph of G, then $cw(H) \leq cw(G)$. So for a given graph G with cutwidth k, we must guarantee that each of its biconnected component has cutwidth at most k. The main approach we deal with the 3-cutwidth problem on a given graph G is to layout $G's$ biconnected components and combine them appropriately to obtain a resulting layout. Clearly, a biconnected graph has cutwidth 2 if and only if it is a simple cycle by the Lemma 5.1. Now we will discuss the characterization of biconnected graph with cutwidth 3. The following lemma will show this characterization.

Lemma 5.2. Let G be a biconnected graph. The cutwidth of G is 3 if and only if :
  (1) All nodes in G have degree at most 4, and
  (2) G is characterized as an outerplanar graph satisfying the "collinear chord" property.

The example of Lemma 5.2 will be showned in Fig. 5.1.

From Lemma 5.2, the algorithm of recognizing the biconnected outerplanar graph [13] can be employed to determine whether a given biconnected graph G has cutwidth 3 or not. This procedure is linear. So the time complexity to solve the 3-cutwidth problem on biconnected graphs is linear.

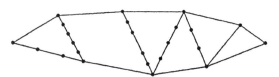

Fig. 5.1. The biconnected graph with cutwidth 3

Now we will design an optimal linear-time algorithm for the 3-cutwidth problem on general graphs. First, we show an important property that will be used in our algorithm.

Definition 5.1. A *cycle separator* of an n-vertex graph is a simple cycle such that when the vertices of the cycle are deleted, the resulting graph has no connected component with more than $n/2$ vertices.

The cycle separator of a graph and the centroid of a tree have the same effect. They respectively divide the original graph and tree into some small components. It is known that every graph has a cycle separator (The cycle separator may be a single vertex, called zero length cycle). Such a separator can be found in $O(n + m)$ sequential time for any graph of n vertices and m edges [1,9]. Now we will give an algorithm to solve the 3-cutwidth problem on general graph. In section 2, our method to solve the small cutwidth problem on trees is recursively finding the basic path. This method can be made a little modification by finding a basic biconnected component chain for the general graph case. The algorithm is shown as follows.

Algorithm 3-MINCUT$_G$

input: A graph $G = (V, E)$ of n vertices.
output: determine whether there exists a Min Cut Linear Arrangement
for G such that its cutwidth is at most 3.

1. If the maximum vertex degree of this graph G are greater than 6, then stop and answer "G has cutwidth larger than 3".
2. Find the biconnected components G_i ($1 \leq i \leq q$ for some q) of G.
3. Determine each biconnected components G_i's cutwidth (lemma 5.2). If there exists one G_i which has cutwidth greater than 3, then stop and the cutwidth of this graph G is greater than 3.
4. Find a cycle separator of G.
5. Find a biconnected chain $C = C_1 C_2 \cdots C_m$ that contains the cycle separator.
6. Determine the cutwidth of subgraph that attached to the chain C. If all of them has cutwidth at most 2, then go to step 8.
7. Recursively find the chain that satisfies the condition in step (6) (The detail is similar to the method used in section 3).
8. Suppose the chain we obtained is BC. Layout the subgraph that are attached to BC into the interval *nearest* to the vertex which it attached to the chain BC. If all subgraphs that are attached to the chain BC can be layout such that the resulting graph has cutwidth ≤ 3. Then answer "G has cutwidth ≤ 3, else answer "G has cutwidth greater than 3".

end of algorithm 3-MINCUT$_G$

Theorem 5.1. The 3-cutwidth problem for a graph G can be solved by algorithm 3-MINCUT$_G$ in $O(n)$ steps.

Proof: Each step in this algorithm can be done in linear time from the above discussions. We omit the detail. So the total time complexity to solve the 3-cutwidth problem on general graph is linear. ∎

6. Conclusion

We present linear time algorithms for the small cutwidth problem on trees and the 3-cutwidth problem on general graph. The main results of the two problems illustrated above are based on the concept of the basic path and the basic chain respectively.

Finally, we will list all the result about the min cut linear arrangement problem as follows.

trees :

(1) Thomas Lengauer gave an $O(n \log n)$ time approximation algorithm that produces a layout with cutwidth at most twice the optimal [10].

(2) M. Chung et al. present an algorithm that solves the problem in time $O(n(\log n)^{d-2})$ while d is the maximum degree of the tree [5].

(3) M. Yannakakis showed that the cutwidth of any tree can be found in $O(n \log n)$ time [17].

(4) A linear time algorithm to recognize the small cutwidth tree is presented in this paper.

graphs :

(1) This k-cutwidth problem (Min Cut Linear Arrangement problem) is NP-complete for graphs with maximum vertex degree three [12].

(2) For fixed k, the problem can be solved in $O(n^k)$ time by dynamic programming technique [8].

(3) A linear time algorithm to recognize the 3-cutwidth graph is presented in this paper.

REFERENCES

[1] Alok Aggarwal, Richard J. Anderson, and Ming Yanf Kao, "Parallel Depth-First search in general directed graphs," *SIAM J. Comput.*, Vol. 19, No. 2, April 1990, pp. 397-409.

[2] C. Berge, *Graphs and Hypergraphs*, North-Holland, Amsterdam 1973.

[3] F. Buckley and F. Harary, *Distance in Graphs*, Addison-Wesley, Reading, 1990.

[4] Fan R. K. Chung, "On the Cutwidth and the Topological Bandwidth of a Tree," *SIAM J. Alg. Disc. Meth.*, Vol. 6, No. 2, April 1985, pp. 268-277.

[5] M. J. Chung, F. Makedon, I. H. Sudborough and J. Tarner, "Polynomial algorithm for the min-cut linear arrangement problem on degree restricted trees," *SIAM J. Comput.*, Vol. 14, No. 1, 1985, pp. 158-177.

[6] A. E. Dunlop and B. W. Kernighan, "A Procedure for Placement of Standard-Cell VLSI Circuits," *IEEE Trans. on Computer-Aided Design of Integrated Circuits and Systems*, Vol. CAD-4, No. 1, Jan. 1985, pp. 92-98.

[7] M. R. Garey, D. S. Johnson, *Computers and Intractability : A Guide to the Theory of NP-Completeness*, Freeman, San Franciso 1979.

[8] E. Gurari, and I. H. Sudborough, "Improved dynamic programming algorithms for bandwidth minimization and min-cut linear arrangement problem," *J. Algorithms*, Vol. 5, 1984, pp. 531-546.

[9] M. Y. Kao, "All graphs have cycle separators and planar directed depth-first search is in DNC," in Proc. 3rd Aegean Workship on Computing, Corfu, Greece, J. H. Reif, ed.; *Lecture Notes in Computer Science 319*, Springer-Verlay, Berlin, New York, 1988, pp. 53-63.

[10] Thomas Lengauer, "Upper and Lower Bounds on the Complexity of the Min-Cut Linear Arrangement Problem on Trees," *SIAM J. Alg. Disc. Meth.*, Vol. 3, No. 1, March 1982, pp. 99-113.

[11] A. D. Lopez, and H. F. S. Law, "A Dense Gate Matrix Layout Method for MOS VLSI," *IEEE Trans. on Electronic Devices*, ED-27, 8, 1980, pp. 1671-1675.

[12] F. S. Makedon, C. H. Papadimitriou and I. H. Sudborough, "Topological Bandwidth," *SIAM J. Alg. Disc. Meth.*, Vol. 6, No. 3, July 1985, pp. 418-444.

[13] Sanda L. Mitchell, "Linear Algorithms to Recognize Outplanar and Maximal Outerplanar Graphs," *Information Processing Letters*, Vol. 9, No. 5, December 1979, pp. 229-232.

[14] T. H. Ohtsuki, H. Mori, E. S. Kuh, T. Kashiwabara and T. Fujiswa, "One Dimensional Logic Gate Assignment and Interval Graphs," *IEEE Trans. Circuits and Systems*, Vol. 26, 1979, pp. 675-684.

[15] A. L. Rosenberg, "The Diogenes Approach to Testable Fault-Tolerant Arrays of Processors," *IEEE Trans. on Computers*, Vol. C-32, No. 10, Oct. 1983, pp. 902-910.

[16] R. E. Tarjan, "An Efficient Parallel Biconnectivity Algorithm," *SIAM J. Comput.*, Vol. 14, No. 4, November 1985, pp. 862-874.

[17] M. Yannakakis, "A Polynomial Algorithm for the Min Cut Linear Arrangement of Trees," *J. ACM.*, Vol. 32, No. 4, 1985, pp. 950-988.

The k-Edge-Connectivity Augmentation Problem
of Weighted Graphs

Toshimasa Watanabe, Toshiya Mashima and Satoshi Taoka

Department of Circuits and Systems, Faculty of Engineering, Hiroshima University
4-1, Kagamiyama 1-chome, Higashi-Hiroshima, 724 Japan
E-mail:watanabe@huis.hiroshima-u.ac.jp

Abstract. The k-edge-connectivity augmentation problem (k-ECA) is the subject of the paper. Four approximation algorithms FSA, FSM, SMC and HBD for k-ECA are proposed, and both theoretical and experimental evaluation are given.

1. Introduction

The subject of the paper is the *k-edge-connectivity augmentation problem* (k-ECA) defined as follows: "Given a complete graph $G=(V,E)$, a subgraph $G'=(V,E')$ with $E' \subseteq E$, and a cost function $c : E \to Z^+$ (nonnegative integer) with $c(e)=0$ if $e \in E'$ and $c(e)>0$ if $e \in E-E'$, find a set of edges $E'' \subseteq E-E'$ of minimum total cost $c(E'')$ such that $G''=(V,E' \cup E'')$ is k-edge-connected." Note that this definition prohibits adding multiple edges to G' and G' is simple. (We can define k-ECA in which adding multiple edges is allowed or G' may have multiple edges; the discussion is similar, and the results will be reported elsewhere. We often denote G'' as $G'+E''$. The *edge-connectivity* $ec(H)$ of a graph H is the minimum number of edges whose deletion disconnect it, and H is *k-edge-connected* if $ec(H) \geq k$. If $c(e)=c(e')$ for e, $e' \in E-E'$ then the problem is called the unweighted version (denoted by UW-k-ECA) and the weighted one (denoted by W-k-ECA) otherwise.

Some known results on W-k-ECA are summarized in the following. (For UW-k-ECA, see [3,5,7,15,19,20,22].) For W-2-ECA, [6] showed that it remains NP-complete even if G' is restricted to a tree with edge costs from $\{1,2\}$. An $O(|V|^2)$ approximation algorithm based on minimum-cost arborescence is presented and it is shown that worst approximation is no more than twice (three times, respectively) the optimum if $ec(G')=1$ (if $ec(G')=0$). For W-3-ECA, [21] proved that the problem remains NP-complete even if G' is restricted to a 2-vertex-connected graph with edge costs chosen from $\{1,2\}$, and an $O(|V|^3)$ approximation algorithm FS that utilizes a minimum-cost arborescence was presented. Four more approximation algorithms with both theoretical and experimental evaluation are given in [23]. [20] mentioned that both W-k-ECA and W-k-VCA with $k \geq 2$ is NP-complete if G' is an empty graph.

The paper mainly considers W-k-ECA and for simplicity, k-ECA means W-k-ECA unless otherwise stated. The problem has application to designing robust networks: a k-edge-connected network can survive k-1 communication lines failure. Let $\lambda=ec(G')$ and $k=\lambda+\delta$. $(\lambda+1)$-ECA, that is, k-ECA with $\delta=k-ec(G')=1$, is mainly considered, and general $(\lambda+\delta)$-ECA is solved by repeatedly finding solutions to $(\lambda+1)$-ECA. If $\lambda=0$ and $\delta=1$ then the problem is optimally solved by using an algorithm for finding a minimum-cost spanning tree. We assume that $\lambda \geq 1$ in the paper.

First it is proved that the recognition version of $(\lambda+1)$-ECA is NP-complete even if edge costs are either 1 or 2. We can provide a unified NP-completeness proof of $(\lambda+1)$-

ECA and $(\lambda+1)$-VCA (the $(\lambda+1)$-vertex-connectivity augmentation problem definde similarly to $(\lambda+1)$-ECA). Then four approximation algorithms FSA, FSM, SMC and HBD are proposed for $(\lambda+1)$-ECA. FSA is based on a minimum-cost arborescence algorithm, and its time complexity is $O(|V|^3)$. FSM, based on a maximum-cost matching algorithm, runs in $O(|V|^3 \log|V|)$ time. SMC is an alternative to FSM and is a greedy algorithm finding an edge set of small total cost in $O(|V|^3)$ time. HBD is a combination of FSM and SMC and runs in $O(|V|^3 \log|V|)$ time.

It is shown theoretically that all of FSA, FSM and HBD produce worst approximation no greater than twice the optimum if all costs are equal. (Note that there is an algorithm that optimally solves k-ECA if all costs $c(e) \in$ E-E' are equal.) For the case where all costs are not equal. It is experimentally observed that FSM gives the best approximation in longest computation time, while SMC and HBD give good solutions in shortest computation time. For each of FSM and SMC there is a case such that worst approximation cannot be bounded by a constant. However, it is experimentally shown that FSM produces a good solution in cases for which SMC gives unbounded solutions, and vice versa. Also observed is that HBD generates sharp approximation in each of these unbounded cases.

For general $(\lambda+\delta)$-ECA, approximate solutions can be obtained by repeatedly using any one of the four algorithms. Experimental results for the case with $1 \le \delta \le 4$ are given. It is also observed that FSM gives best solutions among the four algorithms.

2. Preliminaries

Technical terms not given here can be identified in [4,10,18]. An *(undirected) graph* $G=(V(G),E(G))$, or simply denoted as $G=(V,E)$, consists of a set $V(G)$ of *vertices* and a set $E(G)$ of *undirected edges*, where an edge between u and v is denotes as (u,v). A *directed graph* $\overline{G}=(V(\overline{G}),A(\overline{G}))$, or simply denoted as $\overline{G}=(V,A)$, consists of a set of vertices $V(\overline{G})$ and a set of *directed edges* $A(\overline{G})$. A directed edge from u to v is denoted by $<u,v>$. The *degree* of a vertex v is denoted as $d_G(v)$, or simply $d(v)$: v is often called a degree-$d(v)$ vertex. We consider a pair of multiple edges as a cycle of length 2.

A *leaf* in a tree is a vertex with only one edge incident on it. An *arborescence* is a directed acyclic graph with one specified vertex, called the *root*, having no entering edges, and all other vertices having exactly one entering edge. A minimum-cost arborescence is an arborescence of minimum total cost. A *cactus* is an undirected connected graph in which any pair of cycles share at most one vertex: each shared vertex is a cutpoint. A leaf of a cactus is a vertex v with $d(v)=1$ or v' included in a cycle with $d(v')=2$.

Let $S \subseteq V \cup E$ be any minimal set such that G-S (a graph obtained by deleting all element of S from G) is disconnected. S is called a *separator* of G, or in particular a (u,v)-separator if u and v are disconnected in G-S. A *minimum separator* S of G is a separator of minimum cardinality among those of G, and $|S|$ is the *edge-connectivity* (denoted by $ec(G)$) of G in case $S \subseteq E$; particularly such $S \subseteq E$ is called a *minimum* cut (of G). (For a mimimum separator S, $|S|$ is the vertex-connectivity $v_C(G)$ if $S \subseteq V$.) If $|S|=1$ then the element of S is called a *cutpoint* in case $S \subseteq V$ or a *bridge* in case $S \subseteq E$. A minimum cut $S \subseteq E$ is often denoted as (X,Y), where $X \cup Y$ is a partition of V such that $S=\{(u,v) \in E | u \in X, v \in Y\}$. G is k-edge-connected if $ec(G) \ge k$. For two vertices u, v of G, let $\lambda(u,v;G)$, or simply $\lambda(u,v)$, denote the maximum number of pairwise edge-disjoint paths between u and v. A *k-edge-connected component* (k-ecc for short) of G is a subset $S \subseteq V$ satisfying the following (a) and (b): (a) $\lambda(u,v;G) \ge k$ for any pair $u,v \in S$; (b) S is a maximal set satisfying (a). A 1-ecc is called a *component*.

A *structural graph* F(G) of a given graph G=(V,E) with edge-connectivity ec(G)=λ is a representation of all minimum cuts of G. F(G) is an edge-weighted cactus of $O(|V|)$ nodes and edges such that each tree edge has weight λ and each cycle edge has weight $\lambda/2$. Particularly if λ is odd then F(G) is a weighted tree. It is shown that F(G) can be constructed in $O(|V||E|)$ time [11] or $O(|E|+\lambda^2|V|\log(|V|/\lambda))$ time [7]. Each vertex in G maps to exactly one vertex in F(G), and F(G) may have some other vertices, called *empty vertices*, to which no vertices of G are mapped. Let $\varepsilon(G)\subseteq V(F(G))$ denote the set of all empty vertices of F(G). Let $\rho:V(G)\to V(F(G))-\varepsilon(G)$ denote this mapping, and we represent as $\rho(X)=\{\rho(v)|v\in X\}$ for $X\subseteq V$ and $\rho^{-1}(Y)=\{u\in V|\rho(u)=v, v\in Y\}$ for $Y\subseteq V(F(G))$. F(G) has the following properties: each minimum cut (X,Y) in F(G), with Y=V(F(G))-X, corresponds to a minimum one $(\rho^{-1}(X-\varepsilon(G)),\rho^{-1}(Y-\varepsilon(G)))$ in G, and conversely, for each minimum cut (X,Y) in G, there exists at least one partition of $\varepsilon(G)$ into two sets ε_1 and ε_2 such that $(\rho(X)\cup\varepsilon_1,\rho(Y)\cup\varepsilon_2)$ is a minimum cut of F(G). Note that if λ is even then replacing each tree edge by a pair of multiple edges preserves the properties of structural graphs and makes their handling easy because the resulting graphs have no bridges. This graph and a tree in the case where λ is odd are called *modified cactuses*. In the following F(G) denotes a modified cactus unless otherwise stated. Suppose that $\lambda\geq2$, and let Z be a set of edges connecting vertices of F(G) and such that $Z\cap E(F(G))=\emptyset$. Neglect edge-weights of F(G), and suppose that ec(F(G)+Z)$\geq$2 if λ is odd (that is, F(G) is a tree) or ec(F(G)+Z)$\geq$3 if λ is even (that is, F(G) is a union of some cycles). Then Z is called a *solution* to F(G). For each (u,v)$\in$Z, the set A(u,v)={u,v}$\cup${w$\in$ V(F(G))-{u,v}|w is a cutpoint separating u from v}, is called a (u,v)-*augmenting set* of G. Fig.1 shows an example of graph G_1 with λ=2 and its structural graph F(G_1) as a weighted modified cactus with $\varepsilon(G_1)=\emptyset$ and $(\lambda+1)$-eccs

$\rho^{-1}(a)=\{2\}, \rho^{-1}(b)=\{1,3\}, \rho^{-1}(c)=\{4\}, \rho^{-1}(d)=\{5,7,11\},$
$\rho^{-1}(e)=\{6\}, \rho^{-1}(f)=\{8,10\}, \rho^{-1}(g)=\{9\}, \rho^{-1}(h)=\{12\},$
$\rho^{-1}(i)=\{13\}, \rho^{-1}(j)=\{14\}, \rho^{-1}(k)=\{15\}, \rho^{-1}(l)=\{16\}, \rho^{-1}(m)=\{17\}.$

Let $\lceil x \rceil$ ($\lfloor x \rfloor$, respectively) denote the minimum integer not smaller (the maximum one not greater) than x.

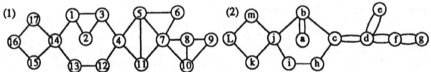

Fig.1. An example of a graph G_1 with ec(G_1)=2 and a structural graph F(G_1):
(1) G_1; (2) F(G_1).

3. Intractability of K-ECA

We can prove that the recognition versions of k-ECA and k-VCA with G' restricted to a (k-1)-edge-connected graphs are NP-complete, by showing a polynomial-time reduction of the 3-dimensional matching problem (3DM), known to be NP-complete [4], to these problems.
(The recognition version of W-$(\lambda+\delta)$-ECA (W-$(\lambda+\delta)$-VCA), respectively)
Instance: A complete graph G=(V,E), a spanning subgraph G'=(V,E') of G with ec(G')=λ

($vc(G')=\lambda$), $c(e)=0$ for $c \in E'$ and nonnegative costs $c(e)>0$ for edges $e \in E-E'$ and nonnegative integers B and δ.

Question: Is there a subset $E'' \subseteq E-E'$, with $c(E'')=\sum_{c \in E''} c(e) \leq B$, such that $ec(G'+E'') \geq \lambda + \delta$ ($vc(G'+E'') \geq \lambda + \delta$)?

We can provide a unified proof of the theorem: the details are omitted due to shortage of space (see [13]).

Theorem 3.1. The recognition versions of W-$(\lambda+\delta)$-ECA and W-$(\lambda+\delta)$-VCA with $\lambda \geq 1$ are NP-complete even if $\delta=1$ and costs $c(e) \in \{1,2\}$ for $e \in E-E'$. ◆

4. Approximation Algorithms for W-K-ECA

We first describe a general scheme as an algorithm EC.

ALGORITHM EC

Input: a complete undirected graph $G=(V,E)$ with a cost function $c:E \to Z^+$ with $c(e)=0$ if $e \in E'$ and $c(e)>0$ if $e \in E-E'$, and a λ-edge-connected spanning subgraph $G'=(V,E')$ of G.

Output: A set of edges, $E'' \subseteq E-E'$, such that $G''=(V,E' \cup E'')$ is $(\lambda+\delta)$-edge-connected.

1. $H \leftarrow G$, $H' \leftarrow G'$.
2. Construct a structural graph $G'_s=(V_s,E_s')(=F(H'))$ of H' as in [11] or [7], and $\theta \leftarrow ec(H')$.
3. if $\theta \geq k$ then stop.
4. Construct a complete graph $G_s=(V_s,E_s)$ and define a new cost function $c':E_s \to Z^+ \cup \{\infty\}$ by $c'(u,v)=\min\{\{\infty\} \cup \{c(x,y)|(x,y) \in E(H)-E(H'), x \in \rho^{-1}(u)$ and $y \in \rho^{-1}(v)\}\}$, where $\rho^{-1}(w)$ is a $(\theta+1)$-ecc of H' and is represented as $w \in V_s$-$\epsilon(H')$. Define a backpointer $b:E_s \to E(H)$ by $b(u,v) \leftarrow (x,y)$ if $c'(u,v)=c(x,y)$ or $b(u,v) \leftarrow (u,v)$ if $c'(u,v)=\infty$.
5. Find an edge set E_s'' of small total cost $c'(E_s'')$ such that E_s'' is a solution to G_s'.
6. Construct a solution $\Gamma=\{b(u,v) \in E(H)-E(H')|(u,v) \in E_s''\}$ (with multiplicity deleted) such that $ec(H'+\Gamma)=\theta+1$.
7. $H' \leftarrow H'+\Gamma$, $E'' \leftarrow E'' \cup \Gamma$, $c(x,y) \leftarrow 0$ for any $(x,y) \in E''$, and goto Step 2. ◆

Constructing G_s', G_s, c' and b can be done in $O(\min\{|V||E'|,\Delta\}+|V|^2 \log|V|)$ time, where $\Delta=|E'|+\lambda^2|V| \log(|V|/\lambda)$. If E_s'' of Step 4 can be found in $O(\xi)$ time then EC runs in $O(\delta(\xi+\min\{|V||E'|,\Delta\}))$ time.

For a subset $E'' \subseteq E-E'$, let $c(E'')=\sum_{(u,v) \in E''} c(u,v)$, and similarly for $c'(E_s'')$. The following lemma shows that it suffices to consider $(\lambda+1)$-ECA for a cactus $G_s'=(V_s,E_s')$.

Lemma 4.1. If $ec(G'+E'') \geq \lambda+1$ for some $E'' \subseteq E-E'$ then there is $E_s'' \subseteq E_s-E_s'$ with $c'(E_s'') \leq c(E'')$ such that E_s'' is a solution to G_s'. For a subset $E_s'' \subseteq E_s-E_s'$, let $b(E_s'')=\{b(u,v)|(u,v) \in E_s''\}$ with multiplicity deleted. If E_s'' is a solution to G_s' then $ec(G'+b(E_s'')) \geq \lambda+1$ and $c'(E_s'') \geq c(b(E_s''))$. ◆

Thus the remaining task is to devise efficient algorithms producing good approximate solutions E_s'' to G_s' in Step 5 of EC for the case where $ec(H')<k$. Four procedures FSA* (Finding Solution by Arborescence), FSM* (Finding Solution by Matching), SMC* (finding solution by Selecting Minimum-Cost edges) and HBD* (combining FSM and SMC) are to be proposed for Step 5 of EC. EC with FSA* is denoted as FSA, and similarly for other procedures.

4.1 Procedure FSA* Based on Minimum-Cost Arborescence

FSA* is based on a minimum-cost arborescence algorithm [1,2,8,17]. The algorithm is outlined as follows. First, G_s' is changed into a simple graph by deleting multiplicity, and then a tree G_b' will be constructed as follows: for each cycle C that is remaining in this simple graph, a new vertex v_C is added. For every cycle C, each vertex on C and v_C are connected by an edge. Then all edges of C are deleted. Let G_b' denote the resulting tree. Next, we choose a degree-1 vertex r of G_b' as the root, and direct every edge of G_b' toward r. Let $\overline{G_b}'$ denote the resulting directed graph. Some modification of costs will be done. Let $\overline{G_b}$ and d' be the complete directed graph and a final cost function, respectively, where $\overline{G_b}'$ is a spanning subgraph of $\overline{G_b}$. We find a minimum-cost arborescence $T=(V_b,A_b")$ of $\overline{G_b}$ with respect to d'. Finally an approximate solution is obtained by means of backpointers.

We first present procedure REMAKE that changes a modified cactus G_s' in the case where λ is even into a spanning tree G_b'.

PROCEDURE REMAKE;
/* Input: $G_s'=(V_s,E_s')$. Output: $G_b'=(V_b,E_b')$. */
1. If λ is odd then $G_b'\leftarrow G_s'$ and stop.
2. Delete multiplicity of edges in G_s', making it simple. Then find all cycles (equal to 2-eccs of G_s') by a depth-first-search.
3. For each cycle C, add a dummy vertex w_C, connect w_C to every vertex on C and delete all edges (u,v) of C. Let $G_b'=(V_b,E_b')$ be the resulting graph. Let $G_b=(V_b,E_b)$ be a complete graph on V_b, and extend the domains E_s of c' and b to E_b as follows: for all dummy vertices w_C and all vertices u of V_b, $c'(w_C,u)\leftarrow-\infty$ and $b(w_C,u)\leftarrow\emptyset$. ◆

We use a distance function $d:E_b\rightarrow Z^+\cup\{\infty\}$. This function was introduced in [6] in order to avoid poor choice of edges in finding a minimum-cost arborescence. In this paper it is defined as $d(u,v)=\min\{c'(x,y)|$ u and v are on a path from x to y in $G_b'\}$.

PROCEDURE DIST;
/* Input: a complete undirected graph $G_b=(V_b,E_b)$ with a cost function $c':E_b\rightarrow Z^+\cup\{\infty\}$, a backpointer b and a spanning subgraph $G_b'=(V_b,E_b')$ of G_b. */
/* Output: A distance function $d:E_b\rightarrow Z^+\cup\{\infty\}$ and a backpointer $b':E_b\rightarrow E_b$ such that $c'(b'(u,v))=d(u,v)$. */
1. For each pair of vertices u and v, compute the number a(u,v) of edges on the path between u and v in G_b', find the vertex s(u,v) adjacent to v (s(v,u) adjacent to u, respectively) on this path, $d(u,v)\leftarrow c'(u,v)$ and $b'(u,v)\leftarrow(u,v)$.
2. Bucketsort the edges (u,v) of E_b-E_b' into nonincreasing order of a(u,v). For each edge (u,v) in E_b-E_b' do the following in its sorted order:
    $d(u,s(u,v))\leftarrow d(u,v)$ and $b'(u,s(u,v))\leftarrow b'(u,v)$ if $d(u,v)<d(u,s(u,v))$;
    $d(s(v,u),v)\leftarrow d(u,v)$ and $b'(s(v,u),v)\leftarrow b'(u,v)$ if $d(u,v)<d(s(v,u),v)$. ◆

DIST correctly computes the distance function d and the backpointer b' in $O(|V|^2)$ time [6]. For two different edges $(u,v),(u',v')\in E_b$, it may happen that $b'(u,v)=b'(u',v')$. Hence, in general, $b'(Z)=\{b'(u,v)|(u,v)\in Z\}$ may be a multiset for a set $Z\subseteq E_b$. However, we assume that b'(Z) denotes the one with multiplicity deleted unless otherwise stated, for notational simplicity. Similar notation will be used for other backpointers to be defined later.

The formal description of FSA* is omitted due to shortage of space (see [13]). The following lemma can be proved, and we obtain Theorem 4.1.

Lemma 4.2. FSA* generates a set of directed edges $A_b"$ such that $(V_b,A_b'\cup A_b")$ is strongly connected. ◆

Theorem 4.1. FSA generates E" with $ec(G'+E'')\geq\lambda+1$ in $O(|V|^2+\min\{|V||E'|,\Delta\})$ time.
◆

4.2 Procedure FSM* Based on Maximum-Cost Matchings

FSM is based on a maximum-cost matching algorithm. The idea is very simple. As a set of $\lceil |L|/2 \rceil$ edges, each connecting a pair of leaves in a cactus G_s', the one E''_0 of minimum total cost is selected by using a maximum-cost matching algorithm in $O(|V_s|^3)$ time [17]. If E_0'' is not a solution to G_s' then similar process will be repeated for $G_s'+E_0''$: repetition is at most $O(\log|V_s|)$ time. The description of FSM is given as follows.

PROCEDURE FSM:*

/* *Input*: a complete undirected graph $G_s=(V_s,E_s)$ with a cost function $c':E_s\to Z^+\cup\{\infty\}$, and a backpointer $b:E_s\to E$, and a spanning subgraph $G_s'=(V_s,E_s')$ of G_s. */

/* *Output*: A set of edges $E_s''\subseteq E_s-E_s'$ that is a solution to G_s'. */

1. $G_0\leftarrow G_s$, $G_0'\leftarrow G_s'$, $c_0\leftarrow c'$, $i\leftarrow 0$.
2. Construct a tree G_b' from G_i' by using REMAKE, and a complete graph G_b from G_i.
3. $H_i\leftarrow G_b$, $H_i'\leftarrow G_b'$.
4. Find a distance function $d_i:E(H_i)\to Z^+\cup\{\infty\}$ with a back pointer $b_i':E(H_i)\to E(G_i)$ by using DIST. For each dummy vertex w_C added within the cycle C of G_s', $d_i(w_C,v)\leftarrow-\infty$ for any vertex v on C.
5. Compute $d_i':E(H_i)\to Z^+\cup\{\infty\}$ and $b_i'':E(H_i)\to E(H_i)$ by finding a shortest path $P(u,v)$ of H_i for each pair u,v of degree-1 vertices of H_i' as follows: for the edge set E_P of $P(u,v)$, $d_i'(u,v)\leftarrow d_i(E_P)$ and $b_i''(u,v)\leftarrow E_P$ (actually $b_i''(u,v)$ is a pointer to the list maintaining E_P).
6. Construct a complete subgraph S_i, with $E(S_i)\subseteq E(H_i)$, on the set of all degree-1 vertices of H_i'.
7. For each cost $d_i'(u,v)$ with $(u,v)\in E(S_i)$, $d_i''(u,v)\leftarrow MAX+1-d_i'(u,v)$, where MAX is the maximum edge-cost of S_i. Find a maximum-cost matching $M_i\subseteq E(S_i)$ of S_i with respect to d_i''.
8. For each edge $(u,v)\in M_i$, insert the corresponding set of edges $b_i'(b_i''(u,v))$ into E_i'' (and then any multiplicity is deleted).
9. If E_i'' is not a solution to G_i' then $\Gamma\leftarrow E_i''$ execute (i)-(vi):
   (i) $G_{i+1}'\leftarrow G_i'$;
   (ii) while $\Gamma\neq\emptyset$ repeat (a) and (b):
        (a) choose $(u,v)\in\Gamma$ and $\Gamma\leftarrow\Gamma-\{(u,v)\}$;
        (b) $G_{i+1}'\leftarrow$ the graph obtained by shrinking the (u,v)-augmenting set A(u,v) of G_{i+1}'; (Note that the resulting G_{i+1}' is also a modified cactus.)
   (iii) let G_{i+1} denote a smaller complete graph on $V(G_{i+1}')$ constructed from G_i by these shrinking;
   (iv) define a new cost function $c_{i+1}:E(G_{i+1})\to Z^+\cup\{\infty\}$ by
        $c_{i+1}'(u,v)=\min\{c_i'(x,y)|(x,y)\in E(G_i), x\in S_u, y\in S_v\}$,
        where S_w denotes the set of vertices of $G_i'+\Gamma$ that are shrunk into $w\in V(G_{i+1})$ and the edge (x,y) is referenced by a backpointer $b_{i+1}(u,v)=(x,y)$, where $b_{i+1}:E(G_{i+1})\to E(G_i)$.
   (v) $i\leftarrow i+1$ and goto step 2.
10. If E_i'' is a solution to G_i' then do the following (i) and (ii):
    (i) $E_s''\leftarrow E_0''$;
    (ii) if $i\geq 1$ then for each j $(j=1, ..., i)$ repeat $E_s''\leftarrow E_s''\cup b_1(...(b_j(E_j''))...)$. ◆

Theorem 4.2. FSM generates E" with $ec(G'+E")\geq\lambda+1$ in $O(|V|^3\log|V|+\min\{|V||E'|,\Delta\})$ time. ◆

4.3 Procedure SMC* Based on Minimum-Cost Edges

SMC is a greedy algorithm that finds approximate solutions without using a maximum-cost matching algorithm. The description of SMC is almost the same as that of FSM. The only difference is that SMC finds a solution to G_s' by choosing, at each leaf v of tree H_i', a minimum-cost edge among those incident upon v. Delete Step 5, replace Step 7 by the following statement in FSM and rewrite $b_i'(b_i"(u,v))$ as $b_i'(u,v)$ in Step 8:

Step 7. For each degree-1 vertex v of H_i', let e_v denote an edge $(u,v)\in E(H_i)$ with $d_i'(u,v)=\min\{d_i'(u',v)|(u',v)\in E(H_i)\}$, and $M_i\leftarrow E(k_i)\{e_v|v$ is a degree-1 vertex of $H_i'\}$ (with multiplicity deleted).

Theorem 4.3 SMC generates E" with $ec(G'+E")\geq\lambda+1$ in $O(|V|^3+\min\{|V||E'|,\Delta\})$ time. ◆

4.4. Procedure HBD* Combination of FSM and SMC

HBD is a combination of FSM and SMC, and is almost the same as that of FSM. The only difference is that a maximum-cost matching algorithm, to be repeatedly used in finding a solution to G_s', is applied to the set $E(K_i)=\{e_v|v$ is a degree-1 vertex of $H_i'\subseteq E(S_i)$ (with multiplicity deleted) instead of $E(S_i)$, where $E(K_i)$ will be constructed similarly to Step 7 of SMC, but is slightly different from it. If we replace Step 7 of FSM by the following statement then the description of HBD is obtained.

Step 7. For each degree-1 vertex v of H_i', let f_v denote an edge$(u,v)\in E(S_i)$ with $d_i'(u,v)=\min\{d_i'(u',v)|(u,v)\in E(S_i)\}$, and $E(K_i)\leftarrow\{f_v|v$ is a degree-1 vertex of $H_i'\}$ (with multiplicity deleted). Let K_i denote the subgraph $(V(S_i),E(K_i))$ of S_i. For each cost $d_i'(u,v)$ with $(u,v)\in E(K_i)$, $d_i"(u,v)\leftarrow MAX+1-d_i'(u,v)$, where MAX is the maximum edge-cost of K_i. Find a maximum-cost matching $M_i\subseteq E(K_i)$ of K_i with respect to $d_i"$.

Theorem 4.4 HBD generates E" with $ec(G'+E")\geq\lambda+1$ in $O(|V|^3\log|V|+\min\{|V||E'|,\Delta\})$ time. ◆

5. Evaluation of Worst Approximation

Worst approximation by proposed algorithms is evaluated theoretically for unweighted cases and experimentally for weighted cases. (Since UW-k-ECA can be solved optimally in polynomial time, evaluation in unweighted cases is easy.) Let OPT or APP denote total cost of an optimum or approximate solution, respectively. The ratio APP/OPT is going to be evaluated concerning $(\lambda+\delta)$-ECA (mainly for the case with $\delta=1$), where EC with FSA* is denoted as FSA, and similarly for other procedures.

5.1. Theoretical Evaluation

For the unweighted $(\lambda+1)$-ECA it is known that QPT=$\lceil q/2\rceil$ [15,20], where q is the total number of leaves of G_s'. Theoretical evaluation for FSA, FSM and HBD are given. It suffices to consider solutions to G_S'.

5.1.1. FSA. Since a minimum arborescence to be found contains exactly q-1 edges, we have APP=q-1 and, therefore, APP/OPT$\leq$(q-1)/(q/2)=2-2/q<2.

5.1.2. FSM and HBD. We can prove that $lim_q\to\infty$(APP/OPT)$\leq$2. (See [13] for the details.)

5.2. Experimental Evaluation

5.2.1. Input data. We briefly explain how input data, G' and c, are constructed (see [13] for the details).
1. The number IVI of vertices is given:
    IVI=10,15,20,40,60,80,100,120,140,160,180,200.
2. Two types of data are provided: type C and type T.
3. Costs on edges are given randomly.

5.2.2. Experimental results. We have tried 1575 data so far. A workstation SUN SPARC station is used. Tables 1 through 3 show a part of our experimental results for data IVI$\leq$200 and IEI$\leq$2055 of type C and for data IVI$\leq$200 and IEI$\leq$1014 of type T. For 360 data with IVI=10,15,20, optimum solutions are sought by exhaustive search, and error (APP/OPT-1)$\times$100 (%) is computed for these 360 data as summarized in Table 1. Table 2 shows results for large data. Both Tables 1 and 2 shows results for δ=1, while Table 3 shows those for δ>1.

Table 1. Total number of data (left) and its ratio (right), for which each algorithm produces solutions with errors (APP/OPT-1)$\times$100(%) falling into the corresponding intervals. λ=4, δ=1 and the total number of data is 360, to each of which an optimum solution is found by exhaustive search.

err.	err.=0%		0%<err.$\leq$5%		5%<err.$\leq$10%		10%<err.$\leq$15%		15%<err.	
FSA	99	27.50%	22	6.11%	43	11.94%	34	9.44%	162	45.00%
FSM	285	79.17%	29	8.06%	31	8.61%	11	3.06%	4	1.11%
SMC	189	52.50%	26	7.22%	52	14.44%	32	8.89%	61	16.94%
HBD	195	54.17%	43	11.94%	70	19.44%	28	7.78%	24	6.67%

Table 2. A part of our experimental results (1575 data in total) for the case where ec(G')=4, δ=1, type C and ζ=IV$_s$I: The columns cost and time show total cost (left) and CPU time with unit time in 1/60 second (right).

#	IVI	IE'I	λ	ζ	FSA		FSM		SMC		HBD	
					cost	time	cost	time	cost	time	cost	time
4091	160	354	4	110	114	936	86	1372	100	707	88	1125
4092	160	350	4	119	116	999	94	1670	99	730	95	1322
4093	160	345	4	133	127	1104	102	1794	110	738	103	1483
4094	160	348	4	123	129	1080	109	1966	118	768	110	1405
4095	160	364	4	107	110	931	86	1280	94	717	87	1164
4101	180	394	4	134	121	1283	92	1947	104	918	92	1592
4102	180	427	4	89	79	1038	66	1240	75	903	66	1141
4103	180	384	4	148	149	1435	112	2445	126	982	118	2011
4104	180	387	4	144	136	1374	109	2515	120	951	111	1850
4105	180	398	4	132	127	1309	96	2006	109	966	100	1693
4111	200	429	4	159	148	1821	111	3144	125	1274	117	2541
4112	200	438	4	146	157	1702	119	2802	134	1193	122	2221
4113	200	493	4	88	78	1348	67	1523	79	1187	68	1434
4114	200	429	4	162	150	1816	110	3105	127	1229	113	2515
4115	200	435	4	155	137	1784	100	2841	115	1238	101	2332

Let time(FSM) and cost(FSM) denote computation time and total cost for FSM, respectively, and similarly for others. Experimental results show the following (1)-(4).

(1) In general, cost(FSM)<cost(HBD)<cost(SMC)<cost(FSA). Especially for data of type T, optimum or nearly optimum solutions are obtained by FSM.

(2) We have 360 data to each of which an optimum solution is found by exhaustive search (see Table 1). For 345 data (95.83%) of them, FSM generates approximate solutions with errors (APP/OPT-1)×100≤10%.

(3) For data of type C, each algorithm produces better solutions for λ that is odd rather than even.

(4) In general, time(SMC)<time(FSA)<time(HBD)<time(FSM). For some data of type C with λ odd, time(HBD) is the longest (and other order is unchanged).

5.2.3. Unbounded approximation. There are examples for which FSA, FSM or SMC generates solutions such that APP/OPT fails to be bounded by constants. It is observed in our experimentation that FSM produces good solutions for data to which SMC gives unbounded solutions, and vice versa. This is why we propose HBD: it produces sharp approximation for these unbounded data.

5.2.4. The general (λ+δ)-ECA. We show a part of our experimental results in Table 3, where λ=1 and 1≤δ≤4.

Table 3. A part of our experimental results on (λ+δ)-ECA with λ=1 and 1≤δ≤4, where the column AP/OP shows the ratio APP/OPT, and other columns are similar to those of Table 2.

#	$|V|$	$|E'|$	λ	δ	FSA cost	FSA AP/OP	FSA time	FSM cost	FSM AP/OP	FSM time	SMC cost	SMC AP/OP	SMC time	HBD cost	HBD AP/OP	HBD time	OPT cost	OPT time
1001	10	11	1	1	43	1	5	43	1	5	43	1	4	43	1	5	43	24
				2	157	1	9	157	1	9	157	1	6	157	1	9	157	522
				3	312	1.065	13	293	1	12	293	1	9	293	1	11	293	6124
				4	537	1.051	17	537	1.051	16	537	1.051	13	522	1.022	16	511	35908
1002	10	10	1	1	30	1	5	30	1	5	35	1.167	3	30	1	5	30	22
				2	104	1.156	9	90	1	10	90	1	6	90	1	9	90	89
				3	236	1.103	12	228	1.065	13	228	1.065	9	228	1.065	13	214	988
				4	459	1.159	16	416	1.051	17	416	1.051	11	434	1.096	17	396	14407
1003	10	11	1	1	37	1	5	37	1	5	40	1.081	3	37	1	4	37	53
				2	103	1	9	120	1.165	8	106	1.029	6	120	1.165	8	103	144
				3	289	1.151	14	251	1	12	261	1.04	8	251	1	12	251	2331
				4	523	1.13	17	471	1.017	18	515	1.112	12	484	1.045	18	463	12041
1004	10	11	1	1	52	1	5	52	1	5	52	1	4	52	1	5	52	69
				2	151	1	9	151	1	8	151	1	7	151	1	9	151	1441
				3	283	1.105	12	283	1.105	11	283	1.105	9	283	1.105	13	256	2412
				4	505	1.107	16	505	1.107	16	555	1.217	12	505	1.107	16	456	9418
1005	10	10	1	1	10	1	5	10	1	6	10	1	4	10	1	6	10	5
				2	134	1.055	9	127	1	10	127	1	7	138	1.087	11	127	1189
				3	310	1.051	13	304	1.031	14	311	1.054	10	310	1.051	14	295	23412
				4	549	1.114	15	507	1.028	19	534	1.083	12	532	1.079	12	493	400030

6. Concluding Remarks

This paper proposed four approximation algorithms FSA, FSM, SMC and HBD for k-ECA, and both theoretical and experimental evaluation of their approximate solutions are given. The following (1) through (3) are left for future research:

(1) theoretical evaluation of HBD, which is being continued (our conjecture is APP/OPT≤2); (2) proposing approximation algorithms of better approximation for k-ECA; (3) providing more experimental results on k-ECA with k=λ+δ and δ≥2.

Acknowledgements

The research of T. Watanabe is partly supported by The Okawa Institute of Information and Telecommunication.

References

[1] P.M.Camerini, L.Fratta and F.Maffioli, A note on finding optimum branchings, Networks, 9, 309-312 (1979).

[2] Y-J Chu and T-H Liu, On the shortest arborescence of a directed graph, SCIENTIA SINICA, 14, 1396-1400 (1965).

[3] K.P.Eswaran and R.E.Tarjan, Augmentation problems, SIAM J.Comput, 5, 653-655 (1976).

[4] S.Even, Graph Algorithms, Pitman, London (1979).

[5] A.Frank, Augmenting graphs to meet edge connectivity requirements, Proc. 31st Annual IEEE Symposium on Foundations of Computer Science, 708-718 (1990).

[6] G.N.Fredericson and J.Ja'ja', Approximation algorithms for several graph augmentation problems, SIAM J.Comput., 10, 270-283 (1981).

[7] H.N.Gabow, Applications of a poset representation to edge connectivity and graph rigidity, Proc. 32nd IEEE Symp. Found. Comp. Sci., 812-821 (1991).

[8] H.N.Gabow, Z.Galil, T.Spencer and R.E.Tarjan, Efficient algorithms for finding minimum spanning trees in undirected and directed graphs, Combinatorica, 6(2), 109-122 (1986).

[9] Z.Galil and G.F.Italiano, Reducing edge connectivity to vertex connectivity, SIGACT NEWS, 22, 57-61 (1991).

[10] M.R.Garey and D.S.Johnson, Computers and Intractability: a Guide to the Theory of NP-Completeness, Freeman, San Francisco (1978).

[11] J.E.Hopcroft and R.E.Tarjan, Dividing a graph into triconnected components, SIAM J. Comput., 2, 135-158 (1973).

[12] A.V.Karzanov and E.A.Timofeev, Efficient algorithm for finding all minimal edge cuts of a nonoriented graph, Cybernetics, 156-162, Translated from Kibernetika, No.2, pp.8-12 (March-April, 1986).

[13] T.Mashima, S.Taoka and T.Watanabe, Approximation Algorithms for the k-Edge-Connectivity Augmentation Problem, IEICE of Japan, Tech. Reserch Rep., COMP92-24, 11-20 (1992).

[14] H.Nagamochi and T.Ibaraki, A linear time algorithm for computing 3-edge-connected components of a multigraph,Tech. Rep. #91005, Dept. of Applied Mathematics and Physics, Faculty of Engineering, Kyoto Univ., Kyoto Japan, 606 (1991).

[15] D.Naor, D.Gusfield and C.Martel, A fast algorithm for optimally increasing the edge-connectivity, Proc. 31st Annual IEEE Symposium on Foundations of Computer Science, 698-707 (1990).

[16] S.Taoka, T.Watanabe and K.Onaga, A linear time algorithm for computing all 3-edge-connected components of an multigraph, Trans. IEICE, E75-3, 410-424 (1991).

[17] R.E.Tarjan, Finding optimum branchings, Networks, 7, 25-35 (1977).

[18] R.E.Tarjan, Data Structures and Network Algorithms, CBMS-NSF Regional Conference Series in Applied Mathematics, SIAM, Philadelphia, PA (1983).

[19] S.Ueno, Y.Kajitani, and H.Wada, The minimum augmentation of trees to k-edge-connected graphs, Networks, 18, 19-25 (1988).

[20] T.Watanabe and A.Nakamura, Edge-connectivity augmentation problems, Journal of Computer and System Sciences, 35, 96-144 (1987).

[21] T.Watanabe, T.Narita and A.Nakamura, 3-Edge-connectivity augmentation problems, Proc. 1989 IEEE ISCAS, 335-338 (1989).

[22] T.Watanabe, M.Yamakado and K.Onaga, A linear-time augmenting algorithm for 3-edge-connectivity augmentation problems, Proc. 1991 IEEE ISCAS, 1168-1171 (1991).

[23] T.Watanabe, S.Taoka and T.Mashima, Approximation algorithms for the 3-edge-connectivity augmentation problem of graphs, IEEE Asia-Pacific Conference on Circuits and Systems 1992, to appear.

Principal Lattice of Partitions of submodular functions on graphs: Fast algorithms for Principal Partition and Generic Rigidity

Sachin Patkar[*]   H. Narayanan[†]

Abstract

In this paper we use a single unifying approach (which we call the Principal Lattice of Partitions approach) to construct simple and fast algorithms for problems including and related to the "Principal Partition" and the "Generic Rigidity" of graphs. Most of our algorithms are at least as fast as presently known algorithms for these problems, while our algorithm for Principal Partition problem (complete partition and the partial orders for all critical values) runs in $O(|E||V|^2 log^2|V|)$ time and is the fastest known so far.

1 Introduction

Many structural properties of a submodular function μ are revealed if we study the partitions of the underlying set E using whose blocks the lower Dilworth Truncation $(\mu - \lambda)_*$ of $(\mu - \lambda)$ (λ is a rational number) can be computed [6].

The partitions Π of E which have the following property: $\sum_{N \in \Pi}(\mu - \lambda)(N) = (\mu - \lambda)_*(E)$, form a lattice which we call the Dilworth Truncation Lattice (DTL) of $(\mu - \lambda)$. Further it can be shown that if $\lambda_1 > \lambda_2$ then the partitions in the DTL of $\mu - \lambda_1$ are finer than every partition in the DTL of $\mu - \lambda_2$. The collection of all such partitions (for all values of λ) is called the *Principal Lattice of Partitions* (PLP) of μ and is made up of the DTLs of $(\mu - \lambda)$ for a finite (and minimal) set of λ's called the critical values of the PLP. The maximum and the minimum partitions in the DTLs are called the critical partitions of the PLP.

The PLP arises naturally in the study of Electric Network analysis by decomposition [6]. For the case of graphs the well known problem of computing the *Principal Partition* [2, 3, 4, 5, 6, 8, 12] can be solved by computing the PLP of the function $-|\mathcal{E}(.)|$ (where $\mathcal{E}(U)$ denotes the set of edges having both their endpoints in the vertex subset U) [6, 8, 9]. (Let $f_k^w(X) = w(X) - k * r(X)$ where $r(.)$ is the rank function of the graph $G(V, E)$, $w(.)$ is a real positive weight assignment on E and k is a positive real number. The *Weighted Principal Partition Problem* is to find for each k, the

---

[*]Dept. of Computer Science and Engg., IIT Bombay, Bombay 400 076, INDIA
[†]Dept. of Electrical Engg., IIT Bombay, Bombay 400 076, INDIA

collection of all sets X which maximize f_k^w.) The problem of testing membership of a real vector $w(.)$ in the polyhedron of the cycle matroid of a graph is equivalent to that of finding a set that maximizes $w(.) - r(.)$ over the subsets of E and hence can be solved by the PLP approach. The DTL of the function $(2|\mathcal{V}(.)| - 3)$ (where $\mathcal{V}(X)$ is the set of endpoints of edge set X) arises naturally in the study of the rigidity of graphs [1, 2, 4, 7, 11]. We study this function as one of a whole class of functions for all of which we are able to provide efficient algorithms. (For want of space, the review of the literature is incomplete. More details may be found in [9].)

In this paper we study the partitions which yield the lower Dilworth Truncation of the functions

1. $(-|\mathcal{E}(.)| - \lambda)$ in order to compute the Principal Partition of a graph.

2. $r(.) - \lambda$ (where $r(.)$ is the rank function of a graph) in order to study generic rigidity and its generalisations.

In the present paper, using the PLP approach, we outline efficient algorithms for problems related to Principal Partition and Generic rigidity (due to lack of space the details are omitted, they may be found in [7, 8, 9]).

1. $O(|E||V|^2 log^2|V|)$ algorithm for the unweighted Principal Partition problem and $O(|E||V|^3 log|V|)$ algorithm for the weighted version of the Principal Partition for a graph (The previous best algorithms were given in [8]).

2. $(|E||V|^2 log|V|)$ algorithm for testing the membership of a real vector in the polyhedron of the cycle matroid of a graph.

3. $O(|E|^2|V| log^2|V|+|E||V|^3 log|V|)$ algorithm for computing the critical partitions in the PLP of the rank function of a graph.

4. $O(|E|^3)$ algorithm for computing the DTLs (in the PLP of the rank function) implicitly.

5. $O(k^3|V|^2)$ algorithms for finding the minimum and the maximum partitions in the DTL of $(k|\mathcal{V}(.)| - (2k - q))$.

As a byproduct of our techniques we get simple and fast algorithms for several sub-problems of the generic rigidity problem.

2   Preliminaries and Notation

We deal throughout with finite sets. The graphs we deal with are simple (no parallel edges and self loops). $\coprod$ denotes disjoint union. Let Π and Π' be two partitions of E. We say that $\Pi \succeq \Pi'$ if every block of Π' is completely contained in some block of Π (*i.e.* no block of Π' is broken in Π). In this case we call Π coarser than Π', or equivalently, Π' is finer than Π. Clearly $(\succeq)$ is a partial order. Using this partial order we define the following lattice operations on the collection of partitions. We denote by $\Pi \bigvee \Pi'$ ($\Pi \bigwedge \Pi'$) the finest (coarsest) partition coarser (finer) than Π and Π'. If a collection of partitions of E is closed under $\bigvee$ and $\bigwedge$ then it forms a lattice.

Let λ be a rational, we define $(f - \lambda)(X) = f(X) - \lambda$.

f is said to be submodular if $f(A) + f(B) \geq f(A \bigcup B) + f(A \bigcap B)....\forall\, A, B \subseteq E$. f is said to be supermodular iff $-f$ is submodular. The lower Dilworth Truncation of a function f is called f_* which is defined as

$$f_*(\emptyset) = 0 \text{ and } f_*(X) = \min \sum_{N \in \Pi} f(N) \; \forall\, X \neq \emptyset.$$

where the minimum is taken over all partitions Π of X. It is clear that a submodular function μ is equal to μ_* if $\mu(\emptyset) = 0$.

Since we are mainly dealing with submodular functions only, we may refer to lower Dilworth truncation by Dilworth Truncation without causing much ambiguity.

For the definition of a matroid and the related concepts we refer the reader to [13]. Let $r(.)$ denote the rank function on the edge set of a given graph $G(V, E)$. $G(V, E) \times Z$ denotes the graph obtained by contracting $E - Z$ from $G(V, E)$. $G(V, E) \bullet Z$ denotes the subgraph of $G(V, E)$ induced by Z. $|\mathcal{V}(.)| : 2^E \longrightarrow R$ is defined as $|\mathcal{V}(X)|$ = the number of vertices of the edges in X. $|\mathcal{E}(.)| : 2^V \longrightarrow R$ is defined as $|\mathcal{E}(U)|$ = the number of edges having both the vertices in U.

The following two theorems [6] speak of the existence of Dilworth Truncation Lattice and PLP.

Theorem 1 *Let $f : 2^E \longrightarrow R$ be submodular. Then the collection of partitions Π of E that minimize $\sum_{N \in \Pi} f(N)$ form a lattice. We call this lattice the DTL of the function f.*

Theorem 2 *Let $f : 2^E \longrightarrow R$ be submodular. Let λ and λ' be two constants and $\lambda > \lambda'$. Let Π and Π' be some partitions in the DTLs of $(f - \lambda)$ and $(f - \lambda')$ respectively, then $\Pi \preceq \Pi'$.*

3 PLP of the rank function of a graph

In this section we study the problem of computing the PLP of the rank function of a graph.

The PLP of rank function $r(.)$ of a graph $G(V, E)$ contains the partition Π_{bicon} of E into the edge sets of the biconnected components of the graph. In fact Π_{bicon} is the minimum partition in the DTL of $r(.)$. Further all the partitions of E, which are coarser than Π_{bicon}, also belong to the PLP of the rank function of the graph. So to compute the PLP of the rank function of a graph, it suffices to solve the problem on the biconnected components of the graph.

The following theorem [6] shows that for a biconnected graph $G(V, E)$, the PLPs of the rank function $r(.)$ of the graph and the function $(|\mathcal{V}(.)| - 1)$ (equivalently, the PLP of $|\mathcal{V}(.)|$) are identical.

Theorem 3 *If $G(V, E)$ is a biconnected graph then the DTLs of the functions $(r - a)$ and $(|\mathcal{V}(.)| - 1 - a)$ are identical.*

It can be shown [7, 9] that to compute the PLP of the rank function of a graph it suffices to compute the DTLs of $(k|\mathcal{V}(.)| - (2k - q))$ (equivalently, the DTLs of $|\mathcal{V}(.)| - \frac{2k-q}{k}$), for integers q and k such that $0 < q < k$.

3.1 Sketch Of The Main Algorithm

In this section we sketch the algorithm to find the maximum partition in the DTL of $(k|\mathcal{V}(.)| - (2k - q))$ or equivalently that of $|\mathcal{V}(.)| - \frac{2k-q}{k}$. The algorithm to find the minimum partition is similar.

Algorithm to compute the maximum partition in the DTL of $(k|\mathcal{V}(.)|-(2k-q))$.

Suppose we are given a partition, say $\Pi = \{P_1, P_2, ..., P_l\}$, which is known to be finer than the maximum partition in the DTL of $(k|\mathcal{V}(.)| - (2k - q))$. Let $\Pi'[i]$ denote the maximum partition in the DTL of the function $(k|\mathcal{V}(.)| - (2k - q))$ restricted to $P_1 \coprod ... \coprod P_i$. Suppose we have already computed $\Pi'[i]$, then we proceed to find $\Pi'[i+1]$ as follows:

Let $\mu(.)$ denote the function $(k|\mathcal{V}(.)| - (2k - q))$.

Consider the function $\sigma : 2^{\Pi'[i] \bigcup \{P_{i+1}\}} \longrightarrow \mathbb{R}$ defined as follows:

$$\sigma(W) = \mu(\coprod_{M_j \in W} M_j) - \sum_{M_j \in W} \mu(M_j) \quad \forall\, W \subseteq \Pi'[i] \bigcup \{P_{i+1}\}$$

Find the maximum Z that minimizes σ over the subsets of $\Pi'[i] \bigcup \{P_{i+1}\}$ containing P_{i+1}. W.l.o.g. let $Z = \{N_1, ..., N_p, P_{i+1}\}$, where $N_1, ..., N_p$ are some of the blocks of $\Pi'[i]$. [Note that p may be 0.] Then the partition, obtained by merging the blocks $N_1, ..., N_p, P_{i+1}$ of the partition $\Pi'[i] \bigcup \{P_{i+1}\}$ into a single block, is the required partition $\Pi'[i+1]$. We continue in the above manner to build the maximum partition in the DTL of $(k|\mathcal{V}(.)| - (2k - q))$.

The crucial subroutine in this algorithm is that of computing the maximum set Z that minimizes the submodular function σ over certain collection of subsets. The set Z is computed by finding the minimum cut with the largest $s - part$ [7, 9] in the flow network defined below.

The node set of the flow network
$= \{source \ (s), \ sink \ (t)\} \coprod (\Pi'[i] \bigcup \{P_{i+1}\}) \coprod \mathcal{V}(P_1 \coprod P_2 ... \coprod P_{i+1})$. And the arcs along with their capacities are as follows:
$cap(s, M) = k|\mathcal{V}(M)| - (2k - q) \ ... \ \forall M \in \Pi'[i]$.
$cap(s, P_{i+1}) = k|\mathcal{V}(P_{i+1})|$.
$cap(u, t) = k \ ... \ \forall u \in \mathcal{V}(P_1 \coprod ... \coprod P_{i+1})$.
$cap(M, u) = \infty \ ... \ \forall M \in \Pi'[i] \bigcup \{P_{i+1}\}$ and $\forall\, u \in \mathcal{V}(M)$.

Thus to find the maximum partition in the DTL of $(k|\mathcal{V}(.)|-(2k-q))$ we need to solve $|\Pi|$ (where $\Pi = \{P_1, ..., P_l\}$) flow maximization problems on the flow networks of the type described above. It can be shown [9] that in each such flow network the number of arcs is bounded by $O(|\Pi| + |V|)$ and the length of any undirected source-sink path is bounded by $O(|V|)$. Thus by [10], the time complexity of computing the maximum partition in the DTL of $(k|\mathcal{V}(.)| - (2k - q))$ is bounded by $O(|\Pi||V|(|\Pi| + |V|)log|V|)$.

An alternative approach to compute the maximum partition is as follows: In the above scheme we can take $\Pi = \{\{e\} \mid e \in E\}$ (i.e. the singleton partition of E). We then need to solve a sequence of flow maximization problems. It may be noted that these networks are obtained from the previous ones via a slight modification. This fact allows us to use the maximum flow in a flow network to get an initial flow in the next flow network. Also, using some technical lemmas we can bound the number of flow augmentations required throughout the algorithm, as well as the number of arcs in such flow networks by $O(k^2|V|)$ and $O(k|V|)$, respectively. Thus we have a different time bound, *viz* $O(k^3|V|^2)$, for computing the maximum partition in the DTL of $(k|\mathcal{V}(.)| - (2k - q))$. Due to lack of space we skip the details, which may be found in [7, 9].

In the next sections we outline our approach for computing the PLP of the rank function of a graph. The details may be found in [7, 9].

3.2 Computation of all critical partitions in the PLP of the rank function of a graph

As noted before, to compute the PLP of the rank function of a graph it suffices to compute the PLP of the function $|\mathcal{V}(.)|$. In this section we outline the scheme which we use to compute all the critical partitions in the PLP of $|\mathcal{V}(.)|$. We can find all the critical partitions in $O(|E|^2|V|log^2|V| + |E||V|^3log|V|)$ time.

We use the following notation:

Notation: Let Π^λ and Π_λ denote the maximum and the minimum partitions in the DTL of $|\mathcal{V}(.)| - \lambda$.

It can be seen that the critical values in the PLP of the rank function of a graph are rationals with the numerator bounded by $O(|E||V|)$ and the denominator is bounded by $O(|E|)$. So there are $O(|E|^2|V|)$ possible critical values of λ.

Let Π be a partition of E in the PLP of $|\mathcal{V}(.)|$ such that $\Pi \leq \Pi_\lambda \leq \Pi^\lambda$. We have already given a sketch of the subroutine which computes Π_λ and Π^λ given a ("*basis*") partition Π which is known to be finer than Π_λ, in time polynomial in $|\Pi|$ and $|V|$. In fact, as remarked above the worst case complexity of this subroutine can be shown to be $O(|\Pi||V|(|\Pi| + |V|)log|V|)$. We will now sketch our method for finding all the critical partitions using the above subroutine.

1. Start with the singleton partition Π_0 of E. Note that $|\Pi_0| = |E|$.

2. Locate the critical value λ such that $|\Pi^\lambda| \leq \frac{|\Pi_0|}{2}$ and $|\Pi_\lambda| > \frac{|\Pi_0|}{2}$, by executing the subroutine that computes Π^λ and Π_λ for a given λ using the singleton partition Π_0 as the basis partition.

3. Find all the critical partitions that are coarser than Π^λ using the same bisection approach and the subroutine for computing the maximum and the minimum partitions with Π^λ as the "basis" partition.

4. Find all the critical partitions that are finer than Π_λ. To solve this, it suffices to find for each block N of Π_λ the critical partitions in the PLP of the function $|\mathcal{V}(.)|$ restricted to N. For each of these subproblems, we use the singleton partitions of the respective blocks as the "basis" partitions.

Observe that the above bisection technique reduces the size of the "basis" partition to half. But the complexity of the basic subroutine that computes the maximum and the minimum partitions also depends on $|V|$, *i.e.*, the size of the vertex set which does not change. So we continue with the above bisection technique as long as the size of the "basis" partition dominates $|V|$, otherwise we make use of a different method (described in [6]) to compute the critical partitions. It can be shown with the help of a few technical lemmas that the complexity of computing all the critical partitions by this method is bounded by $O(|E|^2|V|log^2|V| + |E||V|^3log|V|)$ (The details may be found in [9]).

3.3   Computation of DTLs in the PLP of the rank function of a graph

Recall that Π^λ and Π_λ are the maximum and the minimum partitions in the DTL of $|\mathcal{V}(.)| - \lambda$. And λ is a critical value iff $\Pi^\lambda \neq \Pi_\lambda$. We would like to find all the partitions in the DTL of $|V(.)| - \lambda$ for a critical λ, but there may be exponentially many of them. So we use the idea of building a hypergraph which helps in describing these partitions implicitly. This idea is borrowed from [6]. Our implementation of this idea yields a fast algorithm which computes the whole DTL implicitly in $O(|E|^3)$ time (The details may be found in [9]).

3.4   Some applications: Generic Rigidity

There are several ideas which are useful for solving optimization problems on certain submodular functions of the type μ_*. Also, using the minimum partition in the DTL we can describe the circuits when μ_* is a rank function of a matroid M on E. For the case when $\mu(X) = k|\mathcal{V}(X)| - (2k - q)$ the fundamental functions [2] of the polymatroid μ_* can be computed easily. In particular, when $\mu(.) = k|\mathcal{V}(.)| - (2k - 1)$, it can be easily seen that μ_* is a matroid rank function, and simple (and fast) procedures can be given for computing fundamental circuits, closure of a set, rank etc. based on the following theorems.

Theorem 4 *Let $\mathcal{M}$ be a matroid with the rank function $(k|\mathcal{V}(.)| - (2k - 1))_*$. Let $X \subseteq E$. Let Π be the maximum partition in the DTL of $k|\mathcal{V}(.)| - (2k - 1)$ restricted to X. Then the closure of X in the matroid $\mathcal{M}$, denoted by, $cl(X) = \coprod_{N_i \in \Pi} \mathcal{E}(\mathcal{V}(N_i))$.*

Theorem 5 *Let $\mathcal{M}$ be a matroid with the rank function $(k|\mathcal{V}(.)| - (2k - 1))_*$. Let $X \subseteq E$ be independent in $\mathcal{M}$ and $e \notin X$ be such that $X + e$ is dependent in $\mathcal{M}$. Let Π be the minimum partition in the DTL $k|\mathcal{V}(.)| - (2k - 1)$ restricted to $X + e$. Then the only nonsingleton block of Π is the fundamental circuit (of the matroid $\mathcal{M}$) created by the introduction of e in X.*

Thus several matroids can be studied through PLP of the rank function of a graph. The most important being the generic rigidity case [7, 9].

Theorem 6 *For integral $k \geq 1$, $(k|\mathcal{V}(.)| - (2k - 1))_*$ is a matroid rank function. In particular $(2|\mathcal{V}(.)| - (3))_*$ is a matroid arising out of the generic rigidity problem, and is called a "planar rigidity matroid".*

The problem of generic rigidity [1, 2, 11].

A graph $G(V, E)$ is said to be **rigid** (in a plane) iff it has at least one realization as a bar and joint framework in the plane (A bar and joint framework is a collection of rigid bars connected with universal joints) wherein it resolves all static equilibrium loads (equivalently, a realization in which it has no nontrivial infinitesimal motions). A rigid graph is said to be **redundant rigid** iff for every edge, its removal retains the rigidity. A graph is said to be (generic) **independent** iff there does not exist any redundant rigid subgraph of it.

It has been described in [9] how our techniques result in fast algorithms for the following problems arising in the study of rigidity,
(1) Testing generic independence and minimal rigidity. (2) Testing rigidity and redundant rigidity. (3) Computing the decomposition of a graph into maximal rigid

subgraphs. (4) Computing the decomposition of a graph into connected components with respect to the matroid rank function $(2|V| - 3)_*$. In fact, this completely solves the problem of redundant rigidity. (5) Computing the Henneberg sequence [9] of a minimally rigid graph. (6) Representing the family of all rigid subgraphs of a minimally rigid graph.

Our algorithms for these problems run in $O(|V|^2)$ time except for the problem of representing all the rigid subgraphs of a minimally rigid subgraph (This problem is slightly different from the one considered in [4]), which runs in $O(|V|^3)$ time. These are the fastest algorithms known for the above problems except for the first. (For the details see [7, 9]).

It can also be shown [7, 9] that the PLP of the rank function of a graph has attractive structural properties which provide useful information on the presence of long cycles, cliques and, in general, dense subgraphs [7, 9].

4 Improved time bounds for computation of the Principal Partition of a graph

In this section we improve the time bounds for finding the Principal Partition of a graph $G(V, E)$. We use an approach which is based on finding a suitable partition of vertices rather than a set of edges. In fact, we compute the PLP of the submodular function $-|\mathcal{E}(.)|$, which is defined on the set of vertices. We present an algorithm which computes the Principal Partition in $O(|E||V|^2 log^2|V|)$ time for an unweighted graph and in $O(|E||V|^3 log|V|)$ time for an arbitrarily weighted graph. [The fastest known earlier algorithm [8] has a complexity $O(|E|^2|V|log|V|)$ for the unweighted case and $O(|E|^2|V|^2 log|V|)$ for the weighted case.]

We begin by stating the Weighted Principal Partition Problem:

Let $f_k^w(X) = w(X) - k*r(X)$ where $r(.)$ is the rank function of the graph $G(V, E)$, $w(.)$ is a real positive weight assignment on E and k is positive real number. The *Weighted Principal Partition Problem* is to find for each k, the collection of all sets X which maximize f_k^w. For the sake of convenience, we will discuss our approach only for the unweighted case, i.e. for $w(.) = |.|$. The same approach with obvious modifications will work for the arbitrary weight case.

Let $f_k(X) = |X| - k*r(X)$ where $r(.)$ is the rank function of a graph and k is a positive real. The *Principal Partition Problem* is to find for each k, the collection of all sets X which maximize f_k. Note that f_k is a supermodular function when $k \geq 0$. Let X_k denote the minimum set that maximizes f_k and X^k denote the maximum set that maximizes f_k.

4.1 A scheme for construction of X_k and X^k

In this section we describe a scheme for construction of the sets X_k for a graph $G(V, E)$. A minor modification of this scheme will be adequate for construction of X^k. Also note that it suffices to solve this problem on the connected components of the underlying graph.

For the further discussion in this paper we use the following notation.

Let the vertex set V be (arbitrarily) ordered as $v_1, v_2, ..., v_n$.

Then V^i denotes $\{v_1, v_2, ..., v_i\}$.

$X_k[i]$ denotes the minimum set that maximises f_k for $G(V^i, \mathcal{E}(V^i))$.

$G_k[i]$ denotes the graph $G(V^i, \mathcal{E}(V^i)) \times (\mathcal{E}(V^i) - X_k[i-1])$.

With respect to the function $w(.)-k*r(.)$, the flow graph $FLG(G(U, Z), u, k, w(.))$ corresponding to a graph $G(U, Z)$ and a vertex $u \in U$, is defined as follows:

Let s and t denote the source and sink, respectively.

The node set of the network $= \{s, t\} \coprod U \coprod Z$.

$$\text{The arc set} = \{(s, v) \mid v \in U\} \coprod \{(e, t) \mid e \in Z\}$$
$$\coprod \{(v, e) \mid v \in U, \ e \in Z \text{ and } e \text{ incident on } v \text{ in } G(U, Z)\}$$

The capacities are

$cap(s, v) = k \ ....\forall \ v \neq u. \ cap(s, u) = 0. \ cap(v, e) = \infty. \ cap(e, t) = w(e).$

Any cut partitions the node set of the flow network in two parts. The set of nodes containing the source (sink) is called s-part (t-part) of the cut.

For the unweighted case, we take $w(.) = |.|$.

It was shown in [8] that $X_k[i] = X_k[i-1] \coprod Y$, where Y is the minimum set that maximises $|.| - k * \bar{r}(.)$ ($\bar{r}(.)$ is the rank function of $G_k[i]$) over the subsets of $\mathcal{E}(V^i) - X_k[i-1]$. Also it was shown in [8] that the above set Y can be found using the following lemma.

Lemma 1 $Y \subseteq (\mathcal{E}(V^i) - X_k[i-1])$ *is the minimum set that maximises* $|.| - k * \bar{r}(.)$ *over the subsets of* $\mathcal{E}(V^i) - X_k[i-1]$ *iff* Y *is the minimum set among the sets of the type* $T \bigcap (\mathcal{E}(V^i) - X_k[i-1])$ *where* T *is the t-part of some minimum cut in the flowgraph* $FLG(G_k[i], v_i, k, |.|)$.

Using the above ideas, it is easy to show that (the details may be found in [9]), to compute X_k we need to solve $O(r(E))$ flow maximization problems on flow networks of the type $FLG(G(U, Z), u, k, |.|)$. We use Sleator and Tarjan's [10] implementation of Dinic's layered network algorithm for solving the flow maximisation problem, which requires $O(|A|log|A|)$ (where A denotes the set of arcs in the underlying network) time for every stage of the layered network approach. Note that the number of stages of the layered network algorithm is bounded by $O(r(E))$ in the network $FLG(G(U, Z), u, k, |.|)$. This yields an $O(|E|r(E)^2 log|E|)$ algorithm for computing X_k.

Also, it can be shown that the supernodes of the graph $G_k[n]$ represent the blocks of the *minimum* partition of the vertex set V which minimises the function

$$g(\Pi) = \sum_{U \in \Pi} (k(|U| - 1) - |\mathcal{E}(U)|)$$

over the partitions of V.

4.2   Computation of the critical sets

We now consider the problem of finding X^λ and X_λ for all the critical values λ in the Principal Partition of the given unweighted graph $G(V, E)$. Clearly in the unweighted case the critical values are rationals with the numerator bounded by $|E|$ and the denominator bounded by $r(E)$. We adopt the following approach to find all the critical sets.

Using binary search, locate a λ such that $r(X_\lambda) < \frac{r(E)}{2}$ and $r(X^\lambda) \geq \frac{r(E)}{2}$. This involves solving $O(log(|V|))$ problems of maximizing f_λ $(= |.| - \lambda * r(.))$ for different λ's. Time required for the above computation is $O((log|V|)(|E|r(E)^2log|V|))$.

Now, we split the problem of computing the remaining critical sets into two subproblems each of size not greater than $(\frac{r(E)}{2})$ as follows:

- Contract X^λ and repeat the above procedure of locating an appropriate rational on the contracted graph.

- Repeat the above procedure of locating an appropriate rational on the subgraph on X_λ.

Continue the bisection as described above to get all the X^λs. Delete repetitions to get the critical sets. It can be shown that [9],

Theorem 7 *Total time required for computing all the critical sets is bounded by* $O((log^2|V|) * (|E|r(E)^2))$.

Once we know all the critical sets, the critical values can be found as follows. Order the critical sets under inclusion. Let $Y \subset Z$ be a pair of consecutive critical sets in this order. Find λ such that $|Y| - \lambda * r(Y) = |Z| - \lambda * r(Z)$. Such a λ is a critical value. Also $X_\lambda = Y$ and $X^\lambda = Z$. All the critical values may be obtained in this manner.

4.3 Computation of the Preorder

As is well known [3, 5, 8, 9, 12], all the sets in the Principal Partition of a graph can be represented by suitably defined preorders for each critical value. And the principal ideals of these preorders can be found using simple network flow techniques [8, 9]. Also it can be shown that [9] the time complexity of computing the preorders is dominated by that of computing the critical sets. Thus the complexity of our algorithm to compute the Principal Partition for the unweighted case is dominated by the computation of all the critical sets, which is $O(|E|r(E)^2log^2(|V|))$. Minor modifications in the above technique yield an $O(|E||V|^3log|V|)$ algorithm to construct weighted Principal Partition of a graph.

5 Conclusion

The algorithms outlined in this paper for Principal Partition and rigidity problems are all simple and efficient (Indeed, for Principal Partition of a graph ours is the fastest of known algorithms). They are all based on the PLP approach and involve maintaining maximum or minimum partitions in the DTL of the restrictions of $(\mu - \lambda)$ to subsets of E which enables us to limit the number of steps in the algorithm. Similar ideas can be used in general to bound the number of steps in other algorithms and improve the complexity.

References

[1] Gabow, H.N. and Westermann, H.H.: Forests, Frames and Games: Algorithms for Matroid sums and Applications, *in Proc. 20th STOC*, 1988, pp. 407-421.

[2] Imai, H.: Network flow algorithms for lower truncated transversal polymatroids, *J. of the Op. Research Society of Japan*, vol. 26, 1983, pp. 186-210.

[3] Iri, M. and Fujishige, S.: Use of Matroid Theory in Operations Research, Circuits and Systems Theory, *Int. J. Systems Sci.*,vol. 12, no. 1, 1981, pp. 27-54.

[4] Nakamura, M.: On the Representation of the Rigid Sub-systems of a Plane Link System, *J. Op. Res. Soc. of Japan*, vol. 29, No. 4, 1986, pp. 305-318.

[5] Narayanan, H.: *Theory of Matroids and Network Analysis*, Ph.D. thesis, Department of Electrical Engineering, IIT Bombay, INDIA, 1974.

[6] Narayanan, H.: The Principal Lattice of Partitions of a Submodular function, *Linear Algebra and its Applications*, 144, 1991, pp. 179-216.

[7] Patkar, S. and Narayanan, H.: Principal Lattice of Partitions of the Rank Function of a Graph, *Technical Report VLSI-89-3*, IIT Bombay, INDIA, 1989.

[8] Patkar, S. and Narayanan, H.: Fast algorithm for the Principal Partition of a graph, *in Proc. 11th ann. symp. on Foundations of Software Technology and Theoretical Computer Science (FST & TCS-11), LNCS-560*, 1991, pp. 288-306.

[9] Patkar, S.: *Investigations into the structure of graphs through the Principal Lattice of Partitions approach*, Ph.D. thesis, Dept. of Computer Sci. and Engg., IIT Bombay, INDIA, 1992.

[10] Sleator, D.D. and Tarjan, R.E.: A data structure for dynamic trees, *J. Comp. and System Sci.*, vol. 26, 1983, pp. 362-391.

[11] Sugihara, K.: On Redundant Bracing in Plane Skeletal Structures, *Bulletin of the Electrotechnical Laboratory*, vol. 44, 1980, pp. 376-386.

[12] Tomizawa, N.: Strongly Irreducible Matroids and Principal Partition of a Matroid into Strongly Irreducible Minors (in Japanese), *Transactions of the Institute of Electronics and Communication Engineers of Japan*, vol. J59A, 1976, pp. 83-91.

[13] Welsh, D. J. A.: *Matroid Theory*, Academic Press, New York, 1976.

The Application of the Searching over Separators Strategy to Solve Some NP-Complete Problems on Planar Graphs

R. Z. Hwang and R. C. T. Lee

Department of Computer Science, National Tsing Hua University, Hsinchu, Taiwan, 30043, Republic of China.

Abstract. Recently, we proposed a new strategy for designing algorithms, called the searching over separators strategy. We applied this approach to solve some famous NP-Complete problems in subexponential time such as the discrete Euclidean P-median problem, the discrete Euclidean P-center problem, the Euclidean P–center problem and the Euclidean traveling salesperson problem. In this paper, we further extend this strategy to solve two well known NP-Complete problems, the planar partition–into–clique problem (PCliPar) and the planar steiner tree problem (PStTree). We propose $O(n^{O(\sqrt{n})})$ algorithms for both problems, where n is the number of vertices in the input graph.

1 Introduction

In [8], Johnson listed a summary table of eleven famous problems. Six of them remain NP-complete even when they are defined on planar graphs. The six problems are the planar maximum independent set problem (PIndSet), the planar partition– into–clique problem (PCliPar), the planar chromatic number problem (PChrNum), the planar Hamiltonian circuit problem (PHamCir), the planar maximum dominating set problem (PDomSet) and the planar steiner tree problem (PStTree) [8, 5].

In [10], the authors proposed a planar separator theorem. This method uses a set of vertices to partition a planar graph into two parts. They applied this theorem to solve the PIndSet problem and the PChrNum problem in $O(2^{O(\sqrt{n})})$ time complexity where n is the number of vertices on the planar graph. They also claimed that the similar method can be used to solve the problem which can be formulated as a nonserial dynamic programming problem [4]. Therefore the PDomSet problem can be solved in a similar way.

Miller invented a simple cycle separator theorem which uses a simple cycle to partition a planar graph into two parts [12, 13]. By applying the simple cycle separator theorem, Hwang, Lee and Chang proposed the searching over separators strategy [7]. They solved the Euclidean traveling salesperson problem (ETSP) [9] and the PHamCir problem in $O(n^{O(\sqrt{n})})$ time complexity, and solved the Euclidean P-center problem and the Euclidean discrete P-median problem [5, 11] in $O(n^{O(\sqrt{P})})$ time complexity.

In this paper, we shall show that the PCliPar problem and the PStTree problem can be solved in $O(2^{O(\sqrt{n})})$ and $O(n^{O(\sqrt{n})})$ time complexities, respectively.

In the next section, we will introduce the searching over separators strategy. In Section 3, we will solve the PCliPar problem and the PStTree problem is solved in Section 4. Conclusions are given in the last section.

2 The Searching over Separators Strategy

Before we present the searching over separators strategy, we first introduce the simple cycle separator theorem [12] . For a comprehensive discussion of this topic, consult [13] .

Suppose that we are given a planar graph G with non–negative weights assigned to vertices, faces and edges which sum to 1. For our case, we may simply assume that the weights of faces and edges are all zeros. In other words, weights are assigned only to vertices. For a simple cycle B of G, the size of this cycle is the number of vertices on B. The weight of the interior part (the exterior part) is the sum of weights of vertices in the interior part (the exterior part).

The theorem proved by Miller [12] is now stated as follows:

Theorem 1 [12]. *If G is a 2-connected planar graph with all non-negative weights assigned to vertices which sum to m, there exists a simple cycle, called the simple cycle separator, of size at most $2\sqrt{2\lceil d/2 \rceil n}$, dividing the graph into interior and exterior two parts, such that the sum of the weights in each part is no more than $2m/3$, where d is the maximum face size (The face size is the number of edges contained in the facial cycle.) and n is the number of vertices in G. This cycle can be found in linear time.*

In our case, we are interested in the maximal planar graph [13] , where d is equal to 3. (A maximal planar graph is a planar graph in which every face is of size 3.) All maximal planar graphs are 2-connected. Hence the size of the separator is $2\sqrt{2n} = \sqrt{8n}$.

Now we begin to describe our searching over separators strategy. This strategy is similar to the well known divide–and–conquer strategy for designing efficient algorithms [1, 14, 6, 2, 3] . Many problems can be solved by the divide–and–conquer strategy [6] . It usually yields efficient polynomial algorithms. Unfortunately, not every problem can be solved efficiently by the divide–and–conquer approach. One of the reasons is that we can not easily divide the input data into two unrelated subsets, such that the two subproblems with these two subsets as inputs can be solved independently and the solutions later merged into an optimal solution.

In this paper, we shall point out that there may exist planar graph problems with the following properties:

(1) The problem is an NP-complete problem. Thus it is quite unlikely that it can be solved by the divide–and–conquer strategy directly.

(2) On the other hand, for any optimal solution S and any simple cycle separator B, we can use B to divide the input data into two parts A_d and C_d. Thus the final solution can be obtained by solving the two subproblems with A_d and C_d as inputs and merging the two optimal sub–solutions.

If a problem has the above properties, then although it can not be solved by the divide–and–conquer strategy, it can be solved by the searching over separators strategy.

Let us further assume that our problem is an optimization problem, and we are looking for an optimal solution with the minimum cost. The searching over separators strategy for solving planar graph problems works as follows:

The Searching over Separators Strategy:

Input: A planar graph problem D.

Output: An optimal solution S and its cost C.

Step 1:   Let C:=∞.

Step 2:   Find a maximum planar graph G' of the input planar graph.

Step 3:   Find a simple cycle separator B of G'. For B, do:

Step 4:   Divide Problem D into all possible pairs of two independent subproblems corresponding to B.

Step 5:   For each pair of two independent subproblems (A_d, C_d), do:

Step 5.1:   Recursively solve A_d and C_d respectively. Let the solutions be A_s and C_s respectively.

Step 5.2:   Merge A_s and C_s into S'. Let C' be the cost associated with S'.

Step 5.3:   If $C > C'$ then S:=S', C:=C'.

Step 6:   Return solution S as an optimal solution and C the optimal cost.

In order for the above searching over separators strategy to work. One of the pairs of the subproblems must correspond to an optimal solution. If it does, the searching over separators strategy is guaranteed to have examined an optimal solution and present it as a solution correctly. The above presented strategy is slightly different from that presented in [7], because the problems are now planar graph problems which those in [7] are all Euclidean problems. For more about the searching over separators strategy, consult [7] .

3   The Planar Partition–into–Clique Problem

The planar partition–into–clique problem (PCliPar) is defined as follows:

PCliPar: Given a planar graph $G(V, E)$, find the minimum number K, such that the planar graph G can be partitioned into K disjoint cliques $Q_1, Q_2, \ldots, Q_k$.

For example, for the planar graph in Fig.3.1 , the smallest number of cliques that it can be divided into is four. An optimal solution is shown in Fig.3.2.

In the following, we shall explain why we can apply the searching over separators strategy. Consider Fig.3.1 again. Suppose we construct a maximal planar graph out of the planar graph in Fig.3.1, as that illustrated in Fig.3.3. In Fig.3.3, we also display an optimal solution of the PCliPar problem which was originally displayed in Fig.3.2. Now suppose we have a simple cycle separator B of this maximal planar graph, namely j,k,m,l,d,c,b,j, as shown in Fig.3.4. This separator B will divide the original input vertices into three parts: V_a, V_b, and V_c, where V_a (V_c) is the set of vertices inside (outside) of B and V_b is the set of vertices on B. In the case in Fig.3.4, $V_a = \{f, e, g\}$, $V_b = \{j, k, m, l, d, c, b\}$ and $V_c = \{a, h, i, n\}$. Furthermore, since we have an optimal solution, we may further divide V_b into the interior V_{ba} and the exterior V_{bc} two parts. For each vertex v in V_b which is in the same clique with a vertex in V_c (V_a) in the optimal solution, we assign v to V_{bc} (V_{ba}); otherwise we assign v to either V_{bc} or V_{ba}. Consider vertex m in Fig.3.3. Since it is in the same clique with vertex n and n belongs to V_c, in Fig.3.4, we assign m to V_{bc}. After partitioning all vertices in V_b, in this case, we have $V_{bc} = \{j, k, m, l, d, c, b\}$, and $V_{ba} = \emptyset$. Let $V'_a = V_a \cup V_{bc}$, and $V'_c = V_c \cup V_{bc}$. Now, we have two independent PCliPar subproblems as shown in Fig.3.4. If we solve these two independent subproblems and merge the solutions, it is obvious that we shall obtain the original optimal solution.

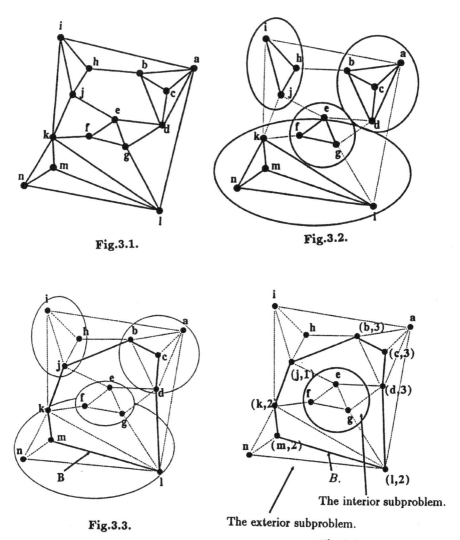

Fig.3.1.

Fig.3.2.

Fig.3.3.

The interior subproblem.

The exterior subproblem.

Fig.3.4.

But we do not have an optimal solution to start with; otherwise, we do not have to work so hard to solve the problem. How can we divide the original problem ? The searching over separators strategy essentially tries all possible methods to partition the data based upon the simple cycle separator and we shall show later that the approach is correct.

In the above partition, we ignore the case that vertex v is in the same clique with vertex V_a and vertex V_c at the same time. In the following theorem, we will show that this case will never happen. Limited by space, we omit this proof.

Theorem 2. *Given a maximal planar graph $G(V, E)$ and a simple cycle separator B of G, if v_1 and v_2 belong to the same clique in a solution of the PCliPar problem, then B will not separate v_1 and v_2 in such a way that v_1 is inside of B and v_2 is outside of B.*

Given a maximal planar graph $G(V, E)$, an optimal solution of the PCliPar problem defined on G and a simple cycle separator B, B divides the vertices in G into three parts, the interior part V_a consisting of all vertices inside B, the exterior part V_c consisting of vertices outside B and V_b the vertices on B. We may use B to divide G according to the following rules:

(1) All vertices in V_a (V_c) are assigned to $G_a(G_c)$.

(2) Let v be a vertex in V_b. Let v belong to a clique q in the optimal solution. If there exists some other vertex in q belonging to $V_a(V_c)$, then assign v to $G_a(G_c)$; otherwise assign v and all vertices in q to G_a.

(3) Assign $\overline{v_1v_2}$ to $G_a(G_c)$ if and only if both v_1 and v_2 belong to $G_a(G_c)$ and $\overline{v_1v_2} \in E$.

Theorem 3. *Given a planar graph $G(V, E)$, an optimal solution and a simple cycle separator B, if we solve two PCliPar subproblems with G_a and G_c as inputs independently, then the union of the two solutions is an optimal solution of the PCliPar problem with G as the input.*

Limited by space, we omit the proof. In the above discussion, we assume that an optimal solution with G as the input is known. Therefore we can distribute the vertices (the edges) in B into either G_a or G_c $(G_a, G_c$ or discarded). However we do not have any optimal solution clique. In the algorithm applying the searching over separators strategy, we generate all possible distributions of the vertices and the edges on B by the following procedures.

Procedure DIST-VERTICES(V_b, Γ)

Input : V_b : a set of vertices.

Output: Γ : a set of pairs of (V_a', V_c'), initialized as empty.

Step 1:   If there is no vertex in V_b, then $\Gamma = \Gamma \cup (V_a', V_c')$ and return.

Step 2:   Select a vertex v from V_b and remove v from V_b, then do:

Step 2.1:   Assign v to V_a' and Call DIST-VERTICES(V_b, Γ).

Step 2.2:   Remove v from V_a'.

Step 2.3:   Assign v to V_c' and Call DIST-VERTICES(V_b, Γ).

Procedure DIST-EDGES(E_b, V_a', V_c', E_a', E_c')
Input : E_b, V_a', V_c' : a set of edges.
Output: E_a', E_c' : two sets of edges, initialized as empty.
Step 1:   If $\overline{v_1 v_2} \in E_b$ and $v_1, v_2 \in V_a'(V_c')$, we assign $\overline{v_1 v_2}$ to E_a (E_c) and remove $\overline{v_1 v_2}$ from E_b.

Now we can solve the PCliPar problem as the following algorithm:

Algorithm A(G, F)
Input :   A planar graph $G(V, E)$.
Output: F: An optimal solution of the PCliPar problem with G as input.
Step 1:   If $|V| < 3$, solve this problem by exhaustive enumeration.
Step 2:   Find a maximal planar graph G' of G, by adding dummy edges.
Step 3:   Find a simple cycle separator B on G'.
Step 4:   Use B to divide the vertices in G into V_a, V_b and V_c, where V_a (V_c) is in the interior (exterior) part of B and V_b is the set of vertices on B.
Step 5:   Use B to divide the edges in G into E_a, E_b and E_c, where E_a(E_c) is in the interior (exterior) part of B and V_b is the set of edges on B.
Step 6:   Call Procedure DIST-VERTICES(V_b, Γ).
Step 7:   For each $\Gamma = (V_a', V_c')$, call Procedure DIST-EDGES(E_b, V_a', V_c', E_a', E_c').
Step 7.1:   Let $G_a = (V_a \cup V_a', E_a \cup E_a')$ and $G_c = (V_c \cup V_c', E_c \cup E_c')$.
Step 7.2:   Call A(G_a, F_a) and A(G_c, F_c).
Step 7.3:   Let F' be the union of F_a and F_c. If the number of cliques in F' is less than that in F, let F be F'.

The correctness of Algorithm A is due to Theorem 3. It can be proved that the above algorithm can be performed in $O(2^{O(\sqrt{n})})$.

4   The Planar Steiner Tree Problem

The PStTree problem is defined as follows:
   PStTree : Given a planar graph $G(V, E)$, a cost $c(e)$ for each $e \in E$ and a subset $R \subset V$, find a subtree $T_s(V_s, E_s)$ of G such that $R \subset V_s$ and the sum of costs of all edges in E_s is minimized.
   In the rest of the paper, we call the vertices in R the fixed vertices. The following simple example illustrates this problem. In Fig.4.1, there is a planar graph G with costs defined on edges. A solution tree of the PStTree problem with G as input is shown in Fig.4.2.
   Now consider Fig.4.3 which shows a simple cycle separator. We may use this simple cycle separator and the optimal solution shown in Fig.4.2 to divide the original problem into two subproblems. There is a difficulty here. Consider the exterior problem instance produced by the simple cycle separator. If a straightforward method is used, it will look like that shown in Fig.4.4. As it can be seen, the graph is not connected any more. But our original PStTree problem is defined on connected planar graphs.

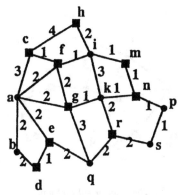

Squares representing the vertices in
$V - R$ and circles representing the
vertices in R.
Fig.4.1.

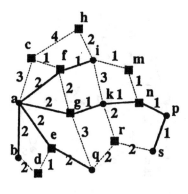

Thick lines representing the edges in
the solution tree.
Fig.4.2.

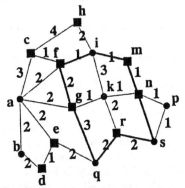

Thick lines representing the edges
on the simple cycle separator B.
Fig.4.3.

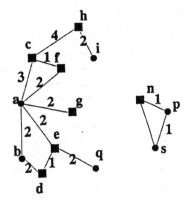

Fig. 4.4.

Our first trick is to preserve the interior (exterior) part when we define the exterior (interior) part. Thus the exterior and interior subproblems produced by using the simple cycle separator will be as shown in Fig.4.5 (a) and Fig.4.5 (b) respectively. Note that for the exterior subproblem, the interior edges, $\overline{fi}, \overline{gk}$ and $\overline{kn}$ must be included. Similarly, for the interior subproblem, the exterior edges, $\overline{af}, \overline{fi}, \overline{ag}, \overline{np}$, $\overline{ps}, \overline{ae}, \overline{eq}$, and $\overline{ab}$ must be included. It is obvious that after we solve these subproblems and merge the subsolutions, we would obtain the original optimal solution.

The above discussion is based upon the assumption that an optimal solution

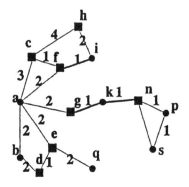

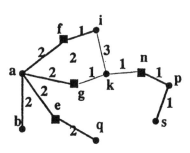

(a) The exterior subproblem.   (b) The interior subproblem.

Fig. 4.5.

is available. In reality, we of course do not have it. Therefore, we have to perform guessing. For each pair of vertices u and v on the simple cycle separator, there are only the following possibilities:

(1) The optimal solution connects u and v and the path of edges connecting u and v is outside of the simple cycle separator.

(2) The optimal solution connects u and v and the path of edges connecting u and v is inside of the simple cycle separator.

(3) The optimal solution connects u and v and the path of edges connecting u and v is on the simple cycle separator.

(4) Vertices u and v are not connected by the above method.

Another important observation is that every solution tree can be condensed into a skeleton tree. A skeleton tree consists of vertices on the simple cycle separator only and each edge is labeled "inside" , "outside" or "on". For instance, the optimal solution tree shown in Fig.4.2 is now represented as a skeleton tree in Fig.4.6.

Our searching over separators strategy eventually guesses all possible skeleton trees. Every skeleton tree corresponds to two subproblems. Given a skeleton tree based upon a simple cycle separator, the rules of producing the subproblems are as follows:

(1) All vertices and edges outside (inside) of the simple cycle separator belong to the exterior (interior) subproblem.

(2) For the exterior (interior) subproblem, remove all vertices and edges inside (outside) of B, all edges on B, and those vertices on B but not on the skeleton tree.

(3) For the exterior (interior) subproblem, two vertices on B are merged into a new vertex if they are connected by an "in" ("out") or "on" edge in the skeleton tree.

(4) All new vertices, produced by rule (3), are fixed vertices for its corresponding subproblem.

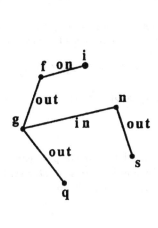

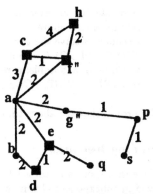

(a) The exterior subproblem.

(b) The interior subproblem.

Fig.4.6.

Fig.4.7.

Let G_a (G_c) denote the input graph in the interior (exterior) subproblem. For instance, the two subproblems corresponding to the skeleton tree in Fig.4.6 are now shown in Fig.4.7 (a) and (b) respectively. After solving the two subproblems, we find the solution trees of the exterior and interior subproblems. A simple way of merging can then be done.

We now give the algorithm to solve the PStTree problembased upon the seaching over separators strategy.

Algorithm SOLVE–STTREE.

Input: A planar graph $G(V, E)$, a set of fixed vertices $R \in V$ and a weight function for each edges on G.

Outout: An optimal solution of the PStTree problem corresponding to the input instance.

Step 1: If $|V| < 3$, then solve this problem with some exhaustive enumerating method, else go to next step.

Step 2: COST $= \infty$.

Step 3: Find a maximal planar graph G' of G, by adding dummy edges onto G'.

Step 4: Find a simple cycle separator B of G'.

Step 5: Generate all possible skeleton trees.

Step 6: For each possible skeleton tree T'_{sk}, do : (Let w be the sum of the costs of "on" edges of the possible skeleton tree.)

Step 6.1 :   Generate the exterior and interior subproblems by using the four rule presented above.

Step 6.2 :   Recursively solve the two subproblems by using this algorithm. Let $COST_c$ and $COST_a$ denote the costs of the solutions of the exterior and interior subproblems respectively.

Step 6.3 :   If $COST > COST_a + COST_c + w$, let $COST = COST_a + COST_c + w$.

It can be proved that the above time complexity is $O(n^{O(\sqrt{n})})$.

5   Concluding Remarks

We do believe that there are still a lot of important NP–complete problems which can be solved in subexponential time by using the simple cycle separator strategy. In the future we will explore more characteristics and more problems.

References

1. A. V. Aho, J. E. Hopcroft, and J. D. Ullman. *The Design and Analysis of Computer Algorithms*. Bell Telephone Laboratories, Inc., New York, 1976.

2. J. L. Bentley. *Divide and Conquer Algorithms for Closest Point Problems in Multidimensional Space*. PhD thesis, Department of Cimputer Science, University of North Carolina, 1976.

3. J. L. Bently. Multidimensional Divide–and–Conquer. *Communications of the ACM*, 23:214–229, 1980.

4. U. Bertele and F. Brioschi. *Nonserial Dynamic Programming*. Academic Press, New York, 1972.

5. M. R. Garey and D. S. Johnson. *Computers and Intractability: a Guide to the Theory of NP–Completeness*. W. H. Freeman and Co., New York, 1979.

6. E. Horowitz and S. Sahni. *Fundamentals of Computer Algorithms*. Computer Science Press, Inc., New York, 1978.

7. R. Z. Hwang, R. C. T. Lee, and R. C. Chang. The Searching over Separators Strategy to Solve Some NP-hard Problems in Subexponential Time. Accepted by *Algorithmica*, 1992.

8. D. S. Johnson. The NP–Completeness Column: an Ongoing Guide. *Journal of Algorithms*, 6:434–451, 1985.

9. E. L. Lawler, J. K. Lenstra, A. H. G. Rinnooy Kan, and D. B. Shmoys. *The Traveling Salesman Problem – A Guided Tour of Combinatorial Optimization*. Wiley-Interscience, New York, 1985.

10. R. J. Lipton and R. E. Tarjan. A Separator Theorem for Planar Graphs. *SIAM Journal on Applied Mathematics*, 36(2):177–189, 1979.

11. N. Megiddo and K. J. Supowit. On the Complexity of Some Common Geometric Location Problems. *SIAM Journal on Computing*, 13(1):182–196, 1984.

12. G. L. Miller. Finding Small Simple Cycle Separators for 2-Connected Planar Graphs. *Journal of Computer and System Sciences*, 32:265–279, 1986.

13. T. Nishizeki and N. Chiba. *Planar Graphs, Theory and Algorithms*. Elsevier Science Publisher B.V., Amsterdam, 1988.

14. F. P. Preparata and M. I. Shamos. *Computational Geometry*. Springer–Verlag, New York, 1985.

Parallel and On-line
Graph Coloring Algorithms

Magnús M. Halldórsson

Japan Advanced Institute of Science and Technology, Hokuriku
Tatsunokuchi, Ishikawa 923-12, Japan

Abstract. We discover a surprising connection between graph coloring algorithms in two orthogonal paradigms: parallel and on-line computing. We present a randomized on-line coloring algorithm with a performance guarantee of $\mathcal{O}(n/\log n)$, an improvement of $\sqrt{\log n}$ factor. Also, from the same principle, we construct a parallel coloring algorithm with the same performance guarantee, for the first such result. Finally, we show how to apply the parallel algorithm to obtain an $\mathcal{NC}$ approximation algorithm for the independent set problem.

1 Introduction

We consider the task of coloring an undirected graph. Each vertex is to be assigned a value, or color, so that no two adjacent vertices are assigned the same color, while minimizing the number of colors used. We investigate this problem in two paradigms. First, we are interested in performing this task in *parallel*, using a polynomial number of processors. Second, we solve it *on-line*, where the graph is revealed to the algorithm node by node and each node must be irrevocably assigned a color before the next vertex is received.

The quality criteria used is the *performance guarantee* of a coloring algorithm, which is the maximum over all inputs of the ratio of the number of colors used to the minimum number of colors needed.

A sequence of results by Johnson [8], Wigderson [16], Berger and Rompel [2], and Halldórsson [6] have pushed the performance guarantee for polynomial-time graph coloring algorithms down to $\mathcal{O}(n(\log\log n)^2/\log^3 n)$. Of particular relevance to this paper is Wigderson's algorithm, which uses $kn^{(k-2)/(k-1)}$ colors on k-colorable graphs.

For on-line graph coloring, the only non-trivial upper bound for deterministic algorithms is a $\mathcal{O}(n/\log^* n)$ performance guarantee due to Lovász, Saks, and Trotter [10]. Vishwanathan [14] was able to improve on that significantly by using randomization. His algorithm uses $\mathcal{O}(k2^k n^{(k-2)/(k-1)}(\log n)^{1/(k-1)})$ expected number of colors on k-colorable graphs, yielding a $\mathcal{O}(n/\sqrt{\log n})$ performance guarantee.

In this paper, we tackle at once graph coloring in two modes of computation: parallel, and on-line. From the same sequential algorithm, we obtain the first parallel coloring algorithm with a non-trivial performance guarantee ($\mathcal{O}(n/\log n)$), as well as a significantly improved performance guarantee for randomized on-line coloring (also $\mathcal{O}(n/\log n)$). Finally, we obtain a $\mathcal{O}(n(\log\log n/\log n)^2)$ performance guarantee for the independent set problem in parallel.

All the results are close to the best we could hope for when compared with the $\Omega(n/\log^2 n)$ approximation lower bound for randomized on-line coloring [7], the $\mathcal{O}(n(\log\log n)^2/\log^3 n)$ [3] ($\Omega(n^\epsilon)$, for some $\epsilon > 0$ [12]) upper (lower) bound, respectively, for sequential graph coloring, and the $\mathcal{O}(n/\log^2 n)$ [6] ($\Omega(n^\epsilon)$, for some $\epsilon > 0$ [1]) upper (lower) bound for the sequential independent set problems.

The following section contains a description of the sequential algorithm and its analysis, with the parallel algorithm covered in section 3, the on-line algorithm described in section 4, and the independent set results closing off in section 5.

2 Approximate Coloring Algorithm

We first focus on the underlying sequential, off-line coloring algorithm that forms the basis of our parallel and on-line algorithms. It can be viewed as a variant of Wigderson's algorithm, or as a deterministic off-line version of Vishwanathan's randomized on-line method.

The algorithm is based on two main principles. The first is based on the observation that the neighborhood of any node in a k-colorable graph must be $k - 1$ colorable. This allows us to break the problem into a collection of presumably easier subproblems. Indeed, when $k = 2$ the problem reduces to bipartite coloring which is fairly straightforward sequentially, in parallel, and on-line.

This suggests a natural heuristic: find a node of maximum degree, recursively color its neighborhood, and iterate this procedure on the remaining graph. This is essentially the method of Wigderson [16]. This iterative process, however, fills the color classes in sequence. We need a method that accumulates many color classes concurrently, rather than searching for individual independent sets, which brings us to the second principle.

A *maximal partial coloring* of a graph is an assignment of some of the nodes into fixed number of "greedy" color classes so that each remaining node is adjacent to some node in each color class. The important thing is that each color class partitions the remaining nodes into subsets of the neighborhoods of the nodes in the color class. If we choose a small class, the remaining uncolored nodes induce relatively few subproblems. The total number of colors used is then the number of greedy color classes, plus the sum of the number of colors used in the subproblems. Maximal partial coloring is easy to achieve sequentially, via the First Fit algorithm which assigns a vertex to the first compatible color class (if one exists).

We now present the main algorithm, illustrated in figure 2.1. The algorithm finds a maximal partial coloring, partitions the remaining nodes around the smallest color class, and recurses on those subproblems. The recursion bottoms out at bipartite graphs, and when the chromatic number is too large, with respect to the size of the graph, we settle on the trivial coloring of one color per vertex.

```
Color(G,k)
begin
  if (k ≤ 2) BipartiteColor(G)
  else if (k > log n) Assign each vertex a different color
  else
    ResidueNodes ← MaximalPartialColor(G,(n/(k − 2))^((k−2)/(k−1)))
    Find the smallest greedy color class, and let w₁, ..., wₚ be its nodes.
    Partition the ResidueNodes into R₁,... Rₚ such that nodes in Rᵢ are adjacent to wᵢ
    for i = 1 to p
      Color(Rᵢ, k − 1)
  endif
end
```

Algorithm 2.1 An approximate coloring algorithm

The precise number of color classes allocated to the MaximumPartialColor (greedy) algorithm was derived analytically to provide the optimal tradeoffs between the greedy and the recursive subproblems. In particular, it yields a stronger performance guarantee than the more straightforward choice of $n^{(k-2)/(k-1)}$.

Theorem 1. *The number of colors used by* Color *on k-colorable graphs is at most*

$$\frac{k-1}{(k-2)^{(k-2)/(k-1)}} n^{(k-2)/(k-1)}.$$

Proof. Let $A(G, n, k)$ denote the number of colors the algorithm uses in total to color a k-colorable graph G on n vertices. Let $\mathcal{A}(n, k)$ denote $((k-1)/(k-2)^{(k-2)/(k-1)})n^{(k-2)/(k-1)}$. The theorem can be restated as: $A(G, n, k) \leq \mathcal{A}(n, k)$

The proof proceeds by induction on k, with the base case, $k = 2$, established by an optimal bipartite algorithm. Hence, assume that $A(G, n, k-1) \leq \mathcal{A}(n, k-1)$ holds.

The number of colors used by our algorithm is the number of greedy colors $s = s(n, k) = (n/(k-2))^{(k-2)/(k-1)}$ plus the number of colors used in the t different subproblems of size m_j each. (To maintain sanity and simplicity of presentation, we ignore all ceilings.) By the inductive assumption, we can assume that number of colors used in the subproblems is bounded by the $\mathcal{A}$ function.

$$A(G, n, k) \leq s + \sum_{j=1}^{t} \mathcal{A}(m_j, k - 1)$$

Since the approximation function is convex, the sum is maximized when the number of subproblems is minimized and their sizes are all equal. If g is the total number of vertices in greedy color classes, then the number of vertices in the smallest class (and thus the number of subproblems) is at most g/s. Let $m = \sum_j m_j$.

$$A(G, n, k) \leq s + \max_g \{\frac{g}{s} \cdot \mathcal{A}(s \cdot \frac{m}{g}, k - 1) \mid n = g + m\}$$

Substitution and simplification gives us

$$A(G, n, k) \leq s + \max_g \{ g \cdot (\frac{n-g}{g})^{(k-3)/(k-2)} \} \cdot [\frac{k-2}{(k-3)^{(k-2)/(k-3)}}] \cdot s^{-1/(k-2)}$$

The function $f(g) = g(\frac{n-g}{g})^{(k-3)/(k-2)}$ has maximum value of $(k-3)^{(k-3)/(k-2)}/(k-2)n$, conveniently simplifying the formula. Now straightforward algebra establishes the claim.

$$\begin{aligned} A(G, n, k) &\leq s + n \cdot s^{-1/(k-2)} \\ &= [(\frac{1}{k-2})^{(k-2)/(k-1)} + (k-2)^{1/(k-1)}] \cdot n^{(k-2)/(k-1)} \\ &= [1 + (k-2)] \cdot (\frac{1}{k-2})^{(k-2)/(k-1)} \cdot n^{(k-2)/(k-1)} \\ &= \mathcal{A}(n, k) \end{aligned}$$
∎

Corollary 2. *The performance guarantee of* Color *is* $\mathcal{O}(n/\log n)$.

Proof. The ratio of the size of the approximation to the chromatic number of the graph is $\mathcal{O}(n^{(k-2)/(k-1)}/k)$, which is maximized when $k \approx \log n$. ∎

3 Parallel Coloring Algorithm

In this section we discuss how to parallelize the steps of the preceding coloring algorithm. The main algorithm can be parallelized easily by executing the loop iterations in parallel. The bipartite graphs can be colored exactly by finding a forest of rooted spanning trees using the Euler tour technique [9], computing the depth of each node using list ranking, and assigning nodes of odd depth one color and nodes of even depth the other color.

The hard part is finding a maximal partial coloring. The idea is to reduce it to the maximal independent set problem which has known $\mathcal{NC}$ solutions. This construction was originally given by Luby [11], but was later independently rediscovered by Vishwanathan [13] when posed by this author.

We create a new graph that consists of many copies of the original graph, with a vertex of one copy adjacent only to the corresponding vertices of the other copies. When the maximal independent set algorithm is applied to this composed graph G', the independent set I found is a maximal partial coloring of G for the following reasons. No vertex of G is contained in I more than once, since all copies of a given vertex of G are adjacent in G'. And if a vertex of G is not contained in I, it must be adjacent to some vertex in each color class, since otherwise some copy of it in G' could have been added to I.

The complexity of the maximal partial coloring algorithm is dominated by the complexity of the maximal independent set algorithm used. The best bounds we are aware of for the latter problem are $\mathcal{O}(\log^3 |V|)$ time and $(|V| + |E|)/\log |V|$ processors, due to Goldberg and Spencer [5]. The number of vertices of the

MaximalPartialColor(G,x)
begin
  Construct the following graph G' on $n \cdot x$ vertices:
    Make x identical copies of G: G^1, G^2, ... G^x.
    $(v_i^j, v_s^t) \in E(G')$ iff $i = s$, or $j = t$ and $(v_i, v_s) \in E(G)$.
  $I \leftarrow$ MIS(G').
  $C_i \leftarrow I \cap G^i$
  ResidueNodes $\leftarrow G - \cup_{i=1}^x C_i$
  return $\{C_i\}$, ResidueNodes
end

Algorithm 3.1 Maximal partial coloring in parallel

composed graph G' is $s \cdot n$, where $s = s(n,k) = (n/(k-2))^{(k-2)/(k-1)}$ is the number of greedy bins allocated. The number of edges is then $sm + n\binom{s}{2} < s(m + ns)$, and thus the processor complexity of the algorithm will be $s(m + ns)/\log n < n^3$.

The complexity of the main algorithm is also dominated by k recursion levels of calls to the maximal partial coloring subroutine, adding a factor of $k = \mathcal{O}(\log n)$ to the time complexity.

Our algorithm assumes that the chromatic number of the graph is given as input. To circumvent that, we can run the algorithm for all natural numbers less than n and retain the best result. However, using binary search for the smallest acceptable value of k is more efficient, adding only a $\log k = \mathcal{O}(\log\log n)$ factor to the time complexity, for a total of at most $\mathcal{O}(\log^4 n \log\log n)$.

4 On-line Coloring Algorithm

The on-line coloring problem is defined as follows. An adversary selects a graph and an ordering of the graph. She feeds us one node at a time along with its edges to the nodes already received. We are to irrevocably assign that node a color consistent with the colors of the previous nodes. We are allowed arbitrary amount of time but we are not given any further information about the graph, including its size and chromatic number.

We shall be using randomization to inhibit the adversary from foreseeing our every move. We also insist that the adversary be *oblivious* in that she must pick the graph and the ordering before we make any choices (rather than constructing the graph adaptively according to our our choices). The performance guarantee of a randomized algorithm is the maximum expected number of colors used on any graph (and any ordering of the graph) divided by the chromatic number of the graph.

We shall show how to convert algorithm 2.1 into an on-line algorithm. The greatest change is that we must process the nodes fully in sequence, rather than processing the chromatic levels in sequence. We must also overcome four difficulties.

One hurdle is that a partitioning class must be selected in advance, before we know its eventual size. The concern is that if the size will be large, many subproblems will be spanned. The solution is to use randomization: pick a uniformly random class as a partitioning class. Since the adversary must choose the ordering of the graph in advance, the expected number of nodes in the class will be sufficiently small.

The second problem is estimating the size of the input, and recursively, the sizes of the subproblems. For this we apply a well known trick in the on-line trade (see e.g. [10]), the *doubling strategy*, which works as follows. Assume, to begin with, that the size of the input, n, equals 2^i, for i initially equal 0. When the actual number of nodes exceeds this estimate, we increment it to the next power of 2, retire the old color classes and start afresh with a new set of color classes. We continue this way, and when the input finally ends, our estimate of n is then never off by more than a factor of 2. Because the function giving our approximation performance is necessarily convex, it follows that the overhead caused by this strategy is no more than a small constant.

It is important to notice that we can apply the same strategy recursively for the subproblems as well. From the perspective of an on-line algorithm, the subgraph induced by the nodes in a residue class (i.e. in a "subproblem") is structurally no different from the original problem, except for the fact that the chromatic number is one less. We shall therefore allocate sets of color classes to this subproblem completely independent of the other subproblems or the parent problem. This strategy is what primarily distinguishes our algorithm from Vishwanathan's.

The third problem – not knowing the chromatic number in advance – can be handled similarly. To begin with we assume the graph is 2-colorable, and as our assumptions are shattered, usually by the bipartite algorithm stumbling upon a non-bipartite subgraph, we increment our estimate by one, and start from scratch with a fresh set of colors.

Finally, we need to find on-line versions of the subroutines called by algorithm 2.1. As stated in [10], there is a simple algorithm for coloring bipartite graphs using no more than $2 \log n$ colors. And a FirstFit algorithm that is restricted to the number of greedy color classes actually available is perfect for performing a maximal partial coloring.

The resulting algorithm, less the mechanisms for updating the estimates of size and chromatic number, is shown in fig. 4.1. The algorithm as presented assigns a color only to the formal variable v (a vertex) in each invocation, while updating a static data structure called a *coloring tree*. This tree is layered into chromatic levels: the graph induced by nodes belonging to some residue bin (subproblem) has a chromatic number strictly less than the parent graph. Each vertex t in the tree contains a set of greedy color classes (and a set of decommissioned classes), links to the subproblems, and a total count of the nodes in the subtree. Coloring a vertex means placing it in a greedy bin at some location in the coloring tree. Either it is placed in the greedy bins at the root, or placed in a residue bin, which means passing it to one of the children of the root. This

On-LineColor (T, k, v)
{ Assign a color to the vertex v. }
{ T is a coloring tree, of colorability k }
begin
  if $(k \leq 2)$
    BipartiteColor(T, v)
  else
    if (FirstFit(T, k, v) is not sufficient)
      Choose some node v_j adjacent to v in the partitioning class
      On-LineColor$(R_j, k-1, v)$
    endif
  endif
end

Algorithm 4.1 A randomized on-line coloring algorithm

recursive behaviour is identical at each level.

UpdateSizeEstimate (T, k, n)
{ n is the new estimate for the size of k-colorable subtree T }
begin
  Retire all color classes in the subtree T
  Allocate $\left(\frac{n}{k-2}\right)^{(k-2)/(k-1)}$ greedy color classes
  Select a randomly uniform color class as a partitioning class.
end

Algorithm 4.2 The size updating procedure

The kind of processing done when the size estimates are updated is indicated in fig. 4.2.

4.1 Analysis

The heart of the analysis argument is contained in the following result, which we shall state as a lemma. The proof resembles the analysis of the parallel approximation, but is more tedious. For reasons of brevity we shall not include it in this abstract.

Lemma 3. *Let* $V_k = \{v_{i+1}, \ldots, v_{i+m}\}$ *be the set of vertices colored during the time that the chromatic estimate* k *remains unchanged. Then no more than*

$$\mathcal{B}(m, k) = \mathcal{O}(m^{(k-2)/(k-1)}(\log m)^{1/(k-1)})$$

expected number of colors are used to color V_k.

Our results follow fairly easily from the lemma.

Theorem 4. *Our on-line algorithm uses at most* $\mathcal{O}(n^{(\lambda(G)-2)/(\lambda(G)-1)}(\log n)^{1/(\lambda(G)-1)})$ *colors, without a priori knowledge of either n or* $\chi(G)$.

Proof. We maintain a lower bounded estimate on the chromatic number of the graph. This estimate is incremented only when the current graph has proven to be non k-colorable.

Consider the nodes colored while the estimate was i. If we denote their count by n_i, we know that On-LineColor used no more than $\mathcal{B}(n_i, i)$ colors to color those nodes. This can be bounded from above by $\mathcal{B}(n_i, k)$, which is a concave function. By Jensen's Inequality the sum of the of these values over all i is maximized when all node counts are equal, for a total of $(k-1)\,\mathcal{B}(\frac{n}{k-1}, k) \leq (k-1)^{1/(k-1)}\,\mathcal{B}(n, k)$ number of colors. ∎

We obtain the same performance guarantee as in the off-line case, with a similar proof.

Corollary 5. *Randomized on-line coloring is* $\mathcal{O}(n/\log n)$ *approximable.*

5 Independent Sets in Parallel

The parallel coloring algorithm can also be used to find cliques or independent sets in parallel. Recall that the algorithm used at most $n^{(k-2)/(k-1)}$ colors, where the graph was provably not $k-1$ colorable since a k-chromatic subgraph had been found. This subgraph is formed by the pivot nodes w_i at each of the $k-2$ recursive levels, along with a non-bipartite subgraph (i.e. an odd cycle) at the end of the recursion. These pivot nodes are, by definition, mutually adjacent, as well as adjacent to the odd cycle, thus producing a k-clique. Since the clique number of the graph is at most the chromatic number, which is at most the number of colors we used, this yields a $n^{(k-2)/(k-1)}/k$ approximation of the maximum clique, for an $\mathcal{O}(n/\log n)$ performance guarantee. Finally, recall that we can obtain an independent set algorithm with the same performance by applying our algorithm to the complement graph. Notice that the algorithm runs in poly-logarithmic time, for all "interesting" values of k (i.e. $k \leq \log^2 n$). This construction is originally due to Wigderson [15].

Another source of a parallel independent set algorithm can be found in the first step of Berger and Rompel's [2] coloring algorithm. It essentially involves choosing polynomial number of random vertex subsets of size $\log_k n$ and testing for non-adjacency. This can be made deterministic via the pigeonhole principle without destroying the independence of the samples, and hence can be parallelized trivially. This method yields a $n/\log_k n$ approximation.

Notice that the two methods complement each other, since our/Wigderson's method performs best for small values of k, while the above random sampling performs relatively better for larger values of k. If we run both methods and retain the better result, we get an algorithm with a performance guarantee of

$$\max_k \min \left(\frac{n^{(k-2)/(k-1)}}{k}, \frac{n}{\log_k n} \right)$$

which is minimized when $k = \log n/(2 \log \log n)$ as $\mathcal{O}(n(\log \log n/\log n)^2)$.

5.1 Graphs with large independent sets

We shall show that given a graph with an independence number at least $(1/3 + c)n$, we can find an independent set of size $\Omega(\sqrt{n})$ in polylogarithmic time.

The method starts with removing a maximal collection of triangles (cliques of size 3) from the graph. Such a collection can be obtained as a maximal independent set in the graph whose vertices are all the triangles in the original graph, with edges between intersecting triangles. Since at most one node in each triangle can be a member of a given independent set, at least $cn = \Omega(n)$ nodes remain after the triangle removal.

TriangleFreeIS(G)
begin
  $I \leftarrow$ MaximalIndependentSet(G) $= (w_1, w_2, \ldots, w_t)$
  $\overline{N}_0 \leftarrow V(G)$
  for $i \leftarrow 1$ **to** t **do**
    $\overline{N}_i \leftarrow$ Non-nbrhd(w_i) $\cap \overline{N}_{i-1}$
    $N_i \leftarrow$ Nbrhd(w_i) $\cap \overline{N}_i$
  od
  return max(N_1, $\{w_1\} \cup N_2$, $\ldots$, $\{w_1, w_2, \ldots, w_{t-1}\} \cup N_t$, $\{w_1, w_2, \ldots, w_t\}$)
end

Algorithm 5.1 Algorithm for finding independent sets in triangle free graphs

Given a triangle-free graph, we apply alg. 5.1 based on a sequential algorithm of [3]. The algorithm finds a maximal independent set with an ordering of the vertices in the set, and finds the common non-neighborhoods $\overline{N}_i$ of any prefix of the vertices, as well the complementary neighborhoods N_i. Each neighborhood must be independent since the graph contains no triangles, and is also contained in an earlier non-neighborhood. The final result is the largest of these independent sets considered, and its size must be at least the smallest t such that $\binom{t}{2} \geq n$, or approximately $\sqrt{2n}$.

Finding the recursive non-neighborhoods $\overline{N}_i$ is a prefix computation and can therefore be done in logarithmic time with linear number of processors. The complexity is therefore asymptotically equivalent to that of the MIS algorithm used.

The above construction can be generalized to find independent sets of size $\Omega(n^{1/(k-1)})$ in k-clique free graphs, providing a constructive parallel proof of the upper bound on Ramsey numbers by Erdős and Szekeres [4]. That also allows us to find such independent sets in graphs with independence number greater than n/k, but, unfortunately, the processor complexity of removing the k-cliques grows as fast as n^k.

Acknowledgements

I would like to thank Ravi Boppana, and Sundar Vishwanathan for many helpful discussions.

References

1. S. Arora, C. Lund, R. Motwani, M. Sudan, and M. Szegedy. Proof verification and intractability of approximation problems. Manuscript, Apr. 1992.
2. B. Berger and J. Rompel. A better performance guarantee for approximate graph coloring. *Algorithmica*, 5(4):459–466, 1990.
3. R. Boppana and M. M. Halldórsson. Approximating maximum independent sets by excluding subgraphs. *BIT*, 32(2):180–196, June 1992.
4. P. Erdős and G. Szekeres. A combinatorial problem in geometry. *Compositio Math.*, 2:463–470, 1935.
5. M. Goldberg and T. Spencer. Constructing a maximal independent set in parallel. *SIAM J. Comput.*, 2(3):322–328, Aug. 1989.
6. M. M. Halldórsson. A still better performance guarantee for approximate graph coloring. Technical Report 90–44, DIMACS, June 1990.
7. M. M. Halldórsson and M. Szegedy. Lower bounds for on-line graph coloring. In *Proc. of the Third ACM-SIAM Symp. on Discrete Algorithms*, pages 211–216, Jan. 1992.
8. D. S. Johnson. Worst case behaviour of graph coloring algorithms. In *Proc. 5th Southeastern Conf. on Combinatorics, Graph Theory, and Computing. Congressus Numerantium X*, pages 513–527, 1974.
9. R. Karp and V. Ramachandran. A survey of parallel algorithms for shared-memory machines. In J. van Leeuwen, editor, *Handbook of Theoretical Computer Science*, volume A. Elsevier Science Publishers B.V., 1990.
10. L. Lovász, M. Saks, and W. T. Trotter. An online graph coloring algorithm with sublinear performance ratio. *Discrete Math.*, 75:319–325, 1989.
11. M. Luby. A simple parallel algorithm for the maximal independent set problem. *SIAM J. Comput.*, 15:1036–1053, 1986.
12. C. Lund and M. Yannakakis. On the hardness of approximating minimization problems. Manuscript, July 1992.
13. S. Vishwanathan. Private communication.
14. S. Vishwanathan. Randomized online graph coloring. In *Proc. 31st Ann. IEEE Symp. on Found. of Comp. Sci.*, pages 464–469, Oct. 1990.
15. A. Wigderson. Personal communications.
16. A. Wigderson. Improving the performance guarantee for approximate graph coloring. *J. ACM*, 30(4):729–735, 1983.

Competitive Analysis of the Round Robin Algorithm

Tsuyoshi Matsumoto

University of Tokyo, Department of Information Science,
Tokyo 113, Japan
e-mail: matumoto@is.s.u-tokyo.ac.jp

Abstract. We investigate on-line algorithms that schedule preemptive tasks on single processor machines when the arrival time of a task is not known in advance and the length of a task is unknown until its termination. The goal is to minimize the sum of the waiting times over all tasks. We formulate an on-line algorithm, RR, which is an ideal version of so-called Round Robin algorithm. It is known that, if all tasks arrive at one time, RR is 2-competitive [W]. We prove that, when tasks may arrive at different times, the competitve ratio of RR is between $2(k-1)/H_{k-1}$ and $2(k-1)$, where k is the maximal number of tasks that can exist at any given time. Our analysis also yields bounds on the sum of response times, and through several criteria we demonstrate the effectiveness of Round Robin algorithm.

1 Introduction

The problem of scheduling tasks is a basic problem in computer science and has been studied for long time by many researchers. Many of scheduling problems are on-line in nature, that is, each task should be processed without having informations on future tasks. For general on-line problems, the "competitive analysis" of an on-line algorithm with respect to the performance of an optimal off-line algorithm is now drawing great attention. Recently, some papers have been published that study the task scheduling problem using the competitive analytic approach ([BKMRRS], [FST], [SWW]). In this paper we study very fundamental on-line scheduling algorithms, such as the so-called Round Robin algorithm, that schedule preemptive tasks on single processor machines, based on the method of competitive analysis.

Our model is as follows. There is only one processor, and only one task can be processed at a time. A task $T_i = (s_i; l_i)$ consists of a starting time s_i and the length l_i. An on-line algorithm can not know the length of an arriving task until the task is processed to completion. Thus the algorithm must decide which task to process without knowing the value of l_i. In general, when a sequence of n tasks $\mathcal{T} = \{T_1, T_2, \ldots, T_n\}$ is given, a scheduling algorithm should decide which task to process at each moment. We allow preemptive scheduling, that is, some task may be interrupted and resumed later.

One criterion of the efficiency of on-line scheduling algorithms is the "response time", the time between the request for a task and its termination. For a designer of operating systems, reducing response times is a critical problem. In this paper, we chose the waiting time as the criterion, subtracting the actual processing time from the response time. For a formal definition, suppose task T_i is terminated at time f_i. We define the cost $c_A(T)$ of the sequence T as $\sum_{i=1}^{n}(f_i - s_i - l_i)$, the total length of the waiting time. We want to reduce the waiting time to zero, but in most cases this is not possible since there is only one processor available in our model.

J.Wein at MIT [W] informed us that the behavior of off-line optimal algorithms for this problem is well known. Let OPT be one of off-line optimal algorithms. Then, OPT processes the shortest job at each moment. For completeness, we provide our proof in this paper.

To evaluate the efficiency of on-line algorithms, we use a measure called "competitive ratio", devised by Sleator and Tarjan [ST] and subsequently generalized by a number of researchers ([BLS], [MMS], [RS]). The competitive ratio of an algorithm A for a certain on-line problem is defined as follows. Let $C_X(\sigma)$ denote the cost incurred by an algorithm X on request sequence σ. On-line algorithm A is c-competitive if there is a constant a such that for any sequence σ,

$$C_A(\sigma) \leq c \cdot C_{OPT}(\sigma) + a.$$

If A is not c-competitive for any constant c, we say A is "not competitive". If an on-line algorithm A is c-competitive and no on-line algorithm has competitive ratio less than c, we say A is an "optimal" on-line algorithm.

As an example, let us consider "first come first serve" algorithm (we call it FCFS). FCFS serves tasks by the coming order. It is easy to see that FCFS is not competitive. Suppose a sequence $T = \{(0; 1000), (1; 1)\}$ which consists of two tasks. OPT incurres the cost of 1, while FCFS incurres the cost of no less than 999. The value 1000 can be made larger, and it is obvious that the ratio of two costs increases infinitely. This means that FCFS is not competitive.

In this paper, we focus our attention on the Round Robin algorithm, which is widely used in practice. Based on the Round Robin algorithm, we define an algorithm RR in the following way: When there are n tasks ready for processing, RR processes all n tasks conceptually in parallel, and the speed of processing each task slows down by the factor of n. RR is a formalization of Round Robin algorithm with its quantum infinitely small and its switching cost zero. Karloff, Vishwanathan and Chandra, all at the University of Cicago, proved that RR is 2-competitive if all jobs begin at the same moment, i.e., if $s_i = s_j$ for any i and j [W]. They have also proved that this is optimal for an on-line algorithm. The proof has not been published yet, so that we provide our proofs in this paper.

We prove that, without the restriction above, RR is $2(k-1)$-competitive, where k is the maximal number of tasks that can exist at any given time. We also prove that the competitive ratio of RR is at least $2(k-1)/H_{k-1}$, where $H_{k-1} = 1/1 + 1/2 + \cdots + 1/(k-1)$, the $(k-1)$th harmonic number.

2 The Off-Line Optimal Algorithm

Off-line algorithms can use the full information of T for scheduling. Let OPT be one of the optimal off-line algorithms. (In fact, there can be several optimal off-line algorithms. We elaborate on this later.)

Theorem 1. *OPT processes the shortest task at each moment. "The shortest task" is the one that has the least amount of remaining job to do at the moment.*

Proof. Our claim is that, at any moment and for any pair of tasks T_i and T_j, if task T_i is longer than task T_j, OPT does not process T_i. So let us suppose, at time t_0, OPT processes T_i and that T_j is shorter than T_i. There are two possibilities:

1. T_i ends later than f_j : We consider another off-line algorithm A which behaves as follows. A is almost the same as OPT, except that A uses T_i's processing time after t_0 in order to process T_j until T_j is terminated. Then T_i uses T_j's time and its own time, terminating at the same moment as OPT. So, using A instead of OPT, T_j's waiting time decreases while waiting times of other tasks, including T_i, does not change, decreasing the total cost. This is a contradiction.
2. T_i ends before f_j : We consider again another off-line algorithm A which behaves as follows. Let f_i be the termination time of T_i. And let Δt denote $f_j - f_i$. Using T_i's time before f_i, A process T_j and make it terminate before f_i. Since T_i is longer than T_j, this is possible. Then, using A instead of OPT, T_j's cost decreases by more than Δt, while T_i's cost increases by no larger than Δt, decreasing the total cost. This is again a contradiction.

This concludes the proof. □

Ambiguity occures when the several shortest task have the same amount of remaining task. But the choice is not substantial because it does not change the cost incurred.

3 Delay Time

We use here the notion of the "delay time" in [MPT], which was independently considered by S.Hasegawa [H].

Definition 2. $D_{i,j}^A$ is the time allocated by algorithm A to job i in the presence of job j. In other words, this is the amount of time by which job i delays job j.

Then, $c_A(T) = \sum_{i \neq j} D_{i,j}^A$.
When the input sequence satisfies certain conditions, the competitive ratio of RR is easily derived.

Theorem 3. *When all tasks have the same starting time, RR is 2-competitive.*

Proof. ([MPT], [H]) Let the input sequence be $T = \{T_1, T_2, \ldots, T_n\}$. By renumbering, we can assume that $l_{i-1} \leq l_i$ for $2 \leq i \leq n$. Then,

$$D_{i,j}^{OPT} = \begin{cases} min(l_i, l_j)(= l_i) & i < j \\ 0 & i > j \end{cases}$$

and

$$D_{i,j}^{RR} = min(l_i, l_j) \qquad \text{for all } i, j.$$

Therefore, $c_{RR} = \sum_{i \neq j} min(l_i, l_j) = 2 \cdot \sum_{i<j} min(l_i, l_j) = 2 \cdot c_{OPT}$ and this means that RR is 2-competitive. □

Also, RR is an optimal on-line algorithm in the case above. Proof is straighforward. Idea is that, if an on-line algorithm A does not process existent tasks uniformly, an adversary loads more work on the task which A puts less priority. Then, OPT can save the cost by processing shorter tasks earlier.

Theorem 4. *When all tasks have the equal length, RR is 2-competitive.*

Proof. Proof is done by induction on the number of tasks. If the sequence consists of only one task, it is clear that $c_{RR} = 2 \cdot c_{OPT}$, since $c_{RR} = c_{OPT} = 0$. Next, let the input sequence be $T = \{T_1, T_2, \ldots, T_{k-1}\}$. By renumbering, we can assume that $s_{i-1} \leq s_i$ for $2 \leq i \leq k-1$. Let T_k be a new task with the equal length and the starting time which satisfies $s_{k-1} \leq s_k$. The proof is done by showing that, when T_k is added to T, c_{RR} increases twice as much as the increase of c_{OPT}.

First, we study RR. For $1 \leq i \leq k-1$, let α_i denote the remaining amount of the job of T_i at time s_k when processed by RR. Note that α_i is smaller than the length of T_k. By adding T_k, $D_{i,j}^{RR}$ does not change for $1 \leq i, j \leq k-1$, and $D_{i,k}^{RR} = \alpha_i$, $D_{k,i}^{RR} = \alpha_i$ for $1 \leq i \leq k-1$. Therefore, the increase of c_{RR} is $2 \cdot \sum_{i=1}^{k-1} \alpha_i$.

Next, we study OPT. For $1 \leq i \leq k-1$, let β_i denote the remaining amount of the job of T_i at time s_k when processed by OPT. Note that β_i is smaller than the length of T_n. By adding T_k, $D_{i,j}^{OPT}$ does not change for $1 \leq i, j \leq k-1$, and $D_{i,k}^{OPT} = \beta_i$, $D_{k,i}^{OPT} = 0$ for $1 \leq i \leq k-1$. Therefore, the increase of c_{OPT} is $\sum_{i=1}^{k-1} \beta_i$.

By the way, $\sum_{i=1}^{k-1} \alpha_i = \sum_{i=1}^{k-1} \beta_i$. This means that, c_{RR} increases twice as much as the increase of c_{OPT}. □

In this situation, FCFS is the optimal on-line algorithm. RR is 2-competitive even in this peculiar situation. This fact reinforces the claim that RR is a stable algorithm.

4 RR in general case

4.1 Upper Bound of the Competitive Ratio of RR

In some cases, we can assume that the maximal number of tasks is bounded from above by some number n, for example due to the restrictions from hardware or operating system. If we restrict the problem by bounding the number of existent tasks at any given time (when the input sequence is processed by RR and OPT) from above by n, the upper bound of the competitive ratio of RR can be evaluated as a function of n. Note that, in this case, the set of admitted input sequences depends on the algorithms RR and OPT. But, if we restrict the problem further by bounding the total number k of tasks, then the set of admitted input sequences becomes independent of the algorithms RR or OPT, and the next theorem still holds for k in place of n (because the number of existent tasks at any given time never exceeds the total number of tasks).

Specifically, let n denote the maximal number of tasks that can exist at any given time (for both RR and OPT). Then, we have the following.

Theorem 5. *RR is $2(n-1)$-competitve.*

Proof. Proof is by induction on the number of tasks k. When $k = 1$, it is clear that $c_{RR} \leq 2(n-1)c_{OPT}$, since $c_{RR} = c_{OPT} = 0$. Let the input sequence be $T = \{T_1, T_2, \ldots, T_{k-1}\}$. We add a new task T_k which satisfies $s_i \leq s_k$ for $1 \leq i \leq k-1$, and examine the increase of the cost.

First, we study RR. For $1 \leq i \leq k-1$, let α_i denote the remaining amount of the job of T_i at time s_k when processed by RR. We may assume $\alpha_{i-1} \leq \alpha_i$ for $2 \leq i \leq k-1$ without loss of generality. Let p be the number that satisfies $\alpha_p \leq l_k < \alpha_{p-1}$. Then, by adding T_k, $D_{i,j}^{RR}$ ($1 \leq i, j \leq k-1$) does not change and

$$D_{i,k}^{RR} = D_{k,i}^{RR} = \begin{cases} \alpha_i & 1 \leq i \leq p \\ l_k & p < i \leq k-1 \end{cases}$$

Therefore, the increace Δc_{RR} of c_{RR} is $2(\sum_{i=1}^{p} \alpha_i + (k-1-p)l_k)$.

Next, we study OPT. For $1 \leq i \leq k-1$, let β_i denote the remaining amount of the job of T_i at time s_k when processed by OPT. We assume $\beta_{i-1} \leq \beta_i$ for $2 \leq i \leq k-1$ without loss of generality. Let q the number that satisfies $\beta_q \leq l_k < \beta_{q-1}$. Then, by adding T_k, $D_{i,j}^{OPT}$ ($1 \leq i, j \leq k-1$) does not change and

$$D_{i,k}^{OPT} = \begin{cases} \beta_i & 1 \leq i \leq q \\ 0 & q < i \leq k-1 \end{cases}$$

and

$$D_{k,i}^{OPT} = \begin{cases} 0 & 1 \leq i \leq q \\ l_k & q < i \leq k-1 \end{cases}$$

Therefore, the increace Δc_{OPT} of c_{OPT} is $\sum_{i=1}^{q} \beta_i + (k-1-q)l_k$.

Next, we compare Δc_{RR} and Δc_{OPT}. Note that $\sum_{i=1}^{k-1} \alpha_i = \sum_{i=1}^{k-1} \beta_i$. We denote this amount by R. Then, if $R \leq l_k$, $p = q = k-1$ and consequently $\Delta c_{RR} = \Delta c_{OPT} = R$. If $l_k < R$, $\Delta c_{RR} = 2(\sum_{i=1}^{p} \alpha_i + (k-1-p)l_k) \leq 2(k-1)l_k \leq 2l_k(n-1)$ and $l_k \leq \Delta c_{OPT}$. This means $\Delta c_{RR}/\Delta c_{OPT} \leq 2 \cdot (n-1)$ and concludes the proof. □

Instead of the sum of waiting times, the sum of response times may be used as a cost measure of on-line algorithms. In this case, the above theorem immediately implies the following corollary.

Corollary 6. *Using response time instead of waiting time as a mesure of cost, RR is still $2(k-1)$-competitve.*

Proof. We define $c'_{OPT}(T)$ and $c'_{RR}(T)$ to be the costs measured by the waiting time. Let L be the total length of all tasks, $\sum_{i=1}^{n} l_i$. Then, $c_{RR}(T) = c'_{RR}(T) + L$, $c_{OPT}(T) = c'_{OPT}(T) + L$. Since $\frac{c'_{RR}(T)}{c'_{OPT}(T)} \leq 2(k-1)$, it is obvious that $\frac{c_{RR}(T)}{c_{OPT}(T)} \leq 2(k-1)$. □

4.2 Lower Bound

Theorem 7. *RR is not c-competitve for any constant that satisfies $c < \frac{2(k-1)}{H_{k-1}}$.*

Proof. For the sequence T described below, RR incurres $\frac{2(k-1)}{H_{k-1}}$ times the OPT's cost.
$T = \{(s_1; l_1), (s_2; l_2), \ldots, (s_k; l_k)\}$, where

$$l_1 = 1, \quad l_i = \frac{l_1}{i-1} \quad \text{for} \quad 2 \leq i \leq k, \text{ and}$$

$$s_1 = s_2 = 0, \quad s_i = l_1 \cdot H_{i-2} \quad \text{for} \quad 3 \leq i \leq k.$$

For this sequence, $c_{OPT}(T) = H_{k-1}$ and $c_{RR}(T) = 2(k-1)$. This implies the theorem. □

5 Conclusion

In this paper, we have proved that the competitive ratio of RR is between $\frac{2(k-1)}{H_{k-1}}$ and $2(k-1)$, where k is the maximal number of running tasks.

We conjecture that the best possible value of RR's competitive ratio matches the lower bound, i.e., $\frac{2(k-1)}{H_{k-1}}$. No larger ratio is ever attained in our experiments. Is it true that RR is optimal among on-line algorithms, and if not, what is the optimal on-line algorithm for this problem? Can the competitive ratio of RR be given as a function of the ratio between the lengthes of the longest task and the shortest task?

Acknowledgement

We are grateful to Hiroshi Imai for suggesting competitive analysis as a fruitful area of research. We are indebted to Sandy Irani and Kazuo Iwano for their lectures on on-line algorithm at IBM TRL. We also thank Susumu Hasegawa, Koji Hakata and Magnus Halldorsson for their useful information and helpful comments.

References

[BKMRRS] S. Baruah, G. Koren, B. Mishra, A. Raghunathan, L. Rosier and D. Shasha. On-line scheduling in the presence of overload. Proc. 32nd IEEE FOCS, pages 100-110, 1991.

[BLS] A. Borodin, N. Linial and M. Saks. An optimal online algorithm for metrical task systems. Proc. 19th ACM STOC, pages 373-382, May 1987.

[FST] A. Feldmann, J. Sgall and S. H. Teng. Dynamic scheduling on parallel machines. Proc. 32nd IEEE FOCS, pages 111-120, 1991.

[H] S. Hasegawa. Oral communication, 1992.

[MMS] M. Manasse, L. A. McGeoch and D. Sleator. Competitive algorithms for on-line problems. Proc. 20th ACM STOC, pages 322-333, May 1988, Chicago.

[MPT] R.Motwani, S.Phillips and E.Torng. Non-Clearvoyant Scheduling. Manuscript, 1992.

[RS] P. Raghavan and M. Snir. Memory versus ramdomization in online algorithms. Proc. 16th ICALP, Italy, July 1989.

[ST] D. D. Sleator and R. E. Tarjan. Amortized efficiency of list update and paging rules. Communications of the ACM, 28:202-208, February 1985.

[SWW] D. B. Shmoys, J. Wein and D. P. Williamson. Scheduling parallel machines on-line. Proc. 32nd IEEE FOCS, pages 131-140, 1991.

[W] J. Wein. Private communication, 1991.

Competitive Analysis of the On-line Algorithms for Multiple Stacks Systems *

Been-Chian Chien   Rong-Jaye Chen
Institute of Computer Science and Information Engineering
National Chiao Tung University

Wei-Pang Yang
Department of Computer and Information Science
National Chiao Tung University
Hsinchu, Taiwan 30050, R.O.C.

Abstract

An on-line problem is one in which an algorithm must handle a sequence of requests, satisfying each request without knowledge of the future requests. A competitive algorithm is an on-line algorithm whose cost is bounded by the cost of any other algorithm, even the algorithm is an optimal off-line algorithm, multipling a constant. This paper discusses the algorithms used to manipulate the multiple stacks problem, which is one of the on-line problems. We find the optimal off-line algorithm first, then show that the Knuth's algorithm is not a competitive algorithm, but Garwick's algorithm is competitive when the number of stacks n is 2. Furthermore, the competitive ratio found here is a low bound if the Garwick's algorithm is also a competitive algorithm for $n \geq 3$.

1   Introduction

An on-line problem is one in which an algorithm must handle a sequence of requests, satisfying each request without knowledge of the future requests. On-line problems are common in computer science. Many on-line problems are discussed of late years, such as *server problems*[5, 6], *metrical task systems*[2] and dynamic data structures[8].

The study of *competitive algorithms* concentrates on comparing on-line algorithms with the optimal off-line algorithm on a sequence of operations. The competitiveness is defined as follows. Let $C_A(\sigma)$ denote the cost incurred by an on-line algorithm A in satisfying a request sequence σ, and $C_B(\sigma)$ denote the cost of any other algorithm B which satisfies the same request sequence σ. The on-line algorithm A is called competitive if

$$C_A(\sigma) \leq c \cdot C_B(\sigma) + a,$$

where c and a are constants, and σ is an arbitrary request sequence. The constant c is called competitive ratio. The algorithm B can be arbitrary algorithms; we usually

---

*This research was partly supported by National Science Council of Taiwan, R.O.C. under contract: NSC81-0408-E009-19(1991).

select the optimal off-line algorithm. More detailed definitions and explanation are given in [5, 6].

In this paper, the multiple stacks problem is fully discussed. *Stack* is a simple and useful data structure; , usually having two operations - *push* and *pop*. The simplest and most natural way to keep a stack inside a computer is to put items in a sequential memory area. It is quite convenient in dealing with only *one* stack. However, system developers frequently encounter programs which involve multiple stacks, each of which has dynamically varying size. In such a situation, keeping multiple stacks in a common area with sequential allocation will cause much trouble. First, developers would hate to impose a maximum size on each stack, since the size is usually unpredictable. Second, to store multiple variable-size stacks in sequential locations of a common memory area, an obstacle called *overflow* must be solved. An overflow situation will cause an 'error'; it means stack is already full, yet there is still more items that ought to be put in. Solution for overflow is *reallocating memory*, making room for the overflowed stacks by taking some space from stacks that are not yet filled. This action needs to move many items to their proper locations in order to assure correctness of *push* operations coming later. However, the movement of items will cause the extra cost of push operations. To minimize the number of item movements, the best way is to know the pushed order of items. That is, the number of item movements is dependent on the push sequence of items. Since the push sequence of items is not known in advance, the traditional off-line algorithm is useless. Under such a situation, an on-line algorithm is a better alternative.

A number of simple solutions based on *move operation* for solving overflow have been available in the literature. Knuth proposed a simple on-line algorithm in re-allocating memory by move operation [4]. Another on-line algorithm was proposed by J. Garwick [4]. Section 2 reviews these two algorithms, whose performances have not been analyzed in a deterministic sense yet. In Section 3, we analyze the on-line property of the two algorithms and show that Knuth's algorithms is not a competitive algorithm while Garwick's is when the number of stacks n is 2. Moreover, the competitive ratio for $n = 2$ is a low bound for Garwick's algorithm. In the last Section, some open questions are outlined.

2   Knuth's algorithm and Garwick's algorithm

The algorithms proposed by Knuth[4] and Garwick[4] are briefly presented in this section. Assume there are n stacks, and the values $BASE[i]$ and $TOP[i]$ represent bottom location and top location of stack i. These stacks all share a common memory area consisting of all locations L with $L_0 < L \le L_\infty$, where L_0 and L_∞ are constants specifying the total number of locations available for use.

Knuth's algorithm starts out with all stacks empty, $BASE[i] = TOP[i] = L_0$, for all i, and sets $BASE[n + 1] = L_\infty$. *Push* and *Pop* algorithms are shown as:

Push :
$TOP[i] \leftarrow TOP[i] + 1;$
if $TOP[i] > BASE[i + 1]$
    then *Overflow*
    else $CONTENTS[TOP[i]] \leftarrow Y;$

Pop :
if $TOP[i] = BASE[i]$
    then *Underflow*
    else begin
        $Y \leftarrow CONTENTS[TOP[i]]$;
        $TOP[i] \leftarrow TOP[i] - 1;$
    end;

When stack i overflows, reallocating strategy will find the smallest k for which $i < k \le n$ or the largest k for which $1 \le k < i$, and k satisfying $TOP[k] < BASE[k+1]$. It then moves the total items between stack $(i + 1)$ and stack k to the right by one entry, if $i < k \le n$; otherwise it moves the total items between stack $(k + 1)$ and stack

i to the left by one entry. We give an example as follows.

Example 1 Assume the lowest address of common memory area is 0, and the number of stacks is 4. Initially, $BASE[i] = TOP[i] = 0$, for $1 \leq i \leq 4$. Let I_i denote an item pushed into stack i and P denote a sequence of push operations. If the *push sequence* P is $I_2I_3I_3I_3I_4I_1$, the configuration of data structure will be shown as Figure 1 (a) after the first 6 push operations finished. As the 7th push operation will cause overflow and shift all items belonging to stack 2, 3 and 4 to the right by one entry, as shown in Figure 1 (b).
□

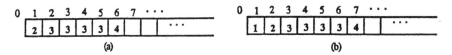

(a)   (b)

Figure 1: Knuth's algorithm: (a) after $I_2I_3I_3I_3I_4$; (b) I_1 .

Garwick's algorithm begins with each stack set to the same size initially except the nth stack. The buttom and top positions of stacks begin with

$$BASE[i] = TOP[i] = \left\lfloor \left(\frac{i-1}{n}\right)(L_\infty - L_0) \right\rfloor + L_0, \quad 1 \leq i \leq n.$$

In addition, let $BASE[n+1] = L_\infty$. If $L_\infty - L_0$ is divisible by n, all stacks have the same number of available spaces. The operations, *push* and *pop*, are similar to Knuth's algorithm. When overflow occurs, the memory reallocation is needed. The reallocation strategy is stated as follows:

1. Find the total amount of available space left.

2. Check whether all the available space is used up.

3. Compute the new $BASE$ address of each stack in accordance with the following principle:

   - Set two reallocation parameters α and β, where $0 \leq \alpha, \beta \leq 1$ and $\alpha + \beta = 1$.
   - Approximately $\alpha \times 100$ percent of available space is shared equally among the n groups, and the other $\beta \times 100$ percent will be reallocated in proportion to the ratio of growth of individual stack since the last overflow. Let x_i be the increasing number of items in stack i since the last overflow and $v = \sum_{i=1}^{n} x_i$. That is, the available spaces remained so far will be allocated to each stack by the following formula:

   $$\alpha \left\lfloor \frac{\text{the remained available space}}{\text{the number of stacks}} \right\rfloor + \beta \left\lfloor \frac{\text{the remained available space}}{v} \right\rfloor \cdot x_i.$$

4. Shift each stack up or down to the accurate position.

5. Adjust related pointers of each stack.

The example of Garwick's algorithm is as follows.

Example 2 Assume the memory size is 16, and the number of stacks is 4. The bottom address of each stack is set to be the same size. As **Example 1**, if the push sequence $I_2I_3I_3I_3I_4I_1$ occurs, we will obtain the configuration shown in Figure 2 (a). Then, if we push an item into stack 3, an overflow will occur. The reallocation shown in Figure 2 (b) sets the parameters $\alpha = 0$ and $\beta = 1$. Thus, after the reallocation, the number of available spaces of stack 1,2 and 4 are $8 \times 1/8$, the number of available spaces in stack 3 is $8 \times 4/8$.
□

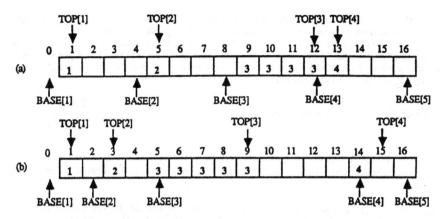

Figure 2: Garwick's algorithm: (a) after $I_2I_3I_3I_3I_3I_4I_1$; (b) I_3.

3   Analysis of the algorithms

An optimal off-line algorithm

Since the performance of an on-line algorithm can be evaluated by the optimal cost of an off-line algorithm, we find the optimal cost for a sequence of m push operations by the following algorithm:

1. Compute the number of items belonging to each stack.

2. Set proper locations to the bottom addresses of all stacks to prevent the occurrence of overflow.

3. Push the m items into stacks; *Push* and *Pop* operations are similar to the operations in Section 2.

The above algorithm takes only $O(m)$ time to finish m push operations. The optimal cost $C_{OPT}(P) = m$, where P is a sequence of m push operations.

Knuth's algorithm

For any push sequence P, except placing data into memory cells, Knuth's algorithm must handle the movements of overflow. For example, the push sequence P contains 6 push operations, which are $I_2I_3I_4I_3I_2I_1$. In that sequence, the first, second and third operations do not cause any overflow. When the 4th item I_3 is pushed into the system, the items belonging to a stacks whose stack number is larger than 3 must be moved. So, the item pushed by the 3th push operation will be moved. If the 5th item I_2 is pushed, the items belonging to a stack whose stack number is larger than 2 must be moved. In total, there are 3 items moved. The cost of Knuth's algorithm $C_K(P)$ is the sum of the number of push operations and the number of item movements. Given a push sequence P having m numbers and n stacks, $\Phi(P)$ is denoted as the number of item movements when the items in P are pushed into stacks system. Since a push sequence P has m items, if there are x_i items belonging to stack i in the push sequence P; $x_1 + \cdots + x_n = m, 1 \leq i \leq n$. Let x_i be fixed and the maximal number of item movements be $\max \Phi(P)$.

Lemma 1

$$\max \Phi(P) = \sum_{i=1}^{n-1} \sum_{j=i+1}^{n} x_j \cdot x_i.$$

Proof. Since the number of items in each stack is known, the maximal movements can be obtained by pushing the items contained in push sequence P in the reverse order of the stack number. First, the x_n items belonging to stack n are pushed; there is no item movement. Then, the x_{n-1} items belonging to stack $n-1$ are pushed, each of the push operation will cause x_n item movements; the number of item movements is $x_n \cdot x_{n-1}$. By this rule, the maximum of $\Phi(P)$ is as follows.

$$\max \Phi(P) = x_n x_{n-1} + (x_n + x_{n-1})x_{n-2} + \cdots + (x_n + \cdots + x_2)x_1 = \sum_{i=1}^{n-1} \sum_{j=i+1}^{n} x_j \cdot x_i.$$

□

Now, if all the x_i, $1 \le i \le n$, are left to be variables, the maximal number of item movements of any push sequence is $\frac{1}{2}(1 - \frac{1}{n})m^2$, which is obtained by Theorem 1 below.

Theorem 1 *A multiple stacks system has n stacks and M sequential memory locations. Total number of item movements caused by Knuth's method is at most $\frac{1}{2}(1 - \frac{1}{n})m^2$ after any sequence of m push operations, where $0 \le m \le M$.*

Proof. Formally, we want to find the solution of the following equations,

$$\max(\sum_{k=1}^{m} \Phi(P)) = \max(\sum_{i=1}^{n-1} \sum_{j=i+1}^{n} x_i \cdot x_j), \quad \text{subject to} \sum_{i=1}^{n} x_i = m, \quad x_i \ge 0. \quad (1)$$

Solving the equation (1), the result is $\max(\Phi(P)) = \frac{1}{2}(m^2 - \frac{m^2}{n}) = \frac{1}{2}(1 - \frac{1}{n})m^2$.

□

In Theorem 1, the worst push sequence is obtained. The number of item movements in the worst push sequence is in proportion to the number of push operations in the push sequence. The cost of Knuth's algorithm is $C_K(P)$, $C_K(P)$ is the value of $\max(\Phi(P))$ in Theorem 1. Thus, the Knuth's algorithm is not a competitive algorithm.

Garwick's algorithm

In Garwick's algorithm, assume the total number of memory cells, M, is equally divided into n parts initially. Each part contains at least $\lfloor \frac{M}{n} \rfloor$ available spaces. So, before the first overflow occurs, there have been at least $\lfloor \frac{M}{n} \rfloor$ items in the stacks systems. Assume the ith overflow is derived from the v_i pushed items since the last overflow was solved. The x_j^i denotes the increasing number of items belonging to stack j since the last overflow. Thus, $x_1^i + \cdots + x_n^i = v_i$. The following relations will exist:

Overflow	The total number of available spaces	The number of available space in individual satck i
Initial	M	$\lfloor \frac{M}{n} \rfloor$
1st	$M - v_1$	$\alpha \lfloor \frac{M - v_1}{n} \rfloor + \beta \lfloor \frac{M - v_1}{v_1} \rfloor \cdot x_j^1$
2nd	$M - v_1 - v_2$	$\alpha \lfloor \frac{M - v_1 - v_2}{n} \rfloor + \beta \lfloor \frac{M - v_1 - v_2}{v_2} \rfloor \cdot x_j^2$
3rd	$M - v_1 - v_2 - v_3$	$\alpha \lfloor \frac{M - v_1 - v_2 - v_3}{n} \rfloor + \beta \lfloor \frac{M - v_1 - v_2 - v_3}{v_3} \rfloor \cdot x_j^3$
$\vdots$	$\vdots$	$\vdots$
kth	$M - v_1 - \cdots - v_k$	$\alpha \lfloor \frac{M - \sum_{i=1}^{k} v_i}{n} \rfloor + \beta \lfloor \frac{M - \sum_{i=1}^{k} v_i}{v_k} \rfloor \cdot x_j^k$

Let t_i be the minimal number of the available spaces after the reallocation of $(i-1)$th overflow, $i \geq 1$; that is,

$$t_1 = \left\lfloor \frac{M}{n} \right\rfloor \quad \text{and}$$

$$t_i = \alpha \left\lfloor \frac{M - v_1 - \cdots - v_{i-1}}{n} \right\rfloor + \beta \left\lfloor \frac{M - v_1 - \cdots - v_{i-1}}{v_{i-1}} \right\rfloor \cdot x_j^{i-1}, \quad 1 \leq j \leq n.$$

Since overflow never occurs when the number of push items is less than the minimal number of available spaces among the n stacks, thus $v_i \geq t_i$. When an overflow occurs, there must be one stack whose available space is used up. All of the stacks ,except the stack 1, must be reallocated. The number of item movements is almost equal to the total number of items at that time. So, if there are m items pushed, the total number of item movements can be evaluated by the sum of the number of items at each time of overflow occurring. For an algorithm, the cost of the optimal off-line algorithm is the criterion of competitiveness. The easiest way to decide whether an algorithm is competitive or not is to find the worst input sequence and compare the cost of the algorithm with the cost of optimal off-line algorithm. If the cost of the algorithm for the worst sequence data is bounded within a constant factor of optimal off-line algorithm, the algorithm will be competitive. However, in Garwick's algorithm we do not know how many overflows will occur in m pushed items, and it is also difficult to find a maximum of the total number of item movement. Here, a push sequence nearly the worst is found. From the push sequence, the times of overflow and the number of items at each time of overflow occurring can be known. Unfortunately, the push sequence shows that the cost of Garwick's algorithm is depentent on the logarithm of the number of push operations but is not a constant factor. The proves are shown as follows.

Suppose the push sequence having m items is arranged as $v_1 = \frac{M}{n}, \ldots, v_k = \frac{M - \sum_{i=1}^{k-1} v_i}{n}$, where the numerator of each v_i is divisible by n. It means that the first overflow will occur after the $\frac{M}{n}$th item is pushed from system is initialized. Then, the second overflow will occur after the $\frac{M-v_1}{n}$th item is pushed since the first overflow was solved; ... and so on. Let $V_k = \sum_{i=1}^{k} v_i$. The values of v_k and V_k are obtained in Lemma 2.

Lemma 2

$$v_k = \frac{M \cdot (n-1)^{k-1}}{n^k} \quad \text{and} \quad V_k = M \cdot \left(1 - \left(\frac{n-1}{n} \right)^k \right).$$

Proof.

$$\text{Since } v_1 = \frac{M}{n}, \qquad\qquad V_1 = v_1 = \frac{M}{n}.$$

$$\vdots \qquad\qquad\qquad\qquad \vdots$$

$$\text{Let } v_{k-1} = \frac{M(n-1)^{k-2}}{n^{k-1}}, \qquad V_{k-1} = M \left(1 - \left(\frac{n-1}{n} \right)^{k-1} \right).$$

$$v_k = \frac{M - V_{k-1}}{n} = \frac{M - M \left(1 - \left(\frac{n-1}{n} \right)^{k-1} \right)}{n} = \frac{M(n-1)^{k-1}}{n^k}.$$

$$V_k = \frac{M}{n} + \frac{M(n-1)}{n^2} + \cdots + \frac{M(n-1)^{k-1}}{n^k}$$

$$= \frac{M}{n^k}(n^{k-1} + (n-1)\cdot n^{k-2} + \cdots + (n-1)^{k-1})$$

$$= \frac{M}{n^k}\left(\binom{k}{1}n^{k-1} - \binom{k}{2}n^{k-2} + \binom{k}{3}n^{k-3} - \cdots + (-1)^{k-1}\right)$$

$$= M\cdot\left(1 - \left(\frac{n-1}{n}\right)^k\right).$$

By induction, this lemma can be proved.
□

Since $t_i = \min(\alpha\lfloor\frac{M-V_{i-1}}{n}\rfloor + \beta\lfloor\frac{M-V_{i-1}}{v_{i-1}}\rfloor \cdot x_j^{i-1})$ and each v_i should be not smaller than t_i, the next lemma exists.

Lemma 3 *Let v_i and V_i be the values in Lemma 2, $v_i \geq t_i$.*

Proof.

Since $v_i = \frac{M(n-1)^{i-1}}{n^i}$ and $t_i = \min(\alpha\lfloor\frac{M-V_{i-1}}{n}\rfloor + \beta\lfloor\frac{M-V_{i-1}}{v_{i-1}}\rfloor\cdot x_j^{i-1})$,

$$v_i - t_i = \frac{M(n-1)^{i-1}}{n^i} - \min(\alpha\lfloor\frac{M-V_{i-1}}{n}\rfloor + \beta\lfloor\frac{M-V_{i-1}}{v_{i-1}}\rfloor\cdot x_j^{i-1}).$$

$$\lfloor\frac{M-V_{i-1}}{n}\rfloor = \lfloor\frac{M(n-1)^{i-1}}{n^i}\rfloor = \lfloor v_i\rfloor.$$

$$\lfloor\frac{M-V_{i-1}}{v_{i-1}}\rfloor = \lfloor\frac{M - M\left(1-\left(\frac{n-1}{n}\right)^{i-1}\right)}{\frac{M(n-1)^{i-2}}{n^{i-1}}}\rfloor = n - 1.$$

Since $v_i = \frac{M(n-1)^{i-1}}{n^i} = \lfloor v_i\rfloor = \lfloor\frac{M(n-1)^{i-1}}{n^i}\rfloor$ and $\alpha + \beta = 1$,

$$v_i - t_i = \frac{M(n-1)^{i-1}}{n^i} - \min(\alpha\lfloor\frac{M(n-1)^{i-1}}{n^i}\rfloor + \beta\cdot(n-1)\cdot x_j^{i-1})$$

$$= \beta[\frac{M(n-1)^{i-1}}{n^i} - \min((n-1)\cdot x_j^{i-1})].$$

The $\min((n-1)\cdot x_j^{i-1})$ should find the maximum of $\min(x_j^{i-1})$ subject to $x_1^{i-1} + \cdots + x_n^{i-1} = v_{i-1}$. Thus,

$$\max\{\min(x_j^{i-1})\} = \frac{v_{i-1}}{n}, \qquad \min(x_j^{i-1}) \leq \frac{v_{i-1}}{n}.$$

$$v_i - t_i \geq \beta(\frac{M(n-1)^{i-1}}{n^i} - (n-1)\cdot\frac{v_{i-1}}{n})$$

$$= \beta(\frac{M(n-1)^{i-1}}{n^i} - \frac{M(n-1)^{i-1}}{n^i})$$

$$= 0.$$

□

Lemma 4 *For Garwick's algorithm, let $v_1 = \frac{M}{n}, \ldots, v_k = \frac{M - V_{k-1}}{n}$, the total number of item movements is at most*

$$M\left[k + \frac{(n-1)^{k+1}}{n^k}\right],$$

where k is a nonnegative integer and $k = \left\lceil \dfrac{\log \frac{M}{M-m}}{\log \frac{n}{n-1}} \right\rceil - 1.$

Proof. Assume the push sequence is the same as our discussion in Lemma 2 and Lemma 3. That is, when overflows occur at $v_1 = \frac{M}{n}, \ldots, v_k = \frac{M - V_{k-1}}{n}$, the total number of item movements is the summation of V_i, $1 \leq i \leq k$, where V_i is the number of items at each time of overflow occurring.

$$\sum_{j=1}^{k} V_j = \sum_{j=1}^{k} M \cdot \left(1 - \left(\frac{n-1}{n}\right)^j\right)$$

$$= M\left[1 - \left(\frac{n-1}{n}\right) + 1 - \left(\frac{n-1}{n}\right)^2 + \cdots + 1 - \left(\frac{n-1}{n}\right)^k\right]$$

$$= M\left[k - (n-1)\left(1 - \left(\frac{n-1}{n}\right)^k\right)\right].$$

What is the value of k? Suppose there are m items in the push sequence and the m pushed items cause k overflow exactly, $k \geq 0$. Since $v_1 + \cdots + v_k = V_k$, the remained items after k overflow will satisfy

$$0 < m - V_k \leq \min\left(\alpha \cdot \left\lfloor \frac{M - V_k}{n} \right\rfloor + \beta \left\lfloor \frac{M - V_k}{v_k} \right\rfloor \cdot x_j^k\right), \quad 1 \leq j \leq n.$$

For $\quad m - V_k > 0, \quad m - M\left(1 - \left(\frac{n-1}{n}\right)^k\right) > 0, \quad \left(\frac{n-1}{n}\right)^k > \frac{M-m}{M}.$

Since $\log \dfrac{n-1}{n} < 0 \quad$ and $\quad \log \dfrac{M-m}{M} \leq 0, \quad k < \dfrac{\log \frac{M}{M-m}}{\log \frac{n}{n-1}}.$ $\qquad$ (2)

For $\quad m - V_k \leq \min\left(\alpha \cdot \left\lfloor \frac{M - V_k}{n} \right\rfloor + \beta \left\lfloor \frac{M - V_k}{v_k} \right\rfloor \cdot x_j^k\right),$

$$m - M \cdot \left(1 - \left(\frac{n-1}{n}\right)^k\right) \leq \alpha \cdot \left(\frac{M(n-1)^k}{n^{k+1}}\right) + \min[\beta \cdot (n-1) \cdot x_j^k].$$

$$m \leq M \cdot \left(1 - \left(\frac{n-1}{n}\right)^k\right) + \alpha \cdot \left(\frac{M(n-1)^k}{n^{k+1}}\right) + \min\left[\beta \cdot (n-1) \cdot x_j^k\right]$$

$$\leq M \cdot \left[1 - \frac{(n-1)^k}{n^{k+1}}(n - \alpha)\right] + \beta \cdot (n-1) \cdot \left(\frac{M(n-1)^{k-1}}{n^{k+1}}\right)$$

$$= M\left(1 - \left(\frac{n-1}{n}\right)^{k+1}\right).$$

Thus, $\quad m \leq M\left(1-\left(\dfrac{n-1}{n}\right)^{k+1}\right), \qquad k+1 \geq \dfrac{\log\left(\frac{M}{M-m}\right)}{\log\left(\frac{n}{n-1}\right)}. \qquad (3)$

In general, $m > n$, $M \geq m$, $n \geq 2$ and $k \geq 0$, from equation (2) and (3)

$$k = \left\lceil \frac{\log \frac{M}{M-m}}{\log \frac{n}{n-1}} \right\rceil - 1.$$

The k should be limited under the condition $M - m \geq n$, for otherwise at least one of the stacks is full, and the overflow occurs every time when the item is pushed. Thus, if $M - m \leq n - 1$, the total number of item movements becomes

$$\sum_{j=1}^{k} V_j + \sum_{j=M-n+1}^{m} j.$$

Under the condition of $0 < M - m \leq n - 1$, we have $m - M \leq -1$ and $M - n \leq m - 1$.

$$\sum_{j=M-n+1}^{m} j = \frac{(m + M - n + 1)(m - (M - n + 1) - 1)}{2}$$

$$\leq \frac{(m + m - 1 + 1)(-1 + n)}{2} = m(n - 1).$$

$$\sum_{j=1}^{k} V_j + \sum_{j=M-n+1}^{m} j \leq M\left[k - (n-1)\left(1 - \left(\frac{n-1}{n}\right)^k\right)\right] + m(n-1)$$

$$\leq M\left[k + \frac{(n-1)^{k+1}}{n^k}\right].$$

□

Theorem 2 *Let m be a fraction of M, $\frac{m}{M} = c$, $0 \leq c < 1$. The Garwick's algorithm is a competitive algorithm when $n = 2$ and $\alpha = 1$, the competitive ratio is $\frac{1}{c} \cdot (k + \frac{1}{2^k}) + 1$, $k = \left\lceil \frac{\log \frac{1}{1-c}}{\log 2} \right\rceil - 1$.*

Proof. In Garwick's algorithm if the number of stacks is 2, the worst sequence P can be found in the following. Initially, the total memory cells, M, is divided into two parts. Each part contains at least $\left\lfloor \frac{M}{2} \right\rfloor$ memory cells. Since the movement only occurs at the second stack, the first overflow must occur after the $\left\lfloor \frac{M}{2} \right\rfloor$ available spaces of stack 2 are used up. After the reallocation of the first overflow, the remaining available spaces are equally divided into two stacks. Then the following pushed items are placed into stack 2 in order to get the maximal number of item movements. Following this rule, the worst push sequence P_w can be found, because any item pushed into stack 2 will decrease the number of item movements. This push sequence P_w is exactly equal to the sequence of Lemma 4. The cost of Garwick's algorithm for such a push sequence P_w is $C_G(P_w)$, which is the sum of the number of pushed items and the total number of item movements. Thus, when $n = 2$

$$C_G(P_w) \leq m + M \cdot \left[k + \frac{1}{2^k}\right] = m\left[\frac{1}{c} \cdot \left(k + \frac{1}{2^k}\right) + 1\right].$$

Since m is a fraction of M, $\frac{m}{M} = c$, and $n = 2$.

$$k = \left\lceil \frac{\log \frac{M}{M-m}}{\log \frac{n}{n-1}} \right\rceil - 1 = \left\lceil \frac{\log \frac{1}{1-c}}{\log 2} \right\rceil - 1.$$

□

4   Conclusions

In this paper, we analyze the algorithm for manipulating multiple stacks. The multiple stacks problem is an on-line problem, it can be applied to many fields. The previous two on-line algorithms, Knuth's algorithm and Garwick's algorithm, have not been analyzed in a deterministic sense until now. This paper shows that the Knuth's algorithm is not a competitive algorithm. The worst push sequence of Garwick's algorithm is difficult to find, but an alternative push sequence is proposed. Fortunately, the push sequence matches the worst push sequence of $n = 2$ when $\alpha = 1$. Therefore, the Garwick's algorithm for $n = 2$ can be proved to be a competitive algorithm. Furthermore, since the push sequence P found here is not the worst sequence for $n \geq 3$, there may exist other push sequences whose total push costs are worse than P's. That is to say that the competitive ratio in Theorem 2 is a low bound for the Garwick's algorithm. If the Garwick's algorithm is a competitive algorithm for all $n \geq 2$, the competitive ratio must be equal to or less than the competitive ratio in Theorom 2. By our analysis, we make a sensible conjecture that the Garwick's algorithm is a competitive algorithm for $n \geq 2$. However, is the conjecture true or false? Does there exist another competitive algorithms for this problem? What is the low bound of the multiple stacks problem? All these questions make this problem the more interesting and worthy of further research.

References

[1] J. L. Bentley and C. C. McGeoch, Amortized Analyses of Self-Organizing Sequential Search Heuristics, *Comm. ACM* 28 (1985) 404-411.

[2] A. Borodin, N. Linial, and M. Saks, An Optimal Online Algorithm for Metrical Task Systems, in *Proceedings 19th Annual ACM Symposium on Theory of Computing*, 1987, pp. 373-382.

[3] B. C. Chien and W. P. Yang, An Amortized Analysis of Linear Hashing, *National Computer Symposium, Taiwan, R.O.C.* 1989, pp. 66-75.

[4] D. E. Knuth, The Art of Computer Programming, Vol. 1: Fundamental Algorithms (Addison-Wesley, reading, MA, 1973).

[5] M. S. Manasse, L. A. McGeoch, and D. D. Sleator, Competitive Algorithms for On-line Problems, in *Proceedings 20th Annual ACM Symposium on Theory of Computing*, 1988, pp. 322-333.

[6] M. S. Manasse, L. A. McGeoch, and D. D. Sleator, Competitive Algorithms for Sever Problems, *J. of Algorithms*, 11 (1990) 208-230.

[7] D. D. Sleator and R. E. Tarjan, Amortized Efficiency of List Update and Paging Rules, *Comm. ACM* 28 (1985) 202-208.

[8] D. D. Sleator and R. E. Tarjan, Self-Adjusting Binary Search Trees, *J. ACM* 32 (1985) 652-686.

Self-Adjusting Augmented Search Trees

Tony W. Lai*

Department of Computer Science, University of Toronto
Toronto, Ontario M5S 1A4, Canada

Abstract. We consider the problem of maintaining a dynamic weighted binary search tree augmented with a secondary search structure. Although we show that partial rebuilding cannot simultaneously achieve optimal search and update times, we introduce a new technique related to partial building called *weighted partial rebuilding*, which supports optimal worst-case search times (within a constant factor) for primary keys, $O(\log n)$ amortized update times, and efficient amortized reweight times. We also give an example application.

1 Introduction

The binary search tree is a famous data structure best known for its ability to efficiently support the operations of *insert*, *delete*, and *member*. By augmenting a binary search tree with an appropriate secondary structure, the tree can support many other operations. Some examples of augmented search trees are the priority search trees of McCreight [7], the persistent search trees of Sarnak and Tarjan [10], the dynamic contour trees of Frederickson and Rodger [4], and the double-ended priority search trees of Klein et al. [5]. In this abstract, we consider the problem of maintaining an augmented search tree that adjusts to a nonuniform query distribution.

The presence of a secondary search structure causes many maintenance problems. For example, in a priority search tree, a rotation may require time proportional to the height of the tree to adjust the secondary structure. Although many efficient schemes for maintaining self-adjusting or adaptive binary search trees have been proposed, such as dynamic binary search [8], splaying [11], biasing [2], weighted randomization [1], partial splaying [6], and deepsplaying [3], none of the schemes can adequately handle augmented trees. Indeed, none of the schemes can ensure an access time of $O(\log n)$ in a priority search tree; they either cannot maintain a worst-case height of $O(\log n)$ or require too many rotations.

Because of the difficulty of modifying an adaptive technique to handle augmented trees, we consider the opposite approach: we take a general technique for maintaining augmented trees and modify it to support adaptation. As our basis, we use Overmars and van Leeuwen's partial rebuilding technique [9]. Unfortunately, although partial rebuilding can maintain an balanced augmented tree

---

* This research was partially supported by an NSERC Postdoctoral Fellowship. Part of this research was performed while the author visited NTT Communication Science Labs, Kyoto, Japan.

efficiently, the technique cannot maintain an adaptive tree efficiently in general. In fact, we prove that it needs an $\Omega(n^{1/l})$ amortized update time to ensure that some fixed external node has depth at most l, for any constant l.

Nevertheless, if our augmented tree can support two operations called *diversions* and *easy deletions* in $O(\log n)$ time, then we can extend partial rebuilding to maintain a weighted tree efficiently. We call this technique *weighted partial rebuilding*. We achieve a worst-case search time of $O(\log\min(n, W/w))$ for the primary keys, where w is the accessed element weight and W is the total tree weight; an amortized update time of $O(\log n)$; and an amortized reweight time of $O(\log\min(n, W/w) + |w' - w|)$, where w' is the new weight of the affected element.

In the following, we assume that all trees are *external-search* trees; that is, all keys are associated with external nodes. We thus require *separating keys* in the internal nodes for guiding searches. We define the *depth* of a node x to be the number of edges on the root-to-x path; the root has depth 0. We define the *size*, $|T|$, of a tree T to be the number of external nodes in T.

We associate a positive integer called a *weight* with each external node. We denote the weight of x by $w(x)$. The *weight of tree* T is the sum of the weights of all external nodes in T.

We define T_L to be the left subtree of T, and T_R to be the right subtree of T. We define T_{LL}, T_{LR}, T_{RL}, and T_{RR} recursively; for example, $T_{LR} = (T_L)_R$. We say that a node x is an *LL* (*LR*, *RL*, *RR*) *node* if there exists a subtree T such that x is the root of T_{LL} (T_{LR}, T_{RL}, T_{RR}, respectively).

2 Partial Rebuilding

Partial rebuilding [9] is based on the following idea: after performing an update, rebuild the subtree rooted at the highest node that is unbalanced with respect to some criterion. By choosing a suitable balance criterion, we can maintain a tree of $O(\log n)$ height in $O(\log n)$ amortized time, provided that we can rebuild any subtree T in $O(|T|)$ time. Unfortunately, partial rebuilding cannot efficiently maintain external nodes near the root, and thus cannot efficiently maintain adaptive external-search trees. To see why, consider a two-player game in which player A inserts nodes into a tree, and player B rebuilds subtrees to perfect balance. Player A can push an external node x downward one level with every insertion, while player B must rebuild large subtrees to bring x upward.

Lemma 1. *For any tree T of size $N \geq 1$ and any constant $k \geq 2$, if player B ensures that x has depth at most l, then player A can perform I insertions with cost greater than $\min(N, I^{l/(l-1)}8^{1-l})$.*

Proof. After an insertion, if player B rebuilds some subtree T', we charge B a cost of $|T'|$; if B does nothing, we charge a cost of 1. Thus, the lemma trivially follows if $I^{l/(l-1)}8^{1-l} < I$, or $I < 8^{(l-1)^2}$.

First, consider the case where $l = 2$. Without loss of generality, assume that x is in T_L. Player A performs I insertions to displace x. Either player B rebuilds the entire tree, implying the cost is at least $N + I - 1$, or B rebuilds T_L after each insertion, implying that the cost is at least $\sum_{i=1}^{I} i > I^2/8$.

To prove the lemma for $l > 2$, we use induction. Suppose the lemma is true for $l = k$, and consider the case where $l = k + 1$. Assume that x is in T_{LL} or is at the root of T_L. Player A first performs $\lfloor I/2 \rfloor$ insertions in T_{LR}. From the induction hypothesis, we know that player A can perform $\lfloor I^{1-1/k} \rfloor$ insertions in T_L with cost at least $\min(\lfloor I/2 \rfloor, \lfloor I^{1-1/k} \rfloor^{k/(k-1)} 8^{1-k})$. Player A repeatedly inserts elements with this strategy until I insertions are performed. We omit the details, but it can be shown that either player B rebuilds the entire tree, implying that the cost is at least $N + I - 1$, or B rebuilds subtrees with a total cost of more than $I^{1+1/k} 8^{-k}$. The lemma follows.

Theorem 2. *Let x be an arbitrary external node. If player B (the rebuilder) ensures that x has depth at most l after each rebuilding, then player A (the adversary) can force player B to use $\Omega(n^{1/l} 8^{-l})$ amortized time per insertion.*

However, by altering separating keys during an insertion, we can prevent an adversary from continually displacing a given external node. We call this key alteration a *diversion*. Assume that an adversary inserts a node with key c that displaces an external node x with key b. If x is an LR or RL node, then we can prevent it from being displaced by changing the separating key of its parent or grandparent; see Fig. 1a. If x is an RR node, though, there are two cases. If $c < b$, then we can still divert c by changing the key of x's parent; see Fig. 1b. Otherwise, if $c > b$, we cannot divert c, but x becomes an RL node after the insertion and cannot be pushed down further. If the node x is an LL node, we can similarly prevent x from being pushed down more than one level.

For efficiency, we also use another operation called a *easy deletion*. As an example, let T be a tree such that its left subtree contains a single external node x. To delete x using partial rebuilding, we must rebuild the entire tree. However, for many types of trees, we can simply delete x by replacing T with T_R or with new tree T' that can be quickly computed from T_R; see Fig. 2. We note that a priority search tree [7] can support both diversions and easy deletions in $O(\log n)$ time if the tree has height $O(\log n)$ in the worst case.

3 The Data Structure

With weighted partial rebuilding, we conceptually maintain a binarized version of a 2-3-4 tree that is balanced with respect to both size and weight. Each internal node of the 2-3-4 tree is represented by a subtree of at most three nodes. To indicate the subtree boundaries, we color the subtree roots black and the other nodes red. Also, for each black node x, we maintain $size(x)$, the total size of the subtree rooted at x; $bigsize(x)$, the maximum permitted value of $size(x)$; and $weight(x)$, the weight of the subtree rooted at x. We also maintain a flag in each external node to indicate whether the node is *normal*, *fixed*, or *semifixed*. Fixed nodes are LR and RL nodes that cannot be displaced downward; semifixed nodes can be displaced at most one level downward. Finally, we maintain a global value

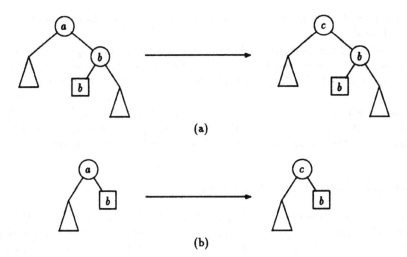

Fig. 1. Insertion and diversion of c, $a < c < b$

Fig. 2. Easy deletion of x

maxweight, which is the maximum weight of the entire tree attained since the last global rebuilding.

Our scheme has three parameters α, β, and γ. We require that $1/2 < \alpha < 2/3 < \beta < 1 < \gamma$. We have six invariants that each node must satisfy.

1. If a node x is black, then $size(x) \leq bigsize(x)$.
2. If a node x is black, then, for any child or grandchild y of x colored black, $size(y) \leq \alpha \cdot size(x)$.
3. If a node x is black, then, for any child or grandchild y of x colored black, $weight(y) \leq \beta^{k+1} maxweight$, where k is the number of proper black ancestors of x.
4. If a node x is red, then x's parent is black.
5. If an external node x is semifixed, then x's parent is black.
6. If an external node x is fixed, then x's parent is red.

If the root of a subtree T inviolates one of the invariants, then we say that T is *unbalanced*.

We also ensure that the total tree weight is at least *maxweight*/2. Otherwise, we rebuild the entire tree. It is straightforward to show that each external node has optimal depth (within a constant factor) with respect to its weight.

Theorem 3. *Let x be an external node of weight w. If all node invariants are satisfied, then the depth of x is at most $2 \cdot \min(\log_{1/\alpha} N, \log_{1/\beta}(W/w)) + O(1)$, where N is the size of the tree, and W is the total weight of the tree.*

3.1 The Construction Algorithm

To construct a tree of size N at most 4, simply build a minimum height tree, label the external nodes normal, and, if $N > 1$, color the root black and the other internal nodes red. In general, to construct a tree T containing elements $x_1, x_2, \ldots, x_N$ with weights $w_1, \ldots, w_N$, for $N > 4$, we consider five cases. Let $W = \sum_{i=1}^{N} w_i$.

1. $w_1 > W/3$.
   Choose the root x and the right child y of x such that T_L is an external node, T_{RL} has size $\lfloor N/2 \rfloor$, and T_{RR} has size $\lceil N/2 \rceil - 1$. Color x black and y red, and label T_L semifixed. Construct T_{RL} and T_{RR} recursively.

2. $w_N > W/3$. Symmetrical to case 1.

3. $w_i > W/3$, where $1 < i \le \lfloor N/2 \rfloor$.
   Choose the root x, the left child y of x, and the right child z of x such that T_{LL} has size $i - 1$, T_{LR} is an external node, T_{RL} has size $\lfloor \frac{N-i}{2} \rfloor$, and T_{RR} has size $\lceil \frac{N-i}{2} \rceil$. Color x black and y and z red, and label T_{LR} fixed. Construct T_{LL}, T_{RL}, and T_{RR} recursively.

4. $w_i > W/3$, where $\lfloor N/2 \rfloor < i < N$. Symmetrical to case 3.

5. $w_i \le W/3$, for all i.
   Choose the root x, the left child y of x, and the right child z of x such that T_L has size $\lfloor N/2 \rfloor$, T_R has size $\lceil N/2 \rceil$, and T_{LL}, T_{LR}, T_{RL}, and T_{RR} have weight at most $2W/3$. Color x black and y and z red. Construct T_{LL}, T_{LR}, T_{RL}, and T_{RR} recursively.

In each case, we have a root x colored black, so we set $size(x)$ to N, $weight(x)$ to W, and $bigsize(x)$ to $\lceil \gamma N \rceil$. If multiple cases are applicable, we may break ties arbitrarily. Observe that we can determine which case is applicable in $O(\log N)$ time if we precompute the partial sums $w_1, w_1 + w_2, \ldots, w_1 + \cdots + w_N$. It is straightforward to show that the above algorithm runs in linear time if the secondary search structure can be rebuilt in linear time.

Theorem 4. *The construction algorithm constructs a tree satisfying all node invariants in $O(N)$ time.*

3.2 The Update Algorithms

To insert an element x, first search for x. Let y be the external node at which the search ends. There are three cases to consider.

1. y is fixed. Divert and insert x.
2. y is semifixed. Let p be y's parent; note p is black. If x's key lies between p's and y's keys, then divert and insert x. Otherwise, insert x naively, color x's new parent red, and relabel y as fixed.
3. y is normal. Insert x naively, and color x's new parent black.

Afterwards, rebuild the largest unbalanced subtree, if any.
  To delete an element x, we consider two cases.

1. x is not fixed and x's parent is red.
   If x's sibling y is fixed, then relabel y as semifixed. Delete x lazily. Rebuild the largest unbalanced subtree, if any.
2. x is fixed or x's parent is black.
   Rebuild the largest unbalanced subtree, if any. If none exists, rebuild the subtree rooted at the lowest black ancestor of x.

Afterwards, if the total tree weight is less than $maxweight/2$, then rebuild the entire tree.
  To increase the weight of an external node x, simply change x's weight, and rebuild the largest unbalanced subtree, if any. To decrease the weight of x, change x's weight, and, if the total tree weight is less than $maxweight/2$, then rebuild the entire tree.

4 Analysis

To analyze the performance of weighted partial rebuilding, we first prove some technical lemmas on subtree rebuilding. There are six possible causes of rebuilding.

A. Invariant 1 is violated at some node.
B. Invariant 2 is violated at some node.
C. Invariant 3 is violated at some node.
D. A normal node with a black parent is deleted.
E. A fixed or semifixed node is deleted.
F. The tree weight is less than $maxweight/2$.

Note that the update algorithms cannot cause violations of invariants 4, 5, or 6.
  Before proving our main theorem, we first bound the number of updates necessary to cause the first four conditions, and we also bound the subtree size when condition E occurs. For brevity, we omit the proofs, as they follow in a straightforward manner from the invariants and the construction algorithm.

Lemma 5. *Let T be a subtree such that T's root violates invariant 1. Then, at least $(1 - 1/\gamma)|T|$ insertions have been performed in T since the last rebuilding of a supertree of T.*

Lemma 6. *Let T be a subtree such that T's root violates invariant 2. Then, at least $(2\alpha - 1)|T|$ updates have been performed in T since the last rebuilding of a supertree of T.*

Lemma 7. *Let T be a subtree such that T's root x violates invariant 3. Then, the weight of T must have increased by at least $(\beta - \frac{2}{3})\beta^k|T|/\alpha^k$ since the last rebuilding of a supertree of T, where k is the number of proper black ancestors of x.*

Lemma 8. *Let T be a subtree that is rebuilt because of condition D. Then, at least $|T|/(4\gamma + 1)$ deletions have been performed in T since the last rebuilding of a supertree of T.*

Lemma 9. *Let T be a subtree that is rebuilt because of condition E, let x be the deleted node, and let M be the maximum weight of x attained since the last global rebuilding. Then, the size of T is at most $\lceil 3\gamma M \rceil$.*

Theorem 10. *Assume that any subtree T can be rebuilt in $O(|T|)$ time, and that a diversion and easy deletion can be performed in $O(h)$ time, where h is the height of the tree. Then, the amortized insertion and deletion time of weighted partial rebuilding is $O(\log N)$, and the amortized time to change the weight of a node from w to w' is $O(\log \min(N, W/w) + \max(0, w' - w))$.*

Proof. To analyze the amortized costs of rebuilding due to conditions A, B, C, and D, we use credits; to analyze the costs due to conditions E and F, we use potential functions. Whenever we change the size of a subtree T, we deposit $\frac{\gamma}{\gamma-1} + \frac{1}{2\alpha-1} + 4\gamma + 1$ credits at T's root. Whenever we increase the weight of T by δ, we deposit $\delta \cdot \alpha^k \beta^{-k}/(\beta - \frac{2}{3})$ credits at T's root, where k is the number of proper black ancestors of T. To analyze the costs of rebuildings due to conditions E and F, we use the potential functions

$$\Phi_E = \sum_{x \in S} \lceil 3\gamma \cdot M(x) \rceil$$

and

$$\Phi_F = \text{maxweight},$$

where S is the set of elements in the tree, and $M(x)$ is the maximum weight x since the last global rebuilding. From all of the above lemmas, it follows that the credits deposited and decreases in potential are sufficient to pay for all subtree rebuildings.

The amortized time to insert a node is no greater than the sum of the search time, which is $O(\log N)$; the time for a diversion, also $O(\log N)$; and the amortized rebuilding time. Since $O(\log N)$ credits are deposited per update, and Φ_E and Φ_F increase by at most a constant, the amortized insertion time is $O(\log N)$.

The amortized time to delete a node is no greater than the sum of the search time, which is $O(\log N)$; the time for an easy deletion, also $O(\log N)$; and the amortized rebuilding time. Since $O(\log N)$ credits are deposited, and Φ_E and Φ_F cannot increase, the amortized deletion time is $O(\log N)$.

The amortized time to change a node's weight is no greater than the sum of the search time, which is $O(\log \min(N, W/w))$, and the amortized rebuilding time. Since at most $O(\max(0, w' - w))$ credits are deposited during a reweight operation, and Φ_E and Φ_F increase by $O(\max(0, w' - w))$, the amortized reweight time is $O(\log \min(N, W/w) + \max(0, w' - w))$.

5 An Application

As an example of how weighted partial rebuilding may be used, we describe an application involving priority search trees [7]. The priority search tree (PST) is an augmented binary search tree that represents a dynamic set S of ordered pairs (x, y), efficiently supporting insertions, deletions, and various queries, including the following *EnumerateRectangle* query.

EnumerateRectangle(x_l, x_h, y_h): Enumerate all pairs (x, y) in S such that $x_l \leq x \leq x_h$ and $y \leq y_h$.

We note that the x-coordinate is the primary key of a PST, and that a PST supports diversions and easy deletions in $O(\log n)$ time and subtree rebuilding in time proportional to the subtree size. For any a, let a^- be the largest x-value in S no greater than a, and a^+ be the smallest x-value in S no less than a. By using weighted partial rebuilding, we can support insertions and deletions in $O(\log n)$ amortized time and *EnumerateRectangle*(x_l, x_h, y_h) queries in $O(\log \min(N, W/w(x_l^-) + W/w(x_h^+)) + k)$ amortized time, where k is the number of points reported.

McCreight observed that a priority search tree can be used to efficiently compute containments in a dynamic set S of linear intervals as follows. We represent each interval $[u, v]$ with the ordered pair (v, u). To determine all intervals that contain $[u, v]$, we enumerate all points whose x-value lies in $[v, \infty]$ and whose y-value is at most u using an *EnumerateRectangle* query.

To obtain an efficient, adaptive structure for computing interval containments, we assign a weight to each interval and maintain the PST structure using weighted partial rebuilding. We also maintain an extra interval $[-\infty, \infty]$, and we increment the weights of $[u, v]$ and $[-\infty, \infty]$ by 1 after each containment query for an interval $[u, v]$ in S. We achieve the following bound.

Theorem 11. *There exists a structure for maintaining a dynamic set S of linear intervals, such that an insertion or deletion requires $O(\log N)$ amortized time, and a containment query for $[u, v] \in S$ requires $O(\log \min(N, W/w) + k)$ amortized time, where w is the number of previous queries for $[u, v]$, W is the sum of N and the total number of queries for all intervals in S, and k is the number of intervals reported.*

Acknowledgements

I would like to thank Derick Wood for his many useful suggestions, including the topic of this paper.

References

1. C. R. Aragon and R. G. Seidel. Randomized search trees. In *Proceedings of the 30th Annual IEEE Symposium on Foundations of Computer Science*, pages 540–545, 1989.
2. S. W. Bent, D. D. Sleator, and R. E. Tarjan. Biased search trees. *SIAM Journal on Computing*, 14:545–568, 1985.
3. V. Estivill-Castro and M. Sherk. Competitiveness and response time in on-line algorithms. In *Proceedings of the 2nd International Symposium on Algorithms*, pages 284–293, 1991.
4. G. Frederickson and S. Rodger. A new approach to the dynamic maintenance of maximal points in the plane. In *Proceedings of the 25th Annual Allerton Conference on Communication, Control, and Computing*, pages 879–888, 1987.
5. R. Klein, O. Nurmi, T. Ottmann, and D. Wood. A dynamic fixed windowing problem. *Algorithmica*, 4:535–550, 1989.
6. T. W. Lai and D. Wood. Adaptive heuristics for binary search trees and constant linkage cost. In *Proceedings of the 2nd Annual ACM-SIAM Symposium on Discrete Algorithms*, pages 72–77, 1991.
7. E. M. McCreight. Priority search trees. *SIAM Journal on Computing*, 14:257–276, 1985.
8. K. Mehlhorn. Dynamic binary search. *SIAM Journal on Computing*, 8:175–198, 1979.
9. M. H. Overmars and J. van Leeuwen. Dynamic multi-dimensional data structures based on quad- and k-d trees. *Acta Informatica*, 17:267–285, 1982.
10. N. Sarnak and R. E. Tarjan. Planar point location using persistant search trees. *Communications of the ACM*, 29:669–679, 1986.
11. D. D. Sleator and R. E. Tarjan. Self-adjusting binary search trees. *Journal of the ACM*, 32:652–686, 1985.

Algorithms for a Class of Min-Cut and Max-Cut Problem

Teofilo F. Gonzalez[1] and Toshio Murayama[2]

[1] Department of Computer Science, University of California,
Santa Barbara, CA 93106, USA
[2] Sony Corporation System LSI Group,
4-14-1 Asahi-cho Atsugi Kanagawa, 243 Japan

Abstract. The *k-Min-Cut (k-Max-Cut)* problem consists of partitioning the vertices of an edge weighted (undirected) graph into k sets so as to minimize (maximize) the sum of the weights of the edges joining vertices in different subsets. We concentrate on the k-Max-Cut and k-Min-Cut problems defined over complete graphs that satisfy the triangle inequality, as well as on d-dimensional graphs. For the one-dimensional version of our partitioning problems, we present efficient algorithms for their solution as well as lower bounds for the time required to find an optimal solution, and for the time required to verify that a solution is an optimal one. We also establish a bound for the objective function value of an optimal solution to the k-Min-Cut and k-Max-Cut problems whose graph satisfies the triangle inequality. The existence of this bound is important because it implies that any feasible solution is a near-optimal approximation to such versions of the k-Max-Cut and k-Min-Cut problems.

1 Introduction

Partitioning a set of objects into equal size subsets is a fundamental problem that arises in many disciplines including CAD (placement of devices) [9] and sparse matrix computations [5]. Our partition problems consist of an integer k, and an edge weighted undirected graph $G = (V, E, W)$, where V is a set of n vertices, E is the edge set, and W is the edge weight function $W : E \rightarrow \Re_0^+$ (the set of non-negative reals). The *k-Min-Cut (k-Max-Cut)* problem consists of partitioning V into k sets so as to minimize (maximize) the sum of the weights of the edges joining vertices in different subsets.

It is well known that the k-Min-Cut and the k-Max-Cut problems are NP-hard even when k is two ([3], and [4]). As a result of this, numerous researchers have studied heuristics, and restricted versions defined over special classes of graphs, e.g. planar graphs, interval graphs, mesh graphs, etc ([5], [6], [9], [10], and [12]). In this paper we concentrate on the k-Max-Cut and k-Min-Cut problems defined over complete graphs that satisfy the triangle inequality, as well as on a restricted class of these graphs called d-dimensional.

We say that a k-Max-Cut (k-Min-Cut) problem is one-dimensional, denoted by *(k,1)-Max-Cut ((k,1)-Min-Cut)*, if the graph is complete, and the set of vertices V can be placed along a straight line in such a way that the weight of each edge is the distance between the two vertices it joins. The (k,d)-*Max-Cut* and (k,d)-*Min-Cut* are defined as the one-dimensional case, but the set of points lie in d-dimensional Euclidean space.

An instance of the k-Min-Cut and k-Max-Cut problems is said to *satisfy the triangle inequality* if the set of edges is complete, and the weight of an edge between any pair of vertices is never larger than the sum of the weights of the edges in any path between such vertices. We shall refer to the problems restricted this way as the (k,t)-*Min-Cut* and (k,t)-*Min-Cut* problems.

In [11] it was established that the (k,t)-Min-Cut and (k,t)-Max-Cut are NP-hard. The proofs are based on the reductions that show that simple Max-Cut is NP-hard [3]. The $(k,2)$-Max-Cut was also shown to be NP-hard in [11]. The proof is involved and follows an approach similar to the one used to establish that a 2-dimensional clustering problem is NP-hard [7].

In this paper we consider the one-dimensional version of our partitioning problems which in our notation are called the $(k,1)$-Max-Cut and $(k,1)$-Min-Cut problems. We present a characterization of optimal solutions to these problems, and present a simple $O(n \log n)$ time algorithms to generate optimal solutions, where n is the number of vertices. Let k be any fixed constant. For the $(k,1)$-Max-Cut, we present a faster algorithm that takes only $O(n)$ time, and for the $(\frac{n}{k},1)$-Max-Cut problem we establish robust $\Omega(n \log n)$ lower bound for the time required to compute an optimal partition. For the $(k,1)$-Min-Cut we discuss an $O(n)$ time algorithm, and for the $(\frac{n}{k},1)$-Min-Cut problem we show that verifying whether or not a solution is an optimal solution requires $\Omega(n \log n)$ time. Of course the lower bounds hold only on certain models of computation.

We also establish a bound for the objective function value of an optimal solution to the (k,t)-Min-Cut and (k,t)-Max-Cut problems, for all k. We give problem instances, for all k, that nearly match the bounds when k is at least 5. For the case when k is two, we give problem instances that match the bound. The existence of these bounds is important because they imply that any feasible solution can be used to approximate the (k,t)-Max-Cut and (k,t)-Min-Cut, and such an approximation is near-optimal for reasonable values of k.

2 One-Dimensional Problems

In this section we study the $(k,1)$-Max-Cut and the $(k,1)$-Min-Cut problems. We present a characterization of optimal solutions to these problems, and present simple $O(n \log n)$ time algorithms to generate optimal solutions. Let k be any fixed constant. For the $(k,1)$-Max-Cut, we present a faster algorithm that takes only $O(n)$ time, and for the $(\frac{n}{k},1)$-Max-Cut problem we establish an $\Omega(n \log n)$ lower bound for the time required to compute an optimal partition. For the $(k,1)$-Min-Cut we discuss an $O(n)$ time algorithm, and for the $(\frac{n}{k},1)$-Min-Cut problem we show that verifying whether or not a solution is an optimal solution requires $\Omega(n \log n)$ time. Of course the lower bounds hold only on certain models of computation.

Let $P_1, P_2, \cdots, P_{km-1}, P_{km}$ be the set of points on a straight line for an instance of the $(k, 1)$-Max-Cut or $(k, 1)$-Min-Cut problems, where $km = n$. We assume without loss of generality that $P_1 \leq P_2 \leq \cdots \leq P_{km-1} \leq P_{km}$. A characterization of optimal solutions to the $(k, 1)$-Min-Cut and $(k, 1)$-Max-Cut problems is given by the following theorem.

Theorem 1. *The partition* $\{S_1, S_2, \cdots, S_k\}$ $(\{S'_1, S'_2, \cdots, S'_k\})$ *is an optimal solution to the* $(k, 1)$-*Min-Cut* $((k, 1)$-*Max-Cut) problem, where*

$$S_i = \{P_i, P_{i+k}, \cdots, P_{i+k(m-2)}, P_{i+k(m-1)}\} \text{ for } 1 \leq i \leq k, \text{ and}$$

$$S'_i = \{P_{(i-1)m+1}, P_{(i-1)m+2}, \cdots, P_{(i-1)m+m-1}, P_{im}\} \text{ for } 1 \leq i \leq k.$$

Proof. We prove this theorem by showing that a partition with smaller (larger) objective function value can be obtained for any partition that differs from the S (S') partition. This is established through the use of interchange arguments. For brevity the proof is omitted.

One can easily show that the partition $\{S'_1, S'_2, \cdots, S'_k\}$ given in Theorem 1 is the only optimal solution to the $(k, 1)$-Max-Cut problem. As pointed out by an anonymous referee, the partition $\{S_1, S_2, \cdots, S_k\}$ is not the only optimal solution to the $(k, 1)$-Min-Cut problem. In this case one can show that any partition in which the points $P_{(i-1)m+1}, P_{(i-1)m+2}, \cdots, P_{(i-1)m+m-1}, P_{im}$, for $1 \leq i \leq k$ belong to distinct sets is an optimal solution. One can show that this represents the sets of all optimal solutions to the $(k, 1)$-Min-Cut problem. Let us now discuss algorithms that generate optimal solutions based on the above characterizations.

Algorithm *sort-it-Min* for the (k,1)-Min-Cut problem reads in the set of points and then sorts them from smallest to largest. The points are then traversed in that order and are assigned to the sets $S_1, S_2, \cdots, S_k, S_1, S_2, \cdots, S_k$, and so on. Clearly, this procedure takes $O(n \log n)$ time. Similarly, algorithm *sort-it-Max* for the (k,1)-Max-Cut problem reads in the points and sorts them from smallest to largest. The smallest m points are assigned to S'_1, the next smallest m points are assigned to S'_2, and so on. This procedure also takes $O(n \log n)$ time. We formalize these results in the following theorem that we state without a proof.

Theorem 2. *Algorithm sort-it-Min (sort-it-Max) solves in $O(n \log n)$ time the (k,1)-Min-Cut ((k,1)-Max-Cut) problem.*

An alternate algorithm, which we call *2-fast-Max*, for the $(2, 1)$-Max-Cut problem first *selects* the smallest $\frac{n}{2}$ points, and assigns them to S'_1. The remaining points are assigned to S'_2. This algorithm takes $O(n)$ time because selection takes linear time [8]. A similar algorithm, *k-fast-Max*, for the $(k, 1)$-Max-Cut takes linear time. Note that k in this case is not an input to the algorithm, i.e., k is a fixed constant in the algorithm (but n is a variable). Thus, we have the following theorem.

Theorem 3. *Algorithm k-fast-Max solves n O(n) time the (k,1)-Max-Cut problem, for any fixed value of k.*

The question now is whether or not a linear time algorithm exists for the other cases? In what follows we show that linear time decision tree algorithms do not exist for the other one dimensional versions of the problem. More specifically, we establish an $\Omega(n \log n)$ lower bound for the $(\frac{n}{2}, 1)$-Max-Cut problems.

By theorem 1 we know that an optimal solution to this problem is of the following form: $\{P_1, P_2\}, \{P_3, P_4\}, \ldots, \{P_{n-1}, P_n\}$. We now show that any algorithm that finds an optimal solution to the $(\frac{n}{2}, 1)$-Max-Cut problem in $T(n)$ time can be used to determine element uniqueness ([2]) of $\frac{n}{2}$ real numbers in $T(n) + O(n)$ time. Such algorithm works as follows. Let $X = (x_1, x_2, \ldots, x_{\frac{n}{2}})$ be the $\frac{n}{2}$ elements in the element uniqueness problem. Let min (max) the smallest number in X. There are two sets of inputs to the $(\frac{n}{2}, 1)$-Max-Cut problem. The first input set is X, and the second input set is X plus the numbers $min - 1$ and $max + 1$. The output to the element uniqueness problem is "yes" if, and only if, the two elements in each of the sets that form an optimal solution to both of the instances of the $(\frac{n}{2}, 1)$-Max-Cut are distinct. Since this checking can be done in $O(n)$ time, we know that any algorithm that solves the $(\frac{n}{2}, 1)$-Max-Cut problem in $T(n)$ time can be used to determine element uniqueness for a set of $\frac{n}{2}$ elements in $T(n) + O(n)$ time. As a result of this, "any" lower bound for element uniqueness also holds for $(\frac{n}{2}, 1)$-Min-Cut problem. Since an $\Omega(n \log n)$ lower bound for element uniqueness is known ([1], and [2]), we also have an $\Omega(n \log n)$ for the $(\frac{n}{2}, 1)$-Min-Cut problem. The following theorem is a trivial extension of these observation.

Theorem 4. *Any algorithm that solves $(\frac{n}{k}, 1)$-Max-Cut problem, for any fixed value of k, in $T(n)$ time can be used to determine element uniqueness for a set of $\frac{n}{2}$ elements in $T(n) + O(n)$ time.*

Let us now discuss very nice $O(n)$ time algorithm for the $(\frac{n}{2}, 1)$-Min-Cut problem which was developed by an anonymous referee. By the comments just after theorem 1 we know that an optimal solution to the $(\frac{n}{2}, 1)$-Min-Cut problem is of the following form: each of the $\frac{n}{2}$ sets contains exactly two points, one from $\{P_1, P_2, \ldots, P_{\frac{n}{2}}\}$ and the other from the remaining points. This characterization suggests a simple $O(n)$ algorithm to construct an optimal partition. First we find the median using the linear time algorithm in [8]. Then we assign each of the first $\frac{n}{2}$ points to distinct sets, the remaining points are assigned similarly. It is simple to see that this linear time algorithm generates an optimal solution to the $(\frac{n}{2}, 1)$-Min-Cut problem, and that it can be easily extended to solve the $(\frac{n}{k}, 1)$-Min-Cut problem, for any fixed value for k.

For the $(2, 1)$-Min-Cut we do not know of any $O(n)$ for its solution, and we do not know of any nontrivial lower bound for it. However, we feel an $\Omega(n \log n)$ lower bound is likely to exist. We base this on the following results for the verification version of these problems. By the verification version of a problem we mean given a problem and a solution, is the solution an optimal solution to the problem. In this case the solution is a partition and we are asked to decide whether or not the partition is an optimal one for the problem.

The $(2,1)$-Max-Cut verification problem can be easily solved in $O(n)$ time by just checking whether or nor all the points in one set of the partition are either located to the left, or to the right, of all the points in the other set. A similar algorithm can be developed for the $(k,1)$-Max-Cut verification problem. The $(\frac{n}{2},1)$-Min-Cut verification problem is a little bit more complex. The problem can be reduced to checking whether or not the sum of the distance between each pair of points in each set in the given solution is equal to the corresponding sum of the sets in an optimal solution, which we know can be computed in $O(n)$ time. A similar algorithm can be developed for the $(\frac{n}{k},1)$-Min-Cut verification problem, for any constant k.

Theorem 5. *The $(\frac{n}{k},1)$-Min-Cut and the $(k,1)$-Min-Cut verification problems, for any fixed value for k, can be solved in $O(n)$ time by the above algorithms.*

Proof. By the above discussion.

For the $(\frac{n}{2},1)$-Max-Cut and the $(2,1)$-Min-Cut verification problems we establish an $\Omega(n\ log\ n)$ lower bound in what follows. Before presenting those results, we need to define a decision problem related to the element uniqueness problem. The *element ϵ-uniqueness* decision problem consists of $\frac{n}{2}$ real numbers, and a real number $\epsilon > 0$. An instance is a yes-instance if, and only if, every two of such numbers are at least ϵ units from each other. It is simple to show that "any" lower bound that can be established for the element uniqueness problem can also be established for the element ϵ-uniqueness problem as long as the bound is derived strictly for the number of "yes" components.

We now show that any algorithm that solves the $(\frac{n}{2},1)$-Max-Cut verification problem in $T(n)$ time can be used to determine element ϵ-uniqueness of $\frac{n}{2}$ real numbers in $T(n) + O(n)$ time. The resuction is defined as follows. Let $X = (x_1, x_2, \ldots, x_{\frac{n}{2}})$ be the $\frac{n}{2}$ elements in the element ϵ-uniqueness problem. The set of inputs to the $(\frac{n}{2},1)$-Max-Cut verification problem is X followed by $x_1 + \epsilon, x_2 + \epsilon, \ldots, x_{\frac{n}{2}} + \epsilon$, and the partition which we want to determine whether or not is an optimal one is: $\{x_1, x_1 + \epsilon\}, \{x_2, x_2 + \epsilon\}, \ldots, \{x_{\frac{n}{2}}, x_{\frac{n}{2}} + \epsilon\}$. With this linear time reduction it is simple to establish the following theorem.

Theorem 6. *Any algorithm that solves $(\frac{n}{k},1)$-Max-Cut verification problem, for any fixed value for k, in $T(n)$ time can be used to determine element ϵ-uniqueness for a set of $\frac{n}{2}$ elements in $T(n) + O(n)$ time.*

Proof. The proof is based on the above reduction.

We now show that any algorithm that solves the $(2,1)$-Min-Cut verification problem in $T(n)$ time can be used to determine element ϵ-uniqueness of $\frac{n}{2}$ real numbers in $T(n) + O(n)$ time. The reduction is defined as follows. Let $X = (x_1, x_2, \ldots, x_{\frac{n}{2}})$ be the $\frac{n}{2}$ elements in the element ϵ-uniqueness problem. The set of inputs to the $(2,1)$-Min-Cut verification problem is X followed by $x_1 +$

$\epsilon, x_2 + \epsilon, \ldots, x_{\frac{n}{2}} + \epsilon$, and the partition which we want to determine whether or not it is an optimal one is: $\{x_1, x_2, \ldots, x_{\frac{n}{2}}\}$ and $\{x_1 + \epsilon, x_2 + \epsilon, \ldots, x_{\frac{n}{2}} + \epsilon\}$. After extending this linear time reduction for any fixed value for k, we can easily establish the following theorem.

Theorem 7. *Any algorithm that solves $(k,1)$-Min-Cut verification problem, or any fixed value for k, in $T(n)$ time can be used to determine element ϵ-uniqueness for a set of $\frac{n}{2}$ elements in $T(n) + O(n)$ time.*

Proof. The proof is based on the above reduction.

3 Triangle Inequality Case

Let us now consider the (k,t)-Min-Cut and (k,t)-Max-Cut problems, i.e., the graph satisfies the triangle inequality. In this section we establish a bound between $f^*_{(k,t)-Min}(I)$ and $f^*_{(k,t)-Max}(I)$, where $f^*_{(k,t)-Min}(I)$ and $f^*_{(k,t)-Max}(I)$ are the objective function values of optimal solutions to the (k,t)-Min-Cut and (k,t)-Max-Cut, respectively.

Theorem 8. *Let I be any instance of the (k,t)-Min-Cut $((k,t)$-Max-Cut) problem. Then,*

$$f^*_{(k,t)-Max}(I) \leq \frac{k+1}{k-1} f^*_{(k,t)-Min}(I), \text{ and } f^*_{(2,t)-Max}(I) \leq 2f^*_{(2,t)-Min}(I).$$

Proof. The proofs are based on elaborate accounting of the cost of transforming from one solution to the other. For brevity the proofs are omitted.

We now show that $f^*_{(2,1)-Max}$ can be arbitrarily close to $2f^*_{(2,1)-Min}$. Consider the example given in Fig. 1. From Theorem 1, we know that an optimal $(2,1)$-Min-Cut partition is

$$S_1 = \{x_1, x_3\} \text{ and } S_2 = \{x_2, x_4\},$$

and an optimal $(2,1)$-Max-Cut partition is

$$S'_1 = \{x_1, x_2\} \text{ and } S'_2 = \{x_3, x_4\}.$$

Then, we know that

$$\lim_{d \to 0} \frac{f^*_{(2,1)-Max}(I)}{f^*_{(2,1)-Min}(I)} = 2.$$

If two vertices are allowed to be at the same location, the above bound becomes exactly 2.

From theorem 2.1 we know that any algorithm which partitions the input into two equal-sized subsets and generates a solution with objective function value $f(I)$ has the property that

$$\frac{|f^*_{(2,1)-Min} - f(I)|}{f^*_{(2,1)-Min}} \leq 1, \text{ and } \frac{|f^*_{(2,1)-Max} - f(I)|}{f^*_{(2,1)-Max}} \leq \frac{1}{2}.$$

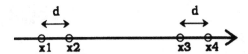

Fig. 1. Example where $f^*_{(2,1)-Max}(I) \approx 2f^*_{(2,1)-Min}(I)$. The distance d is very small.

Thus, any algorithm is a 1-approximation algorithm for the $(2, 1)$-Min-Cut problem and a $\frac{1}{2}$-approximation algorithm for the $(2, 1)$-Max-Cut problem when the input graph satisfies the triangle inequality. These results suggest that there may be other approximation algorithms with a smaller approximation bound.

We know establish that for any positive integer k (≥ 2), there exists an instance I such that

$$f^*_{(k,t)-Max}(I) = \frac{k}{k-1}f^*_{(k,t)-Min}(I).$$

There are nk points partitioned into k subsets $\{S_1, S_2, \cdots, S_k\}$. All the points in each subset have the same coordinate value. Points in different subsets are at a distance d from each other. In this case, the partitions T_1 and T_2 satisfy

$$\frac{t_1}{t_2} = \frac{k}{k-1}$$

where T_1 is the partition $\{S'_1, S'_2, \cdots, S'_k\}$ and T_2 is the partition $\{S_1, S_2, \cdots, S_k\}$, and t_1 and t_2 are the objective function values of T_1 and T_2, respectively. It is simple to show that T_1 is an optimal (k,t)-Max-Cut partition, and T_2 is an optimal (k,t)-Min-Cut partition (the proof is similar to Theorem 1). Therefore, we know that

$$\frac{f^*_{(k,t)-Max}(I)}{f^*_{(k,t)-Min}(I)} = \frac{k}{k-1}.$$

We know from this result that when k is large, the objective function value of any partition is near optimal regardless of n.

4   Discussion

We presented a characterization of optimal solutions to the $(k, 1)$-Max-Cut and the $(k, 1)$-Min-Cut problems. This characterization enabled us to develop a simple $O(n \ log \ n)$ time algorithms to generate optimal solutions. Let k be any fixed constant. For the $(k, 1)$-Max-Cut, we presented a faster algorithm that takes only

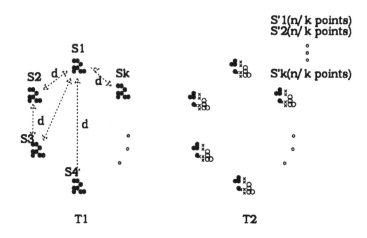

Fig. 2. Possible Upper Bound. T1 (T2) is an optimal solution for the (k,t)-Max-Cut $((k,t)$-Min-Cut) problem.

$O(n)$ time, and for the $(\frac{n}{k}, 1)$-Max-Cut problem we established robust $\Omega(n \ log \ n)$ lower bound for the time required to compute an optimal partition. For the $(k, 1)$-Min-Cut we discussed an $O(n)$ time algorithm developed by an anonymous referee, and for the $(\frac{n}{k}, 1)$-Min-Cut problem we showed that verifying whether or not a solution is an optimal solution requires $\Omega(n \ log \ n)$ time.

We also established a bound for the objective function value of an optimal solution to the (k,t)-Min-Cut and (k,t)-Max-Cut problems. We presented problem instances, for all k, that nearly match the bounds when k is at least 5. For the case when k is two, we presented a problem instance that match the bound. The existence of these bounds is important because they imply that any feasible solution can be used to approximate the (k,t)-Max-Cut and (k,t)-Min-Cut, and such an approximation is near-optimal for reasonable values of k. For a special type of problem instances we can establish tight bounds. For brevity we did not discuss these results.

Several interesting heuristics for our 2-dimensional partitioning problems were developed, and results of an experimental evaluation of these methods is given in [11]. A very interesting open problem is that of finding efficient approximation algorithms for the two dimensional case. A very interesting problem is to determine whether or not the (2,2)-Max-Cut problem is an NP-hard problem. In [11] it was shown that there are problem instances of the (2,2)-Max-Cut problem which do not have an optimal solution in which the two sets in the partition can be separated by k lines, for any fixed value of k.

References

1. M. Ben-Or, "Lower Bounds for Algebraic Computation Trees," Proc. 15th ACM Annual Symposium on Theory of Computing 80 - 86, 1983.
2. D. Dobkin, and R. Lipton, "Multidimensional Searching Problems," *SIAM Journal on Computing*, 15(2), 181 - 186, 1976.
3. M. R. Garey, D. S. Johnson, and L. Stockmeyer, "Some Simplified NP-Complete Graph Problems," *Theoretical Computer Science*, (1), 237 - 267, 1976.
4. M. R. Garey and D. S. Johnson, *"Computers and Intractability,"* Freeman, 1980.
5. A. George and J. W. H. Liu, "An automatic nested dissection algorithm for irregular finite element problems," *SIAM Journal on Numerical Analysis*, 15, 1053 - 1069, 1978.
6. J. R. Gilbert and E. Zmijewski, *"Combinatorial Sparse Matrix Theory,"* Chapter 5, Lecture Notes in Computer Science, UCSB, 1991.
7. T. F. Gonzalez, "Clustering to Minimize the Maximum Intercluster Distance," *Theoretical Computer Science*, 38, 293 - 306, 1985.
8. E. Horowitz and S. Sahni, *"Fundamentals of Computer Algorithms,"* Computer Science Press, 1978.
9. B. W. Kernighan and S. Lin, "An Efficient Heuristic Procedure for Partitioning Graphs," *The Bell System Technical Journal*, 49, 291 - 307, 1970.
10. B. Krishnamurthy, "An improved min-cut algorithm for partition VLSI networks," *IEEE Transactions on Computers*, C(33), 438 - 446, 1984.
11. T. Murayama, "Algorithmic Issues for the Min-Cut and Max-Cut Problems," M.S. Thesis, Department of Computer Science, The University of California, Santa Barbara, December 1991.
12. D. F. Wong, H. W. Leong, and C. L. Liu, *"Simulated Annealing for VLSI Design,"* Kluwer Academic Publishers, 1988.

Algorithms for
Rectilinear Optimal Multicast Tree Problem

Jan-Ming Ho, M. T. Ko, Tze-Heng Ma and Ting-Yi Sung

Institute of Information Science, Academia Sinica
Taiwan, R. O. C.

Abstract. Given a point s called the *signal source* and a set D of points called the *sinks*, a rectilinear multicast tree is defined as a tree $T = (V, E)$ such that $s \in V$, $D \subseteq V$, and the length of each path on T from the source s to a sink t equals the L_1-distance from s to t. A rectilinear multicast tree is said to be *optimal* if the total length of T is minimized. The *optimal multicast tree (OMT) problem* in general is NP-complete [1, 2, 4], while the complexity of the rectilinear version is still open. In this paper, we present algorithms to solve the rectilinear optimal multicast tree (ROMT) problem. Our algorithms require $O(n^{3k})$ and $O(n^2 3^n)$ time, where n denotes $|D|$ and k is the number of dominating layers defined by s.

1 Introduction

In many applications, like VLSI design and message routing in multicomputer networks, sometimes we are asked to find a Steiner tree which connects a source point s and a set of sink points $D = \{t_1, t_2, ..., t_n\}$. A **multicast tree** on (s, D) is a Steiner tree $T = (V, E)$ rooted at s and interconnects all the sinks in D, and the path in T from the source s to a sink t is a shortest path with respect to a given metric or graph topology. This constraint reflects the requirement that the signal propagation delay is minimized. The **optimal multicast tree (OMT) problem** on (s, D) is to find a multicast tree on (s, D) such that the total wire length is minimized. We can consider the OMT problem as a problem of properly selecting a shortest path between each sink and the source, such that the selected shortest paths overlap as much as possible.

The OMT problem is NP-complete if the source and sinks are defined as vertices on a general graph topology [1,2,3]. In this paper, we study the OMT problem where the source and sinks are given as points on the xy-plane. In particular, distances between points are defined by L_1-metric. Since in Euclidean metric, the shortest path between two points is unique; thus, the multicast tree on (s, D) can be trivially obtained.

Whether the **rectilinear optimal multicast tree (ROMT) problem** on (s, D), i.e. the OMT problem in L_1 metric, can be solved in polynomial time is still open. In this paper, we present two algorithms in solving the ROMT problem.

In section 2, we present an $O(n^2 3^n)$ time algorithm based on Dreyer and Wagner's [3] idea in solving the Steiner minimal tree problem. In section 3, we study the ROMT problem by partitioning the sinks into dominating layers with respect to s. An $(O(n^{3k}))$ algorithm is presented, where k is the number of dominating layers. In section 4, we study the special case where there is only one dominating layer, and all sink points in D lies in the first quadrant corresponding to s. We show that an $O(n^2)$ time algorithm can be found to solve this problem.

2   An $O(n^2 3^n)$ Algorithm

A classical algorithm in solving the Steiner minimal tree problem optimally is presented by Dreyer and Wagner [3]. We are going to show in the following that the ROMT problem can also be solved based on a similar idea.

Let's consider an optimal multicast tree T of a given source s and a set of sinks D. First, we notice that a Steiner vertex of T must locates at the intersection, called a natural grid point with respect to $\{s\} \cup D$, of a vertical line L_v and a horizontal line L_h, where L_v and L_h each contains at least a point in $\{s\} \cup D$. We also observe that after the deletion of s from T, T is decomposed into several connected components. Since the degree of s on T is no greater than 4, let's denote the components as $T_0, T_1, T_2,$ and T_3, respectively, where T_i, $0 \leq i \leq 3$ can be empty.

Let's take a closer look at a non-empty component, say T_0, in specific. In other words, T_0 contains at least one vertex in D. Note that the degree of s on T_0 is 1. Let s_0 denote the vertex nearest to s on T_0. Then it can easily be shown that T_0 is an OMT of s_0 and D_0, the set of sinks on T_0. Vertices in T_0 must lie completely in either of the four half spaces defined by either the x- or the y-coordinate. The position of s_0 is also determined uniquely within two candidates.

We can thus present a dynamic programming algorithm for solving the ROMT problem as follows. The algorithms is divided into n iterations, where n is the number of vertices in D. In the ith iteration, the algorithm computes the ROMT $T^{(i)}$ of each subset $D^{(i)}$ of D with exactly i vertices with respect to each grid point p defined by D as the given source, using the smaller ROMT's computed in previous iterations. To compute $T^{(i)}$, we have to enumerate each proper nonempty subset E of points in $D^{(i)}$ such that all the points in E lies completely in either the lower, upper, right, or left half space with respect to p. A grid point q can thus be selected as the given source for feeding the set of points in E. We then compute the sum of the two ROMT's defined by (q, E) and $(p, D^{(i)} - E)$ respectively. The subset $E^* \subseteq D$ giving the minimum sum thus defines the ROMT of $(p, D^{(i)})$.

Note that the number of natural grid points is $O(n^2)$, and that for each subset of D containing i vertices and a given grid point, it takes 2^i time to table-lookup the corresponding ROMT. The time complexity of the dynamic programming algorithm is $O(\sum_{i=1}^{n} \binom{n}{i} n^2 2^i) = O(n^2 3^n)$.

3 An Algorithm by Layer Partition

Before we discuss the general case, we first consider the rectilinear optimal multicast tree problem with sinks in one quadrant.

Given a set of sink points $D = \{t_1, t_2, ..., t_n\}$, let x_*, y_* be the smallest x-coordinate and y-coordinate of points in D respectively. Our aim is to solve the rectilinear optimal multicast tree problem on $(s(D), D)$, $s(D) = (x_*, y_*)$.

Let T be a rectilinear optimal multicast tree. Removing s from T, the tree T is partitioned into two subtrees. The subtrees T_h and T_v denote the subtrees connected to s at a horizontal segment and a vertical segment, respectively. We partition the sink set D into two subsets D_v and D_h according to the subtree each sink belongs to.

Since each sink is connected to the source by a shortest path, in our case, each rectilinear path to s in T increases monotonicly in both x- and y-coordinates. These shortest paths are thus called increasing staircases. Therefore, D_v and D_h can be partitioned by an increasing staircase.

The rectilinear distance between a and b is denoted by $d(a, b)$. The total length of a rectilinear optimal multicast tree on (s, D) is denoted by $\lambda(s, D)$ or $\lambda(D)$ in short.

Lemma 1 *The trees T_h and T_v are rectilinear optimal multicast trees on $(s(D_h), I$ and $(s(D_v), D_v)$, respectively. Furthermore, $\lambda(D) = \lambda(D_v) + \lambda(D_h) + d(s(D_h), s) + d(s(D_v), s)$.*

Lemma 1 can be easily verified. A partition (D_1, D_2) of D is called a **fair partition** if D_1 and D_2 are nonempty sets and separated by an increasing staircase. Let D_1 and D_2 be a fair partition of D, and $\lambda(D_1, D_2)$ be defined as $\lambda(D_1, D_2) = \lambda(D_1) + \lambda(D_2) + d(s(D_1), s(D)) + d(s(D_2), s(D))$. Let D_v and D_h be a fair partition induced by a rectilinear optimal multicast tree. By lemma 1, $\lambda(D) = \lambda(D_h, D_v)$. Since (D_h, D_v) is a fair partition, we can define the following recursive formula for $\lambda(D)$

$$\lambda(D) = \min \{\lambda(D_1, D_2) \mid \text{for all fair partitions } (D_1, D_2) \text{ of } D\}. \quad (1)$$

Therefore, we can use dynamic programming to compute $\lambda(D)$. However, fair partitions are difficult to enumerate. Instead, we will enumerate on a larger set of partitions, called **quasi-fair partitions**. In order to define a quasi-fair partition, we first define a partial order $(D, \geq)$ on the set D. For sinks t_i and t_j, we say $t_i \geq t_j$ if and only if $x_i \geq x_j$ and $y_i \geq y_j$. A sink t_i is said to be **maximal** if there does not exist any sink t_j such that $t_j \geq t_i$. Let $M(D)$ be the set of all maximal sinks of D. This partial order is used to partition D into layers L_i which are given by $L_1 = M(D)$, $L_i = M(D \setminus \cup_{j=1}^{i-1} L_j)$. See Fig. 1 for an illustration. We assume that D is partitioned into k layers.

We define another partial order $(D, \preceq)$ for sinks t_i and t_j. We say $t_i \preceq t_j$ if and only if $x_i \leq x_j$ and $y_i \geq y_j$. Every layer is a chain in the partial order $\preceq$. Let

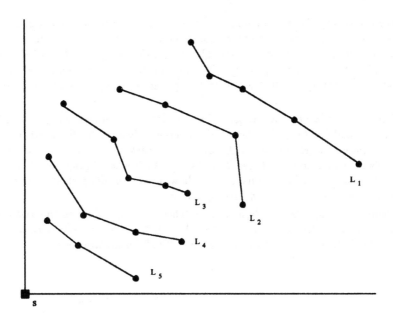

Fig. 1. Layers of sinks in one quadrant

$L_i = \{l_{i1}, l_{i2}, ..., l_{in_i}\}, l_{ip} \preceq l_{iq}$ if $p < q$. We call $\mathbf{L_i[a, b]}=\{l_{ij} | a \leq j < b\}$ an L_i-interval. An L_i-interval can be empty. $L_i[a, b)$ is called an **upper $\mathbf{L_i}$-interval** if $a = 1$; and a **lower $\mathbf{L_i}$-interval** if $b = n_i + 1$.

Lemma 2 *In any fair partition, each layer L_i is partitioned into an upper L_i-interval and the complement, a lower L_i-interval.*

A partition (D_1, D_2) of D, both $D_1 \neq \phi$ and $D_2 \neq \phi$, is said to be a **quasi-fair partition** of D if $D_1 \cap L_i$ is an upper L_i-interval and $D_2 \cap L_i$ is a lower L_i-interval for all $1 \leq i \leq k$. Since a fair partition is quasi-fair, the recursive formula (1) for $\lambda(D)$ can be replaced by

$$\lambda(D) = \min \{\lambda(D_1, D_2) \mid \text{for all quasi-fair partitions } (D_1, D_2) \text{ of } D\}. \quad (2)$$

Let $D(a_1, b_1; \cdots; a_k, b_k)$ denote the set of sinks in $\bigcup_{i=1}^{k} L_i[a_i, b_i)$. To apply the dynamic programming technique, we need to compute $\lambda(D(a_1, b_1; \cdots; a_k, b_k))$, where $1 \leq a_i \leq b_i \leq n_i + 1$ and $i = 1, \cdots, k$. There are $O(\prod_{i=1}^{k} \frac{(n_i+1)(n_i+2)}{2}) \leq O(n^{2k})$ terms to be computed. Any quasi-fair partition of $D(a_1, b_1; \cdots; a_k, b_k)$ is of the form $D(a_1, r_1; \cdots; a_k, r_k)$ and $D(r_1, b_1; \cdots; r_k, b_k)$, where $a_i \leq r_i \leq b_i$ and $r_i \neq a_i$ for some i or $r_i \neq b_i$ for some i. Thus, in evaluating $\lambda(D(a_1, b_1; \cdots; a_k, b_k))$, it takes $O(\prod_{i=1}^{k}(b_i - a_i)) \leq O(\prod_{i=1}^{k}(n_i + 1)) \leq O(n^k)$ time by Equation 2. Thus, to get a rectilinear optimal multicast tree for sinks in one quadrant, it takes $O(n^{3k})$ time, where k is the number of layers in D.

For a general ROMT problem (s, D), we can assume that s is the origin. The plane is then decomposed into four quadrants, $Q_0 = \{(x, y) \mid x \geq 0 \text{ and } y \geq 0\}$, $Q_1 = \{(x, y) \mid x \leq 0 \text{ and } y \geq 0\}$, $Q_2 = \{(x, y) \mid x \leq 0 \text{ and } y \leq 0\}$, and $Q_3 = \{(x, y) \mid x \geq 0 \text{ and } y \leq 0\}$. The sink set D is decomposeded into four subsets $D_i = D \cap Q_i$ for $0 \leq i \leq 3$. In this paper, all indices are taken modulo 4.

Call a point $v = (x, y)$ a *grid point* if $x = x'$ and $y = y''$ for some $p, q \in \{s\} \cup D$, $p = (x', y')$ and $q = (x'', y'')$. The set of grid points, denoted by H, together with s and D induce a grid graph on the plane. It is easy to see that there exists a rectilinear optimal multicast tree T on (s, D) which is a subgraph on that grid graph. Without loss of generality, in the following we consider rectilinear optimal multicast trees contained in the grid graph only.

Define $E_i = Q_i \cap Q_{i+1} \cap H, 0 \leq i \leq 3$. These are the points on the axis that we need to take into consideration. We call c_i, $0 \leq i \leq 3$, the **i-th corner** of T if c_i is the farthest point to s in $T \cap E_i$. See Fig. 2 for illustrations.

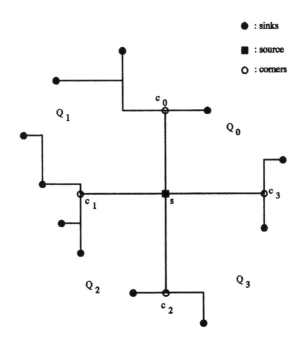

Fig. 2. The corners of an optimal multicast tree.

For a tree T, we use $\ell(T)$ to denote the total length of T. Let $T_i = T \cap Q_i$. The following lemma underlie the applicability of the divide and conquer technique to solve the ROMT problem.

Lemma 3 *Let T be a rectilinear optimal multicast tree. Then T_i is a rectilinear optimal multicast tree on the sink set $D_i \cup \{c_i, c_{i-1}\}$, $0 \le i \le 3$.*

Proof: Let T_i' be a rectilinear optimal multicast tree on the sink set $D_i \cup \{c_i, c_{i-1}\}$. We have $\ell(T_i') \le \ell(T_i)$. Define $T' = \bigcup_{i=0}^3 T_i'$. It follows that $\ell(T') \le \sum_{i=0}^3 \ell(T_i') - \sum_{i=0}^3 d(s, c_i)$. Similarly we have $\ell(T) = \sum_{i=0}^3 \ell(T_i) - \sum_{i=0}^3 d(s, c_i)$. Since $\ell(T_i') \le \ell(T_i)$, it follows that $\ell(T') \le \ell(T)$. Since T is optimal, it follows that $\ell(T') = \ell(T)$ and therefore, $\ell(T_i) = \ell(T_i')$ for all i. The lemma follows. $\square$

By assumption, one of the elements in E_i is the i-th corner of a rectilinear optimal multicast tree for $0 \le i \le 3$. Let $T_i(q_{i-1}, q_i)$, $q_i \in E_i$, denote the rectilinear optimal multicast tree on the sink set $D_i \cup \{q_{i-1}, q_i\}$, $0 \le i \le 3$. By Lemma 3, there exists a quadruple $(q_0, q_1, q_2, q_3) \in E_0 \times E_1 \times E_2 \times E_3$, such that the union of all $T_i(q_i, q_{i-1})$ for $0 \le i \le 3$, is an optimal multicast tree on D with q_0, q_1, q_2, q_3 as corners. To obtain a rectilinear optimal multicast tree, we need $O(n^4)$ time to merge the results if $T_i(q_{i-1}, q_i)$ for all $q_i \in E_i$, $0 \le i \le 3$, are already known.

However, the time complexity of this merging step can be reduced to $O(n^3)$ as shown below. For a given pair (q_1, q_3), we use $T_{01}(q_1, q_3)$ and $T_{23}(q_1, q_3)$ to denote rectilinear optimal multicast trees on the sink set $D_0 \cup D_1 \cup \{q_1, q_3\}$ and $D_2 \cup D_3 \cup \{q_1, q_3\}$, respectively. Then $T_{01}(q_1, q_3)$ can be obtained by searching $T_0(q_3, q_0) \cup T_1(q_0, q_1)$ for all $q_0 \in E_0$ to find the one having the minimum total length. Similarly we can find $T_{23}(q_1, q_3)$. This can be accomplished in $O(n)$ time for each pair (q_1, q_3). A rectilinear optimal multicast tree on the sink set D can be obtained by applying similar process to $T_{01}(q_1, q_3) \cup T_{23}(q_1, q_3)$ for all $(q_1, q_3) \in E_1 \times E_3$. This can also be completed in $O(n^3)$ time.

A straightforward implementation to obtain $T_i(q_{i-1}, q_i)$ for all $q_i \in E_i$, $0 \le i \le 3$, can be done by making $O(n^2)$ calls to the one quadrant algorithm mentioned earlier. The overall complexity would then be $O(n^{3k+2})$. However, we find that there is a more efficient way to obtain $T_i(q_{i-1}, q_i)$ for all $q_i \in E_i$, $0 \le i \le 3$. In the following, we propose a modified dynamic programming algorithm which generates all possible $T_i(q_{i-1}, q_i)$ for one quadrant in $O(n^{3k})$ time. This leads to an improvement from a complexity of $O(n^{3k+2})$ as above to $O(n^{3k})$ for the general ROMT problem. The improvement is significant in the case of $k = 1$.

Let D be a set of sinks in one quadrant with respect to a source s and T be a rectilinear optimal multicast tree of D having p and q at the corners. Similar to the discussion for the case of sink set without corners, removing the root s from T, then T is decomposed into two subtrees, T_v and T_h. Accordingly, the sink set D is also partitioned into two subsets D_v and D_h. Then D_v and D_h also form a fair partition of D.

Consider the subtree T_v in more details. Let s_v be the root (or the source) of T_v. In other words, s_v is the Steiner vertex or sink on the vertical axis which has the smallest y-coordinate. It is obvious that T_v is an optimal multicast tree on $D_v \cup \{p\}$ with source s_v. Removing s_v from T_v, T_v is further decomposed into two subtrees, one with a horizontal segment connecting s_v, denoted by T_{vh} and the other with a vertical segment connecting s_v, denoted as T_{vv}. We partition D_v accordingly into two subsets D_{vh} and D_{vv}. The subtree T_{vh} is a rectilinear optimal multicast tree on D_{vh} with s_{vh} as source. The subtree T_{vv} is a rectilinear optimal multicast tree of D_{vv} with root s_{vv} and sink p as corner on vertical axis. We have the same analysis for T_h. Let (D_1, D_2) be a quasi-fair partition of D and $s_1 = s(D_1 \cup \{p\})$ and $s_2 = s(D_2 \cup \{q\})$. Let $\lambda(D_1, D_2, p, q) = \lambda(D_1 \cup \{p\}) + \lambda(D_2 \cup \{q\}) + d(s, s_1) + d(s, s_2)$, $\lambda(D_1, D_2, p, *) = \lambda(D_1 \cup \{p\}) + \lambda(D_2) + d(s, s_1) + d(s, s(D_2))$ and $\lambda(D_1, D_2, *, q) = \lambda(D_1) + \lambda(D_2 \cup \{q\}) + d(s, s(D_1)) + d(s, s_2)$. Similar to Equation 2, we have

$$\lambda(D \cup \{p, q\}) = \min\{\lambda(D_1, D_2, p, q) \mid \text{ for all quasi-fair partitions } (D_1, D_2) \text{ of } D \}$$

$$\lambda(D \cup \{p\}) = min\{\lambda(D_1, D_2, p, *) \mid \text{ for all quasi-fair partitions } (D_1, D_2) \text{ of } D \}$$

and

$$\lambda(D \cup \{q\}) = min\{\lambda(D_1, D_2, *, q) \mid \text{ for all quasi-fair partitions } (D_1, D_2) \text{ of } D \}$$

Thus, we need to compute $\lambda(D(1, b_1; \cdots; 1, b_k) \cup \{p\})$ and $\lambda(D(a_1, n_1 + 1; \cdots; a_k, n_k + 1) \cup \{q\})$, in addition to $\lambda(D(a_1, b_1; \cdots; a_k, b_k))$, for $1 \le a_i, b_i \le n_i + 1$ and $p \in E_0$ and $q \in E_3$. There are $O(n^{k+1})$ terms to be computed, and each takes $O(n^k)$ time. Thus, we need an additional $O(n^{2k+1})$ time to get all the optimal multicast trees of D with $O(n^2)$ boundary sinks. Putting things together, we get an algorithm of complexity $O(n^{3k} + n^{2k+1} + n^3) = O(n^{3k})$.

4 $O(n^2)$ Algorithm for Sinks of One Layer in One Quadrant

In this section, we further restrict our focus on the special case where the points in D all lie in one quadrant and form a single dominating layer with respect to s. We'll show in the following that this problem can be solved in $O(n^2)$ time. Note that using the algorithm introduced in the previous section, the same problem is solved in $O(n^3)$ time.

Let $D = \{t_1, \cdots, t_n\}$, where $t_i \preceq t_j$ if $i < j$. As before, $D[a, b)$ denotes the set of terminals $\{t_i \mid a \le i < b\}$. Let $s_{a,b}$ denote $s(D[a, b))$ for all $1 \le a < b \le n + 1$. The dynamic programming algorithm presented before is based on a recurrence formula, restated as following:

$$\lambda(D[a, b)) = \min_{a < k < b}\{\lambda(D[a, k)) + \lambda(D[k, b)) + d(s_{a,k}, s) + d(s_{k,b}, s)\} \quad (3)$$

An index k, $a < k < b$, is a **breaking point** of $D[a, b)$ if $\lambda(D[a, b)) = \lambda(D[a, k)) + \lambda(D[k, b)) + d(s_{a,k}, s) + d(s_{k,b}, s)$. A straightforward analysis of this algorithm yields a complexity of $O(n^3)$ since there are $O(n^2)$ terms to be calculated and for each term we have to check on $O(n)$ possible breaking points. A technique similar to that used in constructing optimal binary search trees [5] allows us to search less breaking points and find an $O(n^2)$ time algorithm. The following two lemmas provide the basis for our analysis.

Lemma 4 $\lambda(D[a,b)) + \lambda(D[i,j)) > \lambda(D[a,j)) + \lambda(D[i,b))$, where $a < i < j < b$.

Proof: We will prove by induction on cardinality of $D[a,b)$. It can be verified that Lemma 4 holds when $|D|=4$. Suppose it is true for all $D, |D| \leq b-1-a$.

Let x be a breaking point of $D[a,b)$ and y be a breaking point of $D[i,j)$. Without loss of generality, assume $x \leq y$; it follows that $a < x < j$ and $i < y < b$. See Fig. 3 for an illustration. By the definition of $\lambda(D[a,j))$ and $\lambda(D[i,b))$, we have the following equations.

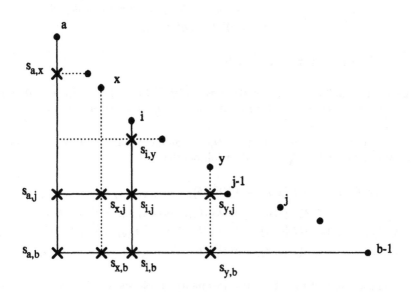

Fig. 3.

$$\lambda(D[a,j)) \leq \lambda(D[a,x)) + \lambda(D[x,j)) + d(s_{a,x}, s_{a,j}) + d(s_{x,j}, s_{a,j}) \quad (4)$$

and

$$\lambda(D[i,b)) \leq \lambda(D[i,y)) + \lambda(D[y,b)) + d(s_{i,y}, s_{i,b}) + d(s_{y,b}, s_{i,b}) \quad (5)$$

Since $|D[x,b)| < |D[a,b)|$, by induction hypothesis,

$$\lambda(D[x,j)) + \lambda(D[y,b)) < \lambda(D[x,b)) + \lambda(D[y,j)) \quad (6)$$

is obtained. It can be verified that

$$d(s_{a,x}, s_{a,j}) + d(s_{x,j}, s_{a,j}) + d(s_{i,y}, s_{i,b}) + d(s_{y,b}, s_{i,b})$$
$$= d(s_{a,x}, s_{a,b}) + d(s_{x,b}, s_{a,b}) + d(s_{i,y}, s_{i,j}) + d(s_{y,j}, s_{i,j}) \tag{7}$$

Joining Equation 4 and Equation 5, with Equation 6 and Equation 7, it follows that

$$\lambda(D[a, j)) + \lambda(D[i, b)) \leq \lambda(D[a, x)) + \lambda(D[i, y)) + \lambda(D[x, j)) + \lambda(D[y, b))$$
$$+ d(s_{a,x}, s_{a,j}) + d(s_{x,j}, s_{a,j}) + d(s_{i,y}, s_{i,b}) + d(s_{y,b}, s_{i,b})$$
$$< \lambda(D[a, x)) + \lambda(D[i, y)) + \lambda(D[x, b)) + \lambda(D[y, j))$$
$$+ d(s_{a,x}, s_{a,b}) + d(s_{x,b}, s_{a,b}) + d(s_{i,y}, s_{i,j}) + d(s_{y,j}, s_{i,j})$$
$$= \lambda(D[a, x)) + \lambda(D[x, b)) + \lambda(D[i, y)) + \lambda(D[y, j))$$
$$+ d(s_{a,x}, s_{a,b}) + d(s_{x,b}, s_{a,b}) + d(s_{i,y}, s_{i,j}) + d(s_{y,j}, s_{i,j})$$
$$= \lambda(D[a, b)) + \lambda(D[i, j))$$

Thus, the lemma is proved by induction. $\qquad\square$

Lemma 5 *Let u and l be breaking points of $D[a, b-1)$ and $D[a+1, b)$ respectively. There exists a breaking point x of $D[a, b)$ such that $u \leq x \leq l$.*

Proof: Let x be a breaking point of $D[a, b)$. Suppose that $x < u$. By the definition of $\lambda(D[a, b))$ and $\lambda(D[a, b-1))$, it follows that

$$\lambda(D[a, b)) = \lambda(D[a, x)) + \lambda(D[x, b)) + d(s_{a,x}, s_{a,b}) + d(s_{x,b}, s_{a,b})$$
$$\leq \lambda(D[a, u)) + \lambda(D[u, b)) + d(s_{a,u}, s_{a,b}) + d(s_{u,b}, s_{a,b})$$
$$\lambda(D[a, b-1)) = \lambda(D[a, u)) + \lambda(D[u, b-1)) + d(s_{a,u}, s_{a,b-1}) + d(s_{u,b}, s_{a,b-1})$$
$$\leq \lambda(D[a, x)) + \lambda(D[x, b-1)) + d(s_{a,x}, s_{a,b-1}) + d(s_{x,b}, s_{a,b-1})$$

Taking the sum of the above two inequality, it follows that

$$\lambda(D[x, b)) + \lambda(D[u, b-1)) \leq \lambda(D[u, b)) + \lambda(D[x, b-1)). \tag{8}$$

Since $x < u < b - 1 < b$, this is a contradiction to Lemma 4. It follows that $x \geq u$. By a similar argument, we can show that $x \leq l$. $\qquad\square$

Based on Lemma 5, for each fixed integer i, $2 \leq i \leq n$, $\lambda(D[a, a+i))$, $a = 1, \cdots, n - i + 1$, can be computed in a total of $O(n)$ time because the index of the breaking point increases monotonicly with a. Thus, it takes $O(n^2)$ time to solve the optimal multicast tree problem where all sink locates in one quadrant and on a single layer with respect to the source.

5 Concluding Remarks

We have studied in this paper the ROMT problem and several special cases. We present two algorithms to solve the rectilinear optimal multicast tree problem, which run in $O(n^2 3^n)$ and $O(n^{3k})$ respectively, where k is the number of dominating layers of sink set with respect to the source. As k is small, the second algorithm is more efficient. Also, in the case where $k = 1$ and all sinks are in one quadrant with respect to the source s, an $O(n^2)$ algorithm is presented, which is an improvement over the general algorithm. Note that the merging step takes $O(n^3)$ time. Thus, our solution to the general rectilinear optimal multicast tree problem with $k = 1$ still takes $O(n^3)$ time.

Although we have proposed several interesting techniques in devising algorithms for solving the ROMT problems, the time complexities of the given algorithms are far from being practical except for several special cases. We conjecture that the rectilinear optimal multicast tree problem is also *NP*-complete as is its general version. To benefit practical applications, efficient approximation algorithms must be pursued in the future.

References

1. H.-A. Choi and A.-H. Esfahanian (1989) "The Complexity of Optimal Distance-Preserving trees," MSU-CPS-ACS-18, Technical Report of Department of Computer Science, Michigan State University.
2. H.-A. Choi, A.-H. Esfahanian and B. C. Houck (1988) "Optimal Communication Trees with Application to Hypercube Multicomputer," Proceedings of the Sixth International conference on the Theory and Application of Graph Theory.
3. S. E. Dreyer and R. A. Wagner, (1972) "The Steiner Problem in Graphs," Networks, vol. 1, pp. 195-207.
4. X. Lin and L. M. Ni, (1989) "Some Theoretical Results on Multicast Communications," MSU-CPS-ACS-17, Technical Report of Department of Computer Science, Michigan State University.
5. D. E. Knuth, (1971) "Optimal Binary Search Trees," Acta Informatica, vol. 1, pp. 14-25.

Approximating Treewidth and Pathwidth of some Classes of Perfect Graphs

Ton Kloks * and Hans Bodlaender **

Department of Computer Science, Utrecht University
P.O.Box 80.089, 3508 TB Utrecht, The Netherlands

Abstract. In this paper we discuss algorithms that approximate the tree-width and pathwidth of cotriangulated graphs, permutation graphs and of cocomparability graphs. For a cotriangulated graph, of which the treewidth is at most k we show there exists an $O(n^2)$ algorithm finding a path-decomposition with width at most $3k + 4$. If $G[\pi]$ is a permutation graph with treewidth k, then we show that the pathwidth of $G[\pi]$ is at most $2k$, and we give an algorithm which constructs a path-decomposition with width at most $2k$ in time $O(nk)$. We assume that the permutation π is given. In this paper we also discuss the problem of finding an approximation for the treewidth and pathwidth of cocomparability graphs. We show that, if the treewidth of a cocomparability graph is at most k, then the pathwidth is at most $O(k^2)$, and we give a simple algorithm finding a path-decomposition with this width. The running time of the algorithm is dominated by a coloring algorithm of the graph. Such a coloring can be found in time $O(n^3)$.
If the treewidth is bounded by some constant, previous results (i.e. [10, 21]), show that, once the approximations are given, the exact treewidth and path-width can be computed in linear time for all these graphs.

1 Introduction

In many recent investigations in computer science, the notions of treewidth and pathwidth play an increasingly important role. One reason for this is that many problems, including many well studied NP-complete graph problems, become solvable in polynomial and usually even linear time, when restricted to the class of graphs with bounded tree- or pathwidth (See e.g. [1, 3, 5, 7]). Of crucial importance for these algorithms is, that a tree-decomposition or path-decomposition of the graph is given in advance. Much research has been done in finding a tree-decomposition with a reasonable small treewidth. Recent results ([25]) show that an $O(n \log n)$ algorithm exists to find a suitable tree-decomposition for a graph with bounded treewidth. However, the constant hidden in the 'big oh', is exponential in the treewidth, limiting the practicality of this algorithm.

For many *special classes* of graphs, it has been shown that the treewidth can be computed efficiently. In this paper we discuss the problem of finding approximate

---

* This author is supported by the foundation for Computer Science (S.I.O.N) of the Netherlands Organization for Scientific Research (N.W.O.), Email: ton@cs.ruu.nl.

** This author is partially supported by by the ESPRIT Basic Research Actions of the EC under contract 7141 (project ALCOM II). Email:hansb@cs.ruu.nl.

tree- and path-decomposition for cotriangulated graphs, permutation graphs and for cocomparability graphs. We also show that for these graphs, if the treewidth is at most k, then the pathwidth is bounded by some polynomial in k.

It has been shown that computing the pathwidth of a triangulated graph is NP-complete ([20]). It is unknown whether there exist good approximations, which can be computed efficiently, for the pathwidth of a triangulated graph. Surprisingly, for cotriangulated graphs, the pathwidth and treewidth are related in a linear fashion, and there is a very simple algorithm that computes an approximate path-decomposition.

Permutation graphs have a large number of applications in scheduling problems. See for example [17], where permutation graphs are used to describe a problem concerning the memory requirements of a number of programs at a certain time (see also [19]). Permutation graphs also arise in a natural way, in the problem of sorting a permutation, using queues in parallel. In [19] it is shown that this problem is closely related with the coloring problem of permutation graphs.

We show that the pathwidth of a permutation graph is at most two times the treewidth of that graph, and we give a linear time algorithm which produces a path-decomposition which is at most two times off the optimal one. If the treewidth of the permutation graph is bounded by a constant, this result, together with earlier results show that an optimal tree- and path-decomposition can be computed in linear time.

Comparability graphs and their complements have a large number of applications (see e.g. [19, 28]). These graph classes, as well as those of the triangulated and of the cotriangulated graphs, are among the largest and most important classes of perfect graphs. Cocomparability graphs properly contain many other well known classes of perfect graphs, like interval graphs, cographs (or P_4-free graphs), permutation graphs, indifference graphs, complements of superperfect graphs etc. (see e.g. [19] or [6, pages 67-96]). Cocomparability graphs are exactly the intersection graphs of continuous functions $F_i : (0, 1) \to R$ (see [29]). Comparability graphs can be recognized in polynomial time (see [24, 19, 18, 30]). Given a comparability graph, a transitive orientation of the edges can be found in $O(n^2)$ time ([30]).

We show that, if the treewidth of a cocomparability graph G is at most k, then the pathwidth is at most $O(k^2)$, and we give a simple algorithm which finds a path-decomposition with this width. Our algorithm uses a transitive orientation of the complement $\overline{G}$, and the *height function* which can be obtained from this orientation in linear time. The algorithm also uses a vertex coloring of the graph G. This vertex coloring, or clique cover in the complement, can be found in $O(n^3)$ time using a flow algorithm, as described in [19].

2 Preliminaries

In this section we start with some definitions and easy lemmas. More backgrounds can be find in [19] and [6]. One of the oldest classes of graphs known to be perfect are the triangulated graphs.

Definition 1. A graph is called *triangulated* (or *chordal* if it has no induced chordless cycle of length at least four. The complement of a triangulated graph is called *cotriangulated*.

Definition 2. An undirected graph $G = (V, E)$ is called a *comparability graph*, or transitively orientable graph, if there exists an orientation (V, F) of the edges satisfying:

$$F \cap F^{-1} = \emptyset \wedge F + F^{-1} = E \wedge F^2 \subseteq F$$

where $F^2 = \{(a, c) \mid \exists_{b \in V} (a, b) \in F \wedge (b, c) \in F\}$. Such an orientation is called a *transitive* orientation. A *cocomparability graph* is the complement of a comparability graph.

For our algorithm we shall need the concept of a height function defined in [19]. Let F be an acyclic orientation of an undirected graph $G = (V, E)$. A *height function* h assigns a non-negative integer to each vertex as follows: $h(v) = 0$ if v is a sink; otherwise, $h(v) = 1 + \max\{h(w) \mid (v, w) \in F\}$. In other words, $h(v)$ is the maximal length of a path from v to a sink. A height function can be assigned in linear time ([19]), and represents a proper vertex coloring of G. If F is a transitive orientation, the coloring by h is optimal (i.e. uses the least possible number of colors) (see e.g. [19]).

We also need a coloring of the cocomparability graph. In [19] a method is described to find an optimal coloring of a cocomparability graph by using a minimum flow algorithm. It follows that this coloring can be found in $O(n^3)$ time.

A permutation π of the numbers $1, \ldots, n$ is denoted as a sequence $\pi = [\pi_1, \ldots, \pi_n]$. We use the notation π_i^{-1} for the position of the number i in this sequence.

Definition 3. If π is a permutation of the numbers $1, \ldots, n$, we can construct an undirected graph $G[\pi] = (V, E)$ with vertex set $V = \{1, \ldots, n\}$, and edge set E:

$$(i, j) \in E \Leftrightarrow (i - j)(\pi_i^{-1} - \pi_j^{-1}) < 0$$

An undirected graph is called a *permutation graph* if there exists a permutation π such that $G \cong G[\pi]$.

In this paper we assume that the permutation π is given, and we show some results on the pathwidth and treewidth of $G[\pi]$, which is sometimes called the *inversion graph* of π. If the permutation π is *not* given, transitive orientations of G and $\overline{G}$ can be computed in $O(n^2)$ time ([30]). Given these orientations, a permutation can be computed in $O(n^2)$ time (see e.g. [19]).

A permutation graph $G[\pi]$ is an intersection graph, which is illustrated by the *matching diagram* of π ([19]).

Definition 4. Let π be a permutation of $1, \ldots, n$. The *matching diagram* of π can be obtained as follows. Write the number $1, \ldots, n$ horizontally from left to right. Underneath, write the numbers $\pi_1, \ldots, \pi_n$, also horizontally from left to right. Draw n straight line segments joining the two 1's, the two 2's, etc.

Notice that two vertices i and j of $G[\pi]$ are adjacent if and only if the corresponding line segments in the matching diagram of π intersect. Matching diagrams are often useful in visualizing certain concepts.

Definition 5. A *tree-decomposition* of a graph $G = (V, E)$ is a pair $D = (S, T)$ with $T = (I, F)$ a tree and $S = \{X_i \mid i \in I\}$ a collection of subsets of vertices of G, one subset for each node in T, such that the following three conditions are satisfied:

1. $\bigcup_{i \in I} X_i = V$.
2. For all edges $(v, w) \in E$ there is a subset $X_i \in S$, such that both v and w are contained in X_i.
3. For each vertex x, the set $\{i \in I \mid x \in X_i\}$ forms a connected subtree of T.

A *path-decomposition* of a graph G is a tree-decomposition (S, T) such that T is a path. We also use the notation $(X_1, X_2, \ldots)$ for a path-decomposition. The *width* of a tree-decomposition (S, T), with $S = \{X_i \mid i \in I\}$, is $\max_{i \in I} (|X_i| - 1)$.

Definition 6. The *treewidth* of G is the minimum width over all tree-decomposition of G. The *pathwidth* of G is the minimum width over all path-decomposition of G.

An alternative way to define the class of graphs with treewidth at most k is by means of *partial k-trees* (see e.g. [22]). There exist linear time algorithms for many NP-complete problems, when restricted to the class of partial k-trees for some *constant* k and when a tree-decomposition with bounded width is given (see e.g. [1, 5, 3, 13, 14, 15]).

Determining whether the treewidth or pathwidth of a given graph is at most a given integer k is NP-complete ([2]). In view of this, the results of Robertson and Seymour on minor closed classes of graph are of great interest. Robertson and Seymour proved that every minor closed class of graphs is recognizable in $O(n^3)$ time ([26]). Since for every fixed k, the class of graphs with treewidth (pathwidth) at most k is minor closed, it follows that for every constant k there is a polynomial algorithm that recognizes graphs with treewidth (pathwidth) at most k. In fact, for these minor closed classes faster algorithms exist. For $k = 2, 3$ there is a linear time algorithm for the treewidth problem using rewrite rules (see e.g. [4] and [23]). For fixed $k \geq 4$ an $O(n \log n)$ algorithm exists which constructs a tree-decomposition with width k ([25], [10] and [21]). If an (approximate) tree-decomposition with bounded width is given, the exact treewidth, and a corresponding tree-decomposition, can be computed in linear time ([10]). In this case, when also the pathwidth is bounded, also the pathwidth and an optimal path-decomposition can be computed in linear time. For an introductory overview of recent results dealing with treewidth and pathwidth, see [7].

3 An approximate path-decomposition for cotriangulated graphs

For triangulated graphs the treewidth is equal to the maximum clique size minus one. It follows that, for triangulated graphs, the treewidth can be computed in linear time. Computing the pathwidth of a triangulated graph is an NP-complete problem, as shown in [20].

It is unknown if the treewidth or pathwidth of cotriangulated graphs can be computed efficiently. We show how to find good approximations for the treewidth and pathwidth of cotriangulated graphs. Let G be a cotriangulated graph with treewidth at most k. We show that the pathwidth of G is at most $3k + 4$ and there exists an $O(n^2)$ algorithm that finds a path-decomposition with this width. The following lemma is easily checked. For a similar result see e.g. [27].

Lemma 7. *Let $G = (V, E)$ be a triangulated graph with n vertices. There is a clique C, such that every component of $G[V - C]$ has at most $\lceil \frac{1}{2}(n - |C|) \rceil$ vertices.*

The following lemma will also be useful (see e.g. [9]).

Lemma 8. *The treewidth of the complete bipartite graph $G = K(m, n)$ is $\min(m, n)$.*

Let C be a clique in $\overline{G}$ as mentioned in the lemma 7. The vertices of $V \setminus C$ can be partitioned in two set A and B such that no vertex of A is adjacent to a vertex of B, and both A and B have at most $\lceil \frac{2}{3}(n - |C|) \rceil$ vertices. Notice that the subgraph induced by A and B has a complete bipartite subgraph in G, since every vertex of A is adjacent to every vertex of B. Since the treewidth of G is at most k and by lemma 8:

$$\lfloor \frac{1}{3}(n - |C|) \rfloor \leq \min(|A|, |B|) \leq k$$

Hence $|A \cup B| \leq 3(k + 1)$. We can triangulate G by adding edges such that $A \cup B$ becomes a clique. Since C is a stable set in G, the result is a splitgraph. For a splitgraph, the following lemma is easily checked.

Lemma 9. *Let G be a splitgraph with clique number c. If G is an interval graph, the pathwidth is $c - 1$, otherwise, the pathwidth is c.*

This proves the following theorem.

Theorem 10. *If G is a cotriangulated graph with treewidth at most k, then the pathwidth is at most $3k + 4$ and there exists an $O(n^2)$ algorithm which produces a path-decomposition with this width.*

As a consequence, we have $O(n^2)$ approximation algorithms for the treewidth and for the pathwidth of cotriangulated graphs with performance ratio $3 + \varepsilon$ for all $\varepsilon > 0$.

4 An approximate path-decomposition for permutation graphs

In this section, let $G[\pi]$ be a permutation graph with n vertices and with treewidth k. We show there exists a path-decomposition of width at most $2k$, and we give a linear time algorithm to compute this. The algorithm outputs a set X_i for $1 \leq i \leq n$. A vertex j is put in all sets X_k with $\pi_j^{-1} \leq k < j$ or $j \leq k < \pi_j^{-1}$. The precise algorithm is given below.

```
Procedure Pathdec (input π; output X)
for i ← 1 to n do Xᵢ ← ∅
for j ← 1 to n do
    if πⱼ⁻¹ = j then Xⱼ ← Xⱼ ∪ {j}
    if πⱼ⁻¹ > j then
        for k ← j to πⱼ⁻¹ - 1 do Xₖ ← Xₖ ∪ {j}
    if πⱼ⁻¹ < j then
        for k ← πⱼ⁻¹ to j - 1 do Xₖ ← Xₖ ∪ {j}
```

The next lemma shows that the constructed sets form indeed a path-decomposition.

Lemma 11. *Let $S = \{X_i \mid 1 \le i \le n\}$ be the subsets of vertices constructed by the algorithm. Let $P = (1, \ldots, n)$ be the path with n vertices. Then (S, P) is a path-decomposition for the permutation graph $G[\pi]$.*

Proof. We first show that each vertex is in at least one subset of S. Consider a vertex i. If $\pi_i^{-1} \ge i$ then i is in the subset X_i. If $\pi_i^{-1} < i$ then i is in the subset X_{i-1}. Notice that the subsets containing i are consecutive. The only thing left to show is that every edge is in at least one subset. Consider again a vertex i and let j be a neighbor of i. Assume without loss of generality that $i < j$. In the matching diagram, the line segment corresponding with j must intersect the line segment of i. Since $i < j$, this implies that $\pi_j^{-1} < \pi_i^{-1}$. We consider the different orderings of i, j, π_i^{-1} and π_j^{-1}. If $i < j \le \pi_j^{-1} < \pi_i^{-1}$, then both i and i are contained in the subset X_j. If $i \le \pi_j^{-1} \le j \le \pi_i^{-1}$ then both are contained in $X_{\pi_j^{-1}}$. If $i \le \pi_j^{-1} < \pi_i^{-1} \le j$, then both are contained in $X_{\pi_j^{-1}}$. If $\pi_j^{-1} \le i \le \pi_i^{-1} \le j$, then both are contained in X_i. Finally, if $\pi_j^{-1} < \pi_i^{-1} \le i < j$, then both i and j must be in $X_{\pi_j^{-1}}$. $\square$

We now show that the width of this path-decomposition is at most $2k$.

Lemma 12. *Each subset produced by the algorithm has at most $2k + 1$ elements.*

Proof. Consider a subset X_i. Notice that $X_i \subset S_1 \cup S_2 \cup \{i\}$ where S_1 and S_2 are defined by: $S_1 = \{j \mid j \le i < \pi_j^{-1}\}$ and $S_2 = \{j \mid \pi_j^{-1} \le i < j\}$. Note that, as π is a permutation, there must be as many lines in the matching diagram with their upper point left of i and their lower point right of i, as lines with their upper point right of i and their lower point left of i. Hence $|S_1| = |S_2|$. Every vertex in S_1 is adjacent to every vertex in S_2, hence the subgraph induced by $S_1 \cup S_2$ contains a complete bipartite subgraph $K(m, m)$, with $m = |S_1|$. By lemma 8, this implies that $k \ge m$. Hence $|X_i| \le |S_1| + |S_2| + 1 \le 2k + 1$. $\square$

Notice that the algorithm can be implemented to run in $O(nk)$ time, since at each step one new element is put into a subset. Hence we have proved the following theorem:

Theorem 13. *If $G[\pi]$ is a permutation graph with treewidth at most k, then the pathwidth of $G[\pi]$ is at most $2k$, and the $O(nk)$ time algorithm Pathdec produces a path-decomposition with width at most $2k$.*

Using results of [10] this shows that, if the treewidth is bounded by a constant, an optimal path-decomposition and tree-decomposition can be computed in linear time. Also, it follows that we have $O(n^2)$ approximation algorithms for pathwidth and treewidth with performance ratio 2.

5 An approximate path-decomposition for cocomparability graphs

In this section, let G be a cocomparability graph with treewidth at most k. We assume that a height function h of the complement $\overline{G}$ with transitive orientation

F, and a coloring of G are given. Let $C_1, \ldots, C_s$ be the color classes of G, with $s = \chi(G)$ is the chromatic number of G. Since a graph with treewidth $\leq k$ can always be colored with $k+1$ colors, we may assume that the number of color classes $s \leq k+1$.

The first step of the algorithm is to renumber the vertices.

Definition 14. Let $G = (V, E)$ be a cocomparability graph with n vertices and let h be a height function of the transitively oriented complement $\overline{G}$. A *height labeling* of G is a bijection $L : V \rightarrow \{1, \ldots, n\}$, such that $h(x) > h(y)$ implies that $L(x) > L(y)$.

A height labeling clearly can be computed in $O(n)$ time, if the height function h is given.

Definition 15. For $i = 1, \ldots, n$ and for $t = 1, \ldots, s$, let $C_t(i)$ be the set of vertices in color class C_t with label at most i: $C_t(i) = C_t \cap \{x \mid L(x) \leq i\}$. Write $C_t(i) = \{x_1, \ldots, x_m\}$, with $i \geq L(x_1) > L(x_2) > \ldots > L(x_m)$. We define for $1 \leq i \leq n$ and $1 \leq t \leq s$:

$$\overline{C_t}(i) = \begin{cases} C_t(i) & \text{if } m \leq k+1 \\ \{x_1, \ldots, x_{k+1}\} & \text{otherwise} \end{cases}$$

Notice that, since C_t is a clique in $\overline{G}$, all heights of vertices in C_t are different. It follows that $\overline{C_t}(i)$ is uniquely determined as the set of $k+1$ vertices of $C_t(i)$ with the largest heights.

Definition 16. For $i = 1, \ldots, n$ and $t = 1, \ldots, s$, define $\mathcal{F}_t(i) = \{x \mid L(x) > i \wedge |Adj(x) \cap C_t(i)| \geq k+1\}$, where $Adj(x)$ is the set of neighbors of x.

We can now define the subsets of the path-decomposition:

Definition 17. For $i = 1, \ldots, n$, let X_i be the set of vertices $X_i = \bigcup_{1 \leq t \leq s}(\mathcal{F}_t(i) \cup \overline{C_t}(i))$.

In the rest of this section we prove that each subset X_i has at most $(2k+1)\chi(G)$ elements and that $(X_1, \ldots, X_n)$ forms indeed a path-decomposition.

Lemma 18. *Each $y \in \mathcal{F}_t(i)$ is adjacent to all vertices of $\overline{C_t}(i)$.*

Proof. Suppose not. Let $y \in \mathcal{F}_t(i)$ be not adjacent to every vertex in $\overline{C_t}(i)$. Let $C_t(i) = \{x_1, \ldots, x_m\}$ with $h(x_1) > h(x_2) > \ldots > h(x_m)$. If $m < k+1$, then by definition $\mathcal{F}_t(i) = \emptyset$. Hence we may assume $m \geq k+1$ and $\overline{C_t}(i) = \{x_1, \ldots, x_{k+1}\}$. Let $x_w \in \overline{C_t}(i)$ (with $1 \leq w \leq k+1$) be not adjacent to y. Consider the complement $\overline{G}$ with the transitive orientation F. $C_t(i)$ is a clique in $\overline{G}$ and $(x_p, x_q) \in F$ for all x_p and x_q in $C_t(i)$ with $p < q$. Since y is adjacent to x_w in $\overline{G}$, we must have $h(y) \neq h(x_w)$. Since $L(y) > L(x_w)$, it follows that $h(y) > h(x_w)$ and hence $(y, x_w) \in F$. Since F is transitive, we find that y is adjacent in $\overline{G}$ to all vertices x_p with $w \leq p \leq m$. So y can have at most $w - 1$ neighbors in $C_t(i)$ in G, hence it can not be in $\mathcal{F}_t(i)$, contradiction. $\square$

Corollary 19. *A vertex y with $L(y) > i$ is in $\mathcal{F}_t(i)$ if and only if it is adjacent to the vertex in $\overline{C_t}(i)$ with the smallest label.*

Theorem 20. *If G is a cocomparability graph and has treewidth at most k, then $|\overline{C_i}(i)| \leq k + 1$ and $|\mathcal{F}_i(i)| < k + 1$.*

Proof. Notice that $|\overline{C_i}(i)| \leq k+1$, by definition. If $|\overline{C_i}(i)| < k+1$, then $C_i(i)$ contains less than $k + 1$ elements, and then, by definition $\mathcal{F}_i(i) = \emptyset$. Now assume that $C_i(i)$ contains at least $k + 1$ vertices. Then $|\overline{C_i}(i)| = k + 1$. By lemma 18 all vertices of $\mathcal{F}_i(i)$ are adjacent to all vertices of $\overline{C_i}(i)$. The treewidth of G is at most k. Hence G can not have a complete bipartite subgraph $K(k + 1, k + 1)$. This implies that $|\mathcal{F}_i(i)| < k + 1$. □

Corollary 21. $\forall_i |X_i| \leq (2k + 1)\chi(G)$.

Corollary 21 shows that, if $(X_1, \ldots, X_n)$ is a path-decomposition of G, then the width of this path-decomposition is at most $2k^2 + 3k$. The next three lemmas show that $(X_1, \ldots, X_n)$ is indeed a path-decomposition.

Lemma 22. *For every vertex x of G: $x \in X_{L(x)}$.*

Proof. Let x be in the color class C_p. By definition, $x \in \overline{C_p}(L(x)) \subseteq X_{L(x)}$. □

Lemma 23. *Let $\{x, y\} \in E$ be an edge of G. Then there is a subset X_i such that x and y are both in X_i.*

Lemma 24. *The subsets X_i containing a given vertex x, are a consecutive subsequence of $(X_1, \ldots, X_n)$.*

By lemmas 22, 23, and 24 the sequence of subsets $(X_1, \ldots, X_n)$ is a path-decomposition for the cocomparability graph G. According to corollary 21, the width of this path-decomposition is at most $(2k + 1)\chi(G) - 1$. In the following theorem we summarize these results.

Theorem 25. *Let G be a cocomparability graph. The sequence $(X_1, \ldots, X_n)$ with X_i defined in definition 17, is a path-decomposition for G. If the treewidth of G is at most k, then the width of this path-decomposition is at most $(2k + 1)\chi(G) - 1 \leq 2k^2 + 3k$.*

Consider the time it takes to compute the sets X_i. We can sort all the color classes according to increasing labels in $O(nk)$ time. Then each set $\overline{C_i}(i)$ can be computed in $O(k)$ time. Now notice that $\mathcal{F}_i(i) \subseteq \mathcal{F}_i(i + 1) \cup \{x \mid L(x) = i + 1\}$. If we use an adjacency matrix to represent G, we can compute each set $\mathcal{F}_i(i)$ in time $O(k)$: An element y in $\mathcal{F}_i(i + 1) \cup \{x \mid L(x) = i + 1\}$ is in $\mathcal{F}_i(i)$ if and only if y is adjacent to the vertex with the smallest label in $\overline{C_i}(i)$ (corollary 19). The sets $\overline{C_i}(i)$ and $\mathcal{F}_i(i)$ have at most $k + 1$ elements and there are at most $k + 1$ of each for each i. Hence we can easily compute each set X_i in time $O(k^2)$.

Corollary 26. *Let G be a cocomparability graph with treewidth k. Assume a vertex coloring of G is given (with at most $k + 1$ colors), and a transitive orientation F of the complement $\overline{G}$. If an adjacency matrix is used to represent G, then a path-decomposition of G with width at most $2k^2 + 3k$ can be computed in time $O(nk^2)$.*

We end this section by remarking that finding approximations for the treewidth of bipartite graphs (within a constant factor away from optimal) is as hard as this problem for general graphs (the transformation uses a subdivision of each edge).

6 Conclusions

In this paper we described very simple and efficient algorithms to approximate pathwidth and treewidth of cotriangulated graphs, permutation graphs and cocomparability graphs. There are classes of graphs for which the exact pathwidth and treewidth can be computed efficiently. For example cographs [9], splitgraphs (lemma 9) and interval graphs. The treewidth can also be computed efficiently for chordal graphs and circular arc graphs [31]. It would be of interest to know, if there exists a fast algorithm which computes the treewidth for permutation graphs or cotriangulated graphs. Finally, in [8] it is shown that there is a polynomial algorithm that finds a tree-decomposition of G with treewidth at most $O(k \log n)$, where k is the treewidth of G and n the number of vertices. We have shown that there are many graph classes for which there is a polynomial algorithm that finds a tree-decomposition with width at most some fixed polynomial of k. It would be very interesting to know if this could be generalized.

Acknowledgements

We like to thank D. Seese, D. Kratsch and B. Reed for valuable discussions.

References

1. S. Arnborg, Efficient algorithms for combinatorial problems on graphs with bounded decomposability — A survey. *BIT* **25**, 2 − 23, 1985.
2. S. Arnborg, D.G. Corneil and A. Proskurowski, Complexity of finding embeddings in a k-tree, *SIAM J. Alg. Disc. Meth.* **8**, 277 − 284, 1987.
3. S. Arnborg, J. Lagergren and D. Seese, Easy problems for tree-decomposable graphs, *J. Algorithms* **12**, 308 − 340, 1991.
4. S. Arnborg and A. Proskurowski, Characterization and recognition of partial 3-trees, *SIAM J. Alg. Disc. Meth.* **7**, 305 − 314, 1986.
5. S. Arnborg and A. Proskurowski, Linear time algorithms for NP-hard problems restricted to partial k-trees. *Disc. Appl. Math.* **23**, 11 − 24, 1989.
6. C. Berge and C. Chvatal, *Topics on Perfect Graphs*, Annals of Discrete Math. **21**, 1984.
7. H.L. Bodlaender, A tourist guide through treewidth, Technical report RUU-CS-92-12, Department of computer science, Utrecht University, Utrecht, The Netherlands, 1992. To appear in: *Proceedings 7th International Meeting of Young Computer Scientists*, Springer Verlag, Lecture Notes in Computer Science.
8. H. Bodlaender, J. Gilbert, H. Hafsteinsson and T. Kloks, Approximating treewidth, pathwidth and minimum elimination tree height, In G. Schmidt and R. Berghammer, editors, *Proceedings 17th International Workshop on Graph-Theoretic Concepts in Computer Science WG'91*, 1 − 12, Springer Verlag, Lecture Notes in Computer Science, vol. 570, 1992.
9. H. Bodlaender and R.H. Möhring, The pathwidth and treewidth of cographs, In *Proceedings 2nd Scandinavian Workshop on Algorithm Theory*, 301 − 309, Springer Verlag, Lecture Notes in Computer Science vol. 447, 1990.
10. H. Bodlaender and T. Kloks, Better algorithms for the pathwidth and treewidth of graphs, *Proceedings of the 18th International colloquium on Automata, Languages and Programming*, 544 − 555, Springer Verlag, Lecture Notes in Computer Science, vol. 510, 1991.

11. A. Brandstädt and D. Kratsch, On the restriction of some NP-complete graph problems to permutation graphs, *Fundamentals of Computation Theory, proc. FCT* 1985, 53−62, Lecture Notes in Comp. Science vol. 199, Springer Verlag, New York, 1985.

12. A. Brandstädt and D. Kratsch, On domination problems for permutation and other perfect graphs, *Theor. Comput. Sci.* **54**, 181 − 198, 1987.

13. B. Courcelle, The monadic second-order logic of graphs I: Recognizable sets of finite graphs, *Information and Computation* **85**, 12 − 75, 1990.

14. B. Courcelle, The monadic second-order logic of graphs III: Treewidth, forbidden minors and complexity issues, Report 8852, University Bordeaux 1, 1988.

15. B. Courcelle, Graph rewriting: an algebraic and logical approach. In J. van Leeuwen, editor, *Handbook of Theoretical Computer Science, Vol. B*, 192 − 242, Amsterdam, 1990. North Holland Publ. Comp.

16. G.A. Dirac, On rigid circuit graphs, *Abh. Math. Sem. Univ. Hamburg* **25**, 71−76, 1961.

17. S. Even, A. Pnueli and A. Lempel, Permutation graphs and transitive graphs, *J. Assoc. Comput. Mach.* **19**, 400 − 410, 1972.

18. Gilmore and Hoffman, A characterization of comparability and interval graphs, *Canad. J. Math.* **16**, 539 − 548, 1964.

19. M.C. Golumbic, *Algorithmic Graph Theory and Perfect Graphs*, Academic Press, New York, 1980.

20. J. Gustedt, Pathwidth for chordal graphs is NP-complete. Preprint TU Berlin, 1989.

21. J. Lagergren and S. Arnborg, Finding minimal forbidden minors using a finite congruence, *Proceedings of the 18th International colloquium on Automata, Languages and Programming*, 532 − 543, Springer Verlag, Lecture Notes in Computer Science, vol. 510, 1991.

22. J. van Leeuwen, Graph algorithms. In *Handbook of Theoretical Computer Science, A: Algorithms an Complexity Theory*, 527 − 631, Amsterdam, 1990. North Holland Publ. Comp.

23. J. Matoušek and R. Thomas, Algorithms Finding Tree-Decompositions of Graphs, *Journal of Algorithms* **12**, 1 − 22, 1991.

24. A. Pnueli, A. Lempel, and S. Even, Transitive orientation of graphs and identification of permutation graphs, *Canad. J. Math.* **23**, 160 − 175, 1971.

25. B. Reed, Finding approximate separators and computing treewidth quickly, To appear in: Proceedings STOC'92, 1992.

26. N. Robertson and P.D. Seymour, Graph minors—A survey. In I. Anderson, editor, *Surveys in Combinatorics*, 153 − 171. Cambridge Univ. Press 1985.

27. N. Robertson and P.D. Seymour, Graph minors II. Algorithmic aspects of treewidth. *J. Algorithms* **7**, 309 − 322, 1986.

28. F.S. Roberts, Graph theory and its applications to problems of society, NFS-CBMS Monograph no. 29 (SIAM Publications, Philadelphia, PA. 1978).

29. M.C. Golumbic, D. Rotem and J. Urrutia, Comparability graphs and intersection graphs, *Discrete Math.* **43**, 37 − 46, 1983.

30. J. Spinrad, On comparability and permutation graphs, *SIAM J. Comp.* **14**, No. 3, August 1985.

31. R. Sundaram, K. Sher Singh and C. Pandu Rangan, Treewidth of circular arc graphs, Manuscript 1991.

Graph Spanners and Connectivity

Shuichi UENO and Michihiro YAMAZAKI

Department of Electrical and Electronic Engineering
Tokyo Institute of Technology, Tokyo 152, Japan

Yoji KAJITANI

School of Information Science
JAIST Hokuriku, Ishikawa 923-12, Japan
Department of Electrical and Electronic Engineering
Tokyo Institute of Technology, Tokyo 152, Japan

Abstract

Given an n-vertex graph or digraph G, a spanning subgraph S is a k-spanner of G if for every $u, v \in V(G)$, the distance from u to v in S is at most k times longer than the distance in G. This paper establishes some relationships between the connectivity and the existence of k-spanners with $O(n)$ edges for graphs and digraphs. We give almost tight bounds of the connectivity of G which guarantees the existence of k-spanners with $O(n)$ edges.

1   Introduction

We denote the vertex set and edge set of a graph G by $V(G)$ and $E(G)$, respectively. $d_G(u, v)$ is the distance from a vertex u to a vertex v in G. Given an integer k, a spanning subgraph S of G is called a *k-spanner* of G if $d_S(u, v) \leq k \cdot d_G(u, v)$ for every $u, v \in V(G)$. We refer to k and $|E(S)|$ as the *stretch factor* and the *size* of S, respectively. A k-spanner S of G is said to be *linear* if the size of S is $O(|V(G)|)$. k-spanners have various applications to distributed systems [3, 17], communication networks [4, 5, 6, 18], biology [7], and the design of approximation algorithms for problems that involve finding shortest distances in the plane [9, 10, 11, 12, 13, 14]. This paper establishes some relationships between the connectivity and the existence of linear k-spanners for graphs and digraphs. We give almost tight bounds of the connectivity of G which guarantees the existence of linear k-spanners in G. We also mention some known results about the weighted graph case for the sake of completeness.

1.1 Graph Case

k-spanners were introduced by Peleg and Ullman [17] in connection with the design of synchronizers. The synchronizer is a simulation methodology introduced by Awerbuch [3] which enables the execution of a synchronous algorithm on an asynchronous network. The k-spanner is the underlying graph structure of the synchronizer, and the stretch factor and size of the k-spanner are closely related to the time and communication complexities of the synchronizer, respectively. k-spanners are also used in the design of succinct routing tables for communication networks. In this case, the stretch factor and size of the k-spanner are related to the length of routes and the total memory required for the table, respectively.

In the above-mentioned applications, the problem is to find k-spanners such that both of the stretch factor and size are as small as possible. Unfortunately, this problem is NP-hard as proved by Peleg and Schäffer [16].

Our interest here is to explore an asymptotic problem on k-spanners. We investigate the existence of linear $O(1)$-spanners which is important from the practical point of view. Unfortunately, linear $O(1)$-spanners do not always exist in general. Peleg and Schäffer [16] proved that for every positive integer k, there exist n-vertex graphs for which every k-spanner requires $\Omega(n^{1+1/(k+2)})$ edges. However, they also proved that for every n-vertex graph and for every integer $k \geq 5$, there exists a k-spanner with $O(n^{1+4/(k-1)})$ edges [16]. Moreover, we can construct such one in polynomial time. Notice that these bounds are almost tight.

Although there exists no linear $O(1)$-spanner in general, there exist linear $O(1)$-spanners for many classes of graphs. Examples are bounded-degree graphs, bounded-diameter graphs, bounded-tree-width graphs, and planar graphs. Although these are rather trivial classes, we also have nontrivial results. Peleg and Ullman proved that hypercubes have linear 3-spanners [17]. Also Peleg and Schäffer proved that chordal graphs have linear 5-spanners [16].

Another trivial example is the complete graph K_n, which has a linear 2-spanner with $n - 1$ edges, namely a star $K_{1,n-1}$. Since the connectivity of K_n is maximal, this leads us the following natural question: What kind of connectivity assures the existence of linear $O(1)$-spanners?

We give in Section 2 an answer to this question. We show that a relatively high connectivity guarantees the existence of linear $O(1)$-spanners. We denote the connectivity of a graph G by $\kappa(G)$. We prove that for any n-vertex graph G, and for any integers $\alpha \geq 1$ and $k \geq 16\alpha + 3$, if $\kappa(G) \geq \frac{1}{\alpha}n^{(8\alpha+2)/(k+1)}$ then G has a linear k-spanner which can be constructed in polynomial time. The bound above cannot be significantly improved, as we also show that for every integer k and infinitely many integers n, there exist n-vertex graphs G with $\kappa(G) \geq \lfloor \frac{1}{8}n^{1/(k+2)} \rfloor$ for which every k-spanner requires $\Omega(n^{1+1/(k+2)})$ edges. This strengthens the result by Peleg and Schäffer [16] mentioned above.

1.2 Digraph Case

The situation for digraphs is generally harder than that for graphs. For every positive integer k, there exist n-vertex digraphs for which every k-spanner requires $\Omega(n^2)$ arcs [16]. It should be noted that for any positive integer k, any n-vertex digraph has a

k-spanner with $O(n^2)$ arcs. Surprisingly, a very high connectivity of $\Omega(n/k^2)$ cannot guarantee the existence of linear $O(1)$-spanners. Peleg and Schäffer [16] implicitly proved that for every positive integers k and $n \geq k$, there exists an n-vertex digraph G with $\kappa(G) \geq \lfloor (n-k)/k^2 \rfloor + 1$ for which every k-spanner requires $\Omega(n^2/k^2)$ arcs.

However, we show in Section 3 that for any integer $k \geq 4$ and n-vertex digraph G, if $\kappa(G) \geq 2(n-1)/(k-2)$ then G has a linear k-spanner which can be constructed in polynomial time.

1.3 Weighted Graph Case

For weighted graphs, the distance between two vertices is the length of the shortest weighted path between the vertices. Weighted k-spanners are important to some practical applications such as the design of routing tables of communication networks and the design of approximation algorithms for problems that involve finding shortest distances in the plane.

The general situation for weighted graphs is quite similar to that for graphs. Althöfer, Das, Dobkin, and Joseph [2] proved that for every positive integer $k \geq 3$ and n-vertex weighted graph, there is a polynomially constructible k-spanner with $O(n^{1+2/(k-1)})$ edges.

However, for the weighted graph case, even the highest connectivity cannot guarantee the existence of linear $O(1)$-spanners. That is, they also mentioned that for infinitely many integers n and k, there exists an n-vertex, weighted complete graph for which every k-spanner requires $\Omega(\frac{2}{k+1}n^{1+1/(k+2)})$ edges [2].

Although the situation is hopeless in general, it has been known that complete graphs with certain restricted edge weights have linear $O(1)$-spanners. There exist linear $O(1)$-spanners for complete graphs whose vertices are points in the plane or space and edge weights are defined by means of some L_i metric [1, 2, 9, 10, 11, 12, 13, 14, 20].

2 Graph Case

Peleg and Schäffer proved the following theorems.

Theorem I ([16]) *Given a graph G and integers k and m, it is NP-complete to decide if G has a k-spanner with m edges.*

Theorem II ([16]) *For every integer $t \geq 3$, there exist n-vertex graphs for which every $(t-2)$-spanner requires $\Omega(n^{1+1/t})$ edges.*

Theorem III ([16]) *For every n-vertex graph and for every positive integer t, there exists a $(4t+1)$-spanner with $O(n^{1+1/t})$ edges which can be constructed in polynomial time.*

Althöfer, Das, Dobkin, and Joseph slightly improved Theorem III as follows.

Theorem IV ([2]) *For every positive integer t and n-vertex graph, there is a $(2t+1)$-spanner with $O(n^{1+1/t})$ edges which can be constructed in polynomial time.*

We prove in this section the following two complementing theorems.

Theorem 1 *For any n-vertex graph G, and for any integers $\alpha \geq 1$ and $k \geq 16\alpha + 3$, if $\kappa(G) \geq \frac{1}{\alpha} n^{(8\alpha+2)/(k+1)}$ then G has a linear k-spanner which can be constructed in polynomial time.*

Theorem 2 *For every positive integer k and for infinitely many integers n, there exist n-vertex graphs G with $\kappa(G) \geq \lfloor \frac{1}{8} n^{1/(k+2)} \rfloor$ for which every k-spanner requires $\Omega(n^{1+1/(k+2)})$ edges.*

2.1 Proof of Theorem 1

Before proving Theorem 1, we need some definitions and notations. $\delta(G)$ is the minimum degree of the vertex of G. $g(G)$ is the girth of G which is the length of the shortest cycle in G. The diameter of G is defined as $diam(G) = max\{d_G(u,v)|u,v \in V(G)\}$. Given a set of vertices $S \subseteq V(G)$, $G[S]$ denotes the subgraph induced by S in G. The j-neighborhood of a vertex $v \in V(G)$ is defined as $N_j(v) = \{u|d_G(u,v) = j\}$. The $(\leq j)$-neighborhood of a vertex $v \in V(G)$ is defined as $N_{(\leq j)}(v) = \{u|d_G(u,v) \leq j\}$. Given $X, Y \subseteq V(G)$, we define that $d_G(X,Y) = min\{d_G(x,y)|x \in X, y \in Y\}$. We denote $d_G(v, X)$ instead of $d_G(\{v\}, X)$ for simplicity. X and Y are said to be independent if $d_G(X,Y) \geq 1$. X and Y are said to be adjacent if $d_G(X,Y) = 1$. A collection $\mathcal{C}$ of subsets of $V(G)$ is said to be independent if any two elements of $\mathcal{C}$ are independent.

Now we are ready to prove Theorem 1. Let α and β be positive integers, and G be an n-vertex graph with $\kappa(G) \geq \frac{1}{\alpha} n^{1/(1+\beta)}$. We are going to prove that G has a linear $(2(4\alpha + 1)(\beta + 1) - 1)$-spanner. If $\alpha \geq diam(G)/2$ then a BFS-tree rooted at a vertex in G is obviously a linear 4α-spanner of G. Thus we assume in the following that $\alpha < diam(G)/2$.

Lemma 1 $|N_{(\leq \alpha)}(v)| \geq n^{1/(1+\beta)}$ *for any* $v \in V(G)$.

Proof. Since G is $\lceil \frac{1}{\alpha} n^{1/(1+\beta)} \rceil$-connected and $\alpha < diam(G)/2$, $|N_j(v)| \geq \frac{1}{\alpha} n^{1/(1+\beta)}$ for any $j(1 \leq j \leq \alpha)$. Thus $|N_{(\leq \alpha)}(v)| = \sum_{j=0}^{\alpha} |N_j(v)| \geq n^{1/(1+\beta)}$. $\square$

From the definition of $N_{(\leq \alpha)}(v)$, we have the following.

Lemma 2 $diam(G[N_{(\leq \alpha)}(v)]) \leq 2\alpha$ *for any* $v \in V(G)$.

Let $\mathcal{C}$ be a maximal independent set of $(\leq \alpha)$-neighborhoods of vertices in G, and $X = \bigcup \{N_{(\leq \alpha)}(v)|N_{(\leq \alpha)}(v) \in \mathcal{C}\}$.

Lemma 3 *For any vertex $w \in V(G) - X$, there exists $N_{(\leq \alpha)}(v) \in \mathcal{C}$ such that $d_G(w, N_{(\leq \alpha)}(v)) \leq \alpha$.*

Proof. If there exists a vertex $u \in V(G) - X$ such that $d_G(u, N_{(\leq \alpha)}(v)) > \alpha$ for any $N_{(\leq \alpha)}(v) \in \mathcal{C}$, then $\mathcal{C} \bigcup \{N_{(\leq \alpha)}(u)\}$ is also independent, contradicting to the maximality of $\mathcal{C}$. $\square$

Let $\mathcal{D}$ be a partition of $V(G)$ obtained from $\mathcal{C}$ by adding each $w \in V(G) - X$ to some $N_{(\leq \alpha)}(v) \in \mathcal{C}$ with $d_G(w, N_{(\leq \alpha)}(v)) \leq \alpha$.

Lemma 4 $|\mathcal{D}| \leq n^{\beta/(1+\beta)}$.

Proof. From Lemma 1 and the definition of $\mathcal{D}$, $|D| \geq n^{1/(1+\beta)}$ for any $D \in \mathcal{D}$. Thus $|\mathcal{D}| \leq n/n^{1/(1+\beta)} = n^{\beta/(1+\beta)}$. □

From Lemma 2 and the definition of $\mathcal{D}$, we have the following.

Lemma 5 $diam(G[D]) \leq 4\alpha$ for any $D \in \mathcal{D}$.

Define the graph H as follows:

$$V(H) = \mathcal{D}; E(H) = \{(D, D')|D, D' \in \mathcal{D}, d_G(D, D') = 1\}.$$

Lemma 6 H has a $(2\beta + 1)$-spanner S_H with $O(n)$ edges.

Proof. From Theorem IV, H has a $(2\beta + 1)$-spanner S_H with $O(|V(H)|^{1+1/\beta})$ edges. The size of S_H is $O(n)$ by Lemma 4. □

We use the following lemma to prove Lemma 7 below.

Lemma I ([16]) A subgraph S of a graph G is a k-spanner of G if and only if $d_S(u, v) \leq k$ for every edge $(u, v) \in E(G)$.

Lemma 7 G has a linear $(2(4\alpha + 1)(\beta + 1) - 1)$-spanner.

Proof. For each edge $(D, D') \in E(S_H)$, choose just one edge $(u, v) \in E(G)$ such that $u \in D$ and $v \in D'$. Such an edge exists because D and D' are adjacent in G. Let A be the set of all such edges $(u, v) \in E(G)$.

Each $D \in \mathcal{D}$ is consisting of one $N_{(\leq \alpha)}(v) \in \mathcal{C}$ and some vertices in $V(G) - X$. Let T_D be a BFS-tree rooted at a vertex in $G[D]$. $diam(T_D) \leq 4\alpha$ by Lemma 5.

Let S be the spanning subgraph of G consisting of the edges in $\bigcup\{E(T_D)|D \in \mathcal{D}\} \bigcup A$. We show that S is a required spanner of G.

Claim 1: $|E(S)| = O(n)$.

$|\bigcup\{E(T_D)|D \in \mathcal{D}\}| = O(n)$ since $\mathcal{D}$ is a partition of $V(G)$ and T_D is a spanning tree in $G[D]$. Also $|A| = O(n)$ by Lemma 6. Thus we have Claim 1.

Claim 2: S is a $(2(4\alpha + 1)(\beta + 1) - 1)$-spanner of G.

Since S_H is a $(2\beta + 1)$-spanner of H and $diam(T_D) \leq 4\alpha$, $d_S(u, v) \leq 4\alpha(2\beta + 2) + 2\beta + 1 = 2(4\alpha + 1)(\beta + 1) - 1$ for any edge $(u, v) \in E(G)$. Thus we have Claim 2 by Lemma I.

Claims 1 and 2 complete the proof of the lemma. □

Spanner S above can be constructed in polynomial time as easily seen. Thus, putting $k = 2(4\alpha + 1)(\beta + 1) - 1$, we have Theorem 1 from Lemma 7.

2.2 Proof of Theorem 2

Theorem II was proved based on the following two results. The first one is due to Bollobás, and second one is due to Peleg and Schäffer.

Theorem V ([8]) *For every integer $t \geq 3$, there exist n-vertex graphs G such that $g(G) \geq t$ and $\delta(G) \geq \frac{1}{2}n^{1/t}$.*

Lemma II ([16]) *For every integers $k \geq 1$ and $t \geq k + 2$ and for every graph G with $g(G) \geq t$, the only k-spanner of G is G itself.*

Theorem 2 can be proved by using these results and the following useful theorem due to Mader (see also [19]).

Theorem VI ([15]) *Every graph G with $\delta(G) \geq 4r$ contains an r-connected subgraph.*

Lemma 8 *For every integer $t \geq 3$ and infinitely many integers n, there exist n-vertex graphs G such that $g(G) \geq t$ and $\kappa(G) \geq \lfloor \frac{1}{8}n^{1/t} \rfloor$.*

Proof. There exists an n-vertex graph G with $g(G) \geq t$ and $\delta(G) \geq \frac{1}{2}n^{1/t}$ by Theorem V. G contains a $\lfloor \frac{1}{8}n^{1/t} \rfloor$-connected subgraph H by Theorem VI. Since H is a subgraph of G, $g(H) \geq t$. Since $\lfloor \frac{1}{8}n^{1/t} \rfloor \leq |V(H)| \leq n$, $\kappa(H) \geq \lfloor \frac{1}{8}|V(H)|^{1/t} \rfloor$, and we have the lemma. $\square$

From Lemmas II and 8, and well-known facts that $\delta(G) \geq \kappa(G)$ and $|E(G)| \geq \delta(G)|V(G)|/2$, we have Theorem 2.

3 Digraph Case

Let $k \geq 1$ and $n \geq k$ be integers, and $N = \{0, 1, 2, \ldots, n-1\}$. Fix $p = \lfloor (n-k)/k^2 \rfloor$. For every $0 \leq l \leq p$, let

$$A_l = \{(i, i + lk + 1) | 0 \leq i \leq n - 1\},$$

where addition is taken modulo n, and take $A = \bigcup_{0 \leq l \leq p} A_l$. Let G_n be the digraph with vertex set N and arc set A.

Peleg and Schäffer proved the following theorem.

Theorem VII ([16]) *The digraph G_n constructed above is $(\lfloor (n - k)/k^2 \rfloor + 1)$-arc-connected, and every k-spanner of G requires $\Omega(n^2/k^2)$ arcs.*

We show that G_n is actually $(\lfloor (n-k)/k^2 \rfloor + 1)$-connected, and prove the following two complementing theorems.

Theorem 3 *For every positive integers k and $n \geq k$, there exists an n-vertex digraph G with $\kappa(G) \geq \lfloor (n - k)/k^2 \rfloor + 1$ for which every k-spanner requires $\Omega(n^2/k^2)$ arcs.*

Theorem 4 *For any integer $k \geq 4$ and n-vertex digraph G, if $\kappa(G) \geq 2(n-1)/(k-2)$ then G has a linear k-spanner which can be constructed in polynomial time.*

3.1 Proof of Theorem 3

By Theorem VII, it suffices to show that $\kappa(G_n) \geq \lfloor(n-k)/k^2\rfloor + 1 = p + 1$. Let $X \subseteq N$ be a set of vertices of G_n such that

(0) $|X| \leq p$.

We are going to show that there exists a dipath from a vertex in $N - X$ to a vertex in $N - X$ consisting of some vertices in $N - X$. We say that a vertex $v \in N$ is reachable from a vertex $u \in N - X$ if there exists a dipath from u to v consisting of some vertices in $N - X$ except the end. A set of vertices S is said to be reachable from a vertex u if every vertex in S is reachable from u. From the definition of G_n, we may assume without loss of generality that $0 \in N - X$, and it suffices to show that $N - X$ is reachable from 0.

For every integer $m \geq 0$, let

$$V(m) = \{mk + j | 0 \leq j \leq k - 1\},$$

and

$$W(m) = \{mk, mk + 1, mk + 2, \ldots, (m + p + 1)k + (k - 1)\},$$

where addition is taken modulo n. Notice that $V(m) \subseteq W(m)$.

From (0), there exist $V(m_0)$ and $V(m_1)$ $(0 \leq m_0 < m_1 \leq p + 1)$ such that $V(m_0), V(m_1) \subseteq N - X$.

(1) $V(m_1)$ is reachable from 0.

Because, $(0, m_0 k + 1) \in A$, since $0 \leq m_0 \leq p$. Also $(m_0 k + (k - 1), m_1 k) \in A$, since $m_0 + 1 \leq m_1 \leq m_0 + p + 1$. Thus we have (1), since $(i, i + 1) \in A$ for any $i \in N$.

(2) If $V(q)$ is reachable from 0 and $V(q) \subseteq N - X$ then $W(q)$ is also reachable from 0.

Because, the vertex $ik + j + 1$ $(q \leq i \leq q + p, 0 \leq j \leq k - 1)$ is reachable from 0, since $(qk + j, ik + j + 1) \in A$. Also the vertex $(q + p + 1)k + j$ $(1 \leq j \leq k - 1)$ is reachable from 0. This can be proved recursively as follows: (i)Vertex $(q+p+1)k+1$ is reachable from 0, since there exists a vertex ik $(q + 1 \leq i \leq q + p + 1)$ which is reachable from 0 and in $N - X$, and $(ik, (q + p + 1)k + 1) \in A$; (ii)If vertex $(q + p + 1)k + j - 1$ $(2 \leq j \leq k - 1)$ is reachable from 0, there exists a vertex $ik + j - 1$ $(q + 1 \leq i \leq q + p + 1)$ which is reachable from 0 and in $N - X$, and since $(ik + j - 1, (q + p + 1)k + j) \in A$, vertex $(q + p + 1)k + j$ is also reachable from 0. Thus we have (2).

It follows from (0) that:

(3) If $W(r)$ is reachable from 0 then there exists $V(s)$ $(r + 1 \leq s \leq r + p + 1)$ such that $V(s)$ is reachable from 0 and $V(s) \subseteq N - X$.

Notice that $s > r$ and $V(s) \subseteq W(r)$.

From (1),(2), and (3), there exists a sequence $W(m_1), W(m_2), \ldots, W(m_h)$ such that $m_1 < m_2 < \ldots < m_h$, $W(m_i) \cap W(m_{i+1}) \neq \phi$ $(1 \leq i \leq h - 1)$, $W(m_j)$

is reachable from 0 $(1 \leq j \leq h)$, and $(m_h + p + 1)k + (k - 1) \geq n + pk$. Then $N \subseteq \bigcup_{1 \leq i \leq h} W(m_i)$, and thus N is reachable from 0. This completes the proof of Theorem 3.

3.2 Proof of Theorem 4

Construct a BFS-tree T_{out} rooted at a vertex v. Also construct a BFS-tree T'_{in} rooted at v of the digraph obtained from G by reversing the direction of each arc. Let T_{in} be the tree of G consisting of the arcs corresponding to the arcs in T'_{in}.

The number of vertices in each level of T_{out} and T'_{in} is at least $2(n-1)/(k-2)$ except the level 0 and the heighest level, since $\kappa(G) \geq 2(n-1)/(k-2)$. Thus the height of T_{out} and T'_{in} is at most $\lceil (k-2)/2 \rceil$. It follows that there exists a dipath of length at most $\lceil (k-2)/2 \rceil$ from v to each vertex in T_{out}, and there exists a dipath of length at most $\lceil (k-2)/2 \rceil$ from each vertex to v in T_{in}. Let S be the subgraph of G consisting of the arcs in T_{out} and T_{in}. Since $diam(S) \leq 2\lceil (k-2)/2 \rceil \leq k$, S is a k-spanner with at most $2(n-1)$ edges.

Acknowledgment. The authors would like to thank Prof. Halldórsson of JAIST Hokuriku for helpful comments and encouragement.

References

[1] I. Althöfer. On optimal realizations of finite metric spaces by graphs. *Discrete and Computational Geometry*, 3:103–122, 1988.

[2] I. Althöfer, G. Das, D. Dobkin, and D. Joseph. Generating sparse spanners for weighted graphs. In *LNCS 447*, pages 26–37, 1990.

[3] B. Awerbuch. Complexity of network synchronization. *J. ACM*, 32:804–823, 1985.

[4] B. Awerbuch, A. Bar-Noy, N. Linial, and D. Peleg. Compact distributed data structures for adaptive routing. In *STOC*, pages 479–489, 1989.

[5] B. Awerbuch, O. Goldreich, D. Peleg, and R. Vainish. A tradeoff between information and communication in broadcast protocols. In *LNCS 319*, pages 369–379, 1988.

[6] B. Awerbuch and D. Peleg. Sparse partitions. In *FOCS*, pages 503–513, 1990.

[7] Bandelt and Dress. Reconstructing the shape of a tree from observed dissimilarity data. *Advances in Appl. Maths.*, 7:309–343, 1986.

[8] B. Bollobás. *Extremal Graph Theory*. Academic Press, 1978.

[9] L. P. Chew. There are planar graphs almost as good as the complete graph. *J. Comput. System Sci.*, 39:205–219, 1989.

[10] G. Das and D. Joseph. Which triangulations approximate the complete graph? In *LNCS 401*, pages 168–192, 1989.

[11] D. P. Dobkin, S. J. Friedman, and K. J. Supowit. Delaunay graphs are almost as good as complete graphs. In *FOCS*, pages 20–26, 1987.

[12] J. M. Keil. Approximating the complete euclidean graph. In *LNCS 318*, pages 208–213, 1988.

[13] J. M. Keil and C. A. Gutwin. Classes of graphs which approximate the complete euclidean graph. *Discrete Comput. Geom.*, 7:13–28, 1992.

[14] C. Levcopoulos and A. Lingas. There are planar graphs almost as good as the complete graphs and as short as minimum spanning trees. In *LNCS 401*, pages 9–13, 1989.

[15] W. Mader. Existenz n-fachen zusammenhängenden teilgraphen in graphen genügend grosser kantendichte. *Abh. Math. Sem. Univ. Hamburg*, 37:86–97, 1972.

[16] D. Peleg and A. A. Schäffer. Graph spanners. *J. Graph Theory*, 13:99–116, 1989.

[17] D. Peleg and J. D. Ullman. An optimal synchronizer for the hypercube. *SIAM J. Comput.*, 18:740–747, 1989.

[18] D. Peleg and E. Upfal. A tradeoff between space and efficiency for routing tables. In *STOC*, pages 43–52, 1988.

[19] C. Thomassen. Paths, cycles and subdivisions. In L. W. Beineke and R. J. Wilson, editors, *Selected Topics in Graph Theory III*, Academic Press, 1988.

[20] P. M. Vaidya. A sparse graph almost as good as the complete graph on points in k dimensions. *Discrete Comput. Geom.*, 6:369–381, 1991.

Randomized Range-Maxima in Nearly-Constant Parallel Time*

(Extended Abstract)

Omer Berkman[1], Yossi Matias[2], and Uzi Vishkin[2]

[1] King's College London***
[2] University of Maryland and Tel Aviv University[†]

Abstract. Given an array of n input numbers, the *range-maxima* problem is to preprocess the data so that queries of the type "what is the maximum value in subarray $[i..j]$" can be answered quickly using one processor. We present a randomized preprocessing algorithm that runs in $O(\log^* n)$ time with high probability, using an optimal number of processors on a CRCW PRAM; each query can be processed in constant time by one processor. We also present a randomized algorithm for a parallel comparison model. Using an optimal number of processors, the preprocessing algorithm runs in $O(\alpha(n))$ time with high probability; each query can be processed in $O(\alpha(n))$ time by one processor. ($\alpha(n)$ is the inverse of Ackermann function.) A constant time query can be achieved by some slowdown in the performance of the preprocessing stage.

1 Introduction

Given an array of n input numbers, the *range-maxima* problem is to preprocess the data so that queries of the type "what is the maximum value in subarray $[i..j]$?" can be answered quickly using one processor. The *prefix-maxima* problem is to compute the maximum value in subarray $[1..j]$ for all $j = 1, \ldots, n$. Algorithms for the range-maxima problem solve the prefix-maxima problem as well. We mention two recent papers that need range minima (or maxima) search: [3] on scaled string matching, and [20] on parallel triconnectivity. Other applications of the algorithm are given in [13].

Using n processors in the deterministic comparison model, the maximum problem, and hence the prefix-maxima and the range-maxima problems, have $\Omega(\log \log n)$ time lower bounds [25]. Reischuk [21] showed that randomization enables computing the maximum in constant time; however, for the prefix-maxima and range-maxima problems with general input, the best deterministic algorithms were so far also the best randomized algorithms: [7] gave a deterministic $O(\log \log n)$ time preprocessing algorithm with linear number of operations for the range-maxima problem, and constant time query retrieval. This improved on Alon and Schieber [2] whose algorithm takes $O(n\alpha(n))$ operations and $O(\log \log n)$ time, but for a more general

---

* Partially supported by NSF grants CCR-9111348 and CCR-8906949.

*** Dept. of Computing, School of Physical Sciences and Engineering, King's College London, Strand, London WC2R 2LS ENGLAND; omer@dcs.kcl.ac.uk.

[†] Institute for Advanced Computer Studies, University of Maryland, College Park, MD 20742, USA. Also with the Department of Computer Science, Tel Aviv University, ISRAEL; {matias,vishkin}@umiacs.umd.edu.

problem. Gil and Rudolph [16] gave an $O(\log \log n)$ time prefix-maxima algorithm using n processors. Schieber [22] gave an $O(\log \log n)$-time algorithm for prefix maxima that uses $n/\log \log n$ processors. As for sequential algorithms, Gabow, Bentley, and Tarjan [13] gave the first linear-time preprocessing algorithm for range maxima that results in constant-time query retrieval. More related work is surveyed in [6].

1.1 Results

We present range-maxima algorithms for two models of parallel computation. (The algorithms also explicitly solve the prefix-maxima problem with respect to the input.) Both algorithms use an optimal number of processors, i.e., the time-processor product is $O(n)$ for the preprocessing stage, and each query is processed by one processor. The preprocessing time is with probability $\geq 1 - 2^{-n^c}$ for some constant $c > 0$ ("almost surely"); the query-processing time is worst case. We have

- *Algorithm for the PRAM model:* Preprocessing in $O(\log^* n)$ time and query retrieval in constant time.
- *Algorithms for the Comparison Model:* Preprocessing in $O(I_m(n))$ time and query retrieval in $O(m)$ time, for $m = 1, \ldots, \alpha(n)$. The function $I_m(\cdot)$ is the inverse of a function at the m'th level of the primitive recursive hierarchy, and the function $\alpha(\cdot)$ is the inverse Ackermann function. In particular, this gives $O(\alpha(n))$ time for both preprocessing and query response.

Thus, the results improve the running times of the best randomized PRAM and comparison model algorithms (for both prefix-maxima and range-maxima) from $O(\log \log n)$ to $O(\log^* n)$ and $O(\alpha(n))$, respectively.

Significance. The $O(\log^* n)$ result is for a problem that is a substantial generalization of the problem of computing the maximum over an array of numbers: The output enables constant-time retrieval of *any* of $O(n^2)$ data rather than providing a single datum. The $O(\alpha(n))$ result is a first example where the star-tree data structure of [8] is used to obtain such result on *unrestricted* input. Also, this is a first example for an $O(\alpha(n))$ *randomized* result. (This latter result is only for the comparison model though.) These results add to a growing knowledge base of "nearly constant-time" parallel algorithms (see [26]).

Postscript. In an independent work, Dietz [12] considered a related problem of building a complete binary heap in parallel for an input array. He gives an $O(\log \log n)$ time deterministic algorithm using $n/\log \log n$ processors (this is optimal), and a randomized $O(\alpha(n))$ time (with high probability) algorithm using $n/\alpha(n)$ processors. Both algorithms are for the comparison model. Note that the heap data structure is insufficient to enable super fast retrieval of range maxima queries.

1.2 Outline

The rest of the paper is organized as follows. In Sect. 2 some definitions, notations, and basic facts are given. A basic randomized constant-time algorithm for range maxima that uses $n \log n$ processors is described in Sect. 3. The main algorithms, both for the comparison model and for the PRAM model, are described in Sect. 4. We conclude with some open problems in Sect. 5. Due to space limitation, some of the algorithms are omitted from this extended abstract and can be found in [6].

2 Basics

2.1 Models of Computation

We use two models of parallel computation.

The PRAM Model. Our main result is given for the concurrent-read concurrent-write (CRCW) PRAM model of computation. We assume the Arbitrary-CRCW model, in which if two or more processors try to write into the same memory cell, one of them succeeds but we do not know in advance which one. It turns out, but not shown in the paper, that our PRAM results actually hold for the weaker Tolerant-CRCW model, in which if two or more processors try to write (even the same value) into the same memory cell, the content of the cell does not change.

Valiant's Comparison Model [25]. Only comparison steps are taken into account. For instance, processors allocation is given free of charge in this model. This model has provided yardsticks for the limits of comparison based parallel algorithms (in [25] for finding the maximum, in [10] for merging and in [5, 1, 9] for sorting.) An upper bound within this model implies that in order to complement it into a PRAM upper bound issues of processor allocation have to be taken into account.

Definition 1. Given an array $A[1..n]$, the *prefix* of $A[i]$, $1 \le i \le n$, is subarray $A[1..i]$; the *suffix* of $A[i]$ is subarray $A[i..n]$. The *prefix maximum* of an element in A is the maximum over its prefix. The *prefix maxima of A* is an array consisting of the n prefix maxima of A's elements. Given an array A, the *prefix maxima problem* is to compute the prefix maxima of A.

Definition 2. Given an array $A[1..n]$, let $[i, j]$ denote the integer interval $[i, i + 1, \ldots, j - 1, j]$, and $\max[i, j]$ denote the maximum of A over the subarray $A[i, j]$; i.e., $\max[i, j] = \max\{A[k] : k \in [i, j]\}$. A *range-maximum query* is given as a pair $\langle i, j \rangle$ and is answered by $\max[i, j]$. The *range-maxima problem* is to preprocess a given array A so that any range-maximum query can be processed within a certain time bound. (Typically, it is required that the processing time is constant or "nearly-constant".)

2.2 Notations for Probabilities

We say that a probability $p(n)$ is *n-polynomial* if $p \ge 1 - n^{-c}$, for some constant $c > 0$. We say that $p(n)$ is *n-exponential* if $p \ge 1 - 2^{-n^c}$, for some constant $c > 0$.

This notation classifies confidence in probabilistic analysis into polynomial and exponential. Precise degrees of polynomials and exponent parameters are of lesser significance. This somewhat coarse notation is convenient for a clearer presentation and analysis as explained below, and in [15].

We use the following bounds on the tail of the binomial distribution.

Fact 3 (Chernoff Bounds [11, 4]). *Let x be random variable with a binomial distribution and let $\bar{x} = \mathbf{E}(x)$. Then, for $0 < \epsilon < 1$ we have $(1 - \epsilon)\bar{x} \le x \le (1 + \epsilon)\bar{x}$ with $(\epsilon^2 \bar{x})$-exponential probability.*

2.3 A Thinning-Out Principle

Suppose that we have n "active" elements and we want to turn them into non-active ones; suppose also that we have at hand a procedure that, using r processors, turns an active element into non-active with r-polynomial probability. The thinning-out principle applies this procedure to each of the active elements simultaneously and *independently*. The number n' of remaining active elements can be upper bounded by a binomial variable with parameters (n, r^{-c}), where $c > 0$ can be any constant. By Chernoff bounds (Fact 3), $n' < \frac{3}{2}n/r^c$ with n/r^c-exponential probability (by taking $\epsilon = 1/2$), and the number of active elements has indeed been thinned-out with high probability. This paradigm, where high success probability is converted into decrease in the number of active elements, was used in fast randomized algorithms [14] and in the log-star paradigm in [19].

In this paper we use the thinning-out principle as follows. We have many subproblems of "small" size, say r, that we try to solve simultaneously. We use two constant time subroutines: a randomized subroutine in which r processors can solve a subproblem with r-exponential probability, and a deterministic subroutine which uses r^b processors, with $b = 2, 3$, or 4 (depending on the subproblem). There are initially n/r subproblems, each allocated with r processors. The randomized subroutine is used for each subproblem, solving it in constant time with r-exponential probability. By using Chernoff bounds, it is shown that the number of unsolved subproblems is at most n/r^4, with appropriately high probability. To each subproblem r^b processors are allocated (with $b = 2, 3$ or 4); they can solve the subproblem deterministically in constant time. In the comparison model the allocation is given for free. To implement the allocation on a PRAM, we map each representative of an unresolved subproblem into a unique cell in an array of size n/r^b, where r^b processors are standing by. [19] showed how to compute such mapping using a *linear approximate compaction* algorithm:

Fact 4 (Linear Approximate Compaction [19]). *Assume that we are given an array of n processors, and a subset $S \subset [1..n]$ of distinguished processors such that $|S| \leq m$. Then, each distinguished processor can be allocated with a private cell from an array B of size $4m$ in $\log^* m$ time with n-exponential probability. If $m \log^{(i)} m \leq n$ and B is of size $m \log^{(i)} m$ then the allocation can be done in $O(i)$ time with n-exponential probability, for any $i = 1, \ldots \log^* n$.*

2.4 Constant-Time Computations of Maximum and of Range-Maxima

The following constant-time algorithms are used as subroutines in our algorithms.

Fact 5 ([21]). *The maximum of n given elements can be computed in constant time, using n processors, with n-exponential probability.*

Given $n^{1+\epsilon}$ processors, for any constant $\epsilon > 0$, the maximum over n given elements can be easily computed deterministically in constant time [23, 25]. This algorithm can be generalized for range-maxima preprocessing:

Fact 6 ([8]). *Assume that we are given an array A of size n, $n^{1+\epsilon}$ processors, and $O(n^{1+\epsilon})$ space, for any constant $\epsilon > 0$. Then, the array A can be preprocessed for range-maxima (deterministically) in constant time. Specifically, the time for preprocessing is $O(1/\epsilon)$. A query can be processed in constant time using a single processor.*

3 The Basic Range-Maxima Algorithm

Lemma 7. *Given n elements and $n \log n$ processors, preprocessing takes constant time, with n-exponential probability. Range-maxima query retrieval will take constant time.*

Assume that n elements $a_1, \ldots, a_n$ are given in an array A. First we present the output of the preprocessing algorithm, and show how the output can be used for processing a range-maximum query. This is followed by the basic preprocessing algorithm.

3.1 Output of the Basic Preprocessing Algorithm

Array A is partitioned into subarrays $A_1, \ldots, A_{n/\log^k n}$ of size $\log^k n$ each, where $A_j = [a_{(j-1)\log^k n+1}, \ldots, a_{j \log^k n}]$, and $k > 0$ is some constant to be determined later. The output consists of local data structures, one for each subarray, and a global data structure for the $n/\log^k n$ subarrays, as follows:

(1) For each subarray A_j we keep a local table. The table enables constant time retrieval of any range-maximum query within the subarray.
(2) Array $B = b_1, \ldots, b_{n/\log^k n}$ consisting of the maximum value in each subarray, i.e., $b_j = \max\{A_j\}$.
(3) A complete binary tree whose leaves are (the values) $b_1, \ldots, b_{n/\log^k n}$. For each internal node v of the tree let $L(v)$ be the subarray consisting of the (values of the) leaves of v. Node v also holds two more arrays: (i) array P_v, consisting of the prefix maxima of $L(v)$; and (ii) array S_v, consisting of the suffix maxima of $L(v)$.
(4) Two arrays of size n each, one contains all prefix-maxima and the other all suffix-maxima with respect to A.

In fact, item (4) is needed only for the recursive statement of the main algorithm in the next section.

3.2 Constant-Time Query Retrieval

Let $\max[i, j]$ be a query to be processed. There are two possibilities:

(i) a_i and a_j belong to the same subarray (of size $\log^k n$): $\max[i, j]$ is computed in $O(1)$ time using the local table that belongs to the subarray.
(ii) a_i and a_j belong to different subarrays. We elaborate below on possibility (ii).

Let right(i) denote the rightmost element in the subarray of a_i and left(j) denote the leftmost element in the subarray of a_j. Then max$[i, j]$ is the maximum among three numbers:

1. max$[i, \text{right}(i)]$, the maximum over the suffix of a_i in its subarray
2. max$[\text{left}(j), j]$, the maximum over the prefix of a_j in its subarray
3. max$[\text{right}(i) + 1, \text{left}(j) - 1]$.

The retrieval of the first and second numbers is similar to possibility (i) above. Denote $i' = \lceil i / \log^k n \rceil + 1$ and $j' = \lceil j / \log^k n \rceil - 1$. Then, a_i is in subarray $A_{i'-1}$ and a_j is in subarray $A_{j'+1}$. We discuss retrieval of the third number. This is equal to finding the maximum over interval $[b_{i'}, \ldots, b_{j'}]$ in B, which is denoted max$_B[i', j']$.

Let LCA(i', j') be the lowest common ancestor of $b_{i'}$ and $b_{j'}$, $v(i')$ be the child of LCA(i', j') that is an ancestor of $b_{i'}$, and $v(j')$ be the child of LCA(i', j') that is an ancestor of $b_{j'}$. Then max$_B[i', j']$ is the maximum among two numbers:

1. max$_B[i', \text{right}(v(i'))]$, the maximum over the suffix of $b_{i'}$ in $v(i')$. We get this from $S_{v(i')}$.
2. max$_B[\text{left}(v(j')), j']$, the maximum over the prefix of $b_{j'}$ in $v(j')$. We get this from $P_{v(j')}$.

It remains to show how to find LCA(i', j'), $v(i')$, and $v(j')$ in constant time. It was observed in [17] that the lowest common ancestor of two leaves in a complete binary tree can be found in constant time using one processor. (The idea is to number the leaves from 0 to $n-1$; given two leaves i' and j', it suffices to find the most significant bit in which the binary representation of i' and j' are different, in order to get the level of their lowest common ancestor.) Thus LCA(i', j'), and thereby also $v(i')$ and $v(j')$, can be found in constant time. Constant time retrieval of max$[i, j]$ query follows.

3.3 The Preprocessing Algorithm

The preprocessing algorithm has four steps:

Step 1. Local Preprocessing: For each sub-array A_j of size $\log^k n$ allocate a team of $\log^{k+1} n$ processors and preprocess the sub-array for range maxima. This can be done in constant time using Fact 6.

Step 2. Take the maximum b_j in each subarray A_j to build array B of size $n/\log^k n$.

Step 3. Global Preprocessing: Preprocess array B for range maxima, using procedure Global Preprocessing below.

Step 4. Compute the prefix maxima for array A: If element a_i is in sub-array A_j, then the prefix-maximum of i is the maximum between the prefix-maximum of i in A_j and the prefix-maximum of b_{j-1} in B. Compute the suffix maxima in a similar manner.

Procedure Global Preprocessing. We first describe the procedure with n-polynomial success probability, and then extend it to n-exponential success probability.

(1) Build a complete binary tree over array B; allocate $\log^{k+1} n$ processors to each element in B ($\log^k n$ per each of its ancestors).

(2) For each internal node v compute the maximum $m(v)$ over its array of leaves $L(v)$ as follows. Replace each leaf in $L(v)$ by $\log^k n$ copies (with respect to v) — one per allocated processor; put these copies into an array $L_1(v)$ (of size $L(v) \cdot \log^k n$) and apply the algorithm of Fact 5 to $L_1(v)$. As is shown in Claim 8 below, making $\log^k n$ copies of each leaf enables this step to run in constant time with n-polynomial probability.

(3) For each internal node v compute the prefix maxima and suffix maxima over its array $L(v)$:

Consider a leaf l in the array of v. Let $LS_v(l)$ be the set of left siblings of the nodes which are on the path from l to v. The leaves in arrays $L(u)$ of nodes u in $LS_v(l)$, together with l itself, are exactly the same leaves as the leaves in the prefix of l in $L(v)$. Therefore, the prefix maximum of l in $L(v)$ is the maximum over $\{m(u) : u \in LS_v(l) \cup l\}$. We allocate $(\log n)^2$ processors to l, and compute its prefix maximum in constant time, using Fact 6. (Note that $|LS_v(l)| \leq \log n$.)

Claim 8. *In Step 2 of Procedure* Global Preprocessing, *all maxima computations terminate in constant time with n-polynomial probability.*

Proof. Consider a level j in the binary tree in which each node v has an array of size L ($= |L(v)| = 2^{j-1}$) and an extended array $L_1(v)$ of size L_1 ($= |L_1(v)| = 2^{j-1} \cdot \log^k n$). In level j there are $n/L \log^k n$ nodes, and using Fact 5 the maximum can be computed for each node in constant time with $(L \log^k n)$-exponential probability. The probability that *all* maxima computations in level j take constant time is therefore $\geq 1 - (n/L \log^k n) \cdot 2^{-(L \log^k n)^\epsilon} \geq 1 - n/\log^k n \cdot 2^{-\log^{k\epsilon} n}$ for some constant $\epsilon > 0$, which is n-polynomial for any constant $k > 1/\epsilon$. This implies that the maxima computations for all nodes in all levels will also be completed in constant time, with n-polynomial probability.

Extensions for n-Exponential Probability. We use the thinning-out principle. Consider a level in the binary tree in which each internal node v has an array of size L. As in the proof of Claim 8, the maximum computation for v was done in Step (2) with $(L \log^k n)$-exponential probability. Let δ be some constant, $1/4 > \delta > 0$. If $L \geq n^\delta$ then the maximum computation with respect to v takes constant time with n-exponential probability. It follows that all maxima computations in this level take constant time with n-exponential probability (since the number of nodes is polynomial). Assume therefore that $L < n^\delta$. By Chernoff bounds (Fact 3) the number of failing nodes (in the level) is at most n/L^3 with (n/L^3)-exponential probability, and hence with n-exponential probability. We therefore add the following step:

(2b) Allocate in constant time L^2 processors to each failing node in the level, and compute the maximum for each allocated node in constant time by comparing each of the $\binom{L}{2}$ possible pairs in parallel. For nodes with $L \leq \log^{k/2} n$, L^2 processors are already allocated (see Step 1). For all other nodes the allocation is done using linear approximate compaction (Fact 4).

Claim 9. *Following Step 2b, all maxima computations terminate with n-exponential probability.*

4 The Main Algorithm

Theorem 10. *There exists a PRAM preprocessing algorithm for the range-maxima problem that, using an optimal number of processors and linear space, takes $O(\log^* n)$ time with n-exponential probability. Subsequently, each query takes constant (worst case) time using one processor.*

In this extended abstract we only sketch an algorithm whose query processing time is $O(\log^* n)$. We later hint on how constant query time can be obtained. Let $T(n)$ be the preprocessing time with n-exponential probability, for a problem of size n. The algorithm has three recursive steps. (1) Array A is partitioned into subarrays of size $\log n$ each. (2) We allocate $\log n$ processors to each subarray and build recursively a local data structure for each subarray in $T(\log n)$ time with $(\log n)$-exponential probability. Using $\log^4 n$ processors for each subarray which failed building its (recursive) data structure in $T(n)$ time, a table consisting of answers to all possible range-maxima queries can be built in constant time. From the thinning-out principle it follows that, with n-exponential probability, we can build for every subarray either a recursive local data structure or a table in $T(\log n) + O(1)$ time. (3) Let B be an array of size $n/\log n$ consisting of the maximum in each subarray. We compute a global data structure over B in constant time with n-exponential probability, using the basic algorithm of Lemma 7. Thus, the preprocessing time is $T(n) = T(\log n) + O(1) = O(\log^* n)$ with n-exponential probability. This preprocessing algorithm can be readily modified to use an optimal number of processors.

Query retrieval is done somewhat similarly to the basic algorithm of Sect. 3, where in our case queries over subarrays of size $\log n$ are computed recursively. Hence we get $O(\log^* n)$ time for query retrieval.

In order to get constant-time query retrieval (and thereby satisfy Theorem 10) we need to circumvent the recursive query retrieval over $\log n$-size subarrays. Indeed, recursive query retrieval can be replaced by prefix- and suffix-maxima queries; these can be precomputed by the preprocessing algorithm. Allocation of processors for this computation is done using a linear approximate compaction algorithm (Fact 4). Finally we note that constant preprocessing time can be obtained by using $n \log^{(i)} n$ processors, for any constant $i > 0$.

To state the result for the comparison model algorithm we need the following definitions.

The Inverse-Ackermann Function. We define the inverse-Ackermann function as in [8]. (For a more standard definition see [24].) Consider a real function f. Let $f^{(i)}$ denote the i-th iterate of f. (Formally, we denote $f^{(1)}(n) = f(n)$ and $f^{(i)}(n) = f(f^{(i-1)}(n))$ for $i > 1$.) Next, we define the $*$ (pronounced "star") functional that maps the function f into another function $*f$: $*f(n) = minimum\{i \mid f^{(i)}(n) \leq 1\}$ (we only consider functions for which this minimum is well defined). (Note that the function $\log^*$ will be denoted $* \log$ using our notation; this change is for notational convenience.)

We define inductively a series I_k of slow-growing functions from the set of integers that are larger than 2 into the set of positive integers. (i) $I_1(n) = \lceil n/2 \rceil$, and (ii) $I_k = *I_{k-1}$ for $k \geq 2$. The first three in this series are familiar functions: $I_1(n) = \lceil n/2 \rceil$, $I_2(n) = \lceil \log n \rceil$ and $I_3(n) = * \log n$. The *inverse-Ackermann function* is $\alpha(n) = minimum \{i \mid I_i(n) \leq i\}$.

Theorem 11. *For any m, $m = 2, \ldots, \alpha(n)$, there exists a preprocessing algorithm for the range-maxima problem that runs in $O(m + I_m(n))$ time using an optimal number of processors on the comparison model with n-exponential probility. Subsequently, each query takes $O(m)$ (worst case) time using one processor.*

The sequence of algorithms that realize Theorem 11 uses the recursive *-tree data structure proposed in [8]. These algorithms are based on a sequence of algorithms that use $nI_m(n)$ processors to get $O(m)$ preprocessing time with n-exponential probability, for $m = 2, \ldots, \alpha(n)$. An algorithm for some $m > 2$ uses recursively the algorithm for $m - 1$ in a way somewhat similar to the way the $O(\log^* n)$ time algorithm above uses the basic algorithm. In each recursive step of this comparison model algorithm the thinning-out principle is used to guarantee that with n-exponential probability the computation is completed in time for all the local sub-problems. Let us note that to get a PRAM algorithm, we must explicitly show how to allocate processors to tasks, when implementing the thinning-out principle. This can be done using linear approximate compaction (Fact 4) leading to an optimal PRAM algorithm for $m = 3$, as in Theorem 10. In fact, the resulting PRAM algorithm is equivalent to the constant query-time PRAM algorithm sketched above.

5 Conclusion and open problems

The constant time randomized algorithm for the maximum problem proved that randomization can beat deterministic lower bounds in parallel algorithms. In this paper we demonstrated that randomization is powerful enough to enable "nearly-constant" time optimal algorithms for problems which are much more general than the maximum problem.

Our results suggest several open problems:

- Give a preprocessing algorithm for the PRAM in $O(\alpha(n))$ expected time or show that $\Omega(\log^* n)$ is optimal.
- Prove $\Omega(\alpha(n))$ expected time lower bound for the comparison model or give an $O(1)$ expected time algorithm.
- Recently, MacKenzie [18] gave a lower bound of $\Omega(\log^* n)$ expected time for the linear approximate compaction problem. This implies that the PRAM algorithm cannot be improved by having a better algorithm for linear approximate compaction — one of the main tools for processor allocation in super fast randomized algorithms. Can one find alternative tools for processors allocations?

References

1. N. Alon and Y. Azar. The average complexity of deterministic and randomized parallel comparison-sorting algorithms. *SIAM J. Comput.*, 17:1178–1192, 1988.
2. N. Alon and B. Schieber. Optimal preprocessing for answering on-line product queries. Technical Report 71/87, Eskenasy Inst. of Comp. Sc., Tel Aviv Univ., 1987.
3. A. Amir, G. M. Landau, and U. Vishkin. Efficient pattern matching with scaling. In *SODA '90*, pages 344–357, 1990.
4. D. Angluin and L. G. Valiant. Fast probabilistic algorithms for hamiltonian paths and matchings. *J. Comput. Syst. Sci.*, 18:155–193, 1979.

5. Y. Azar and U. Vishkin. Tight comparison bounds on the complexity of parallel sorting. *SIAM J. Comput.*, 16:458–464, 1987.
6. O. Berkman, Y. Matias, and U. Vishkin. Randomized range-maxima in nearly-constant parallel time. Technical Report UMIACS-TR-91-161, Institute for Advanced Computer Studies, Univ. of Maryland, 1991. To appear in *j. computational complexity*.
7. O. Berkman, B. Schieber, and U. Vishkin. Some doubly logarithmic parallel algorithms based on finding all nearest smaller values. Technical Report UMIACS-TR-88-79, Univ. of Maryland Inst. for Advanced Computer Studies, Oct. 1988. To appear in *J. Algorithms* as 'Optimal doubly logarithmic parallel algorithms based on finding all nearest smaller values'.
8. O. Berkman and U. Vishkin. Recursive *-tree parallel data-structure. In *FOCS '89*, pages 196–202, 1989. Also in UMIACS-TR-90-40, Institute for Advanced Computer Studies, Univ. of Maryland, March 1991. To appear in *SIAM J. Comput.*
9. R. B. Boppana. The average-case parallel complexity of sorting. *Inf. Process. Lett.*, 33:145–146, 1989.
10. A. Borodin and J. E. Hopcroft. Routing, merging, and sorting on parallel models of computation. *J. Comput. Syst. Sci.*, 30:130–145, 1985.
11. H. Chernoff. A measure of asymptotic efficiency for tests of a hypothesis based on the sum of observations. *Annals of Math. Statistics*, 23:493–507, 1952.
12. P. F. Dietz. Heap construction in the parallel comparison tree model. In *SWAT '92*, pages 140–150, July 1992.
13. H. N. Gabow, J. L. Bentley, and R. E. Tarjan. Scaling and related techniques for geometry problems. In *STOC '84*, pages 135–143, 1984.
14. J. Gil and Y. Matias. Fast hashing on a PRAM—designing by expectation. In *SODA '91*, pages 271–280, 1991.
15. J. Gil, Y. Matias, and U. Vishkin. Towards a theory of nearly constant time parallel algorithms. In *FOCS '91*, pages 698–710, Oct. 1991.
16. J. Gil and L. Rudolph. Counting and packing in parallel. In *ICPP '86*, pages 1000–1002, 1986.
17. D. Harel and R. E. Tarjan. Fast algorithms for finding nearest common ancestors. *SIAM J. Comput.*, 13(2):338–355, 1984.
18. P. D. MacKenzie. Load balancing requires $\Omega(\log^* n)$ expected time. In *SODA '92*, pages 94–99, Jan. 1992.
19. Y. Matias and U. Vishkin. Converting high probability into nearly-constant time—with applications to parallel hashing. In *STOC '91*, pages 307–316, 1991. Also in UMIACS-TR-91-65, Institute for Advanced Computer Studies, Univ. of Maryland, April 1991.
20. V. Ramachandran and U. Vishkin. Efficient parallel triconnectivity in logarithmic parallel time. In *AWOC '88*, pages 33–42, 1988.
21. R. Reischuk. Probabilistic parallel algorithms for sorting and selection. *SIAM J. Comput.*, 14(2):396–409, May 1985.
22. B. Schieber. *Design and analysis of some parallel algorithms*. PhD thesis, Dept. of Computer Science, Tel Aviv Univ., 1987.
23. Y. Shiloach and U. Vishkin. Finding the maximum, merging, and sorting in a parallel computation model. *J. of Alg.*, 2:88–102, 1981.
24. R. E. Tarjan. Efficiency of a good but not linear set union algorithm. *J. ACM*, 22(2):215–225, Apr. 1975.
25. L. G. Valiant. Parallelism in comparison problems. *SIAM J. Comput.*, 4:348–355, 1975.
26. U. Vishkin. Structural parallel algorithmics. In *ICALP '91*, pages 363–380, 1991.

Fault-Tolerant Broadcasting in Binary Jumping Networks

Yijie Han[1], Yoshihide Igarashi[2], Kumiko Kanai[3] and Kinya Miura[2]

[1] Department of Computer Science, University of Hong Kong, Hong Kong
[2] Department of Computer Science, Gunma University, Kiryu, 376 Japan
[3] Research and Development Center, Toshiba Corporation, Kawasaki, 210 Japan

Abstract. The following results are shown. Let N be the number of processors in the network. When N is a power of 2, $\log_2 N + f + 1$ rounds suffice for broadcasting in the binary jumping network if at most f processors and/or links have failed and $f \leq \log_2 N - 1$. For an arbitrary N, $\lceil \log_2 N \rceil + f + 2$ rounds suffice for broadcasting in the binary jumping network if at most f processors and/or links have failed and $f \leq \lceil \log_2 N \rceil - 2$. It is also shown that the vertex connectivity of the binary jumping network with N nodes is $\lceil \log_2 N \rceil$.

1 Introduction

Data broadcasting in a network is a very fundamental operation and has been much studied. As suggested in [1][4], it can be accomplished by the data disseminating process in a network in a way that each processor repeatedly receives and forwards messages without physical broadcast. This fashion of broadcasting is suitable for a synchronized distributed system requiring the condition that any processor can be the source of information and any round can be the start of broadcasting. This type of broadcasting schemes have been studied in [1][3][4][7], and they are obviously different from broadcasting schemes with the fixed starting round such as schemes discussed in [2][6][8][9].

Binary jumping networks have been proposed for interconnecting processors due to their topological richness[1][4]. For any N, there exists a binary jumping network with N nodes, whereas there exists a hypercube with N nodes only if N is a power of 2. In this paper we discuss the fault tolerance of an information disseminating scheme in binary jumping networks. In the network with N processors, each processor is addressed by an integer from $\{0, \cdots, N-1\}$, and there is a link from processor u to processor v if and only if $v - u$ *modulo* N is 2^k for some $0 \leq k \leq \lceil \log_2 N \rceil - 1$. This network is called the binary jumping network with N processsors. The processors in the network are synchronized with a global clock. Han and Finkel have introduced an information disseminating scheme in the binary jumping network[4], although its prototype may be traced back to a scheme by Alon *et al* .[1]. The total number of links in the binary jumping network with N processors is obviously $\Theta(N \log N)$. The Han-Finkel's scheme is known to be time optimal if no processors/links have failed[4]. Information disseminating schemes that require only $\Theta(N)$ links are known[2][5], but these schemes are not time optimal.

Some processors and/or links may be faulty. However, multiple copies of a message can be disseminated through disjoint paths, and the fault tolerance can be achieved by this multiplicity. We consider only the case where faulty processors cannot forward messages, but can receive them, and/or some links may be disconnected. We do not consider cases where a faulty processor alters information and forwards wrong messages. The period for forwarding a message from a processor to one of its neighbors is called a round. We assume that each processor can receive a message and can forward a message in the same round. We also assume that the source processor is always faultless. We evaluate a sufficient number of rounds to complete broadcasting when some processors and/or links have failed. We show that for an arbitrary N, $\lceil \log_2 N \rceil + f + 2$ rounds suffice for broadcasting if f processors and/or links have failed and $f \leq \lceil \log_2 N \rceil - 2$. The case where $f = \lceil \log_2 N \rceil - 1$ is also discussed. It is shown that the vertex connectivity of the binary jumping network with N nodes is $\lceil \log_2 N \rceil$.

2 An Information Disseminating Scheme

The number of processors in the network is denoted by N, and $\lceil \log_2 N \rceil$ is denoted by n. We denote m *modulo* r by $[m]_r$. A directed graph $G = (V, E)$ is called a binary jumping network with N nodes if $V = \{0, \cdots, N - 1\}$ and $E = \{(u, v) | u, v \text{ in } V, \text{ and } [v - u]_N \text{ is } 2^k \text{ for some } 0 \leq k \leq \lceil \log_2 N \rceil - 1\}$.

Throughout this paper we consider the following scheme for disseminating information in a binary jumping network or a network including a binary jumping network as a spanning subgraph. When N is a power of 2, the scheme is the same as the scheme proposed by Alon *et al.*[1], and for a general N it was first proposed by Han and Finkel[4].

procedure *disseminate(N)*
    repeat
        for *round* := 0 **to** $\lceil \log_2 N \rceil - 1$ **do**
            for each processor u send a message
            from u to processor $[u + 2^{round}]_N$ concurrently
    forever

Han and Finkel have showed the next result[4].

Theorem 1 [4]. *Procedure disseminate will broadcast information from any source to all destinations within any consecutive*

1. $\lceil \log_2 N \rceil$ rounds if no processors have failed, and
2. $\lceil \log_2 N \rceil + 2$ rounds if exactly one processor or exactly one link has failed.

For any scheme that can send a message by each processor to one of its neighbors at each round, at most 2^i processors can receive the message from the source within i rounds. Therefore, for such a disseminating scheme it is not possible to complete the broadcast within less than $\lceil \log_2 N \rceil$ rounds. In this sense procedure *disseminate* is time optimal.

3 Message Route Representation

We express a message route from processor u by procedure *disseminate* starting at round s in the form $u(s, a_k \cdots a_1 a_0)$, where a_j is 0 or 1 for each $0 \leq j \leq k$. The message route $u(s, a_k \cdots a_1 a_0)$ is defined by the following procedure as a sequence of processors through which the message moves:

$\alpha := \alpha_0 := u$;
for $i := 0$ **to** k **do**
$\qquad \alpha_{i+1} := [\alpha_i + a_i \times 2^{[s+i]_n}]_N$;
$\qquad$ **if** $a_i \neq 0$ **then** $\alpha := \alpha \alpha_{i+1}$

Representation $u(s, a_k \cdots a_1 a_0)$ denotes not only the message route α but also the message flow timing. If a_j is the g-th nonzero digit of $a_k \cdots a_1 a_0$ from the rightmost, then the message originated from processor u moves from processor α_j to processor α_{j+1} at round $[s + j]_n$ and α_{j+1} is the $g + 1$-st processor in the message route α. Suppose that a_k is the last nonzero digit of $a_k \cdots a_1 a_0$ from the rightmost. Then this message flow from processor u to the last processor of α through the route takes $k + 1$ rounds. For technical convenience we will also use relabelled addresses in the message route. Relabelled addresses can be obtained by replacing $\alpha_{i+1} := [\alpha_i + a_i \times 2^{[s+i]_n}]_N$ by $\alpha_{i+1} := \alpha_i + a_i \times 2^{[s+i]_n}$ in the third line of the message route definition given above. A relabelled address may be greater than $N - 1$. If $x \geq N$, relabelled address x denotes processor $[x]_N$.

Let u and v $(0 \leq u, v \leq N - 1)$ be a pair of processors. For each $0 \leq s, i \leq n - 1$, we define a message route $R_{s,i}(u, v)$ as follows: Let $q_{n-1} \cdots q_1 q_0 = [v - u - 2^{[i+s]_n}]_N$, where $q_h = 0$ or 1 for each $0 \leq h \leq n - 1$. Let $R_{s,i}(u, v) = u(s, c_{n+i} \cdots c_1 c_0)$, where

$$
c_k = \begin{cases} 0 & \text{if } k \leq i - 1 \\ 1 & \text{if } k = i \\ q_{[k+s]_n} & \text{if } i + 1 \leq k \leq n + i \end{cases} \tag{1}
$$

From the above definition $R_{s,i}(u, v)$ denotes a message route from processor u to processor v by procedure *disseminate* starting at round s, and the message located in processor u does not move before round $[s + i]_n$.

Example 1. Suppose $N = 18$. The message flow specified by $R_{1,2}(2, 5) = 2(1, 11010100)$ is shown in Table 1.

The next lemma is immediate.

Lemma 2. *For any $s, i (0 \leq s, i \leq n - 1)$ and any pair of processors u and v, the message from processor u will reach processor v through the route specified by $R_{s,i}(u, v)$ within at most $\lceil \log_2 N \rceil + i + 1$ rounds if no processors and no links in the route have failed.*

Because of the symmetrical property of the scheme, without loss of generality hereafter we may assume that the source of information is processor 0.

Table 1. The message flow specified by $R_{1,2}(2,5) = 2(1, 11010100)$.

		time ⟶							
round		1 (start round)	2	3	4	0	1	2	3
processor	2 (source)	$\boxed{2}$	$\boxed{2}$	10	$\boxed{10}$	11	$\boxed{11}$	15	5 (23)

$\boxed{m}$ indicates that the message remains at processor m.
(r) indicates relabelled address r greater than $N - 1$.

4 Fault Tolerance When $N = 2^n$

Lemma 3. When N is a power of 2, for any $s, i_1, i_2 (0 \leq s \leq n-1, 0 \leq i_1 < i_2 \leq n - 1)$, u and v $(0 \leq u, v \leq N - 1)$, the message routes specified by $R_{s,i_1}(u, v)$ and by $R_{s,i_2}(u, v)$ are node disjoint except for the source and the destination.

Proof. We may assume that each processor is addressed as a binary number of length n and that the source is processor 0. The $[s + i_1]_n + 1st$ bit from the rightmost of the address of any intermediate processor appeared before the second round $[s+i_1]_n$ in $R_{s,i_2}(0, v)$ is different from the corresponding bit of the address of any intermediate processor in $R_{s,i_1}(0, v)$. The message route $R_{s,i_1}(0, v)$ ends at the second round $[s + i_1]_n$ when the destination v has been reached. As for an intermediate processor in the message route $R_{s,i_2}(0, v)$ appeared between the second round $[s + i_1]_n$ and the second round $[s + i_2]_n$, its address is different from the address of any intermediate processor in the message route $R_{s,i_1}(0, v)$ at a bit between the $[s+i_1]_n+1st$ bit and the $[s+i_2]_n+1st$ bit including extremes from the rightmost. Hence, the lemma holds true. □

Theorem 4. Procedure disseminate will broadcast from any source to all destinations within any consecutive $\log_2 N + f + 1$ rounds if N is a power of 2 and at most $f \leq \log_2 N - 1$ processors and/or links have failed.

Proof. We may assume that the source is processor 0. Let f be the number of faulty processors and/or links, where $f \leq \log_2 N - 1$. For each processor v, consider the message routes $R_{s,i}(0, v)$ $(i = 0, \cdots, f)$. From Lemma 3 these $f + 1$ message routes are node disjoint except for the source and the destination. Since we assume that the source is faultless, at least one of them is faultless. Hence, from Lemma 2 the message will reach processor v within at most $\log_2 N + f + 1$ rounds. □

Consider the situation that all f links from processor 0 to processors $2^0, 2^1, \cdots, 2^{f-1}$ are disconnected. Then the message cannot reach processor 2^{f-1} within $\log_2 N + f$ rounds when the source is processor 0 and the starting round is 0. From this observation the bound given in Theorem 4 is tight.

5   Fault Tolerance for an Arbitrary N

We divide the message routes $R_{s,i}(0, v)$, $0 \le s, i \le \lceil \log_2 N \rceil - 1$ into two classes, ascending routes and nonascending routes. A message route $R_{s,i}(0, v)$ is called an ascending route if $2^{[s+i]_n} \le v$, and otherwise it is called a nonascending route. The next lemma is immediate.

Lemma 5. *The sequence of processors in an ascending route is in ascending order.*

For example, $R_{1,2}(0, 10)$ with $N = 18$ is an ascending route since $2^3 \le 10$. Due to Lemma 5, we need to consider only the processors $0, 1, 2, \cdots, v$ for ascending routes. Therefore, when we consider only ascending routes, the sequence of processors in $R_{s,i}(0, v)$ with N processors and that in $R_{s,i}(0, v)$ with $2^{\lceil \log_2 N \rceil}$ processors are the same. Hence, the argument in the proof of Lemma 3 is valid for the class of ascending routes, and the next lemma is immediate.

Lemma 6. *For an arbitrary size of the network all ascending routes are node disjoint except for the source and the destination.*

The relabelled address of processor p is p itself or $N + p$. For example, the message route $(0,8,0,2,6)$ of $R_{1,2}(0, 6)$ with $N = 9$ is expressed as $(0,8,9,11,15)$ by using relabelled addresses. After relabelling, the sequence of the addresses in a nonascending route will be in ascending order. The sequence of relabelled addresses in nonascending route $R_{s,i}(0, v)$ with network size N is exactly the same as that in $R_{s,i}(0, v+N)$ with network size $\frac{3}{2} \cdot 2^{\lceil \log_2 N \rceil}$. Sequence $(0,8,9,11,15)$ is the sequence of the processors in $R_{1,2}(0, 15)$ with $N = 16$. Therefore, all nonascending routes are node disjoint except for the source and the destination. One of the nonascending routes may go through the source processor, where "go through" means "go in and go out". As shown above the route $R_{1,2}(0, 6)$ with $N = 9$ goes through processor 0. It is not possible that more than one nonascending routes go through processor 0. Since we assume that the source processor is not faulty, we have the next lemma.

Lemma 7. *For an arbitrary size of the network, if t nonascending routes are available then they can be used for tolerating $t - 1$ faults.*

We now discuss the interaction between ascending routes and nonascending routes. Let destination v satisfy $2^{k-1} \le v < 2^k < N$. The message route $R_{s,i}(0, v)$ is the same as $R_{k,j}(0, v)$ if $[s+i]_n = [k+j]_n$, though their timing is different. Therefore, for each nonascending route we may assume that the starting round is round k. Then for any nonascending route, each of rounds $0, 1, \cdots, k$ will be executed only once.

Lemma 8. *If a nonascending route did not go through a relabelled processor in $\{N, N+1, \cdots, N+v\}$ at the end of round $k-1$, then it will not go through any of them during the rest of the rounds.*

Proof. Consider a nonascending route satisfying the condition of the lemma. At each round after round $k - 1$, a message will move from a processor p to a processor q (relabelled address) $q \geq p + 2^k$ or the message will remain at the processor p. If p is $N + v$ then the massage has arrived at the destination and there is no need to move the message to processor q. If p is not $N + v$ and $p \geq N$, then $q > N + v$. This is impossible because relabelled processors are in ascending order and must stop at $N + v$. Hence, if p is not $N + v$ then p cannot be any of $N, N + 1, \cdots, N + v - 1$. By the same reason q is either $N + v$ or $q < N$. This argument can be used iteratively to show that the nonascending route does not go through any of $N, N + 1, \cdots, N + v$. $\qquad\qquad\qquad\qquad\qquad\qquad\square$

For a set of processors $P = \{p_1, \cdots, p_u\}$ we define $D(P) = min\{|p_i - p_j| \| i \neq j$ and $p_i, p_j \in P\}$. Let destination v satisfy $2^{k-1} \leq v < 2^k < N$. We consider all nonascending routes $R_{s,i}(0, v)$, $k \leq [s + i]_n \leq n - 1$. We relabel the processors as we did before. Let S be the set of the relabelled addresses of the processors in these nonascending routes immediately after the first round $n - 1$. Then the next lemma is immediate.

Lemma 9. *For any p, q in S, $|p - q|$ is a multiple of 2^k and $D(S) \geq 2^k$.*

The largest relabelled address p_1 in S is less than or equal to $N + v$, the second largest one p_2 in S is less than N(because $v < 2^k$ and from Lemma 9), and the third largest one p_3 in S is less than $N - 2^k$. After the first round $n - 1$, rounds $0, 1, \cdots, k-1$ follow. During these rounds (i.e., rounds $0, 1, \cdots, k-1$), only routes going through p_1 or p_2 could possibly go through a relabelled processor in $\{N, N + 1, \cdots, N + v\}$. Each of the rest of the nonascending routes will end at a processor less than N at the end of round $k - 1$. That is, at the end of round $k - 1$, at most 2 routes among all nonascending routes could possibly go through processors which are gone through in ascending routes. By Lemma 8, among all the nonascending routes at most 2 routes could possibly go through processors which are gone through in ascending routes. From this observation, $\lceil \log_2 N \rceil + f + 3$ rounds suffice for tolerating $f \leq \lceil \log_2 N \rceil - 3$ faults.

We now show that $\lceil \log_2 N \rceil + f + 2$ rounds suffice for tolerating f faults. As we have discussed above, the largest relabelled address p_1 in S is less than or equal to $N + v$. Thus the second largest address p_2 in S is at most $N + v - 2^k$. We note that $p_1 - p_2$ is equal to 2^k or no less than 2^{k+1}. If $p_1 - p_2 \geq 2^{k+1}$, then $p_2 \leq N - 2^k$ and at the end of round $k - 1$ the route going through p_2 ends at a processor less than N. Therefore, from Lemma 8 this route will not go through any processor in $\{N, N + 1, \cdots, N + v\}$. Thus it is sufficient to consider only the case where $p_2 = p_1 - 2^k$. In this case, during rounds $0, 1, \cdots, k - 1$ the route going through p_2 cannot reach $N + v$ because $N + v - p_2 \geq 2^k$, and at the end of round $k - 1$ this route cannot end at a processor p with $N \leq p < N + v$; otherwise during the remaining rounds the message will move to a processor greater than $N + v$ or remain at processor p. Therefore, by Lemma 8 the route going through p_2 will not go through any processor in $\{N, N + 1, \cdots, N + v\}$. Hence, there exists at most one nonascending route which may intersect with ascending routes. Thus we have the next theorem.

Theorem 10. *Procedure disseminate will broadcast from any source to all desti-nations within any consecutive $\lceil \log_2 N \rceil + f + 2$ rounds if at most $f \leq \lceil \log_2 N \rceil - 2$ processors and/or links have failed.*

Example 2. Consider the massage routes $R_{3,i}(0,6)$ $(0 \leq i \leq 5)$ with $N = 37$. These message routes are shown in Table 2. Let S be the set of the relabelled addresses in the nonascending routes immediately after the first round 5. Then $S = \{16, 32, 40\}$. Since $2^2 \leq 6 < 2^3$, the difference of any pair of elements in S is a multiple of 2^3. The nonascending route with the maximum element in S could possibly go through a processor in an ascending route. In fact, processor 4 is included in both $R_{3,0}(0,6)$ and $R_{3,5}(0,5)$. If processors 2, 4, 16 and 32 have failed, only route $R_{3,3}(0,6)$ is not faulty among routes $R_{3,i}(0,10)$ $(0 \leq i \leq 5)$. As shown in Table 2, if the starting round is 3 and the number of faulty processors and/or links is at most 4, then 10 rounds suffice for broadcasting. However, if the starting round is 5 and processors 1, 2, 4 and 32 have failed, 12 rounds are necessary and sufficient for broadcasting. This number of rounds coincides with the bound given in Theorem 10.

Table 2. Message routes $R_{3,i}(0,6)$ $(0 \leq i \leq 5)$ with $N = 37$.

$R_{s,i}(0,v)$	round $\longrightarrow$									
	3 4	5	0	1	2	3	4	5	0 1 2
$R_{3,0}(0,6)$ nonascending	0 8	-	3	4	6				
			(40)	(41)	(43)				
$R_{3,1}(0,6)$ nonascending	0 -	16	-	17	19	-	27	6	
								(43)	
$R_{3,2}(0,6)$ nonascending	0 -	-	32	33	35	-	6		
							(43)		
$R_{3,3}(0,6)$ ascending	0 -	-	-	1	-	5	-	-	- 6
$R_{3,4}(0,6)$ ascending	0 -	-	-	-	2	6			
$R_{3,5}(0,6)$ ascending	0 -	-	-	-	-	4	-	-	- - 6

(r) indicates that r is a relabelled address.

6 The Case of $\lceil \log_2 N \rceil - 1$ Faults

Let v be a destination processor. The distance between the source and the des-tination v is also denoted by v.

Theorem 11. *Let $N = 2^{n-1} + t$ and $2^{k-1} \leq v < 2^k$. If $t + v \geq 2^k$ and the number of faulty processors and/or links $f \leq \lceil \log_2 N \rceil - 1$, then $3\lceil \log_2 N \rceil - k - 1$ rounds suffice for sending a message from processor 0 to processor v.*

Proof. Let $2^k \leq t + v$. For each nonascending route we may assume that the starting round is round k, since we are not concerned with broadcasting time at this stage (it will be considered later). Let $R = R_{k,j}(0, v)$ be the nonascending route that reaches the largest processor p (relabelled address) at the end of the first round $n - 1$. As shown in the proof of Theorem 10, only route R could possibly intersect with an ascending route. It is obvious that p must be a multiple of 2^k.

Case 1: $N + v - p \geq 2^k$.

After the first round $n - 1$ the following round k is needed to send the message in processor p to processor v. Therefore, the message cannot be moved to a processor larger than $N - 1$ from round 0 through round $k - 1$; otherwise round k cannot enforce the message to reach processor v. Hence, in this case route R cannot intersect with any ascending route.

Case 2: $N + v - p < 2^k$.

Since we assume $2^k \leq t + v$ and $N + v - p < 2^k$, $p > 2^{n-1}$. In route R, at the first round $n - 1$ the message in processor $p' = p - 2^{n-1}$ moves to processor p. We now modify the route R and obtain route R'. In route R', the message in processor p' will not move at the first round $n - 1$. We now use rounds $0, ..., k - 1$ following the first round $n - 1$ to send the message in processor p' to processor $p'' = N + v - 2^{n-1}$. This move is possible since $N + v - p < 2^k$. We then use the second round $n - 1$ to move the message in processor p'' to the destination v. In the duration of moving the message from processor p' to processor p'', the route R' does not intersect with any of routes $R_{k,i}(0, v)$ except for route $R = R_{k,j}(0, v)$. At the second round $n - 1$ the message in processor p'' can reach directly the destination v. By replacing $R = R_{k,j}(0, v)$ with R', we obtain a set of n node disjoint routes connecting processor 0 and processor v. Therefore, the network can tolerate $n - 1$ faults for sending the message from the source to processor v.

Even if we change the starting round, each of these n message routes is the same, but the necessary number of rounds for sending the message through the route to the destination v depends on the starting round. For any starting round, $2n$ rounds suffice for sending the message through the routes except for the modified route R'. When the starting round is round $k + 1$, the number of rounds required to send the message through R' is the largest. In the worst case, $3n - k - 1$ rounds may be required to send the message through R'. Hence, the theorem holds. $\qquad\Box$

Example 3. Let $N = 24$, $t = 8$ and the source be processor 0. Consider the destination $v = 7$. Then $2^2 \leq v < 2^3$ and $k = 3$. The condition in Theorem 11, $t + v = 15 > 2^3$ is satisfied. As shown in Table 3, only route $R_{3,0}(0, 7)$ intersects with an ascending route $R_{3,2}(0, 7)$. Route $R' = (0, 8, 9, 11, 15, 7(= [31]_{24}))$ is obtained by modifying $R_{3,0}(0, 7)$. This route is also shown in Table 3. Routes $R_{3,i}(0, 7)$ ($i = 1, 2, 3, 4$) and R' are node disjoint message routes from processor 0 to processor 7. If the starting round is round 4, then 11 rounds are required to send the message to processor 7 through route R'.

Table 3. Message routes $R_{3,i}(0,7)$ $(0 \leq i \leq 4)$ and the modified route R'.

$R_{s,i}(0,v)$	3	4	round $\longrightarrow$ 0	1	2	3	4	0 1 2
$R_{3,0}(0,7)$ nonascending	0 8	0	1	3	7			
		(24)	(25)	(27)	(31)			
$R_{3,1}(0,7)$ nonascending	0 -	16	17	19	23	7		
						(31)		
$R_{3,2}(0,7)$ ascending	0 -	-	1	3	7			
$R_{3,3}(0,7)$ ascending	0 -	-	-	2	6	-	-	7
$R_{3,4}(0,7)$ ascending	0 -	-	-	-	4	-	-	5 7
R' modified route	0 8	-	9	11	15	-	7	
							(31)	

(r) indicates that r is a relabelled address.

7 The Connectivity of Binary Jumping Networks

The vertex connectivity of a graph is the minimum number of nodes whose removal results in a disconnected graph. From Theorem 4, for $N = 2^n$ the vertex connectivity of the binary jumping network with N nodes is exactly $\log_2 N$, but for an arbitrary N it has not been known. In this section we prove that for any N it is exactly $\lceil \log_2 N \rceil$.

Theorem 12. *The vertex connectivity of the binary jumping network with N nodes is $\lceil \log_2 N \rceil$.*

Proof. It is obvious that the vertex connectivity of the binary jumping network with N nodes is not more than $\lceil \log_2 N \rceil$. When $N = 2^n$, from Theorem 4 the vertex connectivity is exactly $\log_2 N$. Let $N = 2^{n-1} + t$, $t \neq 0$ and $2^{k-1} \leq v < 2^k$. From Theorem 11 only the situation we must consider is the connectivity between node 0 and node v such that $t + v < 2^k$. We apply procedure *disseminate(N)* with starting round k, and consider message routes $R_{k,i}(0,v)$ $(0 \leq v \leq n-1)$. Since $t + v < 2^k$, we have $v + N < 2^{n-1} + 2^k$. Hence, among nonascending routes, $R_{k,n-k-1}(0,v)$ has the largest relabelled address at the end of the first round $n-1$. As shown in Section 5, only $R_{k,n-k-1}(0,v)$ could possibly intersect with an ascending route. Note that $R_{k,n-k-1}(0,v)$ first goes through node 2^{n-1}. Since $v \geq 2^{k-1}$ and $t + v < 2^k$, we have $t < 2^{k-1}$. Let $m = \lfloor (2^k - v)/t \rfloor$. Then $m \geq 1$ since $t + v < 2^k$. We construct a route R from node 0 to node v as follows: The first $m + 2$ nodes in R are $0, 2^{n-1}, q_1, q_2, \cdots, q_m$, where $q_j = 2^k - jt$ $(1 \leq j \leq m)$. This construction is possible since $[2^{n-1} + 2^k]_N = 2^k - t$ and $[2^k - jt + 2^{n-1}]_N = 2^k - (j+1)t$ $(1 \leq j \leq m-1)$. In the following rounds 0 to $k-1$ we can move the message in node q_m to node $v + t$ since $v + t - (2^k - mt) < t < 2^{k-1}$. The last link in R is the edge from node $v + t$ to node v. This edge exists since $[v + t + 2^{n-1}]_N = v$. The route R never use nodes from $\{1, \cdots, v-1\}$ nor nodes from $\{2^k, \cdots, N-1\} - \{2^{n-1}\}$. Hence, route R is node disjoint from all ascending routes and nonascending routes except for $R_{k,n-k-1}$ that goes through node 2^{n-1}. Therefore, there exist $\lceil \log_2 N \rceil$ node disjoint routes from node 0 to node v. From the symmetry of the network the vertex connectivity of the network is $\lceil \log_2 N \rceil$. $\qquad\qquad\qquad\qquad\qquad\qquad\qquad\qquad\qquad\qquad\qquad\qquad\quad$ $\square$

8   Concluding Remarks

From our computer experiment, for many values for N the bound given in Theorem 10 is tight. It will be interesting to consider what kind of values for N require $\lceil \log_2 N \rceil + f + 2$ rounds in the worst case. In Theorem 11 and Theorem 12 we have partly solved the fault tolerance of procedure *disseminate* when $\lceil \log_2 N \rceil - 1$ processors and/or links have failed. Note that in general broadcasting through the route R given in the proof of Theorem 12 requires many rounds. We conjecture that for an arbitrary N and $f = \lceil \log_2 N \rceil - 1$, $3\lceil \log_2 N \rceil$ rounds suffice for broadcasting.

References

1. Alon, N., Barak, A., Mauber, U.: On disseminating information reliably without broadcasting. the 7th International Conference on Distributed Computing Systems, 74-81 (1987)
2. Bermond, J-C., Peyrat, C.: Broadcasting in de Bruijn networks. Congressus numerantium **66** (1988) 283-292
3. Carlsson, S., Igarashi, Y., Kanai, K., Lingas, A., Miura, K., Petersson, O.: Information disseminating schemes for fault tolerance in hypercubes. IEICE Trans. on Fundamentals of Electronics, Communications and Computer Science **E75-A** (1992) 255-260
4. Han, Y., Finkel, R.: An optimal scheme for disseminating information. Proceedings of 1988 International Conference on Parallel Processing, Chicago, Illinois, 198-203 (1988)
5. Imase, M., Soneoka, T., Okada, K.: Connectivity of regular directed graphs with small diameters. IEEE Trans. on Computers **34** (1985) 267-273
6. Johnsson, S. L., Ching-Tien Ho: Optimal broadcasting and personalized communication in hypercubes. IEEE Trans. on Computers **38** (1989) 1249-1268
7. Kanai, K., Miura, K., Igarashi, Y.: Fault tolerance of Han-Finkel's scheme for disseminating information. IPSJ Technical Report, Algorithms **91-AL-24-2** (1991)
8. Liestman, A. L.: Fault-tolerant broadcast graphs. NETWORKS **15** (1985) 159-171
9. Ramanathan, P., Shin, K. G.: Reliable broadcasting in hypercube multiprocessors. IEEE Trans. on Computers **37** (1988) 1654-1657

Routing Problems on the Mesh of Buses

Kazuo Iwama[†], Eiji Miyano
Department of Computer Science and Communication Engineering
Kyushu University, Fukuoka 812, Japan

and

Yahiko Kambayashi[‡]
Integrated Media Environment Experimental Lab.
Kyoto University, Kyoto 606, Japan

Abstract. The mesh of buses (MBUSs) is a parallel computation model which consists of $n \times n$ processors, n row buses and n column buses. Upper and lower bounds for the routing problem over MBUSs are discussed. We first show elementary upper and lower bounds, $2n$ and $0.5n$, respectively, for its parallel time complexity. The gap between $2n$ and $0.5n$ is then narrowed to $1.5n$ and n, which is the main theme of the paper. The n lower bound might seem to be trivial but is actually not. Three counter examples will be shown against this kind of easy intuition.

1 Introduction

The two-dimensional mesh is an important structure of physically realizable parallel computers. In this structure, processors are naturally placed at the intersections of the horizontal and vertical grids, while there can be two different types of communication links: The first type is shown in Fig.1. Each processor is connected to its four neighbors and such a system is called *mesh-connected computers* (*MCs* for short). Fig.2 shows the second type: Each processor is connected to a couple of (row and column) buses. The system is then called *mesh-bus computers* (*MBUSs* for short).

There is a large literature on the performance of MCs, in particular, for the sorting and routing problems. Recently tight time-bounds of $3n$ and $4n$ (with respect to minor differences of the model) for sorting have been presented[7, 9]. For routing also, almost optimal deterministic and randomized upper bounds are known [5, 6]. As for MBUSs, [2] first gave nontrivial graph algorithms that run in $O(\log n)$ randomized time, which, a little surprisingly, achieved virtually the same performance as CRCW-PRAMs. MBUSs also have linear time algorithms for sorting [3, 8] but the constant factors are much larger than the above optimal algorithms on MCs.

In this paper, we present upper and lower bounds for routing over MBUSs. The input is given by n^2 data–items that are initially held by the n^2 processors, one for each. Each data–item consists of a destination and an integer. Routing requires that all n^2 such data–items be moved to their destinations. We first show elementary upper and lower bounds, $2n$ and $0.5n$, respectively, for its parallel time complexity. The gap between $2n$ and $0.5n$ is then narrowed to $1.5n$ and n, which is the main theme of the paper. The n lower bound might seem to be trivial but is actually not. Three counter examples will be shown against this kind of easy intuition.

Differences between MCs and MBUSs are clearly based on the differences of their communication facilities. The local communication of MCs raises a trivial lower bound of $2n$ for most problems due to the physical distance between the farthest two processors. Global buses of MBUSs eliminate this kind of distances, which is of course a necessary (but not sufficient) condition for the logarithmic time algorithms mentioned above. Ironically, in proving lower bounds,

---

[†]Supported by Scientific Research Grant, Ministry of Education, Japan, No. 02302047 and No. 04650318.
[‡]Supported by Scientific Research Grant, Ministry of Education, Japan, No. 01302059 and No. 01633010.

this can be a demerit; we can no longer use the physical-distance argument that is quite convenient and is fairly accurate for MCs. Another aspect of the global-bus communication is that it becomes another weak point for highly parallel local communications (buses cannot be "cut"). That is the reason why MBUSs can only offer relatively poor algorithms for sorting compared to MCs. Then one would be naturally lead to the following question: If both local and global communications are equipped, does the performance of such a system increase significantly? The answer is negative. It is known that the bus communication cannot improve sorting over MCs [11, 10] and the local communication in turn cannot improve the logarithmic-time algorithms over MBUSs [2, 4] either.

To narrower the still existing gap (between $1.5n$ and n) is a natural goal of the future research. At this moment we have made little progress to this goal. Trying to raise the lower bound seems to be more promising than to lower the upper bound.

2 Model

Our discussion throughout this paper is based on the following four rules on the model:

(i) We adopt the following common regulation on how to measure the running time of MBUSs: The one-step computation of each processor P consists of (a) reading the current data on both row and column buses P is connected to, (b) executing a constant number of local instructions using the local memory and (c) if necessary, writing data to the row or column bus, or (possibly different data) to both. The written data will be read in the next step.

(ii) Algorithms should be designed so that a collision of data on a single bus will never occur.

(iii) The data written by P on the buses must be the data originally given to P as its input data or one of those so far read by P from the buses. Hence it must be of the form (d, σ) just like the original input data. Thus any kind of data compression is *not* allowed.

(iv) Although data compression is not allowed, each processor P can make use of any information obtained from the buses to determine what P wants to know. Programs on the processors may differ. We do not assume the "oblivious" algorithm[1], i.e., the execution of the programs can be affected not only by the destinations d but also by the integers σ of the input. The local memory can hold everything P knew so far, i.e., which instruction is executed at step i is determined by P's original input data, P's processor number and the data that flowed on the (row and column) buses P is connected to during the $i-1$ steps so far.

Throughout this paper, we use one-dimensional indices rather than two dimensional ones. The input (an *instance* of the problem) of size n^2 is given as $I = ((d_0, \sigma_0), (d_1, \sigma_1), \cdots, (d_{n^2-1}, \sigma_{n^2-1}))$ where data-item $(d_{in+j}, \sigma_{in+j})$, $0 \le i \le n-1$, $0 \le j \le n-1$, is initially held by the processor P_{in+j} standing on the intersection of the ith column and the jth row. We assume that no instance includes two data-items whose destinations are the same. (Note that there is a trivial $n^2/2$ lower bound without this condition, i.e., when all data-items go to the same place.) Therefore the list of n^2 destinations $(d_0, d_1, \cdots, d_{n^2-1})$ is given as a permutation of $N^2 = (0, 1, \cdots, n^2-1)$, where $d_k = in + j$ means the data-item should go to P_{in+j}. The domain D_n of instances I of size n^2 is determined by specifying a set Π of permutations of N^2 and an integer $C \ge 0$ in the following way: $I = ((d_0, \sigma_0), \cdots, (d_{n^2-1}, \sigma_{n^2-1}))$ is in D_n if and only if $(d_0, \cdots, d_{n^2-1})$ is in Π and all σ_i is in $\{0, 1, \cdots, C\}$. Unless otherwise stated, we assume the most general input set, i.e., the case that Π is the set of all permutations of N^2 and as for C no limit depending on n is set. In Section 3, we shall mostly handle restricted domains in which each data-item moves only horizontally and some limit is also set for C.

One can easily think of the following $2n$-step routing algorithm.

Elementary Upper Bound: $2n$ steps are enough for routing on MBUSs.

Proof. During the first n steps we use only row buses. Each processor writes its data-item to the row bus in a regular order, e.g., from left to right and the data moves to its destination with respect to column. For the next n steps, we use column buses and move those data-items to their destinations with respect to row, i.e., the final position. The order of processors writing data should be determined not by the positions of the processors as before but by the data they hold. Namely, the data heading for the first row should be written first, the data heading for the second row next and so on. Note that at most n data can stay in a single processor after the first n steps.　　□

Elementary Lower Bound: $0.5n$ steps are necessary for routing on MBUSs:

Proof. In order to prepare ourselves for the later main theorem, we exhibits the following proof that might seem to be a little lengthy. Let D_n be the general domain of instances and suppose that a routing algorithm A halts in at most $(0.5n - 1)$ steps for every instance in D_n. We shall imply a contradiction by constructing a series of instances, finally the one, denoted by I_f, A cannot manage. For an instance I and a set S of processors, we let $D_n(I, S)$ be the set of instances I' such that for every processor P in S, the data–item of I' initially held by P coincides with the data–item of I held by the same processor P. Another important notion is an *idle bus*. A (row or column) bus is said to be idle at step i if no processor writes data on that bus at that step.

Select an arbitrary instance I_0. If there exists no idle bus during the whole execution of A for I_0, then one can obtain I_f immediately: Recall that we have $2n$ processors and A halts within $(0.5n - 1)$ steps. That means at most $n^2 - 2n$ processors can write data, or at least $2n$ processors cannot write data on any bus during the whole execution of A. Let P be one of such $2n$ processors and select, as I_f, any instance $(\neq I_0)$ in $D_n(I_0, S_{n^2} - \{P\})$. ($S_{n^2}$ is the set of all n^2 processors. I_f is exactly the same as I_0 but only the data–item of processor P is different.) For the new input I_f, P must write data since, if it does not again, then the flow of data on the whole buses should be completely the same for I_0 and I_f, or the routing results for I_0 and I_f should be the same, a contradiction. Suppose therefore that P for the input I_f writes data into a bus, say B, at step i. Then one can see that there must be a collision on bus B at step i since the (other than P) processor, say P', that writes data on B at step i for input I_0, also writes (the same) data on B at step i for I_f. (For P' sees exactly the same flow of information on buses until step $i - 1$ for both instances.)

From now on, we consider that there are hypothetically $2n \times (0.5n - 1)$ buses, called *w-buses* to avoid confusion, in the following sense: Column 0 at step 1 is the first w-bus, column 1 at step 1 is the second, $\cdots$, row 0 at step 1 is the $(n + 1)$st, and so on. Formally, column i at step j is the $(i + 2(j - 1)n + 1)$th w-bus and row i at step j is the $(i + n + 2(j - 1)n + 1)$th. Suppose that for an instance I the kth w-bus that is associated with step j is idle and let S_j be the set of processors that write data at or before step j. Then, it is said that the kth w-bus is *essentially idle* (for the instance I) if for every other instance in $D_n(I, S_j)$ (i.e., whatever data–items of other than S_j are different) it is also idle.

Now suppose that there exist idle buses for the input I_0. Then we seek the earliest w-bus that is idle but is not essentially idle. Let B be such a w–bus and suppose that B is associated with step j. (If there exists no such w–bus, let I_h, defined later, be this I_0.) Then, since B is not essentially idle, one can find an instance I_1 in $D_n(I_0, S_j)$ such that it is different from I_0 and some processor write data on w–bus B. Note that what happens to w–buses earlier than B for I_1 is the same as for I_0. Therefore one can claim that if there is an idle, but not essentially idle, bus for I_1 then it must be later than B in the order defined above. Note that there may be essentially idle w–buses before B. An obvious but important property is that if some w–bus B' earlier than B is essentially idle for I_0 then it is also essentially idle for I_1. Repeating this process, we can construct instances I_2, I_3, etc., and finally I_h that has no w–bus that is idle but is not essentially idle, i.e., if idle w–buses exist then they must be essentially idle.

Now we can apply the previous argument to imply the similar difficulty for A. Since every idle w–bus is essentially idle for I_h, there must happen a collision again for I_f which is constructed from I_h exactly as before. It should be noted that this proof holds also for restricted domains of instances, as restricted as $C = 1$ and Π includes only one (not trivial) permutation.
□

3 Improvements of Bounds

Theorem 1. $1.5n$ steps are sufficient for routing on MBUSs.

Proof. We divide the whole $n \times n$ plane into four $\frac{n}{2} \times \frac{n}{2}$ planes (see Fig.3). Using the first $0.5n$ steps, data–items on the upper-left $\frac{n}{2} \times \frac{n}{2}$ plane are moved horizontally using row buses in the regular order and the same for the lower-right $\frac{n}{2} \times \frac{n}{2}$ plane. Data–items on the upper-right $\frac{n}{2} \times \frac{n}{2}$ plane and on the lower-left $\frac{n}{2} \times \frac{n}{2}$ plane are moved vertically. Thus, using $0.5n$ steps, we move every data–item to its correct place with respect to at least row or column.

In the following n steps, we move those data–items to their final places using both row and column buses. Suppose, for example, that some processor P holds data–item d_1 that is to move to the 1st row, d_2 to the 3rd row and d_3 to the 3rd column. Then at the first step of this second stage, P writes d_1 to its column bus. P sleeps at the second step. At the third step P writes d_2 to its column bus and d_3 to its row bus. Details may be omitted. □

Now we shall show the main result of this paper. Let $Rshift(C)$ be a domain of instances such that an initial data–item (d, σ) on each processor P_{in+j} (the processor at column i and row j) should be moved to $P_{(i+1)n+j}$ if $i \leq n-2$ and to P_{0+j} if $i = n-1$, i.e., should be shifted one position horizontally, and $\sigma \in \{0, 1, \cdots, C\}$.

Theorem 2. Suppose that $C \geq n^{\frac{1-\beta}{\beta}}$ and $0 < \beta \leq \frac{1}{3}$. Then routing for $Rshift(C)$ takes at least $(1 - \beta)n$ steps for sufficiently large n.

Proof will be given after the following observations on this result: Firstly, the condition $C \geq n^{\frac{1-\beta}{\beta}}$ says that σ is up to polynomial in n for a constant β. Since we can make β close to 0 arbitrarily, the theorem gives us the lower bound of roughly n steps.

We next discuss a common question to this kind of lower-bound result. One would have the following intuition: Since every data has to move horizontally, it should be put on row buses at least once for each. The n-step lower bound would then be trivial since we have n^2 data but only n row buses. Against this intuition, we shall give the following algorithm.

Remark 1. There is a $0.97n$ routing algorithm for $Rshift(\frac{n}{10})$. (The algorithm is given in Appendix. Recall that any kind of data compression is not allowed. We can nevertheless exploit a certain kind of coding technique in which column buses play a significant role.)

The following question to Theorem 2 would be more likely to arise: Why is such a restricted domain is adopted? If we would consider more general domain, i.e., each data has to move both horizontally and vertically, then, again intuitively, each data should use both row and column buses at least once for each. Then could we get the same result ($2 \times n^2$ data appearances over $2n$ buses implies n steps) more easily since we now have no extra room in column buses that were exploited to support the efficient communication scheme on row buses in Remark 1? Here is another algorithm to answer this question:

Remark 2. There is a domain of instances for which all data must move both horizontally and vertically and for which there is a $(0.5 + \epsilon)n$, for any small ϵ, routing algorithm. (Note that it is very unlikely that there is such an easy class of inputs if the movement of data is restricted to only horizontally or vertically. Thus, a little surprisingly, the data movement of two directions is, in some case, easier to maintain than the movement of only one direction. See Appendix for details.)

Remark 3. We could calculate the whole amount of information carried by the $2n$ buses for the most general (with respect to the destination) domain: A single bus can carry one of n^2 destinations and C different integers at a time. Hence the number of different patterns of the information flow that appear on the $2n$ buses during αn steps is

$$\left((Cn^2)^{\alpha n}\right)^{2n} = C^{2\alpha n^2}(n^2)^{2\alpha n^2}$$

This value must be greater than the number of all different inputs, which is

$$C^{n^2} \cdot (n^2)!$$

Unfortunately, however, it is hard to get better results than $\alpha > 0.5$ (the same as the elementary lower bound) from this line of brute-force analysis. □

Thus Theorem 2 is much more nontrivial than it looks. Furthermore, it is quite tight in the following sense: We shall say $Rshift(C)$ to be *a fixed-destination domain* since it includes only one permutation in the set Π.

Remark 4. $n + 1$ steps are enough for any (data may move both horizontally and vertically) fixed-destination domain.

Proof. We consider the following bipartite graph $G = (V_1 \cup V_2, E)$: $V_1 = \{r_1, r_2, \cdots, r_n\}$, $V_2 = \{c_1, c_2, \cdots, c_n\}$ and for each data-item (d, σ), if (d, σ) is originally placed on row i and if its destination is on column j, then edge (r_i, c_j) is put into the edge set E. Note that G is a multi-edge graph, i.e., there may be more than one edge between a pair of vertices. Then one can see that G is a regular bipartite graph of degree n.

Now it is not hard to modify the König–Hall Theorem (G has a matching iff $N(S) \geq |S|$ for any subset S of V_1) to work for multi-edge graphs. It then follows that G can be decomposed into n disjoint 1-factors. A movement of n data–items according to a single 1-factor can be realized in two steps, one for using row buses and the other for using column buses. It should be noted that we can overlap the second portion (using column buses) of this single movement with the first portion (using row buses) of another movement represented by another 1-factor. Thus $n + 1$ steps are sufficient to complete n movements (n 1-factors) of n data-items. (Suppose that the input contains such n data-items $(d_1, \sigma_1), \cdots, (d_n, \sigma_n)$ that no two items share the same row or column originally and destinations of no two items do so either. Then $n + 1$ steps can be reduced to n steps.) □

As will be seen in the next section, being able to use a fixed-destination domain gives us several advantages in proving the lower bound, which is not restricted to the present paper solely [7] [9]. Remark 4 could be a major obstacle to improve our lower bound.

4 Proof of Theorem 2

In this section an " instance " means one in $Rshift(C)$. Let $\alpha = 1 - \beta$. Then the statement becomes: For $C \geq n^{\frac{a}{1-a}}$ and $\frac{2}{3} \leq \alpha < 1$, routing for $Rshift(C)$ takes at least αn steps. Let A be an algorithm that completes routing for every instance I in at most αn steps.

Lemma 1. Take an arbitrary (row or column) bus. Then, without depending upon the instance I, there exists at most one fixed processor which can write data into that bus at the first step in algorithm A.

Proof. Since there have been no communications so far, each processor has to decide whether or not it writes data only by the processor number and its original data–item. Suppose that a processor P writes data into the bus for instance $I_1 = (\cdots, (d, \sigma)_P, \cdots)$, where $(d, \sigma)_P$ is the data–item given to P, and a different processor Q writes into the same bus for $I_2 = (\cdots, (d', \sigma')_Q, \cdots)$ where $(d', \sigma')_Q$ is given to Q. Then it turns out that both P and Q write into the (same) bus for instances like $I_3 = (\cdots, (d, \sigma)_P, \cdots, (d', \sigma')_Q, \cdots)$, a contradiction. □

Let a *history of A up to step l for an instance I*, denoted by $H_l(I)$, be a sequence $(D_{1,0}, D_{1,1}, \cdots, D_{1,n-1}, D_{2,0}, \cdots, D_{i,j}, \cdots, D_{l,n-1})$ where $D_{i,j}$ is the data on row j at step i. If no data is on the bus at that step, then $D_{i,j} = nil$. $H_{\alpha n}(I)$ is also written as $H(I)$. The following lemma might make clear why we adopted the restricted domain.

Lemma 2. If $H(I_1) = H(I_2)$, then $I_1 = I_2$. (For different instances I_1 and I_2, the information flow on both column and row buses would be different. However, if the flow on row buses were the same then, whatever happens on the column buses, the results of routing should be the same.)

Proof. Suppose hypothetically that $H(I_1) = H(I_2)$ for different I_1 and I_2. Let $I_1 = (\cdots, (d, \sigma)_P, \cdots)$ and $I_2 = (\cdots, (d, \sigma')_P, \cdots)$ where $(d, \sigma)_P$ and $(d, \sigma')_P$ are both given to the same processor P and $\sigma \neq \sigma'$. Now we consider the third instance I_3 (see Fig.7) such that the whole data on the column, denoted by CL_R, that is right next to P's column are the same as those of I_1 and all the other data are the same as I_2.

Now we prove the validity of the following three propositions by induction on i: For the instance I_3 and during the first i $(1 \leq i \leq \alpha n)$ steps, (i) information flow for I_3 on the row buses is completely the same as I_1 and I_2, (ii) the flow for I_3 on CL_R is the same as I_1 and (iii) the flow for I_3 on all the other columns is the same as I_2.

When $i = 1$: By Lemma 1, which processor writes on each row bus at the first step is completely the same for I_1, I_2 and I_3. Hence $H_1(I_3)$ must be equal to $H_1(I_1)$ since $H_1(I_1) = H_1(I_2)$ by the assumption. (Any processor holding different data for I_1 and I_2 cannot write at this step under the assumption that $H(I_1) = H(I_2)$.) Omitted for (ii) and (iii).

Now suppose that the propositions are true for $i \geq 1$. We observe an arbitrarily chosen row to which some processor, say Q, writes data at step $i + 1$. Case 1: Q belongs to a column other than CL_R. In this case, by the induction hypothesis, the information flow on both row and column buses that Q (for instance I_3) has observed up to step i is the same as Q for instance I_2. The original input data given to Q is also the same for I_2 and I_3. Hence Q for instance I_3 writes the same data on the row bus as Q for I_2 (and also as Q for I_1 by the assumption). Case 2: Q belongs to CL_R. Q for instance I_3 has observed exactly the same data as Q for I_1. Hence Q for I_3 writes the same data on the row bus as Q for I_1 (and I_2). Thus (i) is true for $i + 1$.

The argument for propositions (ii) and (iii) is very similar and may be omitted. As a result, we can conclude that all processors on CL_R for instance I_3 observe exactly the same information flow during the whole steps as the same-positioned processors for I_1. That means the final result on CL_R for I_3 should be equal to the result on CL_R for I_1. This, however, creates a contradiction on the place right next to P's. (P is the processor introduced at the beginning.) □

Lemma 3. Suppose that $H_i(I_1) = H_i(I_2)$. Then, for an arbitrary row bus, which processor writes to that row bus at step $i + 1$ is the same for I_1 and I_2, i.e., it is the same for even different instances if H_i is the same.

Proof. Omitted. □

Now we are ready to analyze the behavior of algorithm A for all different instances in $Rshift(C)$. More specifically, we make an evaluation on how much the information flow on the entire row buses can differ according to those different instances.

At the first step, there are no other choices for each processor P than to write its own data–item. By Lemma 1, which processor does so is uniquely determined for all instances and the destination is the unique by the assumption of $Rshift(C)$. Thus the data on each row bus can change only in its integer part, i.e., the possible amount of differences is $C + 2$ (different $C + 1$ integers plus the case that no processors write) for a row single bus and $(C + 2)^n$ for all row buses.

At the second step: By Lemma 3, which processor writes data on some particular row is unique if we fix one of the $(C + 2)^n$ flows at the first step. That processor, say P, can choose to write (i) its original data–item or the data P took from its row at the first step or (ii) the data P took from its column at the first step. Again remember that the destination part of those data is much restricted. For (i), the amount of differences is $C + 2$ (for the original data or empty) $+1$ (for the previous row data). For (ii), it is $C + 1$ (since which processor wrote data on its column was unique and its destination is fixed). Therefore the total amount is $2C + 3$.

Now we encounter more complicated situation at step 3: As above, (a unique if we fix a particular data flow at steps 1 and 2) P can choose, as the data it writes, say D_3, one among (i) the original data–item or the data P took from its row, (ii) the column-bus data taken at step 1, and (iii) the similar one taken at step 2. (i) and (ii) are exactly the same as the second step, but we have to be careful as for (iii): Let CL_P be the column P belongs to. Recall that all the processors on CL_P commonly knew the data appeared on CL_P at the first step. That means they can decide which processor is to write to CL_P at step 2 without collision due to that previous data. Thus the data D_3 can be any of P's own and those held by the other processors on CL_P at step 2.

For a while, we assume that, in general, a data written to some row bus has never appeared on *any* row bus so far. Then D_3 can only be one of n initial data on CL_P except those written on row buses at steps 1 and 2. Let k be the number of those data written already. Then the amount of differences is $(n-k)C$ per row. Note that at steps 1 and 2 there are up to n processors on CL_P which wrote data on their row buses. Thus k can change from 0 to n. One can see, however, that the average of k is at most 2. It is not hard to see that we can get the maximum value of the differences for all the rows when assuming that those processors writing to row buses at steps 1 and 2 distribute evenly to all the columns and k is the same ($=$ the average value) for those processors. Thus the amount of differences is at most $(n-2)C$ per row. At step i, this value becomes $(n-i+1)C$ per row.

Now we consider the case that P writes data, say D, that has already appeared on some row bus. Note that the row on which D has appeared may be P's own row or one of the others. In the latter case, D should have been transmitted to P via CL_P. Hence, if the current step is i, then the number of those data can be up to $n \cdot i$ or, more roughly, at most n^2 at any step. Thus the amount of differences calculated before should be increased by n^2 at every step. (Note that it is not $n^2 \cdot C$ since, once P selects one of those data, the value of the selected data cannot differ from the one appeared before.) We can conclude that the total amount of differences at step i is at most $(n-i+1)C + n^2$ per row.

Let N be the amount of differences that a single row bus can create during the entire execution of the algorithm A for all different instances in $Rshift(C)$. Recall that $C \geq n^2$ since $\alpha \geq \frac{2}{3}$. (The following analysis is not so tight. The result might be improved by making it more strict.)

$$
\begin{aligned}
N &\leq (C+2)(2C+3)((n-2)C+n^2)\cdots((n-i+1)C+n^2)\cdots((n-\alpha n+1)C+n^2) \\
&\leq C \cdot 3C \cdot (n-1)C \cdots (n-i+2)C \cdots (n-\alpha n+2)C \\
&\leq C^{\alpha n} \cdot n!/((1-\alpha)n+2)!
\end{aligned}
$$

Using the Stirling's approximation $n! = \sqrt{2\pi}n^{n+\frac{1}{2}}e^{-n}f(n)$ where $f(n) = 1 + \frac{1}{12n} + \frac{1}{288n^2} + O(n^{-3})$:

$$
\begin{aligned}
&n!/((1-\alpha)n+2)! \\
&= [\sqrt{2\pi}n^{n+\frac{1}{2}}e^{-n}f(n)]/[\sqrt{2\pi}((1-\alpha)n+2)^{((1-\alpha)n+2.5)} \cdot e^{-((1-\alpha)n+2)} \cdot f((1-\alpha)n+2)] \quad (1)
\end{aligned}
$$

Obviously $f(n) < f((1-\alpha)n+2)$. Hence

$$
\begin{aligned}
(1) &< n^{n+\frac{1}{2}}e^{-n}/((1-\alpha)n+2)^{((1-\alpha)n+2.5)} \cdot e^{-((1-\alpha)n+2)} \\
&< n^{\alpha n-2}e^{-\alpha n+2} \cdot \left(\frac{1}{1-\alpha}\right)^{(1-\alpha)n+2.5} \quad (2)
\end{aligned}
$$

If $n \geq \frac{2-2.5\ln(1-\alpha)}{\alpha+(1-\alpha)\ln(1-\alpha)}$ then $e^{-\alpha n+2} \cdot \left(\frac{1}{1-\alpha}\right)^{(1-\alpha)n+2.5} \leq 1$. Therefore

$$
(2) < n^{\alpha n-2} < n^{\alpha n}
$$

Note that the number of different instances in $Rshift(C)$ is C^{n^2} or C^n per row. We can then conclude that if

$$
C^n \geq C^{\alpha n} \cdot n^{\alpha n} \quad (3)
$$

then $N < C^n$ is met, i.e., the number of different instances in $Rshift(C)$ surpasses $|\{H(I) \mid I \in Rshift(C)\}|$, which contradicts to Lemma 2. It is straightforward to show that inequality

$$
C \geq n^{\frac{\alpha}{1-\alpha}}
$$

is equivalent to (3). □

References

[1] A. Borodin and J. Hoporoft. Routing, merging and sorting on parallel models of computations. In *Proc. 14th ACM Symposium on Theory of Computing*, pp. 338–344, 1982.

[2] K. Iwama and Y. Kambayashi. An $o(logn)$ parallel connectivity algorithm on the mesh of buses. In *Proc. 11th IFIP World Computer Congress*, pp. 305–310, 1989.

[3] K. Iwama, E. Miyano, and Y. Kambayashi. A parallel sorting algorithm on the mesh-bus machine. *Information Processing Society of Japan SIGAL Reports*, No. 18-2,, 1990.

[4] K. Iwama, E. Miyano, and Y. Kambayashi. Mesh-buses vs. mesh-connections. Technical Report 91–C07, Kyushu Univ., Kyushu, Japan, 1991.

[5] D. Krizanc, S. Rajasekaran, and T. Tsantilas. Optimal routing algorithms for mesh-connected processor arrays. *LNCS*, Vol. 319, pp. 411–422, 1988.

[6] M. Kunde. Routing and sorting on mesh-connected arrays. *LNCS*, Vol. 319, pp. 423–433, 1988.

[7] Y. Ma, S. Sen, and I. Scherson. The distance bound for sorting on mesh-connected processor arrays is tight. In *Proc. 27th IEEE Symp. on Foundations of Computer Science*, pp. 255–263, 1986.

[8] K. Nakano, T. Masuzawa, and N. Tokura. Optimal parallel sorting algorithms on processor arrays with multiple buses, 1991. manuscript.

[9] C. Schnorr and A. Shamir. An optimal sorting algorithms for mesh connected computers. In *Proc. 18th ACM Symposium on Theory of Computing*, pp. 255–263, 1986.

[10] Q. Stout. Meshes with multiple buses. In *Proc. 27th IEEE Symp. on Foundations of Computer Science*, pp. 264–273, 1986.

[11] Q. F. Stout. Mesh-connected computers with broadcasting. *IEEE Trans. Comput.*, Vol. C-32, 9, pp. 826–830, 1983.

Appendix

Proof of Remark 1. The whole $n \times n$ plane is divided into left $\frac{4n}{10} \times n$, central $\frac{3n}{10} \times n$ and right $\frac{3n}{10} \times n$ regions as shown in Fig.4. We introduce a *saving zone* on each column of the central region. The saving zone (denoted by a bold line-segment in the figure) is of length $\frac{n}{10}$ and is placed over the top $\frac{n}{10}$ processor at the leftmost column, over the second $\frac{n}{10}$ processors at the second column and so on. It is placed again over the top $\frac{n}{10}$ processors at the 11th column of the central region and so on. It should be noted that, in each row, there are $\frac{3n}{100}$ processors belonging to the saving zone of the central region.

Row buses are used as follows. During the first $\frac{4n}{10}$ steps, processors in the left region write their initial data–items in the regular order. Then, during the following $\frac{3n}{10} - \frac{3n}{100}$ steps, processors in the central region use the buses. Details are given later but the saving-zone processors do not use the bus, i.e., they are skipped. During the last $\frac{3n}{10}$ steps, processors in the right region use the buses in the normal way as those of the left region.

In each column of the central region, the saving zone is followed by a *coding zone* of length $\frac{2n}{10}$ as in Fig.5. Using the first $\frac{4n}{10}$ steps, the central-region processors belonging to the saving and coding zones (in parallel with the left-region processors that are using row buses) put their data–items on the column buses to let each processor on the same zones know all the data–items in the zones. Now consider an arbitrary column of the central region and suppose that the initial data–items over the saving zone and coding zone are $a_0, a_1, \cdots, a_{\frac{n}{10}-1}$ and $b_0, b_1, \cdots, b_{\frac{n}{10}-1}$, $b_{\frac{n}{10}}, \cdots, b_{\frac{2n}{10}-1}$, respectively. Our basic strategy is that the integer part σ_0 of a_0 ($\leq \frac{n}{10}-1$) is not propagated directly using its own row but by putting b_0 on the (σ_0+1)st row in the coding zone. Similarly b_1 is put on the row that is determined by σ_1 (=the integer part of a_1) as follows. If $\sigma_1 \geq \sigma_0$, then b_1 is put on the (σ_1+2)nd row. Otherwise, i.e., if $\sigma_1 < \sigma_0$, then b_1 is put on the (σ_1+1)st row and b_0, once assigned to (σ_0+1)st row, is changed to (σ_0+2)nd row. The general idea is this: Suppose that σ_i (the integer part of a_i) is the k-th smallest (if σ_i's are, say 3, 3, 5, 6, 6, $\cdots$, then we consider, for example, that one of the two 6's is the 4th smallest and the other 5th) among σ_0 to $\sigma_{\frac{n}{10}-1}$. Then the value σ_i is informed by putting b_i on the (r_i+1)st row where $r_i = \sigma_i + k - 1$. After completing the assignment of b_0 through $b_{\frac{n}{10}-1}$, there are "vacant" rows in the coding zone. One can see that the value σ_i is equal to the number of those vacant rows existing above the row where b_i is assigned.

There are several ways to construct such an assignment. The following method might be the simplest: Using the first $\frac{n}{10}$ steps, each processor P in the saving zone writes a data–item, say (d_i, σ_i), in the regular order. At the same time, P evaluates the order of its own σ_i among σ_0 through $\sigma_{\frac{n}{10}-1}$ by counting the smaller ones (and equal σ_j's if $j < i$) flowing on the bus. Then, using the following $\frac{n}{10}$ steps, those saving–zone processors again write their data–items from the smallest one first, then the second smallest one and so on. Using the following $\frac{2n}{10}$ steps, b_0 through $b_{\frac{n}{10}-1}$ are written on the column bus in the regular order. Now one can see how the coding–zone processors computes the assignment described above. After the assignments of b_0 through $b_{\frac{n}{10}-1}$ are set, the rest ($b_{\frac{n}{10}}$ through $b_{\frac{2n}{10}-1}$) are then assigned to the vacant rows, $b_{\frac{n}{10}}$ to the uppermost vacant row, $b_{\frac{n}{10}+1}$ to the second vacant row and so on.

Thus, using the first $\frac{4n}{10}$ steps, each processor P in the central region calculates what data–item it should put on its own row bus. If P is in the coding zone, the calculation is described above. If P is in the saving zone, P has NO data–items to put on. Otherwise, i.e., if P is not in the coding or saving zone, P puts its original input. P should also know when it is supposed to put the data–item actually on the row bus, which is calculated by (P's physical column position) – (the number of code–zone processors exiting in the left side of P).

Recall that we are now assuming a restricted domain, i.e., all data–items is shifted one-position (" one " is not important) to the right. So, pay attention to the right next column of P's. For the data–items sent from the coding zone of P's column, we put those data–items on the column bus in the regular order. Then it is not so hard for the processors that are supposed to receive data–item from the saving zone to "decode" that data–item from those sent from the coding zone. This decoding process (details are omitted) takes place during the final $\frac{3n}{10}$ steps.
□

Proof of Remark 2. The integer part of all input data–items is restricted to 0 or 1. The data movement is shown in Fig.6: Each data–item on upper-left and down-right $\frac{n}{2} \times \frac{n}{2}$ planes moves to its mirror-image position with respect to the diagonal from upper-right corner to down-left corner. Data-item on upper-right and down-left planes move similarly with respect to the other diagonal.

We first apply exactly the same algorithm as in Theorem 1 during $0.5n$ steps. After that, each processor P on the above diagonal has $\frac{n}{2}$ data–items. Those $\frac{n}{2}$ data–items are divided into groups, i.e., the $\log \frac{n}{2}$ data–items that reached first make the first group, the following $\log \frac{n}{2}$ data–items make the second group and so on. It should be noted that since the integer part of all data–items is 0 or 1, such a group of $\log \frac{n}{2}$ data–items can be regarded as a single $\log \frac{n}{2}$-bit integer. That means this group of data–items can be informed by transmitting only one data–item among $\frac{n}{2}$ data–items, held by the diagonal processor, by regarding its destination part as representing the above $\log \frac{n}{2}$-bit integer. (Recall that all different $\frac{n}{2}$ destinations are available in each diagonal processor.) Coding and decoding processes between $\log \frac{n}{2}$ 0/1's and a (single) $\log \frac{n}{2}$-bit integer take place in parallel with the data transmission. Thus the total number of steps we need is

$$\frac{n}{2} + \frac{n}{2} \div \log \frac{n}{2} + \log \frac{n}{2} = \frac{n}{2}\left(1 + \frac{1}{\log \frac{n}{2}} + \frac{2\log \frac{n}{2}}{n}\right)$$

where the last $\log \frac{n}{2}$ term is for the time to decode the $\log \frac{n}{2}$-bit integer sent last. □

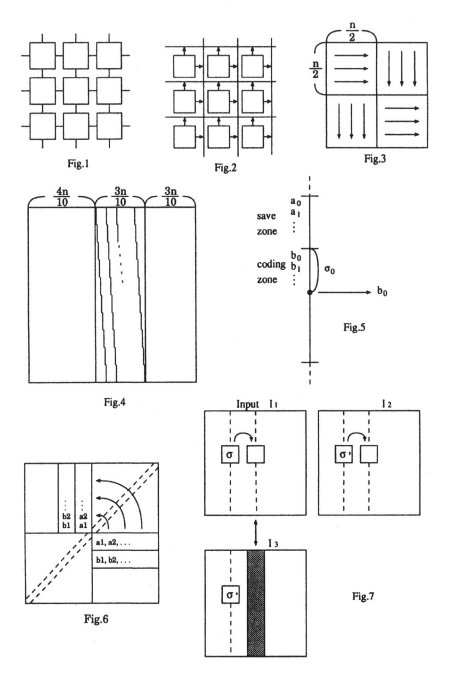

Fig.1

Fig.2

Fig.3

Fig.4

Fig.5

Fig.6

Fig.7

Selection Networks with $8n \log_2 n$ Size and $O(\log n)$ Depth

Shuji Jimbo and Akira Maruoka

Tohoku University, Sendai 980, JAPAN

Abstract. An $(n,n/2)$-selector is a comparator network that classifies a set of n values into two classes with the same number of values in such a way that each element in one class is at least as large as all of those in the other. Based on utilization of expanders, Pippenger[6] constructed $(n,n/2)$-selectors, whose size is asymptotic to $2n \log_2 n$ and whose depth is $O((\log n)^2)$. In the same spirit, we obtain a relatively simple method to construct $(n,n/2)$-selectors of depth $O(\log n)$. We construct $(n,n/2)$-selectors of size at most $8n \log_2 n + O(n)$. Moreover, for arbitrary $C > 3/\log_2 3 = 1.8927 \cdots$, we construct $(n,n/2)$-selectors of size at most $Cn \log_2 n + O(n)$.

1 Introduction

An (n,k)-selector is a comparator network (see Knuth[4]) that classifies a set of n values into two classes with k values and $n - k$ values in such a way that each element in one class is at least as large as all of those in the other class, while an n-sorter is a comparator network that sorts n values into order. Since an n-sorter is obviously an (n,k)-selector, the existence of n-sorters with $O(n \log n)$ size and $O(\log n)$ depth, given in [1][2], immediately implies the existence of (n,k)-selectors with the same size and depth. Main concern in [1][2] seems to prove just the existence of n-sorters of size $O(n \log n)$ and depth $O(\log n)$, so the construction of sorters is of some intricacy, and the constant factors in the depth and size bound are enormous. In fact, in spite of further efforts in [5] to improve the previous construction, the numerical constant in the size bound have not been brought below 1000. Since selectors don't do as much as sorters do, it is natural to ask whether we can obtain much simpler construction of (n,k)-selectors with smaller constant factors. Based on the same fundamental idea, namely utilization of expanders, as their sorters, Pippenger[6] constructed $(n,n/2)$-selectors, whose size is asymptotic to $2n \log_2 n$ and whose depth is $O((\log n)^2)$. In the same spirit, we investigate efficient selectors, obtaining relatively simple $(n,n/2)$-selectors of size at most $8n \log_2 n + O(n)$ and depth $O(\log n)$. Moreover, for arbitrary $C > 3/\log_2 3 = 1.8927 \cdots$, we can construct a family of $(n,n/2)$-selectors of depth $O(\log n)$ and size at most $Cn \log_2 n + O(n)$.

2 Preliminaries

We shall draw comparator networks as in Knuth [4] (See Fig. 1). Each compara-

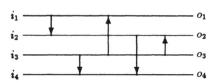

Fig. 1. The representation of a comparator network

tor is represented by vertical connections with arrows between two horizontal lines called registers. Input elements enter at the left and move along the registers. Each comparator causes an interchange of its inputs, if necessary, so that the larger element appears on the register on the side of the head of the comparator after passing the comparator.

Let N be a comparator network. The size of N is the number of comparators belonging to N, and is denoted by size(N). The depth of N is the maximum number of comparators on a path from an input terminal of N to an output terminal of N which passes along the registers and comparators and does not pass from the right to the left on registers. The depth of N is denoted by depth(N).

Definition 1 (A selector). Let k and n be integers with $1 \leq k \leq n-1$. Let S be a totally ordered set. A (n,k)-selector is a comparator network with n registers and designated k output terminals such that, for every input in S^n, each element appearing on the k output terminals is at most as large as all elements appearing on the other output terminals.

In a similar way to the proof of zero-one principle (in [4]), the following statement is proved: If a comparator network with n registers always selects the smallest k elements among arbitrary n elements in $\{0,1\}$, it will always select the smallest k elements among arbitrary n elements in arbitrary totally ordered set. Hence, without loss of generality, we assume that the input set to selectors is $\{0,1\}^n$ rather than S^n.

For $x \in \{0,1\}^n$, $\#_1 x$ denotes the number of x's entries equal to 1 and $\#_0 x$ denotes the number of x's entries equal to 0.

3   Comparator Subnetworks Obtained from Expanders

$(n,n/2)$-selectors we shall construct are composed of several kinds of subnetworks whose structure and behavior will be given in this section.

Definition 2 (An expander). Let $0 < \alpha \leq 1$ and $\beta \geq 1$. Let $G = (A, B, E)$ be a bipartite graph with left vertices A, right vertices B and edges E. G is an (α, β)-expander if every subset $X \subseteq A$ with $|X| \leq \alpha|A|$ satisfies $|\{y \in B \mid (\exists x \in X)((x,y) \in E)\}| \geq \beta|X|$.

Definition 3 (A compressor). Let n, m be positive integers and let α, β be real numbers with $0 < \alpha < 1$ and $\beta > 1$. We shall define two types of comparator networks with $n + m$ registers. Let $x = (x_1, x_2, \ldots, x_{n+m})$ be an input to such a comparator network, and $y = (y_1, y_2, \ldots, y_{n+m})$ denote the output corresponding to x. An (n,m,α,β)-compressor of type 1 is a comparator network with $n + m$ registers satisfying that if $x \in \{0,1\}^{n+m}$ and $\#_1 x \leq \lceil (\beta + 1) \lfloor \alpha n \rfloor \rceil$ then $\beta \#_1(y_1, y_2, \ldots, y_n) \leq \#_1(y_{n+1}, y_{n+2}, \ldots, y_{n+m})$. An (n,m,α,β)-compressor of type 0 is a comparator network with $n + m$ registers satisfying that if $x \in \{0,1\}^{n+m}$ and $\#_0 x \leq \lceil (\beta + 1) \lfloor \alpha m \rfloor \rceil$ then $\beta \#_0(y_{n+1}, y_{n+2}, \ldots, y_{n+m}) \leq \#_0(y_1, y_2, \ldots, y_n)$. For both types of compressors, the registers $1, 2, \ldots, n$ are called *upper registers* and the registers $n + 1, n + 2, \ldots, n + m$ are called *lower registers*.

Definition 4 (An extractor). Let n, m be positive integers and let $0 \leq \gamma \leq 1$. An (n,m,γ)-extractor is a comparator network N with $n+2m$ registers satisfying that for every $x = (x_1, \ldots, x_{n+2m}) \in \{0,1\}^{n+2m}$, $\#_1 x \leq (3/4)n + m$ implies $\#(y_1, \ldots, y_m) \leq \gamma m$ and $\#_0 x \leq (3/4)n + m$ implies $\#_0(y_{n+m+1}, \ldots, y_{n+2m}) \leq \gamma m$, where $(y_1, \ldots, y_{n+2m})$ is the output of N corresponding to x. The registers $1, 2, \ldots, m$, the registers $m + 1, m + 2, \ldots, n + m$ and the registers $n + m + 1, n + m + 2, \ldots, n + 2m$ of an (n,m,γ)-extractor are called *upper registers*, *middle registers* and *lower registers*, respectively.

It is easy to see that an (n,m,α,β)-compressor of type 1 becomes an (m,n,α,β)-compressor of type 0 by interchanging register 1 and register $n + m$, register 2 and register $n + m - 1$, and so on, and keeping the direction of each comparator.

Proposition 5 ([1][2]). *Assume that there exists an (α, β)-expander with n left vertices and m right vertices. Let s denote the number of edges and k the maximum degree of a vertex of the expander. Then there exists a (n,m,α,β)-compressor of type 1 of size at most s and of depth at most k. There also exists a (m,n,α,β)-compressor of type 0 of the same size and depth.*

The following Lemma 6 and Lemma 7 are obtained directly from the results of Bassalygo[3]. Note that $x/H(x) \to 0$ $(x \to 0)$ and, for every $c > 0$, $H(cx)/H(x) \to c$ $(x \to 0)$ where $H(x) = -x \log x - (1 - x)\log(1 - x)$.

Lemma 6 ([3]). *Let $k \geq 2$ and $m \geq 1$ be integers. For every $\varepsilon > 0$ with $m(k - 1 - \varepsilon) > 1$, there exist an integer $n_0 > 0$ and a real $\alpha > 0$ such that for every positive integers $n \geq n_0$ and $l < mn$, there exists a $(mn\alpha/l, k - 1 - \varepsilon)$-expander with l left vertices and n right vertices such that the degree of every left vertex is at most k and the degree of every right vertex is at most mk*

Lemma 7 ([3]). *Let α and β be real numbers with $0 < \alpha < 1/\beta < 1$. There exist positive integers m_0 and k such that for every positive integers $m \geq m_0$ and $n \leq m$, there exists a $(m\alpha/n, \beta)$-expander with n left vertices and m right vertices such that the maximum degree of its vertex is at most k.*

By Proposition 5, Lemma 6 and Lemma 7, the following Lemma 8 and Lemma 9 are proved immediately.

Lemma 8. *For every $0 < \delta < 1$, every $0 < \varepsilon < k - 2$ and every integer $k \geq 3$, there exist a positive integer $n_A(\delta, k, \varepsilon)$, and a real $\alpha_A(\delta, k, \varepsilon)$ with $0 < \alpha_A(\delta, k, \varepsilon) < 1$ such that, for every integers n and m with $n \geq n_A(\delta, k, \varepsilon)$ and $m \geq \delta n$, there exist an $(n, m, \alpha_A(\delta, k, \varepsilon), k - 1 - \varepsilon)$-compressor of type 1 and an $(m, n, \alpha_A(\delta, k, \varepsilon), k - 1 - \varepsilon)$-compressor of type 0 such that those depths are both at most k/δ and those sizes are both at most kn.*

Lemma 9. *For every $0 < \gamma < 1$ and every $0 < \delta < 1/2$, there exist positive integers $n_B(\gamma, \delta)$, $k_B(\gamma, \delta)$ such that for every positive integers m, n with $\delta(m + n) \leq m \leq (1 - \delta)(m + n)$ there exists an (n, m, γ)-extractor of depth at most $k_B(\gamma, \delta)$ and size at most $k_B(\gamma, \delta)m$.*

Note that compressors and extractors constructed by Lemma 8 and 9 are all of constant depth.

Lemma 10. *Let N be a comparator network with n registers. Let i, j and k be integers with $1 \leq i < j < k \leq n+1$. Let $a > 0$ and $0 \leq c \leq 1$ be some constants. Assume that N satisfies the following conditions:*

1. *There is no comparator which connects a register belonging to $\{1, 2, \ldots, i-1\}$ and a register belonging to $\{i, i+1, \ldots, n\}$ and there is no comparator which connects a register belonging to $\{1, 2, \ldots, k - 1\}$ and a register belonging to $\{k, k+1, \ldots, n\}$.*

2. *Let N' denote the comparator subnetwork of N composed of registers $i, i + 1, \ldots, k - 1$ and comparators of N which connect two registers belonging to $\{i, i + 1, \ldots, k - 1\}$. For every $z \in \{0, 1\}^{k-i}$, $\#_1 z \leq a$ implies $\#_1(N'_i(z), \ldots, N'_{j-1}(z)) \leq c\#_1(N'_i(z), \ldots, N'_{k-1}(z)) = c\#_1 z$, where $(N'_i(z), \ldots, N'_{k-1}(z))$ denotes the output of N' corresponding to input z.*

Let $u > 0$ and $v \geq u$ be real numbers with $\lfloor v \rfloor - \lfloor u \rfloor \leq a$. Let $x = (x_1, \ldots, x_n) \in \{0, 1\}^n$ with $\#_1(x_1, \ldots, x_{i-1}) \leq u$ and $\#_1(x_1, \ldots, x_{k-1}) \leq v$. Let $y = (y_1, \ldots, y_n)$ denote the output of N corresponding to input x. Then,

$$\#_1(y_1, y_2, \ldots, y_{j-1}) \leq u + c(v - u) .$$

4 Construction of $(n, n/2)$-selectors

In this section, we give the procedure to construct $(n, n/2)$-selectors and verify its justification. For simplicity, we assume that n is even. In what follows, we take arbitrary real constants ε_C, θ_C and δ_C and an arbitrary integer constant k_C so that $0 < \varepsilon_C$, $0 < \theta_C < 1$, $0 < \delta_C < 1$ and $k_C \geq \dfrac{2(1 + \sqrt{1 - \delta_C})}{\delta_C} + \varepsilon_C$.

Definition 11. Let n be a positive integer. For a positive integer i, $X_n(i)$ denotes the set $\{\lceil (n/2)(1-\theta_C^{i-1})\rceil + 1, \lceil (n/2)(1-\theta_C^{i-1})\rceil + 2, \ldots, \lceil (n/2)(1-\theta_C^i)\rceil\}$ and $Y_n(i)$ denotes the set $f(X_n(i))$, where $f(x) = n+1-x$. If $\lceil (n/2)(1-\theta_C^{i-1})\rceil = \lceil (n/2)(1-\theta_C^i)\rceil$ then $X_n(i) = Y_n(i) = \emptyset$. For an integer $i \geq 0$, $Z_n(i)$ denotes the set $\{1,\ldots,n\} \setminus \left(\bigcup_{j=1}^{i} X_n(j) \cup \bigcup_{j=1}^{i} Y_n(j) \right)$.

Definition 12 (Constants). For ε_C, θ_C, δ_C and k_C, we define real constants d_C, α_C, γ_C and Δ_C and integer constants k_D, n_C and $j_{\max}(n)$ as:

$$d_C = \frac{1+\sqrt{1-\delta_C}}{\delta_C}, \quad k_D = \left\lceil \frac{1}{\delta_C}\left(1 + \frac{2d_C}{k_C - \varepsilon_C}\right) + \varepsilon_C \right\rceil,$$

$$\alpha_C = \min\{\alpha_A(\theta_C/2, k_C, \varepsilon_C), \alpha_A(\theta_C/2, k_D, \varepsilon_C)\},$$

$$\gamma_C = \frac{\theta_C \alpha_C}{2d_C^2}, \quad \Delta_C = \frac{2\theta_C(1-\theta_C)}{3(1+\theta_C)},$$

$$n_C = \max\{n_B(\gamma_C, \Delta_C), n_A(\theta_C/2, k_C, \varepsilon_C), n_A(\theta_C/2, k_D, \varepsilon_C)\},$$

$$j_{\max}(n) = \left\lfloor \frac{1}{\log 1/\theta_C} \log\left(\frac{(d_C\delta_C - 1)\theta_C(1-\theta_C)\alpha_C}{6d_C\delta_C n_C} n \right) \right\rfloor .$$

Lemma 13. *Let i be an integer belonging to $\{1,2,\ldots,j_{\max}(n)\}$. The following inequalities hold.*

$$d_C\delta_C > 1, \quad k_D \leq k_C, \quad |X_n(i)| \geq n_C, \quad \frac{d_C\delta_C - 1}{d_C\delta_C}\alpha_C|X_n(i)| \geq 1,$$

$$\Delta_C(2|X_n(i)| + |Z_n(i)|) \leq |X_n(i)| \leq (1-\Delta_C)(2|X_n(i)| + |Z_n(i)|),$$

$$\frac{1}{\delta_C}\left(\frac{1-\left(\frac{1}{d_C}\right)^i}{1 - \frac{1}{d_C}} - 1 + \frac{1}{d_C} \right) \leq \frac{1-\left(\frac{1}{d_C}\right)^i}{1 - \frac{1}{d_C}}$$

and if $i \leq j_{\max}(n) - 1$ then $(\theta_C/2)|X_n(i)| \leq |X_n(i+1)|$.

First we define a comparator network called a layer (See Fig. 2).

Definition 14. We define compressors $C_n^1(i)$ and $C_n^0(i)$ for $i = 1, 2, \ldots, j_{\max}(n)$ as follows. First, we consider the case of $i = 2$. $C_n^1(2)$ is a $(|X_n(1)|, |X_n(2)|, \alpha_A(\theta_C/2, k_D, \varepsilon_C), k_D - 1 - \varepsilon_C)$-compressor of type 1 on the registers $X_n(1) \cup X_n(2)$. $C_n^0(2)$ is a $(|X_n(2)|, |X_n(1)|, \alpha_A(\theta_C/2, k_D, \varepsilon_C), k_D - 1 - \varepsilon_C)$-compressor of type 0 on the registers $Y_n(2) \cup Y_n(1)$. Next, we consider the case of $i \geq 3$. $C_n^1(i)$ is a $(|X_n(i-1)|, |X_n(i)|, \alpha_A(\theta_C/2, k_C, \varepsilon_C), k_C - 1 - \varepsilon_C)$-compressor of type 1 on the registers $X_n(i-1) \cup X_n(i)$. $C_n^0(i)$ is a $(|X_n(i)|, |X_n(i-1)|, \alpha_A(\theta_C/2, k_C, \varepsilon_C), k_C - 1 - \varepsilon_C)$-compressor of type 0 on the registers $Y_n(i) \cup Y_n(i-1)$. Next, we define extractors $E_n(i)$ for $i = 1, 2, \ldots, j_{\max}(n)$ as follows. If $i \leq j_{\max}(n)-1$ then $E_n(i)$ is a $(|Z_n(i)|, |X_n(i)|, \gamma_C)$-extractor on the registers $X_n(i) \cup Z_n(i) \cup Y_n(i)$. $E_n(j_{\max}(n))$ is some sorting network on the registers $X_n(j_{\max}(n)) \cup Z_n(j_{\max}(n)) \cup Y_n(j_{\max}(n))$. The existence of compressors and extractors above is shown by Definition 12 and Lemma 13. For every positive integer n, a sorting network with n registers is constructed by Batcher's odd-even merge.

Let n be a even positive integer. For each $j = 1, 2, \ldots, j_{\max}(n)$, we inductively define the *layer of rank j with n registers*, denoted by $L_n(j)$, as:

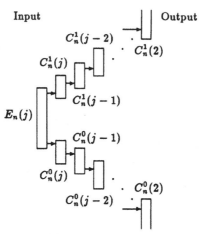

Input   Output

Fig. 2. The structure of the layer $L_n(j)$

1. If $j_{\max}(n) = 0$, then $L_n(1)$ is defined to be some sorting network.
2. If $j_{\max}(n) \geq 1$, then $L_n(1)$ is $E_n(1)$.
3. If $j_{\max}(n) \geq 1$ and $1 < j \leq j_{\max}(n)$ then $L_n(j)$ is obtained by replacing $E_n(j-1)$ in $L_n(j-1)$ with the comparator network constructed from $E_n(j)$, $C_n^1(j)$ and $C_n^0(j)$ by joining the output terminals on the upper registers of $E_n(j)$ to the input terminals on the lower registers of $C_n^1(j)$ and joining the output terminals on the lower registers of $E_n(j)$ to the input terminals on the upper registers of $C_n^0(j)$. When the replacement is made, the output terminals on the upper registers of $C_n^1(j)$ are joined to the input terminals on the lower registers of $C_n^1(j-1)$ and the output terminals on the lower registers of $C_n^0(j)$ are joined to the input terminals on the upper registers of $C_n^0(j-1)$.

Note that $E_n(j_{\max}(n))$ is always $(|Z_n(j_{\max}(n))|, |X_n(j_{\max}(n))|, 0)$-extractor, because $E_n(j_{\max}(n))$ is a sorting network.

Now, we express a procedure to construct $(n,n/2)$-selectors. The input of the procedure is a positive integer n and the output is an $(n,n/2)$-selector $N(n)$. In the procedure, bold-faced letters represent variables of procedure. For comparator networks M and N with the same number, say m, of registers, $M \circ N$ denotes the comparator network with m registers obtained by joining N to the right side of M so that the i-th output terminal of M is joined to the i-th input terminal of N for each $i = 1, 2, \ldots, m$.

Procedure SELECT.

1. Initial setting: $\mathbf{N} \leftarrow L_n(1)$, $\quad \mathbf{j} \leftarrow 1$, $\quad \mathbf{l}(1) \leftarrow \dfrac{d_C \gamma_C}{\delta_C} |X_n(1)|$
2. If $j_{\max}(n) \leq 1$ then terminate the procedure.
3. (a) If $\mathbf{j} < j_{\max}(n)$ and $\mathbf{l(j)} < (d_C \gamma_C / \delta_C)|X_n(\mathbf{j})|$ then $\mathbf{l}(i) \leftarrow \delta_C \mathbf{l}(i)$ for each $i = 1, 2, \ldots, \mathbf{j}$, $\mathbf{l(j+1)} \leftarrow d_C \mathbf{l(j)}$, $\mathbf{j} \leftarrow \mathbf{j} + 1$ and $\mathbf{N} \leftarrow \mathbf{N} \circ L_n(\mathbf{j})$,

(b) else $l(i) \leftarrow \delta_C l(i)$ for each $i = 1, 2, \ldots, \mathbf{j}$ and $\mathbf{N} \leftarrow \mathbf{N} \circ L_n(\mathbf{j})$.

4. If there exists an integer i such that $2 \leq i \leq \mathbf{j}$ and $\sum_{k=1}^{i} l(k) < 1$ then remove the comparator subnetworks $C_n^1(2), \ldots, C_n^1(i_{\max})$ and $C_n^0(2), \ldots, C_n^0(i_{\max})$ from the layer $L_n(\mathbf{j})$ joined to $\mathbf{N}$ in step 3, where $i_{\max} = \max\{i \in \mathbb{Z} \mid \sum_{k=1}^{i} l(k) < 1\}$.

5. If $\mathbf{j} = j_{\max}(n)$ and $\sum_{k=1}^{\mathbf{j}} l(k) < 1$ then terminate the procedure.

6. Go to step 3.

If the execution of Procedure SELECT terminates at step 2 then the value of $\mathbf{N}$ at the termination is an n-sorter, and hence an $(n, n/2)$-selector. In what follows, therefore, we assume that the execution of Procedure SELECT terminates at step 5, namely $j_{\max}(n) > 1$.

Before proceeding to the main part of the proof, we need to define some notations. $k_{\max}(n)$ denotes the total number of times that step 3 is executed in execution of Procedure SELECT. For $\mathbf{v} \in \{\mathbf{j}, l(h), \mathbf{N}\}$, $h = 1, \ldots, j_{\max}(n)$ and an integer $0 \leq k \leq k_{\max}(n)$, $\mathbf{v}[k]$ denotes the value of $\mathbf{v}$ just after step 4 has been executed k times if $0 < k$, and $\mathbf{v}[0]$ denotes the value of $\mathbf{v}$ just before the first execution of step 3. For positive integers k and n, $P_n(k)$ denotes the predicate which takes the value 1 when $\mathbf{j}[k-1] < j_{\max}(n)$ and $l(\mathbf{j}[k-1])[k-1] < (d_C \gamma_C / \delta_C)|X_n(\mathbf{j}[k-1])|$, and 0 otherwise. Thus, $P_n(k)$ indicates that step 3a is executed in the k-th execution of the loop between step 3 and step 6.

Observing Procedure SELECT, we obtain the following two lemmas.

Lemma 15. *Let k and h be integers with $0 \leq k \leq k_{\max}(n)$ and $1 \leq h \leq \mathbf{j}[k]$. If $k < k_{\max}(n)$ then $l(h)[k+1] = \delta_C l(h)[k] > 0$. If $h < \mathbf{j}[k]$ then $l(h+1)[k] = d_C l(h)[k] > 0$.*

Lemma 16. *Let k be an integer with $1 \leq k \leq k_{\max}(n)$. Then*

$$l(\mathbf{j}[k])[k] < \frac{\alpha_C}{\delta^2 d^2} |X(\mathbf{j}[k])| \tag{1}$$

and

$$\mathbf{j}[k] < j_{\max}(n) \text{ implies } d_C \gamma_C |X(\mathbf{j}[k])| \leq l(\mathbf{j}[k])[k] . \tag{2}$$

Let $u \in \{0,1\}^n$ with $\#_1 u = n/2$. Let k and i be integers with $0 \leq k \leq k_{\max}(n)$ and $1 \leq i \leq \mathbf{j}[k]$. Let $v = (v_1, \ldots, v_n)$ denote the output of $\mathbf{N}[k]$ corresponding to the input u. Let $s : \mathbb{N} \to \mathbb{N}$ denote the function such that $s(i) = \sum_{h=1}^{i} |X_n(h)|$ for all $i = 1, 2, \ldots, j_{\max}(n)$ and $t : \mathbb{N} \to \mathbb{N}$ the function such that $t(i) = n + 1 - s(i)$. Let $\tau_1(i, k; u)$ denote $\#_1(v_1, \ldots, v_{s(i)})$ and $\tau_0(i, k; u)$ $\#_0(v_{t_n(i)}, \ldots, v_n)$.

Lemma 17. *Let $u \in \{0,1\}^n$ with $\#_1 u = n/2$. Then, for each $k = 0, \ldots, k_{\max}(n)$ and each $i = 1, 2, \ldots, \mathbf{j}[k]$,*

$$\tau_1(i, k; u) \leq \sum_{h=1}^{i} l(h)[k] \text{ and } \tau_0(i, k; u) \leq \sum_{h=1}^{i} l(h)[k] . \tag{3}$$

Proof. First, we assume that no compressors are removed in step 4.

We shall prove the lemma by induction on k. In view of the initial setting of Procedure SELECT and by the fact that $L_n(1) = E_n(1)$ is an $(|Z_n(1)|, |X_n(1)|, \gamma_C)$-extractor, we derive $\tau_1(i, k; u) \le \gamma_C|X_n(1)| < 1(1)[0]$ and $\tau_0(i, k; u) \le \gamma_C|X_n(1)| < 1(1)[0]$. Thus, when $k = 0$, the inequalities (3) holds for each i with $1 \le i \le j[k] = 1$.

Assume that $0 \le k < k_{\max}(n)$ and the inequalities (3) hold for each $i = 1, \ldots, j[k]$. In the $(k+1)$-th execution of step 3, the layer, $L_n(j[k+1])$ is joined to the comparator network $N[k]$. By the induction hypothesis and $\#_1 u = n/2$, we have $\#_1(u_1, \ldots, u_{t(j[k+1]-1)}) \le (1/2)|Z_n(j[k+1]-1)| + \sum_{h=1}^{j[k+1]-1} 1(h)[k]$ and $\tau_1(j[k+1]-1, k; u) \le \sum_{h=1}^{j[k+1]-1} 1(h)[k]$. Moreover, we have $\left\lfloor (1/2)|Z_n(j[k+1]-1)| + \sum_{h=1}^{j[k+1]-1} 1(h)[k] \right\rfloor - \left\lfloor \sum_{h=1}^{j[k+1]-1} 1(h)[k] \right\rfloor \le \lceil (1/2)|Z_n(j[k+1]-1)| \rceil = (1/2)|Z_n(j[k+1]-1)|$. Therefore, by applying Lemma 10 to $E_n(j[k+1])$, we have $\tau_1(j[k+1], k+1; u) \le \gamma_C|X_n(j[k+1])| + \sum_{h=1}^{j[k+1]-1} 1(h)[k]$. Moreover, since $E_n(j[k])$ is a $(|Z_n(j[k])|, |X_n(j[k])|, \gamma_C)$-extractor, by using Lemma 15 and Lemma 16, we have $\gamma_C|X_n(j[k+1])| + \sum_{h=1}^{j[k+1]-1} 1(h)[k] \le \frac{1}{d_C} 1(j[k+1])[k+1] + \frac{1}{\delta_C} \sum_{h=1}^{j[k+1]-1} 1(h)[k+1]$. Thus, we have $\tau_1(j[k+1], k+1; u) \le (1/d_C\delta_C)1(j[k+1])[k+1] + \frac{1}{\delta_C} \sum_{h=1}^{j[k+1]-1} 1(h)[k+1]$.

From Lemma 13, Lemma 15 and Lemma 16, we have $\left\lfloor (1/d_C\delta_C)1(i+1)[k+1] + (1/\delta_C) \sum_{h=1}^{i} 1(h)[k+1] \right\rfloor - \left\lfloor (1/\delta_C) \sum_{h=1}^{i-1} 1(h)[k+1] \right\rfloor \le (2/d_C\delta_C)1(i+1)[k+1] + 1 \le (2\alpha_C/d_C\delta_C)\gamma|X_n(i)| + 1 \le 2\alpha_C|X_n(i)|$ for each $i = 2, \ldots, j[k+1]-1$. Note that

$$\frac{1}{k_C - \varepsilon_C}\left(\frac{1}{d_C\delta_C}1(i+1)[k+1] + \frac{1}{\delta_C}1(i)[k+1]\right) + \frac{1}{\delta_C}\sum_{h=1}^{i-1} 1(h)[k+1]$$
$$\le \frac{1}{d_C\delta_C}1(i)[k+1] + \frac{1}{\delta_C}\sum_{h=1}^{i-1} 1(h)[k+1] \tag{4}$$

holds for each $i = 2, \ldots, j[k+1]-1$. Therefore, by applying Lemma 10 to $C_n^1(j[k+1])$, $C_n^1(j[k+1]-1)$, $\ldots$ and $C_n^1(3)$, we have $\tau_1(i, k+1; u) \le \frac{1}{d_C\delta_C}1(i)[k+1] + \frac{1}{\delta_C}\sum_{h=1}^{i-1} 1(h)[k+1]$ for each $i = 2, \ldots, j[k+1]-1$. Moreover, by using Lemma 15, we derive that $\frac{1}{d_C\delta_C}1(i)[k+1] + \frac{1}{\delta}\sum_{h=1}^{i-1} 1(h)[k+1] = \frac{1}{\delta_C}\left(\frac{1 - (1/d_C)^i}{1 - (1/d_C)} - 1 + \frac{1}{d_C}\right)1(i)[k+1] \le \frac{1 - (1/d_C)^i}{1 - (1/d_C)}1(i)[k+1] = \sum_{h=1}^{i} 1(h)[k+1]$ holds for each $i = 1, \ldots, j[k+1]$.

Now, we shall consider the case of applying Lemma 10 to $C_n^1(2)$. Let v denote the input to $C_n^1(2)$. By the inequality (4) for i=2, we have $\#_1 v \le (1/\delta_C)1(1)[k+1] + \frac{2}{(k_C-\varepsilon_C)\delta_C}1(2)[k+1]$. Therefore, we have $\tau_1(1, k+1; u) \le$

$\frac{1}{\delta_C(k_D-\epsilon_C)}\left(\mathbf{l}(1)[k+1]+\frac{2}{k_C-\epsilon_C}\mathbf{l}(2)[k+1]\right)\leq\sum_{h=1}^{1}\mathbf{l}(h)[k+1]$. Thus, we conclude that

$$\tau_1(i,k+1;u)\leq\sum_{h=1}^{i}\mathbf{l}(h)[k+1]$$

holds for each $i=1,\ldots,\mathbf{j}[k+1]$. In a similar way, we can also establish that $\tau_0(i,k+1;u)\leq\sum_{h=1}^{i}\mathbf{l}(h)[k+1]$ holds for each $i=1,\ldots,\mathbf{j}[k+1]$.

Next, we assume that some compressors are removed in step 4. From the previous argument and the condition necessary for compressors to be removed, it is easy to see that no impure elements appear on the output terminals to which input terminals of removed compressors were joined. Thus, we conclude the lemma. □

The following lemma is obvious by Lemma 17 and the fact that $E_n(j_{max}(n))$ is a sorting network.

Lemma 18. *For every even integer $n>0$, $N(n)$ is an $(n,n/2)$-selector.*

From Lemma 15, Lemma 16 and the fact that $j_{max}(n)=O(\log n)$, we obtain the following Lemma.

Lemma 19. $\mathrm{depth}(N(n))=O(\log n)$.

Lemma 20. *For every $0<\epsilon_C$, $0<\theta_C<1$, $0<\delta_C<1$ and $k_C\in\mathbf{Z}$ with $k_C\geq\frac{2(1+\sqrt{1-\delta_C})}{\delta_C}+\epsilon_C$,*

$$\mathrm{size}(N(n))\leq F(\epsilon_C,\theta_C,\delta_C,k_C)n\log_2 n+O(n) ,$$

where

$$F(\epsilon_C,\theta_C,\delta_C,k_C)=\frac{(1-\theta_C)\left[\frac{1}{\delta_C}+\frac{2(1+\sqrt{1-\delta_C})}{\delta_C^2(k_C-\epsilon_C)}+\epsilon_C\right]+\theta_C k_C}{\log_2\frac{1}{\delta_C}} .$$

Proof. Let $s_1^+(k,n)$ denote the total size of the compressors belonging to $L_n(\mathbf{j}[k])$. Let $s_1^-(k,n)$ denote the total size of the compressors removed in step 4. Let $s_1(k,n)$ denote $s_1^+(k,n)-s_1^-(k,n)$. Let $s_2(k,n)$ denote the total size of the extractors belonging to $L_n(\mathbf{j}[k])$. Let $s_3(k,n)$ denote the total size of the sorting networks belonging to $L_n(\mathbf{j}[k])$. Clearly $\mathrm{size}(N(n))=\sum_{k=0}^{k_{max}(n)}(s_1(k,n)+s_2(k,n)+s_3(k,n))$ holds. Since $s_3(k,n)=O(1)$, we derive $\sum_{k=0}^{k_{max}(n)}s_3(k,n)=O(\log n)$ from Lemma 19. Recall that, for $k=1,\ldots,k_{max}(n)$, $P_n(k)$ is the predicate such that $P_n(k)=1$ implies that step 3a is executed in the k-th execution of step 3 and $P_n(k)=0$ implies not. Using Lemma 15 and Lemma 16, we can derive the fact that there exists a positive integer c_1 such that, for every integer $k\geq c_1$ with $\mathbf{j}[k]<j_{max}(n)$, $P_n(k-c_1+1)\vee P_n(k-c_1+2)\vee\cdots\vee P_n(k)=1$ holds. Therefore, we have $\sum_{k=0}^{k_{max}(n)}s_2(k,n)\leq\sum_{k=1}^{\infty}c_1 k_B(\gamma_C,\Delta_C)|X_n(i)|\leq(c_1 k_B(\gamma_C,\Delta_C)/2)n=O(n)$. Let $K=\lceil\log_2 n/\log_2(1/\delta_C)\rceil$. By Lemma 15, we have $\sum_{h=1}^{i}\mathbf{l}(h)[k]<$

$1(i)[k] \sum_{h=0}^{\infty} (1/d_C)^h = (1/\sqrt{1-\delta_C}) 1(i)[k]$ for each $k = 0, \ldots, k_{\max}(n)$ and each $i = 1, \ldots, j[k]$. Moreover, from Lemma 15 and $1(1)[0] < n$, we derive $1(1)[K] < 1$. Therefore, by Lemma 15 we have $1(i)[K + j] < d_C^{i-1} \delta_C^j$ for each integers $i \geq 1$ and $j \geq 0$. From Lemma 8 and the fact that, if $1(i)[k] \leq \sqrt{1-\delta_C}$, then $C_n^1(2), \ldots, C_n^1(i)$ and $C_n^0(2), \ldots, C_n^0(i)$ are all removed in the k-th execution of step 4, we derive $\sum_{k=K+1}^{k_{\max}(n)} s_1(k, n) = O(n)$. On the other hand, it is obvious that $\sum_{k=0}^{K} s_1(k, n) \leq (K + 1)\left((1 - \theta_C)k_D + \theta_C k_C\right)n \leq F(\varepsilon_C, \theta_C, \delta_C, k_C)n \log_2 n + 2\left((1 - \theta_C)k_D + \theta_C k_C\right)n.$ □

By taking $\varepsilon_C = 1$, $\theta_C = 1/2$, $\delta_C = 24/49$ and $k_C = 8$, we obtain the following theorem from Lemma 18, Lemma 19 and Lemma 20.

Theorem 21. *There exists a family of $(n, n/2)$-selectors with even n such that its depth is $O(\log n)$ and its size is at most $8n \log_2 n + O(n)$.*

Theorem 22. *For every $C > 3/\log_2 3 = 1.8927 \cdots$, there exists a family of $(n, n/2)$-selectors with even n such that its depth is $O(\log n)$ and its size is at most $Cn \log_2 n + O(n)$.*

Proof. By taking θ_C, δ_C and k_C as $\delta_C = \dfrac{1}{3 - 2\varepsilon_C}$, $k_C = \left\lceil \dfrac{2(1 + \sqrt{1 - \delta_C})}{\varepsilon_C \delta_C^2} + \varepsilon_C \right\rceil \geq \left\lceil \dfrac{2(1 + \sqrt{1 - \delta_C})}{\delta_C} + \varepsilon_C \right\rceil$ and $\theta_C = \dfrac{\varepsilon_C}{k_C}$ for constant ε_C, we have $F(\varepsilon_C, \theta_C, \delta_C, k_C) \leq \dfrac{3 + \varepsilon_C}{\log_2(3 - 2\varepsilon_C)}$. Since $\dfrac{3 + \varepsilon_C}{\log_2(3 - 2\varepsilon_C)} \to \dfrac{3}{\log_2 3}$ ($\varepsilon_C \to 0$), we can take $\varepsilon_C > 0$ so that $F(\varepsilon_C, \theta_C, \delta_C, k_C) < C$. □

References

1. M. Ajtai, J. Komlós, and E. Szemerédi. An $O(n \log n)$ sorting network. In *Proceedings of the 15th Annual ACM Symposium on Theory of Computing*, pp. 1–9, 1983.
2. M. Ajtai, J. Komlós, and E. Szemerédi. Sorting in $c \log n$ parallel steps. *Combinatorica*, Vol. 3 (1), pp. 1–19, 1983.
3. L. A. Bassalygo. Asymptotically optimal switching circuits. *Problemy Peredachi Informatsii*, Vol. 17, pp. 206–211, 1981. English translation in *Problems of Information Transmission*.
4. D. E. Knuth. *The art of computer programming, Sorting and searching*, Vol. 3. Addison-Wesley, 1975.
5. M. S. Paterson. Improved sorting networks with $O(\log n)$ depth. *Algorithmica*, to appear.
6. N. Pippenger. Selection networks. *SIAM Journal on Computing*, Vol. 20 (5), pp. 878–887, August 1991.

Relativizations of the $P =?NP$ and Other Problems: Some Developments in Structural Complexity Theory[*]

Ronald V. Book
Department of Mathematics
University of California
Santa Barbara, CA 93106 USA
(e-mail: book%henri@hub.ucsb.edu)

Abstract

The $P =?NP$ problem has provided much of the primary motivation for developments in structural complexity theory. Recent results show that even after twenty years, contributions to the $P =?NP$ problem, as well as other problems, still inspire new efforts. The purpose of this talk is to explain some of these results to theoreticians who do not work in structural complexity theory.

1 Introduction

In his fundamental 1971 paper, Cook [Coo71] introduced the idea of NP-completeness. In keeping with earlier work that emphasized the importance of polynomial time computation (for example, see [Cob64]), Cook focused on the notion of polynomial-time reductions and called attention to the class NP of decision problems solvable nondeterministically in polynomial time. In 1972, Karp [Kar72] showed that NP-completeness provided a method for relating a large number of problems from combinatorics, operations research, and computer science, all of which appeared to be computationally intractable since their known solutions involved exhaustive search. Independently and essentially simultaneously, Levin [Lev73] studied many of the same problems; see Trakhtenbrot's historical paper [Tra84] for a description of Russian work of the period. (See [GJ79] for a list of several hundred NP-complete problems.)

While the question of whether every problem in NP is in fact tractable (that is, the $P =?NP$ problem) can be viewed as a special case of the longstanding open problem of whether nondeterministic computation is more powerful than deterministic computation,

[*]The preparation of this paper was supported in part by the National Science Foundation under Grant CCR-8913584 and by the Alexander von Humboldt Stiftung while the author visited the Facultät für Informatik, Universität-Ulm, Germany.

the motivation for the study of this problem stems from work in computational complexity theory as well as from the papers of Cook and of Karp. Beyond the study of the $P =? NP$ problem itself, the notion of NP-completeness has influenced a very large part of computer science. One measure of the importance of this work is the fact that many significant theoretical developments have their roots there.

It is the purpose of the present paper to illustrate some of the developments in structural complexity theory by considering results in that theory that are related to the $P =? NP$ problem and some related problems. The notions of reducibilities computed in polynomial time (or space) and of complete sets for complexity classes are among the basic themes of structural complexity theory, and it is investigations of the properties and power of such reducibilities and complete sets that have provided much of the rich development of that theory (see the two-volume text by Balcázar, Díaz, and Gabarró [BDG88, 90]). In this paper results about relativized computation that relate bounded and unbounded reducibilities computed in polynomial time, "sparse" sets used as oracle sets, and randomly selected oracle sets are described in terms of their relationship to $P =? NP$ and other problems.

To understand what is presented here, the reader needs to know very little about structural complexity; however, it is useful to understand the ideas of **determinism** and **nondeterminism** in computations whose running time is bounded. In the normal method for formalizing the idea of a computation, one has a programming language such that at each step of a computation there is exactly one possible next step (until the computation hlats); this is what is meant by "deterministic" computation. The class P is the collection of all decision problems (languages) that are solvable (resp., accepted) by machines whose running time is bounded above by a polynomial in the size of the input; these are problems that are "feasible" in the sense that their solutions can be feasibly computed.

In "nondeterministic" computation there is more that one possible choice of what to do next; a machine running a program in such a language can "guess" what to do at each step and so it is called a "nondeterministic" machine. Since only decision problems are considered here, a computation of a nondeterministic machine that ends in an accepting state is said to have made a "correct" sequence of guesses. To define the class NP, one considers nondeterministic machines with polynomial time bounds; that is, for each machine there is a fixed polynomial such that in each computation the number of steps is bounded by that polynomial evaluated on the size of the input. In this context the sequence of guesses has length bounded by the same polynomial time bound. For any such program there is a fixed bound on the number of different guesses that can be made at any step. Given an input and any sequence of guesses, checking whether that sequence of guesses corresponds to an accepting computation of the program on the given input can be accomplished in time proportional to the length of the sequence of guesses. Thus, a language (or predicate) L is in the class NP if there is a language (or predicate) B in P and a polynomial q with the property that for every input x, x is in L if and only if there exists a y such that the size of y bounded above by q in the size of x and $\langle x, y \rangle$ is in B; here, x is the input to the nondeterministic machine M, y is a record of the sequence of guesses, and $\langle x, y \rangle$ is in B if and only if y encodes an accepting computation of the machine M on input x. Thus, the class NP can be considered to be the class of decision problems (languages) that can be solved affirmatively (resp., accepted) by this "guess-and-check" procedure in polynomial time.

It is useful to think of the inputs encoded as strings of symbols so that a decision problem is encoded as a set of strings, that is, a language. Thus, the size of an input is simply the length of the string. (For a more formal presentation, see Balcázar, Díaz, and Gabarró [BDG88, 90]).

2 NP-Completeness

The idea of solving a problem by reducing it to the solution of another problem is an old idea and one that is well-known to computer scientists. One formalization of this idea was carried out in recursive function theory (see [Rog67]) through the notion of "reducibilities." This formalization was used by Cook. The simplest of the various types of reducibilities is that of "many-one reducibility," which was used by Karp, and that is the starting point in the present paper.

Language A is *many-one reducible to* language B *in polynomial time*, denoted $A \leq_m^P B$, if there exists a polynomial-time computable function f with the property that for all strings x, $x \in A$ if and only if $f(x) \in B$.

Notice that the binary relation $\leq_m^P$ between languages can be specified by means of the collection of functions that map strings to strings and that can be computed in polynomial time. One way of interpreting the notion "A is many-one reducible to B" in polynomial time is that the membership problem for A can be transformed in polynomial time to the membership problem for B.

The condition "$x \in A$ if and only if $f(x) \in B$" in the definition of $A \leq_m^P B$ has two specific features that must be emphasized. First, to determine membership in A, it is sufficient to ask just one question about membership in B. Second, the answer to the question "is x in A?" agrees with the answer to the question "is $f(x)$ in B?"

A language B is $\leq_m^P$-*hard for a class* C if for every language A in C, $A \leq_m^P B$. A language B is $\leq_m^P$-*complete for a class* C if $B \in C$ and B is $\leq_m^P$-hard for C. A language is *NP-complete* if it is $\leq_m^P$-complete for the class NP.

The first language that was shown by Cook to be NP-complete was the collection of Boolean formulas in conjunctive normal form that are satisfiable. This NP-complete set is frequently denoted by SAT. (It is easy to see that SAT is in NP: nondeterministically guess a truth-assignment for the Boolean variables and then check whether that assignment satisfies the given formula. To see that SAT is NP-complete is harder since one must show that for every language B in NP, $B \leq_m^P SAT$.) Subsequently, Karp showed that other languages were NP-complete, languages whose membership problems arise as decision problems in combinatorial mathematics, in operations research, and in computer science.

If a language is NP-complete, then its membership problem is considered to be computationally difficult. This is due to the fact that $P = NP$ if and only if some NP-complete language is in P if and only if every NP-complete language is in P. Most researchers in complexity theory believe that P is not equal to NP.

The following fact may seem surprising at first glance (but is very easy to prove): $P = NP$ if and only if there exists a finite language that is NP-complete if and only if every nonempty finite language is NP-complete.

Suppose that set A is many-one reducible in polynomial time to a finite language F, $A \leq_m^P F$. One can identify this reduction with the notion of "table look-up" in the following sense. Let Π be a program that runs in polynomial time and computes a function g that witnesses $A \leq_m^P F$. Then Π can be transformed into a new program Π' that contains an encoding of the finite language F as a table and on input x, first computes $g(x)$, then determines by table look-up whether $g(x)$ is in F and produces output "yes" if $g(x)$ is in F and output "no" otherwise.

The idea of determining membership in languages like SAT by considering only finite tables seems somewhat unlikely to anyone who has thought about the problem and so one would like to extend the notion of table look-up to infinite tables. But infinite tables as such cannot be stored in finite memory. This consideration led to the notion of "infinite, but slowly growing" tables, and is formalized by the notion of a "sparse" language.

A language S is *sparse* if there exists a polynomial q such that for all $n > 0$, $\| \{x \in S \mid |x| \leq n\} \| \leq q(n)$.

The notion of a sparse language is one way to capture the notion of a language with small density. Every finite language is sparse but there are infinite sparse languages (of arbitrarily large complexity); for example, every language of strings over a one letter alphabet is sparse.

The class of sparse languages plays an important role in the development of structural complexity theory and a prominent role in the present paper. One of the first important results about sparse languages is due to Mahaney [Mah82].

Theorem 1 $P = NP$ *if and only if there exists a sparse language that is $\leq_m^P$-hard for* NP.

Mahaney's work was motivated by the conjecture of Berman and Hartmanis [BH77] that every two NP-complete languages are closely related structurally.

3 Bounded Reducibilities

Recall the comment above about the definition of $A \leq_m^P B$: to determine membership in A, it is sufficient to ask just one query about membership in B; and the answer to the question "is x in A?" agrees with the answer to the query "is $f(x)$ in B?" It is natural to attempt to generalize this, both by allowing more than one query about membership in B and by allowing the answer to the question "is x in A?" to depend on the collection of answers to the various queries about membership in B. There are different ways of doing this but here just one way will be described (and only described informally).

Consider a program Π that gives only 0–1 output. Suppose that in the program Π, there are specific binary branch points, called "query points," and that there is a distinguished variable, called the "query variable." In any computation of Π, the decision of which branch to take when a specific query point is encountered depends on an external source, an "oracle," that gives the correct answer to the question of whether the current value of the query variable, is in a specific language, the "oracle language." If the value of the query variable is in the oracle language, then the oracle says "yes," which is interpreted as instructing the program to take one branch, while if the value of the query variable is not in the oracle language, then the oracle says "no" which is interpreted as instructing the program to take the other branch.

For any language B, consider the language A of all inputs x such that Π's computation on x gives the answer 1 when B is the oracle language. Then A is *Turing reducible to B*, denoted $A \leq_T B$. In addition, if Π runs in polynomial time, then A is *Turing reducible to B in polynomial time*, denoted $A \leq_T^P B$. In addition, if Π runs in polynomial time and there exists a fixed k such that for all x, in Π's computation on x with oracle language B, the oracle is queried at most k times (in other words, in the computation the total number of times the query points are encountered is bounded above by k), then A is *bounded-Turing reducible to B in polynomial time*, denoted $A \leq_{b-T}^P B$.

The notion of bounded Turing reducibility generalizes that of many-one reducibility since more than one query of the oracle language can be made and since the answer that the computation gives depends on the answers to the queries. The notion of being "hard" for a class with respect to some reducibility follows naturally.

A language B is $\leq_{b-T}^P$-*hard for a class* C if for every language A in C, $A \leq_{b-T}^P B$. A language B is $\leq_T^P$-*hard for a class* C if for every language A in C, $A \leq_T^P B$.

A recent result of Ogiwara and Watanabe [OW91] shows that the natural generalization of Theorem 1 to $\leq_{b-T}^P$ does hold.

Theorem 2 $P = NP$ *if and only if there exists a sparse language that is* $\leq_{b-T}^P$-*hard for* NP.

The result of Ogiwara and Watanabe came as a great surprise in structural complexity theory. In developing Theorem 1, Mahaney had built upon the work of others who had been pursuing the Berman-Hartmanis conjecture and had developed a number of results about the structure of NP. While the question of whether Theorem 2 was true was known to many, few results were known about complete languages (for any complexity class) with respect to bounded reducibilities, the principal exception being results by Watanabe [Wat87] who had studied complete languages for the class of languages accepted in exponential time by deterministic machines.

The proof technique used by Ogiwara and Watanabe is extremely different from that used by Mahaney, and it does not appear to shed any light on the Berman-Hartmanis conjecture. It uses a simple ordering of strings that encode accepting computations of a nondeterministic machine on an accepted input. In the case of many-one reducibility, $\leq_m^P$, the new proof technique can be used to give a simple proof of Theorem 1.

There is another important complexity class, the class of decision problems solvable in polynomial space (see [BDG88, 90] for a description). This class is denoted $PSPACE$ since it does not matter whether the deterministic or the nondeterministic modes of computation are considered. While it is clear that P and NP are subclasses of $PSPACE$, it is not known that the inclusions are proper. A recent result of Ogiwara and Lozano [OL91], analogous to Theorem 2, is surprising.

Theorem 3 $P = PSPACE$ *if and only if there exists a sparse language that is* $\leq_{b-T}^P$ *hard for* $PSPACE$.

The reader should keep in mind that for the notion of bounded-Turing reducibility, the bound on the number of queries is a constant that does not depend on the input. The topic of the next section is Turing reducibility in which the number of queries allowed in a computation is bounded only by the running time of the computation.

4   Unbounded Reducibilities

At one time it was thought that the methods of elementary recursive function theory would be sufficient to solve the $P =?NP$ problem (and other problems of complexity theory); while that has not been the case, the attempts to do this have greatly stimulated the development of structural complexity theory. This is particularly true when the notion of Turing reducibility is considered.

For any set A, let $P(B) = \{A \mid A \leq_T^P B\}$ and let $NP(B) = \{A \mid A \leq_T^{NP} B\}$, where $A \leq_T^{NP} B$ is defined analogously to $A \leq_T^P B$ but with nondeterministic polynomial-time programs.

It is reasonable to ask if the answer to the $P =?NP$ problem relativizes: if $P = NP$, is it the case that for all B, $P(B) = NP(B)$? And if $P \neq NP$, is it the case that for all B, $P(B) \neq NP(B)$? While an analogy with elementary recursive function theory would suggest that the answer is "yes," the results of Baker, Gill, and Solovay [BGS75] show that the answer is "no."

Theorem 4 *There exist (recursive) languages A and B such that $P(A) = NP(A)$ and $P(B) \neq NP(B)$.*

What has stimulated much interest is the existence of B such that $P(B) \neq NP(B)$. On one hand, the proof of Baker, Gill, and Solovay says nothing about the existence of a language B such that some NP-complete language is not in $P(B)$. On the other hand, what the proof does do is to describe an oracle set B such that a machine witnessing the separation of $NP(B)$ from $P(B)$ has the property that on certain inputs, the non-deterministic operation of the machine allows the possibility of an exponential number of possible values of the query variable. But for any deterministic machine that runs in polynomial time, for any fixed oracle language, there can be at most a polynomial number of possible values of the query variable.

Theorem 4 led to the investigation (see [BLS84]) of relativizations for which the size of the set of possible values of the query variable was restricted while still leaving the combinatorial difficulties of $P =?NP$ problem.

For any language A, let $NP_B(A)$ denote the collection of all languages $L(M, A)$ where M is an NP oracle machine with the property that for some polynomial q, for all x, the size of $\{y \mid$ there exists a computation of M on x with A as oracle language that queries the oracle about $y\}$ is bounded above by $q(|x|)$. (The subscript "B" refers to the notion that the set of potential query strings is bounded in size.)

This restriction on classes of the form $NP(A)$ retains the combinatorial difficulties of the $P =?NP$ problem but is strong enough to avoid the "Baker-Gill-Solovay" phenomenon.

Theorem 5 *$P = NP$ if and only if for all A, $P(A) = NP_B(A)$ if and only if for all A, some NP-complete set is in $P(A)$.*

Thus, $P \neq NP$ if and only if there exists an A such that $P(A) \neq NP_B(A)$ if and only if there exists an A such that $SAT \notin P(A)$.

When dealing with many-one reducibilities and bounded-Turing reducibilities, interesting results (Theorems 1–3) were obtained by considering reductions to sparse languages.

The same thing is true when considering (unbounded) Turing reductions to sparse languages but in this case the results speak to somewhat different questions, and to state these questions, certain definitions are necessary.

For any language A, let $\Sigma_0^P(A) = P(A)$ and for each $i \geq 0$, let $\Sigma_{i+1}^P(A) = \cup\{NP(B) \mid B \in \Sigma_i^P(A)\}$. Let $PH(A) = \cup_{i \geq 0}\Sigma_i^P(A)$. The structure $\{\Sigma_i^P(A)\}_{i \geq 0}$ is the *polynomial-time hierarchy relative to* A. If A is the empty set (or in P), then write $\{\Sigma_i^P\}_{i \geq 0}$ instead of $\{\Sigma_i^P(A)\}_{i \geq 0}$ and PH instead of $PH(A)$. The structure $\{\Sigma_i^P\}_{i \geq 0}$ is the *polynomial-time hierarchy*.

For any language A, the polynomial-time hierarchy relative to A *is infinite* if for every i, $\Sigma_i^P(A) \neq \Sigma_{i+1}^P(A)$; otherwise, the hierarchy is said to *collapse*.

It is not known whether the polynomial-time hierarchy extends beyond Σ_0^P since it is not known whether P is equal to NP (of course, for any A, $P(A) = NP(A)$ implies $PH(A) = P(A)$). It is clear that for every language B, $PH(B) \subseteq PSPACE(B)$. Yao [Yao85] has shown that there exists a language A such that the polynomial-time hierarchy relative to A is infinite; for any such language A, $PH(A) \neq PSPACE(A)$.

Now the results about Turing reducibilities that are parallel to Theorems 1–3 can be stated.

Theorem 6 *If there exists a sparse language that is $\leq_T^P$-hard for NP, then the polynomial-time hierarchy collapses.*

Theorem 7 *If there exists a sparse language that is $\leq_T^P$-hard for $PSPACE$, then $PH = PSPACE$.*

Theorems 6 and 7 are due to Karp and Lipton [KL82], although their original statements had a different form.

If one considers ways of measuring the information content of a language, then the theory of Kolmogorov complexity (see, for example, [LV90]) suggests that sparse languages have very small information content. This leads to a fairly natural interpretation of Theorems 2, 3, 6, 7: if a language with small information content can be hard for classes such as NP or $PSPACE$ with respect to bounded or unbounded reducibilities computed in polynomial time, then those classes are not as complex as has been thought.

It may be useful to summarize the results of Theorems 2, 3, 6, 7:

(i) $P = NP$ if and only if there exists a sparse language that is $\leq_{b-T}^P$-hard for NP;

(ii) $P = PSPACE$ if and only if there exists a sparse language that is $\leq_{b-T}^P$-hard for $PSPACE$;

(iii) If there exists a sparse language that is $\leq_T^P$-hard for NP, then the polynomial-time hierarchy collapses.

(iv) If there exists a sparse language that is $\leq_T^P$-hard for $PSPACE$, then $PH = PSPACE$.

5   Random Oracles

One theme that has arisen in the study of relativized computation has been the study of properties of complexity classes that hold for "random oracles."

A language A is "randomly selected" if membership in A is decided by a random experiment in which an independent toss of a fair coin is used to decide whether each string is in A. For a class C of languages, write Prob$[C]$ for the probability that $A \in C$ when A is randomly selected. (This can be defined formally in such a way that the result corresponds to the standard Lebesgue measure on the interval $[0, 1]$, and the Kolmogorov 0–1 Law of probability theory holds. See the Appendix for more details.)

If C is a class of languages such that Prob$[C] = 1$, then we say that "almost every language is in C" or "for almost every language A, $A \in C$."

Bennett and Gill [BG81] formulated their "random oracle hypothesis": if a relationship between relativized complexity classes holds for almost every oracle language, then it holds for the unrelativized classes. They established the following result.

Theorem 8 *For almost every language A, $P(A) \neq NP(A) \neq PSPACE(A)$.*

Theorem 8 strengthens Theorem 4. However, the proof by Bennett and Gill says nothing about whether some NP-complete language is in P or whether for almost every A, some NP-complete language is in $P(A)$, and so does not establish the desired relationship between the unrelativized classes.

Cai [Ca89] established a similar separation of PH from $PSPACE$ with respect to randomly selected languagess.

Theorem 9 *For almost every language A, $PH(A) \neq PSPACE(A)$.*

Again, Cai's proof says nothing about whether some language that is complete for $PSPACE$ is in PH or whether for almost every A, some $PSPACE$-complete language is in $PH(A)$.

While Kurtz [Kur83] provided evidence that the random oracle hypotheses fails (also, see [Boo91]), many researchers believe that Theorems 8 and 9 provide strong evidence for the separation of the unrelativized classes. However, the proofs used by Bennett and Gill and by Cai have not been extended to show such separations. But results about randomly selected oracle languages are of interest in the present context and these are developed below.

In the development in Sections 2–4, the use of sparse languages as oracle languages has been described. While sparse languages are considered to have small information content, certain languages with very high information content have some of these same properties when used as oracle languages. The remainder of this section is concerned with a specific class of languages having very high information content, the "algorithmically random languages," and describes properties of such languages used as oracle languages. Unfortunately, the definitions are not well known and so they must be given here; but many are quite technical and so they are relegated to the Appendix.

The collection of algorithmically random languages will be denoted $RAND$. Martin-Löf [ML66] defined the class $RAND$ (with different notation and different motivation) so that Prob$(RAND) = 1$, and showed that no recursively enumerable language is in

RAND. RAND is considered to be the class of those languages essentially having the greatest possible information content in the following sense. For each language A and each n, let $A_{\leq n}$ denote the finite language $x \in \{A \mid |x| \leq n\}$. Then the Kolmogorov complexity of $A_{\leq n}$ is no greater than $2^{n+1} + c$, where c is a fixed constant that does not depend on A on n. Martin-Löf [ML71] proved that for every language A in *RAND*, the Kolmogorov complexity of $A_{\leq n}$ is strictly greater than $2^{n+1} - 2n$ for all but finitely many n.

For some of the results about relativized computation described in the preceding sections, showing that a property holds for almost every oracle language is equivalent to showing that the property holds for languages in *RAND*. This has been made precise by Book, Lutz, and Wagner [BLW92], who showed that from Theorems 8 and 9, one can obtain the following results:

(i) for every language $A \in RAND$, $P(A) \neq NP(A) \neq PSPACE(A)$.

(ii) for every language $A \in RAND$, $PH(A) \neq PSPACE(A)$.

However, for the purposes of the present paper, the following simple corollary of the results of Book, Lutz, and Wagner is more to the point.

Theorem 10

(i) $P = NP$ *if and only if there exists an algorithmically random language that is* $\leq_{b-T}^{P}$*-hard for* NP.

(ii) $P = PSPACE$ *if and only if there exists an algorithmically random language that is* $\leq_{b-T}^{P}$*-hard for* $PSPACE$.

(iii) *If there exists an algorithmically random language that is* $\leq_{T}^{P}$*-hard for* NP, *then the polynomial-time hierarchy collapses.*

(iv) *If there exists an algorithmically random language that is* $\leq_{T}^{P}$*-hard for* $PSPACE$, *then* $PH = PSPACE$.

6 Concluding Remarks

The reader should note the similarity of Theorem 10 to the summary given at the end of Section 4. In Theorem 10, the languages having the greatest possible information content, algorithmically random sets, serve as the oracle sets. In Theorems 2, 3, 6, 7, the sparse languages, languages having very small information content, serve as oracle languages. But the conclusions are the same. One can interpret the results presented in Theorem 10 as indicating that the information in algorithmically random languages is encoded in such a way that little of it is computationally useful from the standpoint of structural complexity theory, since one may as well use a sparse language. This suggests that a theory that relates the information content of oracle languages to the computational power of reducibilities needs to be developed; the results presented here should be viewed as only "first steps."

Appendix

The characteristic sequences of a language A is a (one-way) infinite sequence χ_A on $\{0,1\}$; one can freely identify a language with its characteristic sequence and the class of all languages on the fixed finite alphabet $\{0,1\}$ with the set $\{0,1\}^\omega$ of all such infinite sequences.

If X is a set of strings (that is, a language) and C is a set of sequences (that is, a class of languages), then $X \cdot C$ denotes the set $\{w\xi \mid w \in X, \xi \in C\}$. For each string w, $C_w = \{w\} \cdot \{0,1\}^\omega$ is the *basic open set* defined by w. Note that C_w is the set of all sequences ξ such that w is a prefix of every string in ξ. An *open set* is a (finite or infinite) union of basic open sets, that is, a set of the form $X \cdot \{0,1\}^\omega$ where $X \subseteq \{0,1\}^*$. A *closed set* is the complement of an open set. A class of languages is *recursively open* if it has the form $X \cdot \{0,1\}^\omega$ for some recursively enumerable set $X \subseteq \{0,1\}^*$. A class of languages is *recursively closed* if it is the complement of some recursively open set.

For a class C of languages, write $\text{Prob}[C]$ for the probability that $A \in C$ when A is chosen by a random experiment in which an independent toss of a fair coin is used to decide whether each string is in A. This probability is defined whenever C is measurable in the usual product topology of $\{0,1\}^\omega$. In particular, if C is a countable union or intersection of (recursively) open or closed sets, then C is measurable so $\text{Prob}[C]$ is defined. There are only countably many recursively open sets, so every intersection of recursively open sets can be expressed as a countable intersection of such sets, and hence is measurable; similarly every union of recursively closed sets is measurable. If $\text{Prob}[C] = 1$, then "almost every language" is in C.

A class C is *closed under finite variation* if $A \in C$ whenever $B \in C$ and A and B have finite symmetric difference. The Kolmogorov 0-1 Law says that for every measurable set $C \subseteq \{0,1\}^\omega$ which is closed under finite variation, either $\text{Prob}[C] = 0$ or $\text{Prob}[C] = 1$.

Assume an effective enumeration of the recursively enumerable languages as $W_1, W_2, \ldots$

A *constructive null cover* of a class C of languages is a sequence of recursively open sets $W_{g(1)} \cdot \{0,1\}^\omega$, $W_{g(2)} \cdot \{0,1\}^\omega, \ldots$ specified by a total recursive function g with the properties that for every k,

(i) $C \subseteq W_{g(k)} \cdot \{0,1\}^\omega$, and

(ii) $\text{Prob}[W_{g(k)} \cdot \{0,1\}^\omega] \leq 2^{-k}$.

If a class C has a constructive null cover, then C is a *constructive null set*, so that $\text{Prob}[C] = 0$. Let $NULL$ be the union of all constructive null sets, so that $\text{Prob}[NULL] = 0$ since $NULL$ is a countable union of sets with probability 0. Define $RAND$ as the complement of $NULL$ (with respect to $\{0,1\}^\omega$); a language is *algorithmically random* if and only if it is in $RAND$. Then $\text{Prob}[RAND] = 1$ since $\text{Prob}[NULL] = 0$.

References

[BGS75] T. Baker, J. Gill, and R. Solovay. Relativizations of the $P =?NP$ problem, *SIAM J. Computing* 4 (1975), 431–442.

[BDG88] J. Balcázar, J. Díaz, and J. Gabarró. *Structural Complexity I*, Springer-Verlag, 1988.

[BDG90] J. Balcázar, J. Díaz, and J. Gabarró. *Structural Complexity II*, Springer-Verlag, 1990.

[BG81] C. Bennett and J. Gill. Relative to a random oracle, $P^A \neq NP^A \neq co\text{-}NP^A$ with probability 1, *SIAM J. Computing* 10 (1981), 96–113.

[BH77] L. Berman and J. Hartmanis. On isomorphism and density of NP and other complete sets, *SIAM J. Computing* 6 (1977), 305–322.

[Boo91] R. Book. Some Observations on separating complexity classes, *SIAM J. Computing* 20 (1991), 246–258.

[BLS84] R. Book, T. Long, and A. Selman. Quantitative relativizations of complexity classes, *SIAM J. Computing* 13 (1984), 461–487.

[BLW92] R. Book, J. Lutz, and K. Wagner. On complexity classes and algorithmically random languages, *STACS 92, Lecture Notes in Computer Sci.* 577, Springer-Verlag (1992), 319–328.

[Cai89] J.-Y. Cai. With proability one, a random oracle separates *PSPACE* from the polynomial-time hierarchy, *J. Comput. Systems Sci.* 38 (1989), 68–85.

[Cob64] A. Cobham. The intrinsic computational difficulty of functions, *Prod. 1964 International Congress for Logic, Methodology and Philosophy of Science*, North Holland (1964), 24–30.

[Coo71] S. Cook. The complexity of theorem-proving procedures, *Proc. 3rd ACM Symp. Theory of Computing* (1971), 151–158.

[GJ79] M. Garey and D. Johnson. *Computers and Intractability: A Guide to the Theory of NP-Completeness*, Freeman & Co., 1979.

[Kar72] R. Karp. Reducibility among combinatorial problems, in R. Miller and J. Thatcher (eds.), *Complexity of Computer Computation*, Plenum Press (1972), 85–104.

[KL82] R. Karp and R. Lipton. Turing machines that take advice, *L'Enseignement Mathématique* 28 2nd series (1982), 191–209.

[Kur83] S. Kurtz. On the random oracle hypothesis, *Info. and Control* 57 (1983), 40–47.

[Lev73] L. Levin. Universal sequential search problems, *Probl. Pered. Inform.* IX (1973), 115–116. English translation in *Probl. Information Transmission* 9 (1973), 265–266.

[LV90] M. Li and P. Vitanyi. Kolmogorov complexity and its applications, in J. van Leeuwen (ed.), *Handbook of Theoretical Computer Science*, vol. A, Elsevier Sci. Publishers (1990), 187–254.

[Mah82] S. Mahaney. Sparse complete sets for NP: solution of a conjecture by Berman and Hartmanis, *J. Comput. and Systems Sci.* 25 (1982), 130–143.

[ML66]   P. Martin-Löf. On the definition of random sequences, *Info. and Control* 9 (1966), 602–619.

[ML71]   P. Martin-Löf. Complexity oscillations in infinite binary sequences, *Zeitschrift für Wahrscheinlichkeitstheory und Verwandte Gebiete* 19 (1971), 225–230.

[OL91]   M. Ogiwara and A. Lozano. On one query self-reducible sets, *Proc. 6th IEEE Conference on Structure in Complexity Theory* (1991), 139–151.

[OW91]   M. Ogiwara and O. Watanabe. On polynomial bounded truth-table reducibility of NP sets to sparse sets, *SIAM J. Computing* 20 (1991), 471–483.

[Rog67] H. Rogers. *Theory of Recursive Functions and Effective Computability*, McGraw-Hill, 1967.

[Tra84] B. Trakhtenbrot. A survey of Russian approaches to Perebor (brute-force search) algorithms, *Annals History of Computing* 6 (1984), 384–399.

[Wat87] O. Watanabe, A comparison of polynomial-time completeness notions, *Theoret. Comput. Sci.* 54 (1987), 249–265.

[Yao85] A. Yao. Separating the polynomial-time hierarchy by oracles, *Proc. 26th IEEE Symp. Foundations of Comput. Sci.* (1985), 1–10.

Boolean Circuit Complexity

Mike Paterson

Department of Computer Science
University of Warwick

ABSTRACT

The talk will survey some old, and some recent, results relating to Boolean circuits, and will concentrate particularly on the lower levels of complexity.

Searching a Solid Pseudo 3-Sided Orthoconvex Grid

Antonios Symvonis[1][*] and Spyros Tragoudas[2]

[1] Basser Department of Computer Science, University of Sydney, N.S.W. 2006,
Australia, symvonis@cs.su.oz.au
[2] Department of Computer Science, Southern Illinois University, Faner Hall,
Carbondale, IL 62901, USA, spyros@buto.c-cs.siu.edu

Abstract. In this paper we examine the edge searching problem on pseudo 3-sided solid orthoconvex grids. We obtain a closed formula that expresses the minimum number of searchers required to search a pseudo 3-sided solid orthoconvex grid. From that formula and a rather straight forward algorithm we show that the problem is in P. We obtain a parallel version of that algorithm that places the problem in NC. For the case of sequential algorithms, we derive an optimal algorithm that solves the problem in $O(m)$ time where m is the number of points necessary to describe the orthoconvex grid. Another important feature of our method is that it also suggests an optimal searching strategy that consists of $O(n)$ steps, where n is the number of nodes of the grid.

1 Introduction

The *edge-searching* problem was introduced by Parson in [7]. An undirected graph was given and the objective was to clean its contaminated edges (or, in a different statement of the problem, to capture a fugitive hidden in them). Three kinds of actions were allowed in this cleaning operation:

1. *place(node):* This action places a searcher at the node specified as parameter of the action.
2. *pick(node):* This action picks up a searcher from the node specified as parameter of the action.
3. *move(origin, destination):* This action moves a searcher along the edge that connects the *origin* and the *destination* nodes. For the action to be legal, the two nodes must be connected by an edge and a searcher must be initially located at the *origin* node.

The *search number* of a graph G, denoted by $es(G)$, was defined in [6] as the minimum number of searchers that are required in order to clean the graph (or capture the fugitive that is hidden in its edges). In that paper it was proven that the decision problem *"Given a graph G and an integer k, can G be cleared with k searchers?"* is NP-Hard. The authors also pointed out that the problem

---

[*] Supported by the University of Sydney Research Grant Scheme

would belong in the class NP if it was true that recontamination cannot help in searching a graph. We say that a clean edge is *recontaminated* if it becomes adjacent with a contaminated edge and no searcher is placed at their common node. Recontamination can start when a searcher that is positioned at a node adjacent to a clean edge and at least one contaminated edge, leaves the node (either by a *pick* or *move* action) and allows the clean edge to be contaminated again. We assume that recontamination propagates at an infinite speed, i.e., if recontamination occurs as a result of an action t of the searching, then, before action $t+1$, all edges that can become contaminated again will do so. LaPaugh [4] proved that recontamination does not help to search a graph and thus the edge searching problem was included in the class NP. Besides its theoretical importance, this result is useful in the sense that allows us to assume that there exists a strategy that searches the graph using the minimum number of searchers and never allows recontamination. A consequence is that the graph can be searched in a finite number of actions. After LePaugh's work a great deal of effort was devoted to the searching problem. Most of the results related the searching problem with other combinatorial optimization problems such as *pebbling* [2], *cutwidth* [5] and *graph separators* [1].

In this paper, we concentrate on the searching problem on a special kind of graphs, namely, the pseudo 3-sided solid orthoconvex grids (Section 3.1). We show how to determine the search number of such graphs in optimal time. We do that by defining a modified version of the edge searching problem which we call *modified edge searching* (Section 2). Then, for the modified edge searching problem, by proving that there are searching strategies that possess several properties regarding the way the grid is searched, we are able to obtain a closed formula that expresses the minimum number of searchers required to search a pseudo 3-sided solid orthoconvex grid (Section 3.2). From that formula we derive an algorithm that computes $es()$ in polynomial time (Section 4).

We can also obtain a parallel version of that algorithm that places the problem in NC. For the case of sequential algorithms, we derive an optimal algorithm that solves the problem in $O(m)$ time where m is the number of points necessary to describe the orthoconvex grid. Another important feature of our method is that it also suggests an optimal searching strategy that consists of $O(n)$ steps, where n is the number of nodes of the grid.

Previously, we were able to determine the searching number in optimal time only for the class of trees [6]. We were also able to solve the decision problem *"Given a graph G of n nodes can we search G by using a constant number of k searchers ? "* in $O(n^2)$ [3]. We improve this result for the case of pseudo 3-sided solid orthoconvex grids. Some work has already been done for search problem in rectangular grids [9]. However, in that work the searchers are more powerful than the ones we use (i.e., different actions are assumed) and contamination does not propagate in an infinite speed. Furthermore, the rectangular grid that was assumed as the underlying graph structure belongs in the class of the pseudo 3-sided solid orthoconvex grids.

Because of space limitations, we omit all proofs of lemmata and theorems. Someone interested in them can refer to [8].

2   A New Version of the Searching Problem

In this section, we define a new version of the searching problem which we call *modified edge searching*. The difference between the two searching problems are in the possible actions that can take place during the search.

Definition 1. We say that we have a *modified edge searching problem on a graph* G if we are allowed to search the graph using all 3 actions of the original edge searching problem as well as the additional 4th one:

4. *clean(node1, node2):* This action cleans edge $(node1, node2)$ or the path $(node_1, node_i)$, $(node_i, node_2)$ where $node_i$ is of degree 2. For the action to be legal searchers must be placed on both $node1$ and $node2$.

Kirousis and Papadimitriou [2] defined a similar searching problem which they called *node searching*. In their version of the game only *place, pick* and *clean* actions were allowed and the clean action could clean only one edge.

Definition 2. A *searching strategy* $S(G)$ is a sequence of actions $< a_1; \ldots; a_m >$ such that when applied on a graph G which has all of its edges contaminated has the effect to clean the edges of G. Action $a_i, 1 \le i \le m$, is any action allowed in the searching problem. A searching strategy is said to be *optimum* when there is no other strategy that uses a smaller number of searchers and also searches the graph.

Two searching strategies are *equivalent* if they both search the same graph with the same number of searchers.

We say that a node is *clean* if all of its adjacent edges are clean. A node is *dirty* if all of its adjacent edges are contaminated. If it is neither clean nor dirty we say that it is *partially clean*.

A *move(origin, destination)* action is *useless* if it neither cleans an edge nor results into contamination. A useless *move* action occurs when the searcher moves from a clean node, or, from a partially clean node to a clean one and is not causing recontamination, or, from a dirty node that has no other searcher.

We say that a searcher is *useful for action* a_t (or simply that it is useful at time t) if its removal will cause recontamination, or it will prevent action a_t of happening. Otherwise, we say that the searcher is *useless*.

We can prove the following lemmata:

Lemma 3. *There is an optimum searching strategy for the (modified) edge searching problem on graph G that contains only useful move actions.*   □

Lemma 4. *There is an optimum searching strategy for the modified edge searching problem on graph G that has the property that no useless searcher is on G immediately before any place, move, or clean action.*   □

Corollary 5. *There is an optimum searching strategy for the original edge searching problem on graph G that has the property that no useless searcher is on G immediately before any place, or move action.* ☐

Lemma 6. *If there is a searching strategy that solves the modified edge searching problem on graph G using k searchers then there is a searching strategy that solves the edge searching problem on graph G using either k or k + 1 searchers.* ☐

3 Searching strategies for Pseudo 3-Sided Solid Orthoconvex Grids

3.1 Definitions

Let $G^\infty(V_\infty, E_\infty)$ be the infinite undirected graph whose node set V_∞ consists of all points of the plane with integer coordinates and in which two vertices are connected by an edge in E_∞ if and only if the Euclidean distance between them is equal to one. Let $G_i(V_i, E_i)$ be a finite node-induced subgraph of G.

A *grid graph* $D(V, E)$ is a subgraph of G_i where, $V = V_i$ and $E \subseteq E_i$. In the following discussion we will consider only graphs that are connected and all of their nodes have degree greater than 1. We say that a grid graph $D(V, E)$ is *solid* if it has no holes.

If we color black all unit squares in G^∞ that are subrounded by the edges of a solid grid graph D, we will divide the plane into two regions, one black and one white. A node v that belongs into a solid grid graph and is adjacent to both the black and the white region is said to be a *boundary node*. The set of all boundary nodes of a solid grid D is said to constitute the boundary of D.

Assume a solid grid graph D and its corresponding black and white regions. D is said to be *orthoconvex* if and only if the intersection of any line parallel to any coordinate axis with the black region consists of at most one line segment.

A node v at the boundary of a solid grid graph is said to be a *convex boundary node* if it is of degree 2 and a *concave boundary node* if it is of degree 4. A node v at the boundary of a solid grid graph is said to be a *turning boundary node* if it is either a convex or a concave boundary node. Otherwise, it is called a *simple boundary node*.

From the above definitions it is obvious that a solid grid graph can be completely defined by the coordinates of its turning boundary nodes. In the rest of the paper, we assume that the grid under consideration is represented by its turning boundary nodes given in the order they appear if we traverse its boundary in the clockwise direction. An arbitrary node is selected to be the start of the traversal.

During our traverse of the boundary of a solid grid graph and assuming that we always look towards the next boundary node, we have to make several turns. So, the traversal of a solid grid can be represented by a word which has length equal to the number of turning boundary nodes over an alphabet of two letters namely, L for left and R for right. We call that word the *coding* of the solid orthoconvex grid. Since we can start our traversal of the boundary of the grid from any turning point, it is useful to think of the coding as a circular word where the first and the last characters wrap around.

Since we can return to the point from which we started, it is obvious that we have 4 more R's than L's. Also observe that in an orthoconvex grid we never have two consecutive L's in its coding. By canceling each L and the R that follows it in the coding of a solid orthoconvex grid, we are left with 4 R's. These correspond to 4 convex boundary nodes. These points define the sides of the grid. In that sense, all orthoconvex grids are 4-sided.

However, we can relax that definition for the special case of the solid orthoconvex grids that contain the patterns RRRR or RRRRR. In these two cases, we can combine 2 sides together and thus, consider the orthoconvex to be composed by a *base*, a *rising region* that is immediately after the base in the clockwise direction, and a *falling region* that follows the rising region in the clockwise direction. For this reason, we call all the solid orthoconvex grids that fall into that category *pseudo 3-sided*. In the following we will refer to the boundary nodes that lie between any two sides (or pseudo sides) as *corners*. Figure 1 shows the two types of pseudo 3-sided solid orthoconvex grids along with their codings, corners and sides.

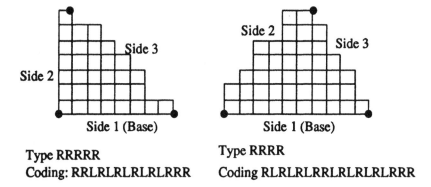

Type RRRRR
Coding: RRLRLRLRLRLRRR

Type RRRR
Coding RLRLRLRRLRLRLRLRRR

Fig. 1. The two types of pseudo 3-sided solid orthoconvex grids and their codings.

A *cord* is any path (possibly a closed one) internal to the grid that consists of grid edges as well as line segments that connect nodes that are of distance 2 and diagonally opposite of each other. If it is not a closed one it must have its endpoints on the boundary.

A *diagonal* is a cord that has its end-nodes on two different sides. We make the convention that a convex boundary node that separates two sides belongs in both of them and, in that sense, it is considered to be a diagonal as well. It is obvious that the nodes that belong into a nontrivial diagonal that touches each of its adjacent side at most once form a cut-set for the solid grid.

During the course of the searching there are regions of the grid that are *clean* and others that are considered to be *contaminated* (or *dirty*).

Definition 7. Assume a searching strategy $S(G)$ on a graph G. Let D_c^t be the graph that is induced by removing all cleaned edges and all nodes that have no incident contaminated edges after t steps of $S(G)$. $Conn(D_c^t)$ is the number of connected components of D_c^t and it denotes also the number of contaminated regions at time t. We say that a graph can be searched in such a way that we have at most μ contaminated regions if and only if $\max_{t>0} Conn(D_c^t) \leq \mu$.

3.2 Searching strategies for Pseudo 3-Sided solid Orthoconvex Grids

In this section we show that there exists an optimum searching strategy for the modified edge searching problem on a pseudo 3-sided solid orthoconvex grid D in which during the course of the searching there exists only one contaminated region. Based on that, we compute $mes(D)$ and we show how from it to derive $es(D)$.

We can prove that[8]:

Lemma 8. *Assume a solid orthoconvex grid and a cord that has its end-nodes on the same side of the grid. We can search the region that is bounded from that side and the cord using the modified edge searching with a number of searchers equal to the cord points and in such a way that one searcher ends up at every cord node.* ☐

Corollary 9. *Assume a solid orthoconvex grid and a cord that has its end-nodes on the same side of the grid. We can search the region that is bounded from that side and the cord using the modified edge searching with a number of searchers equal to the cord points and in such a way that one searcher starts from each cord node.* ☐

Lemma 10. *Assume a pseudo 3-sided solid orthoconvex grid D and a diagonal. We can search the region of D that is bounded from the diagonal and the part of the boundary that contains exactly 1 corner using the modified edge searching with a number of searchers equal to the diagonal points and in such a way that one searcher ends up at every diagonal node.* ☐

Corollary 11. *Assume a pseudo 3-sided solid orthoconvex grid D and a diagonal. We can search the region of D that is bounded from the diagonal and the part of the boundary that contains exactly 1 corner using the modified edge searching with a number of searchers equal to the points of the diagonal and in such a way that one searcher starts from each point of the diagonal.* ☐

Lemma 12. *Any solid grid D that has b boundary nodes can be searched with b searchers in such a way that one searcher starts at every boundary node, or, one searcher ends at every boundary node.* ☐

Using the above lemmas, we can prove [8] the following theorem that will enable us to obtain a closed formula for $mes()$.

Theorem 13. *There is an optimal searching strategy for the modified edge searching problem on a pseudo 3-sided solid orthoconvex grid which has the property that during the search there is only one contaminated area.* □

Let $d(n_1, s)$ be the length of a shortest diagonal from the boundary node n_1 to side s. $d(n_1, s) = 0$ if n_1 is on side s. If node n_1 is a corner then let $s(n_1)$ denote the side which is opposite of it. Let e denote a boundary edge (n_1, n_2) or a path $(n_1, n_i), (n_i, n_2)$ where n_i is of degree 2. Let $S_a(e)$ to denote the side that is following the one that e lies on if we move from n_1 to n_2, and $S_b(e)$ to denote the side that is following that e lies on if we move from n_2 to n_1. In order for $S_a()$ and $S_b()$ to be well defined, e must not be a path of length 2 that contains a corner. In that case (e lies on two sides) we define $S_a(e)$ ($= S_b(e)$) to be the third side of the pseudo 3-sided solid orthoconvex grid.

Based on Theorem 13 we prove:

Theorem 14. *The minimum number of searchers that are required in order to solve a modified edge searching problem on a pseudo 3-sided solid orthoconvex grid D is given by:*

$$mes(D) = \min(corner\_distance, diagonal\_pair\_distance)$$

where, $corner\_distance = \min\{d(c, s(c))\}$ over any corner c (out of 3 possible) and $diagonal\_pair\_distance = \min\{d(n_1, S_b(e)) + d(n_2, S_a(e))\}$ over any boundary edge $e = (n_1, n_2)$ or any path $e = <(n_1, n_i), (n_i, n_2)>$ where n_i is of degree 2. □

Up to now, we have determined $mes(D)$, the minimum number of searchers that are required in order to solve the modified edge searching problem on a pseudo 3-sided solid orthoconvex grid. By Lemma 6 we know that at most $mes(D) + 1$ searchers can solve the original edge searching problem. So, we have a way to approximate $es(D)$ within 1 from the optimum. Unfortunately, as Figure 2 demonstrates, there are grids that can be searched optimally with the same number of searchers on both problems. So, in order to get an algorithm that computes $es(D)$ we have to identify these grids.

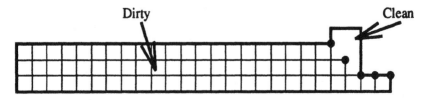

Fig. 2. A grid D for which $mes(D) = es(D) = 5$.

Lemma 15. *If there is a searching strategy that solves the edge searching problem on grid D that contains no path of the form $(n_1, n_i), (n_i, n_j), (n_j, n_2)$ where $degree(n_i) = degree(n_j) = 2$ using k searchers, $k > 2$, then there is a searching strategy that solves the modified edge searching problem on grid D using at most $k - 1$ searchers.* □

Lemma 16. *Assume a pseudo 3-sided solid orthoconvex grid D that contains a path of the form $(n_1, n_i), (n_i, n_j), (n_j, n_2)$ where $degree(n_i) = degree(n_j) = 2$ and n_j is a corner, and the diagonal from n_j to the opposite side has $mes(D)$ points. Then, $mes(D) = es(D)$ if and only if there are two points in the diagonal with the same X or Y-coordinate.* □

Theorem 17. *Assume that the minimum number of searchers that are required to solve the modified edge searching problem on a pseudo 3-sided solid orthoconvex grid D is $mes(D)$. Then, in the case where there is a diagonal from a corner c of D to the opposite side of length $mes(D)$ such that c is next to a degree 2 node and that diagonal has two points with the same X or Y-coordinate, $es(D) = mes(D)$. Otherwise, $es(D) = mes(D) + 1$.* □

4 Algorithms for determining es(D)

In this section we present algorithms to determine $es(D)$. The algorithms are based on Theorems 14 and 17. Through their proofs [8], these theorems also suggest an optimum searching strategy that consists of $O(n)$ actions where n is the number of nodes of the grid D.

It is customary to express the complexity of an algorithm that determines the minimum number of searchers that are required to search a graph as a function of n. However, for a grid we can define two new quantities: the number of boundary nodes b and the number of turning points m. Obviously m turning nodes can completely define the grid and thus it is desired to express the time complexity of the algorithm that determines $es(D)$ as a function of m. Observe that there are grids with $m = o(b)$ and $b = o(n)$.

A quantity that we have to compute is the distance from a boundary node v to some side of the grid. In the rest of the paper we assume that the base of the grid is parallel to the X axis. Informally, we describe how this can be done when the boundary node is on the rising side and we want to compute is distance to the falling side. All other cases are handled in a similar way. We move from node v diagonally up (on a line parallel with $l : x = y$) until we hit the boundary. If we hit the wanted side we are done. If not, we move to the right at the next concave turning point. We then move diagonally up, and so on. From the above discussion, it becomes obvious that we need to compute the distance from any concave turning point to any side.

The following lemma is important for deriving efficient algorithms:

Lemma 18. *Assume a pseudo 3-sided solid grid D that has its base parallel to the X axis. In order to determine $mes(D)$ as Theorem 14 indicates, it is* *sufficient to examine diagonals that start from i) turning nodes, ii) boundary nodes that are adjacent to turning nodes, and iii) the nodes that are at the intersection of the base and all lines that pass from concave turning nodes on the rising side and are parallel with $l : y = -x$.*

It is trivial to compute the minimum diagonal between any boundary node and the base of the grid since one minimum diagonal will be parallel to the Y axis. The nontrivial part is to compute the the minimum diagonal from a concave turning node that lies on the the rising (falling) side to the falling (rising) side. *Algorithm_1* [8] supports the following lemma:

Lemma 19. *Given the side s of a pseudo 3-sided solid orthoconvex grid D that contains m turning nodes, we can compute the length of the minimum diagonals between all turning nodes x and s in $O(m^2)$ steps.* □

We now proceed to construct *Algorithm_2* that computes $es(D)$ based on $mes(D)$ as defined in Theorem 14.

From Lemma 18, we know that we can concentrate only at $O(m)$ nodes of the grid. Recall the definitions from Section 3.2 (following Theorem 13).

Algorithm_2

1. Compute the length of the minimum diagonals between all nodes v indicated in Lemma 18, and any side s.
2. $corner\_distance = \min\{d(c, s(c))\}$ over any corner (out of the 3 possible) c.
3. $diagonal\_pair\_distance = \min\{d(n_1, S_a(e)) + d(n_2, S_b(e))\}$ over any boundary edge $e = (n_1, n_2)$ adjacent to a node specified in Lemma 18, or any path $e =< (n_1, n_i), (n_i, n_2) >$ where n_i is of degree 2 and thus, a convex turning node).
4. $mes(D) = \min(corner\_distance, diagonal\_pair\_distance)$.
5. Determine $es(D)$ from $mes(D)$ based on Theorem 17.

Lemma 20. *Given a pseudo 3-sided solid orthoconvex grid D, $es(D)$ can be determined in $O(m^2)$.* □

Algorithm_2 can be parallelized to yield a parallel algorithm that runs in polylog time using a polynomial number of processors. So we can state:

Theorem 21. *Given a pseudo 3-sided solid orthocnvex grid D, the problem of determining $es(D)$ is in NC.* □

We can modify *Algorithm_1* to compute the length of the minimum diagonals between all turning points v and any side s in $O(m)$ time by a more complicated method that is not easy (if possible) to be parallelized [8]. This improvement leads to an optimum algorithm and together with the fact that Theorem 14 suggests a searching strategy, our final result can be stated as:

Theorem 22. *Given a pseudo 3-sided solid orthocnvex grid D, an optimum edge searching strategy $S(D)$ that consists of $O(n)$ actions can be constructed. It uses $es(D)$ searchers where, $es(D)$ can be computed optimally in $O(m)$ time.*

5 Conclusion

We have shown how to compute the searching number for the class of the pseudo 3-sided solid orthoconvex grids. Our algorithms also suggest an optimal searching strategy. It is not clear how to search any orthoconvex grid. Especially, the property that allowed us to design our algorithms does not hold for every orthoconvex grid. We are able to create orthoconvex grids that cannot be searched by maintaining only one clean area during the searching. However, our conjecture is that there are optimum searching strategies that create at most two dirty areas. This can lead to algorithms for computing the searching number of any solid orthoconvex grid. Other interesting problems are i) how to search general solid grids(non orthoconvexes), and ii) how the existence of holes in the grid can effect the complexity of the problem.

References

1. J.A. Ellis, I.H. Sudbourough and J.S. Turner, "Graph Separation and Search Number".
2. L.M. Kirousis and C.H. Papadimitriou, "Searching and Pebbling", Theoretical Computer Science, 47, 1986, pp. 205-218.
3. M.R. Fellow and M.A. Langston, "Nonconstructive Tools for Proving Polynomial-Time Decidability", Journal of the Association for Computing Machinery, Vol. 35, No. 3, July 1988, pp. 727-739.
4. A.S. LaPaugh, "Recontamination Does Not Help to Search a Graph", Technical Report, Electrical engineering and Computer Science Department, Princeton University, 1983.
5. F. Makedon and I.H. Sudborough, "On Minimizing Width in Linear Layouts", Discrete Applied Mathematics, 23, 1989, pp. 243-265.
6. N. Megiddo, S.L. Hakimi, M.R. Garey, D.S. Johnson and C.H. Papadimitriou, "The complexity of Searching a Graph", Proceedings of the 22nd IEEE Foundations of Computer Science Symposium, 1981, pp. 376-385.
7. T.D. Parsons, "Pursuit Evasion in a Graph", in Theory and Applications of Graphs, Y. Alavi and D.R. Lick, eds., Springer Verlag, 1976, pp.426-441.
8. A. Symvonis and S. Tragoudas, "Searching a Solid Pseudo 3-Sided Orthoconvex Grid", Technical Report 438, Basser Department of Computer Science, University of Sydney, May 1992.
9. K. Sugihara and I. Suzuki, "Optimal Algorithms for a Pursuit-Evasion Problem in Grids", SIAM Journal of Discrete Mathematics, Vol. 2, No. 1, February 1989, pp. 126-143.

An Efficient Parallel Algorithm for Geometrically Characterising Drawings of a Class of 3-D Objects[1]

Nick D. Dendris[2,4], Iannis A. Kalafatis[2,4] and Lefteris M. Kirousis[2,3,4]

Abstract

Labelling the lines of a planar line drawing of a 3-D object in a way that reflects the geometric properties of the object is a much studied problem in computer vision and considered to be an important step towards understanding the object from its 2-D drawing. Combinatorially, the labellability problem is a restricted version of the Constraint Satisfaction Problem and has been shown to be NP-complete even for images of polyhedral scenes. In this paper, we examine scenes that consist of a set of objects each obtained by rotating a polygon around an arbitrary axis. The objects are allowed to arbitrarily intersect or overlay. We show that for these scenes, there is a sequential linear-time labelling algorithm. Moreover, we show that the algorithm has a fast parallel version that executes in $O(\log^4 n)$ time on a EREW PRAM with $O(n^3/\log^2 n)$ processors. The algorithm not only answers the decision problem of labellability, but also produces a legal labelling, if there is one. This parallel algorithm should be contrasted with the techniques of dealing with special cases of the constraint satisfaction problem. These techniques employ an effective, but inherently sequential, relaxation procedure in order to restrict the domains of the variables.

1 Introduction

A planar line drawing (the image) of a 3-D object (the scene) is a planar graph whose lines correspond to a depth or orientation discontinuity in the scene. We assume that the drawing does not carry any information about texture, shadows, lighting, etc. The scene, in general, is assumed to consist of opaque, solid objects. Moreover, as is often done in the literature, we assume the scene to be *trihedral* and *non-degenerate*. Trihedral means that any vertex of the scene is the intersection of at most three surfaces, while non-degenerate means that zero-width solids and cracks are not allowed. The line drawing is obtained by projecting the edges and vertices of the scene onto a plane via an orthographic projection. We assume that the projection plane satisfies the restriction of the *general viewpoint*, i.e., the topology of the image graph will not change if the scene is infinitesimally perturbed.

For scenes as above whose faces are planar (i.e., polyhedra), Clowes [1971] and Huffman [1971] introduced a labelling scheme of the lines of their images that reflect certain geometric properties of the projected scene. For example, the label "+" on a line means that the corresponding edge is convex as seen from the projection plane, "−" means that it is concave and the label "→" means that the corresponding edge reflects a depth discontinuity where one surface occludes another (the direction of the arrow is such as to leave the occluding surface to the right). This labelling scheme was generalized by Malik [1987] to objects bounded by piecewise smooth surfaces. Of course, in images of curved objects, the lines of the image graph are in general not straight lines, but curves. Now, the requirement that the image is the projection of a 3-D object belonging to the class of allowed scenes and not an "impossible object" imposes severe constraints on the possible labellings of the lines that are incident onto a junction of the

---

[1] This research was partially supported by the ESPRIT II Basic Research Actions Program of the EC under contract no. 3075 (project ALCOM).

[2] Department of Computer Engineering and Informatics, University of Patras, P.O. Box 1045, Patras 26110, Greece.

[3] Computer Technology Institute, P.O. Box 1122, Patras 26110, Greece.

[4] E-addresses: {dendris, kalafat, kirousis}@grpatvx1.bitnet

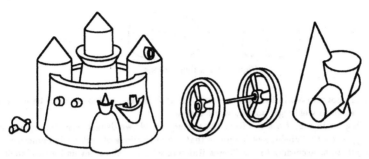

Figure 1: A scene from the permissible class of 3-D objects.

image. Thus the labellability problem becomes a special case of the combinatorial constraint satisfaction problem. The importance of obtaining a legal labelling—i.e. a labelling satisfying the constraints imposed by the geometry of the projected object—towards solving the practically more important problem of fully realizing the projected object has been demonstrated by many researchers (see, e.g. [Malik, 1987]). This importance stems mainly from the fact that many algorithms that realize an object from a line drawing depend on a labelling preprocessing stage (see e.g., [Sugihara, 1984], [Malik and Maydan, 1989]).

However, the labellability problem has been shown to be NP-complete even for images with planar surfaces ([Kirousis and Papadimitriou, 1988]). To overcome this difficulty, one can either use the methods developed in order to deal with the general constraint satisfaction problem or, otherwise, investigate classes of accepted scenes ("worlds") for which efficient labelling algorithms can be designed. The first approach usually makes use of algorithms that do not produce a globally legal labelling, but rather restrict the possible labels of each junction so that any remaining label can participate in a labelling which is only locally consistent with the constraints. This "relaxation" procedure, introduced by Waltz [1975], has been extensively investigated (see e.g., [Mackworth and Freuder, 1985] and [Montanari and Rossi, 1991]). Of course, cases where the relaxation leads to complete solutions have also been studied. However, the relaxation procedure has been shown to be P-complete [Kasif, 1990]. Therefore, algorithms using this method are not amenable to parallelism. The second approach is to restrict the objects that may appear in the scene in a way that leads to efficient algorithms. We believe that this approach is practically important, since artificial vision is usually applied to restricted environments. This method was used in [Kirousis and Papadimitriou, 1988], where a labelling and realizing algorithm for the Manhattan world was obtained (the Manhattan world comprises only of polyhedra whose faces are parallel to one of the Cartesian planes). In [Alevizos, 1991], instead of restricting the possible world, it is assumed that information about the hidden edges of the scene is provided.

In this paper we follow the restricted world approach, but we allow curved objects as well. Moreover, we do not impose any restriction on the orientation of the objects. Specifically, we restrict our scene to comprise of a set of *arbitrarily intersecting or overlaying* objects each obtained by rotating an arbitrary polygon (with edges that are straight-line segments) around an arbitrary axis. The objects do not intersect with the projection plane. We call this permissible universe the **pottery world** (see Figure 1). Following Malik's [1987] approach, in our model, solid or hole cone apices must not belong to any other surface, for example, scenes like the one

Figure 2: An illegal scene.

in Figure 2 are not allowed in the pottery world. In applications, such scenes can be processed in an earlier stage to yield legal pottery world scenes.

In labelling an image of a scene from the pottery world, we depart from the approach of Malik [1987], where the labelling of junctions that are projections of vertices is reduced to the labelling of junctions according to the Clowes-Huffman scheme. We rather find which labellings of such junctions can be realized as projections of scenes from the pottery world. Thus the set of labels of the Clowes-Huffman scheme is restricted even further. Moreover, the propagation of the labels throughout the image graph is not faced as a purely combinatorial problem of combining the set of legal labels at each junction. Instead, the restriction that the image must be realizable as an object in the permissible world is used to exclude combinations of legal labels on various types of components of the image graph. It is this use of the realizability requirement to exclude not only labellings of a junction, but also combinations of labellings on collections of junctions that makes possible the avoidance of the combinatorial explosion. As far as we know, this method was used only in [Kirousis, 1990], but in a very restricted manner. Here we make strong use of it. As far the complexity of our algorithm is concerned, it is proved that it requires linear sequential time. Moreover, we show that it has a fast parallel version that executes in time $O(\log^4 n)$ on a EREW PRAM with $O(n^3/\log^2 n)$ processors. The algorithm not only answers the decision problem but also finds a legal labelling, if there is one. This algorithm makes use of the fast parallel algorithms that produce a maximal independent set of vertices of a graph. Finally, it must be pointed out that our algorithms are easily implementable.

2 Constraint Analysis

2.1 Unary Constraints

The Clowes-Huffman-Malik labelling scheme assigns a label to each line of the image graph according to the way the corresponding edge of the scene is seen from the projection plane: For a **connecting** edge (not belonging to a contour), if it is convex as seen from the projection plane, its projection is labelled by a "+", while if it is concave, its projection is labelled by a "−". For a **contour** edge, where one surface occludes another, its projection is labelled by an "→" (the direction of the arrow is such as to leave the occluding surface to the right). Finally, a **limb** is labelled by a "≫". A limb is not the projection of a "physical" edge. It results when a curved surface occludes itself and the line of sight is tangential to it for all points on the limb (the direction of the "≫" is such as to leave the occluding surface to the right). See Figure 3 for an example of all labels. We also use a generalized notion of a label, a **multi-label**. A multi-label is a set of possible labels which will be processed at a later stage to yield one label from the set (see [Malik, 1987]). The two multi-labels that we introduce are $\{-, \rightarrow\}$ and $\{+, \leftarrow\}$. A line that can be labelled by only one of these multi-labels can get an arrow-label of a uniquely determined direction, whereas one that can be labelled with both multi-labels can get arrow-labels of both directions (this is the reason that the arrows of the two multi-labels are in opposite directions).

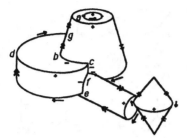

Figure 3: A labelled scene.

Now, the junctions of the image graph are classified as follows (Malik [1987]):

- L-junction: Tangent discontinuity across the junction (e.g. junction e of Fig. 3).

- Curvature-L-junction: Tangent continuity but curvature discontinuity across the junction. (e.g. junction b of Fig. 3).

- T-junction: Two lines have tangent and curvature continuity at the junction. (e.g. junction g of Fig. 3).

- Phantom junction or pseudo-junction: It is not the projection a vertex of the scene and is devised only to indicate the change of label along an line (e.g. junction a of Fig. 3).

- Three-tangent-junction: Three curves with common tangent. Two of them have the same curvature (e.g. junction d of Fig. 3).

- E- or arrow-junction: Three curves with distinct tangents. One angle between two of them is $> \pi$ (e.g. junction c of Fig. 3).

- Y-junction: Three curves with distinct tangents. No angle between them is $> \pi$ (e.g. junction f of Fig. 3).

Figure 4 gives all legal labels on the lines of various types of junctions (modulo rotations in three dimensions). To see that these indeed are all possible labels, we first state three facts that reflect the geometric properties of our restricted world (the proofs of these facts are given in the full paper). The catalogue of legal labels can then be produced by an exhaustive search which exploits these facts to prune the search space.

Fact 1 A line that is a projection of an intersection between two surfaces and is not the projection of a base circle can only be labelled with "−".

Fact 2 A straight line can be assigned only the labels "−" or "➤". Moreover, a limb can occur only in L-, curvature-L- and three-tangent-junctions.

Fact 3 An E-junction in our world, corresponds to an arrangement of volumes either like the ones in Figure 5.a or like the ones in Figure 5.b Thus it can be labelled only in two ways, with $(+, -, +)$, where "−" is the label of the middle line, or with $(-, +, -)$, where "+" is again the label of the middle line.

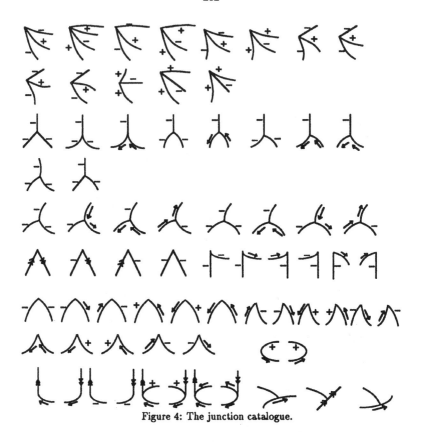

Figure 4: The junction catalogue.

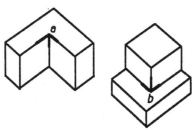

Figure 5: Volume arrangements corresponding to E-junctions.

Figure 6: Junctions that are not included in the component scheme.

Now, by an exhaustive search, we conclude that :

Theorem 1 *Junctions with labels that are not included in the catalogue of Figure 4 cannot be realized as three dimensional objects of the pottery world.*

2.2 Higher Order Constraints

By examining the junction catalogue, one should realize that the problem of labelling a scene of the pottery world cannot be trivially solved in sequential time which polynomially depends on n, the number of junctions of the image graph (since the image graph is planar, this number is of the same order as the number of lines of the image graph).

The labelling problem is a special case of the Constraint Satisfaction Problem (CSP). CSP is usually solved in exponential time using backtracking. Even if a preprocessing stage is applied to the problem, in the way that Mackworth and Freuder [1985] describe, the problem is still prone to exponential explosion, especially when the constraints between the variables are only unary or binary (it is obvious that by requiring two adjacent junctions of the picture graph to share the same label at their common line and by allowing a junction to be labelled according to the junction catalogue, we apply only binary and unary constraints). However, if we take into consideration that the regions of the image graph must be realized as surfaces of the 3-D scene, we get higher order constraints that involve not only adjacent junctions but all lines that delimit a region of the image graph. It is essentially *this fact* that leads to the avoidance of the combinatorial explosion.

To formalize this idea, we introduce the notion of a **component** of an image graph. A component is a subgraph that consists of a maximal connected set of junctions that belong to a certain type. The various types of components are defined below. Our labelling algorithm will at first assign labels to the members of each component separately. However, junctions that can be either uniquely labelled or labelled in more than one way none of which propagates any restriction to the labels of the neighboring junctions need not be (and are not) included in the formation of components. The cases of junctions with a unique labelling can be easily found by inspecting the junction catalogue. The cases of junctions that can be labelled in more than one ways, and yet do not propagate any restriction to neighboring junctions (i.e. junctions with only **local ambiguity**) are depicted in Figure 6 (see full paper for details). Another observation is that a legal labelling of the junction in Figure 7.a, if at all related to the labels of the neighboring junctions, is determined by the neighboring labels to be either one of the labellings in Figure 7.a, or one of the labellings in Figure 7.b. Moreover, the latter case need not be included in the component formation scheme. Finally, observe that phantom junctions do not pose any problem. To see why this is the case, one has to observe that a phantom junction is equivalent to a legal L-junction with a label of $(+, \leftarrow)$ or $(\leftarrow, +)$. In both cases, if such a phantom junction occurs in an L-component, it bears no effect onto the set of possible labels of the component.

We now describe the various types of components.

Figure 7: A special case of an L-junction.

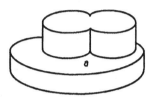

Figure 8: Curvature-L ambiguity.

- Y-component of type 1: Such a component is formed by Y-junctions with exactly one straight line. In these Y-junctions the straight line is always labelled by "−" and the other two can be labelled by either "−" or "→". Thus the legal labellings of such a component are only two and reflect the fact that by looking at a scene like the one in Figure 8, there is an inherent ambiguity that cannot be resolved: do the two cylinders stand on or above the base of the third? This ambiguity will be referred to as the **curvature-L** ambiguity, because it is also observed in curvature-L-junctions.

- Y-component of type 2: Such components are formed by Y-junctions of the type in Figure 9. One should observe that in such a component no label of any line is *a priori* known. However, once a label is found to be an arrow, this arrow will uniquely propagate.

- E-component: Such components include only E-junctions. These, by Fact 3 in the previous subsection, can have two possible labellings: $(+, -, +)$, where "−" is the label of the middle line, or $(-, +, -)$, where "+" is the label of the middle line. Thus, the whole component has two possible labellings; moreover, given the label of a line in the component, the labels of the whole component are uniquely determined.

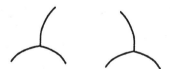

Figure 9: Y-junctions that form Y-components of type 2.

Figure 10: L-junctions that form L-components.

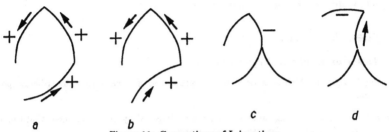

Figure 11: Connections of L-junctions.

- L-components of type 1: This component type is formed by junctions like the ones of Figure 10.a. We use the multi-label $\{+, \leftarrow\}$ for labelling these junctions (recall the case of Figure 7.b).

- L-component of type 2: These components are formed by junctions like the ones in Figure 10.b. By taking into consideration their possible realizations, one can prove that a multi-label (either $\{+, \leftarrow\}$ or $\{-, \rightarrow\}$) on one of their end junctions uniquely propagates to the rest. That is, if the first line in the component can be labelled as a "$+$" or "$\leftarrow$" then these will be the possible labels of the second line and so on. The same applies if the leading junction can be labelled as "$-$" or "$\rightarrow$". So in every case, even if the number of valid labels that can be applied to these components is exponential, the component propagates one multi-label throughout itself (see Figures 11.a and 11.b).

Now, the facts proved above lead to the following rules about the legal labels of various types of components:

- **Rule 1:** Once we know one label of an E-component, the labels of all lines of the component are uniquely determined.

- **Rule 2:** L-components of type 2 connected to an E-component can have two multi-labels if the label of the E-component is not given, or one, if one label in the E-component is given. That is, one can first label the E-component in one of the two possible ways, obtain a label ("$+$" or "$-$") for each line connected to the L-component, and then propagate the apropriate multi-label ($\{+, \leftarrow\}$ or $\{-, \rightarrow\}$, respectively) through the L-component. If a contradiction occurs, one will have to backtrack only once, then re-label the E-component and propagate the new labelling.

- **Rule 3:** Whenever an E-component is connected to a Y-component, the connecting line is labelled as a "$-$". Therefore, the E-component is uniquely labelled.

- **Rule 4:** An L-component of type 2 connected to the same Y-component of type 2 by both its end junctions can have at most two labellings. That is, if the connection is made through a junction like the one in Figure 11.c, the label of the common line is "$-$", else if the connecting junction is like the one in Figure 11.d, the common line is labelled with the multi-label $\{-, \rightarrow\}$. In both cases, the connecting line of the Y-junction is uniquely labelled.

- **Rule 5:** If a Y-component of type 1 is connected to an L-component of type 1, then they both share the curvature-L ambiguity.

3 The Algorithm

3.1 Description of the Algorithm

- *input:* a graph which is an orthographic projection of a scene from the trihedral pottery world.

- *output:* a legal labelling satisfying constraints imposed by the requirement that the image be realized as a scene from the pottery world.

- *step 1:* Disconnect the middle line of all T-junctions from the two lines forming its bar. Apply the following steps separately onto each connected component thus obtained, under the restriction that the two "bar" lines must be labelled by an "$\rightarrow$", if they are curved lines or by a "$\gg$" if they are straight lines.

- *step 2:* Find the various components in the graph as defined in section 2.2.

- *step 3:* Label all junctions that are not part of any component with either a label or a multi-label.

- *step 4:* Propagate the labels of step 3 as far as they can go, by using the label catalogue of Figure 4.

- *step 5:* Label the lines that are shared between Y-components of type 2 on one hand and either E-components or L-components on the other (recall that the lines of Y-components of type 1 are uniquely labelled, except the case of the inherent curvature-L ambiguity). To do this, follow the case analysis below:

  - An E-component connects to a Y-component of type 2. Use Rule 3 to determine the label of the common line.

  - An L-component connects to a Y-component of type 2. Then by use of Rule 4 determine the type of the label of the common line.

- *step 6:* Propagate all labels that were introduced in step 5 of the algorithm, according to the rules and the catalogue.

- *step 7:* By now, all lines shared by two components must be already labelled, except the following case: An E-component is connected with the rest of the graph by means of L-components. Then follow this procedure: give an arbitrary label from the set $\{+, -\}$ to one of the lines of the E-component (say, this is the line a). Thus the whole E-component is uniquely labelled. Propagate the newly introduced labels as far as they can go. If at any stage a contradiction occurs, then label a with the alternative label. Propagate all labels as far as they can go. If any other contradiction is reached then the graph is not labelable. Note that this backtracking will not give rise to an exponential algorithm.

- *step 8:* Label the lines of the Y-components that are not shared with another component and which are not determined by already existing arrows, by arbitrarily giving to the whole set of lines the label "−".

- *step 9:* The final step of the algorithm concerns the labelling of the internal lines of an L-component. In this case, the label "−" can be arbitrarily assigned to the lines of the component, if their multi-label is $\{-, \rightarrow\}$. If on the other hand, their multi-label is $\{+, \leftarrow\}$, then the label "+" is assigned to all lines.

If at any stage of the above algorithm (except in step 7) we come to a label contradiction on any line, then the graph is not labelable and we stop the algorithm.

3.2 The Complexity of the Algorithm

First observe that since the image graph is planar, its number of lines is $O(n)$, where n is the number of its junctions. Therefore, the tasks of finding any type of components can be performed sequentially in time $O(n)$ and in parallel in time $O(\log^2 n)$ on a EREW PRAM using $o(n)$ processors. The labelling of junctions that are not part of components can of course be performed sequentially in linear time and in parallel with a linear number of processors in constant time. The propagation of labels is also performed sequentially in linear time (observe that the bactracking employed in connections of E-components with L-components does not destroy the linearity, since if the label of a junction does not lead to an inconsistency while propagated as far as possible, it will *not* be changed later on during the execution of the algorithm). Parallelwise, these steps can be executed by the standard technique of employing Boolean matrix multiplication in order to find the transitive closure of a directed graph (here the transitive closure corresponds to the propagation of a label). Therefore these steps can be performed in $O(\log^2 n)$ time on a EREW PRAM using $O(M(n))$ processors, where $M(n) = O(n^{2.376})$ (see, e.g., [Karp and Ramachandran, 1990]). However, as observed in [Karp and Wigderson, 1985] for the problem of 2SAT, the transitive closure technique leads only to the solution of the *decision* problem. To find a legal labelling if there is one, we have to use the technique of finding a maximal independent set. For that, we first construct a directed *implication graph*. That graph has a junction for each legal labelling l_e of each line e of the image graph. Also, in the implication graph, there is an edge from l_e to $l'_{e'}$ if the label l_e of line e propagates on the line e' as the label $l'_{e'}$. The implication graph can be constructed in $O(\log^2 n)$ time using $O(M(n))$ processors by the transitive closure technique. What we do next is to construct a *conflict graph*. The conflict graph has the same set of junctions as the implication graph, but an (undirected) edge connects every pair of labels that are incompatible according to the implication graph. Now a maximal independent set of junctions for the conflict graph gives a legal labelling for the image graph (if the maximal independent set has cardinality less than the number of lines in the image graph then the image graph is not labellable). Given the implication graph, the construction of the conflict graph requires constant time with $O(n^3)$ processors. Now, by the fast parallel algorithm in [Goldberg and Spencer, 1989], the construction of a maximal independent set of the conflict graph requires $O(\log^4 n)$ time and $O(n^2)$ processors (the number of processors required by this algorithm is linear in the number of edges). Putting the above together, we get that:

Theorem 2 *The problem of deciding whether a scene from the pottery world is labellable and of finding a legal labelling if there is one can be solved in linear sequential time and in $O(\log^4 n)$ time on a EREW PRAM using $O(n^3/\log^2 n)$ processors.*

Acknowledgment

We thank Paul Spirakis for many illuminating conversations.

References

P. Alevizos, "A linear algorithm for labeling planar projections of polyhedra," Proc. IEEE/RSJ Int. Workshop on Intelligent Robots and Systems (Osaka, Japan, 1991), 595–601.

M.B. Clowes, "On seeing things," *Artifical Intelligence*, 2 (1971), 79–116.

M. Goldberg and T. Spencer, "A new parallel algorithm for the maximal independent set problem, " *SIAM J. Computing*, 18 (1989), 419–427.

D.A. Huffman, "Impossible objects as nonsense sentences," *Machine Intelligence*, 6 (1971), 295–323.

R.M. Karp and V. Ramachandran, "Parallel algorithms for shared-memory machines," in: J. van Leeuwen (ed.), Handbook of Theoretical Computer Science, vol. A (Elsevier, 1990) 869–942.

R.M. Karp and A. Wigderson, "A fast parallel algorithm for the maximal independent set problem," *J. Association of Computing Machinery*, 32 (1985), 762–773.

S. Kasif, "On the parallel complexity of discrete relaxation in constraint satisfaction networks," *Artificial Intelligence*, 45 (1990), 275–286.

L.M. Kirousis and C.H. Papadimitriou, "The complexity of recognizing polyhedral scenes," *Journal of Computer and System Sciences*, 37 (1988) 14–38.

L.M. Kirousis, "Effectively labeling planar projections of polyhedra," *IEEE Transactions on Pattern Analysis and Machine Intelligence* 12 (1990) 123–130.

M. Luby, "A simple parallel algorithm for the maximal independent set problem," *SIAM J. Computing* 15 (1986), 1036–1053.

J. Malik, "Interpreting line drawings of curved objects," *International J. of Computer Vision*, 1 (1987), 73–103.

J. Malik and D. Maydan, "Recovering three-dimensional shape from a single image of curved objects," *IEEE Transactions on Pattern Analysis and Machine Intelligence* 11 (1989) 555–566.

A.K. Mackworth and E.C. Freuder, "The complexity of some polynomial network consistency algorithms for constraint satisfaction problems," *Artificial Intelligence* 25 (1985), 65–74.

U. Montanari and F. Rossi, "Constraint relaxation may be perfect," *Artificial Intelligence*, 48 (1991), 143–170.

K. Sugihara, "A necessary and sufficient condition for a picture to represent a polyhedral scene," *IEEE Transactions on Pattern Analysis and Machine Intelligence*, 6 (1984), 578–586.

D. Waltz, "Understanding line drawings of scenes with shadows," in: P.H. Winston (ed.), *The Psychology of Computer Vision*, (McGraw-Hill, New York, 1975), 19–91.

Topologically Consistent Algorithms Related to Convex Polyhedra

Kokichi Sugihara

Department of Mathematical Engineering and Information Physics
Faculty of Engineering, University of Tokyo
7-3-1 Hongo, Bunkyo-ku, Tokyo 113, Japan

Abstract The paper presents a general method for the design of numerically robust and topologically consistent geometric algorithms concerning convex polyhedra in the three-dimensional space. A graph is the vertex-edge graph of a convex polyhedron if and only if it is planar and triply connected (Steinitz' theorem). On the basis of this theorem, conventional geometric algorithms are revised in such a way that, no matter how poor the precision in numerical computation may be, the output graph is at least planar and triply connected. The resultant algorithms are robust in the sense that they do not fail in finite-precision arithmetic, and are consistent in the sense that the output is the true solution to a perturbed input.

1. Introduction

Rapid development of computational geometry in the last decade has given a tremendous number of "efficient" algorithms for various geometric problems. However, these algorithms are usually designed on the assumption that no numerical error takes place, and hence the straightforward translation into computer language does not necessarily generate practically valid computer programs because numerical errors are inevitable in ordinary floating-point arithmetic; these algorithms often enter an infinite loop, end abnormally, or give nonsense output.

Recently, the importance of numerical robustness has become widely recognized [5], and many approaches has been proposed. The previous approaches can be classified into three categories. The approaches in the first category use high-precision arithmetic [8, 11]. The precision necessary to judge the topological structure of geometric objects explodes as the degree of polynomials representing the objects becomes higher, and hence the approaches in this category are mainly applied to linear objects such as lines and planes. Still the resultant algorithms require costly computation.

The approaches in the second category start with the assumption that a certain fixed precision is available [4, 6, 7]. The results of numerical computation are divided into reliable and unreliable according to the error analysis based on the assumed precision, and the unreliable results are discarded if they contradict the reliable ones. The algorithms thus designed usually have complicated structures; one of the reasons for this is that they have many thresholds due to the assumed precision and consequently have many branches of processing that have no essential relation to the original geometric problems.

The approaches in the third category, which were proposed by the author's group [10, 12], make no assumption on the numerical precision; any numerical result is considered as not necessarily reliable. The basic part of the algorithm is constructed only in terms of combinatorial computations, and numerical results are used as secondary-priority information to resolve ambiguity in choosing the branches of the processing. The algorithms thus designed are simple because, unlike the first category, they have no exceptional branches for degeneracy and, unlike the second category, they have no additional branches due to nonessential thresholding.

An algorithm is said to be *numerically robust* (or *robust* for short) if it always gives an output that belongs to a certain reasonable output set (in other words, if it does not fail on the way of processing). An algorithm is said to be *topologically consistent* (or *consistent* for short) both if it is robust and if the output is a correct solution of a perturbed problem. Algorithms in the first category are consistent, for which, of course, expensive computational cost should be paid. Algorithms in the second and the third categories are robust, but not necessarily consistent. For example, an algorithm for the line arrangement by Milenkovic [7] (which belongs to the second category) outputs a pseudo-line arrangement, but a pseudo-line arrangement can not necessarily be realized by straight lines, and an algorithm for the Voronoi diagrams by Sugihara and Iri [12] (which belongs to the third category) outputs a planar graph whose dual graph is simple, but such a graph is not necessarily isomorphic to a Voronoi diagram [3]. However, it should be noted that the inconsistency of these algorithms is not because they are designed inappropriately, but because it is a hard, and still open, problem to characterize the line arrangement or a Voronoi diagram purely combinatorially. Thus, to construct a consistent algorithm is difficult in general.

In this paper we present a general framework to design various types of consistent algorithms belonging to the third category for a class of geometric problems whose solutions are convex polyhedra. Such problems include, for example, the construction of three-dimensional convex hulls and the construction of two-dimensional Voronoi diagrams with respect to Laguerre distance.

Our design framework is based on Steinitz' theorem, which characterizes convex polyhedra in a combinatorial manner. We design an algorithm in such a way that the output satisfies at least the combinatorial properties given by the theorem, and thus attain the topological consistency.

Hopcroft and Kahn [6] pointed out that Steinitz' theorem is useful for constructing consistent geometric algorithms, and applied it to design an incremental-type algorithm for intersecting two convex polyhedra. Their approach belongs to the second category, and consequently the resultant algorithm as well as its analysis is complicated, although their approach accompanies the bound of the distance between the output and the correct solution. The present approach, on the other hand, gives simpler algorithms with simpler analysis, because we can separate the consistency issue from the evaluation of numerical errors.

An example of topological inconsistency created by a conventional algorithm is given in Section 2, and Steinitz theorem is reviewed in Section 3. In Section 4 we list typical problems that can be solved by the present approach, and in Section 5 two examples of algorithms for solving those problems will be constructed on the same and simple design principle.

2. Topological Inconsistency Due to Numerical Errors

Fig. 1 depicts a simple example in which geometric computation in finite precision arithmetic generates inconsistency. Suppose that we have pentagonal prism Π with top face f having vertices $p_1, p_2, \cdots, p_5$, as shown in (a), and that we want to cut Π by plane α. Imagine the case where f and α are very close and almost parallel

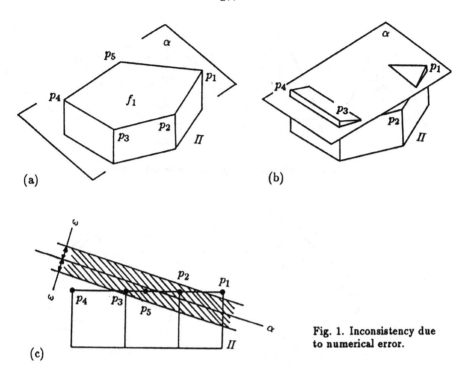

Fig. 1. Inconsistency due to numerical error.

to each other. Numerical errors are involved in the computation to judge whether a vertex is in one side of α or in the other, and hence the judgement is done as if it was at random. Consequently, numerical result may tell that p_1, p_3, p_4 are in one side and p_2, p_5 in the other. Then, we get the configuration shown in Fig. 1(b), which is a contradiction because f and α meet twice although two distinct planes can meet at most once.

A conventional approach to such difficulty is to introduce a positive small number ε, called tolerance, and to judge that p_i is exactly on α if the distance from p_i to α is less than ε. In this approach it might be expected that all the points $p_1, p_2, \cdots, p_5$ are judged being on α. However, if α slants a little as shown in (c), the judgement with tolerance ε tells that p_2, p_3, p_5 are on α while p_1 and p_4 are in the mutually opposite sides of α. This is a contradiction because the three points being on α implies that f is completely contained in α and consequently the other two points should also be on α. Note that such a contradictory situation can happen for any value of ε. Thus, an algorithm that is correct in infinite precision arithmetic is not necessarily valid in finite precision arithmetic.

3. Steinitz' Theorem

Here we will review Steinitz' theorem and prove a slightly extended version of the theorem. These theorems characterize the graph structure of the convex polyhedron, and are used as a fundamental tool for the design of consistent algorithms.

For four reals a, b, c, d satisfying $(a, b, c) \neq (0, 0, 0)$, set $H = \{(x, y, z) \mid ax + by + cz + d \leq 0\}$ is called a *half space* defined by a, b, c, d, and the vector (a, b, c) is

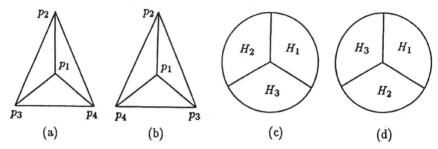

Fig. 2. Vertex-edge graphs of simple polyhedra.

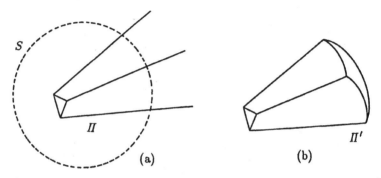

Fig. 3. Representation of the graph structure of an unbounded polyhedra.

called the *outward normal* to the half space. We define $\partial H = \{(x, y, z) \mid ax + by + cz + d = 0\}$. The intersection $H_1 \cap H_2 \cap \cdots \cap H_k$ of a finite number of half spaces $H_1, H_2, \cdots, H_k$ is called a *convex polyhedron* (or *polyhedron* for short). We restrict our consideration to the polyhedron satisfying the following conditions.

(C1) $k \geq 3$, and the set $\{H_1, H_2, \cdots, H_k\}$ includes three half spaces whose outward normals are linearly independent.

(C2) $H_1 \cap H_2 \cap \cdots \cap H_k$ is not a subset of any plane.

The condition (C1) excludes the intersection of two or less half spaces, and also excludes the case where the boundaries (planes) of all the half spaces are parallel to a common line. On the other hand, the condition (C2) excludes the cases where the intersection shrinks to a plane, a line, a point or an empty set. In what follows, by a polyhedron we mean the intersection $H_1 \cap H_2 \cap \cdots \cap H_k$ that satisfies the conditions (C1) and (C2). Hence, in particular, the boundary of a polyhedron has at least one vertex, and all the edges and the vertices form one connected structure.

A polyhedron is said to be *bounded* if there exists a sphere that encloses the polyhedron inside, and *unbounded* otherwise. The vertices and the edges of a bounded polyhedron can be considered as a graph drawn on the boundary of the polyhedron; we call this graph the *vertex-edge graph* of the (bounded) polyhedron. In what follows we draw the vertex-edge graph on the plane in the way we see the polyhedron from outside. Hence, for example, the convex hull of four points p_1, p_2, p_3, p_4 can generate two distinct vertex-face graph, as shown in Fig. 2(a) and (b), depending on the orientation of the four points.

A graph is called *planar* if we can draw the graph on the plane in such a way that the edges do not intersect except at the end vertices. A graph is called *triply connected* both if it has four or more vertices and if the deletion of any two vertices and all the edges incident to them does not make the remaining graph disconnected. A triply connected planar graph is related to a bounded polyhedron by the following theorem.

Theorem 1 (Steinitz' theorem [9]). A graph is the vertex-edge graph of a bounded convex polyhedron if and only if it is planar and triply connected.

This theorem can be easily extended to unbounded polyhedra. Let Π be an unbounded polyhedron. Π has unbounded edges. In order to treat them as edges of an ordinary graph, we introduce the following convention. As shown in Fig. 3(a), consider sphere S that is large enough so as to contain all the vertices of Π inside S, and cut off the portion of Π that is outside S, as shown in (b). Let Π' be the resultant set of points. Π' has a "new (curved) face" that is part of S, and the boundary of this face is a cycle composed of "new edges" and "new vertices"; unbounded edges and unbounded faces of Π are cut off at these new vertices and new edges. Thus, the resultant vertices and edges (including the new vertices and edges) form a graph in an ordinary sense. We call this graph the *vertex-edge graph* of Π. In this graph, the cycle corresponding to the boundary of the "new face" is distinguished from others and is called the *cycle at infinity*; also the "new vertices" and the "new edges", which are on the cycle at infinity, are called *vertices at infinity* and *edges at infinity*, respectively. In what follows we draw the vertex-edge graph of an unbounded polyhedron in such a way that the cycle at infinity is the outermost circle enclosing all the other part of the graph. For example, $H_1 \cap H_2 \cap H_3$ can generate the two distinct vertex-edge graphs as shown in Fig. 2(c) and (d).

Let G be a triply connected planar graph. When we draw G on the sphere without any intersection of edges except at the end vertices, the sphere is partitioned into connected regions bounded by the edges of G. Let us call such a connected region a *face*, and the boundary of a face a *face cycle*. It is known that there are exactly two topologically distinct ways of drawing G on the sphere (if G is planar and triply connected), and that one is the mirror image of the other [13]. Hence, whether a cycle of G is a face cycle or not does not depend on the way of drawing G. A face cycle is said to be of *degree three* if all the vertices on the cycle are of degree three.

Obviously, the vertex-edge graph of an unbounded convex polyhedron is a triply connected planar graph, and the cycle at infinity is a degree-three face cycle of this graph. The converse is also true as stated in the next theorem.

Theorem 2. Let G be a triply connected planar graph and c be a degree-three face cycle of G. Then, there exists an unbounded convex polyhedron whose vertex-edge graph is isomorphic to G and whose cycle at infinity corresponds to c.
Proof. Suppose that G and c are as stated in the theorem. From Theorem 1, there exists a bounded convex polyhedron, say Π, whose vertex-edge graph is isomorphic to G. Let f_0 be the face of Π such that the boundary of f_0 corresponds to c, and let ∂H_0 be the plane containing f_0. Let p be a point outside Π from which f_0 is visible but all the other faces are invisible (see Fig. 4(a)). We transform the whole space by the projective transformation such that the points on ∂H_0 are mapped to

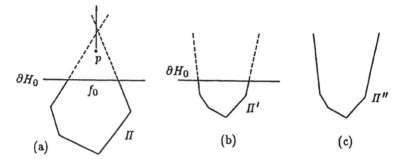

Fig. 4. Transformation from a bounded polyhedron to an unbounded polyhedron.

themselves and p is mapped to a point at infinity in the direction of the outward normal to f_0. Let Π' be the resultant polyhedron. Then, as shown in Fig. 4(b), all the faces of Π' except f_0 are invisible from the point at infinity in the direction of the outward normal of f_0. Let $\Pi' = H_0 \cap H_1 \cap \cdots \cap H_k$ where ∂H_0 contains f_0 and, for any $1 \leq i \leq k$, ∂H_i contains one of faces of Π' (i.e. $\{H_0, H_1, \cdots, H_k\}$ contains no redundant half space). Let Π'' be the polyhedron defined by $\Pi'' = H_1 \cap \cdots \cap H_k$, as shown in (c). Π'' is an unbounded convex polyhedron. Moreover, since c is of degree three, the vertex-edge graph of Π'' is isomorphic to G and whose cycle at infinity corresponds to c. $\qquad\square$

4. Geometric Problems Related to Convex Polyhedra

Here we list geometric problems whose solution set is the set of convex polyhedra.

Problem 1 (Convex hull). Given finite set P of four or more points in the three-dimensional space, construct the convex hull of P.

Problem 2 (Intersection of two convex polyhedra). Given two bounded convex polyhedra, construct their intersection.

Problem 3 (Intersection of half spaces). Given three or more, but a finite number of, half spaces, construct their intersection.

For circle c_i centered at (x_i, y_i) with radius r_i and for point $p = (x, y)$ in the plane R^2, *Laguerre distance*, denoted by $d_L(p, c_i)$, from p to c_i is defined by

$$d_L(p, c_i) = (x - x_i)^2 + (y - y_i)^2 - r_i^2. \qquad (1)$$

Let $C = \{c_1, c_2, \cdots, c_n\}$ be a set of n circles in the plane, and let us define

$$R_L(c_i) = \{p \in R^2 \mid d_L(p, c_i) < d_L(p, c_j), \ j \neq i\}. \qquad (2)$$

The regions $R_L(c_1), \cdots, R_L(c_n)$ determine the partition of the plane; this partition is called the *Voronoi diagram* for C *with respect to Laguerre distance* (or *Laguerre Voronoi diagram* for C, for short).

Problem 4 (Voronoi diagram in Laguerre geometry). Given a finite number of circles in the plane, construct the Laguerre Voronoi diagram for these circles.

It is not obvious that this problem is related to convex polyhedra. The relationship is established in the following manner. For circle $c_i \in C$, we define half space $H(c_i)$ in the three-dimensional space by

$$z \geq 2x_i x + 2y_i y - x_i^2 - y_i^2 + r_i^2 \quad (= -d_L(p, c_i) - x^2 - y^2). \tag{3}$$

Then, the orthogonal projection of $H(c_1) \cap H(c_2) \cap \cdots \cap H(c_n)$ onto the x-y plane coincides with the Laguerre Voronoi diagram for C [1]. Thus the construction of the Laguerre Voronoi diagram can be reduced to Problem 3.

In each of the four problems, the solution is a convex polyhedron, and moreover any convex polyhedron is the solution of some instance of the problem.

The construction of the ordinary Voronoï diagram can also be reduced to the intersection of half spaces [2], but the converse is not true. Since the half spaces are restricted to some special type, an arbitrary unbounded polyhedron does not necessarily correspond to an ordinary Voronoi diagram [3]. Hence, unlike the Laguerre Voronoi diagram, Theorems 1 and 2 are not enough to characterize the ordinary Voronoi diagram.

5. Topologically Consistent Algorithms

Our basic idea for constructing a consistent algorithm is to describe the fundamental part of the algorithm by only combinatorial computations in such a way that the output satisfies the combinatorial conditions given in Theorem 1 or 2. Such a description of processing has ambiguity in the choice of branches, and we use numerical computation in order to choose the branch that is most likely to lead to the correct solution.

First, consider an incremental method for Problem 3. Assume that we already constructed $\Pi_i = H_1 \cap H_2 \cap \cdots \cap H_i$ and now want to construct $\Pi_{i+1} = \Pi_i \cap H_{i+1}$. Let $G = (V, E)$ be the vertex-edge graph of Π_i with the vertex set V and the edge set E, and U be the set of vertices of G that are outside H_{i+1}. For $X \subseteq V$, let $G(X)$ be the subgraph of G composed of the vertices in X and the edges whose both end vertices belong to X. U satisfies the following two conditions.

(U1) Both $G(U)$ and $G(V - U)$ are connected.
(U2) For any face cycle c of G, the vertices and the edges belonging to both c and $G(U)$ form a connected graph.

The condition (U1) comes from that the substructure composed of the vertices and edges in each side of the plane ∂H_{i+1} is connected. The condition (U2) comes from that ∂H_{i+1} can cut off at most one connected portion of each face of Π_i. For any triply connected planar graph G, vertex set U of G is called a π-set if U satisfies (U1) and (U2).

As above, suppose that U is the set of vertices of G that are outside H_{i+1}. When the polyhedron Π_i is cut off by the plane ∂H_{i+1}, the associated vertex-edge graph changes accordingly. The change of the graph is depicted in Fig. 5. Suppose that the vertices represented by solid circles in Fig. 5(a) belong to U. We first generate new vertices (as represented by hollow circles) on the edges connecting vertices in U and vertices outside U, partitioning the edges into two edges. We next generate a cycle (as represented by broken lines) by new edges connecting new vertices, and finally remove the subgraph inside the new circle. The resultant graph is as shown in Fig. 5(b). This way of changing the graph shall be called the *face cut with respect to U*.

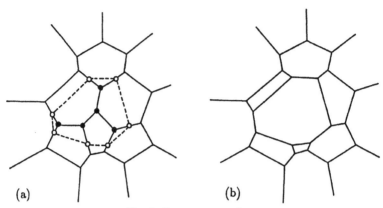

Fig. 5. Face-cut operation.

Now we are ready to describe our first algorithm. In the algorithm, the basic part is described in terms of combinatorial procedure, and numerical procedure to choose one among the possible branches is added in brackets.

Algorithm 1 (Incremental intersection of half spaces).
Input: n half spaces $H_1, H_2, \cdots, H_n$.
Aimed output: Vertex-edge graph of $H_1 \cap H_2 \cap \cdots \cap H_n$.
Procedure:
1. Choose one of the two vertex-edge graphs shown in Fig. 2(c) and (d) [which is the vertex-edge graph of $H_1 \cap H_2 \cap H_3$], and name it G.
2. For $i = 4, 5, \cdots, n$, do 2.1 and 2.2:
    2.1. choose π-set U of G [such that U is the set of vertices that are outside H_i];
    2.2. if U contains all the vertices of G, set $G \leftarrow \emptyset$ and go to 3,
        else if $U \neq \emptyset$, execute the face cut with respect to U.
3. Return G. ☐

Fig. 6 shows an example of the behavior of Algorithm 1. The diagram in (a) is the output of Algorithm 1 in single precise arithmetic for 25 half spaces. When we added artificial errors to all numerical computations, Algorithm 1 gave the output in (b), and when all the numerical computation were replaced by random numbers, Algorithm 1 gave the output in (c). As seen in this example, in any poor-precision arithmetic Algorithm 1 executes the task to the end and gives some output.

Let G be the vertex-edge graph of a polyhedron. Considering G as a graph drawn on the sphere, we say that set F of some faces of G is *simply connected* if the union of the faces in F and their boundaries constitute a simply connected region. Face set F is said to be *nontrivial* if F is nonempty and is a proper subset of the set of all the faces of G. The incremental method for constructing the convex hull is modified in the following way.

Algorithm 2 (Incremental construction of the convex hull).
Input: set P of n (≥ 4) points in R^3.
Aimed output: vertex-edge graph G of CH(P).
Procedure:

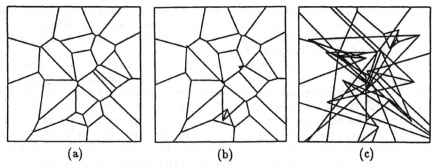

Fig. 6. Behavior of Algorithm 1: (a) output for 25 half spaces computed in single precision; (b) output in the case where artificial numerical errors are added; (c) output in the case where all the numerical computations are replaced by random numbers.

1. Number the points from p_1 to p_n [in the sorted order of the x coordinates].
2. Choose one of the two graphs in Fig. 2(a) and (b) [which is the vertex-edge graph of $CH(\{p_1, p_2, p_3, p_4\})$].
3. For $i = 5, 6, \cdots, n$ do 3.1 and 3.2:
   3.1. choose nontrivial simply connected face set F of G [such that F is the set of faces visible from p_i];
   3.2. add the cone having the base F and the apex p_i to G, and remove F from G.
4. Return G.   □

For both of the algorithms, the basic flow of processing is specified by combinatorial computation and hence they never fail due to numerical errors; thus they are numerically robust. It can be shown that, throughout the procedures of the algorithms, the graph G is planar and triply connected, and consequently from Theorems 1 and 2, both of the algorithms are topologically consistent. If the numerical computations are done in infinite precision and if no degeneracy takes place, these algorithms behave in the same way as the conventional algorithms and consequently they give correct solutions. Moreover, these algorithms are simple in the sense that they have no special branch for degeneracy; this is because degeneracy can not be discerned by computation with errors and therefore degenerate cases need not be considered.

Problem 1 can be solved by Algorithm 2. Problem 2 can be solved by Algorithm 1, where one of the input polyhedra is used as the initial polyhedron and the half spaces defining the other is used as the input half spaces, just in the same manner as [6]. Problem 3 can be solved by Algorithm 1. Problem 4 can be reduced to Problem 3 and can be solved by Algorithm 2.

6. Concluding Remarks.

Steinitz' theorem and its extension are used to construct topologically consistent algorithms for geometric problems related to convex polyhedra. The problems treated here includes the construction of three-dimensional convex hulls, the intersection of half spaces, and the construction of two-dimensional Laguerre Voronoi diagrams.

The algorithms designed here are of the incremental type. The same approach is also applicable to the design of a gift-wrapping algorithm and a divide-and-conquer algorithm for the three-dimensional convex hull; however, they require a little lengthy preparation and will be discussed elsewhere.

The approach taken in this paper belongs to the third category defined in the introduction, which we call the "topology-oriented approach". For example, VORONOI2, the incremental algorithm for the construction of the ordinary Voronoi diagrams [12], is similar to Algorithm 2. The construction of the dual of the Voronoi diagram, i.e., the Delaunay diagram, can be reduced to the construction of the convex hull of points scattered on a sphere [2]. In this connection, VORONOI2 can be interpreted as the procedure for constructing the convex hull. However, there is a great difference between the two; the topological consistency of Algorithm 2 is guaranteed by Steinitz' theorem, while VORONOI2 is not topologically consistent. We have to say that VORONOI2 is consistent only "partially" with respect to some necessary (but not sufficient) topological conditions used in the algorithm. Considering this situation, we can understand that Steinitz' theorem is very important and powerful for the design of consistent algorithms. This is the point the author would like to emphasize most in this paper.

References

[1] F. Aurenhammer: Power diagrams — Properties, algorithms and applications. *SIAM Journal on Computing*, Vol. 16 (1987), pp. 78–96.

[2] K. Q. Brown: Voronoi diagrams from convex hulls. *Information Processing Letters*, Vol. 9 (1979), pp. 223–228.

[3] M. B. Dillencourt: Toughness and Delaunay triangulations. *Discrete and Computational Geometry*, Vol. 5 (1990), pp. 575–601.

[4] L. Guibas, D. Salesin and J. Stolfi: Epsilon geometry — Building robust algorithms from imprecise computations. *Proceedings of the 5th ACM Annual Symposium on Computational Geometry*, Saarbrücken, 1989, pp. 208–217.

[5] C. M. Hoffmann: The problems of accuracy and robustness in geometric computation. *IEEE Computer*, Vol. 22, No. 3 (March 1989), pp. 31–41.

[6] J. E. Hopcroft and P. J. Kahn: A paradigm for robust geometric algorithms. *Technical Report*, TR89-1044, Department of Computer Science, Cornell University, Ithaca, 1989.

[7] V. Milenkovic: Verifiable implementations of geometric algorithms using finite precision arithmetic. *Artificial Intelligence*, Vol. 37 (December 1988), pp. 377–401.

[8] T. Ottmann, G. Thiemt and C. Ullrich: Numerical stability of geometric algorithms. *Proceedings of the 3rd ACM Annual Conference on Computational Geometry*, Waterloo, pp. 119–125, 1987.

[9] E. Steinitz: Polyeder und Raumeinteilungen. *Encyklopädie der mathematischen Wissenschaften*, Band III, Teil 1, 2. Hälfte, IIIAB12, pp. 1–139, 1916.

[10] K. Sugihara and M. Iri: Two design principles of geometric algorithms in finite-precision arithmetic. *Applied Mathematics Letters*, Vol. 2 (1989), pp. 203–206.

[11] K. Sugihara and M. Iri: A solid modelling system free from topological inconsistency. *Journal of Information Processing*, Vol. 12 (1989), pp. 380–393.

[12] K. Sugihara and M. Iri: Construction of the Voronoi diagram for "one million" generators in single-precision arithmetic. *Proceedings of the IEEE* (to appear).

[13] H. Whitney: Congruent graphs and the connectivity of graphs. *American Journal of Mathematics*, Vol. 54 (1932), pp. 150–168.

Characterizing and Recognizing
Visibility Graphs of Funnel-Shaped Polygons*

(extended abstract)

Seung-Hak Choi, Sung Yong Shin, and *Kyung-Yong Chwa*

Department of Computer Science
Korea Advanced Institute of Science and Technology
Kusong-dong 373-1, Yusong-gu, Taejon 305-701, Korea
{shchoi,syshin,kychwa}@gayakum.kaist.ac.kr

Abstract. A funnel, which is notable for its fundamental role in visibility algorithms, is defined as a polygon that has exactly three convex vertices two of which are connected by a boundary edge. In this paper, we investigate the visibility graph of a funnel which we call an F-graph. We first present two characterizations of an F-graph, one of whose sufficiency proof itself is an algorithm to draw a corresponding funnel on the plane in $O(e)$ time, where e is the number of the edges in an input graph. We next give an $O(e)$ time algorithm for recognizing an F-graph. When the algorithm recognizes a graph to be an F-graph, it also reports one of the Hamiltonian cycles defining the boundary of a corresponding funnel. We finally show that an F-graph is weakly triangulated and therefore perfect. This agrees with the fact that many of perfect graphs are related to geometric structures.

1. Introduction

A *simple polygon* (*polygon* for short) is defined as a closed connected region in the plane bounded by a closed chain of line segments with no two non-adjacent line segments intersecting. The chain is called the *boundary* of the polygon, and line segments and their endpoints are called *edges* and *vertices*, respectively. Given a polygon P, two points in P are said to be *visible* if the line segment connecting them is contained in P. Note that if the line segment touches the boundary of P, they are still considered to be visible (in some other definitions, however, they are considered to be invisible [1, 2, 4, 6, 14]). The visibility relationship among the vertices of a polygon can naturally be represented by a graph called a *visibility graph*. The visibility graph, denoted by $VG(P)$, of a polygon P is defined as a graph where the vertices of $VG(P)$ correspond to the vertices of P, and two vertices are adjacent in $VG(P)$ if their corresponding vertices are visible in P. Throughout this paper, a vertex of $VG(P)$ has the same label with the corresponding vertex of P.

The problem of characterizing the visibility graph of a polygon has remained open. Regarding algorithmic problems, it has also remained open to design an algorithm for recognizing the visibility graph. Only partial results have been reported in three directions: restricting the class of graphs, finding necessary conditions,

---

*Research supported by Korea Science and Engineering Foundation (Project No. 91-01-01).

and restricting the class of polygons.

ElGindy [7, 17] first restricted the class of graphs, showing that every maximal outerplanar graph is the visibility graph of a monotone polygon. He also established necessary conditions for a graph to be the visibility graph of a convex fan. Ghosh [8, 9] gave three necessary conditions for a graph G to be the visibility graph of a general polygon with emphasis on the cycles and the invisible vertex pairs of G. Recently, Coullard and Lubiw [6] discovered a new necessary condition in terms of the ordering of the vertices of G.

Everett and Corneil [8] restricted the attention to spiral polygons, i.e., polygons that consist of exactly two subchains of consecutive reflex vertices and convex vertices, respectively (see Fig. 1(a)). They characterized the visibility graph of a spiral polygon as a subclass of interval graphs and also gave a linear time algorithm to recognize such a class of graphs. Furthermore, this algorithm reports one of the Hamiltonian cycles defining the boundary of a corresponding spiral polygon when a graph is recognized as the visibility graph of a spiral polygon. Recently, Abello et al. [1] characterized the visibility graph of a staircase polygon (see Fig. 1(b)).

This paper concerns the visibility graph of a funnel which will be called an *F-graph*. A *funnel* is a polygon that has exactly three convex vertices, two of which are connected by a boundary edge. We will call the third convex vertex *cusp* (see Fig. 2). A funnel arises naturally in many visibility algorithms as an elementary structure [3, 10, 12, 14, 16, 18]. For example, in [10], the visibility graph of a polygon with holes is constructed incrementally by a plane-sweep algorithm. During the plane-sweep process, the algorithm maintains a sequence of funnels for each polygonal edge so that when a vertex v is added, all the vertices that are visible from v and precede v in the sweep order are covered by the current sequences of funnels.

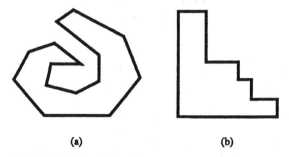

(a)   (b)

Fig. 1. A spiral polygon (a) and a staircase polygon (b).

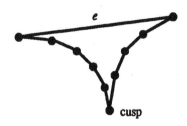

Fig. 2. A funnel.

In Section 2, we introduce some definitions and notation. Two characterizations of an F-graph are provided in Section 3. The sufficiency proof of the first characterization gives an algorithm to draw a corresponding funnel in $O(e)$ time, where e is the number of the edges in the input graph. Section 4 gives an $O(e)$ time algorithm to recognize an F-graph. Like the algorithm for a spiral polygon due to Everett and Corneil [8], our algorithm reports one of the Hamiltonian cycles defining the boundary of a corresponding funnel when a graph turns out to be an F-graph. In Section 5, we show that an F-graph is perfect, and conclusions are finally given in Section 6. Due to lack of space, we omit most of the lemmas and all proofs except one. Interested readers can refer to the full version [5] for details.

2. Definitions

A polygon is generally represented by a sequence of vertices listed in the counterclockwise order along the boundary. As shown in Fig. 2, the boundary of a funnel consists of three parts: an edge e which connects two adjacent convex vertices and two chains of reflex vertices which connect e and the cusp.

A sequence of vertices in brackets denotes a path that consists of the vertices and the edges between consecutive vertices in the sequence. Let $R = [r_1, r_2, \ldots, r_{t-1}, r_t, r_{t+1}]$ be a path of length $t \geq 1$ and $L = [l_1, l_2, \ldots, l_{s-1}, l_s, l_{s+1}]$ be a path of length $s \geq 1$.

Given a graph G, let $\mathcal{F}(G)$ be the set of funnels F such that $VG(F) = G$. Let H be a Hamiltonian cycle of G. If H corresponds to the boundary of a funnel in $\mathcal{F}(G)$, then H is said to be a *boundary cycle* of G. The boundary cycle is represented by $B(R, L)$, where $R = [r_1, r_2, \ldots, r_{t-1}, r_t, r_{t+1}]$, $t \geq 1$, and $L = [l_1, l_2, \ldots, l_{s-1}, l_s, l_{s+1}]$, $s \geq 1$, correspond to two chains of reflex vertices of $F \in \mathcal{F}(G)$. Here, r_1 and l_1 are adjacent convex vertices of F, respectively, and both r_{t+1} and l_{s+1} denote the same convex vertex which is the cusp.

Given a graph G, $V(G)$ and $E(G)$ denote the set of vertices and edges of G, respectively, and a graph G' is a *subgraph* of G if $V(G') \subseteq V(G)$ and $E(G') \subseteq E(G)$. Suppose that W is a nonempty subset of $V(G)$. Then, the subgraph of G, whose vertex set is W and whose edge set consists of the edges of G that have both ends in W, is called the subgraph of G *induced by* W and is denoted by $G[W]$. $G[W]$ is also said to be an *induced subgraph* of G. A cycle C is called a *chordless* cycle of G if C itself is an induced subgraph of G. For a vertex $v \in V(G)$, $\Gamma(v)$ is the set of vertices adjacent to v (not including v).

A Hamiltonian cycle H of a graph G is called a Λ-*cycle* if H has three distinct vertices v_a, v_b, and v_c satisfying that $v_a v_b \in E(H)$, and the path $H - v_a v_b = [v_a, \ldots, v_c, \ldots, v_b]$ can be divided into two paths $P_1 = [v_a, \ldots, v_c]$ and $P_2 = [v_b, \ldots, v_c]$ such that $E(P_1) = E(G[V(P_1)])$ and $E(P_2) = E(G[V(P_2)])$, i.e., each of P_1 and P_2 itself is an induced subgraph of G. Such a cycle H is denoted by $\Lambda(P_1, P_2)$, and v_c is also called the cusp.

3. Necessary and Sufficient Conditions

In this section, we give two characterizations of an F-graph. The first one, which is simple and intuitive, provides an intrinsic nature of an F-graph. More precisely, it is based on the well-known fact that every polygon can be triangulated and the boundary of a funnel consists of two reflex chains connected by an edge. The first characterization shows that these two conditions are just sufficient to characterize an F-graph. However, it hardly leads to a recognition algorithm. Therefore, we give the second characterization which is a little complicated but easy to obtain the algorithm.

Let H be a Hamiltonian cycle of a graph G. Then, a cycle C in G is said to be *ordered with respect to H* if the vertices in C preserve their order in H. Let C_k be a cycle of length k. If G has no chordless C_k ordered with respect to H for any $k \geq 4$, then G is said to be *ordered chordal with respect to H*.

Ghosh [8, 9] showed that the visibility graph of a simple polygon P is ordered chordal with respect to the Hamiltonian cycle corresponding to the boundary of P. This is one of the most fundamental properties of a visibility graph since every ordered cycle of length greater than three induces a subpolygon which has at least one diagonal. This together with the fact that the boundary cycle of an F-graph is a Λ-cycle gives a basic necessary condition which is summarized in the first lemma.

Lemma 1. *If G is an F-graph with the boundary cycle $B(R, L)$, then G is ordered chordal with respect to the Λ-cycle $\Lambda(R, L)$.*

Now we prove the sufficiency. Note that the proof given is a brief outline.

Lemma 2. *If G is an ordered chordal graph with respect to a Λ-cycle $\Lambda(R, L)$, then G is an F-graph with the boundary cycle $B(R, L)$.*

Proof: We prove this lemma by induction on $|V(G)|$. The case $|V(G)| = 3$ is trivial. Now consider the case where $|V(G)| > 3$. Let $R = [r_1, r_2, \ldots, r_{t-1}, r_t, r_{t+1}]$, $t \geq 1$, and $L = [l_1, l_2, \ldots, l_{s-1}, l_s, l_{s+1}]$, $s \geq 1$. Since, $r_1 l_2 \in E(G)$ or $r_2 l_1 \in E(G)$, without loss of generality, assume that $r_1 l_2 \in E(G)$. If G' is the subgraph of G induced by $V(G) - \{l_1\}$, then G' is ordered chordal with respect to the Λ-cycle $\Lambda(R, L - l_1)$. By the induction hypothesis, there exists a funnel $F' \in \mathcal{F}(G')$ with the boundary cycle $B(R, L - l_1)$. Let $h(q_1, q_2)$ denote the closed half-plane which lies to the right as the line passing through two points q_1 and q_2 is traversed from q_1 to q_2. Also let r_k be the last vertex in the path $R' = [r_1, \ldots, r_t]$ such that $r_k l_1 \in E(G)$ when R' is traversed from r_1. Then, it follows that $r_i l_1 \in E(G)$, $1 \leq i \leq k$, and $r_k l_2 \in E(G)$. Thus, if we place l_1 in $h(l_2, l_3) \cap h(l_2, r_{k+1}) \cap h(r_k, l_2)$, the resulting polygon becomes a funnel $F \in \mathcal{F}(G)$ (see Fig. 3). Therefore, G is an F-graph with the boundary cycle $B(R, L)$. □

From Lemma 1 and Lemma 2, an F-graph can be characterized as follows.

Theorem 1. *A graph G is the visibility graph of a funnel if and only if G is ordered chordal with respect to a Λ-cycle.*

Note that given an F-graph G with the boundary cycle $B(R, L)$, the proof of Lemma 2 is itself a recursive algorithm to draw in the plane a funnel $F \in \mathcal{F}(G)$. It can be easily verified that this algorithm takes $O(|E(G)|)$ time, which is summarized as follows.

Lemma 3. *Given an F-graph G with the boundary cycle $B(R, L)$, a funnel $F \in \mathcal{F}(G)$ can be drawn in the plane in $O(|E(G)|)$ time.*

As mentioned at the beginning of this section, the importance of the above characterization is that an F-graph can be characterized by two fundamental geometric properties of a funnel: "ordered chordal" from the fact that every subpolygon has at least one diagonal and "Λ-cycle" from the definition of a funnel.

Now, consider the recognition algorithms for determining whether or not a graph G is an F-graph. If a graph G has no chordless C_k for any $k \geq 4$, then G is said to be *chordal* (or *triangulated*). Since the Hamiltonian cycle problem for a

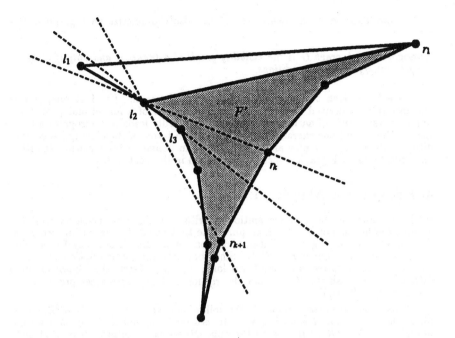

Fig. 3. An illustration for the proof of Lemma 2.

chordal graph is NP-complete [15], determining whether or not a graph G has the Hamiltonian cycle H such that G is ordered chordal with respect to H is an NP-hard problem. Therefore, given a graph G, it would be inefficient to check if G is an F-graph by a two-phase algorithm directly derived from Theorem 1, i.e., the first to find every Hamiltonian cycle H of G such that G is ordered chordal with respect to H and then to check if H is a Λ-cycle. Another two-phase approach may be to find every Λ-cycle H and then to check whether or not G is ordered chordal with respect to H. However, this approach does not seem plausible, either. This motivates us to find another characterization which provides us with a tool for avoiding the two-phase approach.

In order to give the second characterization, the structure of an ordered chordal graph is investigated in the following lemma. We call the conditions in it the *F-conditions*.

Lemma 4. *Suppose that a graph G has a Λ-cycle $H = \Lambda(R, L)$. Then, G is ordered chordal with respect to H if and only if H satisfies the followings:*

(i) $|\Gamma(r_i)| \geq 3$, and $|\Gamma(l_j)| \geq 3$ for $2 \leq i \leq t$ and $2 \leq j \leq s$.

(ii) $G[\Gamma(r_i) \cap V(L)]$ forms a single path on L, $1 \leq i \leq t$, and $G[\Gamma(l_j) \cap V(R)]$ forms a single path on R, $1 \leq j \leq s$.

(iii) If $[l_a, \ldots, l_b] = G[\Gamma(r_i) \cap V(L)]$ and $[l_c, \ldots, l_d] = G[\Gamma(r_{i+1}) \cap V(L)]$, $1 \leq i \leq t-1$, then $a \leq c \leq b \leq d$. If $[r_a, \ldots, r_b] = G[\Gamma(l_j) \cap V(R)]$ and $[r_c, \ldots, r_d] = G[\Gamma(l_{j+1}) \cap V(R)]$, $1 \leq j \leq s-1$, then $a \leq c \leq b \leq d$.

From Theorem 1 and Lemma 4, we can also characterize an F-graph as follows.

Theorem 2. *A graph G is the visibility graph of a funnel if and only if G has a Λ-cycle satisfying the F-conditions.*

We have given two characterizations of an F-graph. The first one, simple and intuitive, is derived from the geometric structure of a funnel and provides a fundamental nature of an F-graph. However, it hardly leads to a recognition algorithm. The second one characterizes an F-graph in terms of three complicated conditions. However, they play an important role in designing the efficient algorithm for recognizing an F-graph, which will be shown in the next section.

4. Recognition Algorithm

In this section, we present a recognition algorithm for deciding, given an arbitrary graph, whether or not it is an F-graph. Note that when the algorithm recognizes a graph G to be an F-graph, it also reports one of the Hamiltonian cycles defining the boundary of a funnel $F \in \mathcal{F}(G)$. Since both the formal description of the algorithm and its correctness proof are long, we only sketch the algorithm in a rather informal fashion. The formal description and the correctness proof can be found in the full version [5].

Given an arbitrary graph G, the following algorithm IS-F-GRAPH recognizes whether or not G is an F-graph. It returns R, L, and the cusp that form a boundary cycle $B(R, L)$ if G is an F-graph. Otherwise, it returns "Not F-graph". Since the boundary cycle of an F-graph is a Λ-cycle, the degree of the cusp is exactly two. Thus, the algorithm checks, for each vertex c of degree two, if c can be the cusp. Moreover, there are at most three vertices of degree two in an F-graph by F-condition (i). Therefore, the number of vertices that are checked by the algorithm is at most three.

Algorithm IS-F-GRAPH

input : A graph G.
output : R, L, and the cusp that form a boundary cycle $B(R, L)$ if G is an F-graph. "Not F-graph" otherwise.

// $deg(v) = |\Gamma(v)|$ is the degree of a vertex v in G.

1. if $^\exists v \in V(G)$ such that $deg(v) \leq 1$ then
2.   return "Not F-graph"
3. if not $(1 \leq$ (number of vertices of degree two) $\leq 3)$ then
4.   return "Not F-graph"
5. for each $c \in V(G)$ such that $deg(c) = 2$ do
6.   if CHECK$(G, c, R, L) = $ "F-graph" then
7.   return (R, L, c)
8. endfor
9. return "Not F-graph"

end Algorithm IS-F-GRAPH

The procedure CHECK takes a graph G along with a vertex $c \in V(G)$ of degree two as the input and checks whether or not G is an F-graph when the cusp is restricted to $c \in V(G)$. It returns "F-graph" as well as R and L that form a boundary cycle $B(R, L)$ if G is an F-graph with the cusp c. Otherwise, it returns "Not F-graph". CHECK consists of three phases called CONSTRUCT-CLUSTERS, FIND-Λ-CYCLE, and CHECK-F-CONDITIONS in sequence.

Procedure CHECK(G, c, R, L)

input : A graph G and a vertex $c \in V(G)$ of degree two.
output : "F-graph" as well as R and L that form a boundary cycle $B(R, L)$ if G is an F-graph with the cusp c. "Not F-graph" otherwise.

1. CONSTRUCT-CLUSTERS
2. FIND-Λ-CYCLE
3. CHECK-F-CONDITIONS

end Procedure CHECK

In the first phase CONSTRUCT-CLUSTERS, we partition G into k subgraphs called clusters for $k \geq 1$. With the vertex c as a seed, we first grow the subgraph containing c incrementally by adding vertices in G as long as we have only one choice for constructing the boundary cycle. This process is based on the local information such as adjacency and the F-conditions. When there occur more than one ways of adding a vertex v to the subgraph, we stop growing it and start a new subgraph with the current vertex v as a new seed. This process is repeated until all vertices of G are added to one of the subgraphs. We call each of these subgraphs as a cluster. Each cluster has exactly two disjoint paths P_i and Q_i, $1 \leq i \leq k$, which are called the *path segments*. The vertices of each path segment in sequence form a subpath of H for all boundary cycles $H = B(R, L)$ of G when G is an F-graph with the cusp c. They also satisfy the condition that if one of P_i and Q_i, $1 \leq i \leq k$, lies on R, then the other lies on L. The following lemma derived from the F-conditions shows how the clusters can be constructed.

Lemma 5. *Let $B(R, L)$ be a boundary cycle of an F-graph G and $W_{ij} = \{r_m \mid 1 \leq m < i\} \cup \{l_k \mid 1 \leq k < j\}$, $1 \leq i \leq t$ and $1 \leq j \leq s$, where $R = [r_1, \ldots, r_t, r_{t+1}]$ and $L = [l_1, \ldots, l_l, l_{s+1}]$. If $|W_{ij}| \geq 2$ and $r_i l_j \in E(G)$, then $X_{ij} = W_{ij} \cap \Gamma(r_i) \cap \Gamma(l_j)$ satisfies following conditions:*

(i) $|X_{ij}| = 1$ or $|X_{ij}| = 2$.

(ii) *If $|X_{ij}| = 1$, then*
$d(r_i) \neq d(l_j)$, *and* $\min\{d(r_i), d(l_j)\} = 1$, *where* $d(r_i) = |\Gamma(r_i) \cap W_{ij}|$ *and* $d(l_j) = |\Gamma(l_j) \cap W_{ij}|$.
Furthermore, $X_{ij} = \{r_{i-1}\}$ if $d(r_i) = 1$, and $X_{ij} = \{l_{j-1}\}$ if $d(l_j) = 1$.

(iii) *If $|X_{ij}| = 2$, then*
$X_{ij} = \{r_{i-1}, l_{j-1}\}$, *and* $r_{i-1} l_{j-1} \in E(G)$.

The second phase FIND-Λ-CYCLE finds a Λ-cycle of G. This can be done by properly combining all path segments under the restriction imposed by the edges lying across the clusters. There may be up to 2^k possibilities for assembling the path segments to form a boundary cycle. However, the structure of a funnel can be exploited to choose the right one efficiently, if any. In the third phase CHECK-F-CONDITIONS, we employ the F-conditions to verify that the finally-assembled sequence of the path segments indeed forms a boundary cycle.

Now, we have the main result of this section.

Theorem 3. *Let G be a graph. Then it can be recognized in $O(|E(G)|)$ time* *whether or not G is an F-graph. Furthermore, one of the boundary cycles of G can be found in $O(|E(G)|)$ time if G is an F-graph.*

We close this section, regarding the algorithm for drawing a funnel representation of a graph if there exist any. This can be directly derived from Theorem 3 and Lemma 3.

Theorem 4. *Given a graph G, we can draw a funnel $F \in \mathcal{F}(G)$, if any, in $O(|E(G)|)$ time.*

5. Combinatorial Properties of F-graphs

In this section, we discuss the relationship among F-graphs and other classes of graphs, showing that an F-graph is contained in the class of weakly triangulated graphs. Let G be a graph. The *complement* of G, denoted by $\bar{G}$, is defined as the graph such that $V(\bar{G}) = V(G)$ and that two vertices are adjacent in $\bar{G}$ if and only if they are not adjacent in G. A graph G is said to be *weakly triangulated* if neither G nor $\bar{G}$ contains chordless C_k for any $k \geq 5$, where C_k denotes a cycle of length k.

Theorem 5. *An F-graph is a weakly triangulated graph.*

Let $\chi(G)$ be the chromatic number of a graph G and $\omega(G)$ the maximum clique size of G. An undirected graph G is called *perfect* if $\chi(A) = \omega(A)$ for every induced subgraph A of G. Perfect graphs attract much attention because some NP-complete problems for general graphs can be solvable in polynomial time. It is also well-known that perfect graphs include many of intersection graphs that represent the intersection relation among geometric objects [11].

Hayward [13] proved that weakly triangulated graphs are perfect, which gives the following.

Corollary 1. *F-graphs are perfect.*

Fig. 4 illustrates the relationship among F-graphs and other classes of graphs. Although visibility graphs are not perfect in general, it confirms the fact that many of perfect graphs have geometric representations.

6. Conclusions

Although it is fundamental to characterize the visibility graph of a simple polygon, no complete characterization is available, yet. In this paper, we give new results on the problem by restricting the class of polygons to funnels. We characterize the visibility graph of a funnel in two ways, one providing its basic nature and the other for the recognition algorithm. Using the second characterization, we present an optimal algorithm for recognizing the visibility graph of a funnel. It is also shown that the visibility graph of a funnel is contained in the class of perfect graphs. This confirms the fact that many of perfect graphs have geometric representations.

It is easy to see that a polygon has at least three convex vertices. An immediate generalization of a funnel is a polygon with exactly three convex vertices since such a polygon can be obtained by relaxing the restriction that two convex vertices are connected by a boundary edge. It is an interesting problem to characterize the visibility graph of this class of polygons. This may give an insight for further investigating the visibility graph of a simple polygon in general.

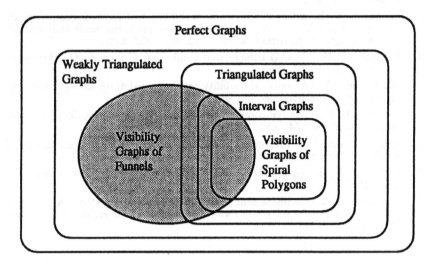

Fig. 4. The relationship among F-graphs and other classes of graphs.

References

1. Abello, J., O. Egecioglu, and K. Kumar, "Characterizing visibility graphs of staircase polygons," manuscript (extended abstract), 1991.

2. Asano, T., T. Asano, L. Guibas, J. Hershberger, and H. Imai, "Visibility of disjoint polygons," *Algorithmica*, vol. 1, pp. 49 - 63, 1986.

3. Avis, D. and G. T. Toussaint, "An optimal algorithm for determining the visibility of a polygon from an edge," *IEEE Trans. Computers*, vol. 30, no. 12, pp. 910 - 914, 1981.

4. Chazelle, B. and L. Guibas, "Visibility and intersection problems in plane geometry," *Discrete and Computational Geometry*, vol. 4, pp. 551 - 581, 1989.

5. Choi, S.-H., S. Y. Shin, and K.-Y. Chwa, "Characterizing and recognizing the visibility graph of a funnel-shaped polygon," Tech. Report CS-TR-92-71, Dept. Comp. Sci., KAIST, Korea, 1992.

6. Coullard, C. and A. Lubiw, "Distance visibility graphs," *Proc. 7th ACM Symp. Computational Geometry*, pp. 289 - 296, 1991.

7. ElGindy, H. A., "Hierarchical decomposition of polygons with applications," Ph.D. dissertation, School of Comp. Sci., McGill Univ., Montreal, 1985.

8. Everett, H. and D. G. Corneil, "Recognizing visibility graphs of spiral polygons," *J. Algorithms*, vol. 11, pp. 1 - 26, 1990.

9. Ghosh, S. K., "On recognizing and characterizing visibility graph of simple polygons," Tech. Report JHU/EECS-86/14, Dept. EECS, The Johns Hopkins Univ., Baltimore, 1986. Also in *Lecture Notes in Computer Science 318*, R. Karlsson and A. Lingas Eds., Springer-Verlag, New York, 1988

10. Ghosh, S. K. and D. M. Mount, "An output-sensitive algorithm for computing visibility graphs," *SIAM J. Comput.*, vol. 20, no. 5, pp. 888 - 910, 1991.

11. Golumbic, M. C., *Algorithmic graph theory and perfect graphs*, Academic Press, New York, 1980.

12. Guibas, L., J. Hershberger, D. Leven, M. Sharir, and R. E. Tarjan, "Linear-time algorithms for visibility and shortest path problems inside triangulated simple polygons," *Algorithmica*, vol. 2, pp. 209 - 233, 1987.

13. Hayward, R. B., "Weakly triangulated graphs," *J. Combinatorial Theory, B*, vol. 39, pp. 200 - 209, 1985.

14. Hershberger, J., "An optimal visibility graph algorithm for triangulated simple polygons," *Algorithmica*, vol. 4, pp. 141 - 155, 1989.

15. Johnson, D. S., "The NP-completeness column: an ongoing guide," *J. Algorithms*, vol. 6, pp. 434 - 451, 1985.

16. Lee, D. T. and F. P. Preparata, "Euclidean shortest paths in the presence of rectilinear barriers," *Networks*, vol. 14, no. 3, pp. 393 - 410, 1984.

17. O'Rourke, J., *Art gallery theorems and algorithms*, Oxford Univ. Press, New York, 1987.

18. Toussaint, G. T., "A linear-time algorithm for solving the strong hidden-line problem in a simple polygon," *Pattern Recognition Letters*, vol. 4, pp. 449 - 451, 1986.

On the Complexity of Composite Numbers

Toshiya Itoh   *Kenji Horikawa*
titoh@ip.titech.ac.jp   holly@ip.titech.ac.jp

Department of Information Processing,
Interdisciplinary Graduate School of Science and Engineering,
Tokyo Institute of Technology
4259 Nagatsuta, Midori-ku, Yokohama 227, Japan

1 Introduction

1.1 Background and Motivation

One of the most important problems in the recent cryptographic research is to design a constant move (blackbox simulation) perfect or statistical (almost perfect) zero-knowledge interactive proof (ZKIP) without any unproven assumption, where in this paper, we use a "move" to mean a message transmission from a verifier (resp. a prover) to a prover (resp. a verifier).

It has been shown so far that for a (presumably intractable) language, there exists a constant move (blackbox simulation) perfect ZKIP without any unproven assumption. Goldwasser, Micali, and Rackoff [7] showed that there exists a four move (blackbox simulation) perfect ZKIP for QNR without any unproven assumption, and Goldreich, Micali, and Wigderson [8] showed that there exists a four move (blackbox simulation) perfect ZKIP for GNI without any unproven assumption. These protocols are *optimal* in the light of the round complexity unless QNR $\in \mathcal{BPP}$ (or GNI $\in \mathcal{BPP}$), because Goldreich and Krawczyk [6] proved that if a language L has a three move blackbox simulation ZKIP, then $L \in \mathcal{BPP}$, i.e., L can be recognized in probabilistic polynomial time by a verifier himself.

On the other hand, Bellare, Micali, and Ostrovsky [4] recently showed in an elegant way that for any random self-reducible [12] language L (e.g., QR, GI, etc.), there exists a five move (blackbox simulation) perfect ZKIP without any unproven assumption. It follows from the result above [4] that QR (or GI, etc.) has a constant move (blackbox simulation) perfect ZKIP for itself and its complement without any unproven assumption.

To the best of our knowledge, every language L, which is known to have a constant move blackbox simulation perfect or statistical ZKIP for L and $\bar{L}$, is random self-reducible. Then

Question: Is it necessary for a language L to be random self-reducible so that L has a constant move blackbox simulation perfect or statistical ZKIP for the language L itself and the complement of the language L without any unproven assumption?

One of the main interests in this paper is to solve the question above (or to find an evidence to the question above), because it enables us to capture the *inherent* property of a language L that has a constant move (blackbox simulation) perfect or statistical ZKIP for L and $\bar{L}$.

1.2 Results

In this paper, we mainly consider two kinds of sets of odd composite numbers, +1MOD4 and +3MOD4. Let +1MOD4 (resp. +3MOD4) be a set of odd composite numbers, each of which consists of prime factors congruent to 1 (resp. 3) modulo 4. In addition, let $\overline{+1\text{MOD4}}$ (resp. $\overline{+3\text{MOD4}}$) be a set of odd composite numbers not in +1MOD4 (resp. +3MOD4).

The main results in this paper are (Theorem 3.1) there exists a four move blackbox simulation perfect ZKIP for $\overline{+1\text{MOD4}}$ without any unproven assumption; (Theorem 3.2) there exists a five move blackbox simulation perfect ZKIP for +1MOD4 without any unproven assumption; (Theorem 3.3) there exists a four move blackbox simulation perfect ZKIP for +3MOD4 without any unproven assumption; and (Theorem 3.4) there exists a five move blackbox simulation statistical ZKIP for $\overline{+3\text{MOD4}}$ without any unproven assumption.

The underlying idea for Theorems 3.1 and 3.3 is almost the same as that of the four move (blackbox simulation) perfect ZKIP for QNR [7] or GNI [8], while the idea behind Theorem 3.2 is almost the same as that of the five move (blackbox simulation) perfect ZKIP for GI [4], i.e., double running process. On the other hand, the protocol in Theorem 3.4 includes a new technique, i.e., how to generate a quadratic residue x modulo N that has no fourth root modulo N without knowing the prime factorization of N.

These results can be regarded as negative evidences for the question in subsection 1.1 if +1MOD4 (or +3MOD4) is not random self-reducible. It is not known whether or not +1MOD4 (or +3MOD4) is random self-reducible, however it is believed that it is *not*.

2 Preliminaries

2.1 Notation and Definitions

For any odd integer N, here we use Z_N^* to denote a set of integers x $(1 \leq x \leq N)$ coprime to N, and QR_N (resp. QNR_N) to denote a set of quadratic residues (resp. quadratic nonresidues) $x \in Z_N^*$ modulo N. Let OC be a set of odd composite numbers. It should be noted that it is easy to determine whether or not $N \in \text{OC}$, because it can be determined in deterministic constant time whether or not N is odd, and it can be determined in zero-error probabilistic polynomial [1], [10], [11] (in $|N|$) time whether or not N is composite.

Definition 2.1: Let N be composite and let $N = p_1^{e_1} p_2^{e_2} \cdots p_k^{e_k}$ be the prime factorization of N. Then $N \in +1\text{MOD4}$ iff $N \in \text{OC}$ and $p_i \equiv 1 \pmod 4$ for each i $(1 \leq i \leq k)$.

Definition 2.2: Let N be composite and let $N = p_1^{e_1} p_2^{e_2} \cdots p_k^{e_k}$ be the prime factorization of N. Then $N \in +3\text{MOD4}$ iff $N \in \text{OC}$ and $p_i \equiv 3 \pmod 4$ for each i $(1 \leq i \leq k)$.

In addition, we define a set $\overline{+1\text{MOD4}}$ of odd composite numbers to be $\overline{+1\text{MOD4}} = \text{OC} \setminus +1\text{MOD4}$ and a set $\overline{+3\text{MOD4}}$ of odd composite numbers to be $\overline{+3\text{MOD4}} = \text{OC} \setminus +3\text{MOD4}$. Although the set OC is easy to recognize (i.e., $\text{OC} \in \mathcal{ZPP}$), the set +1MOD4 (resp. +3MOD4) and its complement seem to be hard to recognize, i.e., they seem to be *not* in $\mathcal{BPP}$.

Let $\langle P, V \rangle$ be an interactive protocol. For the details on this, see [7].

Definition 2.3 [7]: Let $L \subseteq \{0,1\}^*$ be a language, and let $\langle P, V \rangle$ be an interactive protocol. Then $\langle P, V \rangle$ is said to be an interactive proof for a language L iff

- **Completeness:** For each $k > 0$ and for sufficiently large $x \in L$, V halts and accepts x with probability at least $1 - |x|^{-k}$, where the probabilities are taken over all of the possible coin tosses of P and V.

- **Soundness:** For each $k > 0$, for sufficiently large $x \notin L$, and for any P^*, V halts and accepts with probability at most $|x|^{-k}$, where the probabilities are taken over all of the possible coin tosses of P^* and V.

It should be noted that P's resource is computationally unbounded, while V's resource is bounded by probabilistic polynomial (in $|x|$) time.

Informally, an interactive proof $\langle P, V \rangle$ for a language L is said to be *zero-knowledge* if for any $x \in L$, an *honest* prover P releases no additional information other than $x \in L$ to any *cheating* verifier V^*. We can categorize the property of zero-knowledge to perfect, statistical (or almost perfect), and computational zero-knowledge according to the level of its indistinguishability. In addition, we can also categorize the property of zero-knowledge to GMR1, auxiliary input, and blackbox simulation zero-knowledge [9] according to the definition of its simulator.

In this paper, we are interested only in blackbox simulation "perfect" and "statistical" zero-knowledge," because it is the most *secure* and the most *robust* definition of zero-knowledge.

Definition 2.4 [9]: *Let $L \subseteq \{0,1\}^*$ be a language, and let $\langle P, V \rangle$ be an interactive proof for a language L. Then $\langle P, V \rangle$ is said to be blackbox simulation perfect (resp. statistical) zero-knowledge iff there exists an expected polynomial (in $|x|$) time Turing machine (universal simulator) M such that for all $x \in L$ and any cheating verifier V^*, $\{\langle P, V^* \rangle(x)\}$ and $\{M(x; V^*)\}$ are perfectly (resp. statistically) indistinguishable on L, where $\{\langle P, V^* \rangle(x)\}$ and $\{M(x; V^*)\}$ denote families of random variables on $x \in L$, and $M(\cdot; V^*)$ denotes a Turing machine that is allowed to have a blackbox access to the cheating verifier V^*.*

2.2 Technical Lemmas

In this subsection, we provide several technical lemmas for the subsequent discussions. All of the lemmas follow from the propositions below.

Proposition 2.5: *Let N be an odd composite number, and let $N = p_1^{e_1} p_2^{e_2} \cdots p_k^{e_k}$ be the prime factorization of N. Then for any $x \in Z_N^*$, $x \in \mathrm{QR}_N$ iff $x \in \mathrm{QR}_{p_i^{e_i}}$ for each i ($1 \leq i \leq k$).*

Proposition 2.6: *Let p^ℓ be an odd prime power. Then $-1 \in \mathrm{QR}_{p^\ell}$ iff $p \equiv 1 \pmod 4$.*

Proposition 2.7: *For any odd prime power p^ℓ, $\|\mathrm{QNR}_{p^\ell}\| = \|Z_{p^\ell} \setminus \mathrm{QR}_{p^\ell}\|$, where $\|A\|$ denotes the cardinality of a (finite) set A.*

Then we can show the following lemmas from propositions above.

Lemma 2.8: $-1 \in \mathrm{QR}_N$ iff $N \in +1\mathrm{MOD}4$.

Proof: It is immediate from the Propositions 2.5 and 2.6. ∎

Lemma 2.9: *If $N \in +3\mathrm{MOD}4$, then $x \equiv u^4 \pmod N$ for $u \in_R Z_N^*$ always has just a single square root $y \in \mathrm{QR}_N$ modulo N, while if $N \notin +3\mathrm{MOD}4$, then $x \equiv u^4 \pmod N$ for $u \in_R Z_N^*$ always has at least two square roots $y \in \mathrm{QR}_N$ modulo N.*

Proof: Let $x \equiv u^4 \pmod N$ for $u \in_R Z_N^*$ and let $N = p_1^{e_1} p_2^{e_2} \cdots p_k^{e_k}$ be the prime factorization of N. It is easy to see that $x \equiv u^4 \pmod N$ always has a square root $y \in \mathrm{QR}_N$, where $y \equiv u^2 \pmod N$. It should be noted that every square root $\tilde{y}$ modulo N of $x \equiv u^4 \pmod N$ must satisfy $\tilde{y} \equiv \pm u^2 \pmod{p_i^{e_i}}$ for each i ($1 \leq i \leq k$).

It follows from Propositions 2.5 and 2.6 that if $N \in +3\mathrm{MOD}4$, then $x \equiv u^4 \pmod N$ for $u \in_R Z_N^*$ always has just a single square root $y \in \mathrm{QR}_N$ modulo N, where $y \equiv u^2 \pmod N$.

When $N \notin +3\mathrm{MOD}4$, we can assume without loss of generality that for some ℓ ($1 \leq \ell \leq k$), $p_1 \equiv p_2 \equiv \cdots \equiv p_\ell \equiv 1 \pmod 4$. Then it follows from Propositions 2.5 and 2.6 that $x \equiv u^4 \pmod N$ always has 2^ℓ square roots $y \in \mathrm{QR}_N$ modulo N such that $y \equiv \pm u^2 \pmod N$ for each i ($1 \leq i \leq \ell$) and $y \equiv u^2 \pmod N$ for each j ($\ell < j \leq k$). Thus if $N \notin +3\mathrm{MOD}4$, then $x \equiv u^4 \pmod N$ for $u \in_R Z_N^*$ always has at least two square roots $y \in \mathrm{QR}_N$ modulo N. ∎

Lemma 2.10: If $N \in \overline{+3\text{MOD}4}$, then $x \equiv u^2 \pmod{N}$ for $u \in_R Z_N^*$ has no fourth root modulo N with probability at least $1/2$ and $y \equiv v^4 \pmod{N}$ for $v \in_R Z_N^*$ always has a fourth root modulo N, while if $N \notin \overline{+3\text{MOD}4}$, then for some $k > 0$, $x \equiv u^2 \pmod{N}$ for $u \in_R Z_N^*$ and $y \equiv v^4 \pmod{N}$ for $v \in_R Z_N^*$ always have 2^k fourth roots modulo N, respectively.

Proof: Let $x \equiv u^2 \pmod{N}$ for $u \in_R Z_N^*$ and let $y \equiv v^4 \pmod{N}$ for $v \in_R Z_N^*$. In addition, let $N = p_1^{e_1} p_2^{e_2} \cdots p_k^{e_k}$ be the prime factorization of N. It is clear that $y \equiv v^4 \pmod{N}$ always has a fourth root v modulo N. It should be noted that every square root $\tilde{u}$ modulo N of $x \equiv u^2 \pmod{N}$ must satisfy $\tilde{u} \equiv \pm u \pmod{p_i^{e_i}}$ for each i ($1 \leq i \leq k$).

When $N \in \overline{+3\text{MOD}4}$, we can assume without loss of generality that for some ℓ ($1 \leq \ell \leq k$), $p_1 \equiv p_2 \equiv \cdots \equiv p_\ell \equiv 1 \pmod{4}$. It follows from Propositions 2.5, 2.6, and 2.7 that $x \equiv u^2 \pmod{N}$ for $u \in_R Z_N^*$ has no fourth root modulo N with probability $1 - 2^{-\ell}$. Thus if $N \in \overline{+3\text{MOD}4}$, then $x \equiv u^2 \pmod{N}$ for $u \in_R Z_N^*$ has no fourth root modulo N with probability at least $1/2$ and $y \equiv v^4 \pmod{N}$ for $v \in_R Z_N^*$ always has a fourth root modulo N.

When $N \notin \overline{+3\text{MOD}4}$, i.e., $N \in +3\text{MOD}4$, it is easy to see from Proposition 2.6 that for each i ($1 \leq i \leq k$), one of $\pm u \pmod{p_i^{e_i}}$ must be a quadratic residue modulo $p_i^{e_i}$. Then it follows from Proposition 2.5 that $x \equiv u^2 \pmod{N}$ always has just a single square root $z_x \in \text{QR}_N$ modulo N. From Lemma 2.9, this is also the case for $y \equiv v^4 \pmod{N}$, i.e., $y \equiv v^4 \pmod{N}$ always has just a single square root $z_y \in \text{QR}_N$ modulo N if $N \in +3\text{MOD}4$. Note that $w \in Z_N^*$ is a fourth root of $x \equiv u^2 \pmod{N}$ (resp. $y \equiv v^4 \pmod{N}$) modulo N iff $w \in Z_N^*$ is a square root of $z_x \in \text{QR}_N$ (resp. $z_y \in \text{QR}_N$) modulo N and that $z_x \in \text{QR}_N$ (resp. $z_y \in \text{QR}_N$) has 2^k square roots modulo N. Thus if $N \notin \overline{+3\text{MOD}4}$, then $x \equiv u^2 \pmod{N}$ for $u \in_R Z_N^*$ and $y \equiv v^4 \pmod{N}$ for $v \in_R Z_N^*$ always have 2^k fourth roots modulo N, respectively. ∎

Note that the property of Lemma 2.8 plays an important role to prove Theorem 3.1 (subsection 3.1) and Theorem 3.2 (subsection 3.2), while the properties of Lemmas 2.9 and 2.10 are essential to show Theorem 3.3 (subsection 3.3) and Theorem 3.4 (subsection 3.4), respectively.

3 Constant Move Perfect ZKIP

3.1 Four Move Perfect ZKIP for $\overline{+1\text{MOD}4}$

Theorem 3.1: There exists a four move blackbox simulation perfect ZKIP for the complement of $+1\text{MOD}4$ without any unproven assumption.

Proof: The correctness of the protocol below is based on Lemma 2.8 in subsection 2.2.

Four Move Perfect ZKIP for $\overline{+1\text{MOD}4}$

common input: $N \in Z$ ($n = |N|$)

V1-1: V chooses $e_i \in_R \{0, 1\}$ and $c_i \in_R Z_N^*$, and computes $C_i \equiv (-1)^{e_i} c_i^2 \pmod{N}$ for each i ($1 \leq i \leq n$).

V1-2: V chooses $e_{ij} \in_R \{0, 1\}$ and $t_{ij}^0, t_{ij}^1 \in_R Z_N^*$, and computes $T_{ij}^0 \equiv (-1)^{e_{ij}} \left(t_{ij}^0 \right)^2 \pmod{N}$ and $T_{ij}^1 \equiv (-1)^{1-e_{ij}} \left(t_{ij}^1 \right)^2 \pmod{N}$ for each i, j ($1 \leq i, j \leq n$).

$V \to P$: C_i ($1 \leq i \leq n$), $\langle T_{ij}^0, T_{ij}^1 \rangle$ ($1 \leq i, j \leq n$).

P1: P chooses $K_i \subseteq_R \{1, 2, \ldots, n\}$ for each i ($1 \leq i \leq n$).

$P \to V$: K_i ($1 \leq i \leq n$).

V2-1: V assigns $w_{ij} = \langle e_{ij}, (t_{ij}^0, t_{ij}^1) \rangle$ for each $j \in K_i$ $(1 \leq i \leq n)$.

V2-2: V assigns $w_{ij} = \langle f_{ij}, s_{ij} \rangle$, where $f_{ij} = e_i \oplus e_{ij}$ and $s_{ij} \equiv c_i \left(t_{ij}^{f_{ij}} \right)^{-1}$ $(\bmod\ N)$ for each $j \notin K_i$ $(1 \leq i \leq n)$.

$V \to P$: w_{ij} $(1 \leq i, j \leq n)$.

P2-1: P checks that $T_{ij}^0 \equiv (-1)^{e_{ij}} \left(t_{ij}^0 \right)^2$ $(\bmod\ N)$ and $T_{ij}^1 \equiv (-1)^{1-e_{ij}} \left(t_{ij}^1 \right)^2$ $(\bmod\ N)$ for each $j \in K_i$ $(1 \leq i \leq n)$.

P2-2: P checks that $C_i \equiv s_{ij}^2 T_{ij}^{f_{ij}}$ $(\bmod\ N)$ for each $j \notin K_i$ $(1 \leq i \leq n)$.

P2-3: If all of the checks in steps P2-1 and P2-2 are successful, then P continues; otherwise P halts and rejects.

P2-4: For each i $(1 \leq i \leq n)$, if $C_i \in \mathrm{QR}_N$, then P assigns $f_i = 0$; otherwise $f_i = 1$.

$P \to V$: f_i $(1 \leq i \leq n)$.

V3-1: V checks that $f_i = e_i$ for each i $(1 \leq i \leq n)$.

V3-2: If all of the checks in step V3-1 are successful, then V halts and accepts; otherwise V halts and rejects.

Completeness: Immediate.

Soundness: When $N \notin \overline{+1\text{MOD4}}$ (i.e., $N \in +1\text{MOD4}$), any cheating prover P^* can not guess the value of e_i better at random for each i $(1 \leq i \leq n)$ even if P^* is *infinitely* powerful, because from the fact that $-1 \in \mathrm{QR}_N$ if $N \in +1\text{MOD4}$ (see Lemma 2.8.), $C_i \in \mathrm{QR}_N$ regardless of the value of e_i, where $C_i \equiv (-1)^{e_i} c_i^2$ $(\bmod\ N)$ for each i $(1 \leq i \leq n)$. Thus the honest verifier V accepts $N \notin \overline{+1\text{MOD4}}$ with probability at most 2^{-n}.

Perfect Zero-Knowledgeness: It can be shown in a way similar to the four move blackbox simulation perfect ZKIP for QNR [7] or GNI [8].

Thus the protocol above is a four move blackbox simulation perfect ZKIP for the complement of $+1\text{MOD4}$ without any unproven assumption. ∎

3.2 Five Move Perfect ZKIP for +1MOD4

Theorem 3.2: *There exists a five move blackbox simulation perfect ZKIP for $+1\text{MOD4}$ without any unproven assumption.*

Proof: The correctness of the protocol below is also based on Lemma 2.8 in subsection 2.2. Here we use b to denote a square root of -1 modulo N.

Five Move Perfect ZKIP for +1MOD4
common input: $N \in Z$ $(n = |N|)$

P1: P chooses $u_0, u_1 \in_R Z_N^*$ and computes $S_0 \equiv u_0^2$ $(\bmod\ N)$ and $S_1 \equiv u_1^2$ $(\bmod\ N)$.

$P \to V$: $\langle S_0, S_1 \rangle$.

V1: V chooses $e_i \in_R \{0,1\}$ and $t_i \in_R Z_N^*$, and computes $q_i \equiv t_i^2 S_{e_i}$ $(\bmod\ N)$ for each i $(1 \leq i \leq n)$.

$V \to P$: q_i $(1 \leq i \leq n)$.

P2: P chooses $r_i \in_R Z_N^*$, and computes $x_i \equiv r_i^2$ $(\bmod\ N)$ for each i $(1 \leq i \leq n)$.

$P \rightarrow V$: x_i $(1 \le i \le n)$.

V2: V reveals $\langle e_i, t_i \rangle$ for each i $(1 \le i \le n)$.

$V \rightarrow P$: $\langle e_i, t_i \rangle$ $(1 \le i \le n)$.

P3-1: P checks that $q_i \equiv t_i^2 S_{e_i}$ (mod N) for each i $(1 \le i \le n)$.

P3-2: If all of the checks in step P3-1 are successful, then P continues; otherwise P halts and rejects.

P3-3: P assigns $w_i \equiv b^{e_i} r_i$ (mod N) for each i $(1 \le i \le n)$, where b denotes a square root of -1 modulo N.

$P \rightarrow V$: $\langle u_0, u_1 \rangle$, w_i $(1 \le i \le n)$.

V3-1: V checks that $S_0 \equiv u_0^2$ (mod N) and $S_1 \equiv u_1^2$ (mod N).

V3-2: V checks that $w_i^2 \equiv (-1)^{e_i} x_i$ (mod N) for each i $(1 \le i \le n)$.

V3-3: If all of the checks in steps V3-2 and V3-3 are successful, then P halts and accepts; otherwise P halts and rejects.

Completeness: Immediate.

Soundness: In step P1, any cheating prover P^* must choose S_0 and S_1 such that $S_0, S_1 \in QR_N$, because P^* afterward must reveal a square root u_0 of S_0 modulo N and a square root u_1 of S_1 modulo N. Then the cheating prover P^* can not guess the value of e_i better at random for each i $(1 \le i \le n)$ even if P^* is *infinitely* powerful, because $q_i \in QR_N$ for each i $(1 \le i \le n)$. Thus the honest verifier V accepts $N \notin +1MOD4$ with probability at most 2^{-n}.

Perfect Zero-Knowledgeness: It can be shown in a way similar to the five move blackbox simulation perfect ZKIP for GI [4]. The outline of the simulation is as follows: In the *special* mode (mode-1) of simulation, the simulator M chooses $u_0, u_1 \in_R Z_N^*$, and computes $\tilde{S}_0 \equiv u_0^2$ (mod N) and $\tilde{S}_1 \equiv -u_1^2$ (mod N). Note that if $N \in +1MOD4$, then the distribution of $\langle \tilde{S}_0, \tilde{S}_1 \rangle$ is *perfectly* identical to that of $\langle S_0, S_1 \rangle$ in the real interaction between the honest prover P and the cheating verifier V^*, because $-1 \in QR_N$. (see Lemma 2.8.)

The simulator M resets V^* to step V2 by changing the values of x_i $(1 \le i \le n)$ in step P2. If the cheating verifier V^* changes his mind, i.e., the first reveal of e_i is different from the second reveal of $\tilde{e}_i$, then it is lucky for the simulator M to find a square root of -1 modulo N; otherwise the simulator M then changes the special mode of simulation (mode-1) to the *normal* mode (mode-0) of simulation. In the normal mode of simulation, the simulator M computes S_0 and S_1 in completely the same way as the honest prover P does. If the cheating verifier V^* does not change his mind, then it is lucky for the simulator M to correctly answer to the questions made by V^*; otherwise the simulator M then tries again keeping the mode of simulation normal. The way of simulation above is called *double running process* [4], and this process is shown to *perfectly* simulate the real interaction between the honest prover P and the cheating verifier V^* in expected polynomial (in $|x|$) time.

Thus the protocol above is a five move blackbox simulation perfect ZKIP for $+1MOD4$ without any unproven assumption. ∎

3.3 Four Move Perfect ZKIP for $+3MOD4$

Theorem 3.3: *There exists a four move blackbox simulation perfect ZKIP for $+3MOD4$ without any unproven assumption.*

Proof: The correctness of the protocol below is based on Lemma 2.9 in subsection 2.2.

Four Move Perfect ZKIP for +3MOD4
common input: $N \in Z$ $(n = |N|)$

V1-1: V chooses $C_i \in_R Z_N^*$, and computes $B_i \equiv C_i^2 \pmod{N}$ and $A_i \equiv B_i^2 \pmod{N}$ for each i $(1 \leq i \leq n)$.

V1-2: V chooses $c_{ij} \in_R Z_N^*$, and computes $b_{ij} \equiv c_{ij}^2 \pmod{N}$ and $a_{ij} \equiv b_{ij}^2 \pmod{N}$ for each i,j $(1 \leq i,j \leq n)$.

$V \rightarrow P$: A_i $(1 \leq i \leq n)$, a_{ij} $(1 \leq i,j \leq n)$.

P1: P chooses $K_i \subseteq_R \{1, 2, \ldots, n\}$ for each i $(1 \leq i \leq n)$.

$P \rightarrow V$: K_i $(1 \leq i \leq n)$.

V2-1: V assigns $z_{ij} \equiv c_{ij} \pmod{N}$ for each $j \in K_i$ $(1 \leq i \leq n)$.

V2-2: V assigns $z_{ij} \equiv C_i c_{ij} \pmod{N}$ for each $j \notin K_i$ $(1 \leq i \leq n)$.

$V \rightarrow P$: z_{ij} $(1 \leq i,j \leq n)$.

P2-1: P checks that $a_{ij} \equiv z_{ij}^4 \pmod{N}$ for each $j \in K_i$ $(1 \leq i \leq n)$.

P2-2: P checks that $A_i a_{ij} \equiv z_{ij}^4 \pmod{N}$ for each $j \notin K_i$ $(1 \leq i \leq n)$.

P2-3: If all of the checks in steps P2-1 and P2-2 are successful, then P continues; otherwise P halts and rejects.

P2-4: P computes $w_i \in QR_N$ such that $w_i^2 \equiv A_i \pmod{N}$ for each i $(1 \leq i \leq n)$.

$P \rightarrow V$: w_i $(1 \leq i \leq n)$.

V3-1: V checks that $w_i = B_i$ for each i $(1 \leq i \leq n)$.

V3-2: If all of the checks in step V3-1 are successful. then V halts and accepts; otherwise V halts and rejects.

Completeness: Immediate.

Soundness: Assume that $N \notin +3MOD4$. It follows from Lemma 2.9 that for any $C_i \in Z_N^*$ $(1 \leq i \leq n)$, $A_i \equiv C_i^4 \pmod{N}$ always has at least two square roots $\hat{B}_i \in QR_N$ modulo N. In addition, for each $j \notin K_i$ $(1 \leq i \leq n)$, $z_{ij} \equiv C_i c_{ij} \pmod{N}$ *uniformly* distributes over all possible fourth roots of $A_i a_{ij} \pmod{N}$, because the honest verifier V chooses $C_i \in_R Z_N^*$ $(1 \leq i \leq n)$ and $c_{ij} \in_R Z_N^*$ $(1 \leq i,j \leq n)$. Then all powerful prover P^* can not guess better at random which C_i the honest verifier V really possesses for each i $(1 \leq i \leq n)$. This implies that when $N \notin +3MOD4$, any cheating prover P^* can not guess better at random which B_i the honest verifier V really possesses for each i $(1 \leq i \leq n)$ even if P^* is *infinitely* powerful. Thus the honest verifier V accepts $N \notin +3MOD4$ with probability at most 2^{-n}.

Perfect Zero-Knowledgeness: It can be shown in a way similar to the four move blackbox simulation perfect ZKIP for QNR [7] or GNI [8]. For any cheating verifier V^*, the simulator M resets V^* and can find a fourth root $\tilde{C}_i$ of A_i modulo N for each i $(1 \leq i \leq n)$ in expected polynomial (in $|N|$) time. Note that when $N \in +3MOD4$, $\tilde{w}_i \equiv \tilde{C}_i^2 \pmod{N}$ must be equal to the square root $B_i \in QR_N$ of A_i modulo N for each i $(1 \leq i \leq n)$, which is sent by the honest prover P interacting with the cheating verifier V^*, because each A_i $(1 \leq k \leq n)$ always has just a single square root $B_i \in QR_N$ modulo N.

Thus the protocol above is a four move blackbox simulation perfect ZKIP for +3MOD4 without any unproven assumption. ∎

3.4 Five Move Statistical ZKIP for $\overline{+3\text{MOD4}}$

Theorem 3.4: *There exists a five move blackbox simulation statistical ZKIP for the complement of +3MOD4 without any unproven assumption.*

Proof: The correctness of the protocol below is based on Lemma 2.10 in subsection 2.2.

To simplify the description of the protocol below, we use the following notation. Let $a = (a_1, a_2, \ldots, a_k)$ and $b = (b_1, b_2, \ldots, b_k)$ be k dimensional vectors over Z_N^*, i.e., $a_i, b_i \in Z_N^*$ ($1 \le i \le k$). We use $c \equiv a \odot b \pmod N$ to denote a componentwise product of a and b modulo N, i.e., $c_i \equiv a_i b_i \pmod N$ ($1 \le i \le k$), and $d \equiv a^{-1} \pmod N$ to denote a componentwise inverse of a modulo N, i.e., $d_i \equiv a_i^{-1} \pmod N$ ($1 \le i \le k$). In addition, we use $f \equiv a^e \pmod N$ to denote a componentwise e-th power of a modulo N, i.e., $f_i \equiv a_i^e \pmod N$ ($1 \le i \le k$).

Five Move Statistical ZKIP for $\overline{+3\text{MOD4}}$
common input: $N \in Z$ ($n = |N|$)

P1-1: P chooses $u_k^0, u_k^1 \in_R Z_N^*$, and computes $s_k^0 \equiv \left(u_k^0\right)^4 \pmod N$ and $s_k^1 \equiv \left(u_k^1\right)^2 \pmod N$ for each k ($1 \le k \le n$).

P1-2: P assigns $S^0 = (s_1^0, s_2^0, \ldots, s_n^0)$ and $S^1 = (s_1^1, s_2^1, \ldots, s_n^1)$.

$P \to V$: $\langle S^0, S^1 \rangle$.

V1-1: V chooses $e_i \in_R \{0, 1\}$ and $c_i \in_R (Z_N^*)^n$, and computes $C_i \equiv c_i^4 \odot S^{e_i} \pmod N$ for each i ($1 \le i \le n$).

V1-2: V chooses $e_{ij} \in_R \{0, 1\}$ and $t_{ij}^0, t_{ij}^1 \in_R (Z_N^*)^n$, and computes $T_{ij}^0 \equiv \left(t_{ij}^0\right)^4 \odot S^{e_{ij}} \pmod N$ and $T_{ij}^1 \equiv \left(t_{ij}^1\right)^4 \odot S^{1-e_{ij}} \pmod N$ for each i, j ($1 \le i, j \le n$).

$V \to P$: C_i ($1 \le i \le n$), $\langle T_{ij}^0, T_{ij}^1 \rangle$ ($1 \le i, j \le n$).

P2: P chooses $K_i \subseteq_R \{1, 2, \ldots, n\}$ for each i ($1 \le i \le n$).

$P \to V$: K_i ($1 \le i \le n$).

V2-1: V assigns $w_{ij} = \langle e_{ij}, (t_{ij}^0, t_{ij}^1) \rangle$ for each $j \in K_i$ ($1 \le i \le n$).

V2-2: V assigns $w_{ij} = \langle f_{ij}, s_{ij} \rangle$, where $f_{ij} = e_i \oplus e_{ij}$ and $s_{ij} \equiv c_i \odot \left(t_{ij}^{f_{ij}}\right)^{-1} \pmod N$ for each $j \notin K_i$ ($1 \le i \le n$).

$V \to P$: w_{ij} ($1 \le i, j \le n$).

P3-1: P checks that $T_{ij}^0 \equiv \left(t_{ij}^0\right)^4 \odot S^{e_{ij}} \pmod N$ and $T_{ij}^1 \equiv \left(t_{ij}^1\right)^4 \odot S^{1-e_{ij}} \pmod N$ for each $j \in K_i$ ($1 \le i \le n$).

P3-2: V checks that $C_i \equiv s_{ij}^4 \odot T_{ij}^{f_{ij}} \pmod N$ for each $j \notin K_i$ ($1 \le i \le n$).

P3-3: If all of the checks in steps P3-1 and P3-2 are successful, then P continues; otherwise P halts and rejects.

P3-4: For each i ($1 \le i \le n$), if every element C_{ik} ($1 \le k \le n$) of C_i has a fourth root modulo N, then P assigns $f_i = 0$; otherwise $f_i = 1$.

$P \to V$: $\langle u_k^0, u_k^1 \rangle$ ($1 \le k \le n$), f_i ($1 \le i \le n$).

V4-1: V checks that $s_k^0 \equiv \left(u_k^0\right)^4 \pmod N$ and $s_k^1 \equiv \left(u_k^1\right)^2 \pmod N$ for each k ($1 \le k \le n$).

V4-2: V checks that $f_i = e_i$ for each i ($1 \le i \le n$).

V4-3: If all of the checks in steps V4-1 and V4-2 are successful, then V halts and accepts; otherwise V halts and rejects.

Completeness: Assume that $N \in \overline{+3\text{MOD4}}$. It follows from Lemma 2.10 that the k-th element s_k^0 of the vector S^0 always has a fourth root modulo N for each k $(1 \le k \le n)$, while with probability at least $1 - 2^{-n}$, there exists the ℓ-th element s_ℓ^1 of the vector S^1 that has no fourth root modulo N for some ℓ $(1 \le \ell \le n)$. Then the honest prover P *successfully* finds f_i for each i $(1 \le i \le n)$ in step P3-4 with probability at least $1 - 2^{-n}$. Thus the honest verifier V accepts $N \in \overline{+3\text{MOD4}}$ with probability at least $1 - 2^{-n}$.

Soundness: In step P1-1, any cheating prover P^* must choose s_k^0 and s_k^1 such that $s_k^0, s_k^1 \in \text{QR}_N$ for each k $(1 \le k \le n)$, because P^* afterward must reveal a fourth root u_k^0 of s_k^0 modulo N and a square root u_k^1 of s_k^1 modulo N for every k $(1 \le k \le n)$. When $N \notin \overline{+3\text{MOD4}}$, we can assume without loss of generality that $N = p_1^{e_1} p_2^{e_2} \cdots p_h^{e_h}$ is the prime factorization of N and for some ℓ $(1 \le \ell \le k)$, $p_1 \equiv p_2 \equiv \cdots \equiv p_\ell \equiv 1 \pmod 4$. In this case, the cheating prover P^* can not guess the value of e_i better at random for each i $(1 \le i \le n)$ even if P^* is *infinitely* powerful, because each k-th element C_{ik} $(1 \le k \le n)$ of the vector C_i always has its 2^ℓ fourth roots modulo N (see Lemma 2.10.) for each i $(1 \le i \le n)$ regardless of the value of e_i. Thus the honest verifier V accepts $N \notin \overline{+3\text{MOD4}}$ with probability at most 2^{-n}.

Statistical Zero-Knowledgeness: Assume that $N \in \overline{+3\text{MOD4}}$. The simulator M computes S^0 and S^1 in completely the same way as the honest prover P does. It is shown in the proof above of **Completeness** that for $u_k^0, u_k^1 \in Z_N^*$ $(1 \le k \le n)$, the k-th element s_k^0 of the vector S^0 always has a fourth root modulo N for each k $(1 \le k \le n)$, while with probability at least $1 - 2^{-n}$, there exists the ℓ-th element s_ℓ^1 of the vector S^1 that has no fourth root modulo N for some ℓ $(1 \le \ell \le n)$. Then the proof follows in a way similar to the four move blackbox simulation perfect ZKIP for QNR [7] or GNI [8] except for the following cases: (Case 1) if there exists the ℓ-th element s_ℓ^1 of the vector S^1 that has no fourth root modulo N for some ℓ $(1 \le \ell \le n)$ (this happens with overwhelming probability.), then the simulator M *perfectly* simulates the real interaction between the honest prover P and the cheating verifier V^*; and (Case 2) if the k-th element s_k^1 of the vector S^1 has a fourth root modulo N for each k $(1 \le k \le n)$ (this happen with negligible probability.), then there is no guarantee that the simulator M *perfectly* simulates the real interaction between the honest prover P and the cheating verifier V^*. It is not difficult to show that the simulator M *statistically* simulates the real interaction between the honest prover P and the cheating verifier V^*.

Thus the protocol above is a five move blackbox simulation statistical ZKIP for the complement of +3MOD4 without any unproven assumption. ∎

The structure of the protocol above is almost the same as the four move perfect ZKIP for QNR [7] or GNI [8] except for the first move of the prover P. The role of the first move of the prover P is to generate with overwhelming probability a pair of vectors $\langle S^0, S^1 \rangle$ such that every element of S^0 always has a fourth root modulo N but every element of S^1 does *not*. Note that when $N \in \overline{+3\text{MOD4}}$, this procedure can be done in probabilistic polynomial (in $|N|$) time.

4   Concluding Remarks

In this paper, we introduced two kinds of subsets of odd composite numbers, +1MOD4 and +3MOD4, and showed that (1) there exists a four (resp. five) move blackbox simulation perfect ZKIP for $\overline{+1\text{MOD4}}$ (resp. +1MOD4) without any unproven assumption; and (2) there exists a four (resp. five) move blackbox simulation perfect (resp. statistical) ZKIP for +3MOD4 (resp. $\overline{+3\text{MOD4}}$) without any unproven assumption.

These results negatively solve the question in subsection 1.1 unless +1MOD4 (or +3MOD4) is random self-reducible, i.e., it is not necessarily random self-reducible for a language L to have a constant move blackbox simulation perfect or statistical ZKIP for L and $\overline{L}$ without any unproven assumption unless +1MOD4 (or +3MOD4) is random self-reducible.

It is already known some results on the complexity of the perfect ZKIP. Fortnow [5] showed that if a language L has a perfect ZKIP, then $\overline{L} \in \mathcal{AM}$, and Aiello and Håstad [2] showed that if a language L has a perfect ZKIP, then $L \in \mathcal{AM}$, where $\mathcal{AM}$ [3] is a set of languages, each of which has a constant move interactive proof.

However it is known nothing on the complexity of the constant move perfect (or statistical) ZKIP. To the best of our knowledge, if a language L has a constant move (blackbox simulation) perfect or statistical ZKIP without any unproven assumption, then $\overline{L}$ also has a constant move blackbox simulation perfect or statistical ZKIP without any unproven assumption. Then

Open Question: Assume that a language L has a constant move blackbox simulation perfect or statistical ZKIP without any unproven assumption. Then does $\overline{L}$ generally have a constant move blackbox simulation perfect or statistical ZKIP without any unproven assumption?

References

[1] Adleman, L.M. and Huang, M.D.A., "Recognizing Primes in Random Polynomial Time," Proc. of STOC, pp.462-469 (1987).

[2] Aiello, W. and Håstad, J., "Statistical Zero-Knowledge Languages Can Be Recognized in Two Rounds," *JCSS*, Vol.42, No.3, pp.327-345 (1991).

[3] Babai, L. and Moran, S., "Arthur-Merlin Games: A Randomized Proof System, and a Hierarchy of Complexity Classes," *JCSS*, Vol.36, No.2, pp.254-276 (1988).

[4] Bellare, M., Micali, S., and Ostrovsky, R., "Perfect Zero-Knowledge in Constant Rounds," Proc. of STOC, pp.482-493 (1990).

[5] Fortnow, L., The Complexity of Perfect Zero-Knowledge," Proc. of STOC, pp.204-209 (1987).

[6] Goldreich, O. and Krawczyk, H., "On the Composition of Zero-Knowledge Proof Systems," Proc. of ICALP'90, LNCS 443, *Springer-Verlag*, Berlin, pp.268-282 (1990).

[7] Goldwasser, S., Micali, S., and Rackoff, C., "The Knowledge Complexity of Interactive Proof Systems," *SIAM J. Comput.*, Vol.18, No.1, pp.186-208 (1989).

[8] Goldreich, O., Micali, S., and Wigderson, A., "Proofs that Yield Nothing But Their Validity or All Languages in NP Have Zero-Knowledge Proof Systems," *JACM*, Vol.38, No.1, pp.691-729 (1991).

[9] Goldreich, O. and Oren, Y., "Definitions and Properties of Zero-Knowledge Proof Systems," *Tech. Rep.* #610, Technion, Department of Computer Science (1990).

[10] Rabin, M.O., "Probabilistic Algorithm for Primality Testing," *Journal of Number Theory*, Vol.12, pp.128-138 (1980).

[11] Solovay, R. and Strassen, V., "A Fast Monte Calro Test for Primality," *SIAM J. Comput.*, Vol.6, No.1, pp.84-85 (1977).

[12] Tompa, M. and Woll, H., "Random Self-Reducibility and Zero-Knowledge Interactive Proofs of Possession of Information," Proc. of FOCS, pp.472-482 (1987).

On Malign Input Distributions for Algorithms

Kojiro Kobayashi

Department of Information Sciences, Tokyo Institute of Technology
Ookayama, Meguro-ku, Tokyo 152, Japan

Abstract. By a measure we mean a function μ from $\{0,1\}^*$ (the set of all binary sequences) to real numbers such that $\mu(x) \geq 0$ and $\mu(\{0,1\}^*) < \infty$. A malign measure is a measure such that if an input x in $\{0,1\}^n$ (the set of all binary sequences of length n) is selected with the probability $\mu(x)/\mu(\{0,1\}^n)$ then the worst-case computation time $t_A^{wo}(n)$ and the average-case computation time $t_A^{av,\mu}(n)$ of an algorithm A for inputs of length n are functions of n of the same order for any algorithm A. Li and Vitányi found that a priori measures are malign. We show that "a priori"-ness and malignness are different in one strong sense.

1 Introduction

One important problem in the theory of complexity of computation is how the average-case computation time of algorithms is related to the probability distribution of inputs.

Concerning this problem, Li and Vitányi [2] found an interesting result that if inputs to algorithms are selected with the probability that is proportional to a semicomputable measure that is known as an a priori measure [3], then the worst-case computation time and the average-case computation time are roughly the same for any algorithm. A more precise statement of this result is as follows.

By a measure we mean a function μ from $\{0,1\}^*$ (the set of all binary sequences) to real numbers such that $\mu(x) \geq 0$ for any x and $\mu(\{0,1\}^*) < \infty$. For an algorithm A and a measure μ, let $t_A^{wo}(n)$ denote the worst-case computation time of A for inputs from $\{0,1\}^n$ (the set of all binary sequences of length n) and let $t_A^{av,\mu}(n)$ denote the average-case computation time of A for inputs from $\{0,1\}^n$ where a sequence x in $\{0,1\}^n$ is given to A with the probability $\mu(x)/\mu(\{0,1\}^n)$. Li and Vitányi's result says that if μ is an a priori measure then for any algorithm A there exists a constant $c(>0)$ such that $t_A^{wo}(n) \leq ct_A^{av,\mu}(n)$ for any n. Miltersen [4] called measures having this property "malign measures." Then Li and Vitányi's result says that a priori measures are malign.

A priori measures are very complicated mathematical objects and it is difficult to imagine a real situation where inputs are generated with the probability that is proportional to an a priori measure. Hence, if "a priori"-ness and malignness are equivalent, we may regard Li and Vitányi's result as one interesting but pathological phenomenon. In the present paper, we consider the problem of whether the two notions "a priori"-ness and malignness are different or not.

For this problem, we first introduce the notion of strongly malign measures and point out that what Li and Vitányi proved is that a priori measures are strongly malign. Although the intuitive notion of malignness allows several natural variations of definition of malignness other than the one introduced by Miltersen, strong malignness seems to imply all of these variations. Hence we may regard strong malignness as the strongest of these variations. Then we show that there exists a semicomputable strongly malign measure that is not a priori. This means that the two notions "a priori"-ness and malignness are different in one strong sense.

We also show two results concerning a priori measures and Kolmogorov complexity. For a rational number p such that $0 < p < 1$, let $P_p(x)$ denote the probability that a universal prefix-free algorithm A_U outputs a binary sequence x when the input binary sequence to A_U is randomly selected with p as the probability that the symbol 1 is selected. It is known that $P(x) = P_{1/2}(x)$ is an a priori measure. We show that $P_p(x)$ is an a priori measure for each p. Let $H_p(x)$ denote the smallest of $|u|_p$ for binary sequences u such that A_U outputs x when u is given to A_U as an input, where $|u|_p$ denotes the value $(-\log_2(1-p))$(number of 0's in u) + $(-\log_2 p)$(number of 1's in u). The value $H(x) = H_{1/2}(x)$ is known as the (prefix-free algorithm based) Kolmogorov complexity of x. We show that for each p there exists a constant c such that $|H(x) - H_p(x)| \leq c$ for any x.

2 Preliminaries

Let Σ denote the set $\{0,1\}$ of two symbols 0, 1, Σ^n denote the set of all sequences of 0, 1 of length n, and Σ^* denote the set $\Sigma^0 \cup \Sigma^1 \cup \Sigma^2 \dots$. We call an element of Σ^* a *word*. For each word x, let $|x|$ denote its length. Let λ denote the empty word, that is, the sequence of length 0. We use N, Q, R to denote the sets of all natural numbers, all rational numbers, and all real numbers respectively. We assume that some standard one-to-one mapping from N onto Σ^* is fixed and we identify a number with its corresponding word. Throughout in this paper, $\log x$ means $\log_2 x$.

By a *measure* we mean a function μ from Σ^* to R such that $\mu(x) \geq 0$ for any x and $\mu(\Sigma^*) < \infty$ (for $L \subseteq \Sigma^*$, $\mu(L)$ denotes $\Sigma_{x \in L} \mu(x)$). A measure μ is said to be *semicomputable* if there exists a computable function f from $\Sigma^* \times N$ to Q such that $f(x, n) \leq f(x, n')$ for $n \leq n'$ and $\lim_{n \to \infty} f(x, n) = \mu(x)$. For two functions f, g from N to R, we write $f \lesssim g$ if there exists a constant $c (> 0)$ such that $f(n) \leq cg(n)$ for any n. We write $f \approx g$ if both of $f \lesssim g$, $g \lesssim f$ hold true.

By an *algorithm* we mean a deterministic Turing machine. We assume that algorithms accept words as inputs and generate words as outputs if they halt. Let $\phi_A(x)$ denote the output of an algorithm A for an input x. When the algorithm does not halt, $\phi_A(x)$ is undefined. We assume that some standard one-to-one mapping from the set of all (representations of) algorithms onto Σ^* is fixed and we identify an algorithm with its corresponding word.

By a *step counting function*, we mean any partial function $\text{time}_A(x)$ that assigns a natural number to an algorithm A and an input x that satisfies the following two conditions (the "Blum's axioms" [1]): (1) $\phi_A(x)$ is defined $\iff$ $\text{time}_A(x)$ is defined; (2) the predicate "$\text{time}_A(x) \leq t$" on A, x, t is decidable. We understand that the predicate "$\text{time}_A(x) \leq t$" is false when $\text{time}_A(x)$ is undefined.

Suppose that one step counting function is fixed. We define the worst case computation time $t_A^{\text{wo}}(n)$ of an algorithm A for inputs of length n by $t_A^{\text{wo}}(n) = \max_{x \in \Sigma^n} \text{time}_A(x)$. For each measure μ, we define the average case computation time $t_A^{\text{av}, \mu}(n)$ of an algorithm A for inputs of length n by $t_A^{\text{av}, \mu}(n) = \Sigma_{x \in \Sigma^n} (\mu(x)/\mu(\Sigma^n))$

$\text{time}_A(x)$.

Definition 1 ([4]) *A measure μ is malign if it satisfies the following two conditions for any step counting function:* (i) *$\mu(\Sigma^n) > 0$ for any n;* (ii) *for any algorithm A that halts for all inputs there exists a constant c (> 0) such that $t_A^{wo}(n) \leq c\, t_A^{av,\mu}(n)$ for any n.*

3 A Priori Measures

There are several equivalent ways to define a priori measures. The first is to define them as semicomputable measures that are largest among all semicomputable measures.

Definition 2 *An a priori measure is a semicomputable measure μ such that for any semicomputable measure μ' there exists a constant $c(> 0)$ such that $\mu'(x) \leq c\mu(x)$ for any x.*

Existence of such measures is well-known ([3]).

Other ways to define a priori measures use prefix-free algorithms. For two words x, y, y is a *prefix* of x if there exists a word z such that $yz = x$. A set of words L is *prefix-free* if it contain no two different words x, y such that y is a prefix of x. Note that the value $\Sigma_{x \in L} 2^{-|x|}$ is always defined and is at most 1 for any prefix-free set L.

An algorithm A is *prefix-free* if the domain $\text{dom}(\phi_A)$ of the partial function ϕ_A is prefix-free. It is well-known that there exist one prefix-free algorithm A_U and one prefix-free set of words L_U that satisfy the condition: for any prefix-free algorithm A there exists $h \in L_U$ such that $\phi_{A_U}(hx) = \phi_A(x)$ for any x. We call such A_U a *universal prefix-free algorithm* and the word h a *code* of the algorithm A for A_U. We use A_U and L_U to denote some fixed universal prefix-free algorithm and set of codes.

For each word x we define a real number $P(x)$ and a natural number $H(x)$ by $P(x) = \Sigma_y \{2^{-|y|} \mid \phi_{A_U}(y) = x\}$ and $H(x) = \min\{|y| \mid y \in \Sigma^*, \phi_{A_U}(y) = x\}$. (We use the notation $\Sigma_x \{f(x) \mid S(x)\}$ to denote the sum of $f(x)$ for all words x that satisfy the condition $S(x)$.) The value $P(x)$ is the probability that a randomly selected infinite sequence of 0, 1 starts with a word y such that $\phi_{A_U}(y)=x$. The value $H(x)$ is the value that is known as the (prefix-free algorithm based) *Kolmogorov complexity* of x ([3]). The second way to define a priori measures μ is by the condition "μ is semicomputable and $\mu(x) \approx P(x)$" and the third is by the condition "μ is semicomputable and $\mu(x) \approx 2^{-H(x)}$." These three definitions of a priori measures are equivalent. From now on we fix one a priori measure and denote it by $\tilde{\mu}$.

Theorem 1 ([2]) *If μ is an a priori measure then it is malign.*

Proof. The proof in [2] is quite simple. However, here we devide it into two parts to make its logical structure clearer. The proof of the condition $\mu(\Sigma^n) > 0$ is easy and we omit it. Let μ be an a priori measure and let A be an algorithm that halts for all inputs. Suppose that a step counting function is fixed. Let $w(n)$ denote one of the worst inputs of length n.

(1) The proof that there exists a constant $c(> 0)$ such that $\mu(w(n)) \geq c\mu(\Sigma^n)$ for any n. Let A_0 be an algorithm such that

$$\phi_{A_0}(y) = \begin{cases} w(|\phi_{A_U}(y)|) & \phi_{A_U}(y) \text{ is defined,} \\ \text{"undefined"} & \text{otherwise.} \end{cases}$$

This algorithm A_0 is prefix-free. Let h be one of its codes. Then there exist constants c_1, c_2 (> 0) such that we have, for any n, $\mu(w(n)) \geq c_1 P(w(n)) \geq c_1 2^{-|h|} \Sigma_y \{2^{-|y|} \mid \phi_{A_0}(y) = w(n)\} \geq c_1 2^{-|h|} P(\Sigma^n) \geq c_1 2^{-|h|} c_2 \mu(\Sigma^n)$. Denoting the constant $c_1 2^{-|h|} c_2$ (> 0) by c, we have $\mu(w(n)) \geq c\mu(\Sigma^n)$ for any n.

(2) The proof that $t_A^{wo}(n) \leq (1/c) t_A^{av,\mu}(n)$ for any n. For any n, we have $t_A^{av,\mu}(n) = \Sigma_{x \in \Sigma^n} \text{time}_A(x) \, \mu(x) \, / \, \mu(\Sigma^n) \geq \text{time}_A(w(n)) \, \mu(w(n)) \, / \, \mu(\Sigma^n) \geq c \, t_A^{wo}(n)$. $\square$

4 Strongly Malign Measures

We defined malignness by Definition 1. However, there are several other ways to define the intuitive notion of malignness. The variations come from two factors.

First factor is what kind of step counting functions are allowed. In [2], this factor was not clearly stated. One extreme way is to fix one step counting function such as the number of steps performed by Turing machines. This is the approach adopted in [4]. Another extreme way is to allow all step counting functions that satisfy the Blum's axioms. Our definition adopts this approach. There are many other ways that are in between these two extreme approaches. At present, we do not know whether the notions of malignness defined by these two extreme approaches are different or not.

The second factor is how to define the sizes of inputs. In Definition 1, the size of an input x meant its length $|x|$. However, in practice we use the term "size of an input" more loosely. For example, if A is an algorithm for computing the product of two Boolean matrices, we will use "an input of size n" to denote a bit sequence of length $2n^2$ that is the representation of two $n \times n$ bit matrices. Generally, the definition of the size of an input depends on the algorithm that uses them as inputs.

There are several ways to formally define malignness. However, it is interesting to note that Li and Vitányi's result that a priori measures are malign seems to hold true for all of these variations of the definition of malignness. Therefore, a priori measures might have one essential feature from which follows the malignness of a priori measure for all these variations.

Analyzing the proof of Li and Vitányi's result (Theorem 1), we see that what they essentially proved is the following property of a priori measures μ.

(A1) There exists a constant $c(> 0)$ such that $\mu(w(n)) \geq c\mu(\Sigma^n)$ for any n.

The malignness of μ immediately follows from (A1). However, Li and Vitányi's proof can be readily modified to prove the following stronger property.

(A2) For any two partial recursive functions f from N to Σ^* and g from Σ^* to N, there exists a constant $c(> 0)$ such that $\mu(f(n)) \geq c\mu(g^{-1}(n))$ for any n such that $f(n)$ is defined.

We think that the porperty (A2) is a feature of a priori measures that deserves investigation in itself and call measures that have this property (A2) "strongly malign."

Definition 3 *A measure μ is strongly malign if it satisfies the following two conditions:*

   (1) $\mu(\Sigma^*) > 0$;

   (2) *for any algorithms A, B, there exists a constant $c(> 0)$ such that, for any $n \in \mathbb{N}$, if $\phi_A(n)$ is defined then $\mu(\phi_A(n)) \geq c\mu(\phi_B^{-1}(n))$.*

As is expected, a priori measures are strongly malign and strongly malign measures are malign.

Theorem 2 *A priori measures are strongly malign.*

Proof. Let μ be an a priori measure. Suppose that A, B are algorithms. Let A_0 be an algorithm such that $\phi_{A_0}(y) = \phi_A(\phi_B(\phi_{A_U}(y)))$. Then the step (1) of the proof of Theorem 1 becomes a proof of the strong malignness of μ if we replace $w(n)$ with $\phi_A(n)$, $|\phi_{A_U}(n)|$ with $\phi_B(\phi_{A_U}(n))$ and Σ^n with $\phi_B^{-1}(n)$.   □

The proof of the following theorem is easy and we omit it.

Theorem 3 *Strongly malign measures are malign.*

Let AP, SSMAL, SMAL, MAL denote the following classes of measures: AP=the class of all a priori measures, SSMAL=the class of all semicomputable strongly malign measures, SMAL=the class of all strongly malign measures, MAL=the class of all malign measures. The above theorems imply AP $\subseteq$ SSMAL $\subseteq$ SMAL $\subseteq$ MAL. We will show that these four classes are all different. To show SSMAL$\neq$SMAL and SMAL$\neq$MAL is straightforward.

Theorem 4 *There exists a strongly malign measure that is not semicomputable.*

Proof (an outline). For each n let x_n be a word of length n such that $H(x_n) \geq n/2$ and let X be the set $\{x_0, x_1, \ldots\}$. We can relativize the notions of "a priori", "strongly malign" and $H(x)$ with respect to X. Let μ be a measure that is a priori relative to X. Then μ is strongly malign relative to X and is hence strongly malign. Let $\tilde{\mu}$ be an a priori measure. Then we can prove that there exist constants c_1, c_2 such that $\tilde{\mu}(x_n) \leq c_1 2^{-n/2}$ and $\mu(x_n) \geq c_2 2^{-H^X(x_n)} \geq c_2 2^{-2\log n} = c_2/n^2$ for any n ($H^X(x)$ denotes the relativized $H(x)$). This shows that μ is not semicomputable.   □

Theorem 5 *There exists a malign measure that is not strongly malign.*

Proof (an outline). Let $\tilde{\mu}$ be an a priori measure. We can easily show that the measure μ defined by $\mu(x) = \tilde{\mu}(x)/|x|$ is malign and is not a priori.   □

5 Characterizations of a Priori and Strongly Malign Measures

Before proceeding to the proof of AP $\neq$ SSMAL we give some characterizations of the two classes AP and SMAL. These characterizations show an essential difference, if any, of the two classes AP and SSMAL.

Theorem 6 *For a measure μ, the following two conditions are equivalent.*
(1) *μ is a strongly malign measure.*
(2) *μ satisfies the conditions:*
  (i) *$\mu(\Sigma^*) > 0$;*
  (ii) *for any algorithm A there exists a constant $c(> 0)$ such that $\mu(x) \geq c\mu(\phi_A^{-1}(x))$ for any x.*

(I) Proof of (1)$\Rightarrow$(2). Suppose that μ is strongly malign. Let A be an algorithm. Let B, C be algorithms such that $\phi_B(n) = n$, $\phi_C(y) = \phi_A(y)$. Then there exists a constant c (> 0) such that $\mu(\phi_B(n)) \geq c\mu(\phi_C^{-1}(n))$ for any $n \in \mathbb{N}$, and hence $\mu(x) = \mu(\phi_B(x)) \geq c\mu(\phi_C^{-1}(x)) = c\mu(\phi_A^{-1}(x))$ for any $x \in \Sigma^*$.
(II) Proof of (2)$\Rightarrow$(1). Suppose that a measure μ satisfies (i) – (ii). Let A, B be algorithms. Let C be an algorithm such that $\phi_C(y) = \phi_A(\phi_B(y))$ for any $y \in \Sigma^*$. There exists a constant $c(> 0)$ such that $\mu(x) \geq c\mu(\phi_C^{-1}(x))$ for any $x \in \Sigma^*$. Then we have $\mu(\phi_A(n)) \geq c\mu(\phi_C^{-1}(\phi_A(n))) \geq c\mu(\phi_B^{-1}(n))$ for any $n \in \mathbb{N}$ such that $\phi_A(n)$ is defined. $\qquad\square$

Recall that by $\tilde{\mu}$ we denote some fixed a priori measure.

Theorem 7 *For a measure μ, the following two conditions are equivalent.*
(1) *μ is an a priori measure.*
(2) *μ satisfies the conditions:*
  (i) *μ is semicomputable;*
  (ii) *$\mu(\Sigma^*) > 0$;*
  (iii) *there exists a constant $c(> 0)$ such that $\mu(x) \geq c\tilde{\mu}(A)\mu(\phi_A^{-1}(x))$ for any algorithm A and any x.*

(I) Proof of (1) $\Rightarrow$ (2). Let μ be an a priori measure. It is obvious that (i), (ii) hold true. Let B be a prefix-free algorithm such that $\phi_B(uv) = \phi_w(\phi_{A_U}(v))$, with $w = \phi_{A_U}(u)$, and let h be one of its codes. If u is a description of A (that is, a word u such that $\phi_{A_U}(u) = A$) and v is a description of a word in $\phi_A^{-1}(x)$, then we have $\phi_{A_U}(huv) = \phi_B(uv) = x$. Therefore there exist constants $c_1, c_2(> 0)$ such that

$$
\begin{aligned}
\mu(x) &\geq c_1 P(x) = c_1\Sigma_y\{2^{-|y|} \mid \phi_{A_U}(y) = x\} \\
&\geq c_12^{-|h|}(\Sigma_u\{2^{-|u|} \mid \phi_{A_U}(u) = A\})(\Sigma_v\{2^{-|v|} \mid \phi_{A_U}(v) \in \phi_A^{-1}(x)\}) \\
&= c_12^{-|h|}P(A)P(\phi_A^{-1}(x)) \geq c_12^{-|h|}c_2\tilde{\mu}(A)\mu(\phi_A^{-1}(x)).
\end{aligned}
$$

(II) Proof of (2)$\Rightarrow$(1) (An outline). Suppose that a measure μ satisfies (i) – (iii). For each word x_0 let A_{x_0} be an alrogithm such that $\phi_{A_{x_0}}(x) = x_0$ for any x. The algorithm A_{x_0} is determined by x_0 and hence there is a constant c_1 such that $H(A_{x_0}) \leq H(x_0) + c_1$. Hence there is a constant $c_2(> 0)$ such that $\tilde{\mu}(A_{x_0}) \geq c_2\tilde{\mu}(x_0)$, and we have $\mu(x_0) \geq c\tilde{\mu}(A_{x_0})\mu(\phi_{A_{x_0}}^{-1}(x_0)) \geq cc_2\tilde{\mu}(x_0)\mu(\Sigma^*)$. Denoting the constant $cc_2\mu(\Sigma^*)$ (> 0) by c_3, we have $\mu(x_0) \geq c_3\tilde{\mu}(x_0)$ for any x_0. Moreover μ is semicomputable. Hence μ is a priori. $\qquad\square$

Comparing these two theorems, we see that the essential difference between SS-MAL and AP lies in the difference of the behavior of $\mu(x)/\mu(\phi_A^{-1}(x))$ as a function of x. For a measure in SSMAL, this value is bounded from below by a constant c_A (> 0) that may depend on the algorithm A. For a measure in AP, the same is true and moreover the constant c_A can be represented as $c\tilde{\mu}(A)$ with a constant c (> 0) that does not depend on A. Therefore, the problem of whether AP = SSMAL or not is the problem of whether the nonuniform existence of constants for measures in SSMAL can be always changed to uniform existence or not.

6 Semicomputable Strongly Malign Measures that are not a Priori

We show one method for constructing a family of semicomputable strongly malign measures. This family contains some of a priori measures, and also measures that are not a priori.

Let f be a computable function from $\Sigma^* \times \Sigma$ to Q such that $f(u,a) > 0$, $f(u,0)+f(u,1) \leq 1$, and $f(uv,a) \geq f(v,a)$ for any $u,v \in \Sigma^*$ and $a \in \Sigma$. For this f, let π_f be the function from Σ^* to Q defined by $\pi_f(\lambda) = 1$ and $\pi_f(a_1a_2...a_m) = f(\lambda,a_1)f(a_1,a_2)f(a_1a_2,a_3)...f(a_1a_2...a_{m-1},a_m)$ for $m \geq 1$. It is easy to show that $\pi_f(u_1...u_m) \geq \pi_f(u_1)...\pi_f(u_m)$ for any words $u_1,...,u_m$ and that $\Sigma_s\{\pi_f(s) \mid s \in L\} \leq 1$ for any prefix-free set L. Finally, let P_f be the measure defined by $P_f(x) = \Sigma_s\{\pi_f(s) \mid \phi_{A_U}(s) = x\}$.

For two rational numbers p,q such that $0 < p$, $0 < q$, $p+q \leq 1$, let $\pi_{q,p}(u)$ and $P_{q,p}(x)$ denote $\pi_f(u)$ and $P_f(x)$ respectively for the function f defined by $f(u,0) = q$, $f(u,1) = p$. For a rational number p such that $0 < p < 1$, let $\pi_p(u)$ and $P_p(x)$ denote $\pi_{1-p,p}(u)$ and $P_{1-p,p}(x)$ respectively. Moreover, for each $a \in \Sigma$ let $|0|_p$ and $|1|_p$ denote $-\log(1-p)$ and $-\log p$ respectively, for each word $u = a_1...a_m$ let $|u|_p$ denote $|a_1|_p + ... + |a_m|_p$ ($= -\log \pi_p(u)$), and for each word x let $H_p(x)$ denote $\min\{|u|_p \mid u \in \Sigma^*, \phi_{A_U}(u) = x\}$. In these notations, the measure $P(x)$ and the Kolmogorov complexity $H(x)$ introduced in Section 3 are $P_{1/2}(x)$ and $H_{1/2}(x)$ respectively.

In this section we show four results: (1) $P_f(x)$ is semicomputable and strongly malign; (2) if $p+q < 1$ then $P_{q,p}(x)$ is not a priori; (3) $P_p(x)$ is a priori; (4) there is a constant c such that $|H(x) - H_p(x)| \leq c$ for any x. The results (1), (2) imply AP $\neq$ SSMAL.

Theorem 8 *The measure $P_f(x)$ is semicomputable and strongly malign.*

Proof. It is obvious that $P_f(x)$ is semicomputable. We use Theorem 6 to prove that $P_f(x)$ is strongly malign. It is obvious that $P_f(\Sigma^*) > 0$. Let A be an algorithm. Let A_0 be a prefix-free algorithm such that $\phi_{A_0}(s) = \phi_A(\phi_{A_U}(s))$ and let h be one of its codes. Then, for any x we have $P_f(x) = \Sigma_r\{\pi_f(r) \mid \phi_{A_U}(r) = x\} \geq \Sigma_s\{\pi_f(hs) \mid \phi_{A_U}(hs) = x\} \geq \pi_f(h)\Sigma_s\{\pi_f(s) \mid \phi_{A_U}(s) \in \phi_A^{-1}(x)\} = \pi_f(h)P_f(\phi_A^{-1}(x))$. Denoting the constant $\pi_f(h)$ (> 0) by c, we have $P_f(x) \geq cP_f(\phi_A^{-1}(x))$. $\square$

For a word $w = a_1a_2...a_n$ of length n and i such that $0 \leq i \leq n$, let $w|i$ denote the head $a_1a_2...a_i$ of w of length i.

Theorem 9 *If f satisfies the condition*

$$\lim_{n \to \infty} \max_{w \in \Sigma^*} \prod_{i=0}^{n-1}(f(w|i,0) + f(w|i,1)) = 0,$$

then the measure $P_f(x)$ is not a priori.

Proof. Let g be the function from $\Sigma^* \times \Sigma$ to Q defined by $g(u, a) = f(u, a)/(f(u, 0) + f(u, 1))$. Suppose that $P_f(x)$ is a priori. Then there exists a constant $c(> 0)$ such that $P_g(x) < cP_f(x)$ for any x because $P_g(x)$ is semicomputable.

Let n_0 be a value such that $\prod_{i=0}^{n-1}(f(w|i, 0) + f(w|i, 1)) \leq 1/c$ for any $n \geq n_0$ and any word w of length n. For any word $s = a_1 a_2 \ldots a_n$ of length $n \geq n_0$ we have $\pi_f(s) \leq (1/c)\pi_g(s)$ because

$$\pi_f(s) = \prod_{i=1}^{n} f(a_1 \ldots a_{i-1}, a_i) = \pi_g(s) \prod_{i=1}^{n}(f(a_1 \ldots a_{i-1}, 0) + f(a_1 \ldots a_{i-1}, 1))$$

$$= \pi_g(s) \prod_{j=0}^{n-1}(f(s|j, 0) + f(s|j, 1)) \leq (1/c)\pi_g(s).$$

Let x be a word such that there is no s of length less than n_0 such that $\phi_{A_U}(s) = x$. Then we have $P_f(x) = \Sigma_s \{\pi_f(s) \mid \phi_{A_U}(s) = x\} \leq \Sigma_s\{(1/c)\pi_g(s) \mid \phi_{A_U}(s) = x\} = (1/c)P_g(x)$. This contradicts $P_g(x) < cP_f(x)$. □

Corollary 1 *If p, q are rational numbers such that $0 < p$, $0 < q$, $p + q < 1$, then $P_{q,p}(x)$ is a semicomputable strongly malign measure that is not a priori.*

Proof. We have $\lim_{n \to \infty} \max_{w \in \Sigma^*} \prod_{i=0}^{n-1}(f(w|i, 0) + f(w|i, 1)) = \lim_{n \to \infty} (p + q)^n = 0$. □

Now we show the results (3), (4). In the remainder of this section we fix one rational number p such that $0 < p < 1$ and denote $1 - p$ by q. Moreover, we simply write $\pi(u)$ instead of $\pi_p(u)$ $(= \pi_{1-p,p}(u))$.

We define "the interval represented by a word u" as follows. The empty word λ represents the unit interval $(0, 1)$. If a word u represents an interval (s, t), then $u0$ represents the interval $(s, s + q(t - s))$ and $u1$ represents the interval $(s + q(t - s), t)$. It is obvious that the size $t - s$ of the interval (s, t) represented by a word u is $\pi(u)$. Let τ be the function from Σ^* to Q such that the interval represented by u is $(\tau(u), \tau(u) + \pi(u))$. This function τ is defined inductively by $\tau(\lambda) = 0$, $\tau(u0) = \tau(u)$, $\tau(u1) = \tau(u) + \pi(u)q$. It is easy to show that if $|u| = |u'|$ and u' is greater than u by one as a binary number then $\tau(u) + \pi(u) = \tau(u')$. We call a number r in the unit interval $(0, 1)$ p-*rational* if there exists a word u such that $r = \tau(u)$, and p-*irrational* otherwise. We call a subinterval (s, t) of $(0, 1)$ a p-*interval* if it is the interval represented by some word u. It is well-known that any subinterval (s, t) of size d $(= t - s)$ of the unit interval $(0, 1)$ contains a $1/2$-interval (a "binary interval") of size at least $d/4$. This fact is used to prove $2^{-H(x)} \gtrsim P(x)$. We show a similar result for p-intervals.

Lemma 1 *Any subinterval of size d of $(0, 1)$ contains a p-interval of size at least $(d/2) \cdot \min\{q, p\}$.*

Proof (an outline). Let (s, t) be a subinterval of size d of $(0, 1)$. We may assume that $0 < s$ and s, t are p-irrationals. Let u be the shortest word such that $\tau(u)$ is in the interval (s, t). There is a word u' such that $u = u'1$. At least one of

$t - \tau(u)$, $\tau(u) - s$ is at least $d/2$. If $t - \tau(u) \geq d/2$ then let h be the value such that $\tau(u) + \pi(u0^h) \geq t$ and $\tau(u) + \pi(u0^{h+1}) < t$. Then the p-interval $(\tau(u0^{h+1})$, $\tau(u0^{h+1}) + \pi(u0^{h+1}))$ satisfies the condition. If $\tau(u) - s \geq d/2$ then let h be the value such that $\tau(u'01^h) \leq s$, $\tau(u'01^{h+1}) > s$. Then the p-interval $(\tau(u'01^{h+1})$, $\tau(u'01^{h+1}) + \pi(u'01^{h+1}))$ satisfies the condition. □

Theorem 10 *If p is a rational number such that $0 < p < 1$ then there is a constant c such that $H_p(x) \leq -\log P(x) + c$ for any x.*

Proof (an outline). It is obvious that $P_p(x)$ is semicomputable. We show that there exists a constant $c(> 0)$ such that $P(x) \leq cP_p(x)$ for any x.

Let X be the recursively enumerable set $\{(x, m) \mid x \in \Sigma^*, m \in \mathbb{N}, 2^{-m} < P(x)/2\}$ and let (x_0, m_0), (x_1, m_1), (x_2, m_2), ... be the enumeration of all elements of X by some algorithm. We have $\Sigma_{i=0}^{\infty} 2^{-m_i} \leq 1$ and hence the interval $(2^{-m_0} + \ldots + 2^{-m_{i-1}}, 2^{-m_0} + \ldots + 2^{-m_i})$ is a subinterval of size 2^{-m_i} of the unit interval $(0, 1)$ for each $i \geq 0$. Let z_i be (the leftmost of) the largest p-interval contained in this interval. Let this p-interval z_i be represented by a word u_i. By Lemma 1, the size $\pi_p(u_i)$ of the p-interval z_i is at least $(2^{-m_i}/2)\min\{q, p\}$. Let A be the algorithm such that if there exists i such that $u = u_i$ then $\phi_A(u) = x_i$, and $\phi_A(u)$ is undefined otherwise. The domain of ϕ_A is the prefix-free set $\{u_0, u_1, \ldots\}$ and hence A is prefix-free. Let h be one of its codes.

Let x be a word. We have $2^{-\lceil -\log P(x) \rceil - 2} < P(x)/2$. Hence $(x, \lceil -\log P(x) \rceil + 2)$ is in X. Let (x_i, m_i) be this element. Then we have $\phi_{A_U}(hu_i) = \phi_A(u_i) = x_i = x$, and hence $H_p(x) \leq |hu_i|_p = -\log \pi(hu_i) = -\log \pi(h) - \log \pi(u_i) \leq -\log \pi(h) + m_i + 1 - \log(\min\{q, p\}) \leq -\log \pi(h) + 4 - \log(\min\{q, p\}) - \log P(x)$. Denoting the constant $-\log \pi(h) + 4 - \log(\min\{q, p\})$ by c, we have $H_p(x) \leq -\log P(x) + c$. □

The proofs of the following two corollaries are easy if we note that $2^{-H_p(x)}$ is semicomputable and that $P_p(x) \geq 2^{-H_p(x)}$.

Corollary 2 *If p is a rational number such that $0 < p < 1$, then $P_p(x)$ is an a priori measure.*

Corollary 3 *If p is a rational number such that $0 < p < 1$, then there exists a constant c such that $|H(x) - H_p(x)| \leq c$ for any x.*

7 One Condition for Semicomputable Strongly Malign Measures to be a Priori

In this section we show one condition for measures to be a priori under the assumption that they are semicomutable and strongly malign. We expect such a condition to be weaker than the conditions for the case where no assumption on the measures is made.

We call a function F from Σ^* to $\mathbb{R}$ *feasible* if it is semicomputable, $F(s) \geq 0$ for any word s, $F(L) \leq 1$ for any prefix-free set L, and there exists a prefix-free algorithm A such that the measure $\mu(x)$ defined by $\mu(x) = \Sigma_s\{F(s) \mid \phi_A(s) = x\}$ is a priori. The function π_p is an example of feasible functions by Corollary 2. Especially the function $F(s) = 2^{-|s|}$ is feasible. For a measure μ, let $\rho_\mu(s)$ denote $\inf_{y \in \Sigma^*} \mu(sy)/\mu(y)$.

Theorem 11 *Let F be a feasible function and μ be a semicomputable strongly malign measure. Then the following two conditions are equivalent.*

(1) *μ is an a priori measure.*

(2) *For any recursively enumerable prefix-free set L there exists a constant $c(> 0)$ such that $F(s) \leq c\rho_\mu(s)$ for any s in L.*

Proof of (1)$\Rightarrow$(2) (an outline) Let L be a recursively enumerable prefix-free set. Using an enumeration (s_0, m_0), (s_1, m_1), ... of the recursively enumerable set $X = \{(s, m) \mid s \in L, m \in \mathbb{N}, 2^{-m} < F(s)/2\}$, we can show that there is a prefix-free algorithm A_0 such that for any $s \in L$ there exists u such that $\phi_{A_0}(u) = s$ and $|u| \leq -\log F(s) + 5$. Its proof is similar to that of Theorem 10, but we use Lemma 1 with $p = 1/2$.

Let A_1 be a prefix-free algorithm such that $\phi_{A_1}(uv) = \phi_{A_0}(u)\phi_{A_U}(v)$ and let h be one of its codes. Then, for any $s \in L$ and any y we have $H(sy) \leq |h| - \log F(s) + 5 + H(y)$, and hence $\mu(sy) \geq c_1 2^{-H(sy)} \geq c_1 2^{-|h|-5} F(s) 2^{-H(y)} \geq c_1 c_2 2^{-|h|-5} F(s)\mu(y)$ for some constants c_1, c_2 (> 0).

Proof of (2)$\Rightarrow$(1). Let A_0 be a prefix-free algorithm such that the measure $\mu'(x)$ defined by $\mu'(x) = \Sigma_s \{F(s) \mid \phi_{A_0}(s) = x\}$ is a priori. The domain $\mathrm{dom}(\phi_{A_0})$ of ϕ_{A_0} is a recursively enumerable prefix-free set. Let $c_1(> 0)$ be a constant such that $F(s) \leq c_1 \rho_\mu(s)$ for any $s \in \mathrm{dom}(\phi_{A_0})$. Let A_1 be the algorithm such that $\phi_{A_1}(sy) = \phi_w(y)$, with $w = \phi_{A_0}(s)$. By Theorem 6 there exists a constant $c_2(> 0)$ such that $\mu(x) \geq c_2 \mu(\phi_{A_1}^{-1}(x))$ for any x.

For any algorithm A and any word x, we have

$$
\begin{aligned}
\mu(\phi_A^{-1}(x))\mu'(A) &= \mu(\{y \mid y \in \Sigma^*, \phi_A(y) = x\})\Sigma_s\{F(s) \mid \phi_{A_0}(s) = A\} \\
&\leq \Sigma_{s,y}\{\mu(y)F(s) \mid s \in \mathrm{dom}(\phi_{A_0}), \phi_{A_1}(sy) = x\} \\
&\leq \Sigma_{s,y}\{c_1\mu(sy) \mid s \in \mathrm{dom}(\phi_{A_0}), \phi_{A_1}(sy) = x\} \\
&\leq c_1\mu(\phi_{A_1}^{-1}(x)) \leq c_1(1/c_2)\mu(x).
\end{aligned}
$$

This shows that μ is a priori by Theorem 7. □

References

1. M. Blum: A machine-independent theory of the complexity of recursive functions. J. ACM 14, 322-336 (1967)

2. M. Li, P.M.B. Vitányi: A theory of learning simple concepts under simple distributions and average case complexity for the universal distribution. In: Proc. of the 30th FOCS, 1989, pp.34-39

3. M. Li, P.M.B. Vitányi: Kolmogorov complexity and its applications. In: J. van Leeuwen (ed.): Handbook of Theoretical Computer Science, Vol. A, Chap. IV. Elsevier and MIT Press 1990, pp.187-254

4. P.B. Miltersen: The complexity of malign ensembles. In: Proc. of the 6th SICT, 1991, pp.164-171

Lowness and the Complexity of Sparse and Tally Descriptions

V. Arvind*
Department of Computer Science and Engineering
Indian Institute of Technology, Delhi
New Delhi 110016, India

J. Köbler and M. Mundhenk
Abteilung für Theoretische Informatik
Universität Ulm, Oberer Eselsberg
D-W-7900 Ulm, Germany

Abstract

We investigate the complexity of obtaining sparse descriptions for sets in various reduction classes to sparse sets. Let A be a set in a certain reduction class $R_r(\text{SPARSE})$. Then we are interested in finding upper bounds for the complexity (relative to A) of sparse sets S such that $A \in R_r(S)$. By establishing such upper bounds we are able to derive the lowness of A.

In particular, we show that if a set A is in the class $R^p_{hd}(R^p_c(\text{SPARSE}))$ then A is in $R^p_{hd}(R^p_c(S))$ for a sparse set $S \in \text{NP}(A)$. As a consequence we can locate $R^p_{hd}(R^p_c(\text{SPARSE}))$ in the EL^Θ_3 level of the extended low hierarchy. Since $R^p_{hd}(R^p_c(\text{SPARSE})) \supseteq R^p_t(R^p_c(\text{SPARSE}))$ this solves the open problem of locating the closure of sparse sets under bounded truth-table reductions optimally in the extended low hierarchy. Furthermore, we show that for every $A \in R^p_d(\text{SPARSE})$ there exists a sparse set $S \in \text{NP}(A \oplus \text{SAT})/F\Theta^p_2(A)$ such that $A \in R^p_d(S)$. Based on this we show that $R^p_{1\text{-}tt}(R^p_d(\text{SPARSE}))$ is in EL^Θ_3.

Finally, we construct for every set $A \in R^p_c(\text{TALLY}) \cap R^p_d(\text{TALLY})$ (or equivalently, $A \in \text{IC}[\log, \text{poly}]$, as shown in [AHH+92]) a tally set $T \in P(A \oplus \text{SAT})$ such that $A \in R^p_c(T) \cap R^p_d(T)$. This implies that the class $\text{IC}[\log, \text{poly}]$ of sets with low instance complexity is contained in EL^Σ_1.

1 Introduction

Sparse sets play a central role in structural complexity theory. The question of the existence of sparse hard sets for various complexity classes under different sorts

---

*Work done while visiting Universität Ulm. Supported in part by an Alexander von Humboldt research fellowship.

of reducibilities is well studied (see for example [KL80, Mah82, OW91, AHH+92, AKM92]). Besides that much work has been done on other issues concerning the complexity of sets reducible to sparse sets (see for example [Kad87, AH92, SL92, AHH+92]). A central motivation for most earlier work (and also for this paper) can be seen as seeking answers to the following two questions.

1. If a set A is reducible to a sparse set, does it follow that A is reducible to some sparse set that is "simple" relative to A?

2. If a set A is reducible to a sparse set, then how easy is it to access all the relevant information contained in the set A when it is used as oracle?

Question 1 originates in the study of the notions of equivalence and reducibility to sparse sets (see for example [TB91, AHOW, GW91]) and addresses the complexity (relative to A) of sparse descriptions for sets A which are reducible to sparse sets. For the case of Turing reductions, Gavaldà and Watanabe [GW91] established an NP $\cap$ co-NP lower bound by constructing a set $B \in R_T^p(\text{SPARSE})$ (in fact, B is even in $R_c^p(\text{SPARSE})$) that is not Turing reducible to a sparse set in $\text{NP}(B) \cap \text{co-NP}(B)$. For truth-table reduction classes to sparse sets, the first question is also investigated in [AHH+92] where various upper bounds for the relative complexity of sparse descriptions are presented.

The second question concerns lowness properties of sets reducible to sparse sets. Lowness is an important structural tool for systematically classifying sets in different complexity classes according to the power they provide when used as oracles. Building on ideas from recursive function theory, Schöning [Sch83] defined the low hierarchy inside NP. In order to classify sets outside of NP (e.g. sets reducible to sparse sets) Balcázar, Book, and Schöning [BBS86] defined the extended low hierarchy. The extended low hierarchy was subsequently refined by Allender and Hemachandra [AH92] and Long and Sheu [LS91] who showed the optimal location of various reduction and equivalence classes to sparse and tally sets in the (refined) extended low hierarchy. Very recently, Sheu and Long [SL92] proved that the extended low hierarchy is an infinite hierarchy.

In order to investigate the first question, we generalize the definition of sparse set descriptions given in [GW91]. Let $\leq_r$ be a reducibility. A sparse set S is a sparse r-description for a set A if $A \leq_r S$. Similarly, a tally set T is a tally r-description for A if $A \leq_r T$. In particular a sparse set S is called a sparse description for A if $A \leq_T^p S$. We are interested in finding upper bounds for the relative complexity of sparse (tally) r-descriptions for sets in $R_r(\text{SPARSE})^1$ (respectively $R_r(\text{TALLY})$). We refer to a sparse r-description satisfying the established upper bound as a *simple* sparse r-description (with respect to that upper bound), and we refer to the corresponding upper bound result as a *simplicity* result.

In Section 3 we establish simplicity results for various reduction classes to sparse and tally sets. In Section 4 we apply the simplicity results to derive lowness properties. These results reveal a close interconnection between the lowness of sets which are reducible to sparse or tally sets and the complexity of computing small descriptions for these sets. More precisely, the general pattern to prove the lowness of a set A that reduces to a sparse set is the following: first we appropriately bound the complexity of a sparse description for A; based on such a simplicity result, we then

---

[1] For a reducibility $\leq_r$ and a class C of sets, $R_r(C) = \{A \mid A \leq_r B \text{ for some } B \in C\}$.

Name	Notation
many-one	$\leq_m^p$
one truth-table	$\leq_{1-tt}^p$
bounded truth-table	$\leq_b^p$
conjunctive	$\leq_c^p$
disjunctive	$\leq_d^p$
bounded Hausdorff	$\leq_{bhd}^p$
Hausdorff	$\leq_{hd}^p$
np many-one	$\leq_m^{np}$
co-np many-one	$\leq_m^{co-np}$

Table 1: Polynomial-time Reductions

apply a deterministic enumeration technique like that of Mahaney [Mah82] or a census technique similar to that of Hemachandra [Hem87] and Kadin [Kad87] to replace the sparse oracle S (and thus A). Intuitively, we use the simplicity result to extract enough information (a suitable initial segment of S or the census thereof) from the oracle set A in order to avoid further queries to A. Our approach clarifies that the appropriate simplicity result is the basic reason for the set to be low. Using this approach we are able to derive several new and optimal lowness results.

2 Preliminaries and notation

Let A be a set. χ_A denotes the characteristic function of A. $A^{=n}$ ($A^{\leq n}$) denotes the set of all strings in A of length n (up to length n, respectively). The cardinality of A is denoted by $|A|$. A set T is called a tally set if $T \subseteq 0^*$. The census function of a set A is $census_A(1^n) = |A^{\leq n}|$. A set S is called sparse if its census function is bounded above by a polynomial. We use TALLY and SPARSE to represent the classes of tally and sparse sets, respectively.

Let C be a class of sets. A set S is called $P(C)$-printable (see [HY84]) if the function $0^n \mapsto \langle S^{\leq n} \rangle$ can be computed inside $FP(C)$.

The reducibilities discussed in this paper are the standard polynomial time bounded reducibilities defined by Ladner, Lynch, and Selman [LLS75] and the Hausdorff reduction introduced by Wagner [Wag87] (see Table 1 for an overview).

Definition 2.1 [Wag87] *A is Hausdorff-reducible to B ($A\leq_{hd}^p B$), if there exists a function f computable in polynomial time, such that for all x, f computes a tuple $f(x) = \langle y_1, y_2, \ldots, y_{2k} \rangle$ such that*

1. $\chi_B(y_1) \geq \chi_B(y_2) \geq \cdots \geq \chi_B(y_{2k})$, and

2. $\chi_A(x) = \bigvee_{i=1}^{k} [\chi_B(y_{2i-1}) \wedge \neg\chi_B(y_{2i})]$,

We call f a bounded Hausdorff reduction ($A\leq_{bhd}^p B$) if k is a constant.

Notation [AHOW] For any reducibility $\leq_r^\alpha$ where $\alpha \in \{p, np, co\text{-}np\}$ and $r \in \{m, c, d, 1\text{-}tt, b, hd, bhd\}$ and any class C of sets let $R_r^\alpha(C) = \{A \mid A \leq_r^\alpha B$ for some $B \in C\}$.

A class $\mathcal{K}$ of sets that includes $\emptyset$ and Σ^* and is closed under union and intersection is said to be a set ring. The following characterization of the boolean closure of set rings due to Hausdorff plays a key role in some of our results.

Theorem 2.2 [Hau14] *Let $\mathcal{K}$ be a set ring and let $BC(\mathcal{K})$ be the closure of $\mathcal{K}$ under union, intersection, and complement. Then every $A \in BC(\mathcal{K})$ can be represented as $A = \bigcup_{i=1}^{k}(A_{2i-1} - A_{2i})$, where $A_j \in \mathcal{K}$, $1 \leq j \leq 2k$, and $A_1 \supseteq A_2 \supseteq \cdots \supseteq A_{2k}$.*

The above representation for A is called a Hausdorff representation for A over $\mathcal{K}$. We state some useful properties relating boolean closures, bounded truth-table closures, and the bounded Hausdorff closures of language classes.

Lemma 2.3 [KSW87] *Let $\mathcal{K}$ be a class that contains P and is closed under many-one reductions. Then $BC(\mathcal{K}) = R_b^p(\mathcal{K})$.*

Lemma 2.4 *Let $\mathcal{K}$ be a class closed under many-one reductions and under join. Then A has a Hausdorff representation over $\mathcal{K}$ if and only if A is bounded Hausdorff reducible to some set in $\mathcal{K}$.*

The next lemma is obtained by combining Theorem 2.2 and the above two lemmas.

Lemma 2.5 *If $\mathcal{K}$ is a set ring which contains P and is closed under many-one reductions then $BC(\mathcal{K}) = R_{bhd}^p(\mathcal{K}) = R_b^p(\mathcal{K})$.*

Theorem 2.6 *1. $R_{bhd}^p(R_c^p(\text{SPARSE})) = R_b^p(R_c^p(\text{SPARSE}))$*

2. $R_{bhd}^p(R_m^{co\text{-}np}(\text{SPARSE})) = R_b^p(R_m^{co\text{-}np}(\text{SPARSE}))$

For a class C of sets and a class $\mathcal{F}$ of functions from Σ^* to Σ^* let $C/\mathcal{F}$ [KL80] be the class of sets A such that there is a set $B \in C$ and a function $h \in \mathcal{F}$ such that for all $x \in \Sigma^*$,

$$\chi_A(x) = \chi_B(x, h(0^{|x|}))$$

Although Karp and Lipton introduced the notion of advice functions in order to characterize nonuniform complexity classes by imposing a quantitative length restriction on the functions in $\mathcal{F}$, we will consider here complexity restricted advice function classes. We refer the reader to [KT90] for a general study of complexity restricted advice functions.

For a set A the class $\Theta(A)$ [LS91] contains all languages $L(M, A)$ accepted by a deterministic polymomial time bounded oracle machine M asking on inputs of length n at most $O(\log n)$ queries to A. A deterministic polynomial-time oracle machine M as in the definition above is called a Θ machine. The Θ levels of the (relativized) polynomial time hierarchy are defined as $\Theta_k^p(A) = \Theta(\Sigma_{k-1}^p(A))$, $k \geq 1$ [Wag90, LS91].

Similarly, the class $F\Theta(A)$ contains all functions computable by some deterministic polymomial time bounded oracle transducer M asking on inputs of length n at most $O(\log n)$ queries to oracle A, and $F\Theta_k^p(A) = F\Theta(\Sigma_{k-1}^p(A))$, $k \geq 1$.

For further definitions used in this paper we refer the reader to standard books on structural complexity theory (for example, [BDG, Sch86]).

3 Upper bounds for sparse and tally descriptions

In this section we consider the following question: If a set A reduces to a sparse set S via a reduction of a certain type, does it follow that A reduces via a reduction of the same type to some sparse set S' that is "simple" relative to A?

We notice that a simplicity argument was already used in the proof of Mahaney's theorem [Mah82] that if NP has sparse hard sets then P = NP. Mahaney first showed that P = NP under the stronger assumption that NP has a sparse complete set. From that the theorem is derived by observing that if a set $A \in$ NP many-one reduces in polynomial time to a sparse set then it actually many-one reduces to a sparse set in NP. This observation can be formalized as a simplicity result for sets in $R_m^p(\mathrm{SPARSE})$.

Theorem 3.1 [LS91] *If* $A \in R_m^p(\mathrm{SPARSE})$ *then there exists a sparse set* $S \in R_m^{np}(A)$ *such that* $A \leq_m^p S$.

It is easy to see that the same upper bound holds for $R_c^p(\mathrm{SPARSE})$, i.e., every set $A \in R_c^p(\mathrm{SPARSE})$ has a sparse c-description in $R_m^{np}(A)$. On the other hand, for a set A in $R_b^p(\mathrm{SPARSE})$ the only known upper bound for the complexity of sparse b-descriptions is $\Delta_3^p(A)$ (this can be seen by a minor modification of the proof that every set A in P/poly has an advice function computable in $\Delta_3^p(A)$ [Sch86]).

Surprisingly, it turns out (see the following theorem) that each set A in the reduction class $R_{hd}^p(R_c^p(\mathrm{SPARSE}))$, which is larger than $R_b^p(\mathrm{SPARSE})$, has a sparse $hd(c)$-description in NP(A). From this it follows (see Section 4) that $R_{hd}^p(R_c^p(\mathrm{SPARSE}))$, and therefore $R_b^p(\mathrm{SPARSE})$ are in EL_3^Θ.

Theorem 3.2 *For every set* $A \in R_{hd}^p(R_c^p(\mathrm{SPARSE}))$ *there is a sparse set* $\hat{S} \in$ NP(A) *such that* $A \in R_{hd}^p(R_c^p(\hat{S}))$.

Corollary 3.3 *For every set* $A \in R_b^p(R_c^p(\mathrm{SPARSE}))$ *there is a sparse set* $S' \in$ NP(A) *such that* $A \in R_b^p(R_c^p(S'))$.

Next we consider sets in $R_d^p(\mathrm{SPARSE})$ and show that every set A in $R_d^p(\mathrm{SPARSE})$ has a sparse d-description in NP($A \oplus \mathrm{SAT}$)/F$\Theta_2^p(A)$. Furthermore we show that there exists a set in NP(A)/F$\Theta_2^p(A)$ such that every element $\langle 0^n, W \rangle$ in this set encodes a finite set W to which A is disjunctively reducible with respect to strings of length n, i.e., $A^{=n} \leq_d^p W$ via the given disjunctive reduction function (cf. the notion of CIR(A) [KoSc85]). Our proof technique is a refinement and an extension of the one developed in [AHH+92] where it is shown that for every set A in $R_d^p(\mathrm{SPARSE})$ there is a sparse set $S' \in$ P(NP(A)) such that $A \in R_d^p(S')$.

Theorem 3.4 *Let* A *be a set in* $R_d^p(\mathrm{SPARSE})$ *witnessed by the sparse set* S *and the reduction function* f. *Then*

1. *there exists a set* $C \subseteq \{\langle 0^n, W \rangle \mid A \leq_d^p W$ *via* f *for inputs of length* $n\}$ *and a polynomial* p *such that* $C \in$ NP(A)/F$\Theta_2^p(A)$ *and for every* n *there is at least one pair of the form* $\langle 0^n, W \rangle$ *in* $C^{\leq p(n)}$,

2. *there exists a sparse set* $\hat{S} \in$ NP($A \oplus \mathrm{SAT}$)/F$\Theta_2^p(A)$ *such that* $A \leq_d^p \hat{S}$ *via* f.

The following theorem shows in general that a simplicity result for a reduction class $R_r^p(\text{SPARSE})$ can be translated into a simplicity result for the reduction class $R_{1\text{-}tt}^p(R_r^p(\text{SPARSE}))$.

Theorem 3.5 *Let C be a relativizable complexity class closed under join such that $C(A \oplus B) = C(A) = C(\overline{A})$ for all sets $A \subseteq \Sigma^*$ and $B \in P$. If every set L in $R_r^p(\text{SPARSE})$ has a sparse r-description in $C(L)$ then every set A in $R_{1\text{-}tt}^p(R_r^p(\text{SPARSE}))$ has a sparse $1\text{-}tt(r)$-description in $C(A)$.*

Corollary 3.6 *For every set A in $R_{1\text{-}tt}^p(R_d^p(\text{SPARSE}))$ there is a sparse set S' in $\text{NP}(A \oplus \text{SAT})/\text{F}\Theta_2^p(A)$ such that $A \in R_{1\text{-}tt}^p(R_d^p(S'))$.*

Next we consider nondeterministic reduction classes to sparse sets. The notion of $\leq_m^{co\text{-}np}$ reducibility can be seen as a generalization of the deterministic polynomial-time conjunctive reducibility.

Definition 3.7 *A set A is co-np many-one reducible to a set B (denoted $A \leq_m^{co\text{-}np} B$) if there exists a polynomial-time nondeterministic Turing transducer M such that for every $x \in \Sigma^*$, $x \in A$ if and only if all outputs of M on input x are in B.*

Note that $A \leq_m^{co\text{-}np} B$ if and only if $\overline{A} \leq_m^{np} \overline{B}$ where $\leq_m^{np}$ is the more familiar polynomial-time nondeterministic many-one reducibility [LLS75]. Clearly, for every set B, $R_c^p(B)$ is contained in $R_m^{co\text{-}np}(B)$, and $R_m^{co\text{-}np}(B)$ is closed downward under $\leq_c^p$ and $\leq_m^{co\text{-}np}$ reductions.

Theorem 3.8 *1. For every set $A \in R_m^{co\text{-}np}(\text{SPARSE})$ there exists a sparse set $S' \in R_m^{np}(A)$ such that $A \in R_m^{co\text{-}np}(S')$.*

2. For every set $A \in R_{hd}^p(R_m^{co\text{-}np}(\text{SPARSE}))$ there is a sparse set $S' \in \text{NP}(\text{SAT} \oplus A)$ such that $A \in R_{hd}^p(R_m^{co\text{-}np}(S'))$.

3. For every set $A \in R_b^p(R_m^{co\text{-}np}(\text{SPARSE}))$ there is a sparse set $S' \in \text{NP}(\text{SAT} \oplus A)$ such that $A \in R_b^p(R_m^{co\text{-}np}(S'))$.

At the end of this section we consider reduction classes to tally sets. The class $R_c^p(\text{TALLY}) \cap R_d^p(\text{TALLY})$ is of particular interest since it coincides with the class $\text{IC}[\log,\text{poly}]$ of sets containing only strings of low instance complexity [AHH+92]. $\text{IC}[\log,\text{poly}]$ and the notion of instance complexity were introduced by Ko, Orponen, Schöning, and Watanabe (see [OKSW]).

Theorem 3.9 *1. For every set $A \in R_c^p(\text{TALLY}) \cap R_d^p(\text{TALLY})$ there is a tally set $T' \in \text{P}(\text{SAT} \oplus A)$ such that $A \in R_c^p(T') \cap R_d^p(T')$.*

2. For every set $A \in R_m^{np}(\text{TALLY}) \cap R_m^{co\text{-}np}(\text{TALLY})$ there is a tally set $T' \in \text{P}(\text{SAT} \oplus A)$ such that $A \in R_m^{np}(T') \cap R_m^{co\text{-}np}(T')$.

The class $\text{IC}[\log,\text{poly}]$ is known to be closed downward under polynomial time bounded truth-table reductions [OKSW]. It is interesting to note that $\text{IC}[\log,\text{poly}]$ and $R_m^{np}(\text{TALLY}) \cap R_m^{co\text{-}np}(\text{TALLY})$ are also closed downward under polynomial time Hausdorff reductions.

Theorem 3.10 *1. $R_{hd}^p(\text{IC}[\log,\text{poly}]) = \text{IC}[\log,\text{poly}]$*

2. $R_{hd}^p(R_m^{np}(\text{TALLY}) \cap R_m^{co\text{-}np}(\text{TALLY})) = R_m^{np}(\text{TALLY}) \cap R_m^{co\text{-}np}(\text{TALLY})$

4 Lowness

The low and high hierarchies inside NP were introduced by Schöning [Sch83]. The notion of lowness has turned out to be an important structural tool for classifying problems and subclasses of NP not known to be NP-complete or in P. This idea was extended by Balcázar, Book, and Schöning [BBS86] who defined the extended low hierarchy in order to classify decision problems and classes not contained in NP. Allender and Hemachandra [AH92] and Long and Sheu [LS91] refined the extended low hierarchy and proved the optimality of the location of various classes therein.

Definition 4.1 [BBS86, AH92, LS91] *The Σ, Δ, and Θ levels of the extended low hierarchy (denoted EL_k^Σ, EL_k^Δ, and EL_k^Θ, respectively) are defined as below.*

1. $EL_k^\Sigma = \{A \mid \Sigma_k^p(A) \subseteq \Sigma_{k-1}^p(\text{SAT} \oplus A)\}$, $k \geq 1$

2. $EL_k^\Delta = \{A \mid \Delta_k^p(A) \subseteq \Delta_{k-1}^p(\text{SAT} \oplus A)\}$, $k \geq 2$

3. $EL_k^\Theta = \{A \mid \Theta_k^p(A) \subseteq \Theta_{k-1}^p(\text{SAT} \oplus A)\}$, $k \geq 2$

Various classes of sets reducible to sparse and tally sets have been shown to be included in the extended low hierarchy (see for example [BBS86, AH92, LS91]). Using the simplicity results of the previous section we are able to derive lowness results for the reduction classes to sparse sets considered here. Some of our extended lowness proofs contain a census argument similar to that used by Hemachandra [Hem87] and Kadin [Kad87].

Theorem 4.2 1. $R_{hd}^p(R_c^p(\text{SPARSE})) \subseteq EL_3^\Theta$

2. $R_d^p(\text{SPARSE}) \subseteq EL_3^\Theta$.

3. $R_{1-tt}^p(R_d^p(\text{SPARSE})) \subseteq EL_3^\Theta$.

These results are optimal since SPARSE $\not\subseteq EL_2^\Sigma$ as proved in [AH92]. Item 4 of the following corollary answers an open question in [AH92] and also extends the recent result by Long and Sheu [LS91] that $R_{1-tt}^p(\text{SPARSE}) \subseteq EL_3^\Theta$.

Corollary 4.3 1. $R_c^p(\text{SPARSE}) \subseteq EL_3^\Theta$

2. $R_c^p(\text{co-SPARSE}) \subseteq EL_3^\Theta$

3. $R_d^p(\text{co-SPARSE}) \subseteq EL_3^\Theta$

4. $R_b^p(\text{SPARSE}) \subseteq EL_3^\Theta$

5. $E_T^p(R_{hd}^p(R_c^p(\text{SPARSE}))) \subseteq EL_3^\Theta$

6. $E_T^p(R_d^p(\text{SPARSE})) \subseteq EL_3^\Theta$

The following theorem states the generalized Θ_2^p-lowness of NP(NP $\cap$ SPARSE) $\cap$ $R_m^{co-np}(\text{SPARSE})$ and is an improvement of Kadin's result that every sparse NP set is low for Θ_2^p.

Theorem 4.4 *If $A \in \mathrm{NP}(\mathrm{NP} \cap \mathrm{SPARSE})$ and $A \leq_m^{co-np} S$ for a sparse set S then $\Theta_2^p(A) \subseteq \Theta_2^p$, i.e. A is low for Θ_2^p.*

Using this theorem we can improve the result of Lozano and Torán [LT91] that every disjunctive self-reducible set that is many-one reducible to some sparse set is low for Θ_2^p. Since co-NP/log $\subseteq R_m^{co-np}(\mathrm{TALLY})$ [AKM92] it subsumes all previously known Θ_2^p-lowness results (e.g. for NP $\cap$ $R_m^p(\mathrm{SPARSE})$ and for NP $\cap$ co-NP/log, cf. [LS91]) regarding NP sets reducible to sparse or tally sets.

Corollary 4.5 *1. If $A \in \mathrm{NP} \cap R_c^p(\mathrm{SPARSE})$ then $\Theta_2^p(A) \subseteq \Theta_2^p$.*

2. If $A \in \mathrm{NP} \cap R_m^{co-np}(\mathrm{SPARSE})$, then $\Theta_2^p(A) \subseteq \Theta_2^p$.

Although we have simplicity results concerning nondeterministic reduction classes to sparse sets (e.g. Theorem 3.8) extended lowness is not a meaningful measure of lowness for such reduction classes since they contain either NP or co-NP. Nevertheless, the next theorem is a kind of lowness result. We show that the full power of a Θ_3^p computation relative to an oracle in $R_{hd}^p(R_m^{co-np}(\mathrm{SPARSE}))$ is not needed in order to access the information contained in the oracle.

Theorem 4.6 *If $A \in R_{hd}^p(R_m^{co-np}(\mathrm{SPARSE}))$ then $\Theta_3^p(A) \subseteq \Theta_2^p(\Sigma_2^p \oplus A)$.*

Finally we locate IC[log,poly] in the first level of the extended low hierarchy. Since $EL_1^\Sigma = EL_2^\Theta = EL_2^\Delta$ the class IC[log, poly] which equals $R_c^p(\mathrm{SPARSE}) \cap R_d^p(\mathrm{co\text{-}SPARSE})$ is located two levels below $R_c^p(\mathrm{SPARSE})$ in the extended low hierarchy.

Theorem 4.7 *If A has a sparse description that is $P(\mathrm{SAT} \oplus A)$ printable then $A \in EL_1^\Sigma$.*

Since the class APT (almost polynomial time [MP79]) is easily seen to be contained in IC[log, poly] the following corollary subsumes all previously known EL_1^Σ-lowness results (e.g. for APT [LS91], for $E_T^p(\mathrm{TALLY})$ [BB86] and for $R_m^p(\mathrm{TALLY})$ [AH92]) regarding sets reducible to sparse or tally sets.

Corollary 4.8 *1. IC[log, poly] $\subseteq EL_1^\Sigma$*

2. $E_T^p(\mathrm{IC}[\log, \mathrm{poly}]) \subseteq EL_1^\Sigma$

3. $R_m^{np}(\mathrm{TALLY}) \cap R_m^{co-np}(\mathrm{TALLY}) \subseteq EL_1^\Sigma$

We summarize some of our results on simplicity and lowness in the following table.

For A in the reduction class	simplicity	lowness
$R_{hd}^p(R_c^p(\mathrm{SPARSE}))$	NP(A)	EL_3^Θ
$R_{1-tt}^p(R_d^p(\mathrm{SPARSE}))$	NP(SAT $\oplus$ A)/F$\Theta_2^p(A)$	EL_3^Θ
$R_c^p(\mathrm{TALLY}) \cap R_d^p(\mathrm{TALLY})$	P(SAT $\oplus$ A)	EL_1^Σ
$R_m^{np}(\mathrm{TALLY}) \cap R_m^{co-np}(\mathrm{TALLY})$	P(SAT $\oplus$ A)	EL_1^Σ
$R_m^{co-np}(\mathrm{SPARSE}) \cap \mathrm{NP}^{\mathrm{NP} \cap \mathrm{SPARSE}}$	$R_m^{np}(A)$	$\Theta_2^p(A) \subseteq \Theta_2^p$
$R_{hd}^p(R_m^{co-np}(\mathrm{SPARSE}))$	NP(SAT $\oplus$ A)	$\Theta_3^p(A) \subseteq \Theta_2^p(\Sigma_2^p \oplus A)$

Acknowledgments

The first author is grateful to Uwe Schöning for his hospitality and for the research environment provided at Universität Ulm during the year 1991-92.

References

[AH92]   E. Allender and L. Hemachandra. Lower bounds for the low hierarchy. *Journal of the ACM*, 39(1):234-250, 1992.

[AHH⁺92] V. Arvind, Y. Han, L. Hemachandra, J. Köbler, A. Lozano, M. Mundhenk, M. Ogiwara, U. Schöning, R. Silvestri, and T. Thierauf. Reductions to sets of low information content. *Proceedings of the 19th International Colloquium on Automata, Languages, and Programming*, Lecture Notes in Computer Science, #623:162-173, Springer Verlag, 1992.

[AHOW] E. Allender, L. Hemachandra, M. Ogiwara, and O. Watanabe. Relating equivalence and reducibility to sparse sets. *SIAM Journal on Computing*, to appear.

[AKM92] V. Arvind, J. Köbler, and M. Mundhenk. On bounded truth-table, conjunctive, and randomized reductions to sparse sets. To appear in *Proceedings 12th Conference on the Foundations of Software Technology & Theoretical Computer Science*, 1992.

[BB86]   J. Balcázar and R. Book. Sets with small generalized Kolmogorov complexity. *Acta Informatica*, 23(6):679-688, 1986.

[BBS86] J.L. Balcázar, R. Book, and U. Schöning. Sparse sets, lowness and highness. *SIAM Journal on Computing*, 23:679-688, 1986.

[BDG]   J.L. Balcázar, J. Díaz, and J. Gabarró. *Structural Complexity I*. EATCS Monographs on Theoretical Computer Science, Springer-Verlag, 1988.

[BLS92] H. Buhrman, L. Longpré, and E. Spaan. Sparse reduces conjunctively to tally. *Technical Report NU-CCS-92-8*, Northeastern University, Boston, 1992.

[GW91]   R. Gavaldà and O. Watanabe. On the computational complexity of small descriptions. *Proceedings of the 6th Structure in Complexity Theory Conference*, 89-101, IEEE Computer Society Press, 1991.

[Hau14] F. Hausdorff. *Grundzüge der Mengenlehre*. Leipzig, 1914.

[HY84]   J. Hartmanis and Y. Yesha. Computation times of NP sets of different densities. *Theoretical Computer Science*, 34:17-32, 1984.

[Hem87] L. Hemachandra. The strong exponential hierarchy collapses. *Proceedings of the 19th ACM Symposium on Theory of Computing*, 110-122, 1987.

[Kad87] J. Kadin. $P^{NP[\log n]}$ and sparse Turing-complete sets for NP. *Journal of Computer and System Sciences*, 39(3):282-298, 1989.

[KL80]   R. Karp and R. Lipton. Some connections between nonuniform and uniform complexity classes. *Proceedings of the 12th ACM Symposium on Theory of Computing*, 302-309, April 1980.

[KoSc85] K. Ko and U. Schöning. On circuit-size complexity and the low hierarchy in NP. *SIAM Journal on Computing*, 14:41-51, 1985.

[Köb92]   J. Köbler. Locating P/poly optimally in the low hierarchy. Ulmer Informatik-Bericht 92-05, Universität Ulm, August 1992.

[KSW87]   J. Köbler, U. Schöning, and K.W. Wagner. The difference and truth-table hierarchies of NP. *Theoretical Informatics and Applications*, 21(4):419-435, 1987.

[KT90]   J. Köbler and T. Thierauf. Complexity classes with advice. *Proceedings 5th Structure in Complexity Theory Conference*, 305-315, IEEE Computer Society, 1990.

[LLS75]   R. Ladner, N. Lynch, and A. Selman. A comparison of polynomial time reducibilities. *Theoretical Computer Science*, 1(2):103-124, 1975.

[LS91]   T.J. Long and M.-J. Sheu. A refinement of the low and high hierarchies. Technical Report OSU-CISRC-2/91-TR6, The Ohio State University, 1991.

[LT91]   A. Lozano and J. Torán. Self-reducible sets of small density. *Mathematical Systems Theory*, 24:83-100, 1991.

[Mah82]   S. Mahaney. Sparse complete sets for NP: Solution of a conjecture of Berman and Hartmanis. *Journal of Computer and System Sciences*, 25(2):130-143, 1982.

[MP79]   A. Meyer, M. Paterson. With what frequency are apparently intractable problems difficult? Tech. Report MIT/LCS/TM-126, Lab. for Computer Science, MIT, Cambridge, 1979.

[OKSW]   P. Orponen, K. Ko, U. Schöning, and O. Watanabe. Instance complexity. *Journal of the ACM*, to appear.

[OW91]   M. Ogiwara and O. Watanabe. On polynomial-time bounded truth-table reducibility of NP sets to sparse sets. *SIAM Journal on Computing*, 20(3):471-483, 1991.

[Sch83]   U. Schöning. A low hierarchy within NP. *Journal of Computer and System Sciences*, 27:14-28, 1983.

[Sch86]   U. Schöning. *Complexity and Structure*, Lecture Notes in Computer Science, #211, Springer-Verlag, 1985

[SL92]   M.-J. Sheu and T.J. Long. The extended low hierarchy is an infinite hierarchy. *Proceedings of 9th Symposium on Theoretical Aspects of Computer Science*, Lecture Notes in Computer Science, #577:187-189, Springer-Verlag 1992.

[TB91]   S. Tang and R. Book. Reducibilities on tally and sparse sets. *Theoretical Informatics and Applications*, 25:293-302, 1991.

[Wag87]   K.W. Wagner. More complicated questions about maxima and minima, and some closures of NP. *Theoretical Computer Science*, 51:53-80, 1987.

[Wag90]   K.W. Wagner. Bounded query classes. *SIAM Journal on Computing*, 19(5):833-846, 1990.

Honest Iteration Schemes of Randomizing Algorithms

Jie Wang * and Jay Belanger

Wilkes University, Wilkes-Barre PA 18766, USA
Internet: jwang@wilkes1.wilkes.edu and belanger@wilkes1.wilkes.edu

Abstract. This paper further studies Blass and Gurevich's iteration schemes of randomizing algorithms for randomized reductions [BG91]. An iteration scheme of a randomizing algorithm is honest if it passes a bit to a process independently of its value. We show that if a fair iteration scheme is honest, then it is independent. This implies the previous known sufficient condition for independence in [BG91]. On the other hand, we show that for independence, honesty is not necessary.

1 Introduction

Average case complexity theory, initiated by L. Levin [Lev84], has recently been an active research area, for example, see [VL88, BCGL89, IL90, BG91, Gur91, VR92, WB92]. Randomized reductions on search problems have been shown to be strictly more powerful than many-one reductions. For example, the randomized graph-coloring problem is complete for randomized NP problems under randomized reductions but not under many-one reductions unless DTIME(2^{poly}) = NTIME(2^{poly}) ([VL88, Gur91]). Randomized reductions are not as easy to use as many-one reductions. However, as mentioned by Blass and Gurevich [BG91], these randomized reductions are in fact iterations of many-one reductions. They pointed out that the iteration is needed for the resulting search algorithms, in order to ensure that a solution can be found with probability 1 in average polynomial time, but is not needed for the reductions themselves. This motivated Blass and Gurevich to study the theory of randomized reductions by treating reductions separately from iterations.

A randomizing algorithm (coin-flipping) is said to solve a search problem in average polynomial time if for every input x, the algorithm finds a solution with probability $\rho(x)$, in which $1/\rho(x)$ is polynomial on average. Reductions between search problems are defined in such a way that they are transitive, and if a search problem Π is reducible to a search problem Π' which is solvable in average polynomial time, then Π is solvable in average polynomial time. In general, if a solution to Π' can be found with certain probability, then with the help of the reduction, a solution to Π on a given input can be found with some probability which may, however, be very small.

---

* Supported in part by NSF grant CCR-9108899.

A randomized reduction from which an algorithm rarely produces a solution may seem like an unsatisfactory reduction. However, Blass and Gurevich [BG91] showed that it is in fact a practical reduction from which a randomizing algorithm M can be used to produce a solution with probability 1 in average polynomial time by iterating the algorithm in an appropriate manner. So while the algorithm M may be a poor way of finding a solution if used directly, it becomes a good way of finding a solution if iterated. Therefore, it is important to investigate what manner is appropriate to iterate a randomizing algorithm.

The appropriate manner in which to iterate a randomizing algorithm must be polynomially-fair and independent. Blass and Gurevich [BG91] defined the notion of perpetual iteration schemes of randomizing algorithms. They proved that if a perpetual iteration scheme M^∞ of a randomizing algorithm M is polynomially-fair and independent, then M halts in average polynomial time if and only if the "haltable" iteration M^* of M halts in average polynomial time with probability 1.

Independence is an important property which guarantees that the processes (subcomputations) generated by the iteration scheme are independent. Blass and Gurevich [BG91] proved that for a fair perpetual iteration scheme, if it serves random bits to its processes on the "first come first served" basis, then it is independent. However, there are situations in which an iteration scheme does not follow the "first come first served" rule. For example, a request for a bit from a process may be delayed. This motivates us to study a more general condition for independence. We show that for a fair perpetual iteration scheme M^∞, if M^∞ passes its random bits honestly to its processes, i.e., independently of its values, then M^∞ is independent. This result implies the previous known sufficient condition for independence from [BG91], mentioned above.

That an iteration scheme is honest is equivalent to saying that the iteration scheme passes a random bit to its process without looking at its value. This seems a rather weak condition. So it might be a necessary condition for independence as well. However, we show that it is not the case. Honesty is still more than required.

This paper is devoted to studying the notion of independent iteration schemes. We first introduce Blass and Gurevich's definition on iteration schemes of randomizing algorithms in Section 2. We then prove our results in Sections 3 and 4.

2   Iteration Schemes

Let M be a Turing machine which flips a fair coin at certain states in order to determine its next move. Given an input x, run M on x for k times. These iterations can be considered as k independent trials of M on x. Let r_i be the sequence of random bits generated by the i-th iteration. Then r_i is finite if the i-th iteration on input x halts. Otherwise, r_i could be infinite. When r_i is finite, the probability of r_i is $2^{-|r_i|}$. When all r_i's are finite, the probability of these sequences is $2^{-\sum_{i=1}^{k}|r_i|}$. But since r_i could be infinite, one would like to have a methodology to include these infinite sequences of random bits as well. Blass and Gurevich [BG91] suggested studying randomizing algorithms using Lebesgue measure, as described below.

A randomizing algorithm is a Turing machine M with an auxiliary read-only tape, called the *random tape*, containing an infinite sequence r of "random" bits (0's and 1's). The random tape is bounded on the left and unbounded on the right, and its head can move only to the right. The set $\{0,1\}^\infty$ of all possible infinite sequences is endowed with the Lebesgue measure λ on the interval $[0,1]$ by corresponding r to $0.r$. λ can also be regarded as a probability measure. For each finite binary string t, we define its measure to be $\lambda\{r \in \{0,1\}^\infty : t$ is a prefix of $r\}$ which is equal to $2^{-|t|}$. A randomizing algorithm M can be informally described as an algorithm that is allowed to flip a fair coin at various stages of its computation with these coin flips independent. When M needs to read a bit from its random tape, we say that M is at the "random" state.

Fix a randomizing algorithm M with a fixed input x. For each $r \in \{0,1\}^\infty$, let Read(r) (or Read$_{M,x}(r)$ if necessary for clarity) be the initial segment of r that is actually read during the computation. If the computation halts, then Read(r) is finite; if not, then Read(r) may be finite or equal to r. As the computation, with M and x fixed, depends only on Read(r), it is clear that, if $r' \in \{0,1\}^\infty$ has Read(r) as an initial segment, then Read$(r') =$ Read(r). Let R$(x) =$ R$_M(x) =$ $\{$Read$_{M,x}(r) : r \in \{0,1\}^\infty\}$, then no member of R$(x)$ is a proper initial segment of another. Clearly, each R(x) is the disjoint union of the sets RF$_M(x)$ and RI$_M(x)$ consisting of the finite and infinite strings in R(x), respectively. Every $r \in \{0,1\}^\infty$ either is in RI$_M(x)$ or has an initial segment Read(r) in RF$_M(x)$.

A perpetual iteration M^∞ of M is a randomizing scheme such that on input x and random sequence $r \in \{0,1\}^\infty$, M^∞ starts many processes, where a process is M in running on the same input x and random bits obtained from r. M^∞ will always start a new process if all the current processes terminate. So a perpetual iteration never halts. A "haltable" iteration M^* of M acts like M^∞ except that M^* halts whenever a process halts.

Let (M, i) denote the i-th process of an iteration scheme of M. Fix an M and x. Let Send$_i(r)$ denote the subsequence of r that the iteration scheme sends to the process (M, i). So Send$_i(r)$ may only depend on the iteration scheme. Notice that an iteration scheme may discard a bit it reads without passing it to any process. Also, a bit that is passed to a process may not be read by the process if it is not at the random state. So for a process, Read(s) may not be the initial segment of s. Also, s could be finite. For simplicity, we assume that a process will read whatever is passed on to it when it is at a random state; and a bit passed to it will be ignored if it is not at a random state. We also assume that if a process asks for a bit, it will wait until it gets a bit. Let $s_i(r)$ denote the subsequence of r passed by the iteration scheme to process (M, i) which is actually read by (M, i) on x. So $s_i(r) =$ Read$($Send$_i(r))$.

It is easy to see that all processes of an iteration scheme are independent if and only if all $s_i(r)$'s are independent. For any k, all $s_i(r)$'s are independent for $i = 1, 2, ..., k$ if and only if the probability of these strings is equal to $2^{-\sum_{i=1}^{k}|s_i(r)|}$. Since $s_i(r)$ may be infinite, we will consider initial segments of $s_i(r)$. To formalize this idea, a list of finite binary strings $t_1, t_2, ..., t_k$ is considered. The $s_i(r)$'s are independent for $i = 1, ..., k$ if and only if for any such list, $\lambda\{ r : t_i$ is a prefix of $s_i(r)$ for all $i = 1, ..., k \}$ is $2^{-\sum_{i=1}^{k}|t_i|}$ if no t_i has a proper initial segment in $\mathrm{RF}_M(x)$, and is 0 otherwise. This is the notion of "careful iteration" defined in [BG91].

Definition 1. [BG91] Let M be a randomizing algorithm and x be an input. M^∞ is *independent* on x if for any k and any list $t_1, \ldots, t_k$ of finite binary strings, the probability $\lambda\{ r : t_i$ is a prefix of $s_i(r)$ for all $i = 1, \ldots, k \}$ is $2^{-\sum_{i=1}^{k}|t_i|}$ if no t_i has a proper initial segment in $\mathrm{RF}_M(x)$, and is 0 otherwise. M^∞ is *independent* if it is independent on all x.

In many situations, we are only interested in a "fair" perpetual iteration scheme which will start infinitely many processes and will also try to finish a process once it starts it.

Definition 2. A perpetual iteration scheme is *fair* if it starts infinitely many processes and each process will be executed until it halts.

So if a process does not halt in a fair perpetual iteration scheme, then it will be executed for infinitely many steps. By attaching a polynomial time bound to a fair iteration scheme, we have the following definition [BG91].

Definition 3. [BG91] A perpetual iteration scheme is *polynomially fair* (*p-fair*, in short) if it is fair and there is a polynomial p such that the n-th step of the i-th process will be executed within $p(i + n)$ steps.

Blass and Gurevich [BG91] proved that if a perpetual iteration scheme M^∞ is p-fair and independent, then M halts in average polynomial time if and only if M^* halts in average polynomial time and is almost total. So independence plays an important role in an iteration scheme. However, the definition of "independence" may not be easy to use. So an easier to use sufficient condition would be useful. Blass and Gurevich [BG91] proved that if a perpetual iteration scheme M^∞ is fair and passes a bit to a process on a "first come first serve" basis, then M^∞ is independent. However, there are iteration schemes which may not follow the first-come-first-serve rule. For example, a request for a bit from a process may be delayed. So a more general condition is needed. We show in Section 3 that a fair perpetual iteration scheme M^∞ is independent as long as it passes a bit independently of its value. That is, M^∞ is "honest" in the sense that it will not do the following: pass a bit to process (M, i) if the bit is 0 and not pass the bit to process (M, i) if the bit is 1 (or conversely). In other words, an honest iteration scheme means that it passes a random bit to a process without looking at its value.

Definition 4. A perpetual iteration scheme M^∞ is *honest* if it passes a bit independently of its value.

At the first glance, it seems that honesty should also be a necessary condition of an independent perpetual iteration scheme since honesty is a natural and a rather weak condition. However, we show that it is not the case. Independence does not imply honesty. We show in Section 4 that there is a perpetual iteration scheme which is independent but not honest. So passing a bit without looking at its value is more than required.

3   Honest Perpetual Iterations

In this section, we will show that if a fair perpetual iteration scheme M^∞ is honest, then M^∞ is independent. This is done by modifying a proof of Blass and Gurevich [BG91]. As a corollary, we obtain Blass and Gurevich's result which states that a fair perpetual iteration scheme M^∞ is independent if it serves random bits on the "first come first served" basis.

Theorem 5. *Let M^∞ be a perpetual iteration scheme. If M^∞ is fair and honest, then M^∞ is independent.*

Proof. Since M^∞ is fair, each process that requests a bit will eventually get it from M^∞. For any k and list of finite binary strings $t_1, t_2, ..., t_k$, let $n = \sum_{i=1}^{k} |t_i|$, and

$$E = \{r : t_i \text{ is a prefix of } s_i(r) \text{ for all } i = 1, ..., k\}.$$

It is easy to see that if some t_i has a proper initial segment in $\mathrm{RF}_M(x)$, then t_i cannot be a prefix of s_i, so $\lambda(E) = 0$. Suppose no t_i has a proper initial segment in $\mathrm{RF}_M(x)$. Then s_i cannot be a proper initial segment of t_i for all $i = 1, ..., k$. Let q be a prefix of r. q is *pertinent* if for all $i = 1, ..., k$, either $s_i(q)$ is a prefix of t_i or t_i is a prefix of $s_i(q)$. Let $s_i'(q)$ be the shorter string of t_i and $s_i(q)$. Let $m(q) = \sum_{i=1}^{k} |s_i'(q)|$. Clearly, $0 \le m(q) \le n$. A pertinent string q is *perfect* if $\lambda[E \mid q \text{ is a prefix of } r] = 2^{-n+m(q)}$. Otherwise, q is called *defective*. We will prove that if q is defective then one of $q0$ and $q1$ must be defective as follows.

Case 1. The next bit after q that M^∞ reads is sent to (M, i) and is read by (M, i) with $i > k$ or with $i \le k$ and $s_i'(q) = t_i$. Then $s_i'(q0) = s_i'(q1) = s_i'(q)$ for all $i = 1, ..., k$. So both $q0$ and $q1$ are pertinent, and $m(q) = m(q1) = m(q0)$. Since q is defective, we have $\lambda[E \mid q \text{ is a prefix of } r] \ne 2^{-n+m(q)}$. Since M^∞ is honest, $\lambda[E \mid q \text{ is a prefix of } r] = \frac{1}{2}\lambda[E \mid q0 \text{ is a prefix of } r] + \frac{1}{2}\lambda[E \mid q1 \text{ is a prefix of } r]$. So at least one of $q0$ and $q1$ must be defective.

Case 2. The next bit after q that M^∞ reads is sent to (M, i) and is read by (M, i) with $i \leq k$ and $s_i'(q) \neq t_i$. That is, $s_i'(q)$ is a proper initial segment of t_i. Then $q0$ (resp. $q1$) will be pertinent exactly when the next bit in t_i after s_i' is 0 (resp. 1). Without loss of generality, let $q0$ be pertinent. Then $s_i'(q0) = s_i'(q)0$ and $s_j'(q0) = s_j'(q)$ for all $j \in \{1, ..., k\} - \{i\}$. Since $q1$ is not pertinent, $\lambda[E \mid q1$ is a prefix of $r] = 0$. So $\lambda[E \mid q$ is a prefix of $r] = \frac{1}{2}\lambda[E \mid q0$ is a prefix of $r]$. Therefore $\lambda[E \mid q0$ is a prefix of $r] \neq 2^{-n+m(q0)}$.

Case 3. The next bit after q that M^∞ reads is discarded or is sent to (M, i) but is not read by (M, i). Then $s_i'(q0) = s_i'(q1) = s_i'(q)$ for all $i = 1, ..., k$. So the same reasoning of Case 1 applies to this case.

Now we will show that the empty string e is perfect and so $\lambda(E) = 2^{-n}$ which is what we wanted to prove. Suppose e is defective, then as shown above, we can obtain a sequence of defective strings, each a one-bit extension of the previous one. Let r' be the limit of these strings. Since M^∞ is fair, each process (M, i) either halts or is infinite on r'. We have shown earlier that each t_i must be an initial segment of $s_i(r')$ for $i = 1, ..., k$. Since all finite initial segments of r', being defective, are pertinent, for a sufficiently long finite initial segment q of r', each t_i is an initial segment of $s_i(q)$, and therefore $s_i'(q) = t_i$ and $m(q) = n$. Therefore, E contains all strings with prefix q, and so $\lambda[E \mid q$ is a prefix of $r] = 1 = 2^{-n+m(q)}$. But then q is not defective. This contradiction completes the proof. □

It is easy to see that if M^∞ serves random bits on the "first come first served" basis, then M^∞ is honest. So we have

Corollary 6. [BG91] *Let M^∞ be a fair perpetual iteration scheme. If M^∞ serves random bits on the "first come first served" basis, then M^∞ is independent.*

From Theorem 5, we know that fairness with honesty guarantees independence. Does independence imply fairness or honesty? We will show in the next section that independence does not imply honesty. However, we can show that independence does imply that a perpetual iteration scheme must start infinitely many processes. On the other hand, it is easy to see that independence does not imply the second condition of fairness. That is, an independent perpetual iteration scheme does not need to try to finish a process. For example, consider a randomizing algorithm which possesses a certain internal state exactly when it has received its last random bit. A perpetual iteration which stalls on any process which is in this state may well be independent, but will not meet the second condition of fairness.

Theorem 7. *Let M^∞ be an independent perpetual iteration scheme. Then, for any input x on which M does not act deterministically, with probability 1 M^∞ will start infinitely many processes.*

Proof. For any i, the probability that M^∞ begins the process (M, i) is

$$\lambda\{r : 0 \text{ is a prefix of } s_i(r)\} + \lambda\{r : 1 \text{ is a prefix of } s_i(r)\} = 2^{-1} + 2^{-1}$$

because of independence, and this is just 1. So the probability that M^∞ starts infinitely many processes is at least

$$1 - \sum_{i=1}^{\infty} \lambda\{M^\infty \text{ does not begin the process } (M, i)\} = 1 - 0 = 1. \qquad \square$$

4 Non-Honest Independent Iterations

In this section, we will show that there is an independent iteration scheme which is not honest. In other words, passing a bit without looking at its value is not necessary for independence.

Given a randomizing Turing machine M, let M^∞ be the following perpetual iteration scheme:

On any input (x, r), every process will be "marked" with a Λ, 0 or 1; and every process will be marked with a Λ when it is started. The k-th stage of M^∞ will attempt one step each of the first k processes in the order of process 1, process 2, ..., process k, according to the following rules for each step of each process:

1. If no bit is requested from the process, then next bit of r is sent to (although not accepted by) the process, and the step is carried out.
2. If a process asks for a bit while marked with a Λ, and the next bit of r is b ($b = 0$ or 1), then the process is marked with a b, this bit of r is discarded, and M^∞ goes to the next process.
3. If a process asks for a bit while marked with a b ($b = 0$ or 1), and the next bit of r is b, then this bit is passed to the process, and the process is marked with a Λ.
4. If a process asks for a bit while marked with a b and the next bit of r is not b, then that bit of r is discarded and M^∞ goes to the next process.

It is easy to see that this perpetual iteration scheme is not honest from rules 3 and 4. On the other hand, rules 2, 3, and 4 guarantee that each random bit sent to a process has an equal chance of being a 0 or a 1. From the rules we set up, we also know that M^∞ reserves certain positions of the random tape for a process and whether a bit on those reserved positions will be sent to the process or discarded depends on its value. So what M^∞ sends to one process has nothing to do with what it sends to any other process. This is what is needed to prove that the iteration scheme is independent.

It will be useful to consider the set, for any finite binary string s, $Q(s) = \{$ finite binary strings $q : \text{Read}(q) = s$, $\text{Read}(q') \neq s$ where $q = q'0$ or $q = q'1\}$.

So s will be a prefix of $\mathrm{Read}(r)$ exactly when r has a prefix $q \in Q(s)$, and this q will be unique. We will then have

$$\lambda\{r : s \text{ is a prefix of } \mathrm{Read}(r)\} = \sum_{q \in Q(s)} \lambda\{r : q \text{ is a prefix of } r\} = \sum_{q \in Q(s)} 2^{-|q|}.$$

Theorem 8. *Let M^∞ be above. Then M^∞ is independent but not honest.*

Proof. That M^∞ is not honest is easy to see. We prove that M^∞ is independent. We first prove:

Lemma 9. *For any process of M^∞, for any finite binary string t,*

$$\lambda\{r : t \text{ is a prefix of } \mathrm{Read}(r)\} = \begin{cases} 0 & \text{if } t \text{ has a proper prefix in } \mathrm{RF}_M(x), \\ 2^{-|t|} & \text{otherwise.} \end{cases}$$

Proof. (By induction on $|t|$.) This is clearly true for $|t| = 0$, since then $t = e$. So assume the lemma is true for $|t'| < n$, and let $|t| = n$. If t has a proper prefix in $\mathrm{RF}_M(x)$, then the process could never read all of t, so $\lambda\{r : t \text{ is a prefix of } \mathrm{Read}(r)\} = 0$. So assume that t has no proper prefix in $\mathrm{RF}_M(x)$. Write $t = t'b$, where $b = 0$ or 1, and $|t'| = n - 1$. Then

$$\lambda\{r : t' \text{ is a prefix of } \mathrm{Read}(r)\} = \sum_{q \in Q(t')} 2^{-|q|}$$

is 2^{-n+1} by our inductive assumption. Then, for any $q \in Q(t')$, since after reading the random bits q, the process has an equal chance of reading a 0 or a 1 from rules 2 and 3,

$$\lambda\{r' : t \text{ is a prefix of } \mathrm{Read}(qr')\} = 1/2.$$

So we have,

$$\lambda\{r : t \text{ is a prefix of } \mathrm{Read}(r)\}$$
$$= \sum_{q \in Q(t')} \lambda\{r : t \text{ is a prefix of } \mathrm{Read}(r) \text{ and } q \text{ is a prefix of } r\}$$
$$= \sum_{q \in Q(t')} \lambda\{r : t \text{ is a prefix of } \mathrm{Read}(r) | r = qr'\} \lambda\{r : q \text{ is a prefix of } r\}$$
$$= \sum_{q \in Q(t')} \frac{1}{2} \lambda\{r : q \text{ is a prefix of } r\}$$
$$= \frac{1}{2} 2^{-n+1}$$
$$= 2^{-n},$$

which is what we wanted to show. $\qquad\square$

Notice that this lemma implies that for any finite binary string t and process (M, i),

$$\lambda\{r : t \text{ is a prefix of } s_i(r)\} = 2^{-|t|}.$$

To see this, we have

$$\lambda\{r : t \text{ is a prefix of } s_i(r)\} = \sum_{q \in Q(t)} \lambda\{r : q \text{ is a prefix of } \text{Send}_i(r)\}.$$

Similarly to the lemma, we can show $\lambda\{r : q \text{ is a prefix of } \text{Send}_i(r)\} = 2^{-|q|}$. So $\sum_{q \in Q(t)} \lambda\{r : q \text{ is a prefix of } \text{Send}_i(r)\} = \sum_{q \in Q(t)} \lambda\{r : q \text{ is a prefix of } r\}$, which is $2^{-|t|}$ by the lemma.

To finish the proof, notice that from the earlier discussion, the values of $\text{Send}_i(r)$ are independent in the sense that what M^∞ sends to one process has nothing to do with what it sends to any other process. Now, for any finite binary strings $t_1, \ldots, t_k$, let

$$E_i = \{s \in \{0, 1\}^\infty : t_i \text{ is a prefix of } \text{Read}(s)\}.$$

Then we have

$$\lambda\{r : t_i \text{ is a prefix of } s_i(r) = \text{Read}(\text{Send}_i(r)), i = 1, \ldots, k\}$$
$$= \lambda\{r : \text{Send}_i(r) \in E_i, i = 1, \ldots, k\}$$
$$= \prod_{i=1}^{k} \lambda\{\text{Send}_i(r) \in E_i\}$$
$$= \prod_{i=1}^{k} \lambda\{r : t_i \text{ is a prefix of } s_i(r))\}$$
$$= \begin{cases} 0 & \text{if some } t_i \text{ has a proper prefix in } \text{RF}_M(x), \\ \prod_{i=1}^{k} 2^{-|t_i|} & \text{otherwise} \end{cases}$$
$$= \begin{cases} 0 & \text{if some } t_i \text{ has a proper prefix in } \text{RF}_M(x), \\ 2^{-\Sigma_{i=1}^{k} |t_i|} & \text{otherwise} \end{cases}$$

So M^∞ is independent. $\qquad\square$

Acknowledgement. We are grateful to Yuri Gurevich for his comments.

References

[BCGL89] S. Ben-David, B. Chor, O. Goldreich, and M. Luby, *On the theory of average case complexity*, Proc. 21st Annual ACM Symposium on Theory of Computing, 1989, pp.204-216.

[BG91] A. Blass and Y. Gurevich, *Randomizing reductions of search problems*, Proceedings of Foundations of Software Technology and Theoretical Computer Science (Invited Talk), New Delhi, India, 1991, pp. 10-24.

[Gur91]   Y. Gurevich, *Average case completeness*, J. of Computer and System Sciences, 42(1991), pp. 346-398.

[IL90]   R. Impagliazzo and L. Levin, *No better ways to generate hard NP instances than picking uniformly at random*, Proc. 31th IEEE Symposium on Foundations of Computer Science, 1990, pp. 812-821.

[Lev84]   L. Levin, *Average case complete problems*, SIAM J. on Computing, 15(1986), pp.285-286. Extended abstract appeared in Proc. 16th ACM Symposium on Theory of Computing, 1984, p. 465.

[VL88]   R. Venkatesan and L. Levin, *Random instances of a graph coloring problem are hard*, Proc. 20th ACM Symposium on Theory of Computing, 1988, pp. 217-222.

[Ven91]   R. Venkatesan, *Average Case Intractability*, Ph.D. thesis, Computer Science Department, Boston University, 1991.

[VR92]   R. Venkatesan and S. Rajagopalan, *Average case intractability of Diophantine and matrix problems*, Proc. of 24th ACM Symposium of Theory of Computing, May 1992, pp. 632-642.

[WB92]   J. Wang and J. Belanger, *On average P vs. average NP*, Proc. of the 7th IEEE Conference on Structure in Complexity Theory, June 1992, pp. 318-326.

Approximating Vertices of a Convex Polygon with Grid Points in the Polygon*

H. S. Lee[†]

Department of Information Management,
Ming Chuan College,
250 Chungshan N. Rd., Sec. 5,
Taipei, Taiwan, 11120, Republic of China
E-mail: hslee@felix.os.nctu.edu.tw

R. C. Chang

Institute of Computer and Information Science,
National Chiao Tung University,
Hsinchu, Taiwan 30050, Republic of China

Abstract

In this paper, we consider the problem of approximating a vertex of a convex polygon by an integer point in the polygon. We show that the nearest grid point in a convex polygon to a vertex can be found if it exists, or decided to be nonexistent, in time $O(n + \log l)$, where l is the diameter of the polygon and n is the number of the polygon's vertices. The underlying technique used is the continued fraction expansion.

1 Introduction

Continued fraction expansions have been shown to be efficient techniques for many problem [2,3,1,6]. In this paper, we show that continued fraction expansions provide an efficient solution to the problem of finding a grid point which is closest to a vertex of a given convex polygon, under the constraint that the nearest grid point must lie in the polygon. Let n be the number of the vertices of the polygon and let l be the diameter of polygon [7]. In time $O(n + \log l)$, the nearest grid point to a specific vertex can be found if there is any grid point lying in the polygon; otherwise, in the same time, it can be concluded that there is no solution in the polygon.

The problem posed above is an extension of the following problem posed by Nagy et al. [6]: Given two lines, find a grid point that is closest to the intersection of two lines, under the constraint that the approximating point lies in a specified sector formed by the two lines. This problem can be equivalently stated as follows. Given a wedge which is formed by two border rays originating from the same point, find a grid point which is closest to the vertex of wedge, under the constraint that the grid point lies in the wedge (Figure 1). The wedge is open on one end. For simplicity, this problem shall be called the vertex approximation of an open wedge (VAOW). As shown in Figure 2, if a convex chain is put on the open end of the wedge and the closest gird point is restricted in the close region, then this is equivalent to the problem we pose. Integer approximation to vertices of a convex polygon can thus be regarded as the following problem: Given a close wedge formed by two rays and a convex chain on the open end, find a grid point in the close wedge such that the grid point is closest to the vertex of wedge. Similarly, this problem shall be called the vertex approximation of a close wedge (VACW). A severe situation for VACW is that there may be no solution at all, but a VAOW always has a solution.

Rest of this paper is organized as follows. In Section 2, we present a new algorithm for VAOW. Though the algorithm runs in the same time as the that proposed by Nagy et al [6], our algorithm is more adaptable. Based on the philosophy of this algorithm, a more elaborated algorithm for VACW is introduced. Section 3 presents an algorithm for deciding whether there are grid points in a convex

---

*This research work was partially supported by the National Science Council of the Republic of China under grant No. NSC80-0408-E009-03.

[†]To whom all the correspondences should be sent.

polygon; if there are any, report any of them. Section 4 deals with the vertex approximation of close wedge. A summary is given in Section 5.

2 Integer Approximation to the Vertex of an Open Wedge

Some results in number theory are introduced first. In approximating a real number α with rational number, a classical method provides an efficient way to find a good approximation: this is the so-called continued fraction expansion of α [5]. One can write α uniquely in the form

$$\alpha = a_0 + \cfrac{1}{a_1 + \cfrac{1}{a_2 + \cdots}}$$

where $a_0 = \lfloor \alpha \rfloor$ and $a_1, a_2, \ldots$ are positive integers. In fact, $a_1, a_2, \ldots$ can be defined by the recurrence

$$\alpha_0 = \alpha, a_0 = \lfloor \alpha \rfloor, \alpha_{k+1} = \frac{1}{\alpha_k - a_k}, a_{k+1} = \lfloor \alpha_{k+1} \rfloor.$$

For an irrational number α, this expansion is infinite; for a rational number α, it is finite but may be arbitrarily long. If we stop after the k-the a_k, i.e. determine the rational number

$$\frac{g_k}{h_k} = a_0 + \cfrac{1}{a_1 + \cfrac{1}{\ddots_{a_{k-1} + \frac{1}{a_k}}}}$$

then we obtain the k-the convergent of α. It is known that $\frac{g_k}{h_k} \to \alpha$, as $k \to \infty$; in fact,

$$|\alpha - \frac{g_k}{h_k}| \leq \frac{1}{h_k h_{k+1}} \tag{1}$$

and h_k grows exponentially fast, so the right hand side of (1) tends to 0.

The continued fraction expansion has many nice properties. The numbers g_k and h_k satisfy the recurrence

$$g_{k+1} = a_{k+1} g_k + g_{k-1}, \qquad h_{k+1} = a_{k+1} h_k + h_{k-1} \tag{2}$$

which makes them easily computable. It follows that there are at most $O(\log d)$ convergents with numerators or denominators no greater than d. They satisfy the important identity

$$g_{k+1} h_k - g_k h_{k+1} = (-1)^k. \tag{3}$$

Moreover, if k is even, then

$$\frac{g_{k-2}}{h_{k-2}} < \frac{g_k}{h_k} \leq \alpha < \frac{g_k + g_{k-1}}{h_k + h_{k-1}} < \frac{g_{k-1}}{h_{k-1}} < \frac{g_{k-3}}{h_{k-3}}; \tag{4}$$

while the inequality signs are reversed when k is odd. This is shown in [4]. For convenience, the convergents can be represented in vector form $v_k = (h_k, g_k)$. Then recurrence (2) becomes $v_{k+1} = a_{k+1} v_k + v_{k-1}$, where $v_{-1} = (0,1)$, and $v_0 = (1, a_0)$. By equation (3), two consecutive convergents v_k and v_{k+1} constitute a basis of integer lattice Z^2 [5]; in other words, the set $\{c_1 v_k + c_2 v_{k+1} | c_1, c_2 \in Z\}$ is equal to the integer lattice Z^2.

Now, let us return to the VAOW. We say that a direction "goes into" an open wedge if this direction is between the directions of two border rays of the wedge; that is, a ray in this direction originating from wedge vertex lies in the wedge. Otherwise, a direction is said to "go across" the wedge; i.e. a ray in this direction will go out the wedge eventually. Let x^+ denote the direction parallel with x-axis toward positive infinity and let x^- denote the direction toward negative infinity. Let y^+ and y^- be defined similarly. A simple case of VAOW is that one of the four directions x^+, x^-, y^+, y^- goes into the wedge.

Without loss of generality, assume it is the direction y^+ that goes into the wedge (Figure 3). In this case, the solution can be easily calculated. Let $q = (q_x, q_y)$ be the vertex of the wedge. The solution must lie on the grid line $L_1 : x = \lfloor q_x \rfloor$ or $L_2 : x = \lceil q_x \rceil$. Let $(\lfloor q_x \rfloor, y_1)$ be the intersection between L_1 and border rays incident to q. Let $(\lceil q_x \rceil, y_2)$ be the intersection between L_2 and border rays of the wedge. Let $p_1 = (\lfloor q_x \rfloor, \lceil y_1 \rceil)$ and $p_2 = (\lceil q_x \rceil, \lceil y_2 \rceil)$. Since border rays incident to q are upward, the nearest grid point in the wedge to q is either p_1 or p_2. Furthermore, either p_1 or p_2 is the lowest among grid points in the wedge. These are summarized in the following.

Lemma 2.1 *If the y^+ direction goes into the wedge, then either p_1 or p_2 is the nearest grid point in the wedge to q. Also one of p_1 and p_2 is the lowest among grid points in the wedge.*

The thorny case of VAOW is that none of the x^+, x^-, y^+, y^- directions go into the wedge (Figure 4). Without loss of generality, we shall assume that the wedge is in the first quadrant. Otherwise, the wedge can be transformed by a simple translation and maybe an additional mirror operation (with respect to origin or axes). In the following, we shall introduce an integer transformation that transform the wedge into the simple case, i.e., y^+ direction goes into the wedge. By an integer transformation, we mean a linear transformation that map grid points onto grid points and vice versa. After transformation, the leftmost grid point among the lowest grid points in the wedge is the solution. Before the transformation is introduced, some observations are made. The first is the following.

Lemma 2.2 *Assume the wedge lies in the first quadrant. Then r is the nearest grid point in the wedge to wedge vertex q if and only if r is the lowest and the leftmost grid point.*

Proof: Assume r is the nearest grid point but not the leftmost grid point in the wedge. As shown in Figure 5, there must be a grid point in the second quadrant of r. Let it be r'. However, this is impossible for ,otherwise, r'' would be the nearest grid point. So r is the leftmost grid point in the wedge. In a similar way, we can show r is also the lowest grid point in the wedge. Conversely, if r is the lowest and the leftmost grid point, there is no grid point in the third quadrant of r. Hence r is the nearest grid point. □

Based on the observation, the following can be obtained.

Lemma 2.3 *Let r be the nearest grid point in the wedge to wedge vertex q. If the wedge lies in the first quadrant and the wedge also lies in the first quadrant after an integer transformation, then r is also the nearest grid point to q in the new coordinate system.*

Proof: As shown in Figure 6, let L_1 be parallel with x-axis of the new coordinate system and let L_2 be parallel with y-axis of the new coordinate system. Then we claim that there is no grid point in the shaded area. This can be concluded by a argument similar to that in the proof of Lemma 2.2. Hence r is the leftmost and the lowest grid point in the new coordinate system. By Lemma 2.2, r is also the nearest grid point in the new coordinate system. □

The final observation is:

Lemma 2.4 *Let r be the nearest grid point in the wedge to vertex q. Assume the wedge lies in the first quadrant and there is an integer transformation such that the x-axis of the new coordinate system is the same but the y^+ direction of the new coordinate system goes into the wedge. Then, in the new coordinate system, r is the leftmost of the lowest grid points in the wedge.*

Proof: Since the x-axis remains unchanged, r is also one of the lowest grid points in the new coordinate system. Moreover, in the new coordinate system, r is the leftmost among the lowest grid points, for r is the leftmost grid point in the original coordinate system. □

As usual, let $v_k = (h_k, g_k)$ be convergents of real number α. Assume the wedge lies in the first quadrant. Let α be a real number between the slopes of two border rays of the wedge. Since the interval bounded by a pair of adjacent convergents with larger indices is contained in the interval bounded by a pair of two adjacent convergents with lesser indices, there is an integer i for which v_i goes into the wedge but v_{i-2} and v_{i-1} do not. Let r be the solution. Recall that any two adjacent convergents constitute a basis of the integer lattice. We shall follow the convention that the first component of a basis is used as the unit vector of x axis and the second component is used as the unit vector of y axis for new coordinate system. By Lemma 2.3, r is also the nearest grid point in the coordinate system using $\{v_{i-1}, v_{i-2}\}$ as basis. By Lemma 2.4, r is therefore the leftmost of the lowest grid points in the wedge using $\{v_{i-1}, v_i\}$ as basis. To find r, we first find one of the lowest grid points in the coordinate system using $\{v_{i-1}, v_i\}$ as basis. This can be easily done according to Lemma 2.1. Let the grid point found be t. Assume now we are in the coordinate system using $\{v_{i-1}, v_i\}$ as basis. Let $t = (t_x, t_y)$. Let (x_t, t_y) be the intersection between line $L : y = t_y$ and the wedge border. Then $(\lceil x_t \rceil, t_y)$ is the coordinate of r in the coordinate system using $\{v_{i-1}, v_i\}$ as basis.

Now the algorithm is straightforward. If one of the four directions x^+, x^-, y^+ and y^- goes into the wedge, then the nearest grid point can be found according to Lemma 2.1. Otherwise, transform the wedge by a simple integer translation and mirror operation so that the wedge lies in the first quadrant. Choose α such that the vector $(1, \alpha)$ goes into the wedge, and find an integer i for which v_i goes into the wedge but v_{i-2} and v_{i-1} do not. Then find the leftmost among lowest grid points within the wedge in the coordinate system using v_{i-1} as x unit vector as v_i as y unit vector. This is the solution.

The last question is how much time it would take to find such i. In fact, it doesn't matter how we choose α provided that $(1, \alpha)$ goes into the wedge. Let m_1 and m_2 be the slopes of the border rays. Assume $m_1 \leq \alpha < \alpha' \leq m_2$. We can show that it makes no difference in efficiency to choose α or α'. Let v_j be first convergent of α that goes into the wedge. Likewise, let $v'_{j'}$ be the first convergent of α' that goes into the wedge. We claim that $|j - j'| \leq 1$. Let j_0 be the integer for which $v_k = v'_k$ for $k \leq j_0$ and $v_{j_0+1} \neq v'_{j_0+1}$. List the convergents of α an α' in ascending order. Assume j_0 is even. According to Lemma 4, we have

$$\frac{g_0}{h_0} < \cdots < \frac{g_{j_0}}{h_{j_0}} < \alpha < \frac{g_{j_0+1}}{h_{j_0+1}} < \alpha' < \frac{g'_{j_0+1}}{h'_{j_0+1}} < \frac{g_{j_0-1}}{h_{j_0-1}} < \cdots < \frac{g_1}{h_1}.$$

Note that $\alpha < \frac{g_{j_0+1}}{h_{j_0+1}} < \alpha'$ can be proved by contradiction. If this is not the case, i.e., $\alpha' \leq \frac{g_{j_0+1}}{h_{j_0+1}}$, then according to Equation 4, it would result in a contradiction that $v'_{j_0+1} = v_{j_0+1}$. Moreover $(j_0 + 2)$-th convergent of α', $\frac{g'_{j_0+2}}{h'_{j_0+2}}$, falls between $\frac{g_{j_0+1}}{h_{j_0+1}}$ and α', i.e.

$$\frac{g_0}{h_0} < \cdots < \frac{g_{j_0}}{h_{j_0}} < \alpha < \frac{g_{j_0+1}}{h_{j_0+1}} \leq \frac{g'_{j_0+2}}{h'_{j_0+2}} \leq \alpha' < \frac{g'_{j_0+1}}{h'_{j_0+1}} < \frac{g_{j_0-1}}{h_{j_0-1}} < \cdots < \frac{g_1}{h_1}.$$

If m_1 falls in $[\frac{g_{j_0}}{h_{j_0}}, \alpha]$ and m_2 falls in $[\alpha', \frac{g'_{j_0+1}}{h'_{j_0+1}}]$, then $j = j_0 + 1$ and $j' = j_0 + 2$, i.e. $|j - j'| = 1$. In other cases, j would equal to j'.

If the slopes of the border rays are rationals and the denominators of the rationals are less than d, then the time to find v_i are bounded by $O(\log d)$. If the solution space is an $N \times N$ lattice, then time to find v_i is $O(\log N)$, for there are at most $O(\log N)$ convergents with denominators less than N. Hence, we have the following conclusion:

Theorem 2.1 *The problem of finding a nearest grid point in the wedge to wedge vertex can be solved in $O(\log N)$ time if the problem is considered in an $N \times N$ lattice.*

3 Finding a Grid Point in a Convex Polygon

Finding a grid point is a crucial step of our algorithm for VACW. If there is no grid point in the polygon, we can stop searching for the nearest grid point in the polygon; otherwise, the searching procedure can proceed based on the grid found. Let l be the diameter of the convex polygon P. We present an $O(n + \log l)$ algorithm for this problem, where n is number of the vertices.

Our strategy is as follows. If the polygon is wide enough, we determine a grid point in the polygon in a simple way. If the polygon is so narrow that it is difficult to determine a grid point in the polygon in a simple way, we devise an integer transformation called "rectification" to transform the "bad" polygon into a "good" one. After the transformation, either the polygon becomes wide enough so that a gird point in the polygon can be found in a simple way, or the polygon is still narrow but it can be covered by a constant number of horizontal or vertical grid lines. Hence, it would be easy to determine whether there is a grid point in the polygon.

At first, we define a measurement for the wideness of a convex polygon. Let $\{a, b\}$ be a diametral pair of a convex polygon P[7]; i.e. the distance between a and b is equal to the diameter of the polygon. Draw two supporting lines parallel to the line through a and b such that P is bounded by these two lines. Then distance between these two lines is defined to be the width of P. Though different width may be obtained for different choosing of diametral pair, widths vary within a ratio of 2. This kind of measurement is good enough for our purpose.

Let c be the furthest vertex of P to the line passing through a and b. If the width measured with respect to diametral pair $\{a, b\}$ is no less than 6, then the nearest grid point to the centroid of triangle abc lies in P. We therefore find a grid point in P. This can be justified as follows. If the width is no less than 6, then the nearest distance from the centroid to sides of triangle abc is no less than 1. The nearest grid point to the centroid therefore lies in the triangle, i.e. in P.

If the width of P is less than 6, the nearest grid point to the centroid is no longer guaranteed to lie in P. We can not find a grid point in P by the simple method above. We apply the rectification transformation to the P. Let P' be the image of P under the rectification. If width of P' is no less than 6, then a grid in P' can be found by the method above. Otherwise, P' can be covered by a constant number of horizontal or vertical grid lines.

The rectification transformation is constructed by finding a new basis for integer lattice. The new basis is found as follows. Let l be the diameter of P and w be the width of P with respect to the diametral pair we choose. Let α be the slope of the line passing through the diametral pair of P chosen by us. Compute convergents v_k's of α. Let i be the largest integer such that $h_i \leq min\{\sqrt{\frac{l}{w}}, l\}$. If $\frac{g_i}{h_i} = \alpha$ or $h_i h_{i+1} > \frac{l}{w}$ then the basis is $\{v_i, v_{i-1}\}$; otherwise the basis would be $\{v_i, v_{i+1}\}$. In the following, we shall show that such a transformation satisfies our needs.

The rectification transformation has the following property: the area of the smallest rectangle containing P' with sides parallel to axes is roughly equal to that of P'. Let $abcd$ denote the smallest rectangle containing P' with sides parallel to axes. Assume that side ab is parallel to x axis and side bc is parallel to y axis. If the rectification transformation uses v_i as x unit vector of new coordinate system, length of bc can be measured as follows. In the original coordinate system, draw two support lines parallel with vector v_i such that P is bounded by these two lines. Let D_{P,v_i} denote the distance between these two lines. Let D_{v_i} denote the distance between two adjacent grid lines parallel with vector v_i. The length of bc is equal to $\frac{D_{P,v_i}}{D_{v_i}}$. It can be easily verified that $D_{v_i} = \frac{1}{\|v_i\|}$. Let θ_{v_i} denote the angle between the vector v_i and the line passing through the diametral pair. Then $D_{P,v_i} \leq l sin\theta_{v_i} + w cos\theta_{v_i}$, where l is the diameter of P and w is the width of P. Therefore length of bc is less than $(l sin\theta_{v_i} + w cos\theta_{v_i})\|v_i\|$ which is in turn less than $(l sin\theta_{v_i} + w)\|v_i\|$. Likewise, length of ab is less than $(l sin\theta_{v_{i-1}} + w)\|v_{i-1}\|$ if v_{i-1} is used as y unit vector of new coordinate system; or it is less than $(l sin\theta_{v_{i+1}} + w)\|v_{i+1}\|$ if v_{i+1} is used as y unit vector of new coordinate system. Now we are ready to show $|\overline{ab}| \times |\overline{bc}|$ is roughly equal to the area of P'.

Lemma 3.1 Let A denote the area of P'. Then $|\overline{ab}| \times |\overline{bc}| \leq 12A$.

Proof: At first we shall show that $|\overline{ab}| \times |\overline{bc}| < 6lw$. Let $d_1 = (l sin\theta_{v_i} + w)\|v_i\|$, $d_2 = (l sin\theta_{v_{i-1}} + w)\|v_{i-1}\|$, and $d_3 = (l sin\theta_{v_{i+1}} + w)\|v_{i+1}\|$. Let θ_{v_{i-1}, v_i} denote the angle between v_i and v_{i-1}. Since α lies between $\frac{g_{i-1}}{h_{i-1}}$ and $\frac{g_i}{h_i}$, $\theta_{v_i} \leq \theta_{v_{i-1}, v_i}$. Also, $sin\theta_{v_i, v_{i-1}} = \frac{\|v_i \times v_{i-1}\|}{\|v_i\|\|v_{i-1}\|}$ if these vectors are treated as three dimensional vectors, where $\times$ denote the cross product. Since $\|v_i \times v_{i-1}\| = |g_{i-1}h_i - g_i h_{i-1}| = 1$, $sin\theta_{v_i, v_{i-1}} = \frac{1}{\|v_i\|\|v_{i-1}\|}$. Then,

$$
\begin{aligned}
d_2 &= (l sin\theta_{v_{i-1}} + w)\|v_{i-1}\| \\
&\leq (l sin\theta_{v_{i-1}, v_i} + w)\|v_{i-1}\| \\
&= \frac{l}{\|v_i\|} + w\|v_{i-1}\|
\end{aligned}
$$

$$< \frac{l}{h_i} + w\sqrt{2}h_{i-1}$$

Likewise, if $\frac{l_i}{h_i} = \alpha$ then $d_1 < \sqrt{2}wh_i$; otherwise,

$$d_1 < (l\sin\theta_{v_i,v_{i+1}} + w)\|v_i\| < \frac{l}{h_{i+1}} + \sqrt{2}wh_i.$$

Similarly, if v_{i+1} is the last convergent, then $d_3 = w\|v_{i+1}\| < \sqrt{2}wh_{i+1}$; otherwise,

$$d_3 < (l\sin\theta_{v_{i+1},v_{i+2}} + w)\|v_{i+1}\| < \frac{l}{h_{i+2}} + \sqrt{2}wh_{i+1}$$

Hence, if v_{i+1} is the last convergent then $d_1d_3 < \sqrt{2}lw + 2w^2h_ih_{i+1}$; otherwise,

$$d_1d_3 < \frac{l^2}{h_{i+1}h_{i+2}} + \sqrt{2}lw + \sqrt{2}lw\frac{h_i}{h_{i+2}} + 2w^2h_ih_{i+1} < 4lw + 2w^2h_ih_{i+1}.$$

In other words, $d_1d_3 < 4lw + 2w^2h_ih_{i+1}$. If $\frac{l_i}{h_i} = \alpha$, $d_1d_2 < \sqrt{2}lw + 2w^2h_ih_{i-1} < 4lw$. If $\frac{l_i}{h_i} \neq \alpha$ and $h_ih_{i+1} > \frac{l}{w}$,

$$d_1d_2$$
$$< \frac{l^2}{h_ih_{i+1}} + \sqrt{2}\frac{h_{i-1}}{h_{i+1}}lw + \sqrt{2}lw + 2w^2h_ih_{i-1}$$
$$< \frac{l^2}{h_ih_{i+1}} + 5lw < 6lw.$$

If $\frac{l_i}{h_i} \neq \alpha$ and $h_ih_{i+1} \leq \frac{l}{w}$, $d_1d_3 < 4lw + 2w^2h_ih_{i+1} \leq 6lw$. Hence we can conclude that $|\overline{ab}| \times |\overline{bc}| \leq 6lw$. The area of P is also A. Since $lw \leq 2A$, $|\overline{ab}| \times |\overline{bc}| < 6lw \leq 12A$. $\square$

Let l' denote the diameter of P' and w' denote the width of P'. Then $A \leq l'w' \leq 2A$. Since $|\overline{ab}| \leq |\overline{bc}| \leq 12A$, we have

$$|\overline{ab}| \times |\overline{bc}| \leq 12l'w'. \tag{5}$$

Assume $|\overline{ab}| \leq |\overline{bc}|$. Then

$$l' \leq \sqrt{2}|\overline{bc}| \tag{6}$$

Combining Equations (5) and (6), obtain $|\overline{ab}| \leq 12\sqrt{2}w'$. We therefore have the following lemma:

Lemma 3.2 *Let w' be the width of P'. Then $\min\{|\overline{ab}|,|\overline{bc}|\} \leq 12\sqrt{2}w'$.*

If the width of P', w' is less than 6, by the lemma above, P' can be covered by a constant number of horizontal or vertical grid lines and therefore searching for the grid points in P' can be done in $O(n)$ time. If w' is no less than 6, the a grid point in P' can be found by the simple method described previously. It takes linear time to compute the diameter of a convex polygon. And the time to find the rectification transformation is $O(\log l)$. We can therefore conclude our results of this section in the following theorem.

Theorem 3.1 *Given a convex polygon P with n vertices, it can be determined whether there is a grid point in P in $O(n + \log l)$ time, where l is the diameter of P.*

4   Integer Approximation to the Vertex of a Close Wedge

In this section, we consider the following problem. Given a convex polygon P and a specific vertex q of P, find a nearest grid point in P to the vertex q. We shall show that this problem can be solved in $O(n + \log l)$ time, where n is the vertex number and l is the diameter of P.

Approximating a vertex of a convex polygon with a grid point in the polygon can be considered as a VACW problem. Let $w(q)$ denote the open wedge formed by extending two edges of P incident to q. Then convex polygon P can be considered as the boundary of the convex region surrounded by the wedge $w(q)$ and a convex chain. The nearest grid point in $w(q)$ to q may not be in P; that is, it may be excluded by the convex chain. This makes VACW problem more difficult than VAOW. In the following, we show similar observations for VACW can be obtained and only a slight modification of the procedure for VAOW is needed to have a procedure for VACW.

In VAOW, a simple case is that one of the four directions x^+, x^-, y^+, y^- goes into the $w(q)$. However, this may not hold for VACW. The condition of the simple case for VACW is much stronger. Without loss of generality, assume direction y^+ goes into $w(q)$. Let $q = (q_x, q_y)$. The two adjacent vertical grid lines surrounding q are $L_1 : x = \lfloor q_x \rfloor$ and $L_2 : x = \lceil q_x \rceil$. Let $(\lfloor q_x \rfloor, y_1)$ and $(\lceil q_x \rceil, y_2)$ be the intersections between lines L_1 L_2 and the edges incident to q if they exist. Let $p_1 = (\lfloor q_x \rfloor, \lceil y_1 \rceil)$ and $p_2 = (\lceil q_x \rceil, \lceil y_2 \rceil)$. The additional condition is that either p_1 or p_2 must be in P. A result similar to Lemma 2.1 can be obtained.

Lemma 4.1 *If y^+ goes into $w(q)$ and either p_1 or p_2 lies in P, then the nearest grid point in P to q is either p_1 or p_2. Also, either p_1 or p_2 is the lowest among grid points in P.*

As mentioned above, the nearest grid point and the lowest grid point can be computed by rounding up the intersections between the grid lines L_1 L_2 and the edges incident to q. However, it will not work when y^+ goes into $w(q)$ but there are no grid points in P that lie on the grid lines L_1 and L_2. Fortunately, in this case either some part of the polygon P can be discarded without affecting the solution. Let q_1 be the intersection between L_1 and the convex chain, and let q_2 be the intersection between L_2 and the convex chain. We have the following:

Lemma 4.2 *If y^+ goes into $w(q)$ and there are no grid points in P that lie on L_1 or L_2, then part of P, either to the left of line qq_2 or to the right of line qq_1, can be discarded.*

Proof: Either the convex chain form q to q_1 or the convex chain form q_2 to q is monotonic with respect to y-axis. Assume it is the convex chain form q to q_1 to be monotonic with respect to y-axis. Then there would be no grid point in P to the left of L_1. The part of P to the left of line qq_2 can therefore be discarded. □

Properties similar to those of VAOW can be obtained. We have a result similar to Lemma 2.2:

Lemma 4.3 *Assume the wedge $w(q)$ lies in the first quadrant. Then r is the nearest grid point in the P to q if and only if r is the lowest and the leftmost grid point in P.*

The proof of the lemma above is similar to that of Lemma 2.2. Based on this lemma, a result similar to Lemma 2.3 can be obtained. The proof is similar to that of Lemma 2.3.

Lemma 4.4 *Let r be the nearest grid point in P to q. If $w(q)$ lies in the first quadrant and $w(q)$ also lies in the first quadrant after an integer transformation, then r is also the nearest grid point to q in the new coordinate system.*

A result similar to Lemma 2.4 for VACW is stated in the following. The argument is similar to that of Lemma 2.4.

Lemma 4.5 *Let r be the nearest grid point in P to q. Assume $w(q)$ lies in the first quadrant and there is an integer transformation such that the x-axis of the new coordinate system is the same but the y^+ direction of the new coordinate system goes into $w(q)$. Then, in the new coordinate system, r is the leftmost of the lowest grid points in P.*

In the case mentioned in Lemma 4.5, if there are grid points in P that lie on the two adjacent grid lines surrounding q which are parallel with y^+ direction of the new coordinate system, then r can be found by first finding one of the lowest grid points in P in the new coordinate system according to Lemma 4.1. Once one of the lowest grid points in P in the new coordinate system is known, r can be easily computed.

Based on the properties derived previously, an algorithm for a VACW can be obtained. First of all, apply the integer translation and mirror operation to the polygon so that either $w(q)$ lies in the first quadrant or y^+ direction goes into $w(q)$. If y^+ goes into $w(q)$, then either there are grid points in P that lie on the two adjacent vertical grid lines surrounding q or there are no grid points in P that lie on the two adjacent vertical grid lines. In the first case, the solution can be computed according to Lemma 4.1. In the second case, according to Lemma 4.2, we can obtain a new polygon such that none of x^+, x^-, y^+ and y^- go into $w'(q)$ where $w'(q)$ is the new wedge with respect to new polygon.

From now on, we shall concentrate on the case where none of x^+, x^-, y^+ and y^- go into the wedge associated with q. Without loss of generality, we shall assume the wedge lies in the first quadrant. The first thing to do is to find a grid point in the polygon. According to Theorem 3.1, a grid point in P can be found if there is any. Let s be the grid point found. Let α be the slope of the line joining q and s. Let i be the integer for which v_i goes into $w(q)$ but v_{i-1} and v_{i-2} do not. By Lemma 4.4, the nearest grid point r is also the nearest grid in coordinate system using $\{v_{i-1}, v_{i-2}\}$ as basis. Let L_1 and L_2 be the two adjacent vertical grid lines surrounding q in the coordinate system using $\{v_{i-1}, v_i\}$ as basis. Let q_1 be the intersection between L_1 and the convex chain, and let q_2 be the intersection between L_2 and the convex chain. By Lemma 4.5, if there are grid points in P that lie on L_1 or L_2, r can be computed in $O(n)$ time.

If there are no grid points in P that lie on L_1 or L_2, according to Lemma 4.2, the part of P either to the left of line qq_2 or to the right of line qq_1 can be discarded. If i is even, then s is on the left of L_2. Otherwise, s is on the right of L_1. Hence, part of P to the right of line qq_1 can be discarded if i is even and part of P to the left of line qq_2 can be discarded otherwise. This is shown in Figure 7. That is, a smaller polygon can be obtained; the new wedge associate q in more narrow so that none of the four directions x^+, x^-, y^+ and y^- go into the wedge. We therefore face the same problem as that to be solved originally; that is, the problem can be solved by recursion.

The only problem remained is when to stop the recursion. The condition to stop is that either v_i goes into the wedge and there are grid points on the two adjacent grid lines parallel with v_i that surround q, or the first component of v_{i-1} is larger than l, where l is the diameter of P. Goal of our algorithm is to find the first integer i for which the stoping condition is satisfied. If the first part of the stoping condition is satisfied, according to Lemma 4.5, the nearest grid point can be found. Otherwise, if the second part is satisfied, then s is the nearest grid point. This can be justified by contradiction. Assume this is not the case, i.e., s is not the nearest grid point. Let s' be the nearest grid point. According to the stoping condition, either v_i do not go into the wedge or v_i goes into the wedge but there are no grid points on the two adjacent grid lines parallel with v_i that surround q. In the latter case, we have new smaller wedge such that the both v_i and v_{i-1} go across the new wedge. We can therefore assume that v_i and v_{i-1} go across the wedge. In this case, s' is also the nearest grid point in the coordinate system using $\{v_i, v_{i-1}\}$ as basis. Let $\langle s', s\rangle$ denote the vector from point s' to point s. Since s is on the first quadrant of s' in the coordinate system using $\{v_i, v_{i-1}\}$ as basis, $\langle s', s\rangle = av_i + bv_{i-1}$ where both a and b are non-negative integers and either a or b is greater than 1. Since either $\|\langle s', s\rangle\| \geq \|v_i\|$ or $\|\langle s', s\rangle\| \geq \|v_{i-1}\|$, and $\|v_i\| > \|v_{i-1}\| > l$, we have $\|\langle s', s\rangle\| > l$. It is impossible to have both s and s' in P, for l is the diameter of P. A contradiction is thus obtained.

Let i_0 be the first integer for which v_{i_0} satisfies the stoping condition. It follows that $i_0 - 2 \leq \log_{\frac{1+\sqrt{5}}{2}} l$. Note that each time v_i goes into the wedge but there are no grid points on the two adjacent grid lines parallel with v_i that surround q, a part of P can be pruned away. The time spent in all pruning steps is bounded by $O(n)$, where n is the vertex number of P. It takes $O(n)$ time to compute the diameter of l. Hence, the time to find the nearest gird point is bounded by $O(n + \log l)$. We can conclude our result in the following theorem.

Theorem 4.1 *Given a convex polygon P and a vertex q, the nearest grid point to q in P can be found in time $O(n + \log l)$ if it exists.*

5 Conclusion

In this paper, integer approximation to a vertex of a convex polygon is considered. This problem is closely related to the integer programming. We have shown that continued fraction expansions provide an efficient solution for this problem. On the issue of determining a grid point in a convex polygon, a transformation called "rectification" has been devised. The rectification transformation was also shown to be useful in [3]. One of the related topics is to apply the rectification transformation to other problems; the other is to extend our results to higher dimension.

References

[1] D. H. Greene and F. F. Yao, Finite-resolution computational geometry, *Proc. 27th Symposium on Foundations of Computer Science*, pp. 143-152, 1986.

[2] H. S. Lee and R. C. Chang, Hitting Grid Points in a Convex Polygon with Straight Lines, *Lecture Notes in Computer Science (ISA'91 Algorithms)*, Vol. 557, pp. 176-189, Springer-Verlag, 1991.

[3] H. S. Lee and R. C. Chang, A Fast Algorithm for Weber Problem on a Grid, Techn. Rep., Science and Info. Engineering, NCTU, Taiwan, 1991.

[4] W. J. LeVeque, *Fundamentals of Number Theory*, Addison-Wesley, 1977.

[5] L. Lovasz, *An Algorithmic Theory of Numbers, Graphs and Complexity, Society for Industrial and Applied Mathematics, Pennsylvania*, 1986.

[6] S. Mehta, M. Mukherjee and G. Nagy, Constrained integer approximation to 2-D line intersections, *Second Canadian Conf. on Computational Geometry*, Ottawa, Canada, August 1990, 302-305.

[7] F. P. Preparata and M. I. Shamos, *Computational Geometry: An Introduction*, Springer-Verlag, New York, 1985.

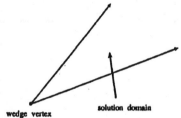

Figure 1: The solution domain for vertex approximation of an open wedge is an open wedge.

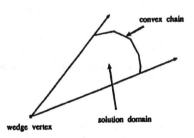

Figure 2: The solution domain for vertex approximation of a close wedge is a close wedge with one end bounded by a convex chain.

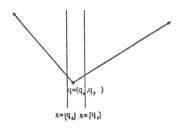

Figure 3: A simple case of VAOW: vertical grid lines intersect the wedge border at one point.

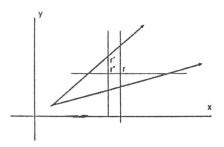

Figure 5: r is the leftmost grid point if r is the nearest grid point.

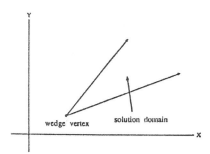

Figure 4: The thorny case of VAOW

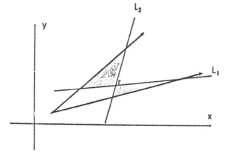

Figure 6: r is also the nearest grid point in the new coordinate system

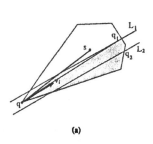

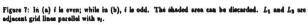

(a)   (b)

Figure 7: In (a) i is even; while in (b), i is odd. The shaded area can be discarded. L_1 and L_2 are adjacent grid lines parallel with v_i.

Algorithms for Determining the Geometrical Congruity in Two and Three Dimensions

Tatsuya Akutsu

Mechanical Engineering Laboratory
1-2 Namiki, Tsukuba, Ibaraki, 305 Japan

Abstract. This paper discusses the problem of determining the geometrical congruity. $O(n \log n)$ time algorithms for point sets in three dimensions and for graphs whose vertices are points in three dimensions are presented, respectively. They are based on Sugihara's algorithm for determining the congruity of polyhedra. Moreover, an $O(\log n)$ time $O(n)$ processors parallel algorithm for graphs in two dimensions which works on a CRCW PRAM is presented. In this algorithm, the original problem is transformed into the string matching problem. All the algorithms are optimal and can be modified for computing the canonical forms.

1   Introduction

In this paper, we discuss on the problem of determining whether two point sets (or two graphs) are congruent or not in two and three dimensional Euclidean space. Objects such as mechanical parts in mechanical CAD systems and three dimensional structures of chemical compounds in chemical database systems are represented with data structures like graphs whose vertices have positions. Thus, determining the congruity of two such graphs is practically important.

Several studies have been done for the related problems. Manacher showed an $O(n \log n)$ time algorithm for determining the congruity of two polygons [6] by applying a linear time string matching algorithm. Sugihara showed an $O(n \log n)$ time algorithm for determining the congruity of two polyhedra [8] based on the algorithm for planar graph isomorphism by Hopcroft and Tarjan [4]. Atkinson showed an $O(n \log n)$ time algorithm for determining the congruity of two point sets in three dimensional space [2].

In this paper, we show the following three algorithms, all of which are optimal as shown in chapter 5:

(A1) An $O(n \log n)$ time algorithm for determining whether two point sets are congruent or not in three dimensional space,

(A2) An $O(n \log n)$ time algorithm for determining whether two graphs are geometrically congruent or not in three dimensional space,

(A3) An $O(\log n)$ time $O(n)$ processors parallel algorithm for determining whether two graphs (as well as two point sets) are geometrically congruent or not in the plane.

Although Atkinson's algorithm is too complicated, Algorithm A1 is (conceptually) very simple. Although the shapes of polyhedra are restricted in Sugihara's algorithm, Algorithm A2 can be applied to almost all the types of polyhedra and other types of objects. As far as we know, no optimal parallel algorithm for two point sets in the plane was known. Moreover, each algorithm can be modified for computing the canonical form.

In this paper, we adopt a RAM (random access machine) as a model of sequential computation and a CRCW (concurrent read exclusive write) PRAM as a model of parallel computation. In both models, we assume that each arithmetic operation including square root and trigonometric functions for real numbers is carried out in one step.

2 Preliminaries

Let U denote a two or three dimensional Euclidean space. An *object* is a (finite or infinite) set of points in U. In this paper, two types of objects are considered. One is a finite set of points. The other is an undirected graph in which each vertex is a point in U and each edge is a straight line segment connecting two vertices. We assume that each graph satisfies a condition that, for each edge e, no vertex except its endpoints is on e. We call this condition as the condition (C1).

We define the congruence of objects in a similar way as Ref.[8]. A mapping T of U onto itself is said to be *isometric* if $|\overline{PQ}| = |\overline{T(P)T(Q)}|$ for any two points P and Q. For any object X, we define $T(X) = \{ Q \mid (\exists P \in X)(Q = T(P)) \}$. Let X and Y be two objects. If there exists an isometric mapping T which satisfies $Y = T(X)$, X and Y are said to be *congruent*. Moreover, such T is called a *congruent mapping* of X onto Y. It is well known that T is composed from translations, rotations and reflections [3].

For searching a database which consists of objects, canonical forms are useful. A data structure $C(X)$ representing an object X is called a *canonical form* of X if it satisfies the condition that $C(X) = C(Y)$ holds if and only if X and Y are congruent. Once canonical forms are computed, the congruity can be tested only by comparing them. Moreover, such technique as binary search can be used to search the database for an identical object.

A *polyhedron* and its *faces*, *edges* and *vertices* are defined in the usual way [8]. The size of a polyhedron is defined as the total number of faces, edges and vertices. Sugihara showed that whether two polyhedra are congruent or not can be determined in $O(n \log n)$ time if polyhedra satisfy some conditions, where n denotes the total size of input polyhedra [8]. Although the conditions are not described here, they are satisfied for all the polyhedra treated in this paper. See Ref.[8] for the details of the conditions. Although only the orientation-preserving mappings (i.e. mappings which does not include reflections) are considered in Ref.[8], it is easy to modify his algorithm for mappings including reflections.

Sugihara's algorithm can be modified for computing the canonical form of a polyhedron. We briefly describe the method of the modification. In his algorithm, edges are firstly partitioned into blocks $B(1), \cdots, B(k)$, and then the blocks are subdivided until each block consists of indistinguishable edges. If the final blocks are totally ordered where the order is invariant with rotations and translations, the canonical form of a polyhedron can be obtained. It is easy to see that the initial blocks $B(1), \cdots, B(k)$ can be sorted according to an appropriate order. When $B(j)$ is divided into $B(j)$ and $B(l)$, they are ordered as $B(j-1) < B(j) < B(l) < B(j+1)$. Since the final blocks are totally ordered in this way, the canonical form of a polygon of size n can be computed in $O(n \log n)$ time.

3 Congruity of Point Sets

In this chapter, we describe a simple algorithm (Algorithm A1) for determining the congruity of two point sets in space. The method is conceptually very simple. Before describing it in detail, we briefly overview it. For each set, points are projected to the surface of the sphere centered at the centroid. Then, a convex hull is constructed for each set of projected points. Lastly, Sugihara's algorithm is applied to convex hulls.

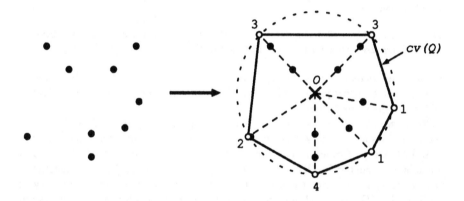

Fig. 1. Construction of $cv(Q)$ (in two dimensions). Black dots denote P_is. White dots denote Q_is. The number shown near Q_i denotes $label(Q_i)$.

Let $P = \{P_1, \cdots, P_n\}$ and $P' = \{P'_1, \cdots, P'_n\}$ be the sets of input points. Note that P and P' are not congruent if $|P| \neq |P'|$. Let O (resp. O') be the centroid of P (resp. P'). Let $R = \max_{P_i \in P} |\overline{OP_i}|$. S denotes the surface of the sphere of radius R centered at O. We assume w.l.o.g. (without loss of generality) that no P_i coincides with O. If P_i coincides with O, we handle $P - \{P_i\}$ with keeping the information that P_i coincides with O. $Q = \{Q_1, \cdots, Q_m\}$ denotes a set of the projected points of P on S (i.e. $(\forall Q_i)(\exists P_j)(\exists \alpha > 0)(\overrightarrow{OQ_i} = \alpha \overrightarrow{OP_j})$). $S(Q_i) = \{q^i_1, q^i_2, \cdots, q^i_{n_i}\}$

$(q_1^i < q_2^i < \cdots < q_{n_i}^i)$ denotes an ordered set of real numbers such that, for each q_k^i, there exists $P_j \in P$ projected to Q_i and $q_k^i = |\overline{OP_j}|$ holds. Q_i and Q_j are said to be equivalent if $S(Q_i) = S(Q_j)$ holds. Next, we define the relation '$\prec$' as follows. For two points Q_h and Q_k in Q, $Q_h \prec Q_k$ holds if and only if

$$(|S(Q_h)| < |S(Q_k)|) \vee$$
$$((|S(Q_h)| = |S(Q_k)|) \wedge (\exists j > 0)((\forall i < j)(q_i^h = q_i^k) \wedge (q_j^h < q_j^k)))$$

holds. $R', Q', S(Q_i'), \ldots$ are defined for P' in the same way. Moreover, the relation '$\prec$' and the equivalence relation are defined between $S(Q_k)$ and $S(Q_h')$ in the same way.

A set $Q \cup Q'$ is partitioned into equivalence classes. This task is done by sorting $Q \cup Q'$ according to '$\prec$'. It is easy to see that the sorting can be done in $O(n \log n)$ time. After the sorting, each number in $[1 \ldots K]$ is associated with different class where K is the number of the equivalence classes. $label(Q_i)$ denotes the number associated with the class to which Q_i belongs. Next, a convex hull $cv(Q)$ (resp. $cv(Q')$) of a set of points Q (resp. Q') is constructed (see Fig.1). Lastly, whether $cv(Q)$ and $cv(Q')$ are congruent (with considering labels) or not is determined by using Sugihara's algorithm. Although labels of vertices are not considered in Sugihara's algorithm, the modification for introducing labels can be made easily.

Theorem 1. *Algorithm A1 determines whether two point sets in three dimensional space are congruent or not in $O(n \log n)$ time.*

Proof. The correctness of the algorithm is almost obvious. The only thing to note is that each projected point is a vertex of a convex hull since it is on the surface of the same sphere.

Next, we consider the time complexity. Computing the centroids and projecting the points are trivially performed in $O(n)$ time. Computing and partitioning Q and Q' are performed in $O(n \log n)$ time. Since a convex hull of N points in space is constructed in $O(N \log N)$ time [7], $cv(Q)$ and $cv(Q')$ are constructed in $O(n \log n)$ time. Since the sizes of $cv(Q)$ and $cv(Q')$ are $O(n)$, congruity of $cv(Q)$ and $cv(Q')$ is tested in $O(n \log n)$ time. Therefore, Algorithm A1 works in $O(n \log n)$ time. □

By using the modified Sugihara's algorithm mentioned in chapter 2, the canonical form of a point set can be computed in $O(n \log n)$ time.

4 Geometrical Congruity of Graphs

In this chapter, we describe Algorithm A2, that is, an algorithm for determining whether two undirected graphs $G(V, E)$ and $G'(V', E')$ are geometrically congruent or not. Let $G(V, E)$ and $G'(V', E')$ be undirected graphs such that $V = \{P_1, \cdots, P_n\}$, $V' = \{P_1', \cdots, P_n'\}$ and $|E| = |E'| = m$. Note that $G(V, E)$ and $G'(V', E')$ are not congruent if $|V| \neq |V'|$ or $|E| \neq |E'|$. We assume w.l.o.g.

that $G(V, E)$ and $G'(V', E')$ are connected graphs. Otherwise, they can be made connected by applying a similar method as shown in the previous chapter and by adding $O(n)$ edges.

Algorithm A2 is conceptually very simple. Each input graph is transformed into a polyhedron. Let $poly(G)$ (resp. $poly(G')$) denotes the polyhedron transformed from $G(V, E)$ (resp. $G'(V', E')$). Then, whether $poly(G)$ and $poly(G')$ are congruent or not is determined by using Sugihara's algorithm. $G(V, E)$ is transformed by means of the following two steps. Although only the transformation for $G(V, E)$ is described, $G'(V', E')$ is transformed in the same way. In the following, ε_1 and ε_2 denote very small real numbers. However, their explicit values are not used in the algorithm. They are used as symbols which denote very small numbers.

STEP 1: Making each vertex fat (see Fig.2).

For each vertex $P_i \in V$, P_i is replaced by a convex hull $cv(P^i)$ as follows.

Let $E^i = \{ P_j \,|\, \{P_i, P_j\} \in E \}$ and $P^i = \{ P_j^i \,|\, \overrightarrow{P_i P_j^i} = \varepsilon_1 \dfrac{\overrightarrow{P_i P_j}}{|P_i P_j|}$ where $P_j \in E^i \}$.

Then, $cv(P^i)$ denotes a convex hull of P^i. Let F^i be a set of edges of $cv(P^i)$ and let $F = \{ \{P_j^i, P_i^j\} \,|\, \{P_i, P_j\} \in E \}$. Then, $G(V, E)$ is replaced by $\hat{G}(\hat{V}, \hat{E})$ where $\hat{V} = P^1 \cup P^2 \cup \cdots \cup P^n$ and $\hat{E} = F^1 \cup F^2 \cup \cdots \cup F^n \cup F$.

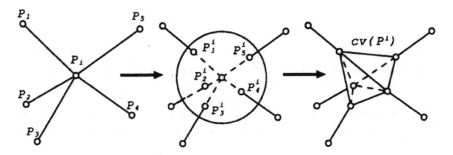

Fig. 2. Making a vertex fat.

STEP 2: Construction of bridges (see Fig.3).

For each edge $\{P_j^i, P_i^j\} \in F$, it is replaced by a *bridge* BR_{ij}. BR_{ij} and BR_{ji} are identical and it is constructed as follows. From P_j^i (resp. P_i^j), a column K_j^i (resp. K_i^j) is spanning towards P_i^j (resp. P_j^i). K_j^i is constructed as follows. Let H_j^i denote a plane perpendicular to $\overline{P_i P_j^i}$ such that the distance between H_j^i and P_j^i is ε_2. For each edge $\{P_j^i, P_h^i\} \in F^i$, compute the intersection between $\overline{P_j^i P_h^i}$ and H_j^i. Let $\{Q_1, \cdots, Q_k\}$ be a set of the intersections. Since all the Q_is are on the same plane perpendicular to $\overline{P_i P_j}$, the convex hull of $\{Q_1, \cdots, Q_k\}$ becomes a convex polygon. The polygon becomes the bottom of K_j^i, and K_j^i is spanning towards P_i^j. K_i^j is constructed in a similar way. K_j^i and K_i^j hit each other at the midpoint of $\overline{P_i P_j}$. The union of K_j^i and K_i^j is BR_{ij}. The hit region is partitioned into polygons. It is easy to see that this partition can be done in the time proportional to the sum of the size of K_j^i and the size of K_i^j. Although this construction can not be applied if a convex hull is degenerated, it can be extended for such cases easily.

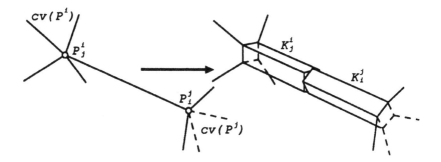

Fig. 3. Construction of a bridge.

Theorem 2. *Whether two graphs $G(V, E)$ and $G'(V', E')$ in three dimensional space that satisfy the condition (C1) are geometrically congruent or not is determined in $O(n \log n)$ time where $n = \max(\{|V|, |E|, |V'|, |E'|\})$.*

Proof. It is easy to see that $poly(G)$ and $poly(G')$ are congruent if and only if $G(V, E)$ and $G'(V', E')$ are geometrically congruent. Thus, Algorithm A2 is correct.

Next, we consider the time complexity. In STEP 1, a convex hull $cv(P^i)$ is constructed for each P_i. Since $cv(P^i)$ is computed in $O(deg(P_i) \log(deg(P_i)))$ time[7] where $deg(P_i)$ denotes the vertex degree of P_i in G, the total time required for STEP 1 is

$$\sum_{P_i \in V} deg(P_i) \log(deg(P_i)) = O(n \log n).$$

In STEP 2, a bridge BR_{ij} is constructed for each edge $\{P_i, P_j\} \in E$. The size of K_j^i is proportional to $deg(P_j^i)$ where $deg(P_j^i)$ denotes the vertex degree of P_j^i in $cv(P^i)$. Since

$$\sum_{P_j^i \in P^i} deg(P_j^i) = O(|F^i|) = O(deg(P_i))$$

holds, the total size of all the K_j^is are $O(n)$. Since each K_j^i is computed in the time proportional to its size, the total time required for STEP 2 is $O(n)$. Since the size of $poly(G)$ (as well as $poly(G')$) is $O(n)$, the congruity of $poly(G)$ and $poly(G')$ is tested in $O(n \log n)$ time by using Sugihara's algorithm. Therefore, the time complexity is $O(n \log n)$. ☐

By using the modified Sugihara's algorithm, the canonical form can be computed in $O(n \log n)$ time as well.

Remark. In Algorithm A2, faces are not included in inputs. However, it is easy to handle the case where there are faces. The problem with faces is reduced to the problem of this chapter as follows. For each face (polygon) F which contains vertices $P_1, \cdots, P_k$, a dummy vertex P_F is created at the centroid of $\{P_1, \cdots, P_k\}$ and P_F is labeled so as to indicate that P_F is a dummy vertex for representing a face. Moreover, P_F is connected to each of $\{P_1, \cdots, P_k\}$ with an edge. Since the size increases at most constant factor by this transformation, the time complexity is $O(n \log n)$.

5 Parallel Algorithm in the Plane

In this chapter, we show an optimal parallel algorithm (Algorithm A3) for determining the geometric congruity of two graphs (as well as two point sets) in the plane. We reduce the problem to the string matching problem. Manacher used a similar idea to obtain an $O(n \log n)$ time sequential algorithm for determining the congruity of two polygons [6]. However, the method presented in this chapter is more general.

Let $G(V, E)$ and $G'(V', E')$ be graphs in the plane. We assume w.l.o.g. that $|V| = |V'| = n$ and $|E| = |E'| = m$. The case of point sets can be treated by considering graphs with no edges. We assume w.l.o.g. that congruent mappings do not contain reflections.

First, we show that the lower bound of the complexity of the problem is $\Omega(n \log n)$. The lower bound of the complexity for determining the equality of two sets of numbers was shown to be $\Omega(n \log n)$ under the algebraic decision tree model [2, 7]. Thus, if there is no edge in graphs, the lower bound is trivially $\Omega(n \log n)$. We show that the same lower bound holds even if input graphs are connected. We reduce the problem for set equality to the problem for graphs as follows. Let a set of numbers be $\{a_1, \cdots, a_n\}$ (we assume $n > 2$). We construct $G(V, E)$ such that $V = \{P_0, P_1, \cdots, P_n\}$ and $E = \{\{P_0, P_i\} | 1 \leq i \leq n\}$, where $P_0 = (0, 1)$ and $(\forall 1 \leq i \leq n)(P_i = (a_i, 0))$. $G'(V', E')$ is constructed in the same way. These constructions are done in $O(n)$ time. It is easy to see that two sets of numbers are equivalent if and only if $G(V, E)$ and $G(V', E')$ are geometrically congruent. Therefore, the lower bound is $\Omega(n \log n)$. Moreover, it follows that the lower bounds for the problems of Algorithm A1 and Algorithm A2 are also $\Omega(n \log n)$.

Before describing a parallel algorithm, we show a sequential algorithm. Let $V = \{P_1, \cdots, P_n\}$ and $V' = \{P_1', \cdots, P_n'\}$. First, the centroid O of V is computed.

For each P_i, θ_i denotes the angle $\angle P_1 O P_i$. The angles are defined in such a way as the range of the angles is $[0, 2\pi)$. We define $P_i < P_j$ as follows:

$$(P_i < P_j) \quad \leftrightarrow \quad ((\theta_i < \theta_j) \vee ((\theta_i = \theta_j) \wedge (|\overline{OP_i}| < |\overline{OP_j}|))).$$

By sorting V according to '$<$', P_is are renumbered so that $P_1 < P_2 < \cdots < P_n$ holds. P_i's are renumbered in the same way.

Let $A = \{\angle P_{i-1} P_i P_{i+1}\} \cup \{\angle P_{i-1}' P_i' P_{i+1}'\}$ and $L = \{|\overline{P_i P_{i+1}}|\} \cup \{|\overline{P_i' P_{i+1}'}|\}$ where indices are computed in modulo n. We sort A and L, respectively. Let sorted lists be $(\alpha_1, \alpha_2, \cdots, \alpha_h)$ and $(l_1, l_2, \cdots, l_k)$, respectively. Let $\alpha(P_i)$ denote j such that $\alpha_j = \angle P_{i-1} P_i P_{i+1}$. Let $l(P_i)$ denote j such that $l_j = |\overline{P_i P_{i+1}}|$. $\alpha(P_i')$ and $l(P_i')$ are defined in the same way.

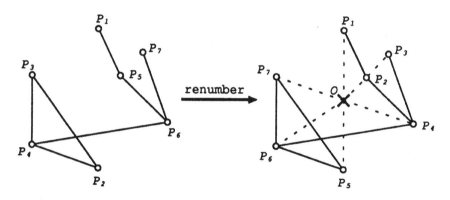

Fig. 4. Construction of $str(G)$. $str(P_2) =$"#7:1:2:2:6#" where $\alpha(P_2) = 7$ and $l(P_2) = 1$. $str(P_6) =$"#4:3:3:1:5:6#" where $\alpha(P_6) = 4$ and $l(P_2) = 3$.

We construct a string $str(G)$ as follows (see Fig.4). Let I_j denotes $j + n - i$ modulo n. For each P_i, $\{P_{n_1}, P_{n_2}, \cdots, P_{n_j}\}$ is a set of the adjacent vertices of P_i where $I_{n_1} < I_{n_2} < \cdots < I_{n_j}$ holds. For two strings s_1 and s_2, $s_1 s_2$ denotes a concatenation of s_1 and s_2. For each P_i, we define a string $str(P_i)$ as follows:

$$str(P_i) = "\#" \ \alpha(P_i) \ ":" \ l(P_i) \ ":" \ j \ ":" \ I_{n_1} \ ":" \ I_{n_2} \ ":" \ \cdots \ ":" \ I_{n_j} \ "\#" \ .$$

$str(G)$ is defined as $str(G) = str(P_1) str(P_2) \cdots str(P_n)$. $str(G')$ is defined in the same way. Then, $G(V, E)$ and $G'(V', E')$ are geometrically congruent if and only if $str(G')$ is a substring of $str(G) str(G)$ (note that the lengths of $str(G)$ and $str(G')$ are the same). Next, we analyze the time complexity. Since it is easy to see that the other parts can be done in $O(n + m)$ time, we consider the time required for the sorting and the substring matching. The length of $str(G)$ (resp.$str(G')$) is $O(n + m)$ where we assume that the size of each index is $O(1)$. It is easy to see that the total number of elements to be sorted is

$O(n+m)$. Since substring matching can be done in linear time and sorting of N elements can be done in $O(N \log N)$ time, the time complexity (the total work) is $O((n+m) \log(n+m))$.

Next, we consider a parallel implementation. Sorting and substring matching are critical for a parallel implementation. Sorting of N elements is done in $O(\log N)$ time using $O(N)$ processors on a CRCW PRAM (as well as EREW PRAM) [1]. Substring matching of length N and M ($N > M$) for general alphabet is done in $O(\log M)$ time using $O(\log N/\log M)$ processors on a CRCW PRAM [9]. Therefore, the sorting and the substring matching can be done in $O(\log(n+m))$ time using $O(n+m)$ processors. Since it is easy to see that the other parts can be done in $O(\log(n+m))$ time using $O(n+m)$ processors, the following theorem holds.

Theorem 3. *Whether two graphs $G(V,E)$ and $G'(V',E')$ in two dimensional space are geometrically congruent or not is determined in $O(\log n)$ time using $O(n)$ processors on a CRCW PRAM where $n = \max\{|V|, |E|, |V'|, |E'|\}$.*

The canonical form of G is computed in the following way. $a_j \cdots a_N a_1 \cdots a_{j-1}$ is said to be the canonical form of a (circular) string $a_1 \cdots a_N$ if the following condition is satisfied[5]:

$$(1 \leq \forall i \leq N)(a_j \cdots a_N a_1 \cdots a_{j-1} \leq a_i \cdots a_N a_1 \cdots a_{i-1}).$$

It is easy to see that the concatenation of $(\alpha_1, \cdots, \alpha_h)$, $(l_1, \cdots, l_k)$ and the canonical form of $str(G)$ becomes the canonical form of G. Since the canonical form of a circular string of length N over an alphabet of size $O(N)$ can be computed in $O(\log N)$ time using $O(N)$ processors on a CRCW PRAM [5], the canonical form of G can be computed in $O(\log n)$ time using $O(n)$ processors on a CRCW PRAM.

6 Remarks

The followings are the remarks common to all the algorithms in this paper.

Remark 1. Although we assumed a real RAM or a real PRAM as a computational model in which each arithmetic operation including square root and trigonometric functions for real numbers is carried out in one step. However, square root and trigonometric functions are not necessarily required since we need not explicitly compute such values as longitudes, latitudes and positions of projected points.

Remark 2. In practical situations, positions of input data usually contain errors. This problem is discussed in Ref.[8] and a similar method can be applied for cases of this paper. Of course, the method can not be applied to the cases of computing canonical forms.

Remark 3. All the algorithms presented can be modified for determining whether two objects are similar in the sense that they differ only in their sizes. This problem is discussed in Ref.[8] and a similar method can be applied for our cases. For that purpose, it is enough that the size of each input object is normalized by expanding or shrinking it according to the length between the centroid and the farthest vertex from it.

References

1. Ajitai, M., Komós, J., Szemerédi, E.: Sorting in $c\log(n)$ parallel steps. Combinatorica **3** (1983) 1–19
2. Atkinson, M. D.: An optimal algorithm for geometrical congruence. J. Algorithms **8** (1987) 159–172
3. Coxeter, H. S. M.: Regular Complex Polytopes. Cambridge University Press (1974)
4. Hopcroft, J. E., Tarjan, R. E.: A $V\log V$ algorithm for isomorphism of triconnected planar graphs. J. Computer and System Sciences **7** (1973) 323–331
5. Iliopoulos, C. S., Smyth, W. F.: Optimal algorithms for computing the canonical form of a circular string. Theoretical Computer Science **92** (1992) 87–105
6. Manacher, G.: An application of pattern matching to a problem in geometrical complexity. Information Processing Letters **5** (1976) 6–7
7. Preparata, F. P., Shamos, M. I.: Computational Geometry - An Introduction. Springer-Verlag, New York (1985)
8. Sugihara, K.: An $n\log n$ algorithm for determining the congruity of polyhedra. J. Computer and System Sciences **29** (1984) 36–47
9. Vishkin, U.: Optimal parallel pattern matching in strings. Information and Control **67** (1985) 91–113

On the Relationships among Constrained Geometric Structures

Esther Jennings   Andrzej Lingas

Department of Computer Science, Lund University
Lund, Sweden

Abstract

In this paper, we show the inclusion relation among several constrained geometric structures. In particular, we examine the constrained relative neighborhood graph in relation with other constrained geometric structures such as the constrained minimum spanning tree, constrained Gabriel graph, straight-line dual of bounded Voronoi diagram and the constrained Delaunay triangulation. We modify a linear time algorithm for computing the relative neighborhood graph from the Delaunay triangulation and show that the constrained relative neighborhood graph can also be computed in linear-time from the constrained Delaunay triangulation in the $(\Re^2, L_p)$ metric space.

1 Introduction

For many problems in clustering, pattern recognition, perception and terrain interpolation, suitable geometric structures are needed to extract the internal shape of a set of points. This need extends even to those cases where a set of non-intersecting edges (connecting points in the point set) is imposed as constraints.

Toussaint [14] introduced the relative neighborhood graph $RNG(V)$ for a point set V and discussed its applications in pattern recognition. Gabriel and Sokal [4] presented the Gabriel graph for geographic variation analysis of patterns of variation and covariation of characteristics of organisms distributed over an area. Delaunay triangulation has been studied extensively for its applications in the closest point problem, finite element method, stress analysis of two-dimensional continua and interpolation.

The above geometric structures have been generalized to take a planar straight-line graph $PSLG$ G as input instead of a point set V. Let G denote a given $PSLG$ throughout this paper. In geographical applications, the edges of G may represent extra information in addition to the given sites. These edges are important and should be preserved during the computation of various structures within G. Therefore, several constrained geometric structures such as the constrained relative neighborhood graph $CRNG(G)$, constrained Gabriel graph $CGG(G)$, and constrained Delaunay triangulation $CDT(G)$ were introduced [2, 7, 12]. In [12], it is shown that $CRNG(G)$ and $CGG(G)$ are both subgraphs of $CDT(G)$. We show that these structures, together with the constrained unique nearest neighbor graph $CUNNG(G)$, constrained

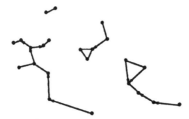

Figure 1: *A directed edge from a to b indicates that b is a nearest neighbor of a while an undirected edge $\overline{ab}$ indicates that a is a nearest neighbor of b and vice versa.*

minimum spanning tree $CMST(G)$ and straight-line dual of bounded Voronoi diagram $SBV(G)$, fit into a nice inclusion relation, i.e. $CUNNG(G) \subseteq CMST(G) \subseteq CRNG(G) \subseteq CGG(G) \subseteq SBV(G)$. Furthermore, Wang and Schubert [15] showed that $SBV(G) \subseteq CDT(G)$.

Given G on n vertices, the lower bound for the $CRNG$ problem is $\Omega(n\,log\,n)$ [12]. However, the $CRNG$ problem is usually solved in two phases. In phase one, $CDT(G)$ is constructed. Then, edges from $CDT(G) - CRNG(G)$ are eliminated in phase two. The best known algorithm for computing $CRNG(G)$ for $\Re^2$ space with L_2 norm is by Su and Chang [12] which takes $O(n\,log\,n)$ time for each phase. In this paper, we improve the time bound for phase two to $\mathcal{O}(n)$ by modifying the algorithm of Jaromczyk et al. [6] to work for the constrained structures in $(\Re^2, L_p)$ metric-space with $1 < p < \infty$.

This paper is organized as follows. In Section 2, we give the definitions of the different geometric structures used in this paper. Geometric properties of the constrained relative neighborhood graph and its relationship to other constrained geometric structures are discussed in Section 3. In Section 4, we show that several geometric properties of the relative neighborhood graph are preserved under constraints (prespecified edges). Establishing these results allows us to apply an algorithm presented in [6] to compute the constrained relative neighborhood graph in linear time from the constrained Delaunay triangulation. Section 5 summarizes this paper.

2   Preliminaries

Let $\Re^2$ denote the two-dimensional plane. The metric L_p is defined by a distance function d_p where $d_p(a, b) = \left(\sum_{j=1}^{2} |x_j(a) - x_j(b)|^p\right)^{\frac{1}{p}}$. Let V be a finite vertex set. The nearest neighbor of $a \in V$ is a subset of $V - a$ containing vertices that are closest to a. The nearest neighbor graph $NNG(V)$ for a vertex set V is a directed graph such that a directed edge from a to b, denoted by $\widehat{ab}$, is in $NNG(V)$ if and only if $a, b \in V$ and $d_p(a, b) = min\{d_p(a, c) : c \in V - a\}$. Note that the nearest neighbor relation is neither symmetric nor one-to-one (see Fig. 1). We introduce the *unique nearest neighbor graph* $UNNG(V)$ for a vertex set V as a subgraph of $NNG(V)$ such that $\widehat{ab}$ is in $UNNG(V)$ if and only if a has exactly one nearest neighbor, which is b.

A *minimum spanning tree* $MST(V)$ for a vertex set V is a tree which contains all vertices of V such that the total edge length of the tree is minimum. The *lune* of a and b, denoted by $lun(a, b)$, is defined as the intersection of the L_p-circles drawn by using a, b as centers and with radii equal to $d_p(a, b)$. An edge $\overline{ab}$ is in $RNG(V)$ if and only if there are no other vertices of V in $lun(a, b)$ [5, 6, 14].

Let $disk(a, b)$ denote the L_p-circle with $\overline{ab}$ as its diameter. The *Gabriel graph* $GG(V)$ for a vertex set V is a graph such that an edge $\overline{ab}$ is in $GG(V)$ if and only if there are no other vertices of V lying inside or on $disk(a, b)$ [4, 6].

A *PSLG* (planar straight-line graph) G is a pair (V, E) where V is the set of vertices and E is the set of open pairwise disjoint straight-line segments whose end-points are in V. A *triangulation* of a vertex set V is a maximal *PSLG* whose vertex set is V. There are many ways to triangulate V. We will only discuss the Delaunay triangulation in this paper. The Delaunay triangulation $DT(V)$ in L_p metric of a vertex set V is a triangulation of V such that the L_p circumcircle of any finite triangular face (i.e. empty triangle) does not contain any other points. This triangulation is unique if no four points in V are co-circular.

A *diagonal* of a *PSLG* G is an open straight-line segment which does not intersect any edge of E nor contain any vertex of V except at its end-points which are in V [8]. By considering all possible diagonals, $MST(V)$ on n vertices in the plane can be computed in quadratic time while computing $RNG(V)$ and $GG(V)$ in this way could require cubic time. However, knowing that $MST(V) \subseteq RNG(V) \subseteq GG(V) \subseteq DT(V)$ [1, 10, 14], optimal $\mathcal{O}(n \log n)$ time algorithms for the computation of $MST(V)$, $RNG(V)$ and $GG(V)$ have been designed. These algorithms first compute $DT(V)$ in $\mathcal{O}(n \log n)$ time and then remove edges that violate the requirements of $MST(V)$, $RNG(V)$ or $GG(V)$ in linear time. By using $DT(V)$, only $\mathcal{O}(n)$ Delaunay edges are examined instead of the $\mathcal{O}(n^2)$ possible diagonals.

In [7], Lee and Lin introduced the notion of a *generalized Delaunay triangulation* (also known as constrained Delaunay triangulation, $CDT(G)$ [11]) of a *PSLG* G. The generalized Delaunay triangulation $CDT(G)$ complements G with a partial triangulation T such that $G \cup T$ is a (full) triangulation of the vertices of G and the L_p circumcircle of each finite triangular face does not contain in its interior any other point which is visible from all three vertices of the triangle.

Wang and Schubert [15] introduced the notion of Voronoi diagram with barriers ($BV(G)$ for short, also known as bounded Voronoi diagram) for a *PSLG* G. $BV(G)$ is a net which together with G partitions the plane into regions $P(v), v \in V$ such that a vertex q is in $P(v)$ if and only if (q, v) is the shortest straight-line segment connecting q with a vertex in V that does not intersect any edge in E.

Analogous to the constrained Delaunay triangulation, we extend the domain of the unique nearest neighbor graph on a vertex set to a planar straight-line graph. Let $G = (V, E)$ denote a *PSLG* whose edges are considered as visibility barriers. Two vertices a and b of a *PSLG* are visible to each other if $\overline{ab}$ is an edge of E or if $\overline{ab}$ does not properly intersect any edge of E. The "distance" of the end-points of a prespecified edge in G is zero while the "distance" of two vertices not visible from each other is ∞. The distance function d for a pair of vertices is defined as follows:

$$d(u, v) = \begin{cases} 0 & \text{if } \overline{uv} \in E \\ d_p(u, v) & \text{if } \overline{uv} \notin E \\ & \text{and } u, v \text{ visible from each other} \\ \infty & \text{otherwise} \end{cases}$$

We define the distance of a vertex to an edge as follows. Let c be a vertex and $\overline{ab}$ an edge. The distance is defined as

$$dist(c, \overline{ab}) = \begin{cases} inf\{d_p(c, x) : x \in \overline{ab}\} & \text{if } \overline{cx} \cap E = \emptyset \\ \infty & \text{otherwise} \end{cases}$$

We define the constrained unique nearest neighbor graph $CUNNG(G)$ of a $PSLG$ G as a $PSLG$ such that a directed edge $\widehat{ab}$ is in $CUNNG(G)$ if b is the unique nearest visible neighbor of a and a is not incident to any edge of E. A *constrained minimum spanning tree* $CMST(G)$ of a $PSLG$ G is a $PSLG$ on the same set of vertices as G such that $G \cup CMST(G)$ is connected and the total edge length (a function of d) of $CMST(G)$ is minimum. Chew [2] mentioned that the $CDT(G)$ could be used to compute $CMST(G)$. Recently, Su and Chang [12] defined the *constrained relative neighborhood graph* $CRNG(G)$ for a $PSLG$ G such that a diagonal $\overline{ab}$ is in $CRNG(G)$ if and only if there are no other points in $lun(a, b)$ visible from both a and b. In the same article, they introduced the *constrained Gabriel graph* $CGG(G)$ for a $PSLG$ such that a diagonal $\overline{ab}$ is in $CGG(G)$ if and only if there are no other points visible from both a and b inside or on $disk(a, b)$.

Previous work has shown that $CMST(G) \subseteq CDT(G)$ [2], $CRNG(G), CGG(G) \subseteq CDT(G)$ [12], and $SBV(G) \subseteq CDT(G)$ [15] in $(\Re^2, L_2)$ metric space.

3 Subgraphs and Supergraphs of CRNG

Note that if u is the unique nearest visible neighbor of v then $lun(u, v)$ cannot contain any other points in its interior visible from both u and v. We obtain directly that $CUNNG(G) \subseteq CRNG(G)$. Obviously, $CUNNG(G)$ is a forest and every node in $CUNNG(G)$ has at most out-degree 1. Furthermore, we observe that any cycle in $CUNNG(G) \cup G$ is composed entirely of edges of G. The proof is omitted due to space limitation. Recall that G is a given $PSLG$, we show that $CUNNG(G) \subseteq CMST(G)$.

Theorem 1 : $CUNNG(G) \subseteq CMST(G)$ for $(\Re^2, L_p)$, $1 < p < \infty$.

Proof: Let $G = (V, E)$ and let T be a $CMST(G)$. We show that $CUNNG(G) \subseteq CMST(G)$. Suppose there exists e in $CUNNG(G)$ which is not in T, then $\{e\} \cup T \cup G$ contains a cycle C passing through e. Let $e = \overline{vw}$ where w is the unique nearest neighbor of v. Let e' be the other edge on C incident to v. Note that $e' \notin G$ because if so, neither of the end-points of e' could have any outgoing edges, a contradiction. Thus e' is in T. If $d(e') = d(e)$ then w is not an unique nearest neighbor of v, and if $d(e') < d(e)$ then w is not a nearest neighbor of v. Thus, $d(e') > d(e)$. By exchanging e for e' in T, we obtain a new spanning tree T' whose total length is shorter than T, a contradiction. □

Next, we show that $CMST(G) \subseteq CRNG(G)$.

Theorem 2 : Given G, then $CMST(G) \subseteq CRNG(G)$ for $(\Re^2, L_p)$, $1 < p < \infty$.

Proof: Consider a diagonal $\overline{ab}$ in a $CMST(G)$ T. Suppose $\overline{ab} \notin CRNG(G)$, i.e. a vertex c visible from both a and b is in the interior of $lun(a, b)$. Then $d(a, c) < d(a, b)$ and $d(b, c) < d(a, b)$. By removing $\overline{ab}$ from T, $G \cup T$ is split into two connected components A and B where $a \in A$ and $b \in B$. If $c \in A$ then add $\overline{bc}$ to $(T - \overline{ab})$, otherwise add $\overline{ac}$ to $(T - \overline{ab})$ to obtain a new constrained spanning tree T'. Note that the total length of T' is less than T in either case which leads to a contradiction. □

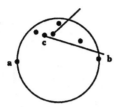

Figure 2: *There exists a closest vertex c not hidden by an edge of E in disk(a,b) if disk(a, b) is not empty and there does not exist a prespecified edge passing through disk(a,b).*

We now show that $CRNG(G) \subseteq CGG(G)$.

Observation 1 : $CRNG(G) \subseteq CGG(G)$ for $(\Re^2, L_p), 1 < p < \infty$.

Proof: It follows from the fact that $disk(a, b) \subset lun(a, b)$. □

Recall that $SBV(G)$ is the straight-line dual of the bounded Voronoi diagram, we establish the following lemma.

Lemma 1 : *Let $\overline{ab}$ be a diagonal of G. If $disk(a, b)$ is not free from vertices in $V-\{a, b\}$ and there does not exist a prespecified edge e which passes through $disk(a, b)$ hiding all of the vertices from $\overline{ab}$ then there exists a vertex c within $disk(a, b)$ visible from both a and b.*

Proof: Let $G = (V, E)$ and $E' = \{e : e \cap disk(a, b) \neq \emptyset\}$. If $E' = \emptyset$ and $disk(a, b)$ is not free from vertices of $V - \{a, b\}$, then there exists a vertex c closest to $\overline{ab}$ visible from both a and b. Otherwise, since there does not exist a prespecified edge passing through $disk(a, b)$ hiding all of the vertices from $\overline{ab}$, every edge of E' has at least one end-point in $disk(a, b)$. No matter how these end-points are hidden from a or b, the end-point closest to $\overline{ab}$ cannot be hidden by any edge of E' (see Fig. 2). □

Corollary 1 : *Let $\overline{ab}$ be a diagonal of G such that $\overline{ab} \in CGG(G)$. If there exists a vertex v in $disk(a, b)$ then there exists at least one prespecified edge that passes through $disk(a, b)$ which hides v from a and b.*

Theorem 3 : $CGG(G) \subseteq SBV(G)$ for $(\Re^2, L_p), 1 < p < \infty$.

Proof: Let a and b be two arbitrary points of G such that $\overline{ab}$ is a diagonal of G. Let $P(a)$ and $P(b)$ denote the bounded Voronoi regions of a and b respectively. If $\overline{ab}$ is an edge of $CGG(G)$ and not an edge of $SBV(G)$, then there exists an interior point m of $\overline{ab}$ which is in the region $P(c)$ of a vertex c different from a and b. If c is outside of $disk(a, b)$ then the bisectors of $\overline{ac}$ and $\overline{bc}$ cannot contain an interior point of $\overline{ab}$ (see Fig. 3). Therefore, c must be within $disk(a, b)$ and if c is visible from one end-point of $\overline{ab}$, then from lemma 1 there exists a vertex c' visible from both a and b, contradicting $\overline{ab}$ being an edge of $CGG(G)$. If c is hidden by an edge e from both a and b then no point in $\overline{ab}$ is contained in $P(c)$, a contradiction. Thus $P(a)$ and $P(b)$ must share a boundary edge. □

In [8, 15] it is shown that $SBV(G) \subseteq CDT(G)$ in L_2 metric. This result can be generalized to the L_p metric by similar proofs which are omitted. The above theorems combined with previously known results give us the following theorem.

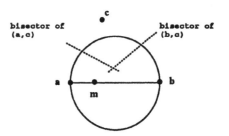

Figure 3: *This illustrates in L_2 metric that if c is outside of disk(a, b), then the bisectors of $\overline{ac}$ and $\overline{bc}$ do not intersect at any point on $\overline{ab}$.*

Theorem 4 : $CUNNG(G) \subseteq CMST(G) \subseteq CRNG(G) \subseteq CGG(G) \subseteq SBV(G) \subseteq CDT(G)$ for $(\Re^2, L_p)$, $1 < p < \infty$.

4 An Optimal CRNG Algorithm

In this section, we modify the algorithm by Jaromczyk, Kowaluk and Yao [6] to obtain $CRNG(G)$ from $CDT(G)$ in $(\Re^2, L_p)$ metric space in $\mathcal{O}(n)$ time. The key observations behind the algorithm by Jaromczyk et al. are the following: (1) if $\triangle abc$ is a triangle of $DT(V)$ and $v \in V$, then v eliminates at most two edges of $\triangle abc$; (2) if $v, w \in V$ eliminate $\overline{ab}$ where $\overline{ab} \in DT(V)$ and v, w are on the opposite side as c w.r.t. the line passing through $\overline{ab}$, then v, w do not eliminate both $\overline{ac}$ and $\overline{bc}$; and (3) let $\overline{ab} \in DT(V)$ and v be a closest point (w.r.t. d_p) of V eliminating $\overline{ab}$. Then v eliminates all edges (whose end-points are neither a nor b) of $DT(V)$ that lies between $\overline{ab}$ and v. By using the above properties, they avoid sweeping along sorted coordinates as the method of [13]. Instead, they define the elimination path for a vertex as an ordered list of edges eliminated by that vertex. The union of elimination paths forms the elimination forest. Since the elimination forest contains exactly those edges in $DT(V) - RNG(V)$, by elimination all the edges in the forest from $DT(V)$, $RNG(V)$ is obtained. A suitable data structure (static *UNION-FIND*) is chosen to represent the elimination forest in order to achieve the desired linear-time performance of the algorithm.

4.1 Geometrical properties of $CRNG(G)$ in $\Re^2$

In this section, we extend some geometrical properties which are essential for the design of the main algorithm from $RNG(V)$ to $CRNG(G)$.

Lemma 2 : *Given $G = (V, E)$, if $\triangle abc$ is a triangle of $CDT(G)$ and $v \in V$ then v eliminates at most two edges of $\triangle abc$.*

Proof: We consider the following two cases. (1) $\triangle abc$ does not contain a prespecified edge. The proof is the same as lemma 2.2 of [5] with the substitution of $CDT(G)$ for $DT(V)$. (2) $\triangle abc$ contains at least one prespecified edge. Clearly a vertex $v \in V$ could only eliminate the (at most two) diagonal(s) of $\triangle abc$. □

Let $v \in V$ eliminate a diagonal $\overline{ab}$. If $\triangle abc$ is a triangle of $CDT(G)$ and v does not lie on the same side as c w.r.t. the line passing through a, b, then v is external w.r.t. $\overline{ab}$.

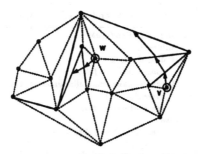

Figure 4: *This illustrates two elimination paths w.r.t. vertices v and w. The solid edges are vertices of the elimination paths. The directed edges show the predecessor/successor relation among adjacent edges of the elimination paths. Dashed edges are edges of the CDT(G), and bold solid edges are the prespecified edges.*

Lemma 3 : *Given $G = (V, E)$. Let $v, w \in V$ eliminate $\overline{ab}$ where $\overline{ab} \in CDT(G)$ and v, w are external to $\triangle abc$ w.r.t. $\overline{ab}$. Then v, w do not eliminate both $\overline{bc}$ and $\overline{ac}$.*

Proof: Again, we consider the following two cases. (1) $\triangle abc$ does not contain a prespecified edge. The proof is identical to lemma 2.3 of [5] with the substitution of $CDT(G)$ for $DT(V)$. (2) $\triangle abc$ contains at least one prespecified edge, then v, w could eliminate at most one of $\overline{bc}$ or $\overline{ac}$ whichever is a diagonal. □

Lemma 4 : *Given $G = (V, E)$. Let $\overline{ab} \in CDT(G)$ and v be a closest vertex (w.r.t. dist) of V eliminating $\overline{ab}$. Then v eliminates all edges $e \in CDT(G)$ (whose end-points are neither a nor b) such that $e \cap \triangle abv \neq \emptyset$.*

Proof: If $\triangle abv \cap E = \emptyset$, see lemma 2.5 of [5]. Let $e \in E$ such that $e \cap \triangle abv \neq \emptyset$ and there is not another edge $e' \in E$ in $\triangle abv$ such that e' separates e from $\overline{ab}$. We show by contradiction that e cannot be a prespecified edge. Since v is the closest point to $\overline{ab}$ (w.r.t. dist) which eliminates $\overline{ab}$, then v must be visible from both a and b. This implies that both end-points of e are in $\triangle abv$. Hence v is not the closest vertex of V which eliminates $\overline{ab}$. □

In Fig. 4, we show two elimination paths. Adjacent edges in an elimination path induced by a vertex always belong to a common triangle. By lemma 2, no three edges in an elimination path belong to the same triangle. Let e_i and e_j be two edges which belong to a common triangle and both edges belong to an elimination path induced by a vertex v. If e_i separates e_j from v, e_i is the immediate predecessor of e_j and there exists a directed edge from e_i to e_j in the elimination path. An elimination path for a vertex v ends if one of the following is true: a prespecified edge is encountered, an edge not eliminated by v is reached or an edge on the convex hull is reached. By lemma 3, if two elimination paths share a common edge with the same direction, then starting from this edge, one of the paths is completely included in the other. The last edge of each elimination path is a root. Thus the union of elimination paths is a forest of rooted and oriented trees. Each diagonal can appear twice in the elimination forest. The following algorithm is a modification of the algorithm by Jaromczyk et al. to compute the elimination path for a vertex $v \in V$ and an edge $\overline{ab}$ where $\triangle abc \in CDT(G)$.

```
Algorithm  Build_ElimPath(v, ab)
        path ← ∅;
        if  ab ∈ E  then
            can_continue ← false;
        if  can_continue and v eliminates ab  then
            current_e ← ab;
            can_continue ← true
        else
            can_continue ← false;

        while  can_continue  do
            path ← append(path, current_e);
            new_triangle ← triangle adjacent to current_e
            on the other side of current_e w.r.t. v;
            next_e ← an edge of new_triangle
            which is the successor of current_e;
            if  (∃ next_e) and (next_e ∉ E) and
                (v eliminates next_e)  then
                current_e ← next_e
            else
                can_continue ← false
        endwhile
        path(v, ab) ← path
end  Build_ElimPath
```

As a consequence of lemma 4, every edge in $CDT(G) - CRNG(G)$ belongs to a certain elimination path. Note that there are $\mathcal{O}(n)$ paths each of which may be $\mathcal{O}(n)$ long. To avoid a quadratic algorithm, static *UNION-FIND* is used to merge the elimination paths achieving a linear-time algorithm. Establishing lemmas 3 and 4 allows us to apply the techniques of [3, 6] for RNG to achieve an optimal $CRNG$ algorithm.

4.2   An optimal algorithm for $CRNG(G)$ in $\Re^2$

In [3], Gabow and Tarjan presented a linear-time algorithm for a special case of the disjoint set union problem in which the structure of the unions (static union tree) is known in advance. For an intermixed sequence of m union and find operations on n elements, the algorithm takes $\mathcal{O}(m + n)$ time and $\mathcal{O}(n)$ space.

Here is an overview of the $CRNG$ algorithm. A static union tree is obtained in linear time from a directed graph $CDT^*(X, A)$ where X is the vertex set and A is the set of directed edges. Furthermore, a vertex of X represents a diagonal of $CDT(G)$. There is a directed edge in A from $e_1 \in X$ to $e_2 \in X$ if and only if e_1 and e_2 belong to the same triangle in $CDT(G)$.

As a consequence of lemma 3, we have the following. Let $\triangle abc$ be a triangle of $CDT(G)$ such that $\overline{ab}$ is a diagonal. If a vertex v external to $\triangle abc$ eliminates $\overline{ab}$, then the only other edge of $\triangle abc$ which v may eliminate is: (1) the only remaining diagonal (if $\triangle abc$ contains exactly one prespecified edge) or (2) the longest of the remaining diagonals. Similar to the method used in [6], to obtain a static union tree, we construct a subgraph $H(X, B)$ of $CDT^*(X, A)$ such that each vertex in $H(X, B)$ has out-degree 0 or 1. For each vertex in X (which is a diagonal of $CDT(G)$ denoted as $\overline{ab}$), if $\triangle abc \in CDT(G)$ and only one of $\overline{ac}$, $\overline{bc}$ is a diagonal then include in B the directed edge from $\overline{ab}$ to $\overline{ac}$ or $\overline{bc}$ whichever is the diagonal. Otherwise, include in B the directed edge from $\overline{ab}$ to the (strictly) longest edge of $\overline{ac}$ and $\overline{bc}$. Again, by lemma 3, every elimination path is a path in $H(X, B)$.

As [6] pointed out, each connected component in $H(X, B)$ may contain at most one cycle so one cannot use $H(X, B)$ as the underlying static union tree. Instead, if a cycle exists in a connected component of $H(X, B)$, it must be broken. Jaromczyk et al. proposed a method of breaking the existing cycles by removing an arbitrary edge on each cycle. The removed edges are recorded. Let $T(X, B)$ denote the static union tree obtained by breaking cycles of $H(X, B)$. To compute $H(X, B)$ from $CDT^*(X, A)$ takes linear time. Breaking the cycles in $H(X, B)$ takes no more than linear time. When the static union tree $T(X, B)$ is obtained, the algorithm continues in two phases. In phase one, the linear-time disjoint union algorithm [3] is applied to $T(X, B)$. Note that the vertex set of $T(X, B)$ is also a superset of the edges of $CDT(G) - CRNG(G)$. Let $E' = CDT(G) - CRNG(G)$. In phase two, nodes of $T(X, B)$ which are edges of E' are traversed and marked. During the traversal, compensating actions are taken for the removed edges of $H(X, B)$. This is done in the following way. For each $\triangle abc \in CDT(G)$ such that b eliminates $\overline{ac}$, edges of $T(X, B)$ starting from node $\overline{ac}$ are traversed upwards, and the edges which b eliminates are marked. This is essentially the building of elimination paths. Let $\overline{e_1 e_2}$ denote a directed edge which was removed from $H(X, B)$ during cycle breaking. During the traversal of $T(X, B)$ and marking of nodes, if a marked node v is traversed and $v \neq e_1$ then make a union of the current elimination path with the previously computed elimination path (or tree) containing v. Otherwise, there are two cases to consider. If $v = e_1$ and the elimination path needs to be extended from $v = e_1$ to e_2, then find e_2 and continue. On the other hand, if $v = e_1$ is the root of the elimination tree containing e_2, there is no need to continue on the current path because e_2 was already examined. When the nodes of $T(X, B)$ which represent the edges of E' are marked, $CRNG(G)$ is obtained by $CDT(G) - E'$.

We note that the only modifications needed for the algorithm in [6] to work for $CRNG(G)$ are in the preprocessing of $H(X, B)$. Since a prespecified edge cannot appear on an elimination path, excluding them from $H(X, B)$ would not affect the correctness of the algorithm. Our contribution is to show that the geometric properties of $RNG(V)$ needed for computing $CRNG(G)$ from $CDT(G)$ in linear time are preserved under the constraint of prespecified edges.

5 Summary

In this paper, we have shown the inclusion relation among some specific planar constrained geometric structures in L_p metric. Specifically, we proved that $CUNNG(G) \subseteq CMST(G) \subseteq CRNG(G) \subseteq CGG(G) \subseteq SBV(G) \subseteq CDT(G)$. We then establish the fact that certain geometric properties of the relative neighborhood graph are preserved under the constraint of a set of prespecified edges. This allows us to extend the result of Jaromczyk et al. to the constrained relative neighborhood graph in $(\Re^2, L_p)$ metric space.

We could further refine the above inclusion relations by introducing the constrained version of the so called β-skeleton, $1 \leq \beta \leq 2$, between $CRNG(G)$ and $CGG(G)$. A β-skeleton for a finite point set V is by definition a supergraph of $RNG(V)$ and a subgraph of $GG(V)$ which respectively equals $RNG(V)$ for $\beta = 2$ and $GG(V)$ for

$\beta = 1$. See [6, 9] for details. As Jaromczyk et al. extended their algorithm in [6], we could analogously extend our algorithm to compute the constrained β-skeleton in L_p metric from the constrained Delaunay triangulation in linear time if $\beta = 2$ or $p = 2$.

Recently, Lingas has developed a simple linear-time algorithm for constructing $RNG(V)$ and (more general, for β-skeletons) from $DT(V)$ which does not use any advanced data structure [9]. This algorithm can also be generalized to compute $CRNG(G)$ from $CDT(G)$.

References

[1] F. Aurenhammer, *"Voronoi Diagrams - A Survey of a Fundamental Geometric Data Structure"*, ACM Comp. Surveys, Vol. 23, No. 3, Sep. 1991, pp.345-406.

[2] L. P. Chew, *"Constrained Delaunay Triangulations"*, Algorithmica 4 (1989), pp. 97-108.

[3] H. N. Gabow and R. E. Tarjan, *"A Linear-Time Algorithm for a Special Case of Disjoint Set Union"*, Journal of Computer and System Sciences 30, 1985, pp. 209-221.

[4] K. R. Gabriel and R. R. Sokal, *"A New Statistical Approach to Geographic Variation Analysis"*, Systematic Zoology 18 (1969), pp. 259-278.

[5] J. W. Jaromczyk and M. Kowaluk, *"A Note on Relative Neighborhood Graphs"*, Proceedings of the 3rd Annual Symposium on Computational Geometry, Waterloo, Canada, 1987, pp. 233-241.

[6] J. W. Jaromczyk, M. Kowaluk and F. F. Yao, *"An Optimal Algorithm for Constructing β-skeletons in L_p Metric"*, (manuscript, 1992) to appear in the SIAM Journal of Computing.

[7] D. T. Lee and A. K. Lin, *"Generalized Delaunay Triangulation for Planar Graphs"*, Discrete and Computational Geometry, 1 (1986), pp. 201-217.

[8] A. Lingas, *"Voronoi Diagrams with Barriers and the Shortest Diagonal Problem"*, Information Processing Letters 32 (1989), pp. 191-198.

[9] A. Lingas, *"A Simple Linear-Time Algorithm for Constructing the Relative Neighborhood Graph from the Delaunay Triangulation"*, 1992, (submitted).

[10] F. P. Preparata and M. I. Shamos, *"Computational Geometry, An Introduction"*, Springer-Verlag, 1985.

[11] R. Seidel, *"Constrained Delaunay Triangulations and Voronoi Diagrams with Obstacles"*, In Rep. 260, IIG-TU, Graz, Australia, 1988, pp. 178-191.

[12] T. H. Su and R. C. Chang, *"Computing the Constrained Relative Neighborhood Graphs and Constrained Gabriel Graphs in Euclidean Plane"*, Pattern Recognition, Vol. 24, No. 3, 1991, pp. 221-230.

[13] K. J. Supowit, *"The Relative Neighborhood Graph, with an Application to Minimum Spanning Trees"*, J. of Assoc. for Comput. Mach. Vol. 30, No. 3, July 1983, pp. 428-448.

[14] G. Toussaint, *"The Relative Neighborhood Graph of a Finite Planar Set"*, Pattern Recognition, Vol. 12 (1979), pp. 261-268.

[15] C. A. Wang and L. Schubert, *"An Optimal Algorithm for Constructing the Delaunay Triangulation of a Set of Line Segments"*, Proceedings of the 3rd Annual Symposium on Computational Geometry, June 8-10, 1987, Waterloo, Ontario, Canada, pp. 223-232.

Generating Small Convergent Systems Can Be Extremely Hard[*]

- Extended Abstract -

Klaus Madlener[1], Friedrich Otto[2], Andrea Sattler-Klein[1]

[1] Fachbereich Informatik, Universität Kaiserslautern,
6750 Kaiserslautern, West Germany

[2] Fachbereich Mathematik/Informatik, Gesamthochschule Kassel,
3500 Kassel, West Germany

Abstract. A sequence $(R_{n,m})_{n,m\in\mathbb{N}}$ of normalized string-rewriting systems on some finite alphabet Σ is constructed such that, for all $n, m \in \mathbb{N}$,
- $R_{n,m}$ contains 44 rules, it is of size $O(n + m)$, and it is compatible with a length-lexicographical ordering $>$ on Σ^*, but
- given the system $R_{n,m}$ and the ordering $>$ as input, the Knuth-Bendix completion procedure will generate more than $A(n, m)$ intermediate rules before a finite convergent system $S_{n,m}$ of size $O(n + m)$ is obtained, where A denotes Ackermann's function.

1 Introduction

Given a finite string-rewriting system R on some alphabet Σ and a total well-founded ordering $>$ on Σ^* as input, the Knuth-Bendix completion procedure [3] tries to transform the system R into an equivalent finite system S such that S is convergent and compatible with the given ordering. This procedure is correct and complete in the sense that it will succeed if and only if there exists a finite system S having these properties. Since it is undecidable in general whether or not such a system exists [6], termination of the completion procedure is undecidable as well. Thus, there is no recursive function that can serve as an apriori upper bound for the running time of the completion procedure for those inputs $(R, >)$ that lead to termination. Hence, for each recursive function $f : \mathbb{N} \to \mathbb{N}$, there exists a sequence of finite string-rewriting systems $(T_n)_{n\in\mathbb{N}}$ such that, given the system T_n and the ordering $>$ as input, the completion procedure will terminate eventually, but it will perform more than $f(size(T_n))$ steps until then. One obvious reason for this phenomenon to occur is the fact that the resulting finite convergent system S_n can be extremely large. However, here we illustrate the fact that, even if the resulting finite convergent system S_n is only of the size of the input system T_n, the completion procedure may make extremely many steps. In fact, we construct a sequence $(R_{n,m})_{n,m\in\mathbb{N}}$ of finite normalized string-rewriting systems on a fixed finite alphabet Σ such that, for each $n, m \in \mathbb{N}$, there exists a finite convergent system $S_{n,m}$ of approximately the same

---

[*] The results presented here where obtained while the second author was visiting at the Fachbereich Informatik, Universität Kaiserslautern during his sabbatical 1991/92.

size as the system $R_{n,m}$ such that $S_{n,m}$ is equivalent to $R_{n,m}$, and $S_{n,m}$ is compatible with a length-lexicographical ordering $>$ on Σ^*, but given the system $R_{n,m}$ and the ordering $>$ as input, the Knuth-Bendix completion procedure will generate more than $A(n,m)$ intermediate rules before the system $S_{n,m}$ is obtained. Here A denotes Ackermann's function.

The secret behind our construction is the fact that each system $R_{n,m}$ presents the trivial monoid, i.e., for all $n, m \in \mathbb{N}$ and all $w \in \Sigma^*$, we have $w \leftrightarrow^*_{R_{n,m}} \lambda$, where λ denotes the empty string, and $\leftrightarrow^*_{R_{n,m}}$ is the Thue congruence on Σ^* generated by $R_{n,m}$. Thus, for all $n, m \in \mathbb{N}$, $S_{n,m}$ is simply the system $S_{n,m} = \{a \rightarrow \lambda \mid a \in \Sigma\}$. Since it is undecidable in general whether or not a finite string-rewriting system R on Σ defines the trivial monoid, there is no recursive function that can serve as an apriori upper bound for the running time of the completion procedure applied to finite string-rewriting systems presenting the trivial monoid. Our examples illustrate this fact by showing explicitly how the completion procedure may become very inefficient in this case.

In Section 2 we present the necessary definitions and notation in short. Section 3 contains the development of our example systems. We first describe, for each $n \in \mathbb{N}$, a sequence $(R^1_{n,m})_{m \in \mathbb{N}}$ of finite string-rewriting systems on some finite alphabet Σ_n such that $R^1_{n,m}$ defines the trivial monoid, but given the system $R^1_{n,m}$ and a syllable ordering $>_{1,syl}$ as input, the Knuth-Bendix completion procedure will generate more than $A(n,m)$ intermediate rules. This sequence differs from our intended example in three ways. First of all, we do not have a fixed finite alphabet. Observe that it is really in the first argument that Ackermann's function is so rapidly increasing. Secondly, the systems $R^1_{n,m}$ are not normalized, and, while the system $R^1_{n,m}$ is of size $O(n + m)$, the resulting normalized system is much larger. Finally, $R^1_{n,m}$ contains length-increasing rules, and hence, it is not based on a length-lexicographical ordering. Now a syllable ordering may admit very long sequences of reductions, while a length-lexicographical ordering induces a simple exponential upper bound on the length of reduction sequences.

Subsequently, the example systems are refined in several steps to yield our intended systems $(R_{n,m})_{n,m \in \mathbb{N}}$. Each time we give the motivation behind the intended refinements and we shortly explain the technique used. We hope that in this way the reader will gain confidence in our construction.

2 Convergent String-Rewriting Systems

Here we restate the necessary definitions in short and establish notation. For additional information regarding the notions introduced the reader is asked to consult the literature, e.g., [1, 2].

Let Σ be a finite alphabet. Then Σ^* denotes the set of strings over Σ including the empty string λ. A *string-rewriting system* R on Σ is a subset of $\Sigma^* \times \Sigma^*$. Its *size* is defined as $size(R) = \sum_{(\ell \rightarrow r) \in R} (\mid \ell \mid + \mid r \mid)$, where $\mid u \mid$ denotes the length of the string u. For all $u, v \in \Sigma^*$ and $(\ell \rightarrow r) \in R$, $u\ell v \rightarrow_R urv$, i.e., $\rightarrow_R$ is the *single-step reduction relation* induced by R. Its reflexive transitive closure $\rightarrow^*_R$ is the *reduction relation* induced by R. A string u is *irreducible* if $u \rightarrow_R v$ does not hold for any string v. By $IRR(R)$ we denote the set of all irreducible strings.

The smallest equivalence relation $\leftrightarrow_R^*$ on Σ^* containing the relation $\rightarrow_R$ is called the *Thue congruence* induced by R. The set of congruence classes forms a monoid $M_R = \Sigma^* / \leftrightarrow_R^*$, and the ordered pair $(\Sigma; R)$ is called a (*monoid-*) *presentation* of this monoid.

In their seminal paper [3] Knuth and Bendix show how certain properties of finite (string-) rewriting systems yield a syntactically simple algorithm for deciding the word problem, and in addition, they show how to attempt to construct a system that has these properties from a given system. To discuss this approach for solving the word problem, we need the following additional notions.

A string-rewriting system R on Σ is called *noetherian* if there is no infinite sequence of the form $u_0 \rightarrow_R u_1 \rightarrow_R u_2 \rightarrow_R \ldots$, *confluent* if, for all $u, v, w \in \Sigma^*$, $u \rightarrow_R^* v$ and $u \rightarrow_R^* w$ imply that v and w have a common descendant, and *convergent* if it is noetherian and confluent. If a string-rewriting system R is convergent, then each congruence class $[u]_R$ contains a unique irreducible string $\hat{u}$, and $u \leftrightarrow_R^* v$ holds if and only if $\hat{u} = \hat{v}$. Thus, the word problem for R can simply be solved by reduction.

If $(\ell_1 \rightarrow r_1)$ and $(\ell_2 \rightarrow r_2)$ are two (not necessarily distinct) rules of R such that $\ell_1 = x\ell_2 y$ for some $x, y \in \Sigma^*$ or $x\ell_1 = \ell_2 y$ for some $x, y \in \Sigma^*$, $|x| < |\ell_2|$, then (r_1, xr_2y), respectively (xr_1, r_2y), is called a *critical pair* of R. A noetherian system R is confluent if and only if, for each critical pair (z_1, z_2) of R, z_1 and z_2 have a common descendant.

If the system R is not confluent, then certain critical pairs do not have a common descendant. Knuth and Bendix propose a *completion procedure* [3] that attempts to construct a convergent system S that is equivalent to R by adding rules to R. Here two systems on the same alphabet are called *equivalent* if they induce the same Thue congruence. Since new rules may result in additional critical pairs, this process may not terminate. Further, when adding new rules one must be careful to ensure that the new system is still noetherian. For this a total ordering $>$ on Σ^* is chosen that is *admissible* (i.e., $u > v$ implies $xuy > xvy$ for all $u, v, x, y \in \Sigma^*$), *well-founded* (i.e., there is no infinite descending sequence $u_0 > u_1 > u_2 > \ldots$), and *compatible* with R (i.e., $\ell > r$ holds for all rules $(\ell \rightarrow r) \in R$). If (z_1, z_2) is a critical pair of R such that z_1 and z_2 do not have a common descendant, then both these strings are reduced to some irreducible descendant $\hat{z}_i$ ($i = 1, 2$), and then the rule $(\hat{z}_1 \rightarrow \hat{z}_2)$ is added to R, if $\hat{z}_1 > \hat{z}_2$, or the rule $(\hat{z}_2 \rightarrow \hat{z}_1)$ is added to R, if $\hat{z}_2 > \hat{z}_1$. In addition, the actual system is kept as small as possible by *interreduction*. Here a string-rewriting system R is called *interreduced* or *normalized* if, for each rule $\ell \rightarrow r$ of R, $r \in IRR(R)$ and $\ell \in IRR(R - \{\ell \rightarrow r\})$. If $>$ is an admissible well-ordering, then there exists a unique (possibly infinite) normalized convergent system that is compatible with this ordering and that is equivalent to R.

In this paper we consider two types of orderings. Let $\Sigma = \{a_1, a_2, \ldots, a_k\}$ be a finite alphabet, and let $>$ be a total ordering on Σ called a *precedence*. Then the induced *length-lexicographical ordering* $>_{\ell\ell}$ on Σ^* is defined as follows:

$$u >_{\ell\ell} v \text{ iff } |u| > |v| \text{ or } (|u| = |v| \text{ and } u >_{lex} v),$$

where $>_{lex}$ denotes the lexicographical ordering on Σ^* induced by $>$.

Let $max(u)$ denote the largest letter with respect to the precedence $>$ that occurs in u. Then the induced *syllable ordering* $>_{syl}$ is defined as follows:

$$u >_{syl} v$$

$$\text{iff}$$
$$| u |_{max(uv)} > | v |_{max(uv)} \text{ or}$$
$$(max(uv) = a, \ | u |_a = | v |_a = n, \ u = u_1 a \cdots u_n a u_{n+1}, \ v = v_1 a \cdots v_n a v_{n+1},$$
and $\exists i \in \{1, \ldots, n+1\} : u_i >_{syl} v_i$ and $u_j = v_j$ for all $j \in \{i+1, \ldots, n+1\}$).

This syllable ordering corresponds to the well-known recursive path ordering for monadic terms [8].

The Thue congruence generated by a string-rewriting system R on Σ is called *trivial* if $u \leftrightarrow_R^* v$ holds for all strings $u, v \in \Sigma^*$. In this situation the monoid $\mathbf{M}_R$ presented by $(\Sigma; R)$ contains the identity element only, i.e., it is the *trivial monoid*. In this case the finite convergent string-rewriting system $S = \{a \to \lambda \mid a \in \Sigma\}$ is equivalent to R. Since S is compatible with every admissible well-ordering on Σ^*, the Knuth-Bendix completion procedure generates the trivial system S given the system R and an ordering $>_u$ or $>_{syl}$ as input.

3 Presentations of the Trivial Monoid

We use the following variant of Ackermann's function [5]:
$$A(0, y) = y + 1, \ A(x + 1, 0) = A(x, 1), \ A(x + 1, y + 1) = A(x, A(x + 1, y)).$$
For all $j \in \mathbb{N}$, we define the function A_j through $A_j(y) = A(j, y)$ $(y \in \mathbb{N})$. Now, for each $n \in \mathbb{N}$, we present a sequence of finite string-rewriting systems $(R^1_{n,m})_{m \in \mathbb{N}}$ on some alphabet Σ_n such that the Thue congruences $\leftrightarrow_{R^1_{n,m}}^*$ are trivial, and the systems $R^1_{n,m}$ are compatible with a syllable ordering $>_{1,syl}$ on Σ_n^*. Thus, $R^1_{n,m}$ is equivalent to the finite convergent system $S_n = \{a \to \lambda \mid a \in \Sigma_n\}$. However, on input the system $R^1_{n,m}$ and the ordering $>_{1,syl}$ the Knuth-Bendix completion procedure generates a sequence of rules that simulates the process of counting from 1 up to $A_n(m)$. Only when this value is reached, i.e., when the corresponding rule is obtained, a critical pair is generated that results in a rule, which in turn is crucial for producing the trivial system S_n by interreduction.

Definition 1. Let $n \in \mathbb{N}$, let $\Sigma_n = \{A_0, \ldots, A_n, w, V, i, X, a, b, c, d, r, u, o, O\}$, and let $>_{1,syl}$ be the syllable ordering on Σ_n^* that is induced by the following precedence:

$$w > V > A_n > \cdots > A_1 > A_0 > i > X > a > b > c > d > r > u > o > O.$$

The system $R^1_{n,m}$ $(m \geq 0)$ consists of the following three groups of rules:

(a)
1. $ub^m a \to o$
2. $wu \to rVA_n$
3. $Vb \to bV$
4. $Va \to cd$

5. $A_0 b \to bb$
6. $A_0 a \to ba$
7. $A_j b \to A_{j-1} A_j$ $(j = 1, \ldots, n)$
8. $A_j a \to A_{j-1} ba$ $(j = 1, \ldots, n)$

9. $wo \to o$

(b)
1. $irbc \to O$
2. $cX \to bc$
3. $OX \to O$

(c)
1. $Od \to \lambda$
2. $iob \to \lambda$
3. $wb \to \lambda$

4. $wa \to \lambda$
5. $Oa \to \lambda$

6. $Oob \to \lambda$
7. $ra \to \lambda$

The system $R^1_{n,m}$ has only two critical pairs:

- $(rVA_n b^m a, wo)$ from group (a), since $rVA_n b^m a \leftarrow wub^m a \to wo$, and

- $(OX, irbbc)$ from group (b), since $OX \leftarrow irbcX \rightarrow irbbc$.

The rules (a)5-8 are chosen such that they simulate the computation of the function A_n.

Lemma 2. $A_i b^j a \rightarrow^*_{R^1_{n,m}} b^{A_i(j)} a$ for all $i \leq n$ and $j \in \mathbb{N}$.

Since $wo \rightarrow o$, and since $Vb^k a \rightarrow^* b^k cd$, the first critical pair yields the rule $rb^{A_n(m)}cd \rightarrow o$. Since $OX \rightarrow O$, the second critical pair results in the rule $irb^2 c \rightarrow O$, from which we get another critical pair: $(OX, irb^3 c)$, since $OX \leftarrow irb^2 cX \rightarrow irb^3 c$. Thus, the rules $irb^k c \rightarrow O$, $k = 2, \ldots, A_n(m)$, are added to $R^1_{n,m}$ one by one. When the last of these rules has been added, a critical pair of a different form results: (Od, io), since $Od \leftarrow irb^{A_n(m)}cd \rightarrow io$. Since $Od \rightarrow \lambda$, this yields the rule $io \rightarrow \lambda$. Now interreduction replaces the rule $iob \rightarrow \lambda$ by the rule $b \rightarrow \lambda$, and further interreduction results in the trivial system S_n. We summarize the important properties of $R^1_{n,m}$.

Theorem 3. For all $n, m \in \mathbb{N}$,

(a) $size(R^1_{n,m}) = m + 9n + 55$,

(b) the Thue congruence $\leftrightarrow^*_{R^1_{n,m}}$ is trivial,

(c) $R^1_{n,m}$ is compatible with the syllable ordering $>_{1,syl}$, but

(d) given $R^1_{n,m}$ and $>_{1,syl}$ as input, the completion procedure generates at least $A_n(m) + 1$ intermediate rules from overlaps before the trivial system S_n is obtained.

The system $R^1_{n,m}$ is not normalized, since, for each $j \in \{1, \ldots, n\}$, the right-hand side of rule (a8) is reducible. In fact, $A_{j-1}ba \rightarrow^*_{R^1_{n,m}} b^{A_j(0)}a$ by Lemma 2. By replacing each rule $A_j a \rightarrow A_{j-1}ba$ by the rule $A_j a \rightarrow b^{A_j(0)}a$ we obtain a finite normalized system $\hat{R}^1_{n,m}$ that still has the above properties (b) to (d). However, $size(\hat{R}^1_{n,m}) = m + 7n + 55 + \sum_{j=1}^{n} A_j(0) \notin O(n+m)$.

To get around this difficulty we replace each of the rules $A_j a \rightarrow A_{j-1}ba$ by a rule of the form $A_j a \rightarrow w_{j-1}$, where w_{j-1} is an irreducible string of length 3.

Definition 4. Let $n \in \mathbb{N}$, let $\Delta_n = \Sigma_n \cup \{a_0, a_1, \ldots, a_{n-1}, z, Z, B\}$, and let $>_{2,syl}$ be the syllable ordering on Δ_n^* that is obtained by extending the precedence on Σ_n as follows:

$$A_n > a_{n-1} > A_{n-1} > \cdots > a_0 > A_0 > z > Z > B > i.$$

The system $R^2_{n,m}$ ($m \geq 0$) is obtained by replacing the rules (a8) $A_j a \rightarrow A_{j-1}ba$, $j = 1, \ldots, n$, with the following group of rules:

(d)	1. $A_j a \rightarrow z a_{j-1} B$ $(j = 1, \ldots, n)$		4. $Vz \rightarrow VZ$
	2. $A_j z \rightarrow A_j Z$ $(j = 0, 1, \ldots, n)$		5. $ZB \rightarrow ba$
	3. $Z a_j \rightarrow A_j Z$ $(j = 0, 1, \ldots, n-1)$		6. $Za \rightarrow \lambda$

When completing the system $R^1_{n,m}$ the rules $A_j a \rightarrow A_{j-1}ba$ $(j = 1, \ldots, n)$ are used to reduce the string $V A_n b^m a$ to $b^{A_n(m)}cd$. However, in this context these rules can be simulated in $R^2_{n,m}$.

Lemma 5. *Let* $\alpha \in \{V, A_0, A_1, \ldots, A_n\}$. *Then* $\alpha A_j a \rightarrow^*_{R^2_{n,m}} \alpha A_{j-1} ba$ *for all* $j = 1, \ldots, n$.

Thus, the completion of $R^2_{n,m}$ is essentially parallel to the completion of $R^1_{n,m}$. Observe that the rule $Za \rightarrow \lambda$ guarantees that $Z \leftrightarrow^* \lambda$, i.e., that the system $R^2_{n,m}$ still defines the trivial monoid. The system $R^2_{n,m}$ is normalized, and $size(R^2_{n,m}) = m + 17n + 69$.

The systems $R^1_{n,m}$ and $R^2_{n,m}$ are not compatible with a length-lexicographical ordering. To use such an ordering, we need to simulate the computation of the value $A_n(m)$ from m in a different way. The idea is to use a sequence of intermediate rules that are generated by the completion procedure and that encode this computation.

Definition 6. Let $n \in \mathbb{N}$, and let Δ_n be the alphabet from the previous definition. However, here we choose the following different precedence on Δ_n:

$$w > V > z > Z > B > A_n > A_{n-1} > \cdots > A_1 > A_0 > a_{n-1} > \cdots > a_1 > a_0 >$$
$$i > X > a > c > b > d > r > u > o > O.$$

The system $R^3_{n,m}$ ($m \geq 0$) is obtained from the system $R^2_{n,m}$ by replacing the rules (a2) $wu \rightarrow rVA_n$ and (d1) $A_j a \rightarrow za_{j-1}B$, $j = 1, \ldots, n$, with the following group of rules:

> (e)   1. $w^2 u \rightarrow rVA_n$ 2. $A_j a^3 \rightarrow za_{j-1}B$ $(j = 1, \ldots, n)$ 3. $oa \rightarrow o$

The system $R^3_{n,m}$ is compatible with the induced length-lexicographical ordering $>_{1,\ell\ell}$. It contains all the rules of group (b), and hence, the sequence $irb^k c \rightarrow O$, $k = 2, 3, \ldots, A_n(m)$, is generated during the completion of $R^3_{n,m}$.

Lemma 7. *When applied to the system* $R^3_{n,m}$ *and the ordering* $>_{1,\ell\ell}$ *the completion procedure will eventually generate the rule* $rb^{A_n(m)}cd \rightarrow o$.

Proof. Instead of giving a formal proof we just consider a simple example. Let $n = m = 2$, i.e., $A_n(m) = 7$. From (e1) and (a1) we obtain a critical pair:

$$rVA_2 b^2 a \leftarrow w^2 ub^2 a \rightarrow w^2 o .$$

Since $A_2 b^2 a \rightarrow^* A_1^2 A_2 a$ and $w^2 o \rightarrow^* o$, this gives the rule $rVA_1^2 A_2 a \rightarrow o$. With the corresponding rule of (e2) this rule yields the next critical pair:

$$oa^2 \leftarrow rVA_1^2 A_2 aa^2 \rightarrow rVA_1^2 za_1 B.$$

Since $oa^2 \rightarrow^* o$ and $A_1 za_1 B \rightarrow^* A_1 A_0 A_1 a$, this gives the rule $rVA_1^2 A_0 A_1 a \rightarrow o$. Continuing in this way the following rules are generated one after another:

$$rVA_1 A_0^3 A_1 a \rightarrow o, \quad rVA_0^5 A_1 a \rightarrow o, \quad rb^7 cd \rightarrow o.$$

Thus, each of these rules encodes an intermediate step of the computation of $A_2(2)$ until finally the rule $rb^{A_2(2)}cd \rightarrow o$ is obtained.   □

We can summarize the properties of $R^3_{n,m}$ as follows.

Theorem 8. *For all* $n, m \in \mathbb{N}$, $R^3_{n,m}$ *is a normalized system such that*
(a) $size(R^3_{n,m}) = m + 19n + 73$,
(b) *the Thue congruence* $\leftrightarrow^*_{R^3_{n,m}}$ *is trivial,*
(c) $R^3_{n,m}$ *is compatible with the length-lexicographical ordering* $>_{1,\ell\ell}$, *but*

(d) given $R_{n,m}^3$ and $>_{1,\ell\ell}$ as input, the completion procedure generates more than $A_n(m)$ intermediate rules from overlaps before the trivial system is obtained.

With the parameter n also the number of letters used and the number of rules of the system $R_{n,m}^3$ increase, whereas we would rather have a fixed finite alphabet and a fixed number of rules. To this end we shall encode the rules

$$A(0, y) \to y + 1, \quad A(x + 1, 0) \to A(x, 1), \quad A(x + 1, y + 1) \to A(x, A(x + 1, y))$$

in the completion process. At each step in the evaluation of $A(n, m)$, we have a term of the form $A(n_1, A(n_2, \ldots, A(n_k, n_{k+1}) \ldots))$, and the next step involves the last two arguments n_k and n_{k+1}. Since we would rather work from the left, we consider the function $\hat{A}(y, x) := A(x, y)$. Then we have intermediate terms of the form $\hat{A}(\ldots \hat{A}(\hat{A}(n_1, n_2), n_3), \ldots, n_k)$, which we encode as $ub^{n_1}ab^{n_2}a \cdots ab^{n_k}ae$.

Definition 9. Let $\Sigma = \{B, C, F, K, O, S, U, X, a, b, c, d, e, f, i, o, r, u, w\}$, and let $>$ be the following precedence on Σ:

$$w > i > f > U > B > K > C > X >$$
$$b > a > e > F > S > d > c > r > u > o > O.$$

For $n, m \in \mathbb{N}$, let the system $R_{n,m}^4$ contain the following three groups of rules:

(a)
1. $ub^m ab^n ae \to o$		7. $faaa \to Sbaa$		13. $Ka \to aa$	
2. $fb \to Bf$		8. $faae \to Fbcd$		14. $Cb \to bC$	
3. $BS \to Sb$		9. $faab \to Sbab$		15. $Ca \to ab$	
4. $BF \to Fb$		10. $Ufab \to uba$		16. $wo \to o$	
5. $UF \to r$		11. $Bfab \to SabK$		17. $wu \to Uf$	
6. $US \to u$		12. $Kb \to bCK$			

(b)
1. $irbc \to O$		2. $cX \to bc$		3. $OX \to O$	

(c)
1. $iob \to \lambda$		4. $Oa \to \lambda$		7. $wS \to \lambda$	
2. $wb \to \lambda$		5. $Oob \to \lambda$		8. $UX \to \lambda$	
3. $wa \to \lambda$		6. $Od \to \lambda$		9. $ra \to \lambda$	

It is easily verified that the system $R_{n,m}^4$ is normalized, and that it is compatible with the syllable ordering $>_{3,syl}$ induced by the given precedence. There are two overlaps - one between rules (a1) and (a17), and one between rules (b1) and (b2). As in the examples before, the latter gives the rule $irb^2c \to O$, and by iteration we get the sequence of rules $irb^kc \to O$, $k = 1, 2, \ldots$. From the other overlap the critical pair $(Ufb^m ab^n ae, wo)$ results. Now $wo \to o$, and

$$Ufb^m ab^n ae \to^* \begin{cases} rb^{m+1}cd & : \quad n = 0 \\ ubab^{n-1}ae & : \quad m = 0 \text{ and } n > 0 \\ ub^{m-1}ab^n ab^{n-1}ae & : \quad m > 0 \text{ and } n > 0. \end{cases}$$

Thus, the result of reducing $Ufb^m ab^n ae$ encodes the sequence of arguments of $\hat{A}$ after one step of the calculation process. Accordingly, the rule $rb^{m+1}cd \to o$ is generated, if $n = 0$, or one of the rules $ubab^{n-1}ae \to o$ or $ub^{m-1}ab^n ab^{n-1}ae \to o$ is generated. In the former case, the value $\hat{A}(m, 0) = m + 1$ has been computed, and no further overlaps occur in group (a). In the latter case, the new rule overlaps with rule (a17), and the resulting critical pair yields a rule that encodes the sequence of arguments of $\hat{A}$ after the next step in the evaluation process. By iteration we finally obtain the rule $rb^{\hat{A}(m,n)}cd \to o$.

This rule overlaps with the rule $irb^{\hat{A}(m,n)}c \to O$ of the sequence of rules generated from group (b). The resulting critical pair is (Od, io), which gives the rule $io \to \lambda$. As in the previous examples this rule now is responsible for the fact that by interreduction the trivial system $S = \{a \to \lambda \mid a \in \Sigma\}$ is obtained.

Theorem 10. *For all $n, m \in \mathbb{N}$, $R_{n,m}^4$ is a normalized system such that*

(a) $size(R_{n,m}^4) = m + n + 118$, and $R_{n,m}^4$ contains 29 rules,

(b) the Thue congruence $\leftrightarrow_{R_{n,m}^4}^*$ is trivial,

(c) $R_{n,m}^4$ is compatible with the syllable ordering $>_{3,syl}$, but

(d) given $R_{n,m}^4$ and $>_{3,syl}$ as input, the completion procedure generates more than $A(n,m)$ intermediate rules from overlaps before the trivial system is obtained.

The system $R_{n,m}^4$ contains the length-increasing rule (a12) $Kb \to bCK$, and hence, it is not compatible with a length-lexicographical ordering. In fact, if $i > 0$ and $j > 0$, then $Ufb^iab^jae \to^* ub^{i-1}ab^jab^{j-1}ae$ is the reduction used to simulate one step of the evaluation process of $\hat{A}(m,n)$, and this reduction consists essentially of the copying of the syllable $b^{j-1}a$. Hence, this technique is not applicable when a length-lexicographical ordering is used. Thus, this copying must then also be simulated by a sequence of overlaps resulting in further intermediate rules.

Definition 11. Let Σ be the following 25-letter alphabet

$$\Sigma = \{A, B, C, F, G, K, O, P, R, S, T, U, X, a, b, c, d, e, f, g, i, o, r, u, w\},$$

and let $>$ be the following precedence:

$$w > f > g > A > C > T > B > S > R > F > G > K > U > P >$$
$$u > a > c > b > X > d > r > i > e > o > O.$$

For $n, m \in \mathbb{N}$, the system $R_{n,m}$ contains the following three groups of rules:

(a)
1. $wwu \to Uff$	11. $TTKb \to SbKC$	21. $BG \to Gb$
2. $ffb \to Bff$	12. $TTKa \to Sa$	22. $UG \to ub$
3. $ffa \to Agg$	13. $ggb \to BTT$	23. $BS \to Sb$
4. $ggae \to Fd$	14. $TTb \to BTT$	24. $AS \to Sa$
5. $AF \to Fc$	15. $TTa \to RRa$	25. $US \to u$
6. $BF \to Fb$	16. $BRR \to RRb$	26. $Cb \to bC$
7. $UF \to rb$	17. $BARRb \to SabKP$	27. $Ca \to ab$
8. $ggaa \to Ga$	18. $Pb \to bP$	28. $wo \to o$
9. $ggab \to Gb$	19. $ub^mab^nae \to o$	29. $Pa \to aa$
10. $AG \to Ga$	20. $UARRb \to uba$	

(b) The rules of group (b) from the system $R_{n,m}^4$.

(c) The rules of group (c) from the system $R_{n,m}^4$, where the rule (c9) is replaced by the following rules:

9. $fS \to \lambda$	11. $TS \to \lambda$	12. $wR \to \lambda$
10. $gS \to \lambda$		

Again it is easily verified that the system $R_{n,m}$ is normalized, and that it is compatible with the length-lexicographical ordering $>_{2,u}$ induced by the above precedence. From the rules of group (b) completion produces the sequence of rules $irb^k c \to O$, $k = 1, 2, \ldots$. From the rules of group (a) the rule $rb^{\hat{A}(m,n)}cd \to o$ is generated as in the system $R^4_{n,m}$; however, from the rule $ub^m ab^n ae \to o$ $(n, m > 0)$ the rule $ub^{m-1}ab^n ab^{n-1}ae \to o$ is not generated in a single step but through a sequence of steps. Let us consider the example that $m > 0$ and $n = 2$. We first get the critical pair $(wwo, Uffb^m ab^2 ae)$ from overlapping the rules (a1) and (a19). Since $wwo \to^* o$ and $Uffb^m ab^2 ae \to^* ub^{m-1}abKbaae$, we obtain the new rule $ub^{m-1}abKbaae \to o$. This rule overlaps with the rule (a1) giving the critical pair $(wwo, Uffb^{m-1}abKbaae)$. Since $Uffb^{m-1}abKbaae \to^* ub^{m-1}ab^2 Kabae$, we obtain the rule $ub^{m-1}ab^2 Kabae \to o$. Finally, this rule gives the critical pair $(wwo, Uffb^{m-1}ab^2 Kabae)$, and since $Uffb^{m-1}ab^2 Kabae \to^* ub^{m-1}ab^2 abae$, we obtain the intended rule $ub^{m-1}ab^2 abae \to o$.

Theorem 12. *For all $n, m \in \mathbb{N}$, $R_{n,m}$ is a normalized system such that*
(a) $size(R_{n,m}) = m + n + 189$, and $R_{n,m}$ contains 44 rules,
(b) the Thue congruence $\leftrightarrow^*_{R_{n,m}}$ is trivial,
(c) $R_{n,m}$ is compatible with the length-lexicographical ordering $>_{2,u}$, but
(d) given $R_{n,m}$ and $>_{2,u}$ as input, the completion procedure generates more than $A(n, m)$ intermediate rules before the trivial system is obtained.

4 Concluding Remarks

The above examples nicely illustrate the fact that even for finite normalized string-rewriting systems that admit equivalent finite and convergent systems of a small size the completion procedure can be very inefficient. The reason for this inefficiency is here the fact that the completion procedure, which works in a purely syntactic way, does not realize that it is dealing with a presentation of the trivial monoid before the rules $a \to \lambda$ $(a \in \Sigma)$ are generated. However, if along with the system $R_{n,m}$ and the ordering $>_{2,u}$ the information was provided that the monoid presented by $R_{n,m}$ is trivial, then the convergent system S that is equivalent to $R_{n,m}$ could be produced immediately. Thus, this additional information would speed-up the process of completing $R_{n,m}$ enormously. This observation leads to the following interesting question.

Let R be a finite string-rewriting system on Σ such that the monoid presented is finite. Then the process of completing R is guaranteed to succeed for any admissible well-ordering on Σ^*. How much additional information on the finite monoids presented would be needed to obtain a variant of the completion procedure that runs efficiently on all finite presentations of finite monoids? How much does the information help that the monoid presented by R is indeed finite? How much would it help to know the order of the monoid presented by R? Would it be helpful to know that the finite monoid is indeed a finite group?

Observe that in our examples the process of completion could be sped-up considerably if the additional information was provided that the string-rewriting systems considered present groups. Consider e.g. the system $R^1_{n,m}$ of Definition 1. The rule

$wo \to o$ would imply $w \leftrightarrow^* \lambda$ by the cancellation laws for groups, and in fact, $g \leftrightarrow^* \lambda$ could then easily be derived for all letters $g \in \Sigma_n$. However, we can modify this example to get systems $(\tilde{R}^1_{n,m})_{n,m \in \mathbb{N}}$ such that the following properties are also met:

1. for all $(\ell \to r) \in \tilde{R}^1_{n,m}$, ℓ and r have neither a common nonempty prefix nor a common nonempty suffix,
2. for all $(\ell_1 \to r_1), (\ell_2 \to r_2) \in \tilde{R}^1_{n,m}$, if $r_1 = r_2$ and $\ell_1 \neq \ell_2$, then ℓ_1 and ℓ_2 have neither a common nonempty prefix nor a common nonempty suffix.

Hence, the cancellation laws do not apply to the rules of the modified systems $\tilde{R}^1_{n,m}$, $n, m \in \mathbb{N}$. Modifications of this form can also be applied to our other examples - however, in this way the example systems get technically even more involved.

Obviously, the above questions can also be asked for group-presentations instead of monoid-presentations, i.e., for string-rewriting systems that provide inverses of length one for all generators. Also they can be asked for other types of algorithms dealing with finite structures like e.g. the Todd-Coxeter method of coset enumeration for subgroups of finite groups [4]. In this context we would like to point out that in [7] Sims discusses some examples of finite presentations of groups for which the Knuth-Bendix completion procedure seems to be more useful than coset enumeration. We hope that our examples and this short discussion will contribute to focusing attention on the important question of how to measure the complexity of algorithms that deal with finite structures.

References

1. R.V. Book. Thue systems as rewriting systems. *Journal Symbolic Computation* 3, 39-68 (1987)
2. M. Jantzen. *Confluent String Rewriting.* EATCS Monographs on Theoretical Computer Science 14. Berlin: Springer 1988
3. D.E. Knuth, P. Bendix. Simple word problems in universal algebras. In: J. Leech (ed.): *Computational Problems in Abstract Algebra.* New York: Pergamon 1970, pp. 263-297
4. R.C. Lyndon, P.E. Schupp. *Combinatorial Group Theory.* Berlin: Springer 1977
5. R. McNaughton. *Elementary Computability, Formal Languages, and Automata.* Englewood Cliffs: Prentice-Hall 1982
6. C. O'Dunlaing. Undecidable questions related to Church-Rosser Thue systems. *Theoretical Computer Science* 23, 339-345 (1983)
7. C. Sims. The Knuth-Bendix procedure for strings as a substitute for coset enumeration. *Journal Symbolic Computation* 12, 439-442 (1991)
8. J. Steinbach. *Comparing on Strings: Iterated Syllable Ordering and Recursive Path Orderings.* SEKI Report SR-89-15. Kaiserslautern: Universität 1989

Chew's theorem revisited
- uniquely normalizing property of
nonlinear term rewriting systems -

Mizuhito Ogawa

NTT Basic Research Laboratories
3-9-11 Midori-cho Musashino-shi Tokyo 180 Japan
mizuhito@ntt-20.ntt.jp

Abstract. This paper gives a purely syntactical proof, based on proof normalization techniques, of an extension of Chew's theorem. The main theorem is that a weakly compatible TRS is uniquely normalizing. Roughly speaking, the weakly compatible condition allows possibly nonlinear TRSs to have nonroot overlapping rules that return the same results. This result implies the consistency of CL-pc which is an extension of the combinatory logic CL with parallel-if rules.

1 Introduction

The Church-Rosser (CR) property is one of the most important properties for term rewriting systems (TRSs). When a TRS is nonterminating, a well-known condition for CR is Rosen's theorem, which states that a left-linear weakly nonoverlapping TRS is CR - or, simply, that a left-linear nonoverlapping TRS is CR[11, 13]. (A pair of reduction rules is said to be overlapping if their applications interfere with each other (i.e., they are unified at some nonvariable position), and a TRS is said to be nonoverlapping if none of its rules are overlapping (except that a same rule overlaps itself at the root). A TRS is said to be weakly nonoverlapping if applications of an overlapping pair of rules return the same result. Without the assumption of linearity, on the other hand, CR for a nonoverlapping TRS is not guaranteed for the following two reasons:

(i) A pair of nonoverlapping rules may overlap modulo equality. For instance, R_1 has a sequence $d(2,2) \rightarrow d(2, f(2)) \rightarrow d(f(2), f(2)) \rightarrow d(f(2), f^2(2)) \rightarrow \cdots$ s.t. $d(2,2)$, $d(f(2), f(2))$, $\cdots$ are reduced to 0, and $d(2, f(2))$, $d(f(2), f^2(2))$, $\cdots$ are reduced to 1. Because 0 and 1 are normal forms, R_1 is not CR[11].

(ii) A reducible expression (redex) for a nonlinear rule may not be recovered after some reduction destroys the identity of nonlinear variables. For instance, R_2 has a sequence $1 \rightarrow f(1) \rightarrow d(1, f(1)) \rightarrow d(f(1), f(1)) \rightarrow 0$. Thus, $1 \overset{*}{\rightarrow} 0$ and $1 \rightarrow f(1) \overset{*}{\rightarrow} f(0)$. Since 0 is a normal form and $f(0)$ simply diverges to $d(0, d(0, d(\cdots)))$, R_2 is not CR[3].

$$R_1 = \begin{cases} d(x,x) & \rightarrow 0 \\ d(x, f(x)) \rightarrow 1 \\ 2 & \rightarrow f(2) \end{cases} \qquad R_2 = \begin{cases} d(x,x) \rightarrow 0 \\ f(x) & \rightarrow d(x, f(x)) \\ 1 & \rightarrow f(1) \end{cases}$$

Chew and Klop have shown the sufficient condition for the uniquely normalizing (UN) property instead of CR - that is, a strongly nonoverlapping TRS is UN[3, 7]. A TRS is said to be strongly nonoverlapping if its linearization (i.e., the renaming of repeated variables with fresh individual variables) is nonoverlapping. Chew also states, more generally that a compatible TRS is UN[7]. In the compatible case, however, Chew's proof is hard to recognize, and its journal version has not yet been published[4].

Several trials to get a new proof have been reported[3, 15], and they show partial answers. Their main technique is to first transform a nonlinear TRS to a linear TRS (either unconditional or conditional) and use its weakly nonoverlapping property to prove it CR. Then, they translate the CR property of the linearized TRS to the UN property of the original nonlinear TRS[7, 8]. De Vrijer uses a similar technique to show that CL-pc (combinatory logic with parallel-if) is UN[15] and the key of his proof is the consistency check (i.e., $T \neq F$) accomplished by constructing a model of CL-pc. Finally the consistency shows that an application of $Czzz \to z$ may overlap modulo equality with an application of either $CTxy \to x$ or $CFxy \to y$, but not both.

$$\text{CL-pc} = \text{CL} \cup \left\{ \begin{array}{l} CTxy \to x \\ CFxy \to y \\ Czzz \to z \end{array} \right\} \quad \text{where} \quad \text{CL} = \left\{ \begin{array}{l} Sxyz \to xz(yz) \\ Kxy \to x \\ Ix \to x \end{array} \right\}$$

This paper shows a purely syntactical proof for an extension of Chew's theorem. This extension states that a weakly compatible TRS is UN. Roughly speaking, this weakly compatible condition allows a pair of nonroot overlapping rules if they return the same results (whereas a compatible condition allows only root overlapping rules). The main technique of this syntactical proof is an equational proof normalization called E-normalization that shows UN directly (not by way of the CR of a linearized TRS). This normalization technique differs those described in [2] and [9] in that it can be used even for TRSs that are nonterminating.

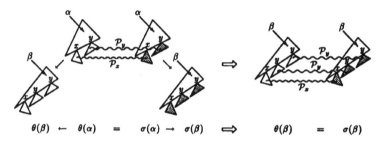

$$\theta(\beta) \leftarrow \theta(\alpha) \quad = \quad \sigma(\alpha) \to \sigma(\beta) \quad \Longrightarrow \quad \theta(\beta) \quad = \quad \sigma(\beta)$$

Fig. 1. Elimination of a reduction peak modulo equality on substitutions (Basic E-normalization)

Intuitively speaking, E-normalization is an elimination of reduction peaks modulo equality on substitutions (strongly overlapping cases, Fig. 1) and their variations (weakly compatible cases). Section 2 reviews the basic notation and terminology used in this paper, and it reviews results related to CR. Section 3 introduces a weight for proof structures that guarantees the termination of the E-normalization procedure. This weight is principally an extension of the parallel steps of an objective proof. Section 4 introduces E-normalization rules and their properties. Subsection 5.1 shows that any equality in a weakly compatible TRS has a proof in which there is no reduction peaks modulo equality (not restricted on substitutions. That is, two redexes are combined with equality that does not touch on their root symbols). Lemma 31 in subsection 5.2 then shows that a weakly compatible TRS is UN.

2 Term rewriting systems

Assuming that the reader is familiar with the basic TRSs concepts in [1] and [6], we briefly explain notations and definitions.

A term set $T(F, V)$ is a set of terms where F is a set of function symbols and V is a set of variable symbols. 0-ary function symbols are also called *constants*. The term set $T(F, V)$ may be abbreviated by simply T. A substitution θ is a map from V to $T(F, V)$. To avoid confusion with equality, we will use $\equiv$ to denote the syntactical identity between terms.

Definition 1. A position $position(M, N)$ of a subterm N in a term M is defined by

$$position(M, N) = \begin{cases} \epsilon & \text{if } M \equiv N. \\ i \cdot u & \text{if } u = position(N_i, N) \quad \text{and} \quad M \equiv f(N_1, \cdots, N_n). \end{cases}$$

Let u and v be positions. We denote $u \prec v$ if $\exists w \neq \epsilon$ s.t. $v = u \cdot w$, we denote $u \preceq v$ if either $u = v$ or $u \prec v$, $u \perp v$ if neither $u \preceq v$ nor $u \succeq v$, and we denote $u \, / \, v$ if either $u \preceq v$ or $u \succeq v$.

For a set U of positions and a position v, we will denote $U \perp v$ if $u \perp v$ for $\forall u \in U$, we denote $U \not\perp v$ if not $U \perp v$, and we denote $v \prec U$ if $v \prec u$ for $\forall u \in U$.

The subterm N of M at position u is noted by M/u (i.e., $u = position(M, N)$). We say that u is a nonvariable position if M/u is not a variable. We denote a set of all positions in M by $pos(M)$, a set of all non-variable positions in M by $pos_F(M)$, and a set of positions of variables in M by $pos_V(M)$. A set of variables in M is denoted as $Var(M)$. A variable x is *linear* (in M) if x appears at most once in M. A variable x is *nonlinear* if it is not linear. A *replacement* of T/u with T', where u is a position in T, is denoted by $T[u \leftarrow T']$.

Definition 2. A finite set $R = \{\alpha_i \rightarrow \beta_i\}$ of ordered pairs of terms is said to be a *term rewriting system* (TRS) if each α_i is not a variable and $Var(\beta_i) \subseteq Var(\alpha_i)$. A binary relation called reduction is defined to be $M \rightarrow N$ if there exist a position u and a substitution θ s.t. $M/u \equiv \theta(\alpha_i)$ and $N \equiv M[u \leftarrow \theta(\beta_i)]$. A subterm $M/u \equiv \theta(\alpha_i)$ in M is said to be a *redex*. A *normal form* is a term that contains no redex. A set of normal forms of R is noted as $NF(R)$.

The symmetric closure of $\rightarrow$ is noted as $\leftrightarrow$. If a reduction $M \rightarrow N$ or $M \leftrightarrow N$ occurs at a position u, we will note $M \underset{u}{\rightarrow} N$ or $M \underset{u}{\leftrightarrow} N$. The reflexive transitive closure of $\rightarrow$ is noted as $\overset{*}{\rightarrow}$. An equality noted $=$ is the reflexive symmetric transitive closure of $\rightarrow$. A structure with an equality $=$ is said to be an equational system associated to a TRS R. We will use the default notations R for a TRS and E for an associated equational system.

Definition 3. A TRS R is *Church-Rosser* (CR) if $M = N$ implies $M \downarrow N$ (i.e., $\exists P$ s.t. $M \overset{*}{\rightarrow} P$ and $N \overset{*}{\rightarrow} P$). A TRS R is said to be *uniquely normalizing* (UN) if each set of equal terms has at most one normal form (i.e., $M = N$ and $M, N \in NF(R)$ imply $M \equiv N$).

Definition 4. A reduction rule is said to be *left linear* if any variable in its lhs is linear. A TRS is said to be *left linear* if all its reduction rules are left linear. We say a TRS is *nonlinear* if it is not left linear.

Definition 5. A pair of reduction rules $\alpha_i \rightarrow \beta_i$ and $\alpha_j \rightarrow \beta_j$ is said to be *overlapping* if there exist both a nonvariable position u in α_i and substitutions θ, σ s.t. $\theta(\alpha_i)/u \equiv \sigma(\alpha_j)$ (i.e., α_i/u and α_j are unifiable). We also say $\sigma(\alpha_j)$ overlaps with $\theta(\alpha_i)$ at u. A TRS R is said to be *nonoverlapping* if no pair of rules in R are overlapping except for trivial cases (i.e., $i = j \wedge u = \epsilon$), and a TRS R is said to be *strongly nonoverlapping* if R is nonoverlapping after renaming its nonlinear variables with fresh individual variables. A TRS R is said to be *weakly nonoverlapping* if any overlapping pair of rules returns same result (i.e., $\theta(\beta_i) \equiv \theta(\alpha_i)[u \leftarrow \theta(\beta_j)]$).

Theorem 6 (Rosen[13]). *A left-linear weakly nonoverlapping TRS is* CR.

Corollary 7 ([11]). *A left-linear nonoverlapping TRS is* CR.

This theorem and its corollary intuitively rely on the commutativity of reductions, but nonlinear TRSs have more complex situations. Firstly, the commutativity of reductions is lost because it requires *synchronous* applications of a reduction on each occurrences of a variable (such as those applied in *term graph rewriting*[5]). Moreover, consider a variable x on the lhs of a rule: which x on the lhs is inherited to an occurrence of x's on the rhs? The following compatible conditions guarantee that if a pair of linearizations of rules is overlapping it returns same result under suitable combinations of variable-inheritances. Note that when a TRS is left-linear, a strongly nonoverlapping condition is the same as a nonoverlapping condition, and a weakly compatible condition is the same as a weakly nonoverlapping condition.

Definition 8 ([15]). (i) Let $\alpha \rightarrow \beta$ be a rewriting rule. A *linearization* of a term M is a term M' in which all the nonlinear variables in M are renamed to fresh individual variables. A *cluster* of a rewrite rule $\alpha \rightarrow \beta$ is $\{\alpha' \rightarrow \beta'_1, \cdots, \alpha' \rightarrow \beta'_n\}$, where α' is a linearization of α and each β'_i is a term obtained from β by replacing each nonlinear variable in α with one of the corresponding variables.

(ii) Let $\alpha_1 \to \beta_1$ and $\alpha_2 \to \beta_2$ be two rules of a TRS R. Let $\{\alpha_1' \to \beta_{11}', \cdots, \alpha_1' \to \beta_{1n}'\}$ and $\{\alpha_2' \to \beta_{21}', \cdots, \alpha_2' \to \beta_{2m}'\}$ be the two clusters corresponding to $\alpha_1 \to \beta_1$ and $\alpha_2 \to \beta_2$. We say that $\alpha_1 \to \beta_1$ and $\alpha_2 \to \beta_2$ are *weakly compatible* if the following holds:

If $\sigma(\alpha_2')$ overlaps with $\theta(\alpha_1')$ at a position u, then the two clusters have a common instance wrt a context $C[\ ] \equiv \alpha_1'[u \leftarrow \Box]$. That is,

$$\{(\theta(\alpha_1') \to \theta(\beta_{1i}')) \mid i = 1, \cdots, n\} \cap \{(\sigma(C[\alpha_2']) \to \sigma(C[\beta_{2j}'])) \mid j = 1, \cdots, m\} \neq \phi.$$

A TRS R is *weakly compatible* if all pairs of its reduction rules are weakly compatible.

(iii) Let notations be the same as in (ii), and let $\alpha_1 \to \beta_1$ and $\alpha_2 \to \beta_2$ be reduction rules. We say that $\alpha_1 \to \beta_1$ and $\alpha_2 \to \beta_2$ are *compatible* if they are weakly compatible and α_1' and α_2' may overlap only at the root (i.e., $u = \epsilon$). A TRS R is *compatible* if all pairs of reduction rules are compatible.

Example 1. Regarding a product xy in CL-pc as $apply(x, y)$, CL-pc is a compatible TRS. CL-sp (CL with surjective pairing[12, 15]) is not weakly compatible.

Theorem 9 (Chew[7]). *A compatible TRS is UN.*

Corollary 10 (Klop[3]). *A strongly nonoverlapping TRS is UN.*

3 Equational proof: structure and weight

3.1 Proof structure

Definition 11. A sequence of terms combined by the $\leftrightarrow$ relation is said to be a *proof structure*. A *proof* is a pair consisting of an equality and its proof structure, and it is denoted by $M_0 \underset{u_1}{\leftrightarrow} M_1 \underset{u_2}{\leftrightarrow} \cdots \underset{u_{n-1}}{\leftrightarrow} M_{n-1} \underset{u_n}{\leftrightarrow} M_n \Rightarrow M_0 = M_n$, where a reduction $M_{i-1} \underset{u_i}{\leftrightarrow} M_i$ occurs at a position u_i. We will omit positions u_i if they are clear or unspecified from the context.

Instead of writing whole terms, we will use proof structure variables $\mathcal{P}, \mathcal{P}', \mathcal{P}_1, \mathcal{P}_2, \cdots$ for proof structures. For proofs $\mathcal{P}_1 \Rightarrow M_0 = M_1, \mathcal{P}_2 \Rightarrow M_1 = M_2, \cdots, \mathcal{P}_n \Rightarrow M_{n-1} = M_n$, a concatenation $\mathcal{P}_1, \mathcal{P}_2, \cdots, \mathcal{P}_n$ is denoted by $conc(\mathcal{P}_1, \cdots, \mathcal{P}_n)$, and the resulting proof is denoted by $conc(\mathcal{P}_1, \cdots, \mathcal{P}_n) \Rightarrow M_0 = M_n$. For a proof structure $\mathcal{P}$ of $M_1 \leftrightarrow M_2 \leftrightarrow \cdots \leftrightarrow M_n$ and a context $C[\ ]$, we will denote $C[M_1] \leftrightarrow C[M_2] \leftrightarrow \cdots \leftrightarrow C[M_n]$ by $C[\mathcal{P}]$.

To emphasize the specific structure of a proof structure, we will also introduce two abbreviated notations: a sequence and a collection. A sequence of proofs $[\mathcal{P}_1 \Rightarrow M_0 = M_1; \cdots; \mathcal{P}_n \Rightarrow M_{n-1} = M_n]$ is a proof structure for $M_0 = M_n$. It emphasizes the specific intermediate terms $M_0, \cdots, M_n$. Its proper form is obtained by the unfolding rule (S):

(S) *A sequence $[\mathcal{P}_1 \Rightarrow M_0 = M_1; \cdots; \mathcal{P}_n \Rightarrow M_{n-1} = M_n]$ is a proof structure for $M_0 = M_n$. It is unfolded to $conc(\mathcal{P}_1, \cdots, \mathcal{P}_n) \Rightarrow M_0 = N_n$.*

A collection of proofs $[\![\mathcal{P}_1 \Rightarrow M_1 = M_1', \cdots, \mathcal{P}_n \Rightarrow M_n = M_n']\!]$ is a proof structure for $C[M_1, \cdots, M_n] = C[M_1', \cdots, M_n']$. Consider an example of a proof structure $f(g(a), h(y)) \underset{1}{\to} f(h(a), h(y)) \underset{1}{\leftarrow} f(h(a), g(y))$ for a TRS $\{g(x) \to h(x)\}$. Parallel subproofs $g(a) \to h(a)$ and $h(y) \leftarrow g(y)$ induce an equality $f(g(a), h(y)) = f(h(a), g(y))$. A collection of proofs emphasizes such a situation. Its proper form is obtained by the unfolding rule (C):

(C) *Let u_i be disjoint positions in $pos(M) \cap pos(N)$. A collection $[\![\mathcal{P}_1 \Rightarrow M/u_1 = N/u_1, \cdots, \mathcal{P}_n \Rightarrow M/u_n = N/u_n]\!]$ is a proof structure for $M = N$. It is unfolded to $conc(C_1[\mathcal{P}_1], \cdots, C_n[\mathcal{P}_n]) \Rightarrow M = N$, where $C_i[\Box] \equiv M[u_j \leftarrow N/u_j$ for $\forall j < i; u_i \leftarrow \Box]$.*

Our technique for transforming an equational proof to a *simpler* one is E-normalization. This normalization (introduced in Section 4) eliminates reduction peaks modulo equality on substitutions and their variations. For instance, consider a TRS $\{f(x,y) \rightarrow g(x,y,y), \cdots\}$. Assume there exist an equality $g(\theta(x), \theta(y), \theta(y)) = g(\sigma(x), \sigma(y), \sigma(y))$ that is proved by $g(\theta(x), \theta(y), \theta(y)) \leftarrow f(\theta(x), \theta(y))$, a collection of proofs $[\mathcal{P}_x \Rightarrow \theta(x) = \sigma(x), \mathcal{P}_y \Rightarrow \theta(y) = \sigma(y)] \Rightarrow f(\theta(x), \theta(y)) = f(\sigma(x), \sigma(y))$, and $f(\sigma(x), \sigma(y)) \rightarrow g(\sigma(x), \sigma(y), \sigma(y))$. Then a (basic) E-normalization (Fig. 1) transforms it to a *simpler* proof $[\mathcal{P}_x \Rightarrow \theta(x) = \sigma(x), \mathcal{P}_y \Rightarrow \theta(y) = \sigma(y), \mathcal{P}_y \Rightarrow \theta(y) = \sigma(y)] \Rightarrow g(\theta(x), \theta(y), \theta(y)) = g(\sigma(x), \sigma(y), \sigma(y))$.

To show the termination of E-normalization, the proof weight will be introduced in Subsection 3.2. This normalization decreases a parallel step of a transformed part in a proof structure, but when that part may be properly included in a whole proof, the number of parallel steps of a whole proof may not decrease. We therefore need an extension that is more sensitive to a parallel step of its proper substructure: a proof weight. Lemma 19 in Subsection 3.2 will show the termination of E-normalization procedure (Theorems 21 and 25).

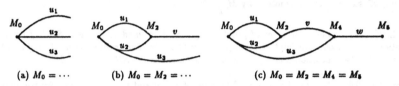

(a) $M_0 = \cdots$ (b) $M_0 = M_2 = \cdots$ (c) $M_0 = M_2 = M_4 = M_6$

Fig. 2. Construction of E-graph for $M_0 \underset{u_1}{\leftrightarrow} M_1 \underset{u_2}{\leftrightarrow} M_2 \underset{u_3}{\leftrightarrow} M_3 \underset{v}{\leftrightarrow} M_4 \underset{w}{\leftrightarrow} M_6$.

The graphical intuition of a proof weight is sketched by a notion of an E-graph[17]. Let $M_0 \underset{u_1}{\leftrightarrow} M_1 \underset{u_2}{\leftrightarrow} M_2 \underset{u_3}{\leftrightarrow} M_3 \underset{v}{\leftrightarrow} M_4 \underset{w}{\leftrightarrow} M_6 \Rightarrow M_0 = M_6$. Assume that u_1, u_2, and u_3 are disjoint, and that $v \prec u_1, u_2$, $v \perp u_3$, and $w \prec u_3, v$. Then, the E-graph is generated as follows. Since u_1, u_2, and u_3 are disjoint, the edges corresponding to u_1, u_2, and u_3 stem in parallel from a vertex M_0 (Fig. 2(a)). Since $v \prec u_1, u_2$, the edges labeled with u_1 and u_2 are collected to a vertex M_2 and an edge labeled v stems from M_2 (Fig. 2(b)). Finally, since $w \prec u_3, v$ and $u_3 \perp v$, the edges labeled u_3 and v are collected to a vertex M_4 and an edge labeled w stems from M_4 to a vertex M_6 (Fig. 2(c)). A set of proof strings (Subsection 3.2) is a set of all cycle-free paths from M_0 to M_6, such as (u_1, v, w), (u_2, v, w), and (u_3, w). The proof weight of $\mathcal{P}$ is a multiset of their lengths $\{3, 3, 2\}$.

3.2 Weight for proofs

Definition 12. Let $(u_1, u_2, \cdots, u_n)$ be a sequence of positions and v be a position. An operation $(u_1, u_2, \cdots, u_n) \odot v$ returns a set defined by

$$\begin{cases} \{(u_1, u_2, \cdots, u_n, v)\} & \text{if } v \not\perp u_n. \\ \{(u_1, u_2, \cdots, u_n), (u_1, u_2, \cdots, u_i, v)\} & \text{if } v \perp u_j \text{ for } \forall j > i \text{ and } v \not\perp u_i. \\ \{(u_1, u_2, \cdots, u_n)\} & \text{if } v \perp u_i \text{ for } \forall i. \end{cases}$$

Definition 13. Let S be a set of sequences of positions, and v be a position. An operation $S \oplus u$ is defined to be

$$\begin{cases} \bigcup_{\bar{u} \in S} \bar{u} \odot v & \text{if } \exists \bar{u} \in S \text{ s.t. } \bar{u} \not\perp v, \text{ and} \\ S \cup \{(v)\} & \text{otherwise.} \end{cases}$$

Here $\bar{u} \perp v$ if $u_i \perp v$ for $\forall u_i \in \bar{u} = (u_1, u_2, \cdots, u_n)$, and $\bar{u} \not\perp v$ if not $\bar{u} \perp v$.

Definition 14. Let $M_0 \underset{u_1}{\leftrightarrow} M_1 \underset{u_2}{\leftrightarrow} \cdots \underset{u_{n-1}}{\leftrightarrow} M_{n-1} \underset{u_n}{\leftrightarrow} M_n \Rightarrow M_0 = M_n$ be a proof. A *proof string* is an element of a set $S(M_0 \underset{u_1}{\leftrightarrow} M_1 \underset{u_2}{\leftrightarrow} \cdots \underset{u_{n-1}}{\leftrightarrow} M_{n-1} \underset{u_n}{\leftrightarrow} M_n)$ inductively defined to be

$$\begin{cases} \phi & \text{if } n = 0, \text{ and} \\ S(M_0 \underset{u_1}{\leftrightarrow} M_1 \underset{u_2}{\leftrightarrow} \cdots \underset{u_{n-1}}{\leftrightarrow} M_{n-1}) \oplus u_n & \text{otherwise.} \end{cases}$$

For two sets S_1, S_2 of sequences of positions, $S_1 \otimes S_2$ is defined to be $\cup_{(v_1,\cdots,v_n) \in S_2} S_1 \oplus v_1 \oplus \cdots \oplus v_n$. Thus $S(conc(\mathcal{P}_1, \mathcal{P}_2)) = S(\mathcal{P}_1) \otimes S(\mathcal{P}_2)$.

Definition 15. Let $\mathcal{P} \Rightarrow M = N$ be a proof. A *proof weight* $w(S(\mathcal{P}))$ is a multiset of string-lengths of proof strings in $S(\mathcal{P})$ where a string-length $l(\bar{u})$ is a number n for $\bar{u} = (u_1, u_2, \cdots, u_n)$. The ordering on proof weights is defined to be the multiset extension of the usual ordering on natural numbers[10].

Definition 16. Let $\mathcal{P} \Rightarrow M = N$ be a proof. A *boundary* $\partial \mathcal{P}$ is a set of positions defined by $min(\{v \mid v \text{ appears in } \bar{u} \in S(\mathcal{P})\})$ (i.e., a set of minimum positions appeared in each proof string in $S(\mathcal{P})$). If a proof structure $\mathcal{P}$ satisfies $\forall v \in \partial \mathcal{P}$ s.t. $w \prec v$, we denote $\mathcal{P}$ by $M \underset{w \prec}{=} N$. If $\forall v \in \partial \mathcal{P}$ s.t. $w \preceq v$, we denote $\mathcal{P}$ by $M \underset{w \preceq}{=} N$.

Definition 17. Let S be a set of proof strings. S^u is defined to be $\{\bar{v} \in S \mid \bar{v} \not\!\!\int u\}$ for a position u. For a set of disjoint positions U, S^U is $\cup_{u \in U} S^u$. For positions v, w s.t. $v \preceq w$, $S^{v/w}$ is $S^v - S^w$. For a position v and a set of disjoint positions W, $S^{v/W}$ is $S^v - S^W$.

Definition 18. Let $\bar{w} = (w_1, \cdots, w_m)$ be a proof string and u be a position. We will denote $(w_1, \cdots, w_i)$ by $\bar{w} \triangleleft u$ if $w_i \not\!\!\int u$ and $w_j \perp u$ for $\forall j > i$, and we will denote $(w_i, \cdots, w_n)$ by $u \triangleright \bar{w}$ if $w_i \not\!\!\int u$ and $w_j \perp u$ for $\forall j < i$. For a set S of proof strings, $S \triangleleft u$ is $\{\bar{w} \triangleleft u \mid \bar{w} \in S\}$ and $u \triangleright S$ is $\{u \triangleright \bar{w} \mid \bar{w} \in S\}$.

Lemma 19. Let an equality $S = T$ have two proof structures $\mathcal{P}_1$, $\mathcal{P}_2$, and let u, v be positions s.t. $u \not\!\!\int v$. Assume $\mathcal{P}_1$ has a form $S \underset{u}{\leftrightarrow} S' \underset{max(u,v) \prec}{=} T' \underset{v}{\leftrightarrow} T$ and $\mathcal{P}_2$ has a form $S \underset{max(u,v) \prec}{=} T$. Then $max(w(S(\mathcal{P}_1))) > max(w(S(\mathcal{P}_2)))$ implies $w(S(\mathcal{P}_1^*)) \gg w(S(\mathcal{P}_2^*))$, where

(i) $\mathcal{P}_1^*$ is $[\mathcal{P} \Rightarrow M = S; \mathcal{P}_1 \Rightarrow S = T; \mathcal{P}' \Rightarrow T = N]$,
(ii) $\mathcal{P}_2^*$ is $[\mathcal{P} \Rightarrow M = S; \mathcal{P}_2 \Rightarrow S = T; \mathcal{P}' \Rightarrow T = N]$, and
(iii) $\gg$ is the multiset extension of the ordering $>$ on natural numbers.

Proof Without loss of generality, we can assume $u \preceq v$ and $max(u, v) = v$. We remark that if positions u, v satisfy $u \preceq v$, then, for any set S of proof strings, $S^u \supseteq S^v$, $max(w(S \triangleleft u)) \geq max(w(S \triangleleft v))$, and $max(w(u \triangleright S)) \geq max(w(v \triangleright S))$.

We will estimate the maximum length in $w(S(\mathcal{P}_1^*) - S(\mathcal{P}_1^*) \cap S(\mathcal{P}_2^*))$ and $w(S(\mathcal{P}_2^*) - S(\mathcal{P}_1^*) \cap S(\mathcal{P}_2^*))$. The former will be shown to be greater than the latter, and this will show that $w(S(\mathcal{P}_1^*)) \gg w(S(\mathcal{P}_2^*))$.

(1) A proof string in $S(\mathcal{P}_1^*) = S(\mathcal{P}) \otimes S(\mathcal{P}_1) \otimes S(\mathcal{P}')$ has one of following forms:

(1-i) $(w_1, \cdots, w_i, u, u_1, \cdots, u_s, v, w'_j, \cdots, w'_n)$ for $(w_1, \cdots, w_i) \in S(\mathcal{P})^u \triangleleft u$
$(u, u_1, \cdots, u_s, v) \in S(\mathcal{P}_1)$
$(w'_j, \cdots, w'_n) \in v \triangleright S(\mathcal{P}')^v$

(1-ii) $(w_1, \cdots, w_i, u, w'_j, \cdots, w'_n)$ for $(w_1, \cdots, w_i) \in S(\mathcal{P})^u \triangleleft u$
$(w'_j, \cdots, w'_n) \in u \triangleright S(\mathcal{P}')^{u/v}$

(1-iii) $(w_1, \cdots, w_i, w'_j, \cdots, w'_n)$ for $(w_1, \cdots, w_i) \in S(\mathcal{P})^{i/u} \triangleleft w'_j$
$(w'_j, \cdots, w'_n) \in w_i \triangleright S(\mathcal{P}')^{i/u}$

These sets are respectively denoted $S_{(1-i)}$, $S_{(1-ii)}$, and $S_{(1-iii)}$. The maximum length of $S_{(1-i)}$ and $S_{(1-ii)}$ are estimated as

$max(w(S_{(1-i)})) = max(w(S(\mathcal{P}) \triangleleft u)) + max(w(S(\mathcal{P}))) + max(w(v \triangleright S(\mathcal{P}')))$ and
$max(w(S_{(1-ii)})) = max(w(S(\mathcal{P}) \triangleleft u)) + 1 + max(w(u \triangleright S(\mathcal{P}')^{u/v}))$.

(2) A proof string in $S(\mathcal{P}_2^*) = S(\mathcal{P}) \otimes S(\mathcal{P}_2) \otimes S(\mathcal{P}')$ has one of following forms:

(2-i) $\quad (w_1, \cdots, w_i, u_1', \cdots, u_s', w_j', \cdots, w_n')$ for $(w_1, \cdots, w_i) \in S(\mathcal{P})^{u_s'} \lhd u_1'$
$\qquad\qquad\qquad\qquad\qquad\qquad\qquad\qquad\qquad (u_1', \cdots, u_s') \in w_i \rhd S(\mathcal{P}_2) \lhd w_j'$
$\qquad\qquad\qquad\qquad\qquad\qquad\qquad\qquad\qquad (w_j', \cdots, w_n') \in u_s' \rhd S(\mathcal{P}')^{u_s'}$

(2-ii) $\quad (w_1, \cdots, w_i, w_j', \cdots, w_n')$ for $(w_1, \cdots, w_i) \in S(\mathcal{P})^{\epsilon/\partial \mathcal{P}_2} \lhd w_j'$
$\qquad\qquad\qquad\qquad\qquad\qquad\qquad\qquad\qquad (w_j', \cdots, w_n') \in w_i \rhd S(\mathcal{P}')^{\epsilon/\partial \mathcal{P}_2}$

These sets are denoted $S_{(2-i)}$ and $S_{(2-ii)}$. Then $S(\mathcal{P}_1^*) \cap S(\mathcal{P}_2^*) = S_{(1-iii)}$ and $S_{(2-i)} \cap S_{(1-iii)} = \phi$. A set $S_{(2-ii)'} = S_{(2-ii)} - S_{(1-iii)}$ consists of elements of the form

(2-ii)' $\quad (w_1, \cdots, w_i, w_j', \cdots, w_n')$ for $(w_1, \cdots, w_i) \in S(\mathcal{P})^{u/\partial \mathcal{P}_2} \lhd w_j'$
$\qquad\qquad\qquad\qquad\qquad\qquad\qquad\qquad\qquad (w_j', \cdots, w_n') \in w_i \rhd S(\mathcal{P}')^{u/\partial \mathcal{P}_2}$.

Since $u \rhd S(\mathcal{P}')^{u/\partial \mathcal{P}_2} = u \rhd S(\mathcal{P}')^{u/v} \cup u \rhd S(\mathcal{P}')^{v/\partial \mathcal{P}_2}$, the maximum lengths $max(w(S_{(2-i)}))$ and $max(w(S_{(2-ii)'}))$ are estimated as follows:

$$max(w(S_{(2-i)})) \leq max(\; w(S(\mathcal{P})^{\partial \mathcal{P}_2} \lhd u)) + max(w(S(\mathcal{P}_2))) + max(w(v \rhd S(\mathcal{P}')^{\partial \mathcal{P}_2}))$$
$$< max(\; w(S(\mathcal{P}) \lhd u)) + max(w(S(\mathcal{P}_1))) + max(w(v \rhd S(\mathcal{P}')))$$
$$= max(\; w(S_{(1-i)}))$$
$$max(w(S_{(2-ii)'})) \leq max(\; w(S(\mathcal{P})^{u/\partial \mathcal{P}_2} \lhd u) + max(w(u \rhd S(\mathcal{P}')^{u/\partial \mathcal{P}_2}))$$
$$\leq max(\; max(w(S(\mathcal{P})^{u/v} \lhd u) + max(w(u \rhd S(\mathcal{P}')^{u/v})),$$
$$max(w(S(\mathcal{P})^{v/\partial \mathcal{P}_2} \lhd u) + max(w(u \rhd S(\mathcal{P}')^{v/\partial \mathcal{P}_2}))\;)$$
$$< max(\; max(w(S_{(1-i)})), max(w(S_{(1-ii)}))\;)$$

Thus $w(S(\mathcal{P}_1^*)) \gg w(S(\mathcal{P}_2^*))$. ∎

4 E-normalization

4.1 Basic E-normalization

Definition 20. Let $\alpha \to \beta$ be a reduction rule, and let θ and σ be substitutions. Then, the following transformation rule (from the upper column to the lower column) is said to be a *basic E-normalization* rule:

$$\begin{bmatrix} \theta(\beta) \leftarrow \theta(\alpha) \;\Rightarrow\; \theta(\beta) = \theta(\alpha) \;; \\ [\![\mathcal{P}_u \;\Rightarrow\; \theta(\alpha)/u = \sigma(\alpha)/u \text{ for } \forall u \in pos_V(\alpha)]\!] \;\Rightarrow\; \theta(\alpha) = \sigma(\alpha) \;; \\ \sigma(\alpha) \to \sigma(\beta) \;\Rightarrow\; \sigma(\alpha) = \sigma(\beta) \end{bmatrix} \Rightarrow \theta(\beta) = \sigma(\beta)$$

$$\overline{[\![\mathcal{P}_v \;\Rightarrow\; \theta(\beta)/v = \sigma(\beta)/v \text{ for } \forall v \in pos_V(\beta)]\!] \qquad\qquad\qquad \Rightarrow \theta(\beta) = \sigma(\beta)}$$

if a proof structure $\mathcal{P}_v$ in the lower column is arbitrarily selected from a set of $\mathcal{P}_u$'s in the upper column s.t. $\alpha/u \equiv \beta/v \in Var(\beta)$.

Theorem 21. *Each step of a basic E-normalization decreases the proof weight. Thus a basic E-normalization procedure for any proof always terminates.*

4.2 E-normalization for a weakly compatible TRS

Definition 22. Let $\alpha_1 \to \beta_1$ and $\alpha_2 \to \beta_2$ be reduction rules, let θ and σ be substitutions, and let $C[\;] \equiv \theta(\alpha_1)[u \leftarrow \square]$. Assume that $\theta(\alpha_1)$ and $C[\sigma(\alpha_2)]$ are combined by a proof structure $\theta(\alpha) \underset{u \lhd}{=} C[\sigma(\alpha_2)]$. If $\theta(\theta(\alpha_1) \underset{u \lhd}{=} C[\sigma(\alpha_2)]) \cap pos_F(\alpha_1) \cap u \cdot pos_F(\alpha_2) = \phi$, then $\theta(\alpha_1)$ and $C[\sigma(\alpha_2)]$ are said to be *quasi-E-normalizable*.

Assume S and S' to be quasi-E-normalizable. The next lemma shows that if a TRS is weakly compatible, then any pair of reduction $T \leftarrow S$ and $S' \to T'$ in a proof can be eliminated by an E-normalization.

Lemma 23. Let a TRS R be weakly compatible. Assume that $\alpha_1 \to \beta_1$ and $\alpha_2 \to \beta_2$ are rules in R, that θ and σ be substitutions, and that a position $u \in pos_F(\alpha_1)$ s.t.

$[\![\mathcal{P}_w \Rightarrow \theta(\alpha_1)/u \cdot w = \sigma(\alpha_2)/w$ for $\forall w \in min(pos_V(\alpha_1/u) \cup pos_V(\alpha_2))]\!] \Rightarrow \theta(\alpha_1) = \sigma(C[\alpha_2])$

for $C[\ ] \equiv \alpha_1[u \leftarrow \square]$. Then linearizations of α_1/u and α_2 overlap, and

$[\![\mathcal{P}_{w'} \Rightarrow \theta(\beta_1)/u \cdot w' = \sigma(\beta_2)/w'$ for $\forall w' \in min(pos_V(\beta_1/u) \cup pos_V(\beta_2))]\!] \Rightarrow \theta(\beta_1) = \sigma(C[\beta_2])$

Proof Let $\{\alpha'_1 \to \beta'_{11}, \cdots, \alpha'_1 \to \beta'_{1n}\}$ and $\{\alpha'_2 \to \beta'_{21}, \cdots, \alpha'_2 \to \beta'_{2m}\}$ be clusters of $\alpha_1 \to \beta_1$ and $\alpha_2 \to \beta_2$. Let $W = min(pos_V(\alpha_1/u) \cup pos_V(\alpha_2))$, and let $C'[\ ]$ be a linearization of $C[\ ]$. Set substitutions θ' for α'_1 and σ' for $C'[\alpha'_2]$ as follows:

For a variable x in α'_1 s.t. $w = position(\alpha'_1, x)$,

$$\theta'(x) \equiv \begin{cases} \sigma(C[\alpha_2]/w) & \text{if } w \in u \cdot (W \cap pos_F(\alpha_2)), \text{ and} \\ \theta(\alpha_1/w) & \text{otherwise.} \end{cases}$$

For a variable y in $C'[\alpha'_2]$ s.t. $w = position(C'[\alpha'_2], y)$,

$$\sigma'(y) \equiv \begin{cases} \theta(\alpha_1/w) & \text{if } w \in u \cdot (W - pos_F(\alpha_2)), \text{and} \\ \sigma(C[\alpha_2]/w) & \text{otherwise.} \end{cases}$$

Then linearizations α'_1 and α'_2 overlap at a position u (i.e., $\theta'(\alpha'_1) \equiv \sigma'(C[\alpha'_2])$). Since R is weakly compatible, there exists the intersection $(\theta'(\alpha'_1) \to \theta'(\beta'_{1s})) \equiv (\sigma'(C[\alpha'_2]) \to \sigma'(C[\beta'_{2i}]))$ between two clusters (i.e., $\theta'(\beta'_{1s}) \equiv \sigma'(C[\beta'_{2i}])$). Thus a set of proofs $\theta(\alpha_1/u \cdot w) = \sigma(\alpha_2/w)$ for $\forall w \in min(pos_V(\alpha_1/u) \cup pos_V(\alpha_2))$ also combines $\theta(\beta_1)$ and $\sigma(C[\beta_2])$ as $\theta(\beta_1) = \theta'(\beta'_{1s}) \equiv \sigma'(C[\beta'_{2i}]) = \sigma(C[\beta_2])$. ∎

Definition 24. Let $\alpha_1 \to \beta_1$ and $\alpha_2 \to \beta_2$ be weakly compatible. Assume α'_1/u and α'_2 to be unified for $u \in pos_F(\alpha'_1)$, where α'_1 and α'_2 are the linearizations of α_1 and α_2. Let $C[\ ]$ be a context that has a hole $\square$ at u s.t. $C[\alpha'_2] \equiv \alpha'_1$, and let θ and σ be substitutions. The following transformation rules are said to be the *E-normalization* rules:

$$\frac{\begin{bmatrix} \theta(\beta_1) \leftarrow \theta(\alpha_1) \Rightarrow \theta(\beta_1) = \theta(\alpha_1) \ ; \\ \begin{bmatrix} \begin{bmatrix} \mathcal{P}_w \Rightarrow \theta(\alpha_1)/u \cdot w = \sigma(\alpha_2)/w \\ \text{for } \forall w \in min(pos_V(\alpha_1/u) \cup pos_V(\alpha_2)) \end{bmatrix} \end{bmatrix} \Rightarrow \theta(\alpha_1) = \sigma(C[\alpha_2]); \\ \sigma(\alpha_2) \to \sigma(\beta_2) \Rightarrow \sigma(C[\alpha_2]) = \sigma(C[\beta_2]) \end{bmatrix} \Rightarrow \theta(\beta_1) = \sigma(C[\beta_2])}{\begin{bmatrix} \begin{bmatrix} \mathcal{P}_{w'} \Rightarrow \theta(\beta_1/u) \cdot w' = \sigma(\beta_2)/w' \\ \text{for } \forall w' \in min(pos_V(\beta_1/u) \cup pos_V(\beta_2)) \end{bmatrix} \end{bmatrix} \quad \Rightarrow \theta(\beta_1) = \sigma(C[\beta_2])}$$

and its inverted form

$$\frac{\begin{bmatrix} \sigma(\beta_2) \leftarrow \sigma(\alpha_2) \Rightarrow \sigma(C[\beta_2]) = \sigma(C[\alpha_2]) \\ \begin{bmatrix} \begin{bmatrix} \mathcal{P}_w \Rightarrow \sigma(\alpha_2)/w = \theta(\alpha_1)/u \cdot w \\ \text{for } \forall w \in min(pos_V(\alpha_1/u) \cup pos_V(\alpha_2)) \end{bmatrix} \end{bmatrix} \Rightarrow \sigma(C[\alpha_2]) = \theta(\alpha_1); \\ \theta(\alpha_1) \to \theta(\beta_1) \Rightarrow \theta(\alpha_1) = \theta(\beta_1) \end{bmatrix} \Rightarrow \sigma(C[\beta_2]) = \theta(\beta_1)}{\begin{bmatrix} \begin{bmatrix} \mathcal{P}_{w'} \Rightarrow \sigma(\beta_2)/w' = \theta(\beta_1/u) \cdot w' \\ \text{for } \forall w' \in min(pos_V(\beta_1/u) \cup pos_V(\beta_2)) \end{bmatrix} \end{bmatrix} \quad \Rightarrow \sigma(C[\beta_2]) = \theta(\beta_1)}$$

Here a proof structure $\mathcal{P}_{w'}$ in the lower column is arbitrarily selected from a set of $\mathcal{P}_w$ in the upper column s.t. $\theta(\alpha_1)/u \cdot w \equiv \theta(\beta_1)/u \cdot w'$ and $\sigma(\alpha_2)/w \equiv \sigma(\beta_2)/w'$.

Theorem 25. *Let R be a weakly compatible TRS. Each step of an E-normalization decreases the proof weight. Thus the E-normalization procedure for any proof always terminates.*

Example 2. For a product xy such as $apply(x, y)$, CL-pc is a compatible TRS. CL-pc has overlapping rules such as $(CTxy \to x, Czxx \to x)$ and $(CFxy \to x, Czxx \to x)$. The following rule is an example of an E-normalization rule in addition to the basic E-normalization rules:

$$\frac{\begin{bmatrix} \theta(x) \leftarrow C \ T \ \theta(x) \ \theta(y) \Rightarrow \theta(x) = C \ T \ \theta(x) \ \theta(y) \ ; \\ \begin{bmatrix} \begin{bmatrix} \mathcal{P}_1 \Rightarrow T = \sigma(z'), \\ \mathcal{P}_2 \Rightarrow \theta(x) = \sigma(x'), \\ \mathcal{P}_3 \Rightarrow \theta(y) = \sigma(x') \end{bmatrix} \end{bmatrix} \Rightarrow C \ T \ \theta(x) \ \theta(y) = C \ \sigma(z') \ \sigma(x') \ \sigma(x') \ ; \\ C \ \sigma(z') \ \sigma(x') \ \sigma(x') \to \sigma(x') \Rightarrow C \ \sigma(z') \ \sigma(x') \ \sigma(x') = \sigma(x') \end{bmatrix} \Rightarrow \theta(x) = \sigma(x')}{\mathcal{P}_2 \Rightarrow \theta(x) = \sigma(x')}$$

for substitutions θ and σ.

5 Sufficient condition for the UN property

5.1 E-overlapping pair

Definition 26. Let $\mathcal{P} \Rightarrow M_0 = M_n$ be a proof. Assume $\mathcal{P}$ has the form
$$M_0 \leftrightarrow M_1 \leftrightarrow \cdots \leftrightarrow M_{i-1} \underset{u}{\leftarrow} M_i \underset{max(u,v)\prec}{=} M_j \underset{v}{\rightarrow} M_{j+1} \leftrightarrow \cdots \leftrightarrow M_{n-1} \leftrightarrow M_n.$$
Then a pair of terms (M_i, M_j) is said to be an E-overlapping pair.

Definition 27. A proof is E-normal if no E-normalization rule is applicable. An equational proof is said to be E-overlapping-pair-free if there are no E-overlapping pairs in it. A TRS is E-nonoverlapping if every E-normal proof is E-overlapping-pair-free.

Theorem 28. *If a TRS R is weakly compatible, then R is E-nonoverlapping.*

Proof Let $\mathcal{P} \Rightarrow M = N$ be an E-normal proof. Assume there exists an E-overlapping pair in $\mathcal{P}$. Let (S, T) be an innermost E-overlapping pair in $\mathcal{P}$. Then (S, T) is quasi-E-normalizable (otherwise there is an inner E-overlapping pair between S and T), and E-normalization can be applicable to (S, T) from Lemma 23. This contradicts the E-normal assumption on $\mathcal{P}$. ∎

5.2 The UN property for an E-nonoverlapping TRS

Definition 29 [11]. A term-length $\Delta(M)$ for a term M is defined to be
$$\begin{cases} \Delta(x) = 1 & \text{for a variable } x, \text{ and} \\ \Delta(f(M_1, \cdots, M_n)) = 1 + \sum_{i=1}^{n} \Delta(M_i) & \text{otherwise.} \end{cases}$$

Proposition 30. Assume that a TRS R is E-nonoverlapping. If $\mathcal{P} \Rightarrow M = N$ is an E-normal proof and $M \not\equiv N$, then there exist a position $u \in \partial\mathcal{P}$, a substitution σ, and a rule $\alpha \rightarrow \beta$ s.t. $\mathcal{P}$ is either $M = N[u \leftarrow \sigma(\beta)] \underset{u}{\leftarrow} N[u \leftarrow \sigma(\alpha)] \underset{u\prec}{=} N$, or $M \underset{u\prec}{=} M[u \leftarrow \sigma(\alpha)] \underset{u}{\rightarrow} M[u \leftarrow \sigma(\beta)] = N$.

Proof $\mathcal{P}$ is a nontrivial proof because $M \not\equiv N$. If $\mathcal{P}$ holds for neither of the cases specified in Proposition 30, there exist a position $u \in \mathcal{P}$, a substitution σ', and a rule $\alpha' \rightarrow \beta'$ s.t. $\mathcal{P}$ includes $M/u \underset{\epsilon\prec}{=} \sigma(\beta) \underset{\epsilon}{\leftarrow} \sigma(\alpha) = \sigma'(\alpha') \underset{\epsilon}{\rightarrow} \sigma'(\beta') \underset{u\prec}{=} N/u$. Then there must exist an E-overlapping pair in $\sigma(\beta) \underset{\epsilon}{\leftarrow} \sigma(\alpha) = \sigma'(\alpha') \underset{\epsilon}{\rightarrow} \sigma'(\beta')$. This is contradiction. ∎

Lemma 31. Assume that a TRS R is weakly compatible. Then

(i) If $M \underset{\epsilon\prec}{=} \theta(\alpha) \rightarrow \theta(\beta)$ is an E-normal proof for some rule $\alpha \rightarrow \beta \in R$ and a substitution θ, then $M \notin NF(R)$.

(ii) If $N_1 = N_2$ and $N_1, N_2 \in NF(R)$, then $N_1 \equiv N_2$.

Proof Since R is weakly compatible, Theorem 28 implies that R is E-nonoverlapping. Assume that (i) of Lemma 31 holds, then Theorem 25 and Proposition 30 imply that if $N_1 = N_2$ and $N_1 \not\equiv N_2$, then either N_1 or N_2 is not a normal form. Thus if (i) holds for $\Delta(M) < n$, (ii) also holds for $\Delta(N_1), \Delta(N_2) < n$.

The proof of (i) is due to the induction on $n = \Delta(M)$. If $n = 1$, M must be a constant or a variable. Then $M \underset{\epsilon\prec}{=} \theta(\alpha)$ implies $M \equiv \theta(\alpha)$, and M is a redex. Let $n > 1$ and $M \underset{\epsilon\prec}{=} \theta(\alpha) \rightarrow \theta(\beta) \Rightarrow M = \theta(\beta)$ be an E-normal proof.

Assume $M \in NF(R)$. Thus $M \not\equiv \theta(\alpha)$ and according to Proposition 30 there are a position $u \in \partial(M \underset{\epsilon\prec}{=} \theta(\alpha))$, a substitution σ, and a rule $\alpha' \rightarrow \beta' \in R$ s.t. $M \underset{\epsilon\prec}{=} \theta(\alpha) \rightarrow \theta(\beta)$ has one of the following forms:
$$\begin{cases} (1) \quad M \underset{u\prec}{=} M[u \leftarrow \sigma(\alpha')] \underset{u}{\rightarrow} M[u \leftarrow \sigma(\beta')] = \theta(\alpha) \rightarrow \theta(\beta) \\ (2) \quad M \underset{u\prec}{=} \theta(\alpha)[u \leftarrow \sigma(\beta')] \underset{u}{\leftarrow} \theta(\alpha)[u \leftarrow \sigma(\alpha')] \underset{u\prec}{=} \theta(\alpha) \rightarrow \theta(\beta) \end{cases}$$

Form (1) satisfies $\Delta(M/u) < \Delta(M)$, and $M/u \underset{\overset{*}{\smile}}{=} \sigma(\alpha') \rightarrow \sigma(\beta')$ is E-normal (because a sub-proof of an E-normal proof is E-normal). Thus $M/u \notin NF(R)$ from the induction hypothesis. This implies $M \notin NF(R)$.

Let us divide Form (2) into two cases: (a) $u \in pos_F(\alpha)$ and (b) $u \notin pos_F(\alpha)$.

(a) Since $u \in pos_F(\alpha)$, $(\theta(\alpha)[u \leftarrow \theta(\alpha')], \theta(\alpha))$ is an E-overlapping pair. Every E-normal proof in R is E-overlapping-pair-free. Thus this is a contradiction.

(b) There exist a nonlinear variable x in α and positions $v_1, v_2 \in position(\alpha, x)$ s.t. $M/v_1 \not\equiv M/v_2$ and $M/v_1 = M/v_2$. (Otherwise, M is a redex for $\alpha \rightarrow \beta$.) If $M \in NF(R)$, then $M/v_1, M/v_2 \in NF(R)$ and applying the induction hypothesis to (ii) shows $M/v_1 \equiv M/v_2$. This is a contradiction. Thus $M \notin NF(R)$. ∎

Theorem 32. *If a TRS R is weakly compatible, then R is UN.*

Corollary 33. *Let R be a weakly compatible TRS. If R has at least two distinct normal forms, then an associated equality system E is consistent.*

Acknowledgments

I thank Dr. Yoshihito Toyama (NTT Communication Science Lab.), Dr. Aart Middeldorp (Hitachi Advanced Research Lab.), and Professor Michio Oyamaguchi (Mie Univ.) for valuable comments and discussions. I also thank my supervisor Hirofumi Katsuno for his careful reading, Professor Touru Naoi (Gifu Univ.) for informing me of references, and Dr. Satoshi Ono (NTT Software Lab.) for suggestions.

References

1. J.van Leeuwen (eds.), "Handbook of Theoretical Computer Science, Vol.B Formal models and Semantics", The MIT Press/Elsevier (1990)
2. L.Bachmair, "Canonical Equational Proofs," Birkhäuser (1991)
3. J.W.Klop, "Combinatory Reduction Systems", Mathematical Centre Tracts 127, CWI Amsterdam (1980)
4. D.S.Johnson (eds.), "STOC/FOCS Bibliography (Preliminary version)", ACM SIGACT (1991)
5. Barendregt,H.P., M.C.J.D.Eekelen, J.R.W.Glauert, J.R.Kennaway, M.J.Pasmeijer, and M.R.Sleep, "Term Graph Rewriting", Proc. Parallel Architectures and Languages Europe Vol.2, pp.141-158, Springer LNCS 259 (1987)
6. J.A.Bergstra,and J.W.Klop, "Conditional Rewrite Rules : Confluence and Termination," Journal of Computer and System Science, Vol.32, pp.323-362 (1986)
7. P.Chew, "Unique Normal Forms in Term Rewriting Systems with Repeated Variables",Proc. the 13th ACM Sympo. on Theory of Computing, pp.7-18 (1981)
8. P.L.Curien,and G.Ghelli, "On Confluence for Weakly Normalizing Systems," Proc. the 4th Int. Conf. on Reduction Techniques and Applications, pp.215-225, Springer LNCS 488 (1991)
9. N.Dershowitz,and M.Okada, "Proof-theoretic techniques for term rewriting theory," Proc. the 3rd IEEE Sympo. on Logic in Computer Science, pp.104-111 (1988)
10. N.Dershowitz, and Z.Manna, "Proving Termination with Multiset Orderings," Comm. of the ACM, Vol.22, No.8, pp.465-476 (1979)
11. G.Huet, "Confluent reductions : Abstract properties and applications to term rewriting systems", Journal of the ACM, Vol.27,No.4, pp.797-821 (1980)
12. J.W.Klop, and R.C.de Vrijer, "Unique normal forms for lambda calculus with surjective pairing," Information and Computation, Vol.80, pp.97-113 (1989)
13. B.K.Rosen, "Tree-manipulating systems and Church-Rosser theorems", Journal of the ACM, Vol.20, No.1, pp.160-187 (1973)
14. R.C.de Vrijer, "Extending the Lambda Calculus with Surjective Pairing is Conservative," Proc. the 4th IEEE Sympo. on Logic in Computer Science, pp.204-215 (1989)
15. R.C.de Vrijer, "Unique Normal Forms for Combinatory Logic with Parallel Conditional, a case study in conditional rewriting," Technical Report, Free University, Amsterdam (1990)
16. M.Ogawa,and S.Ono,"On the uniquely converging property of nonlinear term rewriting systems," IEICE Technical Report COMP89-7, pp.61-70 (1989)
17. M.Oyamaguchi, *Private Communications* (April 1992)

Higher Order Communicating Processes with Value-Passing, Assignment and Return of Results *

Dominique BOLIGNANO and Mourad DEBABI

Bull Corporate Research Center,
Rue Jean Jaurès,
78340 Les Clayes-Sous-Bois,
FRANCE
Dominique.Bolignano@frcl.bull.fr, Mourad.Debabi@frcl.bull.fr

Abstract. Our intent in this paper is to present a denotational model that supports both data and concurrency description. Data can be pure (concurrency free data such as literals) or processes. Concurrency is supported through processes that may communicate data and thus possibly processes through channels. Processes are thus said: *higher order communicating* processes. Functions are considered as processes that take their parameters, access some store, communicate on some channels and return some result. The model can be viewed as an extension of the VPLA language (Value-Passing Language with Assignment) proposed initially by Hennessy, thus an extended CCS without τ's version for handling input, output, communication, assignment and return of results. Furthermore a semantics for a useful set of combinators is defined. A significant subgoal is to investigate the algebraic properties of the model.

1 Introduction

Lately, a great deal of interest has been expressed in concurrency and data description. By description we mean either specification or programming. This interest is motivated by the awareness, that a mature description language, for real, thus complex, distributed systems must provide a framework for describing the two aspects of such a system: control and data.

This gave rise to a class of languages which are intended to satisfy such a requirement. One proposal is based on the combination of a process algebra such as CCS [15] [16] or CSP [12], together with a formalism supporting data abstraction such as ACT-ONE [5], VDM [1] or programming languages such as ML [18]. Good representatives of this approach are LOTOS [3], RSL [7], CML [19] and FACILE [6].

Most of these languages except RSL (RAISE Specification Language) are formally defined through an operational semantics. RSL comes with a denotational semantics [14] that is used in the extraction and the correction of its proof theory (axiomatic semantics). The work reported in this paper is relevant to the models

---

* This work is supported in part by Esprit project LACOS.

underlying the RSL denotational description and their foundations. Our interest in these models is motivated by the fact that RSL, as a specification language, is quite expressive, thus the corresponding models are quite sophisticated.

A great deal of work has been reported, in the literature, on semantic theories for "pure processes" i.e., processes whose communication form is restricted to just pure synchronization. But very little seems to have been developed about processes which may also perform transfer of data of possibly complex types between processes. Recently Hennessy and Ingólfsdóttir [11], inspired by the original RSL denotational description [13], proposed a language called VPLA (Value-Passing Language with Assignment) which takes into account the latter requirement. In fact VPLA is an extension of a CCS without τ's version [17] to handle input, output, communication and assignment. VPLA is based on the well behaved acceptances model [9] and provide a CCS semantics for CSP-like combinators as it will be detailed in the sequel of this paper.

Inspired by our work on the foundations of RSL, the models that will be presented in this paper can be considered as an extension of the VPLA language. Indeed higher order processes are allowed in our framework, in other words, functions that are treated as a special kind of processes are allowed to be communicated along the channels. This feature, as we will see later in more details, complicates even more the construction of the process space. More accurately we will see that the classical methods for construction of reflexive domains are unavailing in this case. We will consider for that a method proposed by R.E.Milne [14] that relies on a combination of the static and the dynamic semantics. Notice also that our model, on the opposite of VPLA, enables one to write expressions of any type, that may return some results after their execution. The model described in this document is endowed with a useful set of polymorphic concurrency combinators whose semantics is inspired by [14]. This latter set is more rich than the VPLA one since it provides a sequencing operator rather than simple prefixing and a variant of the parallel combinator called "interlocking" used in implicit concurrency specification i.e., processes are specified through their interaction with some test processes.

This article is organized as follows. The acceptances model is briefly recalled in the section 2. Then a first extension of the previous model is presented in the section 3, to handle input, output, communication, assignment and return of results. The existence of this concurrency model as well as its algebraic properties are then investigated. Section 4 is devoted to the semantics of some useful concurrency combinators. In section 5, the model is extended to allow higher order processes. The method is then illustrated and the algebraic properties investigated. Finally a discussion about related works and further research is ultimately sketched as a conclusion.

Along this presentation $\mathcal{P}_f$ will stand for the " finite" powerset (the set of finite sets), $\mathcal{P}_\infty$ for the infinite powerset (the set of finite and infinite sets), $A \rightarrowtail B$ for the set of all finite mappings (maps for short) from A to B, $dom(m)$ for the domain of the mapping m and $m \dagger m'$ for the overwriting of the map m with the map m'. We also adopt the VDM notation $[x \mapsto y, ...]$ for defining maps in extension. If x is a tuple then $\pi_i(x)$ stands for the ith projection of x.

2 The acceptances semantics

The acceptances model has been introduced in [9] and further explained in [10], as a denotational model for CSP-like process algebras. In what follows, we will recall the abstract model of acceptances. Then an extension for covering input and output is presented as well as an adaptation for supporting imperative aspects.

Hereafter, the intention is to recall briefly the general model of acceptances. Let Σ be a set of events which the processes can perform. In the rest of this section , we need the notion of *saturated sets*. Thus a set $\mathcal{A} \subseteq \mathcal{P}_f(\Sigma)$, is said to be saturated if it satisfies the following closures:

1. $\bigcup \mathcal{A} \in \mathcal{A}$ .... union closure.
2. $A, B \in \mathcal{A}$ and $A \subseteq C \subseteq B$ implies $C \in \mathcal{A}$ .... convex closure.

The set of all saturated finite subsets of $\mathcal{P}_f(\Sigma)$, all saturated sets over Σ, is denoted by $sat(\Sigma)$. Let $c(\mathcal{A})$ be the saturated closure of a set $\mathcal{A} \subseteq \mathcal{P}_f(\Sigma)$, defined as the least set which satisfies:

1. $\mathcal{A} \subseteq c(\mathcal{A})$
2. $\bigcup \mathcal{A} \in c(\mathcal{A})$
3. $A, B \in c(\mathcal{A})$ and $A \subseteq C \subseteq B$ implies $C \in c(\mathcal{A})$

The process space is then defined as:

$$Processes = \{(m, S) \mid (m, S) \in ((\Sigma \xrightarrow{m} Processes) \times Sat(\Sigma)) \bullet$$
$$dom(m) = \bigcup S\}$$
$$\cup$$
$$\{\bot\}$$

Thus *Processes* is a set of pairs that represent processes where the first component is an association (a map) between finitely many members of Σ and members of *Processes* whilst the second one is an acceptance set.

Now, let us come back to the underlying intuition. A process can be modeled as a pair (m, S) which stipulates that the process is waiting to engage in one event, say e, chosen from $dom(m)$, and once chosen, the process continues progressing as the process $m(e)$. The choice of the event is governed by the corresponding acceptance set. In fact S stands for the set of the possible internal states that can be reached non-deterministically. The actions in a set A from S, are those which the process can perform when in that state. The special element $\bot$ corresponds to the most non-deterministic process, known in the literature as **chaos**. In other words it models divergent processes. We consider here a bounded non-determinism, so at any given time, a process can be waiting to participate in one of only finitely many events. That it is why $dom(m)$ and the relevant internal states of an acceptance set are restricted to be finite.

3   Adapting the model

Our aim in this section, is to present the accommodation of the acceptances model, for handling input, output, communication and imperative aspects. Indeed, in a recent work, Hennessy and Ingólfsdóttir [11], inspired by the original RSL denotational description [13], consider also an extension that takes into account the previously mentioned features. It should be noted that the model presented here is more general than Hennessy's model VPLA. In fact our concurrency model, on the opposite of VPLA, allows one to write expressions that may return results of any type. Also it comes with more concurrency combinators, for instance the interlocking operator skipped in this presentation only for space reasons. Also our model is equipped with a polymorphic sequencing operator rather than simple prefixing.

Processes are able to perform one-to-one communications through unidirectional channels. For these reasons, the accommodated model make distinction between events in order to handle input and output. Thus, events are not simply assumed ranging over an alphabet Σ of actions, but are split into two categories according to their direction. So input and output events are distinguished and correspond respectively to inputs and outputs from certain channels. We also introduce a special event denoted by "$\sqrt{}$" (pronnounced tick) to signal the successful termination of a process. When such an event occurs the process immediately terminates and then returns some value.

Another adaptation, due to communication and sequencing handling, consists in two kinds of sequel processes (or just sequels). In fact sequels are not simply identified, as previously, with processes, but are also classified according to the nature of the relevant event that just occurred. The sequel attached to an input communication is modeled as a λ-abstraction that takes a value of the associated type of the channel, to some process that represents the continuation process, whilst an output sequel is modeled as a map (that associates some continuation processes to the values that may be communicated).

As process execution may affect some accessed variables of a particular store, we will consider an outcome of a process execution to be a pair composed of a value and a store (as usual, a store can be viewed as a map from a set of locations to their corresponding values). Because typically processes may be non-deterministic, we will let them have sets of possible outcomes as results, in other words results become sets of pairs. More formally, the process space of the adapted model can be written as shown in figure 1.

Notice that our process space is reflexive, in other words recursively defined. Hereafter we will prove its existence and investigate its algebraicity. From this point on, in order to lighten notations, we will use the notations S, V, D instead of $STORE$, $Value$ and $Processes$. Hence, more abstractly our space can be written as:

$$
\begin{aligned}
D = \{ \, (R, I, O, A) \, | R &\in \mathcal{P}_f(S \times V), \\
I &\in \Sigma_{in} \underset{m}{\rightarrow} (V \rightarrow Processes), \\
O &\in \Sigma_{out} \underset{m}{\rightarrow} (V \underset{m}{\rightarrow} Processes), \\
A &\in \mathcal{P}_f(\mathcal{P}_f(\Sigma_{in}) \times \mathcal{P}_f(\Sigma_{out})) \bullet condition \} \; \cup \; \{\bot\}
\end{aligned}
$$

$Processes = \{ (Result\_Set,$
$\qquad\qquad Input\_Map,$
$\qquad\qquad Output\_Map,$
$\qquad\qquad Acceptance\_Set$
$\qquad\quad) \mid$
$\qquad\quad Result\_Set \in \mathcal{P}_f(STORE \times Value),$
$\qquad\quad Input\_Map \in \Sigma_{in} \, \overline{m} \, (Value \rightarrow Processes),$
$\qquad\quad Output\_Map \in \Sigma_{out} \, \overline{m} \, (Value \, \overline{m} \, Processes),$
$\qquad\quad Acceptance\_Set \in \mathcal{P}_f(\mathcal{P}_f(\Sigma_{in}) \times \mathcal{P}_f(\Sigma_{out}) \times \mathcal{P}_f(\{\surd\}))$
$\qquad\quad \bullet$
$\qquad\quad Acceptance\_Set \neq \phi \wedge$
$\qquad\quad Sat(Acceptance\_Set) \wedge$
$\qquad\quad dom(Input\_Map) = \bigcup\{\pi_1(z) \mid z \in Acceptance\_Set\} \wedge$
$\qquad\quad dom(output\_Map) = \bigcup\{\pi_2(z) \mid z \in Acceptance\_Set\} \wedge$
$\qquad\quad \surd \in \bigcup\{\pi_3(z) \mid z \in Acceptance\_Set\} \Leftrightarrow Result\_Set \neq \phi \wedge$
$\qquad\quad \surd \in \bigcup\{\pi_3(z) \mid z \in Acceptance\_Set\} \Rightarrow (\phi, \phi, \{\surd\}) \in Acceptance\_Set$
$\qquad \}$
$\qquad \cup$
$\qquad \{\perp\}$

Where:

$\qquad \Sigma = \Sigma_{in} \cup \Sigma_{out} \cup \{\surd\}$ is the set of input and output events
$\qquad STORE$ stands for the set of possible stores.
$\qquad Value$ stands for the set of possible values.
$\qquad Sat(E)$ stands in this model for the following closure condition:
$\qquad\qquad \forall x \in E\bullet$
$\qquad\qquad\quad \forall y \in \mathcal{P}_f(\Sigma_{in}) \times \mathcal{P}_f(\Sigma_{out}) \times \mathcal{P}_f(\{\surd\})\bullet$
$\qquad\qquad\quad (\, \pi_1(y) \subseteq \bigcup\{\pi_1(z) \mid z \in E\} \wedge$
$\qquad\qquad\qquad \pi_2(y) \subseteq \bigcup\{\pi_2(z) \mid z \in E\} \wedge$
$\qquad\qquad\qquad \pi_3(y) \subseteq \bigcup\{\pi_3(z) \mid z \in E\} \wedge$
$\qquad\qquad\qquad \pi_1(x) \subseteq \pi_1(y) \wedge$
$\qquad\qquad\qquad \pi_2(x) \subseteq \pi_2(y) \wedge$
$\qquad\qquad\qquad \pi_3(x) \subseteq \pi_3(y)$
$\qquad\qquad\quad) \Rightarrow y \in E$

Fig. 1. The process space

Hereafter, we will assume that the reader is familiar with domain theory [8] [20] and the general theory of [21]. We have a recursive domain specification of the form $D = F(D)$. It is straightforward that it is enough to prove the existence of the superspace obtained by dropping the predicate *"condition"*. In order to establish the existence as well as the algebraicity of this superspace, we can show that our "tree" constructor F (in the process domain equation: $D = F(D)$) can be turned into a continuous functor in the category of cpo's CPO and then define the process domain as the least fixpoint of the functor F (i.e. the initial object in the category of F-algebras). Afterwards, we have to prove that F preserve algebraicity. An alternative method, is to use the inverse limit construction of Scott [8] [20]. The construction is standard up to some technical details and the interested reader can refer to [2] for the detailed proof.

Let us consider the following order on D:

1. $\forall P \in D \bullet \bot \sqsubseteq P$
2. $(R, I, O, A) \sqsubseteq (R', I', O', A')$ iff:
   (a) $R' \subseteq R$ and $A' \subseteq A$
   (b) $dom(I') \subseteq dom(I)$ and $dom(O') \subseteq dom(O)$
   (c) $\forall x \in dom(I') \bullet I(x) \leq I'(x)$
      $ie: \quad \forall v \in V \bullet I(x)(v) \sqsubseteq I'(x)(v)$
   (d) $\forall y \in dom(O') \bullet O(y) \leq O'(y)$
      $ie: \quad dom(O'(y)) \subseteq dom(O(y)) \wedge \forall v \in dom(O'(y)) \bullet O(y)(v) \sqsubseteq O'(y)(v)$

In the following some propositions whose detailed proofs are given in [2]

Corollary 1. $(D, \sqsubseteq)$ *is a complete partial order (cpo).*

The following result is about the compact (or finite or isolated) elements of D.

Proposition 2. *Let:*

$$K = [\, \mathcal{P}_f(S \times V) \times$$
$$\Sigma_{in} \rightarrowtail (V \rightarrowtail K) \times$$
$$\Sigma_{out} \rightarrowtail (V \rightarrowtail K) \times$$
$$\mathcal{P}_f(\mathcal{P}_f(\Sigma_{in}) \times \mathcal{P}_f(\Sigma_{out}))$$
$$] \cup \{\bot\}$$

where $\rightarrowtail$ *stands for the total function constructor which satisfies:* $f \in A \rightarrowtail B \Leftrightarrow \{x | f(x) \neq \bot\} \in \mathcal{P}_f(A)$. *Let* $C(D)$ *denote the elements of finite depth of* K. $C(D)$ *is then the set of compact elements of* D.

Proposition 3. D *is an algebraic cpo.*

Notice that provided that V is countable, one can prove that D is a Scott-domain, since D is an algebraic cpo and the set of the compact elements is then countable.

4 Concurrency combinators semantics

The semantic technique that will be presented hereafter, takes its origin in De Nicola and Hennessy's work on "CCS without τ's" [17]. In the following we need some semantic functions presented in figure 2. The function *process_set* takes a process to a set of deterministic processes, thus with only one acceptance state, such that the application of $\sqcap$ on this set yields the original process. Now let us call *effect* any function that takes a store to a particular process, and *action* any function that maps a value to a certain effect.

The $\gg$ function in figure 3 has been introduced to model value and store passing. Thus the expression *process* $\gg$ *action* means execute *process* and use then the resulting value and store to form the sequel process. If *process* is not divergent then it can either perform internal or external communications and the choice between both is governed by a particular strategy that will be explained below. The internal communications are those produced by passing a possible available result to the action in order to form the corresponding sequel process. Result passing is considered as a silent move since no visible event is performed. Another behavior consists of not returning any result and performing some input or output events before result passing. This corresponds to *external_process* in figure 3. The last function in figure 2, *merge_store*, takes as a parameter a process and returns an action which executes the process above and combines the resulting store with the one taken as a parameter.

Now let us turn to the combinators semantics (figure 3). First the divergent process **chaos** is mapped to the least element of the domain *Processes*. The constant combinator **stop** stands for deadlock and is represented as a process that refuses either to perform any event or to yield any result. **skip** is the process that immediately terminates successfully, it does not communicate, does not alter the store, and returns the unit value "()". The expression $P; Q$ corresponds to the sequential composition of P and Q. The two combinators $\lceil$ and $\lceil$ stand respectively for internal and external choices. In the two cases only one process is selected and this selection depends on whether there is an influence of the environment or not.

The semantics of the concurrency operator $\parallel$ takes its origin in the work of De Nicola and Hennessy [17]. The semantics presented in figure 4 is an accommodation for handling input, output, communication and assignment. RSL parallel operator is similar to the CCS one up to the accommodation mentioned previously, since it provides both synchronizations and interleavings. More accurately, the parallel composition of two processes P and Q stands for the following behavior: first the two processes are split up into their deterministic sub-processes. The whole concurrent composition amounts to the one of the sub-processes. Now given two sub-processes, each from a different process, we will consider as silent moves any result passing or synchronization between them. The external communications are made of the possible interleavings of the two sub-processes.

5   Extending the adapted model: passing processes along channels

It should be noted that the adapted model of section 3 is more powerful than the VPLA space. More precisely, our space allows one to write expressions that may return results, and also provides more concurrency combinators.

However this space is not powerful enough to cover all the expressiveness of RSL. Indeed, in section 3, we have assumed the universe of all possible values V, as a parameter that does not depend on the process space D. Thus it does not intervene in the recursion. This implies that only values that are not processes can be transmitted through channels. But we do not want this limitation to hold in our model. Thus, in particular, processes can be communicated through channels. The remark above complicates the construction much more, since now we have to deal with two mutually recursive equations of reflexive domains. Let us focus in the following on the structure of the domain of values. In fact, up to some syntactical details V can be written as:

$$V = \mathcal{P}_\infty(V) \oplus V \rightarrowtail V \oplus V \rightarrow V \oplus D \oplus \dots$$

It is clear that the domain equation above has no solution since it makes reference to an infinite powerset constructor and one can show easily that there is no isomorphism between a set and its infinite powerset. Moreover, the functions we allow are not necessarily continuous or monotonic and also maps are allowed to be infinite. All that makes the usual methods unavailing.

In order to get round this difficulty, R.E.Milne in [14] proposed a solution which consists of making a dependence between the static and the dynamic semantics of the language. Making the dynamic domains depend on the static domains means that dynamic domains are henceforth typed according to the hierarchy laid down by the static domains. Thus, dynamic spaces are restricted such that, for example, dynamic value domains are not obliged to contain as elements, functions that take members of these domains as parameters. In addition to that, the static constraints relevant to the well-formedness of an expression will be used in the dynamic semantics. This ensures that the dynamic domains are all that are needed.

In what follows, the intention is to present and discuss the existence and the algebraic properties of the process space. But since the latter and the value space are mutually dependent, then the two spaces will be constructed using the technique mentioned above. For that let us introduce some notation. The expression: $A == constructor(\dots, destructor_i : B_i, \dots)$ denotes the tagged expression $A = \{constructor\} \times (\dots \times B_i \times \dots)$ together with the function definition:

$$destructor_i \quad : A \rightarrow B_i$$
$$destructor_i(x) = \text{let } x = (constructor, (\dots, b_i, \dots)) \text{ in } b_i \text{ end}$$

Also in the sequel of this section, and for the sake of readability, some types will be accompanied with some axiomatic constraints. The "pure" set equivalent form can be trivially derived.

Since the aim hereafter is to illustrate the techniques used and to establish theoretical results that are independent from syntactic sugaring, and in order to avoid distraction and unnecessary complication, we will restrict the space of possible values to contain reals (as an example of literal values), infinite sets (as an example of infinite structures: sets, lists and maps) and of course the set of functions and thus processes.

$$SV == s\_real \oplus S\_Set \oplus S\_Action \dots\dots\dots all\ static\ values$$
$$S\_Set == s\_set(SV) \dots\dots\dots\dots\dots\dots\dots type\ for\ sets$$
$$S\_Action == s\_action(par:SV, eff: S\_Effect)\ type\ for\ functions$$
$$S\_Effect == s\_effect(\ rd: S\_LocSet, \dots\dots\ type\ for\ processes$$
$$wr: S\_LocSet,$$
$$in: S\_PosSet,$$
$$out: S\_PosSet,$$
$$res: SV)$$

$S\_LocSet = \mathcal{P}_f(S\_Loc)$ .... *type for variable sets*
$S\_PosSet = \mathcal{P}_f(S\_Pos)$ .... *type for channel sets*
$S\_Loc == s\_loc(\text{typ: SV, cop: IN})$ *type for variables*
$S\_Pos == s\_pos(\text{typ: SV, cop: IN})$ *type for channels*

which satisfies:

$\forall\ loc \in S\_Loc \bullet not\_cyclic\_def(loc)$
$\forall\ pos \in S\_Pos \bullet not\_cyclic\_def(pos)$
$\forall\ eff \in S\_Effect \bullet rd(eff) \supseteq wr(eff)$

A type expression will be mapped, in the static semantics, to a static value. The set SV above represents the set of all static values. S_Set represents the set of all static values that are types for dynamic set values. The type of a function is called a static *action*. It consists of the type of the parameter of the function and a static value called *effect* that contains the description of the accesses and the result, in other words the types of the read and write variables, the types of input and output channels and the type of the result. Notice that variables (with their access modes) and channels (with their directions) are explicitly declared in the signature of an RSL function. We will call a location the denotation of a variable and a position the denotation of a channel. A static location has two attributes: a type and a copy. The first attribute corresponds to the type of the variable whilst the second is a tag that ensures that there is a finite set of static locations of the same type (id for positions). The function $not\_cyclic\_def(a\_static\_value)$ ensures that a_static_value is not a cyclic definition (thus it is a finite term). Let SV_f be the set of finite terms that are members of SV.

The dynamic domains are then constructed as follows:

$$V = \cup_{x \in SV_f} V_x$$
$$V_{s\_real} = \mathbb{R}$$
$$V_{s\_set(x)} = \mathcal{P}_\infty(V_x)$$
$$V_{s\_action(x,y)} = V_{par(s\_action(x,y))} \rightarrow S_{rd(s\_effect(s\_action(x,y)))} \rightarrow$$
$$D_{eff(s\_action(x,y))}$$
$$S_{locset} = (\cup LSet_{locset}) \xrightarrow{m} (\cup_{x \in locset} V_{typ(x)})$$

With the following invariant on the dynamic stores:

$$\forall d\_store \in S_{locset} \bullet dom(d\_store) = \bigcup LSet_{locset} \land$$
$$(\forall d\_loc \in dom(d\_store) \bullet d\_store(d\_dloc) \in dtyp(d\_loc))$$

$$LSet_{locset} = \mathcal{P}_\infty(\cup_{x \in locset} L_x)$$
$$PSet_{posset} = \mathcal{P}_\infty(\cup_{x \in posset} P_x)$$
$$L_x == dloc(sloc : \{x\}, dtyp : V_{typ(x)})$$
$$P_x == dpos(spos : \{x\}, dtyp : V_{typ(x)})$$

V_x, S_x, P_x, L_x, $PSet_x$, $LSet_x$ stand respectively for dynamic values, stores, positions, locations, position sets and location sets of a type x. The dynamic process domain is defined as: $D = \bigcup_{x \in S\_Effect} D_x$ where:

$$D_x = \begin{cases} (\mathcal{P}_f(S_{wr(x)} \times V_{res(x)}) \times \\ \bigcup_{y \in in(x)} P_y \text{ } \overrightarrow{m} \text{ } \bigcup_{y \in in(x)} (V_{typ(y)} \to D_x) \times \\ \bigcup_{y \in out(x)} P_y \text{ } \overrightarrow{m} \text{ } \bigcup_{y \in out(x)} (V_{typ(y)} \text{ } \overrightarrow{m} \text{ } D_x) \times \\ \mathcal{P}_f(\mathcal{P}_f(\bigcup_{y \in in(x)} P_y) \times \mathcal{P}_f(\bigcup_{y \in out(x)} P_y) \times \mathcal{P}_f(\{\sqrt{}\}))) \\ \cup \{\bot\} \end{cases}$$

Notice that $V_{typ(y)}$, when unwound according to $typ(y)$, cannot contain a subterm D_x because otherwise x would be an infinite term which is a contradiction with the fact that x is in SV_f. Hence $V_{typ(y)}$ does not intervene in the recursion. Thus, we have exactly the same expression as the one in the previous section: $D_x = F(D_x)$. Hence D_x exists and is an algebraic cpo. Then it is straightforward that the same property holds for D.

Corollary 4. $\forall x \in S\_Effect \bullet (D_x, \sqsubseteq_{D_x})$ *is an algebraic cpo.*

Proposition 5. $D = \bigcup_{x \in S\_Effect} D_x$, *ordered by* $\sqsubseteq = \bigcup_{x \in S\_Effect} \sqsubseteq_{D_x}$ *is an algebraic cpo.*

6 Conclusion

Inspired by the original RSL denotational description [14], and the VPLA language [11], we have reported in this paper a concurrency model for higher order communicating processes with value-passing, that is an extension of VPLA and a part of the foundations (type universe) of RSL. This presentation has been done in two steps. A first model has been presented to extend the acceptances model for handling input, output, communication and imperative aspects. Its existence has been proved using an inverse limit construction and we have shown that it constitutes an algebraic complete partial order. A second model has been presented to allow the communication of processes (processes are values) along channels, as well as the communication of a wide range of possible values (infinite sets, infinite lists ,...). But since the inverse limit construction alone is unavailing in this case, we have borrowed the technique proposed by R.E.Milne in [14], that consists of typing the dynamic domains using the static semantics. Thus a dependence between the static and the dynamic semantics is created. The existence proof of the generalized domain relies in part, but essentially on the first existence proof. The algebraic properties have been also lifted to the generalized model.

As a future research, we intend to investigate, as in [11], the relationship between the equivalence induced by this denotational model and the operational intuitions. More precisely, we plan to define an operational semantics for our processes similar to the one reported in [11], to formalize then the notion of MUST testing equivalence and try to establish a comparison with the equivalence on the denotational model.

References

1. D. Bjørner and C.B. Jones. *Formal Specification and Software Development.* Prentice-Hall, 1982.
2. D. Bolignano and M. Debabi. *On the Foundations of the RAISE Specification Language.* Technical Report RAD/DMA/92013, Bull-ORDA, May 1992.
3. T. Bolognesi and E. Brinksma. *The Formal Description Technique LOTOS.* North Holland, 1989.
4. M. Debabi and D. Bolignano. *Comparative Concurrency and Denotational Semantics.* Technical Report RAD/DMA/92002, Bull-ORDA, February 1992.
5. H. Ehrig and B. Mahr. *Fundamentals of Algebraic Specification 1: Equations and Initial Semantics.* Springer Verlag, 1985.
6. A. Giacalone, P. Mishra, and S. Prasad. Facile: a symmetric integration of concurrent and functional programming. *International Journal of Parallel Programming,* 18(2):121–160, April 1989.
7. RAISE Language Group. *The RAISE Specification Language.* Prentice-Hall, 1992.
8. C.A. Gunter and D.S. Scott. Semantic domains. *Handbook of Theoretical Computer Sscience,* 31(3):560–599, July 1990.
9. M. Hennessy. Acceptance trees. *Journal of the ACM,* 32:896–928, October 1985.
10. M. Hennessy. *Algebraic Theory of Process.* MIT Press, 1988.
11. M. Hennessy and A. Ingólfsdóttir. *Communicating Processes with Value-Passing and Assignments.* Technical Report, University of Sussex - Draft, June 1991.
12. C.A.R. Hoare. *Communicating Sequential Processes.* Prentice-Hall, 1985.
13. R.E. Milne. *Concurrency Models and Axioms.* Technical Report, RAISE/CRI/-DOC/4/V1, CRI, 1988.
14. R.E. Milne. *Semantic Foundations of RSL.* Technical Report RAISE/CRI-/DOC/4/V1, CRI, 1990.
15. A.J.R.G. Milner. A calculus of communicating systems. In *Lecture Notes in Computer Science 92,* pages 281–305, Springer-Verlag, 1980.
16. A.J.R.G. Milner. *Communication and Concurrency.* Prentice-Hall, 1989.
17. R. De Nicola and M. Hennessy. Ccs without τ's. In *Lectures Notes in Computer Science 250,* pages 294–305, Springer-Verlag, 1987.
18. M. Tofte R. Milner and R. Harper. *The definition of standard ML.* MIT Press, 1990.
19. J.H. Reppy. Cml: a higher-order concurrent language. In *Proceedings of the ACM SIGPLAN '91 Conference on Programming Language design and Implementation,* pages 294–305, SIGPLAN Notices 26(6), 1991.
20. D.A. Schmidt. *Denotational Semantics.* Allyn and Bacon,inc, 1986.
21. M.B. Smyth and G.D. Plotkin. The category-theoretic solution of recursive domain equations. *SIAM Journal of Computing,* 11(4):761–783, November 1982.

Appendix

$[\![$ chaos $]\!]\sigma = \bot$

$[\![$ stop $]\!]\sigma = (\{\}, [\,], [\,], \{(\phi, \phi, \phi)\})$

$[\![$ skip $]\!]\sigma = (\{(\,\sigma, ()\,)\}, [\,], [\,], \{(\phi, \phi, \{\sqrt{}\})\})$

$[\![$ channel_name? $]\!]\sigma = (\,\{\,\},$
$\qquad\qquad\qquad\qquad [channel\_name \mapsto \lambda v \in V \bullet (\{(\,\sigma, v\,)\}, [\,], [\,], \{(\phi, \phi, \{\sqrt{}\})\})],$
$\qquad\qquad\qquad\qquad [\,],$
$\qquad\qquad\qquad\qquad \{(\{channel\_name\}, \phi, \phi)\}$
$\qquad\qquad\qquad)$

$[\![$ channel_name ! a_value $]\!]\sigma = (\,\{\,\},$
$\qquad\qquad\qquad\qquad\qquad [\,],$
$\qquad\qquad\qquad\qquad\qquad [channel\_name \mapsto [\![\ a\_value\ ]\!]\sigma \mapsto (\{(\,\sigma, ()\,)\}, [\,], [\,], \{(\phi, \phi, \{\sqrt{}\})\})],$
$\qquad\qquad\qquad\qquad\qquad \{(\phi, \{channel\_name\}, \phi)\}$
$\qquad\qquad\qquad\qquad)$

$[\![$ var_name := a_value $]\!]\sigma = (\{(\,\sigma \dagger [var\_name \mapsto [\![\ a\_value\ ]\!]\sigma], ()\,)\}, [\,], [\,], \{(\phi, \phi, \{\sqrt{}\})\})$

$[\![$ P;Q $]\!]\sigma = [\![$ P $]\!]\sigma \gg \lambda v \in V \bullet \lambda\sigma' \in S \bullet [\![$ Q $]\!]\sigma'$

$[\![$ P $\sqcap$ Q $]\!]\sigma =$ case $[\![$ P $]\!]\sigma$, $[\![$ Q $]\!]\sigma$ of
$\qquad\qquad (R, I, O, A), (R', I', O', A') \longrightarrow (R \cup R', I \sqcap I', O \sqcap O', A \sqcap A')$
$\qquad\qquad$ else $\bot$
$\qquad\qquad$ end

$[\![$ P $\square$ Q $]\!]\sigma =$ case $[\![$ P $]\!]\sigma$, $[\![$ Q $]\!]\sigma$ of
$\qquad\qquad (R, I, O, A), (R', I', O', A') \longrightarrow (R \cup R', I \sqcap I', O \sqcap O', A \square A')$
$\qquad\qquad$ else $\bot$
$\qquad\qquad$ end

$where :$

$I \sqcap I' = [\, e \mapsto I(e) \mid e \in dom(I)] \;\dagger\; [e \mapsto I'(e) \mid e \in dom(I')] \;\dagger$
$\qquad\quad [\, e \mapsto \lambda v \in V \bullet I(e)(v) \sqcap I'(e)(v) \mid e \in dom(I) \cap dom(I')]$

$O \sqcap O' = [\, e \mapsto O(e) \mid e \in dom(O)] \;\dagger\; [e \mapsto O'(e) \mid e \in dom(O')] \;\dagger$
$\qquad\quad [\, e \mapsto O(e) \sqcap O'(e) \mid e \in dom(O) \cap dom(O')]$

$O(e) \sqcap O'(e) = [\, v \mapsto O(e)(v) \mid v \in dom(O(e))] \;\dagger\; [v \mapsto O'(e)(v) \mid v \in dom(O'(e))] \;\dagger$
$\qquad\qquad\quad [\, v \mapsto O(e)(v) \sqcap O'(e)(v) \mid v \in dom(O(e)) \cap dom(O'(e))]$

$A \sqcap A' = \{x \mid x \in \mathcal{P}_f(\Sigma_{in}) \times \mathcal{P}_f(\Sigma_{out}) \times \mathcal{P}_f(\{\sqrt{}\}) \bullet \pi_1(x) \subseteq \cup\{\pi_1(y) \mid y \in A \cup A'\} \wedge$
$\qquad\quad \pi_2(x) \subseteq \cup\{\pi_2(y) \mid y \in A \cup A'\} \wedge \pi_3(x) \subseteq \cup\{\pi_3(y) \mid y \in A \cup A'\} \wedge$
$\qquad\quad (\exists y \in A \cup A' \bullet \pi_1(x) \supseteq \pi_1(y) \wedge \pi_2(x) \supseteq \pi_2(y) \wedge \pi_3(x) \supseteq \pi_3(y)) \}$

$A \square A' = \{x \mid x \in \mathcal{P}_f(\Sigma_{in}) \times \mathcal{P}_f(\Sigma_{out}) \times \mathcal{P}_f(\{\sqrt{}\}) \bullet \pi_1(x) \subseteq \cup\{\pi_1(y) \mid y \in A \cup A'\} \wedge$
$\qquad\quad \pi_2(x) \subseteq \cup\{\pi_2(y) \mid y \in A \cup A'\} \wedge \pi_3(x) \subseteq \cup\{\pi_3(y) \mid y \in A \cup A'\} \wedge$
$\qquad\quad ((\exists(s, t) \in A \times A' \bullet \pi_1(x) = \pi_1(s) \cup \pi_1(t) \wedge \pi_2(x) = \pi_2(s) \cup \pi_2(t) \wedge \pi_2(x) = \pi_3(s) \cup \pi_3(t))$
$\qquad\quad \vee(\sqrt{} \in \pi_3(x)))\}$

Fig.3. The semantics of some concurrency combinators

$$process\_set(R, I, O, A) = \{\ (\ R,$$
$$[e \mapsto I(e)\ |\ e \in dom(I) \cap \pi_1(x)\ ],$$
$$[e \mapsto O(e)\ |\ e \in dom(O) \cap \pi_2(x)\ ],$$
$$\{x\}$$
$$)\ |\ x \in A\}$$

$process \succ action = $ case $process$ of
$$\bot \longrightarrow \bot$$
$$(R, I, O, A) \longrightarrow (R, I, O, A) \succ action$$
end

$(R, I, O, A) \succ action =$
$\sqcap\ \{$ let
$$internal\_process\_set = \text{if } \sqrt{} \in \pi_3(A_1) \text{ then } R_1 \succ action \text{ else } \phi,$$
$$external\_process = (\ \{\}, I_1 \succ action, O_1 \succ action, (\pi_1(A_1), \pi_2(A_1), \phi))$$
in
$$\text{if } internal\_process\_set = \phi \text{ then } external\_process$$
$$\text{else } (external\_process \sqcap (\sqcap internal\_process\_set)) \sqcap (\sqcap internal\_process\_set)$$
end
end $|\ (R_1, I_1, O_1, A_1) \in process\_set(R, I, O, A)\}$

$R \succ action = \{r \succ action\ |\ r \in R\}$
$r \succ action = action(\pi_1(r))(\pi_2(r))$
$I_1 \succ action = [e \mapsto \lambda v \in V \bullet I_1(e)(v) \succ action\ |\ e \in dom(I_1)]$
$O_1 \succ action = [e \mapsto [v \mapsto O_1(e)(v) \succ action\ |\ v \in dom(O_1(e))]\ |\ e \in dom(O_1)]$

$merge\_store(process)(\sigma) = \lambda v_1 \in V \bullet \lambda \sigma_1 \in S \bullet$
$\quad process \succ \lambda v_2 \in V \bullet \lambda \sigma_2 \in S \bullet$
let
$$\sigma_3 = \sigma_1 \dagger [loc \mapsto \sigma_2(loc)\ |\ loc \in dom(\sigma_2) \cap dom(\sigma) \bullet \sigma(loc) \neq \sigma_2(loc)]$$
in $(\{((\sigma_3, v_2)), [\ ], [\ ], \{(\phi, \phi, \{\sqrt{}\})\}))$
end

Fig. 2. Some useful functions

$[\ P\ \|Q\ ]\sigma = $ case $[\ P\ ]\sigma\ ,\ [\ Q\ ]\sigma$ of
$$(R, I, O, A), (R', I', O', A') \longrightarrow (R, I, O, A)\|\sigma\|(R', I', O', A')$$
else $\bot$
end

$(R, I, O, A)\ \|\ \sigma\ \|\ (R', I', O', A') =$
$\sqcap\ \{$
let
$$(R_1, I_1, O_1, A_1) = process_1,\ (R_2, I_2, O_2, A_2) = process_2,$$

$internal\_process\_set =$
$$R_1 \succ merge\_store(process_2)(\sigma)\ \cup R_2 \succ merge\_store(process_1)(\sigma)\ \cup$$
$$I_1\ \|\sigma\|\ O_2\ \cup I_2\ \|\sigma\|\ O_1,$$

$external\_process =$
$$(\{\},\ I_1\ \|\sigma\|\ process_2,\ O_1\ \|\sigma\|\ process_2,\ A_1)\sqcap$$
$$(\{\},\ I_2\ \|\sigma\|\ process_1,\ O_2\ \|\sigma\|\ process_1,\ A_2)$$
in
$$\text{if } internal\_process\_set = \{\} \text{ then } external\_process$$
$$\text{else } (external\_process\ \sqcap\ (\sqcap\ internal\_process\_set))\ \sqcap\ (\sqcap\ internal\_process\_set)$$
end
end $|$
$(process_1, process_2) \in process\_set(R, I, O, A) \times process\_set(R', I', O', A')\}$

where:

$$I_i\ \|\sigma\|\ O_j = \bigcup\{I_i(e)\|\sigma\|O_j(e)\ |\ e \in dom(I_i) \cap dom(O_j)\}$$
$$I_i(e)\|\sigma\|\ O_j(e) = \{\ I_i(e)(v)\|\sigma\|O_j(e)(v)\ |\ v \in dom(O_j(e))\}$$

$$I_i\ \|\sigma\|\ process_j = [e \mapsto \lambda v \in V \bullet I_i(e)(v)\|\sigma\|\ process_j\ |\ e \in dom(I_i)]$$

$$O_i\|\sigma\|\ process_j = [e \mapsto O_i(e)\|\sigma\|\ process_j\ |\ e \in dom(O_j)]$$
$$O_i(e)\|\sigma\|\ process_j = [v \mapsto O_i(e)(v)\|\sigma\|process_j\ |\ v \in dom(O_i(e))]$$

Fig. 4. Parallel combinator semantics

Searching informed game trees
(Extended abstract)[1]

Wim Pijls & Arie de Bruin

Erasmus University Rotterdam

wimp@cs.eur.nl

Abstract

Well-known algorithms for the evaluation of the minimax function in game trees are alpha-beta [Kn] and SSS* [St]. An improved version of SSS* is SSS-2 [Pij1]. All these algorithms don't use any heuristic information on the game tree. In this paper the use of heuristic information is introduced into the alpha-beta and the SSS-2 algorithm. The subset of nodes which is visited during execution of each algorithm is characterised completely.

1 Introduction

Game trees are related to two person games with perfect information like Chess, Checkers, Go, Tic-tac-toe, etc. Each node in a game tree represents a game position. The root represents a position of the game for which we want to find the best move. The children of each node n correspond to the positions resulting from one move from the position given by n. The terminals in the tree are positions in the game for which a real valued evaluation function f exists giving the so called game value, the pay-off of that position.

We assume that the two players are called MAX and MIN. A node n is marked as max-node or min-node if in the corresponding position it is max's or min's move respectively. We assume that MAX moves from the start position.

The evaluation function can be extended to the so called minimax function, a function which determines the value for each player in any node. The definition is:

$$f(n) = \max \ \{f(c) \mid c \text{ a child of } n\}, \text{ if } n \text{ is a max node,}$$
$$\min \ \{f(c) \mid c \text{ a child of } n\}, \text{ if } n \text{ is a min node.}$$

We adopt the convention that the minimax value of a game tree T, denoted by $f(T)$, is the minimax value of the root of this tree. In Figure 1 an example of a game tree is shown labeled with its f-values. The bold lines in this figure define a so called solution tree, which is to be defined in Section 3.

The value $f(n)$ in any node n (n not necessarily a max node) indicates the highest attainable pay-off for MAX in the position n, under the condition that both players will play optimally in the sequel of the game. In any node n the move for each player to optimize the pay-off is the transition to a child node c such that $f(c) = f(n)$. In this way, MAX tries to maximize and MIN tries to minimize the profit of MAX. Therefore, an optimal play will proceed along a *critical path*, which is defined as a path from the root to a leaf such that $f(n)$ has the same value for all nodes n in the

---

[1]The full paper, including all proofs, is registered as [Pij2]

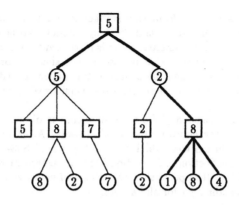

Figure 1: A game tree with some f-values.

path. All nodes in this path have a game value equal to the game value of the root.

For some game trees, heuristic information on the minimax value $f(n)$ is available for any node. This information can be expressed as a pair $H = (U, L)$ where U and L are heuristic functions mapping the nodes of the game tree into the real numbers, such that $U(n) \geq f(n) \geq L(n)$ for any n. The heuristic functions thus denote an upper bound and a lower bound respectively of the minimax value. A heuristic pair $H = (U, L)$ is called *consistent* if for every child c of a given max node n $U(c) \leq U(n)$, and for every child c of a given min node $L(c) \geq L(n)$. From now, we assume that an input instance consists of a pair (G, H), called an informed game tree, where G denotes a game tree and H a pair of consistent heuristic functions. If heuristic information is discarded or is not available at all, we define $U(n) = +\infty$ and $L(n) = -\infty$ for any inner node n and $U(n) = f(n) = L(n)$ in a terminal.

In order to compute the minimax value of a game tree, several algorithms have been developed. The brute force approach would compute the minimax function in each node of the game tree according to the definition. Each feasible algorithm has its own method to avoid examining the entire tree. The oldest algorithm is the so-called alpha-beta algorithm [Kn]. Another important algorithm is called SSS* [St]. The working of this algorithm is rather opaque. In [Pij1] an algorithm, called SSS-2, is presented, which traverses the game tree in the same order as SSS*. However, the underlying paradigm is much more perspicuous than in the case of SSS*. All these algorithms do not take into account any heuristic information. Ibaraki [Ib] introduced the idea to exploit heuristic information for improving the efficiency of game tree algorithms. In this paper alpha-beta and SSS-2 are generalised in the sense that a heuristic pair H features in the algorithm. Furthermore, a complete characterisation is given of the set of nodes visited during execution. For the situation that no heuristic information is taken into account, such a characterisation has been found for alpha-beta and SSS* by Baudet [Ba] and Pearl [Prl] respectively. However, when our characterisation is restricted to such a situation, the results are simpler. Because of the extensive use of recursion, our proofs differ completely from theirs. Moreover, when the proofs in this paper are reduced to the case without heuristic estimates, we obtain improved versions with respect to those in [Pij1].

In Section 2, the extended alpha-beta algorithm is discussed and the characterisation is given for the nodes visited by alpha-beta. After introducing in Section 3 the notion of a solution tree, which plays a key role in SSS-2, the SSS-2 algorithm itself is discussed in Section 4. In Section 5 the characterisation is given for the nodes examined by SSS-2. It appears that SSS-2's set of nodes visited during execution is a subset corresponding set for alpha-beta.

In this paper, recursion is often used in describing algorithms. To prove that a recursive procedure is correct, i.e., the procedure satisfies the specifications, so-called recursion induction is used. In a proof by recursion induction, first, the correctness of the procedure is proven for the case that no recursive calls are performed (the basic step). Next, the correctness is proven under the assumption that the all the inner recursive calls are correct (the induction step).

Several theorems in this paper have the form: for any node n in a given game tree, $P(n) \Leftrightarrow Q(n)$, where P and Q denote propositions. Such theorems will be proved by so-called *path induction*. A proof by path induction consists of three steps. First it is proved that, if n is not the root of the game tree, $P(n) \Rightarrow P(m)$ and $Q(n) \Rightarrow Q(m)$ where $m = father(n)$. Second the implication $P(r) \Leftrightarrow Q(r)$ is proved, where r denotes the root of the given game tree. In the third step we assume $P(n) \wedge Q(n)$ for some n and we prove that $P(c) \Leftrightarrow Q(c)$ for any child c of n.

Now we give a complete list of all definitions in discussing the properties of algorithms:

Definition 1 *For each node n the following quantities are defined (we assume* $\max(\emptyset)$ *$= -\infty$ and $\min(\emptyset) = \infty$):*

$ANC(n) \quad = \quad \{\ x \mid x$ *is a proper ancestor of n* $\}$
$AMAX(n) = \quad \{\ x \mid x \in ANC(n)$ *and x is a max node* $\}$
$AMIN(n) = \quad \{\ x \mid x \in ANC(n)$ *and x is a min node* $\}$
$\bar{L}(n) \quad = \quad \max\{L(x) \mid x \in AMAX(n)\} \qquad \hat{L}(n) \quad = \quad \max\{L(x) \mid x \in ANC(n)\}$
$\bar{U}(n) \quad = \quad \min\{U(x) \mid x \in AMIN(n)\} \qquad \hat{U}(n) \quad = \quad \min\{U(x) \mid x \in ANC(n)\}$
$\hat{f}(n) \quad = \quad \max\{f(x) \mid x \in ANC(n)\}$
A node x is called a left sibling of a given node n, if x is a child of a node m with $m \in ANC(n)$, and x is older than m', where m' denotes the child of m on the path from m to n.
$\alpha(n) = \quad \max\{f(x) \mid x$ *a left sibling of n and $father(x)$ is max node*$\}$
$\beta(n) = \quad \min\{f(x) \mid x$ *a left sibling of n and $father(x)$ is min node*$\}$

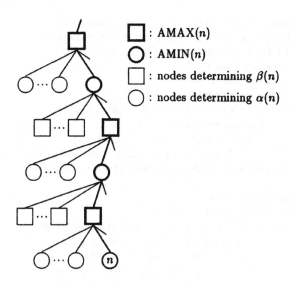

$\square$: AMAX(n)

$\bigcirc$: AMIN(n)

$\square$: nodes determining $\beta(n)$

$\bigcirc$: nodes determining $\alpha(n)$

Figure 2: Illustrating the definitions.

2 The alpha-beta algorithm

In this section we present the generalisation of the alpha-beta algorithm, using a heuristic pair. The algorithm consists of one central recursive procedure, see Figure 3. For a game tree with root r, the minimax value is computed by the call *alphabeta(r, alpha, beta, f)*, where *alpha* and *beta* are real numbers such that *alpha* $\leq f(r) \leq$ *beta*.

Specification of the procedure *alphabeta*.
The input parameters are n, a node in a game tree, and *alpha* and *beta*, two real numbers. There is one output parameter f, a real number.
pre: *alpha* $<$ *beta*,
post: *alpha*$<$$f$$<$*beta* $\Rightarrow$ $f = f(n)$,
          $f \leq$ *alpha* $\Rightarrow$ $f(n) \leq f \leq$ *alpha*,
          $f \geq$ *beta* $\Rightarrow$ $f(n) \geq f \geq$ *beta*.

By recursion induction we can prove that the procedure *alphabeta* meets its specification. By path induction we can prove the following theorems, in which the set of nodes visited by alpha-beta is characterised.

Theorem 1 *If a node n is parameter in a nested call alphabeta(n, alfa, beta,f) anywhere in the recursion during the execution of the call alphabeta(r, $-\infty$, $+\infty$, f) with r the root of a game tree, then* alpha$=$ $\max(\alpha_L(n), \bar{L}(n))$ *and* beta $= \min(\beta(n), \bar{U}(n))$.

Theorem 2 *Suppose the call alphabeta(r, $-\infty$, $+\infty$, f) is executed where r denotes the root of a game tree (G, H). A node n in G is parameter in a nested call alphabeta(n, alpha, beta, f) if and only if* $\max(\alpha(n), \hat{L}(n)) < \min(\beta(n), \hat{U}(n))$.

```
procedure alphabeta(in:  n, alpha, beta; out:  f);
if alpha≥U(n) or L(n)≥beta or U(n)=L(n) then
   [ if L(n) ≥ beta then f:=L(n) else f:=U(n);
     exit procedure; ]
if type(n) = max then
  [ alpha':= max(alpha, L(n));
     for c := firstchild(n) to lastchild(n) do
        [ alphabeta(c, alpha', beta, f');
           if f'>alpha' then alpha':=f';
           if f'≥ min(beta, U(n)) then exit for loop; ]
     f:=maximum of the intermediate f'-values;
  ]
if type(n)=min then
  [ beta':=min(beta, U(n));
     for c := firstchild(n) to lastchild(n) do
        [ alphabeta(c, alpha, beta', f');
           if f'< beta' then beta':=f';
           if f' ≤ max(alpha, L(n)) then exit for loop; ]
     f:=minimum of the intermediate f'-values;
  ]
```

Figure 3: The procedure alphabeta.

3   Solution trees

In the SSS-2 algorithm to be discussed in Section 4, the notion of a solution tree plays a central role. In figure 1 the bold edges generate a solution tree.

Definition 2 *Given an informed game tree* (G, H), *a solution tree* S *is a subtree of* G *with the properties:*
   - *for a max node* n, *either all children of* n *are included in* S *or no child is included;*
   - *for a min node, either exactly one child is included in* S, *or no child is included.*
A node in S *which has no children in* S *is called a tip node of* S.

The set of all solution trees rooted in a node n is denoted by $\mathcal{M}(n)$. For a given game tree (G, H) the set of all max solution trees with the same root as G is denoted by $\mathcal{M}_G$.

If S is any solution tree and m is any node in S, then $S(m)$ denotes the subtree of S, rooted in m. For a node n in a solution tree, the minimax function $g(n)$ is defined as:

$$g(n) \;=\; U(n), \qquad\qquad\qquad if \; n \; is \; a \; tip \; node,$$
$$=\; \max \{g(c) \mid c \; a \; child \; of \; n\}, if \; n \; is \; an \; inner \; max \; node,$$
$$=\; g(c) \qquad\qquad\qquad if \; n \; is \; an \; inner \; min \; node \; and$$
$$c \; is \; the \; single \; child \; of \; n.$$

```
r:=root of the game tree;
expand(r, ∞, g2, T);
repeat
   [ g1:= g2;
     diminish(r, g1, g2);
   ]
until g1= g2;
```

Figure 4: The SSS-2 algorithm.

Similar to the minimax function f in a game tree G, we identify the minimax value $g(S)$ of a solution tree S with the minimax value of the root of S.

Theorem 3 *Let (G, H) be a game tree. Then $f(n) = \min \{g(S) \mid S \in \mathcal{M}(n)\}$.*

The solution tree, which achieves the minimum referred to in Theorem 3, is a solution tree which contains a critical path of the game tree.

In order to investigate the set of solution trees, we introduce the following linear ordering *'older'*, denoted by $\gg$, on this set. (In any non-terminal of a game tree a fixed order for the child nodes is assumed.)

Definition 3 *For two solution trees S and S' in $\mathcal{M}(n)$ with n a node in a game tree (G, H), the relation $\gg$ is defined recursively as follows:*
  - *if n is a tip node in S and n is not a tip node in S', then $S \gg S'$;*
  - *if n is a max node and n is not a tip node in S or in S', then consider the oldest child m of n, such that the subtrees $S(m)$ and $S'(m)$ are different (because $S \neq S'$, such a subtree must exist); if $S(m) \gg S'(m)$ then $S \gg S'$.*
  - *if n is a min node and n is not a tip node in S or in S', then consider m and m', the children of n in S and S' respectively; we define $S \gg S'$, if either m is older than m' or $m = m'$ and $S(m) \gg S'(m)$.*

In the SSS-2 algorithm we pay special attention to those solution trees T in $\mathcal{M}(n)$ such that $g(T) < g(T')$ for any element $T' \in \mathcal{M}(n)$ with $T' \gg T$. A solution tree with this property is called a *milestone* in $\mathcal{M}(n)$. The following Lemma gives a characterisation of a milestone.

Lemma 1 *For an informed game tree (G, H) and a node $n \in G$, a solution tree $S \in \mathcal{M}(n)$ is the oldest solution tree with g-value $\leq g_0$, if and only if one of the following statements holds:*
  - *a) n is a tip node in S (i.e., S consists only of node n), and $U(n) \leq g_0$;*
  - *b) n has at least one child in S, $U(n) > g_0$ and $S(c)$ is the oldest solution tree in $\mathcal{M}(c)$ with g-value $\leq g_0$, for every child c of n in S; moreover, in case n is a min node, $f(c') > g_0$ for every older brother c' of the single child c in G.*

A twin lemma of the last lemma is obtained by replacing $\leq$ by $<$, and by replacing $>$ by $\geq$.

4 The SSS-2 algorithm

Due to Theorem 3, the minimax value $f(G)$ of a game tree (G, H) is equal to the smallest g-value of the trees in the set $\mathcal{M}_G$. In this section we shall develop the SSS-2 algorithm, which computes $f(G)$ by determining the solution tree in $\mathcal{M}_G$ with the smallest g-value. First of all, SSS-2 constructs the oldest solution tree in $\mathcal{M}_G$. Then a loop is set up, in which in each iteration the next younger milestone is determined, which is stored in the global variable T.

The algorithm will be built around two procedures, *diminish* and *expand*. Both procedures have an input parameter n, a node in the game tree, another input parameter g_1, a real number, and an output parameter g_2, a real number, which denotes the value of a solution tree. The *expand* procedure has a second output parameter, called S, which denotes a solution tree. For the specification, we assume that a global variable T contains a solution tree of $\mathcal{M}_G$ and this tree is a milestone.

Specification of the procedure $diminish(n, g_1, g_2)$.
The solution tree in the global variable T on call is denoted by T_1 and the solution tree on exit is denoted by T_2.
Pre: $g(T_1(n)) = g_1$ and $T_1(n)$ is the oldest solution tree in $\mathcal{M}(n)$
    with g-value $\leq g_1$.
Post: $g_1 \geq g_2$;
    $g_1 > g_2 \Rightarrow T_2(n)$ is the oldest solution tree in $\mathcal{M}(n)$ with g-value $< g_1$,
        $g(T_2(n)) = g_2$, and $T_1(n) \gg T_2(n)$;
    $g_1 = g_2 \Rightarrow f(n) = g_1 = g_2$ and $T_1(n) \gg T_2(n)$ or $T_1(n) = T_2(n)$.

Specification of the procedure $expand(n, g_1, g_2, S)$
Post: $g_1 > g_2 \Rightarrow g(S) = g_2$ and
        S is the oldest solution tree in $\mathcal{M}(n)$ with g-value $< g_1$,
    $g_1 \leq g_2 \Rightarrow f(n) \geq g_2 \geq g_1$ (and S is undefined).

The code of *diminish* and *expand* can be found in Figure 5 and Figure 6 respectively. The correctness proof for both procedures can be given by recursion induction. The code of the main program can be found in Figure 4. In the main program of SSS-2, the call $expand(r, +\infty, g_2, S)$ generates the oldest solution tree in $\mathcal{M}_G$. In each iteration of the main loop, the *diminish* call constructs the next younger solution tree with a smaller g-value, if any. If this construction fails, then by the postcondition of *diminish* $f(r) = g_1$

If $L(n) = -\infty$ for all non-terminals n, then the code of both *diminish* and *expand* can be simplified. In that case the solution tree in T can be implemented, using only the tip nodes of T, as shown in [Pij2]. A description is obtained which manipulates pairs. This description is close to that of SSS* which manipulates triples.

5 The nodes, visited during SSS-2

Now we give the theorems, expressing the necessary and sufficient condition respectively for nodes to be visited by the SSS-2 algorithm. The first two theorems in this

```
procedure diminish(n, g₁, g₂);
if L(n) ≥ g₁   then [ g₂:=g₁;
                           exit procedure; ]
if n is a tip node then [ expand(n, g₁, g₂', S);
                           if g₁> g₂' then [ attach S to n in T;
                                            g₂:=g₂'; ]
                           else g₂:=g₁;
                           exit procedure; ]
if type(n)=max then
   [ for c:= firstchild(n) to lastchild(n) do
       [ if g₁=g(T(c)) then diminish(c, g₁, g₂');
         if g₁=g₂' then exit for loop; ]
     g₂:= the maximum of g-values of all children;
   ]
if type(n)=min then
   [ c := the single child of n in T;
     diminish (c, g₁, g₂);
     if g₁= g₂then
         for b:=nextbrother(c) to lastbrother(c) do
           [ expand(b, g₁, g₂', S);
             if g₁> g₂' then
                 [ detach in T from n the subtree rooted in c
                         and attach S to n in T;
                   g₂:= g₂';
                   exit for loop; ]
           ]
   ]
```

Figure 5: The procedure diminish

section are proved by path induction.

Theorem 4 *Suppose a call* $\text{expand}(r', g_1, g_2, S)$ *is applied to a game tree* (G', H) *with root* r'. *A descendant node* n *of a node* r' *is a parameter in a nested expand call anywhere in the recursion if and only if* $\min(\beta(n), \hat{U}(n)) \geq g_1 > \max(\alpha(n), \hat{L}(n))$.

Theorem 5 *A node* n *is parameter in a call* $\text{diminish}(n, g_1, g_2)$ *with* $S = T(n)$ *during SSS-2 if and only if, there exists a milestone* $S \in \mathcal{M}(c)$ *with* $g(S) = g_1$ *and,* $\min(\beta(n), \hat{U}(n)) > g_1 > \max(\alpha(n), \hat{L}(n))$ *and* $g_1 \geq \hat{f}(n)$.

Definition 4 *Let* n *be a node in a game tree, which is not the root. Let* m *be a node in* $ANC(n)$ *such that* $U(m) = \min(\beta(n), \hat{U}(n))$ *or such that for a child* m' *(which is in* $AMIN\text{-}LC(n)$*)* $f(m') = \min(\beta(n), \hat{U}(n))$. *The node closest to the root with this property is called the* β-*ancestor of* n.

```
procedure expand(n, g1, g2, S);
if U(n) < g1  or L(n) ≥ g1   then
     [ if U(n) < g1
          then S := the tree consisting only of node n;
          if U(n) < g1 then g2:= U(n) else g2:=L(n);
          exit procedure; ]
if type(n)=max then
   [ for c := firstchild(n) to lastchild(n) do
        [ expand (c, g1, g2', S');
          if g2'≥ g1  then
             [ g2:= g2';
                exit for loop; ]
        ]
     g2:= max of all intermediate values of g2';
     if g2< g1  then S := the tree composed by attaching
                          all intermediate values of S' to n;
   ]
if type(n)=min then
   [ g2:=g1;
     for c := firstchild(n) to lastchild(n) do
        [ expand (c, g1, g2', S');
          if g2'< g1  then
             [ S := tree with S' attached to n;
               g2:= g2';
               exit for loop; ]
        ]
   ]
```

Figure 6: The procedure expand

Theorem 6 *A node n is parameter in a call expand(n, g_1, g_2, S) during SSS-2 with n not a tip node in T, if and only if $\min(\beta(n), \hat{U}(n)) = g_1 > \max(\alpha(n), \hat{L}(n))$ and $g_1 \geq \hat{f}(m)$, $m = \beta$-ancestor(n)*

In the proof of *only-if* part, m is defined as the node parameter of the *diminish* call, in which a transition to an *expand* call takes place. It is proved that m is the β-ancestor of every node which is visited by this *expand* call. In the proof of *if* part, we show that m is parameter in a *diminish* call, in which a transition to an *expand* call takes place.

Theorem 7 *Let S_1 denote the set of nodes visited by the global alpha-beta algorithm applied to a game tree (G_1, H_1). Let S_2 denote the set of nodes visited by the SSS-2 algorithm applied to a game tree (G_2, H_2). Then $S_2 \subseteq S_1$, if for every node n, $U_2(n) \leq U_1(n)$ and $L_2(n) \geq L_1(n)$ or otherwise $f(n) > L_1(n)$.*

This theorem follows from Theorems 2 and 6. It follows from Theorem 2 that alpha-beta visits less nodes as the heuristic pair becomes tighter. However, this property does not hold for SSS-2, as is shown by a counterexample in [Pij2].

References

[Ba]   G. M. Baudet, *On the branching factor of the alpha-beta pruning algorithm.* Artificial Intelligence 10 (1978), pp 173-199.

[Ib]   T. Ibaraki, *Generalization of Alpha-Beta and SSS* Search Problems,* Artificial Intelligence 29 (1986), pp 73-117.

[Kn]   D.E.Knuth and R.W.Moore, *An Analysis of Alpha-Beta Pruning,* Artificial Intelligence 6 (1975), pp 293-326.

[Prl]   I.Roizen and J. Pearl, *A Minimax Algorithm Better than Alpha-Beta? Yes and No,* Artificial Intelligence 21 (1983), pp 199-220.

[Pij1] W. Pijls and A. de Bruin, *Another view on the SSS* algorithm,* in: Algorithms, Proceedings International Symposium SIGAL '90, Tokyo, August 1990.

[Pij2] W. Pijls and A. de Bruin, *Searching informed game trees,* Report EUR-CS-92-02, Department Computer Science, Erasmus University Rotterdam.

[St]   G.C. Stockman, *A Minimax Algorithm Better than Alpha-Beta?,* Artificial Intelligence 12 (1979), pp 179-196.

How to Generate Realistic Sample Problems for Network Optimization

Masao IRI

Department of Mathematical Engineering and Information Physics
Faculty of Engineering, University of Tokyo
Bunkyo-ku, Tokyo 113, Japan

Abstract. Experimental tests are probably the most direct way of evaluating the performance of computer algorithms and of appealing to their users. However, there seems to be no apparent agreement regarding how to generate sample problems for the test. Commonly used random generation, for example, has several problems since that solely produces those instances which follow truly even distribution and then which do not look like "real" samples we actually encounter in the world of actuality. In this paper, we discuss what kind of properties are significant for proper generation of sample problems from a practical-mathematical point of view. Our attention is focussed mainly on the network optimization problems.

1 Introduction

This is not a paper to report a new result on the subject of the title nor to present a novel observation on it, but long-discussed difficult issues on the subject will be repeated again and the ad-hoc measures which the author's group has been taking for these more than ten years will be shown as raw materials in the hope that somebody else might develop, based on those raw materials, a methodology with which to resolve the issues.

The advent of the concept of computational complexity was epoch-making. It gave us a measure with which to evaluate the quality of an algorithm. But, through the twenty-year experience, we have now come to the recognition that the worst-case order-of-magnitude argument on the complexity of an algorithm does not often reflect the practical performance of the algorithm, although it is practically useful enough for some types of algorithms. The most notorious may be that on the so-called "fast matrix multiplication" (see, e.g., [10]). In fact, it is one of the best examples for demonstrating how insufficient the mere order-of-magnitude argument is to prove the efficiency of an algorithm.

There are problems of the kind for which the worst case rarely occurs in practical situations. A typical case is the simplex method for linear programming. In spite of the famous Klee-Minty examples for which the ordinary simplex methods will take exponential time, practical people still rely on the simplex method because they know that it solves many real-world linear programming problems in a reasonable time and because, for other new-born polynomial-time algorithms such as interior-point methods (not to mention Khachiyan's ellipsoid method), the theoretical complexity analysis has practically nothing to do with their actual behaviour.

If one is not contented with the worst-case argument, one may naturally tend to study the average case. However, we are aware that it is difficult, too, to define the class of problems and to introduce a proper probability distribution in it, proper in the sense that it reflects the occurrences of instances in practical situations. That is probably why the recent main stream in the probabilistic theory of algorithms and complexity is to randomize the algorithm itself instead of introducing a probability distribution in the problem class.

There is another practically important point requiring our due attention. Unless the problems under consideration are purely combinatorial, i.e., non-numerical, we must be careful enough for the numerical stability of an algorithm, i.e., for the influence of rounding errors in numerical computation on the performance of the algorithm. Sometimes, algorithms for linear computation or polynomial evaluation having smaller complexity are known to be less stable numerically. What is more serious is that, for geometrical problems such as the Voronoi diagram construction, rounding errors often give rise to topological inconsistency in current data structures so that some algorithms may crash due to the inconsistency.

Geometrical and/or graphical algorithms will afford us the most appropriate examples to investigate from the above-mentioned points of view. It seems to the author that, in the years when J. E. Hopcroft and R. E. Tarjan first began to cultivate the new field of modern graphical algorithms, all of us were happy because newly invented algorithms had at the same time a nice theoretical complexity bound and a nice practical performance. However, as progress goes on in the field and "theoretical" improvements are done for various algorithms, we have come to see from time to time some theoretically better algorithms not better behaved practically. Then, we are forced to face the difficult problem of mathematically clarifying what "practical situations" really mean.

2 Desirable Properties of Sample Problems for Network Algorithms

When we speak about the generation of sample problems, we must bear in mind at least the following properties to be satisfied by the generation process.

(a) There should be as many sample problems as possible to be generated.

(b) The size as well as other parameters to characterize the problem instances should be freely controllable.

(c) The performance of an algorithm for the sample problems should reflect well that for the "real-world problems".

Obviously, the last property (c) is the most difficult to mathematically define and to practically realize.

For graphical problems, one may naturally remind oneself of the "random graphs" initially studied by P. Erdös and A. Rényi in their epoch-making papers [2, 3, 4] and subsequently by a number of people from the theoretical standpoint as well as from the standpoint of application. (Chapter 8 of H. Frank and I. T. Frisch's book [5] contains an excellent survey of the definitions and the properties of many different kinds of random graphs.) And one may expect that those random graphs might happen to be of use as sample problems.

Although the results expounded in those literatures are highy exquisite and extremely interesting theoretically, one would quickly find those random graphs looking far different from road maps, circuit diagrams, control-system diagrams, programme charts, etc. which we encounter in our daily life. No alternative proposal seems to have been done and examined successfully.

3 Empirical Examples

Under such circumstances as we stated in section 2 we have to give up considering to generate sample problems purely mathematically from scratch but to resort to a more empirical approach. Namely, we shall try to collect a large number of real graphs or networks which we regard as being typical in our world and to modify them in various plausible ways to multiply the number sufficiently. This sort of approach is more or less similar to the so-called "benchmark test", but the former is different from the latter in intention to generate expectedly infinitely many problems with well-controlled parameters.

In the following some of the examples will be shown chosen from among those which the author's research group has dealt with.

3.1 Voronoi diagrams in the plane

In order to test the efficiency of an algorithm for constructing the Voronoi diagrams, especially those in the plane, the randomly distributed points (generators) seem to be good enough. With those sample problems we were able to test and compare the existing algorithms and to improve them [9]. The problem size, which is the number of generators, is easily controllable. Effects of rounding errors are visible in those sample problems. Fig. 1 is an example of such random Voronoi diagrams by means of which we succeeded in developing a numerically very robust algorithm [11].

Generating sample problems with generators subject to a nonuniform distribution is easy, and so is controlling the relevant parameters of the distribution, as is illustrated in Fig. 2.

3.2 Point-location problem in computational geometry

Geographically meaningful partitions of the surface of the globe may make good sample problems, such as the borders of nations, states and administrative sectors and equi-altitude contour lines, for this kind of problems (Fig. 3). Random Voronoi diagrams are of use to some extent, too. However, those partitions seem to be too well behaved for the purpose of evaluating the practical efficiency of algorithms, so that we had to use artificially made more sinister-looking figures as well as those which are unfavorable to particular algoritms [1].

3.3 Drawing a map or a diagram

The problem of optimizing the pen movement of a mechanical plotter to draw a complicated figure on a paper can be reduced to the problem of minimum-weight Euclidean matching for points in the plane followed by the Euler-path problem on a graph, and that of optimizing the movement of a drill to bore many holes on an iron plate, to the travelling-salesman problem for points in the plane. Since the number of points is very large in practical applications, we must be contented with an approximate algorithm if its time-complexity is linear. A quadratic, cubic or even $n \log n$ algorithm is too much time-consuming [8]. Randomly distributed point sets in the plane seem to be good samples for the matching and travelling-salesman problems. However, for the Euler-path problem we need to use "realistic" sample graphs. For this purpose, a real road network was useful (see Fig. 4). VLSI mask patters were used, too. Those diagrams were used not only in their original forms but also with some modification or surgery, where "surgery" means in this case random deletion of existing edges and addition of fictitious edges, the deleted and added edges being relatively small in number for fear that the resulting graphs might deviate from the realistic graphs too far. By cutting out part of the original graph we may increase the number of sample problems and controlling the problem size.

3.4 Maximum flows and shortest paths

The maximum-flow problem and the shortest-path problem are the most fundamental problems in network flow theory in the sense that they are not only of importance themselves in practical applications but also many network-flow type problems utilize them as subproblems.

Algorithms for them are still being "improved" a little by little. It was the principal aim of the papers [6] and [7] to investigate how far the improvement in theoretical complexity goes hand in hand with that in practical efficiency. For that kind of investigation good "realistic" sample problems are indispensable. We made use of real road networks of Fig. 4 and of Figs. 5 and 6 either with or without surgery for sample topologies. For capacities and lengths (costs) of edges, we employed real data of road widths and lengths and their perturbations with the degree of perturbation controlled appropriately. By choosing the origin and destination of the two-terminal problem at random from the vertex set of a sample network we could have many problems from a single network. Needless to say, graphs of more regular structures such as rectangular grid networks with uniform or perturbed capacities/lengths were also employed.

It was interesting to observe that, in the sequence of maximum-flow algorithms in the decreasing theoretical complexity, that by Dinic or Karzanov displayed the best practical performance [6] and that, for the shortest-path problems, rather primitive bucket algorithms behaved themselves quite satisfactorily [7].

3.5 Continuum approximation of a dense network

Finally let us append a rather strange approach to network-flow type problems, i.e., the approximation of discrete flows on a dense network by flow fields on a nonuniform anisotropic continuum (not the approximation of a continuum by a discretum as in finite-element method) [12]. In this case, too, good sample problems were indispensable. We made the continuum approximations of the networks of Figs. 6 and 7 in addition to that of Fig. 5, and analyzed the flow problems on continua and on discrete-continuous mixed systems by attaching "express ways" to those continua.

References

1. T. Asano, M. Edahiro, H. Imai, M. Iri and K. Murota: Practical use of bucketing techniques in computational geometry. In: G. T. Toussaint (ed.): *Computational Geometry*, Elsevier (North Holland), pp.153–195 (1985)

2. P. Erdös and A. Rényi: On the evolution of random graphs. *Publications of the Mathematical Institute of the Hungarian Academy of Sciences*, Vol.5A, pp.17–61 (1960)

3. P. Erdös and A. Rényi: On random graphs I. *Publicationes Mathematicae (Debrecen)*, Vol.6, pp.290–297 (1959)

4. P. Erdös and A. Rényi: On the strength of connectedness of a random graph. *Acta Mathematica*, Vol.12, pp.261–267 (1961)

5. H. Frank and I. T. Frisch: *Communication, Transmission, and Transportation Networks*. Addison-Wesley, Reading, Massachusetts (1971)

6. H. Imai: On the practical efficiency of various maximum flow algorithms. *Journal of the Operations Research Society of Japan*, Vol.26, pp.61–83 (1983)

7. H. Imai and M. Iri: Practical efficiencies of existing shortest-path algorithms and a new bucket algorithm. *Journal of the Operations Research Society of Japan*, Vol.27, pp.43–58 (1984)

8. M. Iri, K. Murota and S. Matsui: An approximate solution for the problem of optimizing the plotter pen movement. In: R. F. Drenick and F. Kozin (ed.): *System Modeling and Optimization* (Proceedings of the 10th IFIP Conference on System Modeling and Optimization, New York, 1981), Lecture Notes in Control and Information Science 38, Springer-Verlag, Berlin, pp.572–580 (1982)

9. T. Ohya, M. Iri and K. Murota: Improvements of the incremental method for the Voronoi diagram with computational comparison of various algorithms. *Journal of the Operations Research Society of Japan*, Vol.27, pp.306–337 (1984)

10. V. Pan: *How to Multiply Matrices Fast*. Lecture Notes in Computer Science 179, Springer-Verlag, Berlin (1984)

11. K. Sugihara and M. Iri: Construction of the Voronoi diagram for "one million" generators in single-precision arithmetic. To appear in: G. T. Toussaint (ed.): *Proceedings of IEEE* — Special Issue on Computational Geometry (1992)

12. A. Taguchi and M. Iri: Continuum approximation to dense networks and its application to the analysis of urban networks. *Mathematical Programming Study*, Vol.20, pp.178–217 (1982)

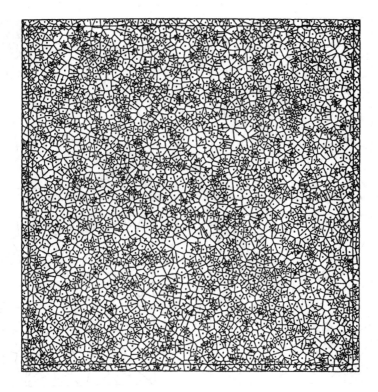

Fig. 1. A Voronoi diagram for uniformly randomly distributed points
— 1/256 part of a one-million-point diagram ([11], Fig. 10(c))

348

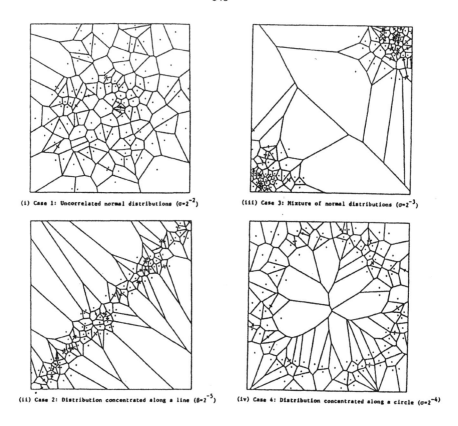

(i) Case 1: Uncorrelated normal distributions ($\sigma=2^{-2}$)

(iii) Case 3: Mixture of normal distributions ($\sigma=2^{-3}$)

(ii) Case 2: Distribution concentrated along a line ($\beta=2^{-5}$)

(iv) Case 4: Distribution concentrated along a circle ($\sigma=2^{-4}$)

Fig. 2. Voronoi diagrams for nonuniformly distributed points ([9], Fig. 11)

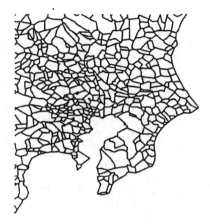

Fig. 3. Map of the administrative sectors of Kanto district ([1], Fig. 4.5)

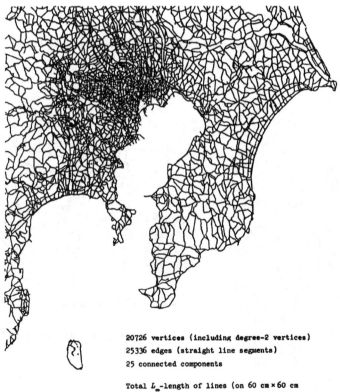

20726 vertices (including degree-2 vertices)
25336 edges (straight line segments)
25 connected components

Total L_∞-length of lines (on 60 cm × 60 cm
sheet): 44.47 m

Fig. 4. Road network of Kanto district of Japan ([8], Fig. 1)

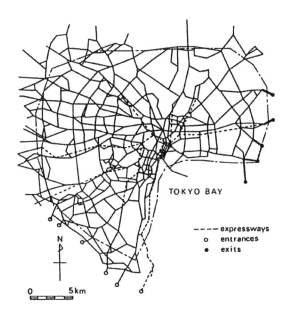

Fig. 5. Road network of Tokyo metropolitan area ([12], Fig. 9)

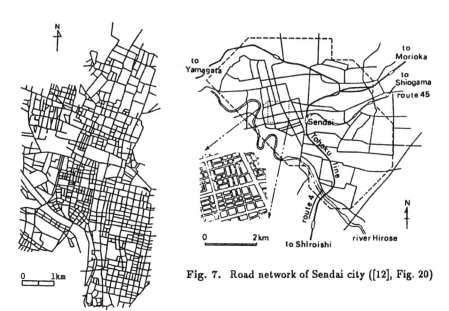

Fig. 7. Road network of Sendai city ([12], Fig. 20)

Fig. 6. Road network of Kofu city
([7], Fig. 5.4)

Generalized Assignment Problems

Silvano Martello
Dipartimento di Informatica, Università di Torino, Italy

Paolo Toth
DEIS, Università di Bologna, Italy

Abstract. We consider generalized assignment problems with different objective functions: min-sum, max-sum, min-max, max-min. We review transformations, bounds, approximation algorithms and exact algorithms. The results of extensive computational experiments are given.

1 Introduction

Given n *items* and m *units*, with

$p_{ij} = penalty$ of item j if assigned to unit i,
$r_{ij} = resource\ requirement$ of item j if assigned to unit i,
$a_i = resource\ availability$ of unit i,

the generalized assignment problems consist in assigning each item to exactly one unit so that the total amount of resource required at any unit does not exceed its availability, and a given penalty function is minimized. We will consider the most natural cases of objective function and describe the main results obtained on relaxations, approximation algorithms and exact algorithms. All the cases considered are NP-hard in the strong sense, since even the feasibility question is so (see, e.g., Martello & Toth (1990), Ch.7).

1.1 Min Sum and Max Sum Generalized Assignment

In the *Min Sum Generalized Assignment Problem* (MINGAP) the objective function is the total penalty corresponding to the assignment. Formally,

$$\text{minimize} \sum_{i=1}^{m} \sum_{j=1}^{n} p_{ij} x_{ij} \tag{1}$$

$$\text{subject to} \sum_{j=1}^{n} r_{ij} x_{ij} \le a_i, \qquad i \in M = \{1,\ldots,m\}, \tag{2}$$

$$\sum_{i=1}^{m} x_{ij} = 1, \qquad j \in N = \{1,\ldots,n\}, \tag{3}$$

$$x_{ij} = 0 \text{ or } 1, \qquad i \in M, j \in N, \tag{4}$$

where

$$x_{ij} = \begin{cases} 1 & \text{if item } j \text{ is assigned to unit } i; \\ 0 & \text{otherwise.} \end{cases}$$

The maximization version of the problem (GAP) is encountered frequently in the literature: by defining v_{ij} as the *value* obtained by assigning item j to unit i, GAP is

$$\text{maximize } z = \sum_{i=1}^{m} \sum_{j=1}^{n} v_{ij} x_{ij} \tag{5}$$

$$\text{subject to} \quad (2), (3), (4).$$

GAP and MINGAP are equivalent. Setting $p_{ij} = -v_{ij}$ (or $v_{ij} = -p_{ij}$) for all $i \in M$ and $j \in N$ immediately transforms one version into the other. If the numerical data are restricted to positive integers (as frequently occurs), the transformation can be obtained as follows. Given an instance of MINGAP, define any integer value t such that

$$t > \max_{i \in M, j \in N} \{p_{ij}\} \tag{6}$$

and set

$$v_{ij} = t - p_{ij} \qquad \text{for } i \in M, \ j \in N. \tag{7}$$

From (1) the solution value of MINGAP is then

$$t \sum_{j=1}^{n} \sum_{i=1}^{m} x_{ij} - \sum_{i=1}^{m} \sum_{j=1}^{n} v_{ij} x_{ij},$$

where, from (3), the first term is independent of (x_{ij}). Hence the solution (x_{ij}) of GAP also solves MINGAP. The same method transforms any instance of GAP into an equivalent instance of MINGAP (by setting $p_{ij} = \hat{t} - v_{ij}$ for $i \in M$, $j \in N$, with $\hat{t} > \max_{i \in M, j \in N} \{v_{ij}\}$).

In the following sections we will always consider the maximization version GAP. A more detailed treatment can be found in Martello & Toth(1990), Ch. 7.

The best known special case of generalized assignment problem is the *Linear Min–Sum Assignment Problem* (or *Assignment Problem*), which is a MINGAP with $n = m, a_i = 1$ and $r_{ij} = 1$ for all $i \in M$ and $j \in N$ (so, because of (3), constraints (2) can be replaced by $\sum_{j=1}^{n} x_{ij} = 1$ for $i \in M$). The problem can be solved in $O(n^3)$ time through the classical Hungarian algorithm (Kuhn (1955), Lawler (1976); efficient Fortran codes can be found in Carpaneto, Martello & Toth (1988)). The assignment problem, however, is not used in general as a subproblem in algorithms for the generalized case.

Another special case arises when $r_{ij} = r_j$ for all $i \in M$ and $j \in N$. Implicit enumeration algorithms for this case have been presented by De Maio & Roveda (1971) and Srinivasan & Thompson (1973).

Facets of the GAP polytope have been studied by Gottlieb & Rao (1990a, 1990b).

1.2 Min Max and Max Min Generalized Assignment

In the *Min Max* (or *Bottleneck*) *Generalized Assignment Problem* (BGAP) the objective is to minimise the maximum penalty incurred in the assignment. Formally,

$$\text{minimise} \quad z = \max_{i,j}\{p_{ij}x_{ij}\} \tag{8}$$

$$\text{subject to} \quad (2), (3), (4).$$

We note that the max-min version of the problem,

$$\text{maximize} \quad v = \min_{i,j}\{v_{ij} : x_{ij} = 1\} \tag{9}$$

$$\text{subject to} \quad (2), (3), (4),$$

can easily be transformed into BGAP. Indeed, (9) can be replaced by

$$\text{maximize} \quad v$$

$$\text{subject to} \quad v \leq v_{ij} + K(1 - x_{ij}), \qquad i \in M, \ j \in N,$$

where K is a large positive number. Given any value $t > \max_{i,j}\{v_{ij}\}$, by defining $p_{ij} = t - v_{ij}$, we obtain by substitution

$$\text{minimize} \quad (t - v)$$

$$\text{subject to} \quad (t - v) \geq p_{ij} - K(1 - x_{ij}),$$

that is, objective function (8) with $z = t - v$.

Mathematical models and a technique for transforming BGAP into GAP have been given by Mazzola & Neebe(1988). The results presented in this survey derive from a paper by Martello & Toth (1991).

1.3 Assumptions

We will suppose, as is usual, that the resource requirements r_{ij} are positive integers. Hence, without loss of generality, we will also assume that

$$p_{ij}, v_{ij} \text{ and } a_i \text{ are positive integers,} \tag{10}$$

$$|\{i : r_{ij} \leq a_i\}| \geq 1 \text{ for } j \in N, \tag{11}$$

$$a_i \geq \min_{j \in N}\{r_{ij}\} \qquad \text{for } i \in M. \tag{12}$$

If assumption (10) is violated, (a) fractions can be handled by multiplying through by a proper factor; (b) units with $a_i \leq 0$ can be eliminated; (c) for each item j of a GAP instance having $\min_{i \in M}\{v_{ij}\} \leq 0$, we can set $v_{ij} = v_{ij} + |\min_{i \in M}\{v_{ij}\}| + 1$ for $i \in M$ and subtract $|\min_{i \in M}\{v_{ij}\}| + 1$ from the resulting objective function value. (The situation can be handled in a similar way for an instance of BGAP.) Instead, there is no easy way of transforming an instance so as to handle negative resource requirements, but all our considerations easily extend to this case too. If an item violates assumption (11) then it cannot be assigned, so the instance is infeasible. Units violating assumption (12) can be eliminated from the instance.

2 Max Sum Generalized Assignment Problem

2.1 Relaxations and Upper Bounds

The continuous relaxation of GAP, $C(GAP)$, given by (5), (2), (3) and

$$x_{ij} \geq 0, \qquad i \in M, j \in N, \tag{13}$$

is rarely used in the literature since it does not exploit the structure of the problem and tends to give solutions a long way from feasibility.

Relaxation of the Resource Requirements

Ross & Soland (1975) have proposed the following upper bound for GAP. First, constraints (2) are relaxed to

$$r_{ij} x_{ij} \leq a_i, \quad i \in M, j \in N,$$

and the optimal solution $\hat{x}$ to the resulting problem is obtained by determining for each $j \in N$,

$$i(j) = \arg \max \{ v_{ij} : i \in M, r_{ij} \leq a_i \}$$

and setting $\hat{x}_{i(j),j} = 1$ and $\hat{x}_{ij} = 0$ for all $i \in M \setminus \{i(j)\}$. The resulting upper bound, of value

$$U_0 = \sum_{j=1}^{n} v_{i(j),j}, \tag{14}$$

is then improved as follows. Let

$$
\begin{aligned}
N_i &= \{ j \in N : \hat{x}_{ij} = 1 \}, & i \in M, \\
d_i &= \sum_{j \in N_i} r_{ij} - a_i, & i \in M, \\
M' &= \{ i \in M : d_i > 0 \}, \\
N' &= \cup_{i \in M'} N_i.
\end{aligned}
$$

Given a set S of numbers, we denote with $\max_2 S$ (resp. $\min_2 S$) the second maximum (resp. minimum) value in S, and with $\arg \max_2 S$ (resp. $\arg \min_2 S$) the corresponding index. Since M' is the set of those units for which the relaxed constraint (2) is violated,

$$q_j = v_{i(j),j} - \max_2 \{ v_{ij} : i \in M, r_{ij} \leq a_i \}, \quad j \in N'$$

gives the minimum penalty that will be incurred if an item j currently assigned to a unit in M' is reassigned. Hence, for each $i \in M'$, a lower bound on the loss of penalty to be paid in order to satisfy constraint (2) is given by the solution to the 0-1 single knapsack problem in minimization form KP_i^1 $(i \in M')$, defined by

$$
\begin{aligned}
\text{minimize } w_i &= \sum_{j \in N_i} q_j y_{ij} \\
\text{subject to } &\sum_{j \in N_i} r_{ij} y_{ij} \geq d_i, \\
&y_{ij} = 0 \text{ or } 1, \qquad j \in N_i,
\end{aligned}
$$

where $y_{ij} = 1$ if and only if item j is removed from unit i. The resulting Ross & Soland (1975) bound is thus

$$U_1 = U_0 - \sum_{i \in M'} w_i. \tag{15}$$

This bound can also be obtained by dualizing constraints (3) in a Lagrangian fashion.

Relaxation of the Semi-Assignment Constraints

Martello & Toth (1981) obtained an upper bound for GAP by removing constraints (3). It is immediately seen that the resulting relaxed problem decomposes into a series of 0-1 single knapsack problems KP_i^2 $(i \in M)$, of the form

$$\text{maximize } z_i = \sum_{j=1}^{n} v_{ij} x_{ij}$$
$$\text{subject to} \quad \sum_{j=1}^{n} r_{ij} x_{ij} \leq a_i,$$
$$x_{ij} = 0 \text{ or } 1, \quad j \in N.$$

In this case too, the resulting upper bound, of value

$$\overline{U}_0 = \sum_{i=1}^{m} z_i, \tag{16}$$

can be improved by computing a lower bound on the penalty to be paid in order to satisfy the violated constraints. Let

$$N^0 = \{j \in N : \sum_{i \in M} x_{ij} = 0\},$$
$$N^> = \{j \in N : \sum_{i \in M} x_{ij} > 1\}$$

be the sets of those items for which (3) is violated, and define

$$M^>(j) = \{i \in M : x_{ij} = 1\} \quad \text{for all } j \in N^>;$$

we can compute, using any upper bound for the 0-1 knapsack problem (see, e.g., Martello & Toth (1990), Ch. 2),

$$u_{ij}^0 = \text{upper bound on } z_i \text{ if } x_{ij} = 0, \quad j \in N^>, i \in M^>(j),$$
$$u_{ij}^1 = \text{upper bound on } z_i \text{ if } x_{ij} = 1, \quad j \in N^0, i \in M$$

and determine, for each item $j \in N^0 \cup N^>$, a lower bound l_j on the penalty to be paid for satisfying (3):

$$
l_j = \begin{cases}
\min_{i \in M}\{z_i - \min(z_i, u_{ij}^1)\} & \text{if } j \in N^0; \\
\sum_{i \in M^>(j)}(z_i - \min(z_i, u_{ij}^0)) \\
\quad - \max_{i \in M^>(j)}\{z_i - \min(z_i, u_{ij}^0)\} & \text{if } j \in N^>.
\end{cases}
$$

The improved upper bound is thus

$$
U_2 = \overline{U}_0 - \max_{j \in N^0 \cup N^>}\{l_j\}. \tag{17}
$$

The Multiplier Adjustment Method

Fisher, Jaikumar & Van Wassenhove (1986) have developed an upper bound, based on the Lagrangian relaxation $L(GAP, \lambda)$ obtained by dualizing constraints (3):

$$
\text{maximize} \quad \sum_{i=1}^{m}\sum_{j=1}^{n}(v_{ij} - \lambda_j)x_{ij} + \sum_{j=1}^{n}\lambda_j
$$
$$
\text{subject to } (2),(4),
$$

which again decomposes into m independent knapsack problems. Obviously, the continuous and integer solutions of a knapsack problem may differ, so (see Fisher (1981)) for the *optimal* Lagrangian multiplier λ^*,

$$
z(L(GAP, \lambda^*)) \leq z(C(GAP));
$$

there is no analitical way, however, to determine λ^*. One possibility is the classical subgradient optimization approach. The novelty of the Fisher–Jaikumar–Van Wassenhove bound consists of a new technique (*multiplier adjustment method*) for determining "good" multipliers. The method starts by setting $\lambda_j = \max_2 \{v_{ij} : i \in M, r_{ij} \leq a_i\}$, $j \in N$: the corresponding Lagrangian relaxation produces the value U_1 of the Ross - Soland bound. Note, in addition, that, with this choice, we have, for each $j \in N$, $\tilde{v}_{ij}(= v_{ij} - \lambda_j) > 0$ for at most one $i \in M$, so there is an optimal Lagrangian solution for this λ which satisfies $\sum_{i=1}^{m} x_{ij} \leq 1$ for all $j \in N$. If some constraint (3) is not satisfied, it is, under certain conditions, possible to select a j^* for which $\sum_{i=1}^{m} x_{ij^*} = 0$ and decrease λ_{j^*} by an amount which ensures that in the new Lagrangian solution $\sum_{i=1}^{m} x_{ij^*} = 1$, while $\sum_{i=1}^{m} x_{ij} \leq 1$ continues to hold for all other j. This phase is iterated until either the solution becomes feasible or the required conditions fail.

The upper bound value is then determined as

$$
U_3 = \sum_{i=1}^{m} z_i + \sum_{j=1}^{n}\lambda_j \,, \tag{18}
$$

where z_i denotes the solution value of the i-th 0-1 knapsack problem into which $L(GAP, \lambda)$ decomposes.

It is possible to prove that no dominance exists between the Fisher - Jaikumar - Van Wassenhove bound (U_3) and the Martello - Toth bounds $(\overline{U}_0$ and $U_2)$, or between the Ross - Soland $(U_0$ and $U_1)$ and Martello - Toth bounds. The only dominances among these bounds are $U_3 \leq U_1 \leq U_0$ and $U_2 \leq \overline{U}_0$.

Guignard & Rosenwein(1989) have proposed a variation of the multiplier adjustment method, by allowing violation of condition $\sum_{i=1}^{m} x_{ij} \leq 1$ and dualizing into the objective function the surrogate constraint $\sum_{j=1}^{n} \sum_{i=1}^{m} x_{ij} \leq n$ (or $\geq n$).

The Variable Splitting Method

Jörnsten & Näsberg (1986) have introduced a new way of relaxing GAP in a Lagrangian fashion. (A general discussion on this kind of relaxation can be found in Guignard & Kim (1987).) By introducing extra binary variables y_{ij} $(i \in M, j \in N)$ and two positive parameters α and β, the problem is formulated, through *variable splitting*, as

$$\text{maximize } \alpha \sum_{i=1}^{m} \sum_{j=1}^{n} v_{ij} x_{ij} + \beta \sum_{i=1}^{m} \sum_{j=1}^{n} v_{ij} y_{ij} \tag{19}$$

$$\text{subject to } \sum_{j=1}^{n} r_{ij} x_{ij} \leq a_i, \quad i \in M, \tag{20}$$

$$\sum_{i=1}^{m} y_{ij} = 1, \quad j \in N, \tag{21}$$

$$x_{ij} = y_{ij}, \quad i \in M, j \in N, \tag{22}$$

$$x_{ij} = 0 \text{ or } 1, \quad i \in M, j \in N, \tag{23}$$

$$y_{ij} = 0 \text{ or } 1, \quad i \in M, j \in N. \tag{24}$$

We denote problem (19)–(24) by XYGAP. It is immediate that XYGAP is equivalent to GAP in the sense that the corresponding optimal solution values, $z(XYGAP)$ and $z(GAP)$, satisfy

$$z(XYGAP) = (\alpha + \beta) \, z(GAP). \tag{25}$$

The new formulation appears less natural than the original one, but it allows a relaxation of constraints (22) through Lagrangian multipliers (μ_{ij}). The resulting problem, $L(XYGAP, \mu)$,

$$\text{maximize } \alpha \sum_{i=1}^{m} \sum_{j=1}^{n} v_{ij} x_{ij} + \beta \sum_{i=1}^{m} \sum_{j=1}^{n} v_{ij} y_{ij}$$

$$+ \sum_{i=1}^{m} \sum_{j=1}^{n} \mu_{ij} (x_{ij} - y_{ij}) \tag{26}$$

subject to (20), (21), (23), (24),

keeps both sets of GAP constraints, and immediately separates into two problems, one, $XGAP(\mu)$, in the x variables and one, $YGAP(\mu)$, in the y variables. The former,

$$\text{maximize } z(XGAP(\mu)) = \sum_{i=1}^{m} \sum_{j=1}^{n} (\alpha v_{ij} + \mu_{ij}) x_{ij}$$

$$\text{subject to } (20), (23),$$

has the same structure as $L(GAP, \lambda)$, hence separates into m 0-1 single knapsack problems, whose solution values will be denoted by $\hat{z}_i$, $i = 1, \ldots, m$; the latter

$$\text{maximize } z(YGAP(\mu)) = \sum_{i=1}^{m} \sum_{j=1}^{n} (\beta v_{ij} - \mu_{ij}) y_{ij}$$

$$\text{subject to } (21), (24),$$

has the same structure as the initial Ross - Soland relaxation, hence its optimal solution is

$$y_{ij} = \begin{cases} 1 & \text{if } i = i(j) \text{ ;} \\ & \quad\quad\quad\quad \text{for } j \in N, \\ 0 & \text{otherwise ,} \end{cases}$$

where

$$i(j) = \arg \max \{ \beta v_{ij} - \mu_{ij} : i \in M, \ r_{ij} \leq a_i \}.$$

The solution to $L(XYGAP, \mu)$ then has the value

$$z(L(XYGAP, \mu)) = \sum_{i=1}^{m} \hat{z}_i + \sum_{j=1}^{n} (\beta v_{i(j),j} - \mu_{i(j),j}), \tag{27}$$

and the upper bound is

$$U_4 = \lfloor z(L(XYGAP, \mu))/(\alpha + \beta) \rfloor.$$

For $\alpha + \beta = 1$ and for the *optimal* Lagrangian multipliers λ^*, μ^*,

$$z(L(XYGAP, \mu^*)) \leq z(L(GAP, \lambda^*)).$$

Good multipliers μ^* are obtained through subgradient optimization.

2.2 Exact Algorithms

The most commonly used methods in the literature for the exact solution of GAP are depth-first branch-and-bound algorithms.

In the Ross & Soland (1975) scheme, upper bound U_1 is computed at each node of the branch-decision tree. The branching variable is selected through the information determined for computing U_1. In fact, the variable chosen to separate, $x_{i^*j^*}$, is the one, among those with $y_{ij} = 0$ $(i \in M', \ j \in N')$ in the optimal solution to problems KP_i^1 $(i \in M')$, for which the quantity

$$\frac{q_j}{r_{ij}/(a_i - \sum_{k=1}^{n} r_{ik} x_{ik})}$$

is a maximum. This variable represents an item j^* which is "well fit" into unit i^*, considering both the penalty for re-assigning the item and the residual resource availability of the unit. Two branches are then generated by imposing $x_{i^*j^*} = 1$ and $x_{i^*j^*} = 0$.

In the Martello & Toth (1981) scheme, upper bound $\min(U_1, U_2)$ is computed at each node of the branch-decision tree. In addition, at the root node, a tighter upper bound on the global solution is determined by computing $\min(U_3, U_2)$. The information computed for U_2 determines the branching as follows. The separation is performed on item

$$j^* = \arg \max \{l_j : j \in N^0 \cup N^>\},$$

i.e. on the item whose re-assignment is likely to produce the maximum decrease of the objective function. If $j^* \in N^0$, m nodes are generated by assigning j^* to each unit in turn; if $j^* \in N^>$, with $M^>(j^*) = \{i_1, i_2, \ldots, i_{\overline{m}}\}$, $\overline{m} - 1$ nodes are generated by assigning j^* to units $i_1, \ldots, i_{\overline{m}-1}$ in turn, and another node by excluding j^* from units $i_1, \ldots, i_{\overline{m}-1}$. With this branching strategy, m single knapsack problems KP_i^2 must be solved to compute the upper bound associated with the root node, but only one new KP_i^2 for each other node of the tree. In fact if $j^* \in N^0$, imposing $x_{kj^*} = 1$ requires only the solution of problem KP_k^2, the solutions to problems KP_i^2 $(i \neq k)$ being unchanged with respect to the generating node; if $j^* \in N^>$, the strategy is the same as that used in the Martello & Toth (1980) algorithm for the 0-1 multiple knapsack problem for which the solution of $\overline{m}$ problems KP_i^2 produces the upper bounds corresponding to the $\overline{m}$ generated nodes.

The execution of the above scheme is preceded by a preprocessing which: (a) determines an approximate solution through a procedure, MTHG, outlined in the next section; (b) reduces the size of the instance, through reduction procedures. At each decision node, a partial reduction is performed by searching for unassigned items which can currently be assigned to only one unit. The listing of the Fortran implementation of the resulting algorithm (MTG) is included in the diskette accompanying Martello & Toth (1990).

In the Fisher, Jakumar & Van Wassenhove (1986) scheme, upper bound U_3 is computed at each node of the branch-decision tree. The branching variable is an $x_{i^*j^*}$ corresponding to an $r_{i^*j^*}$ which is maximum over all variables that have not been fixed to 0 or 1 at previous branches. Two nodes are then generated by fixing $x_{i^*j^*} = 1$ and $x_{i^*j^*} = 0$.

2.3 Approximation Algorithms

As previously mentioned, determining whether an instance of GAP (or MINGAP) has a feasible solution is NP-complete. It follows that, unless $\mathcal{P} = \mathcal{NP}$, these problems admit no polynomial-time approximation algorithm with fixed worst-case performance ratio, hence also no polynomial-time approximation scheme.

A polynomial-time approximation algorithm, MTHG, was presented by Martello & Toth(1981). Let f_{ij} be a measure of the "desirability" of assigning item j to unit i. The algorithm iteratively considers all the unassigned items, and determines the item j^* which has the maximum difference between the largest and the second largest f_{ij} ($i \in M$); j^* is then assigned to the unit for which f_{ij^*} is a maximum. In the second part of the algorithm the current solution is improved through local exchanges. The algorithm can be implemented efficiently by initially sorting, in decreasing order, for each item j, the values f_{ij} ($i \in M$) such that $r_{ij} \leq a_i$. This requires $O(nm \log m)$ time, and makes immediately available, at each iteration, the pointers to the maximum and the second maximum f_{ij}. The overall time complexity of MTHG is $O(nmlogm + n^2)$. Computational experiments have shown that good results can be obtained using the following choices for f_{ij}:

(1) $f_{ij} = v_{ij}$;

(2) $f_{ij} = v_{ij}/r_{ij}$;

(3) $f_{ij} = -r_{ij}$;

(4) $f_{ij} = -r_{ij}/a_i$.

The listing of a Fortran implementation of MTHG, which determines the best solution obtainable with choices (1) - (4) for f_{ij}, is included in the diskette accompanying Martello & Toth (1990). A more complex approximation algorithm, involving a modified subgradient optimization approach and branch-and-bound, can be found in Klastorin (1979).

Mazzola (1989) has derived from MTHG an approximation algorithm for the generalization of GAP that arises when constraints (2) are nonlinear.

2.4 Computational Experiments

Tables 1 to 4 compare the exact algorithms of Section 2.2 on four classes of randomly generated problems. For the sake of uniformity with the literature (Ross & Soland (1975), Martello & Toth (1981), Fisher, Jaikumar & Van Wassenhove (1986)), all generated instances are *minimization* problems of the form (1), (2), (3), (4). All the algorithms we consider except the Ross & Soland (1975) one, solve maximization problems, so the generated instances are transformed through (7), using for t the value

$$t = \max_{i \in M, j \in N} \{p_{ij}\} + 1.$$

The classes are

(a) r_{ij} uniformly random in $[5, 25]$, $i \in M$, $j \in N$,
   p_{ij} uniformly random in $[1, 40]$, $i \in M$, $j \in N$,
   $a_i = 9(n/m) + 0.4 \max_{i \in M}\{\sum_{j \in N_i} r_{ij}\}$, $i \in M$
     (where N_i is defined as in Section 2.1);

(b) r_{ij} and p_{ij} as for class (a),
   $a_i = 0.7(9(n/m) + 0.4 \max_{i \in M}\{\sum_{j \in N_i} r_{ij}\})$, $i \in M$;

(c) r_{ij} and p_{ij} as for class (a),
   $a_i = 0.8 \sum_{j=1}^{n} r_{ij}/m$, $i \in M$;

(d) r_{ij} uniformly random in $[1, 100]$, $i \in M$, $j \in N$,
   p_{ij} uniformly random in $[r_{ij}, r_{ij} + 20]$, $i \in M$, $j \in N$,
   $a_i = 0.8 \sum_{j=1}^{n} r_{ij}/m$, $i \in M$.

Table 1. GAP. Class (a). HP 9000/840 seconds.
Average times / Average numbers of nodes over 10 problems.

m	n	RS		MTG		FJV		MTGFJV		MTGJN	
		time	nodes	time	nodes	time	nodes	time	nodes	time	nodes
2	10	0.003	3	0.014	1	0.010	2	0.022	1	0.054	1
	20	0.020	10	0.038	2	1.792	113	0.042	2	0.152	2
	30	0.018	5	0.060	2	10.089(9)	54	0.086	2	0.283	2
3	10	0.006	2	0.016	1	0.005	1	0.018	1	0.019	1
	20	0.035	18	0.124	16	5.927	293	0.100	5	1.004	15
	30	0.045	15	0.075	1	0.124	6	0.071	1	0.069	1
5	10	0.017	10	0.067	7	0.056	3	0.074	3	0.387	7
	20	0.057	19	0.258	41	6.799	320	0.190	7	1.837	41
	30	0.044	10	0.224	12	0.117	6	0.150	3	1.108	12

Table 2. GAP. Class (b). HP 9000/840 seconds.
Average times / Average numbers of nodes over 10 problems.

m	n	RS		MTG		FJV		MTGFJV		MTGJN	
		time	nodes	time	nodes	time	nodes	time	nodes	time	nodes
2	10	0.089	93	0.123	32	0.416	15	0.312	14	0.793	26
	20	13.906	9907	14.827	4012	79.951(5)	714	49.628(8)	630	69.573(6)	2231
	30	time limit		time limit		time limit		time limit		time limit	
3	10	0.157	165	0.228	54	0.925	69	0.539	31	1.312	41
	20	48.804(6)	32264	32.487(7)	5982	time limit		42.530(7)	489	39.231(7)	853
	30	93.695(1)	56292	63.842(6)	6824	time limit		90.755(2)	1003	82.991(3)	1108
5	10	0.314	259	0.454	94	2.094	110	0.565	33	2.739	74
	20	17.036(9)	9194	19.536	2311	92.831(3)	2043	24.755	551	75.723(3)	944
	30	74.580(3)	33854	68.075(4)	5950	81.932(2)	438	80.615(2)	925	80.587(2)	693

Table 3. GAP. Class (c). HP 9000/840 seconds.
Average times / Average numbers of nodes over 10 problems.

m	n	RS		MTG		FJV		MTGFJV		MTGJN	
		time	nodes	time	nodes	time	nodes	time	nodes	time	nodes
2	10	0.110	121	0.176	50	0.575	22	0.507	25	1.156	44
	20	17.608	13637	18.717	4788	96.525(4)	1242	48.471(7)	760	44.267(7)	1614
	30	98.142(1)	68620	83.363(2)	13362	time limit		96.024(1)	853	86.580(2)	1985
3	10	0.245	240	0.243	43	1.648	92	0.576	28	0.959	19
	20	47.476(6)	32727	18.215(9)	4106	time limit		33.769(8)	534	31.446(8)	826
	30	92.670(1)	52612	47.820(7)	4924	time limit		91.459(2)	951	67.374(5)	830
5	10	0.474	393	0.794	226	4.028	195	1.708	70	6.881	216
	20	61.614(6)	31328	15.698	1332	time limit		47.882	514	59.436(7)	524
	30	time limit		time limit		time limit		time limit		time limit	

Table 4. GAP. Class (d). HP 9000/840 seconds.
Average times / Average numbers of nodes over 10 problems.

m	n	RS		MTG		FJV		MTGFJV		MTGJN	
		time	nodes	time	nodes	time	nodes	time	nodes	time	nodes
2	10	0.150	162	0.168	30	0.254	30	0.364	25	0.760	14
	20	50.575	42321	14.344	2275	86.809(5)	3362	62.123(7)	1405	21.322	290
	30	time limit		79.582(3)	8230	time limit		97.681(1)	1475	95.702(1)	1076
3	10	0.541	575	0.350	48	0.966	80	0.870	41	2.244	38
	20	time limit		16.890	1587	time limit		95.024(2)	876	91.181(3)	801
	30	time limit		97.000(1)	6287	time limit		time limit		time limit	
5	10	0.810	697	0.498	73	1.677	108	1.244	55	3.481	63
	20	time limit		21.203(3)	7722	time limit		time limit		time limit	
	30	time limit		time limit		time limit		time limit		time limit	

Problems of class (a) have been proposed by Ross & Soland (1975) and generally admit many feasible solutions. Problems of classes (b), (c) and (d) have tighter constraints on the resource availabilities; in addition, in problems of class (d) a correlation between penalties and resource requirement (often found in real–world applications) has been introduced.

The entries in the tables give average running times (expressed in seconds) and average numbers of nodes generated in the branch–decision tree. A time limit of 100 seconds was imposed on the running time spent by each algorithm for the solution of a single instance. For data sets for which the time limit occurred, the corresponding entry gives, in brackets, the number of instances solved within 100 seconds (the average values are computed by also considering the interrupted instances). The cases where the time limit occurred for all the instances are denoted as "time limit". The following algorithms have been coded in Fortran IV and run on an HP 9000/840 computer, using option "-o" for the Fortran compiler:

RS   = Ross & Soland (1975) ;

MTG   = Martello & Toth (1981);

FJV   = Fisher, Jaikumar & Van Wassenhove (1986);

MTGFJV   = MTG with $\min(U_2, U_3)$ computed at each node;

MTGJN   = MTG with $\min(U_1, U_2, U_4)$ computed at each node.

For all the algorithms, the solution of the 0–1 single knapsack problems was obtained using the Fortran program MT1 included in the diskette accompanying Martello & Toth (1990).

The tables show that the fastest algorithms are RS for the "easy" instances of class (a), and MTG for the harder instances (b), (c), (d). Algorithms MTGFJV and MTGJN generate fewer nodes than MTG, but the global running times are larger (the computation of U_3 and U_4 being much heavier than that of U_1 and U_2), mainly for problems of classes (b), (c) and (d).

Table 5 gives the performance of the Fortran IV implementation of approxi- mation algorithm MTHG on large–size instances. The entries give average running times (expressed in seconds) and, in brackets, upper bounds on the average per- centage errors. The percentage errors were computed as $100(U - z^h)/U$, where $U = \min(U_1, U_2, U_3, U_4)$. Only classes (a), (b) and (c) are considered, since the com- putation of U for class (d) required excessive running times. The table shows that the running times are quite small and, with few exceptions, practically independent of the data set. The quality of the solutions found by MTHG is very good for class (a) and clearly worse for the other classes, especially for small values of m. However, it is not possible to decide whether these high errors depend only on the approximate solution or also on the upper bound values. Limited experiments indicated that the error computed with respect to the optimal solution value tends to be about half that computed with respect to U.

Table 5. GAP. Algorithm MTHG. HP 9000/840 seconds.
Average times / Average percentage errors over 10 problems.

m	n	Class (a) time	(a) % err.	Class (b) time	(b) % err.	Class (c) time	(c) % err.
5	50	0.121	0.18	0.140	5.43	0.136	6.82
	100	0.287	0.06	0.325	4.75	0.318	5.73
	200	0.887	0.03	0.869	4.55	0.852	6.15
	500	2.654	0.01	3.860	5.68	3.887	6.15
10	50	0.192	0.02	0.225	3.43	0.240	6.24
	100	0.457	0.02	0.521	5.16	0.550	5.91
	200	1.148	0.00	1.271	4.80	1.334	5.19
	500	3.888	0.01	5.139	5.70	5.175	5.55
20	50	0.393	0.06	0.399	1.23	0.438	6.48
	100	0.743	0.00	0.866	1.19	0.888	5.19
	200	1.693	0.01	2.011	2.14	2.035	4.54
	500	2.967	0.00	7.442	3.45	7.351	4.37
50	50	0.938	0.00	0.832	0.13	0.876	2.02
	100	0.728	0.01	1.792	0.18	2.016	4.04
	200	3.456	0.00	3.849	0.30	4.131	3.25
	500	2.879	0.00	12.613	0.52	12.647	3.20

3 Min Max Generalized Assignment Problem

In this section we consider problem BGAP ((8), (2), (3), (4)) defined in Section 1.2.

3.1 Relaxations and Lower Bounds

We consider lower bounds obtained by relaxing constraints (2), either directly (by decreasing the resource requirements) or through surrogate techniques. In any case, we never allow an item j to be assigned to a unit i if, in the non-relaxed instance, $r_{ij} > a_i$.

Relaxation of the Resource Requirements

An immediate lower bound can be obtained by eliminating constraints (2). It is then evident that the resulting problem can be exactly solved, in $O(nm)$ time, by determinig

$$i_1(j) = \text{arg } \min_i \{p_{ij} : r_{ij} \le a_i\} \quad (j \in N) \tag{28}$$

and computing the lower bound value

$$L_0 = \max_j \{p_{i_1(j),j}\}. \tag{29}$$

If the corresponding solution (obtained by assigning each item j to unit $i_1(j)$) satisfies constraints (2), then this is clearly the optimum. Otherwise, a better bound L_1 can be obtained by imposing one of the violated constraints (as shown in Martello & Toth (1991)). The time complexity for the computation of L_0 and L_1 is $O(nm)$.

Surrogate Relaxation

For a given vector (π_i) of non-negative multipliers, we define the *surrogate relaxation* of BGAP, $S(BGAP, \pi)$, as

$$\text{minimize} \quad \hat{z}(\pi) = \max_{i,j} \{p_{ij}x_{ij}\}$$

$$\text{subject to} \quad \sum_{i=1}^{m} \pi_i \sum_{j=1}^{n} r_{ij}x_{ij} \leq \sum_{i=1}^{m} \pi_i a_i,$$

$$(3), (4).$$

For any non-negative vector (π_i),

$$L_2(\pi) = \hat{z}(\pi)$$

is then a lower bound for BGAP. It is proved in Martello & Toth (1991) that, for any vector (π_i) of multipliers, lower bound $L_2(\pi)$ can be computed in $O(nm)$ time.

3.2 Approximation Algorithms

A feasible solution to BGAP of value not greater than a given threshold ϑ can be heuristically found through a procedure which considers the items according to decreasing values of the difference between second smallest and smallest resource requirement for a feasible assignment, and assigns the considered item to the unit having the smallest resource requirement. If an item is found for which no feasible assignment is possible, the procedure returns no solution; otherwise it returns the feasible solution found. This procedure, which can be implemented to run in $O(nm\log m + n^2)$ time, can be used to determine an approximate solution to BGAP by searching for the lowest value ϑ for which a feasible solution is returned. If this is done through binary search, the overall time complexity of the resulting approximation algorithm is $O(nm\log m + \beta(nm + n^2))$, where β denotes the number of bits required to encode $\max_{i,j}\{p_{ij}\}$.

Other approximation algorithms are described in Martello & Toth (1991).

3.3 A Branch-and-Bound Algorithm

The results of the previous sections have been used to obtain a branch-and-bound algorithm for the exact solution of BGAP.

Branching Scheme

The algorithm consists of a depth-first search in which, at each level, an item j^* is selected for branching and assigned, in turn, to all feasible units.

The branching item is selected according to the following criterion. Let z denote the best incumbent solution value, L the current lower bound value, $\bar{a}_i$ ($i = 1, \ldots, m$) the amount of resource currently available for unit i, and U the set of currently unassigned items. For each $j \in U$, $M_j = \{i \in M : r_{ij} \leq \bar{a}_i \text{ and } p_{ij} \leq L\}$ is the set of units to which j can be assigned without increasing the lower bound. Hence

$$r_j = \frac{\sum_{i \in M_j}(r_{ij}/\bar{a}_i)}{|M_j|}$$

represents the average percentage resource requirement of item j, while

$$\delta_j = \min{}_2\{r_{ij} : i \in M_j\} - \min\{r_{ij} : i \in M_j\}$$

is the minimum additional resource requirement if item j is not assigned to the unit with minimum requirement. Since the higher r_j or δ_j , the more critical is item j, the branching item is selected through

$$j^* = \arg \max_{j \in U}\{r_j(1 + \delta_j)\}.$$

Now let $\overline{M}_j = \{i \in M : r_{ij} \leq \bar{a}_j \text{ and } p_{ij} < z\}$ denote the set of feasible units for item $j \in U$. $|\overline{M}_{j^*}|$ son nodes are generated by assigning j^* to all $i \in \overline{M}_{j^*}$, according to increasing values of r_{ij^*}, and the search is resumed from the first of these nodes.

Fathoming Decision Nodes

Consider a decision node generated, say, by assigning item $j_a \in U$ to unit i_a. Before computing the corresponding lower bound value, the following dominance criterion is applied. The node can be fathomed if there exists an item $j_b \notin U$, currently assigned to a unit $i_b \neq i_a$, such that by interchanging the assignments the current lower bound and resource requirements do not increase.

Initialization Phase

At the root node of the branch-decision tree, a lower bound L^* on the optimal solution value is first computed. The approximation procedure of Section 3.2 is then applied, obtaining a first incumbent solution of value z. If $z > L^*$, the first branching item is determined, and the enumeration process begins.

3.4 Computational Experiments

The branch-and-bound algorithm of the previous section was coded in Fortran and computationally tested on a Digital VAX station 3100.

The computational experiments were performed on four classes of randomly-generated problems. Table 6 gives, for different values of n and m, the average number of decision nodes and the average CPU times (expressed in seconds) computed over ten problem instances. For each instance, the execution was halted as soon as the number of nodes reached 10^6. For such cases, we give (in brackets) the number of solved problems and compute the average values over them.

Class (1) is very similar to class (a) used for GAP:

(1) r_{ij} uniformly random in [5,25], $i \in M$, $j \in N$,
$\quad p_{ij}$ uniformly random in [10,50], $i \in M$, $j \in N$,
$\quad a_i = 9(n/m) + 0.4 \max_{k \in M} \{\sum_{j:i_1(j)=k} r_{kj}\}, i \in M$.

The results show that the problems of this class are very easy for BGAP too. Most of the instances were solved by the initialization phase. The computing time increases almost linearly with both n and m.

More difficult problems can be obtained by decreasing the a_i values:

(2) r_{ij} and p_{ij} as for Class (1),
$\quad a_i =$ 60% of the value obtained for Class (1), $i \in M$.

The computational results show indeed a considerable increase in the computing times, especially for $m=5$ and $m=10$. Most of the instances of this class admit no feasible solution for $m=2$ or 3, and very few feasible solutions for $m=5$ or 10.

For both Classes (1) and (2) the range of the penalty values is very limited. In order to test the behaviour of the algorithm when the optimal solution value must be found in a larger range, we considered the following class:

(3) r_{ij} uniformly random in [1,1000], $i \in M$, $j \in N$,
$\quad p_{ij}$ uniformly random in [1,1000], $i \in M$, $j \in N$,
$\quad a_i = 0.6 \sum_{j \in N} r_{ij}/m, i \in M$.

The results show a satisfactory performance of the algorithm. For almost all values of m the difficulty of the instances increases for n going from 10 to 100.

The last class was obtained by introducing a correlation between penalties and resource requirements:

(4) r_{ij} uniformly random in [1,800], $i \in M$, $j \in N$,
$\quad p_{ij}$ uniformly random in [1,1000-r_{ij}], $i \in M$, $j \in N$,
$\quad a_i = \sum_{j \in N} r_{ij}/m, i \in M$.

The computational results show good behaviour of the algorithm for these problems too, with comparatively higher computing times for $n \le 50$.

In Table 7 we analyse the performance of the approximation algorithm of Section 3.2. For the same instances as Table 6 we give the average CPU time (expressed in seconds) and the average percentage error $100(z^a-z)/z$, where z^a is the approximate value, and z the optimal solution value or the lower bound value computed in the initialization phase (for the instance not solved exactly). The results show very good behaviour of the approximation algorithm, both for running time and the quality of the solutions found.

Table 6. BGAP exact solution. VaxStation 3100 seconds.
Average times / Average numbers of nodes over 10 problems.

m	n	Class (1) time	nodes	Class (2) time	nodes	Class (3) time	nodes	Class (4) time	nodes
2	10	0.01	1	0.01	0	0.01	0	0.01	2
	25	0.03	3	0.01	0	0.01	0	0.01	0
	50	0.02	0	0.02	0	0.02	0	0.03	0
	100	0.03	0	0.03	0	0.03	0	0.03	0
3	10	0.01	0	0.07	13	0.10	6	0.07	5
	25	0.03	3	0.09	27	0.34	67	0.14	16
	50	0.02	0	0.30	76	0.21	19	0.09	4
	100	0.04	0	0.05	0	3.02	404	0.04	0
5	10	0.01	0	0.11	18	0.10	3	0.10	7
	25	0.02	0	5.69	1856	1.58	275	0.60	129
	50	0.03	0	8.35	1950	5.92	1162	0.04	0
	100	0.06	0	70.99	13078	10.25	1557	0.06	0
10	10	0.04	1	0.07	2	0.02	1	0.04	1
	25	0.03	0	1.21	177	0.76	74	2.77	491
	50	0.05	0	41.96	5951	7.46	786	367.40	54926
	100	0.11	0	338.05	41311	0.22(9)	0	0.12	0

Table 7. BGAP approximate solution. VaxStation 3100 seconds.
Average times / Average percentage errors over 10 problems.

m	n	Class (1) time	% err.	Class (2) time	% err.	Class (3) time	% err.	Class (4) time	% err.
2	10	0.01	0.00	0.01	0.00	0.01	0.00	0.02	0.00
	25	0.02	0.00	0.01	0.00	0.01	0.00	0.01	0.00
	50	0.02	0.00	0.02	0.00	0.02	0.00	0.02	0.00
	100	0.03	0.00	0.03	0.00	0.03	0.00	0.03	0.00
3	10	0.01	0.00	0.05	0.00	0.09	0.00	0.06	0.00
	25	0.03	0.00	0.03	0.00	0.16	0.12	0.09	0.00
	50	0.02	0.00	0.04	0.00	0.12	0.00	0.06	0.00
	100	0.04	0.00	0.04	0.00	0.41	0.00	0.04	0.00
5	10	0.01	0.00	0.07	0.00	0.10	0.00	0.08	0.00
	25	0.02	0.00	0.21	0.31	0.28	1.03	0.19	0.65
	50	0.03	0.00	0.25	0.26	0.51	0.29	0.04	0.00
	100	0.06	0.00	0.41	1.09	0.26	0.02	0.06	0.00
10	10	0.04	0.00	0.06	0.00	0.02	0.00	0.04	0.00
	25	0.03	0.00	0.30	1.54	0.40	0.48	0.37	0.62
	50	0.06	0.00	0.20	0.00	0.95	1.02	0.22	0.00
	100	0.11	0.00	0.31	0.00	0.54	1.34	0.12	0.00

References

G. Carpaneto, S. Martello, P. Toth (1988). Algorithms and codes for the assignment problem. In B. Simeone, P. Toth, G. Gallo, F. Maffioli, S. Pallottino (eds.). *Fortran Codes For Network Optimization, Annals of Operations Research* 13, 193 - 223.

A. De Maio, C. Roveda (1971). An all zero-one algorithm for a certain class of transportation problems. *Operations Research* 19, 1406 - 1418.

M.L. Fisher (1981). The Lagrangian relaxation method for solving integer programming problems. *Management Science* 27, 1 - 18.

M.L. Fisher, R. Jaikumar, L.N. VanWassenhove (1986). A multiplier adjustment method for the generalized assignment problem. *Management Science* 32, 1095-1103.

E.S. Gottlieb, M.R. Rao (1990a). The generalized assignment problem : valid inequalities and facets. *Mathematical Programming* 46, 31-52.

E.S. Gottlieb, M.R. Rao (1990b). (1,k)-configuration facets for the generalized assignment problem. *Mathematical Programming* 46, 53-60.

M.M. Guignard, S. Kim (1987).Lagrangean decomposition: A model yielding stronger Lagrangean bounds. *Mathematical Programming* 39, 215 - 228.

M.M. Guignard, M.B. Rosenwein (1989).An improved dual based algorithm for the generalized assignment problem. *Operations Research* 37, 658-663.

T.D. Klastorin (1979). An effective subgradient algorithm for the generalized assignment problem. *Computers and Operations Research* 6, 155 - 164.

N.W. Kuhn (1955). The hungarian method for the assignment problem. *Naval Research Logistics Quarterly* 2, 83 - 97.

E.L. Lawler (1976). *Combinatorial Optimization: Networks and Matroids*, Holt, Rinehart and Winston, New York.

K. Jörnsten, M. Näsberg (1986).A new Lagrangian relaxation approach to the generalized assignment problem. *European Journal of Operational Research* 27, 313-323.

S. Martello, P. Toth (1980). Solution of the zero-one multiple knapsack problem. *European Journal of Operational Research* 4, 276 - 283.

S. Martello, P. Toth (1981).An algorithm for the generalized assignment problem. In J.P. Brans (ed.), *Operational Research'81*, North-Holland, Amsterdam, 589-603.

S. Martello, P. Toth (1990).*Knapsack Problems: Algorithms and Computer Implementations*, Wiley, Chichester.

S. Martello, P. Toth (1991).The bottleneck generalized assignment problem. Research report DEIS OR/5/91, University of Bologna.

J.B. Mazzola (1989). Generalized assignment with nonlinear capacity interaction. *Management Science* 35, 923 - 941.

J.B. Mazzola, A.W. Neebe (1988).Bottleneck generalized assignment problems. *Engineering Costs and Production Economics* 14, 61-65.

G.T. Ross, R.M. Soland (1975).A branch and bound algorithm for the generalized assignment problem. *Mathematical Programming* 8, 91-103.

V. Srinivasan, G.L. Thompson (1973). An algorithm for assigning uses to sources in a special class of transportation problems. *Operations Research* 21, 284 - 295.

Recognizing an Envelope of Lines in Linear Time

Eric Guévremont*
School of Computer Science
McGill University

Jack Snoeyink†
Department of Computer Science
University of British Columbia

Abstract

The *envelope polygon* of a set of lines, L, is the polygon consisting of the finite length segments that bound the infinite faces of the arrangement $\mathcal{A}(L)$. Given an envelope polygon, we show how to sort its edges by slope in linear time. Using this result, we can determine whether a given polygon P is an envelope in linear time.

1   Introduction

A *polygonal chain* is a circular sequence of points or *vertices* $p_0, p_1, \ldots, p_n$ in which adjacent pairs are joined by *edges*. A chain is *simple* if no consecutive triple of vertices is colinear and the only intersections between edges are adjacent edges meeting at their common vertex. The information that a collection of vertices form a simple chain sometimes allows more efficient algorithms than are possible for general point sets. The most important examples of this are convex hulls [8], Jordan sorting [9] (sorting the intersections of a simple polygon with a line), and triangulation [2].

In this paper, we consider a problem in which simplicity is not enough—the problem of sorting the edges of a polygon by slope. (A *polygon P* is a closed region whose boundary is simple chain.) The comb-like construction of figure 1 illustrates that one can reduce integer sorting to sorting the slopes of the edges of a polygon. Thus, $\Omega(n \log n)$ comparisons

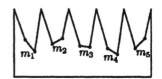

Figure 1: $\Theta(n \log n)$ to sort slopes

are required to sort the edges. In contrast, we develop a linear-time algorithm to sort edges of *envelope polygons*, which are defined in the next section. Section 3 describes the sorting algorithm and section 4 combines it with Keil's algorithm [10], which constructs an envelope given a sorted set of lines, to verify that a polygon is an envelope in linear time.

---

*Currently at Simon Fraser University
†Supported in part by an NSERC Research Grant

2   Definitions and Previous Results

The *envelope of a set of lines L*, shown in figure 2, is a simple polygon consisting of maximal but finite length line segments that bound the infinite faces of the arrangement $A(L)$. One can imagine taking the arrangement $A(L)$, removing segments that extend to infinity, and tracing around the outer boundary to form a polygon in which adjacent segments that belong to the same line are collapsed into a single edge. The convex vertices of the envelope come from the vertices of the arrangement with two or more infinite rays that are consecutive in slope order. (Note that we do not assume general position.) The concave vertices come from vertices on infinite faces having no infinite rays. Vertices of the arrangement with one infinite ray disappear from the envelope because their two adjacent segments are colinear.

We define the *lines of a polygon P*, denoted $L(P)$, to be the minimal set of lines that contain the edges of P. A polygon P is an *envelope polygon* if it is the envelope of its lines.

In previous work, Ching and Lee [3] established an $\Omega(n \log n)$ algebraic computation tree lower bound for the problem of computing the envelope of a set of lines. Eu, Guévremont and Toussaint [6] showed that this bound holds even for the special class of "sail" arrangements, which have only three convex vertices. (This is despite the fact that they can compute the

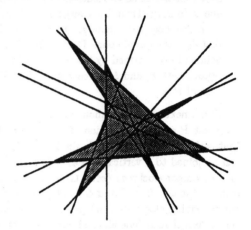

Figure 2: Envelope of a set of lines

convex vertices of a sail arrangement in linear time.) Cole et al. [4] showed that computing the leftmost intersection of a set of lines also take $\Omega(n \log n)$ time.

Suri (unpublished, 1985), Vegter [12], and Keil [10] have all given algorithms for constructing envelopes that achieve the $O(n \log n)$ bound. Edelsbrunner noted to Suri in 1985 (private communication) that one can use the horizon trees from topological sweeping [5] to compute the envelope of L in linear time if the lines of L are sorted by slope. Keil's algorithm does the same. In this work, we show how to obtain the lines of an envelope polygon in slope order and use Keil's algorithm to construct and verify the envelope.

There are three properties of envelope polygons that we shall find useful in the next section. The second of these comes from the close relationship between the envelope polygon P of L and the zone or horizon of the line at infinity in the arrangement $A(L)$.

Lemma 1 *Let P be an envelope polygon.*

    a. *No intersection of two lines of P lies outside of P.*

    b. *The number of vertices/edges of P is at most $3.5|L(P)|$.*

    c. *At most two distinct lines of $L(P)$ are parallel.*

Proof: *a.* From the definition, we can see that the envelope polygon P is the closure of all bounded regions of the arrangement of $L(P)$. Now, suppose that lines ℓ_1 and ℓ_2 of P intersect. Since there is a path in P from an edge on ℓ_1 to an edge on ℓ_2, there is a bounded region in the arrangement with the point $\ell_1 \cap \ell_2$ on its boundary. Therefore, the intersection of ℓ_1 and ℓ_2 is in P.

    b. The recent tight bounds of Bern et al. [1] imply that the zone of the line at infinity has at most $9.5|L(P)|$ edges. Each of the $2|L(P)|$ infinite rays of this zone contributes its two sides to this complexity and contributes one edge on the line at infinity. Thus, the contribution from P is bounded by $3.5|L(P)|$.

    c. Suppose that there are three distinct parallel lines $\ell_1, \ell_2, \ell_3 \in L(P)$ and that ℓ_2 is the middle one. Then there is some bounded edge e on ℓ_2. Then endpoints of e are formed by intersection with two other lines, which form bounded regions with ℓ_1 and ℓ_2 on each side of e. This contradicts the fact that e is an edge of P. ∎

In the next section we show how to sort the lines of P. Our strategy will be to represent P as the intersection of four unbounded regions including two, R and L, that satisfy a property that is similar to edge-visibility: every vertex of R (or L) can be joined to a vertical line at infinity by a line segment within R (or L). We give a counter-clockwise (ccw) Graham scan procedure [7], ccwScan(), that sorts some of the the lines R and L by slope and show that running ccwScan() and the corresponding clockwise (cw) procedure cwScan() produces a constant number of sorted lists that include all the lines of P. Toussaint and Avis [11] have shown that a Graham scan gives an easy algorithm to triangulate an edge-visible polygon; our algorithm in the next section is closely related.

3 Sorting edge slopes

Given an envelope polygon P, this section shows how to obtain the sorted list of the lines of P in linear time. We begin by forming four polygonal regions whose intersection is P.

Lemma 2 *In linear time one can find regions U and D, bounded by concave chains, and polygonal regions L and R with the same lines as P such that $P = U \cap D \cap L \cap R$ and each edge of R (or of L) can be extended to a ray to infinity in R (or L).*

Proof: Choose a line $\sigma \in L(P)$. Let $\sigma' \in L(P)$ be the distinct line parallel to σ if such a line exists, otherwise $\sigma' = \sigma$. (Lemma 1c states that there are at most two distinct parallel lines.) Assume, without loss of generality, that σ is vertical and right of σ'. Let R be the union of P with the halfplane on and left of σ and L be the union of P with the halfplane on and right of σ'. (R is illustrated in figure 3.) Since P is an envelope polygon, the region between σ and σ' is bounded by concave chains above and below; let U be the regions bounded above by the upper chain and let D be the region bounded below by the lower chain.

We know that L and R contain an infinite ray for each line of P. Since, by Lemma 1a, the intersection of any pair of lines of P is contained in L (and R), the lemma holds. ∎

The lines of U and D, if any, are already sorted. We can concentrate on sorting edges of the polygon R, which looks like P to the right of the vertical line σ—the problem of sorting lines of L is symmetric. Later, we can merge the resulting lists in linear time. From now on, we assume that σ corresponds to the y-axis.

From lemmas 1 and 2 we can observe two properties:

(1) The extension of an edge e of R toward the y-axis remains in R and

(2) The extension of e away from the y-axis crosses out of R and then intersects no lines of R.

The first of these properties is similar to edge visibility (from the line at $x = -\infty$).

Input: A polygonal region R constructed according to lemma 2. R satisfies properties (1) and (2) and has vertices $r_0 = (0, -\infty)$, $r_1, \ldots, r_{m-1}, r_m = (0, \infty)$ in ccw order.
Data Structure: A vertex stack with pointers *top* and *next* to the top and next elements.
Output: Lines of R that are visited ccw, sorted by slope.

Procedure ccwScan(R)
    Push(r_0)
    Push(r_1)   /* *Initial stack:* top $= r_1$ *and* next $= r_0$ */
  1. for $i = 2$ to m do
      2. while *next, top, r_i* is a left turn
          3. if (*next, top*) is an edge of P then
                  Output the line through (*next, top*)
             Pop() /* *Remove* top *vertex from hull* */
          end while
      4. Push(r_i)   /* *Push* r_i *on the hull* */

Algorithm 1: The ccw Scan() procedure

Suppose the vertices of R are $r_0 = (0, -\infty)$, $r_1, \ldots, r_{m-1}, r_m = (0, \infty)$ in ccw order. To sort the lines of R we use two Graham scans [7]: one counter-clockwise beginning from r_0 and the other clockwise beginning from r_m. Each Graham scan computes a stack that contains the vertices on the convex hulls of prefixes of the boundary chain of R. We say that an edge $e = (r_i, r_{i+1})$ of R is *visited* by the

Graham scan if r_i and r_{i+1} are ever consecutive on top of the stack. We shall show that visited edges are popped from the stack in sorted order. Algorithm 1 gives the details of the ccw scan procedure, ccwScan(). The cwScan() is similar.

Figure 3 shows the polygon R that derives from figure 2 and the orders that the lines are output by the ccw and cw scans: ccwScan() outputs lines 1, 2, 3, 4, 5 and cwScan() outputs a, b, c, d.

The correctness of the sorting procedure follows from lemmas 5 and 6, which show that every line of R is visited by either the cw or ccw scan and that the visited lines are output in order of increasing slope. First, we recall the lemma establishing the running time of Graham scan.

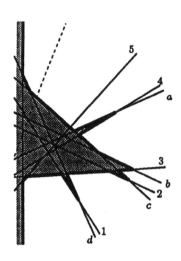

Figure 3: Results of scanning R

Lemma 3 (Graham [7]) ccwScan() *performs* $O(m)$ *steps given any polygon of m vertices.*

Proof: The test in line 3 can be implemented in constant time by checking whether the indices of the vertices *next* and *top* are consecutive. Each time the test is applied, a point is either pushed or popped from the stack; since each point is pushed once and popped at most once, the total time is linear in m. ■

Next, we consider an edge e of R that has the interior of R locally above ℓ_e: the line containing e. In other words, $e = (r_i, r_{i+1})$ with the x-coordinates $r_i.x < r_{i+1}.x$. We show that the boundary of R crosses above ℓ_e only once. If R were locally below ℓ_e, then a symmetric argument shows that the boundary of R crosses below only once.

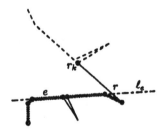

Figure 4: Vertices below ℓ_e

Lemma 4 *Let* $e = (r_i, r_{i+1})$ *be an edge of R with* $r_i.x < r_{i+1}.x$ *and let* r_k *be the first vertex above* ℓ_e. *Vertices that precede e,* $r_0, r_1, \ldots, r_i$, *lie on or below* ℓ_e. *Vertices that follow* r_k *lie above* ℓ_e.

Proof: Let r be a point where the boundary of R crosses above the line ℓ_e from on or below ℓ_e. Property (1) implies that r cannot have smaller x-coordinate than those of e; property (2) implies that there is only one candidate for r with

x-coordinate greater than those of e. Because e must preceed r on the boundary of R, vertices preceeding e lie on or below ℓ_e. And, because e is popped when the vertex r_k is discovered above ℓ_e, r_k and all successive vertices follow r and lie above ℓ_e. ∎

Now we are ready to establish correctness of the sorting procedure.

Lemma 5 *Every edge of R is visited by either the* ccwScan() *or* cwScan().

Proof: Consider an edge $e = (r_i, r_{i+1})$ where the x-coordinate $r_i.x$ is less than $r_{i+1}.x$. Lemma 4 implies that vertices r_0, r_1, ..., r_i, r_{i+1} lie on or below ℓ_e. Thus, edge e is on the convex hull of these vertices. ccwScan() constructs this hull (a fact proved by (Toussaint and Avis's [11]) and visits e. A similar argument shows that, when $r_i.x > r_{i+1}.x$, the edge e is visited by cwScan(). ∎

Lemma 6 *The lines of R output by* ccwScan() *are ordered by increasing slope; those output by* cwScan() *are by decreasing slope.*

Proof: We consider ccwScan() of algorithm 1 and show that all edges popped after a given edge e have greater slope than e. First, consider the edges visited before e and popped after. Such an edge e' must be on the stack with e. Since the stack stores a convex chain from $(0, -\infty)$ to e, the slope of e' is greater than that of e.

Now, consider the edges visited after e is popped. Suppose, for the sake of deriving a contradiction, that after e is popped, an edge e' of smaller slope than e is visited, as depicted in figure 5. By lemma 4, e' is above ℓ_e. The intersection point of ℓ_e and $\ell_{e'}$ is contained in the polygon R, therefore some vertex of the

Figure 5: e' is visited after e has been popped

boundary between e and e' is above the line $\ell_{e'}$. This, however, contradicts lemma 4 for edge e': that everything before e' is visited lies below $\ell_{e'}$. Again, the slope of e' is greater than that of e. ∎

We conclude with theorem 7.

Theorem 7 *One can sort the lines of an envelope polygon P by slope in linear time.*

Proof: Constructing the regions U, D, L and R according to lemma 2 is a simple matter after computing the intersection of P with the chosen lines. Executing the ccwScan() and cwScan() for each of L and R takes linear time by lemma 3. Lemma 5 states that every edge of L and R is visited by one of these scans, that is, every edge is pushed on the stack. Since each scan ends with the stack containing the vertices that lie on the y-axis, every edge except the vertical edges is also popped and every non-vertical line of P is output in some scan. Lemma 6 says that each scan outputs its lines in sorted order; merging the resulting lists by slope and eliminating duplicates is easily accomplished in linear time. ∎

4 Verifying an envelope

Given a polygon P, one can use the sorting procedure of the previous section together with Keil's algorithm for constructing an envelope to determine whether P is an envelope polygon in linear time.

Theorem 8 *One can test whether P is an envelope polygon in linear time.*

Proof: Run the sorting procedure of the previous section to obtain a "sorted" list $\mathcal{L}$ of lines of $L(P)$. If P is an envelope, $\mathcal{L}$ will be correctly sorted. Even if P is not an envelope, this first step runs in linear time by lemma 3.

Verify that the list of lines $\mathcal{L}$ is sorted and run Keil's algorithm [10] to construct the envelope polygon $E(\mathcal{L})$ in linear time. Then check whether $P = E(\mathcal{L})$ by walking around the two polygons and comparing vertices. This takes linear time overall. ∎

Acknowledgement

We thank Godfried Toussaint for introducing us to this problem.

References

[1] Marshall Bern, David Eppstein, Paul Plassmann, and Francis Yao. Horizon theorems for lines and polygons. In Jacob E. Goodman, Richard Pollack, and William Steiger, editors, *Discrete and Computational Geometry: Papers from the DIMACS Special Year*, pages 45–66. American Mathematical Society, Providence, RI, 1991.

[2] Bernard Chazelle. Triangulating a simple polygon in linear time. In *Proceedings of the 31st IEEE Symposium on Foundations of Computer Science*, pages 220–230, 1990.

[3] Y. T. Ching and D. T. Lee. Finding the diameter of a set of lines. *Pattern Recognition Letters*, 18:249–255, 1985.

[4] R. Cole, J. Salowe, W. Steiger, E. Szemerédi An optimal-time algorithm for slope selection SIAM J. Comp. 18(4), pages 792–810, 1989.

[5] Herbert Edelsbrunner and Leonidas J. Guibas. Topologically sweeping an arrangement. In *Proceedings of the 18th Annual ACM Symposium on Theory of Computing*, pages 389–403, 1986.

[6] D. Eu, E. Guévremont, and G. T. Toussaint. On classes of arrangements of lines. Accepted to the Fourth Canadian Conference on Computational Geometry, 1992.

[7] R. Graham. An efficient algorithm for determining the convex hull of a finite planar set. *Information Processing Letters*, 1:132–133, 1972.

[8] R. L. Graham and F. F. Yao. Finding the convex hull of a simple polygon. *Journal of Algorithms*, 4:324–331, 1983.

[9] Kurt Hoffmann, Kurt Mehlhorn, Pierre Rosenstiehl, and Robert E. Tarjan. Sorting Jordan sequences in linear time. In *Proceedings of the ACM Symposium on Computational Geometry*, pages 196–203, 1985.

[10] M. Keil. A simple algorithm for determining the envelope of a set of lines. *Information Processing Letters*, 39:121–124, 1991.

[11] G. T. Toussaint and D. Avis. On a convex hull algorithm for polygons and its application to triangulation problems. *Pattern Recognition*, 15(1):23–29, 1982.

[12] G. Vegter. Computing the bounded region determined by finitely many lines in the plane. Technical Report 87–03, Rijksuniversiteit Groningen, 1987.

Approximation of Polygonal Curves with Minimum Number of Line Segments[+]

W.S. Chan and F. Chin

Department of Computer Science
University of Hong Kong

Abstract. We improve the time complexity to solve the polygonal curve approximation problem formulated by Imai and Iri from $O(n^2 \log n)$ to $O(n^2)$. If the curve to be approximated forms part of a convex polygon, we show that the time complexity can be further reduced to $O(n)$.

[+] This research is partially supported by a UPGC Research Grant

1 Introduction

Under various situations and applications, images of a scene have to be represented at different resolutions. For pixel images, a number of multi-resolution techniques[1, 3, 4, 8, 10] have been devised to generate lower resolution images from the higher ones. If the image consists of sequences of line segments, the problem becomes approximating the original figure with fewer line segments. A number of algorithms[5, 7, 9, 11] have been devised to solve polygonal approximation problems under different constraints and approximation criteria[5, 6, 7]. In this paper, we consider the problem of approximating a piecewise linear curve or polygonal curve by another whose vertices are a subset of the original's. We improve the results by Imai and Iri[7] and Melkman and O'Rourke[9] and show that this problem can be solved in $O(n^2)$ time instead of $O(n^2 \log n)$ time.

Formally, let $P = (p_0, p_1, \ldots, p_{n-1})$ be a *piecewise linear curve* or *polygonal curve* on a plane, i.e., $p_0, p_1, \ldots, p_{n-1}$ is a sequence of points on a plane and each pair of points p_i and p_{i+1}, $i = 0, 1, \ldots, n-2$, are joined by a line segment. Note that in general the line segments may intersect. We can approximate part of a polygonal curve, $(p_r, p_{r+1}, \ldots, p_s)$ by a line segment $\overline{p_r p_s}$ and the *error of a line segment* can be defined as the maximum distance between the segment $\overline{p_r p_s}$ and each point p_k between p_r and p_s, i.e., $r \leq k \leq s$. The distance $d(\overline{p_r p_s}, p_k)$ between a line segment $\overline{p_r p_s}$ and a point p_k is defined to be the minimum distance between $\overline{p_r p_s}$ and p_k, i.e., $d(\overline{p_r p_s}, p_k) = \min_{x \in \overline{p_r p_s}} \{d(x, p_k)\}$ where $d(x, p_k)$ is the Euclidean distance between p_k and x. Thus, the error of a line segment $\overline{p_r p_s}$, $e(\overline{p_r p_s}) = \max_{r \leq k \leq s} \{d(\overline{p_r p_s}, p_k)\}$. Similarly, we say $P' = (p_{i_0}, p_{i_1}, \ldots, p_{i_m})$ is an *approximate curve* of P if $p_{i_0}, p_{i_1}, \ldots, p_{i_m}$ is a subsequence of $p_0, p_1, \ldots, p_{n-1}$ where $i_0 = 0$, $i_m = n-1$ and $0 \leq m < n$. The *error of an approximate curve* P' is defined as the maximum error of each line segment in P', i.e., $e(P') = \max_{0 \leq k < m} \{e(\overline{p_{i_k} p_{i_{k+1}}})\}$. We say that the *error of P' is within ϵ* if $e(P') \leq \epsilon$, i.e., for each line segment $\overline{p_{i_j} p_{i_{j+1}}}$,

$j = 0, 1, \ldots, m-1$, its error is less than or equal to ϵ. Note that $d(\overline{p_{i_j}p_{i_{j+1}}}, p_k) \leq \epsilon$ if p_k lies within the shaded area as shown in Figure 1. With this error measurement, we can guarantee that P' will lie totally within the "band" of width 2ϵ running alongside P if $e(P') \leq \epsilon$ (Figure 2).

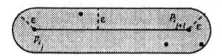

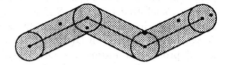

Fig. 1. Definition of being within ϵ of a line segment

Fig. 2. "Band" running alongside a polygonal curve

There are two types of optimization problems associated with the curve approximation problem.

min-# problem Given $\epsilon \geq 0$, construct an approximate curve with error within ϵ and having the minimum number of line segments.

min-ϵ problem Given m, construct an approximate curve consisting of at most m line segments with minimum error.

In this paper, we show that the min-# problem for an open polygonal curve can be solved in $O(n^2)$ time which improves the previous result of $O(n^2 \log n)$ time. We further show that if the polygonal curve forms part of a convex polygon, the min-# problem can be solved in $O(n)$ time for both open and closed polygonal curve. The min-ϵ problem will be presented in a forthcoming paper.

2   The Min-# Problem of a General Polygonal Curve

Given a polygonal curve P and an error bound ϵ, this problem can be solved in two steps. The first step is to construct a *directed* graph $G = (V, E)$, where each vertex v_r in V represents a point p_r in P and each edge (v_r, v_s) is in E if and only if the error of the line segment is within ϵ, i.e., $V = \{v_0, v_1, \ldots, v_{n-1}\}$ and $E = \{(v_r, v_s) \mid r < s \text{ and } e(\overline{p_r p_s}) \leq \epsilon\}$. Note that a directed graph is used to model the problem so that it can later be extended to the closed curve case. The second step is to find the shortest path in G from v_0 to v_{n-1} with each edge of unit length. Thus, the path length will correspond to the number of line segments in the approximate curve. Moreover, the shortest path from v_0 to v_{n-1} also reveals the subset of points of P used in the approximate curve P', i.e., if $(v_{i_0}, v_{i_1}, \ldots, v_{i_m})$ is the shortest path with $i_0 = 0$ and $i_m = n - 1$, then the corresponding $P' = (p_{i_0}, p_{i_1}, \ldots, p_{i_m})$ will be the approximate curve with error $\leq \epsilon$ and the minimum number of line segments.

The brute-force method of constructing G is to check for each pair of points, p_r and p_s, whether the error of $\overline{p_r p_s}$ is within ϵ, i.e., $(v_r, v_s) \in E$. There are $O(n^2)$ pairs of points and checking the error for a line segment, corresponding to a pair of points, takes $O(n)$ time. This brute-force method takes $O(n^3)$ time. Since finding the shortest path in G takes no more than $O(n^2)$ time, the min-# problem can be solved in $O(n^3)$ time.

The critical part of the algorithm is the construction of G. Melkman and O'Rourke [9] described an $O(n^2 \log n)$ algorithm to construct G. For each pair of points, p_r and p_s, their algorithm maintains a data structure which represents a region W_{rs} such that $(v_r, v_s) \in E$ if and only if $p_s \in W_{rs}$. Their data structure allows checking of whether $p_s \in W_{rs}$ and updating W_{rs} to $W_{r,s+1}$ in $O(\log n)$ time. Thus G can be constructed in $O(n^2 \log n)$ time and the time complexity of solving the min-# problem is improved to $O(n^2 \log n)$. In this paper, we improve the complexity of constructing G to $O(n^2)$. As a result, the time complexity of the algorithm for solving the min-# problem is $O(n^2)$, an improvement from $O(n^2 \log n)$. In the following, we shall discuss how G can be constructed in $O(n^2)$ time.

Before we proceed with the algorithm, let us discuss some *necessary* conditions for $d(\overline{p_r p_s}, p_k) \leq \epsilon$ when $p_r \neq p_s$ and $r < k < s$, i.e., $(v_r, v_s) \in E$.

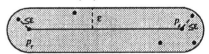

Fig. 3. Conditions B and C

Condition A $d(\overleftrightarrow{p_r p_s}, p_k) \leq \epsilon$ where $\overleftrightarrow{p_r p_s}$ denotes the infinite line extending at both ends of $\overline{p_r p_s}$.

Condition B If $\angle p_k p_r p_s > \pi/2$, $d(p_k, p_r) \leq \epsilon$, where $\angle p_k p_r p_s$ denotes the *convex* angle between line segments $\overline{p_k p_r}$ and $\overline{p_r p_s}$ (Figure 3).

Condition C If $\angle p_k p_s p_r > \pi/2$, $d(p_k, p_s) \leq \epsilon$ (Figure 3).

Note that condition A is intuitively straightforward, while conditions B and C ensure that if p_k lies outside the "range" of $\overline{p_r p_s}$, p_k would not be too far away from p_r or p_s, i.e., in the hemispheres as shown Figure 3. Moreover, one can easily argue that it is not possible to have both $\angle p_k p_r p_s$ and $\angle p_k p_s p_r$ greater than $\pi/2$, and satisfying these three conditions is also sufficient to guarantee that $d(\overline{p_r p_s}, p_k) \leq \epsilon$.

Now we are ready to discuss the construction of G. Basically, G is constructed in two phases, in which $G' = (V, E')$ and $G'' = (V, E'')$ are formed. Note that G, G' and G'' have the same vertex set V and we shall show that $E = E' \cap E''$. G' is a graph generated based on conditions A and B, while G'' on conditions A and C.

Let $\overrightarrow{p_r p_s}$ denote the "ray" emanating from p_r and passing through p_s. If $p_r = p_s$, let $\overrightarrow{p_r p_s} = p_r$. $(v_r, v_s) \in E'$ if and only if $r < s$ and for all $r \leq k \leq s$, $d(\overrightarrow{p_r p_s}, p_k) \leq \epsilon$, i.e., p_k lies in the area as shown in Figure 4(a). Similarly, $(v_t, v_u) \in E''$ if and only if $t < u$ and for all $t \leq k \leq u$, $d(\overrightarrow{p_u p_t}, p_k) \leq \epsilon$, i.e., p_k lies in the area as shown in Figure 4(b).

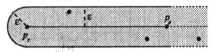

(a) Area for which p_k lies within ϵ of $\overrightarrow{p_r p_s}$

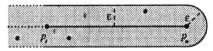

(b) Area for which p_k lies within ϵ of $\overrightarrow{p_u p_t}$

Fig. 4. Relationship between $\overline{p_r p_s}$ and its rays

Since segment $\overline{p_r p_s}$ is part of both $\overrightarrow{p_r p_s}$ and $\overrightarrow{p_s p_r}$, it is easy to observe that $E \subseteq E'$, $E \subseteq E''$ and $E = E' \cap E''$. Instead of of constructing E directly, we construct E' and E'' first. E can then be generated from E' and E''.

Now we describe the construction of E', E'' can be constructed similarly. Consider a particular vertex v_r in V. We want to determine whether $(v_r, v_s) \in E'$ for some $r < s$. Let $r < k < s$. If $d(p_r, p_k) > \epsilon$, let a_{rk} and b_{rk} be the two rays emanating from p_r, one at each side of p_k and at a distance ϵ from p_k, i.e., $d(a_{rk}, p_k) = d(b_{rk}, p_k) = \epsilon$, and let D_{rk} be the convex region bounded by a_{rk} and b_{rk}, including a_{rk} and b_{rk} but excluding point p_r(see Figure 5). If $d(p_r, p_k) \le \epsilon$, let D_{rk} be the whole plane. The following lemma shows that $d(\overrightarrow{p_r p_s}, p_k) \le \epsilon$ as long as $p_s \in D_{rk}$.

Lemma 1. *Assume $r \le k \le s$. $p_s \in D_{rk}$ if and only if $d(\overrightarrow{p_r p_s}, p_k) \le \epsilon$.*

Proof. The proof is rather straightforward and thus omitted. Basically all the cases when $d(p_r, p_k) > \epsilon$, $d(p_r, p_k) \le \epsilon$, $p_r = p_s$ and $p_r \ne p_s$ have to be considered. □

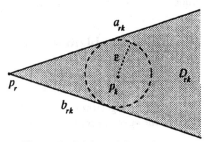

Fig. 5. Definition of D_{rk} when $d(p_r, p_k) > \epsilon$

In order to determine whether $(v_r, v_s) \in E'$, it is necessary and sufficient to check whether $p_s \in D_{rk}$ for all $r \le k < s$ (from Lemma 1). Define $W_{rs} = \bigcap_{r \le k < s} D_{rk}$ for $r < s$.

Lemma 2. *Assume $r < s$. $(v_r, v_s) \in E'$ if and only if $p_s \in W_{rs}$.*

Proof. $(v_r, v_s) \in E'$ iff $d(\overrightarrow{p_r p_s}, p_k) \le \epsilon$ for all $r \le k < s$ (from definition)
iff $p_s \in D_{rk}$ for all $r \le k < s$ (Lemma 1)
iff $p_s \in W_{rs} = \bigcap_{r \le k < s} D_{rk}$ (from definition) □

From Lemma 2, we can determine whether $(v_r, v_s) \in E'$ by testing whether $p_s \in W_{rs}$. For any r, the algorithm would consider all pairs of vertices (v_r, v_s) with s varying from $r+1$ to n. The key part of the algorithm is to determine whether a point is in W_{rs} and to update W_{rs} from $W_{r,s-1}$ efficiently.

Since W_{rs} is an intersection of the D_{rk}'s, the D_{rk}'s and the W_{rs} are either the whole plane or a cone-shape region bounded by the two rays emanating from

```
procedure e1
begin
    for r = 0 to n - 2 do
    begin
        s := r + 1
        W := the whole plane  { W = W_rs }
        while W ≠ ∅ and s < n do
        begin
            if p_s ∈ W
            then output (v_r, v_s) ∈ E'
            if d(p_r, p_s) > ε
            then W := W ∩ D_rs
            s := s + 1;
        end { while }
    end { for }
end
```

p_r. As long as we can keep track of these two rays(boundaries) of W_{rs}, one can easily check whether $p_s \in W_{rs}$ and update W_{rs} in constant time. Checking whether $p_s \in W_{rs}$ is equivalent to checking whether p_s lies within the two rays. As for updating W_{rs}, if $d(p_r, p_s) \le \epsilon$, $W_{rs} = W_{r,s-1}$, otherwise $W_{rs} = W_{r,s-1} \cap D_{r,s-1}$. In fact, if $W_{rs} = \phi$, we can conclude immediately that $(v_r, v_k) \notin E'$ for all $k \ge s$. Thus we can determine all the edges in E' incident from v_r in $O(n)$ time and all edges in E' in $O(n^2)$ time. The pseudo-code of the algorithm is shown in Procedure $e1$.

Theorem 3. $G = (V, E)$ *can be constructed in* $O(n^2)$ *time.*

Proof. In our previous discussion, we can construct E' in $O(n^2)$ time. As E'' can be constructed similarly, $E = E' \cap E''$ can be generated with another n^2 operations. Thus constructing G takes no more than $O(n^2)$ time. □

Corollary 4. *The min-# problem for an open polygonal curve can be solved in* $O(n^2)$ *time.* □

3 Min-# Problem of a Convex Polygonal Curve

We call a polygonal curve $P = (p_0, p_1, \ldots, p_{n-1})$ *convex* if the polygon formed by P with the line segment $\overline{p_{n-1}p_0}$ is convex.

Before the algorithm for the min-# problem of a convex curve is described, we study some special properties of G for a convex polygonal curve.

Lemma 5. *Let $G = (V, E)$ be the corresponding graph defined as above on a convex polygonal curve $P = (p_0, p_1, \ldots, p_{n-1})$. If $r+1 < s$ and $(v_r, v_s) \in E$, then (v_{r+1}, v_s), $(v_r, v_{s-1}) \in E$.*

Proof. By definition, since $(v_r, v_s) \in E$, $d(\overline{p_r p_s}, p_k) \le \epsilon$ for all $r \le k \le s$. We want to show that $d(\overline{p_{r+1}p_s}, p_k) \le \epsilon$ for all k, $r+1 \le k \le s$, and $d(\overline{p_r p_{s-1}}, p_k) \le \epsilon$ for all k, $r \le k \le s - 1$. The proofs for these two cases are similar and so only the former case will be shown here.

Let x be the point on $\overline{p_r p_s}$ which is nearest to p_k. Since P is convex, it can be proved easily that the line $\overline{p_k x}$ has to intersect line $\overline{p_{r+1}p_s}$ at some point y(Figure 6). Thus $d(\overline{p_{r+1}p_s}, p_k) \le d(y, p_k) \le d(x, p_k) = d(\overline{p_r p_s}, p_k) \le \epsilon$. □

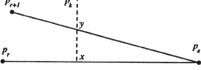

Fig. 6. p_k is closer to $\overline{p_{r+1}p_s}$ than to $\overline{p_r p_s}$

The following corollary can be proved by applying Lemma 5 repeatedly.

Corollary 6. *If $(v_r, v_s) \in E$, $(v_{r'}, v_{s'}) \in E$ for all $r \le r' < s' \le s$.* □

It follows from Corollary 6 that for each vertex v_r where $r < n - 1$, there is a vertex v_s such that $(v_r, v_k) \in E$ for all $k, r < k \le s$ and $(v_r, v_k) \notin E$ for $k, s < k < n$. We call v_s the *furthest vertex reachable from* v_r. We denote this index s by $f(r)$. Thus G can be completely characterized if $f(r)$ is defined for each $r = 0, 1, \ldots, n-2$. From Corollary 6, it can be shown easily that if $r \le s$, $f(r) \le f(s)$. So, intuitively, one would like to find $f(r)$ iteratively, i.e., to search for $f(r+1)$ starting from $f(r)$. With this property and the following lemmas about the distance between p_k and $\overleftrightarrow{p_r p_s}$ for $r < k < s$, condition A (defined in the previous section) can be checked efficiently.

Lemma 7. *Let $(p_r, p_{r+1}, \ldots, p_s)$ be a convex polygonal curve and $r < k < s$. The function $d(\overleftrightarrow{p_r p_s}, p_k)$ is unimodal with respect to k.*

Proof. This fact is well-known and a proof can be found in [2]. □

Based on the unimodal property, condition A can be checked for all the points, p_k, $r < k < s$, with respect to $\overrightarrow{p_r p_s}$ efficiently as long as the index, h_{rs}, of the point which is furthest away from $\overrightarrow{p_r p_s}$ is known. Let h_{rs} be the smallest index l larger than r such that $d(\overrightarrow{p_r p_s}, p_l) = e(\overrightarrow{p_r p_s})$. The following lemma gives an important property of h_{rs} which allows checking condition A for all h_{rs} in $O(n)$ time.

Lemma 8. *Let $(p_0, p_1, \ldots, p_{n-1})$ be a convex polygonal curve.*

1. $h_{rs} \leq h_{r,s+1}$ for $0 \leq r < s < n - 1$,

2. $h_{rs} \leq h_{r+1,s}$ for $0 < r + 1 < s \leq n - 1$.

Proof. The proof of $h_{rs} \leq h_{r,s+1}$ and that of $h_{rs} \leq h_{r+1,s}$ are similar and so only the former case will be shown here. Refer to Figure 7 (a) and (b), since $d(\overrightarrow{p_r p_s}, p_k) < d(\overrightarrow{p_r p_s}, p_{h_{rs}})$, it can be shown easily that $d(\overrightarrow{p_r p_{s+1}}, p_k) < d(\overrightarrow{p_r p_{s+1}}, p_{h_{rs}})$ for all $r < k < h_{rs}$, proving that $h_{rs} \leq h_{r,s+1}$. □

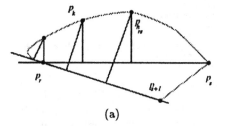 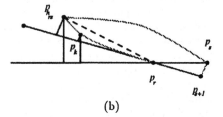

| (a) | (b) |

Fig. 7. Relationship between h_{rs} and $h_{r,s+1}$

In order to find $f(r)$, points p_s with $s > r$, have to be checked sequentially whether $e(\overrightarrow{p_r p_s}) \leq \epsilon$ (condition A), i.e., $d(\overrightarrow{p_r p_s}, h_{rs}) \leq \epsilon$. Thus we have to find h_{rs} efficiently. The above lemmas allow the search of h_{rs} to start from the point previously stopped, i.e., $h_{r-1,s}$.

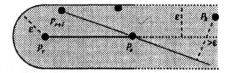

Fig. 8. p_k may not be closer to $\overrightarrow{p_{r+1} p_s}$ than to $\overrightarrow{p_r p_s}$

As for the checking of conditions B and C, one might try to construct G with G^1 and G^2 as in the previous section. Unfortunately, this approach would take more than $O(n)$ time as the property described in Corollary 6 cannot be carried forward to the construction of G^1 and G^2. Let us consider the following example (Figure 8). $d(\overrightarrow{p_r p_s}, p_k) \leq \epsilon$ for all k, $r \leq k \leq s$ but $d(\overrightarrow{p_{r+1} p_s}, p_k) > \epsilon$ for some k', $r + 1 < k' < s$. Checking these p_k's, whether they are more than distance ϵ

away from the endpoints(conditions B and C) by brute-force in order to define $f(r+1), f(r+2), \ldots$, etc., would take $O(n^2)$ time. In the following we shall show how $f(r)$ can be found in $O(n)$ time.

The main problem is to check conditions B and C efficiently. Instead of checking whether every point p_k is distance ϵ away from p_r or p_s, we shall keep track of two indices $\alpha(r) > r$ and $\beta(r) < r$ for each point p_r such that all points p_k with $r < k < \alpha(r)$ (or $\beta(r) < k < r$) would have satisfied condition B (or C) already and $p_{\alpha(r)}$ (or $p_{\beta(r)}$) would be the most likely candidate which would violate condition B (or C). Note that $d(p_r, p_{\alpha(r)})$ and $d(p_r, P_{\beta(r)})$ must be greater than ϵ.

With $\alpha(r)$ defined for $0 \le r < n-1$ and $\beta(r)$ for $0 < r \le n-1$ respectively, we can start sketching the algorithm for constructing $f(r)$. Starting with $r = 0$, $s = 1, 2, \ldots$, are tested sequentially as possible candidates for $f(r)$. For each s, condition A is tested, i.e., whether $e(\overrightarrow{p_0 p_s}) \le \epsilon$, by sequential search for h_{0s} (Lemma 7) starting from $h_{0,s-1}$ (Lemma 8). At the same time, condition B can be checked in constant time by testing whether $\angle p_s p_0 p_{\alpha(0)} \le \pi/2$ based on the property that condition B has already been satisfied for all the points $p_1, p_2, \ldots, p_{\alpha(0)-1}$ and $d(p_0, p_{\alpha(0)}) > \epsilon$. Condition B could not be satisfied if $\angle p_s p_0 p_{\alpha(0)} > \pi/2$. Similarly, condition C can be checked by testing whether $\angle p_0 p_s p_{\beta(s)} \le \pi/2$. Candidates for $f(0)$ are checked sequentially until any one of these conditions A, B or C is not satisfied. Thus finding $f(0)$ takes $O(f(0))$ time. The searching of $f(1)$ can be started from $f(0)$ as well as h_{1s} from h_{0s} (Lemma 8) and this process will take no more than $O(f(1) - f(0))$ time. Similarly, $f(r)$, for $r = 2, 3, \ldots, n-2$, can be found sequentially in $O(f(r) - f(r-1))$ time. Thus given $\alpha(r)$, $0 \le r < n-1$, and $\beta(r)$, $0 < r \le n-1$, respectively, all the $f(r)$'s, and thus G, can be defined in no more than $O(n)$ time. The algorithm for defining $f(r)$ is shown by procedure $Compute\_f(n)$.

```
procedure Compute_f(n)
begin
  r := 0; s := α(0); h := 0
  while r < n - 1 do
  begin
    h := Peak(p_r p_s, h)   { h is h_rs }
    while s < n and d(p_r p_s, p_h) ≤ ε and
      not (α(r) < s and ∠p_α(r) p_r p_s > π/2) and
      not (β(s) > r and ∠p_β(s) p_s p_r > π/2) do
    begin
      s := s + 1
      h := Peak(p_r p_s, h)
    end
    output f(r) = s - 1
    r := r + 1
  end
end
function Peak(L, h)
begin
  k := h
  while d(L, p_k) < d(L, p_{k+1}) do k := k + 1
  return k
end
```

Now we describe the method for computing $\alpha(r)$ for all $0 \le r < n-1$. $\beta(s)$, $0 < s \le n-1$, can be computed similarly. $\alpha(0)$ is chosen to be the smallest index s such that $d(p_0, p_s) > \epsilon$. With $\alpha(0)$ computed, we want to find $\alpha(1)$. If $d(p_1, p_{\alpha(0)}) > \epsilon$, then let $\alpha(1)$ be $\alpha(0)$. In general, we can find $\alpha(r)$ by searching sequentially from $\alpha(r-1)$ for the smallest index s such that $s \ge \alpha(r-1)$ and $d(p_r, p_s) > \epsilon$. If such s does not exist, i.e., $d(p_r, p_s) \le \epsilon$ for all $\alpha(r-1) \le s \le n-1$, $\alpha(r)$ is set to n. The algorithms to compute $\alpha(r)$ and $\beta(s)$ are shown by procedures $Compute\_\alpha(n)$ and $Compute\_\beta(n)$ respectively.

The following lemma shows that all points p_k, $r < k < \alpha(r)$, satisfy condition B for any line segment $\overline{p_r p_s}$.

```
procedure Compute_α(n)                procedure Compute_β(n)
begin                                 begin
    r := 0                                s := n - 1
    s := 1                                r := s - 1
    while r < n - 1 do                    while s > 0 do
    begin                                 begin
        while s < n and d(p_r, p_s) ≤ ε do    while r ≥ 0 and d(p_r, p_s) ≤ ε do
        begin                                 begin
            s := s + 1                            r := r - 1
        end                                   end
        output α(r) = s                       output β(s) = r
        r := r + 1                            s := s - 1
    end                                   end
end                                   end
```

Lemma 9. *For any k, $r < k < \alpha(r)$, and s, $k < s \leq n - 1$, if $\angle p_k p_r p_s > \pi/2$, $d(p_r, p_k) \leq \epsilon$.*

Proof. The lemma is true for $r = 0$ since $d(p_0, p_k) \leq \epsilon$ for all $k < \alpha(0)$. For $0 < r < k < \alpha(r)$ and $k < s \leq n - 1$, consider Figure 9. From the algorithm, since $\alpha(r) > k$, there must exist some vertex $p_{r'}$ where $0 \leq r' \leq r$ with $d(p_{r'}, p_k) \leq \epsilon$. Without loss of generality, assume $r \neq r'$. With the assumption stated in the lemma, $\angle p_k p_r p_{r'} \geq \angle p_k p_r p_s > \pi/2$. It is easy to see that $d(p_r, p_k) \leq \epsilon$. □

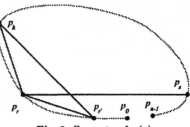

Fig. 9. Property of $\alpha(r)$

Note that $d(p_r, p_{\alpha(r)}) > \epsilon$ if $\alpha(r) \neq n$. The following lemma shows that condition B for $\overline{p_r p_s}$ is not satisfied by all points p_k such that $r < k < s$ if and only if $r < \alpha(r) < s$ and $\angle p_{\alpha(r)} p_r p_s \geq \pi/2$.

Lemma 10. *Assume $0 \leq r < s < n$. $d(p_r, p_k) > \epsilon$ for some k such that $r < k < s$ and $\angle p_k p_r p_s > \pi/2$ if and only if $r < \alpha(r) < s$ and $\angle p_{\alpha(r)} p_r p_s > \pi/2$.*

Proof. It is easily to see the *if* part is true since from the definition of $\alpha(r)$, as $\alpha(r) \neq n$, $d(p_r, p_{\alpha(r)}) > \epsilon$.

The *only if* part can be proved by contrapositive. Assume $\alpha(r) \geq s$ or $\angle p_{\alpha(r)} p_r p_s \leq \pi/2$. It follows from Lemma 9 that $d(p_r, p_k) \leq \epsilon$ for all those points p_k with $\angle p_k p_r p_s > \pi/2$ and $r < k < \alpha(r)$. Since $r < k < \alpha(r)$ covers the case with $\alpha(r) \geq s$, what remains to be shown is for the assumption when $\angle p_{\alpha(r)} p_r p_s \leq \pi/2$ with $\alpha(r) < s$. Since the given polygon P is convex, $\angle p_k p_r p_s \leq \angle p_{\alpha(r)} p_r p_s \leq \pi/2$ for all points p_k with $\alpha(r) \leq k < s$. Thus the lemma is proved. □

After G of a convex polygonal curve has been constructed, we can apply a greedy approach to find the shortest path from v_0 to v_{n-1}. For each step, we go as "far" as possible until v_{n-1} is reached, i.e., the path is $R(n - 1) = (v_{i_0}, v_{i_1}, \ldots, v_{i_m})$ where $v_{i_0} = v_0, v_{i_m} = v_{n-1}, i_{j+1} = f(i_j)$ for $0 \leq j \leq m - 1$. We shall prove in the following lemma that $R(n - 1)$ is a shortest path from v_0 to v_{n-1}.

Lemma 11. *Let $G = (V, E)$ be the corresponding directed graph constructed for a convex polygonal curve $P = (p_0, p_1, \ldots, p_{n-1})$. For any $1 \leq k \leq n-1$, $R(k) = (v_0 = v_{i_0}, v_{i_1}, \ldots, v_{i_m} = v_k)$ with $i_{j+1} = f(i_j)$ for $0 \leq j < m-1$ and $i_m \leq f(i_{m-1})$, is a shortest path from v_0 to v_k.*

Proof. By induction on k. Clearly the lemma is true for $k = 1$. Assume that the lemma is true for all $k' \leq k-1$ with $k \leq n-1$. For the case of k: If $(v_0, v_k) \in E$, $R(k) = (v_0, v_k)$ and the lemma holds trivially. Otherwise any shortest path from v_0 to v_k has at least two edges. Let $R^*(k) = (v_0 = v_{j_0}, v_{j_1}, \ldots, v_{j_{l-1}}, v_{j_l} = v_k)$ be a shortest path from v_0 to v_k. Clearly $(v_{j_0}, v_{j_1}, \ldots, v_{j_{l-1}})$ is a shortest path from v_0 to $v_{j_{l-1}}$. By induction hypothesis, $R(j_{l-1})$ is also a shortest path from v_0 to $v_{j_{l-1}}$. Thus, the new $R^*(k)$ with $(v_{j_0}, v_{j_1}, \ldots, v_{j_{l-1}})$ in $R^*(k)$ replaced by $R(j_{l-1})$ is also a shortest path from v_0 to v_k. Let $R(j_{l-1}) = (v_0 = v_{i_0}, v_{i_1}, \ldots, v_{i_{l-1}} = v_{j_{l-1}})$. Note that $f(i_{l-2}) < k$; otherwise $(v_0, v_{i_1}, \ldots, v_{i_{l-2}}, v_k)$ would be a path from v_0 to v_k with length shorter than that of $R^*(k)$, contradicting the fact that $R^*(k)$ being the shortest. Since $i_{l-1} \leq f(i_{l-2}) < k$ and $(v_{i_{l-1}}, v_k) \in E$, $(v_{f(i_{l-2})}, v_k) \in E$ by Corollary 6. Thus $(v_0 = v_{i_0}, v_{i_1}, \ldots, v_{i_{l-2}}, v_{f(i_{l-2})}, v_k) = R(k)$ as specified in the lemma with $m = l$ is also a feasible path from v_0 to v_k and has the same length as $R^*(k)$. Thus, the lemma is true for k. □

Theorem 12. *Given $\epsilon > 0$, and a convex polygonal curve $P = (p_0, p_1, \ldots, p_{n-1})$, the min-# problem can be solved in $O(n)$ time.* □

4 Closed Polygonal Curves

A polygonal curve is *closed* when there is an edge joining p_{n-1} and p_0, i.e., $p_n = p_0$. The min-# problem for a closed polygonal curve can be solved by considering n separate open curve problems by breaking up the closed curve at each point p_i. Since each min-# problem can be solved in $O(n^2)$ time, the closed curve min-# problem can be solved in $O(n^3)$ time. Similarly, the min-# problem for a closed convex polygonal curve can also be solved in $O(n^2)$ time. We shall show in the following how the problem for the convex curve can be solved in $O(n)$ time.

First of all, we construct the associated G for a closed polygonal curve P, assuming the vertices wrap-around at modulo n. The construction of $f(r), \alpha(r)$ and $\beta(r)$, $0 \leq r \leq n-1$ are similar except that at most $2n$ points have to be considered. Thus the associated G for a closed polygonal curve P can be constructed in $O(n)$ time.

Our next step is to apply greedy approach to construct a sequence of vertices $(v_{i_0} = v_0, v_{i_1}, v_{i_2}, \ldots, v_{i_m})$ with the property that $i_j = f(i_{j-1})$ for $j = 1, 2, \ldots, m$ and $i_{m-1} \leq n-1 < i_m$. From the result in the previous section, we can argue that the optimal approximate curve P_0' for the open curve $P_0 = (p_0, p_1, \ldots, p_{n-1}, p_0)$ consists of m line segments. In fact we can also conclude that the optimal approximate curve P_c' for the closed curve $P_c = (p_0, p_1, \ldots, p_{n-1})$ has at least $m-1$ line segments. Assume the contrary that P_c' consists of only $m-2$ line segments. If p_0 is in P_c, we would break up P_c' at p_0 and arrive at an approximate curve for P_0, which has only $m-2$ line segments. If p_0 is not in P_c', we can replace the line segment $\overline{p_i p_{i+1}}$ in

P'_c, which "covers" p_0, by two line segments $\overline{p_{i_j}p_0}$ and $\overline{p_0 p_{i_{j+1}}}$ and can also arrive at another approximate curve for P_0 which has $m-1$ line segments. Both cases lead to the contradiction that P'_0 must consists of m line segments. Thus we can conclude that P'_0 has either $m-1$ or m line segments. In addition, if P_c is broken up at any point into an open curve, the approximate curve for that open curve has at most $m+1$ and at least $m-1$ line segments.

Given the approximate curve $P'_0 = (p_{i_0} = p_0, p_{i_1}, \ldots, p_{i_m})$ for P_0, it can be shown that for any line segment $\overline{p_{i_j} p_{i_{j+1}}}$ in P'_0, $0 \leq j < m-1$, that the optimal approximate curve P'_c must consist some vertex p_k with $i_j \leq k \leq i_{j+1}$. Thus we can select the line segment with minimum difference $(i_{j+1} - i_j)$ and find the approximate curve for each open curve $(p_k, p_{k+1}, \ldots, p_{n-1}, p_0, p_1, \ldots, p_{k-1}, p_k)$ where $i_j \leq k \leq i_{j+1}$. Note that one of these approximate curves would be the optimal polygonal curve P'_c for P_c. Since each approximate curve has at most $m+1$ edges, it takes $O(m)$ time to search G to find the shortest path and the approximate curve. As the minimum difference $(i_{j+1} - i_j) \leq n/(m-1)$, the total time to find P'_c would be no more than $O(m\frac{n}{m}) = O(n)$ time.

Theorem 13. *Given $\epsilon > 0$ and a closed polygonal curve $P_c = (p_0, p_1, \ldots, p_{n-1})$, the min-# problem for P_c can be solved in $O(n^3)$ time and in $O(n)$ time when P_c is convex.* □

References

1. P.J. Burt. Fast filter transforms for image processing. *Computer Graphics and Image Processing 16*, pp. 20-51,1981.
2. B.M. Chazelle. *Computational Geometry and Convexity*. Ph.D Thesis, Yale Univ., 1980.
3. F. Chin, A. Choi and Y. Luo. Optimal generating kernels for image pyramids by Linear fitting. In *Proc. 1989 Int. Symp. on Computer Architecture and Digital Signal Processing* (Hong Kong, Oct. 11-14, 1989), 612-617.
4. F. Chin, A. Choi and Y. Luo. Optimal generating kernels for image pyramids by Piecewise fitting. *IEEE Trans. Pattern Anal. Machine Intell.* (to appear).
5. L. Guibas, J. Hershberger, J. Mitchell and J.S. Snoeyink. Approximating Polygons and Subdivisions with Minimum Link Paths. *Technical Report 92-5*, University of British Columbia, March 1992.
6. S. L. Hakimi and E. F. Schmeichel. Fitting polygonal functions to a set of points in the plane. *CVGIP, Graphical Models and Image Processing*, Vol. 53 (1991), pp. 132-136.
7. H. Imai and M. Iri. Polygonal Approximations of a Curve-Formulations and Algorithms. In G. T. Toussaint, editor, *Computational Morphology*. North Holland, 1988
8. P. Meer, E.S. Baughter and A. Rosenfeld. Frequency domain analysis and synthesis of image pyramid generating kernels. *IEEE Trans. Pattern Anal. Machine Intell.*, vol. PAMI-9, pp. 512-522, July 1987.
9. A. Melkman and J. O'Rourke. On Polygonal Chain Approximation. In G. T. Toussaint, editor, *Computational Morphology*. North Holland, 1988
10. A. Rosenfeld. (ed.) *Multiresolution Image Processing and Analysis*. New York, Springer-Verlag, 1984.
11. G. T. Toussaint. On the Complexity of Approximating Polygonal Curves in the Plane. In *Proc. IASTED, International Symposium on Robotics and Automation, Lugano, Switzerland, 1985*

Wiring Knock-Knee Layouts:
A Global Approach*

Majid Sarrafzadeh[1], Dorothea Wagner[2], Frank Wagner[3], Karsten Weihe[2]

[1] Department of Electrical Engineering and Computer Science, Northwestern University, Evanston, IL 60208, USA, majid@epsilon.eecs.nwu.edu
[2] Fachbereich Mathematik, Technische Universität Berlin, Straße des 17. Juni 136, W-1000 Berlin 12, Germany, dorothea@combi.math.tu-berlin.de resp. karsten@combi.math.tu-berlin.de
[3] Institut für Informatik, Fachbereich Mathematik, Freie Universität Berlin, Arnimallee 2-6, W-1000 Berlin 33, Germany, wagner@tcs.fu-berlin.de

Abstract. We present a global approach to solve the three-layer wirability problem for knock-knee layouts. In general, the problem is $\mathcal{NP}$-complete. Only for very restricted classes of layouts polynomial three-layer wiring algorithms are known up to now. In this paper we show that for a large class of layouts a three-layer wiring can be constructed by solving a path problem in a special class of graphs or a two-satisfiability problem, and thus may be wired efficiently. Moreover, it is shown that a minimum stretching of the layout into a layout belonging to this class can be found by solving a clique cover problem in an interval graph. This problem is polynomially solvable as well. Altogether, the method also yields a good heuristic for the three-layer wirability problem for knock-knee layouts.

1 Introduction

Routing is an important problem encountered in the design of integrated circuits. After placement and global routing, in the detailed routing phase the course of the wires connecting the cells is determined. Since layouts containing *crossings* or *knock-knees* (points where two wires bend) cannot be realized in a single plane, normally the routing is carried out in two steps, the *layout* and the layer assignment called *wiring*. In case of knock-knee layouts the wiring, i.e., the conversion of a layout in the plane to an actually three dimensional configuration of wires to avoid contacts between different wires, is a non-trivial task.

There have been made several contributions to this problem [1], [2], [8], [9], [11], [16], in general all based on a systematic approach developed by Lipski & Preparata in [15] and [12] for grid based layouts. This combinatorial framework is derived from the observation, that any wiring induces a partition of the layout into a two-colorable

---

* Part of this work was done while Majid Sarrafzadeh, Dorothea Wagner and Frank Wagner were with the LEONARDO FIBONACCI INSTITUTE for the Foundations of Computer Science, Trento, Italy. Majid Sarrafzadeh also acknowledges the National Science Foundation for supporting this research in part under grant MIP-8921540. Dorothea Wagner and Karsten Weihe acknowledge the Deutsche Forschungsgemeinschaft for supporting this research under grant Mö 446/1-3.

map containing diagonal partition lines induced by the knock-knees and additional vertical and horizontal partition lines. Conversely, for wirability in two, three, four or more layers, equivalent conditions for the corresponding partitions are given. One consequence of this approach is that it is $\mathcal{NP}$-complete to decide if a given layout is wirable in three layers [11]. But every layout is easily wirable in four layers [1], [16]. Only very restricted layouts are two-layer wirable. Hence, wirability in three layers is the best one can expect for a given layout in general.

Essentially, there are two different approaches to attack this problem. One possibility is to go one step back within the design process and consider the routing problem itself, i.e. to aim at layouts that are provably three-layer wirable. For channel routing problems, algorithms that guarantee three-layer wirable layouts are given in [7], [10], [14], [15] and [18]. The proof of three-layer wirability strongly depends on the very special structure of these layouts, i.e. they contain at most two knock-knees per vertical line, and in case of two knock-knees these even lie in opposite directions.

The corresponding three-layer partitions contain only vertical partition edges in addition to the diagonals induced by knock-knees.

Such special layouts cannot be expected for more general routing problems and in most cases do not even exist.

The second approach is to transform a layout into a three-layer wirable layout by appropriate *stretchings* [2], [8], [9]. Stretching a layout increases the area required for the layout. Obviously, the problem to find a minimum stretching for three-layer wirability is $\mathcal{NP}$-hard even within only one dimension.

In this paper we consider the problem if for a given layout there is a three-layer partition containing only vertical or only horizontal partition edges (in addition to the diagonal edges induced by the knock-knees). This problem can be formulated as a path problem in a certain graph or as a two-satisfiability problem, and may be decided in time at most linear in the layout area. Local layout modifications, such as those applied in [15] and [10], and the local use of horizontal edges within the vertical approach, or vice versa, are contained as well. Moreover, the method yields the minimum number of vertical lines (resp. tracks) to be added to transform a layout into a stretched layout that admits a legal partition for three-layer wirability with only vertical (resp. horizontal) additional partition edges. In the most general case, this problem is equivalent to the minimum clique cover problem in interval graphs, which is solvable in time linear in the size of the graph [6].

2 Preliminaries

In this section we review the basic definitions and results from [12] concerning the wiring of knock-knee layouts.

Consider a layout in a rectilinear grid graph. A *conducting layer*, or simply *layer* is a graph isomorphic to the layout grid. Conducting layers $L_1, \ldots, L_k$ are assumed to be stacked on top of each other, with L_1 on the bottom and L_k on the top. A contact between two layers, called a *via*, can be placed only at a grid vertex.

A correct *layer assignment* or *wiring* of a given layout is a mapping of each edge of a wire to a layer, such that:

1. No two different wires share a vertex on the same layer.
2. If adjacent edges of a wire are assigned to different layers, a via is established between these layers at their common vertex.
3. If a via connects L_h and L_j ($h < j$), then layers $L_i, h < i < j$, are not used at that vertex by any other wire.

Observe, that a correct wiring can be interpreted as *vertex-disjoint* paths through a three-dimensional grid.

To determine a correct wiring of a knock-knee layout only those grid vertices where two different wires share a vertex, i.e. cross or form a knock-knee, are of relevance. Denote the part of a layout induced by the unit squares around these grid vertices as the *core* of the layout. Then the following lemma holds.

Lemma 1. *[12] A layout is wirable in k layers iff each connected component of its core is.*

The basic idea now is, that any correct wiring of a layout in a fixed number of layers induces a partition of the layout area into the following two types of regions.

- *V-region*: The region where vertical wire edges lie above horizontal ones.
- *H-region*: The region where horizontal wire edges lie above vertical ones.

By this partition, the wiring can be viewed as a *two-colored map*. Obviously, the entire unit square around a crossing in the layout belongs to one color region. Since two wires that form a knock-knee cannot change their relative position through their common grid vertex in a correct wiring, the unit square around a knock-knee must belong to both regions. So, consider the unit square around a knock-knee and the diagonal of the unit square that crosses both bends of the knock-knee. Then the "triangle" above this diagonal belongs to the V-region, and the triangle below to the H-region, or vice versa. Consequently, the set of partition edges P contains all diagonals through knock-knees and, in addition, "appropriate" vertical and horizontal edges of the dual grid. The following properties of two-colorable maps state necessary conditions how a partition inducing a correct wiring has to look like.

Lemma 2. *[12] A set P of diagonals and dual grid edges is a two-colorable map iff*

1. *each interior vertex (of the dual of the grid graph) is incident with an even number of edges of P,*
2. *each connected component of the boundary of the layout core is incident with an even number of edges of P.*

The main result concerning the three-layer wirability problem can then be stated as follows.

Theorem 3. *[12] The core of a layout is three-layer wirable iff there exists a two-colorable map containing none of the eight patterns shown in Figure 1.*

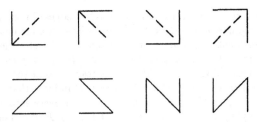

Fig. 1. Forbidden patterns for a three-layer partition. Broken lines denote the absence of edges.

In the following, we always consider the core of the layout. By a *vertical elementary stretching* along a column $\vec{i}$ (between vertical line i and $i + 1$) we denote the operation that modifies the layout by cutting it along $\vec{i}$, moving its pieces horizontally one unit apart and inserting a new horizontal wire edge where a wire was crossing $\vec{i}$. This means that all wires crossing $\vec{i}$ are "stretched" as well. Accordingly, in the routing grid a new vertical line is added. A *horizontal elementary stretching* is defined analogously. Any sequence of horizontal and vertical elementary stretching operations is simply called a *stretching*.

3 The New Wiring Approach

3.1 Constructing a legal partition

Consider the following problem.

- **Given** A layout.
- **Problem** Construct a legal three-layer partition containing only vertical (horizontal) partition edges additional to the diagonals through knock-knees. We call such a partition a *V-partition* (resp. *H-partition.*)

If an edge of the grid belongs to a component of the core, we say that it is *covered* by that core component. We call a connected core component *V-convex* if the set of vertical edges of each vertical grid line covered by the core component forms one connected interval. (H-convex may be defined analogously.) For simplicity, let us first consider only layouts whose core is connected and of V-convex shape.

A canonical approach to construct a V-partition is to scan the layout from left to right and add vertical partition edges in the dual grid depending on the knock-knees of the layout such that a two-colorable map without forbidden patterns arises. (An analogous method can be applied horizontally.)

The following lemma yields an efficient procedure to add vertical partition edges correctly.

Lemma 4. *Let P be a two-colorable map of a layout containing only vertical additional partition edges. Then for any vertical edge e of the grid dual to the layout grid, the state of e, i.e. $e \in P$ or $e \notin P$, uniquely determines the state of all vertical edges e' lying on the same vertical grid line as e.*

Proof. Consider a vertical line L (as a set of vertical edges) in the grid dual to the layout, and the diagonals induced by knock-knees that are incident to an edge of L. Call a vertex incident with one or three diagonals *odd*, all other vertices *even*. Then the set of odd vertices belonging to an edge of L partitions L into intervals $I_1,\ldots,I_k$ say, of edges between two odd vertices resp. an odd vertex and the upper or lower boundary. For a vertical edge e of L, $e \in P$ induces that the interval, say I_j, containing e is completely contained in P, as well as every second interval above and below I_j. Precisely, for a vertical edge e' of L, $e' \in P$ iff e' belongs to an odd indexed interval in case j is odd, resp. to an even indexed interval in case j is even.

Thus, for any vertical line of the grid dual to the layout, there are only two possibilities which vertical edges belong to a V-partition. The decision which possibilities can be chosen for one vertical line only depends on the state of its two neighboring lines. (Look at the forbidden patterns in Figure 1.)

The problem of constructing a V-partition can be formulated as a *path problem* in a directed graph $G_p = (V, E_p)$ as follows. For each vertical line i in the grid dual to the layout, we introduce two vertices v_i^0, v_i^1 corresponding to the two possible states (i.e. v_i^1 corresponding to "the uppermost edge belongs to P"). There are edges only between vertices corresponding to neighboring vertical lines, where v_i^j, $v_{i+1}^k \in E_p$ iff there is no forbidden pattern between line i and line $i+1$ in case i has state j and $i+1$ has state k, for $j, k \in \{0, 1\}$. In addition, we introduce a source s and a target t, and edges from s to the two vertices corresponding to the leftmost vertical line resp. from the two vertices corresponding to the rightmost vertical line to t. We call G_p the *path graph* of the layout. See Figure 2.

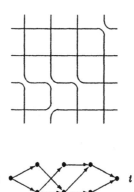

Fig. 2. A layout core and its path graph.

Corollary 5. *The problem to find a V-partition for a V-convex layout core is equivalent to finding an s-t-path in the corresponding path graph.*

The time complexity for constructing the path graph is linear in size of the layout area. In case the knock-knee positions are given in an appropriate order, the time complexity is only linear in the number of knock-knees. Obviously, the time complexity for solving the path problem is linear in the spread of the layout.

Now, consider the general case that the layout core consists of more than one component of arbitrary shape, possibly containing holes. First we informally describe the situation. Observe that by Lemma 1, the components of the layout core may be considered independently of each other. For a vertical line covered by a component C, the edges contained in C are partitioned into maximal vertical intervals. In principle, there are two possible choices of vertical partition edges for each of these maximal intervals. Moreover, forbidden patterns can only appear between neighboring intervals that overlap a common track. By that, the path graph for C arises canonically as a directed graph containing a set of pairs of vertices for each vertical line covered by C and possible edges between vertices corresponding to neighboring intervals that overlap a common track. The problem to find a V-partition for C transforms to several path problems in the path graph that have to be solved simultaneously.

It is more convenient to formulate the simultaneous solvability of these path problems as a *satisfiability problem*, where each clause consists of two literals (2SAT).

Let us first formulate the 2SAT problem corresponding to the simple path problem induced by a layout core of V-convex shape. We introduce a variable x_i for each pair of vertices v_i^0, v_i^1 (resp. vertical line i). For the set of possible edges from a v_i-vertex to a v_{i+1}-vertex (resp. state combinations), we have a set of at most four V-clauses containing x_i and x_{i+1}, negated or non-negated. Actually, we get one clause for each missing edge (resp. forbidden state combination). See Figure 3. Then each s-t-path corresponds to a satisfying truth assignment for the set of all corresponding clauses.

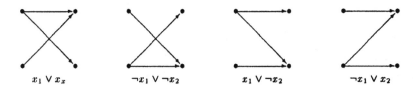

$x_1 \vee x_x$ $\qquad\qquad \neg x_1 \vee \neg x_2$ $\qquad\qquad x_1 \vee \neg x_2$ $\qquad\qquad \neg x_1 \vee x_2$

Fig. 3. The clauses corresponding to state combinations.

In the general case of a core component C of non-convex shape, for example with holes, we have one variable for each maximal interval of a vertical line that belongs to C. The clauses correspond to all forbidden state combinations between neighboring intervals overlapping a common track. Then the existence of a V-partition of C is equivalent to a truth satisfying all corresponding clauses.

2SAT can be solved in time linear in the number of clauses [3]. The number of clauses corresponding to one column of the layout is linear in the number of tracks occupied by the layout. Thus, the total number of clauses for a layout is linear in the layout size. We obtain the following result.

Theorem 6. *For a layout, the existence of a V-partition can be decided in time linear in the layout size. If yes, a V-partition can be determined in linear time as well.*

3.2 Layout Modifications

There are even layouts containing at most two knock-knees per column, where the knock-knees are of opposite direction, as those guaranteed in [15] resp. [10], that do not admit a V-partition.[4]

In such a case however, in [15] resp. [10] V-partitions are constructed for slightly modified layouts. Such layout modifications can also be included into our approach.

A *local layout modification* can be applied to a layout, whenever two nets touch twice in the same column, or in two neighboring columns. Then the layout can be transformed into an equivalent layout by replacing knock-knees by crossings and crossings by knock-knees. See Figure 4.

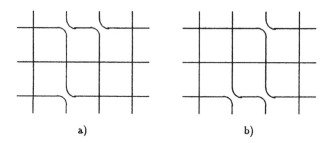

a)   b)

Fig. 4. A layout modification.

Such a layout modification in one resp. two neighboring columns induces a new partition of each of the related vertical lines into intervals between odd vertices. Thus, a possible layout modification can be introduced in the path graph as follows. For vertical lines $i, i+1$ (resp. $i, i+1, i+2$) involved in the modification join additional vertices $u_i^0, u_i^1, u_{i+1}^0, u_{i+1}^1, (u_{i+2}^0, u_{i+2}^1)$ and appropriate edges from v_{i-1}-vertices to u_i-vertices, from u_i-vertices to u_{i+1}-vertices and from u_{i+1}-vertices to v_{i+2}-vertices (resp. from u_{i+1}-vertices to u_{i+2}-vertices and from u_{i+2}-vertices to v_{i+3}-vertices). See Figure 5. In a similar way, the local use of horizontal partition edges may be involved.

---

[4] These examples have been found by applying the implemented method to layouts generated by the algorithms from [15] resp. [10]. Indeed, these examples are fairly large (spread 200). Because of lack of space, they are omitted.

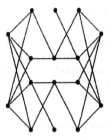

Fig. 5. Part of the path graph corresponding to the layout modification in figure 4. The upper part corresponds to layout a) and the lower part to the modified layout shown in b).

3.3 Stretching for wirability

The problem of finding a minimum stretching of a layout such that the stretched layout is three-layer wirable is $\mathcal{NP}$-hard, even stretching within only one dimension. In this section we give a polynomial time algorithm for the following problem.

- **Given** A layout.
- **Problem** Find a minimum stretching such that the stretched layout admits a V-partition (H-partition).

First, we restrict ourselves to the case that the core of the layout consists of only one component of V-convex shape. Then the method given in Section 3.1 can be applied to this problem too. Consider the path graph $G_p = (V, E_p)$ for a layout. Then for the layout induced by an elementary vertical stretching along column $\overrightarrow{i}$, consider the path graph $G'_p = (V, E'_p)$ obtained by adding all possible edges between v_i-vertices and v_{i+1}-vertices, i.e.

$$E'_p = E_p \cup \{(v_i^j, v_{i+1}^k) : j, k \in \{0, 1\}\}.$$

Thus, any directed path in G terminating with v_i^0 or v_i^1 is in G' connected to any directed path in G leaving v_{i+1}^0 or v_{i+1}^1. Generally, a stretching that guarantees a V-partition corresponds to transforming the path graph G to a new graph G', where directed paths in G are combined to an s-t-path in G' containing at least one v_i-vertex for all vertical lines i of the layout. It follows that a minimum stretching to guarantee a V-partition is equivalent to a minimum number of directed paths in G that can be combined to an s-t-path containing v_i^0 or v_i^1 for all vertical lines i of the original layout.

By scanning the path graph G from s to t, the minimum number of appropriate directed paths, as well as the paths themselves can be determined in time linear in the spread of the layout.

Remark The problem can also be formulated as a shortest path problem in the *weighted path graph* $G_{WP} = (V, E_{WP}; w)$, where again V corresponds to the possible states of the vertical lines, but for all i, $(v_i^0, v_{i+1}^0), (v_i^0, v_{i+1}^1), (v_i^1, v_{i+1}^0), (v_i^1, v_{i+1}^1) \in E_{WP}$, as well as the edges from s and to t. The *weight* w is defined as $w(e) := 0$ for all $e \in E_{WP} \cap E_P$, and $w(e) := 1$ for $e \in E_{WP} \backslash E_P$. Then any shortest s-t-path (with respect to w) in G_{WP} is equivalent to a minimum stretching to guarantee a V-partition of the corresponding layout.

In case the core consists of different components or is not V-convex, the main problem is that for a minimum stretching different components, respectively two parts of the same component that are horizontally separated by a hole, must not be considered independently of each other. The following lemma helps to overcome this problem.

Lemma 7. *Consider m consecutive vertical grid lines in a V-convex layout core that do not admit a V-partition. If there are columns $\vec{i}$ and $\vec{j}$, $1 \leq i < j < m$, such that the layout part between 1 and m stretched along $\vec{i}$ or stretched along $\vec{j}$ admits a V-partition, then for any column $\vec{k}$, $i \leq k \leq j$, the layout stretched along $\vec{k}$ admits a V-partition.*

Proof. Look at the path graph of the layout. There is a directed path from some v_1-vertex to v_j^0 or v_j^1, and a directed path from v_i^0 or v_i^1 to some v_m-vertex. These paths are possibly vertex-disjoint, but overlap in the interval $[i, j]$ in the sense that both of them contain a v_k-vertex for any $k \in [i, j]$. Now, a stretching of the layout along an arbitrary column $\vec{k}$ between $\vec{i}$ and $\vec{j}$ transforms the path graph to a new path graph where these two paths are connected, and induce an s-t-path in the new path graph.

It can be proved:

Theorem 8. *A minimum stretching to admit a V-partition is equivalent to a minimum clique cover of the corresponding interval graph.*

Sketch of the proof: In the path graph of a V-convex layout core of spread n, for any $i, 1 \leq i \leq n$, there exist at most two disjoint paths overlapping at i. We can easily determine all maximal intervals where two disjoint paths overlap. These intervals are pairwise disjoint. Then, with Lemma 7, any choice of a set $S \subseteq \{1, \ldots, n\}$ containing exactly one element of each of these intervals induces a minimum stretching to admit a V-partition. Using this fact, the problem for a layout core consisting of V-convex components can be formulated directly as a minimum *clique cover problem* in *interval graphs*, which is solvable in time $\mathcal{O}(|V| + |E|)$ [6]. In case the interval representation is given, the problem can be solved even in time $\mathcal{O}(|V|)$.

A layout containing core components of arbitrary shape is partitioned into maximal V-convex subcomponents, say of rectangle shape. Then a column which lies just to the right of more than one V-convex subcomponent is called *critical*. For columns that are not critical, the problem is solved by applying the usual minimum clique cover algorithm.

So, consider a critical column $\vec{i}$. Let $I_r, \ldots, I_t$ be those intervals of grid line i which share some track with the same interval J of $i + 1$. Then the usual minimum clique cover algorithm for interval graphs must be applied such that the existence of a legal state combination is guaranteed for all pairs of intervals $(I_j, J), r \leq j \leq t$, simultaneously.

References

1. M. L. Brady, D. J. Brown. VLSI routing: Four layers suffice. In: *Advances in Computing Research 2 (VLSI Theory) (ed. F. P. Preparata) JAI Press* (1984) 245-257
2. M. L. Brady, M. Sarrafzadeh. Stretching a knock-knee layout for multilayer wiring. *IEEE Trans. Comput. 39* (1990) 148-152
3. S. Even, A. Itai, A. Shamir. On the complexity of time table and multicommodity flow problems. *SIAM J. Comput. 5* (1976) 691-703
4. M. Formann, D. Wagner, F. Wagner. Routing through a dense channel with minimum total wire length. *Proc. of the Second Ann. ACM-SIAM Symposium on Discrete Algorithms* (1991) 475-482
5. A. Frank. Disjoint paths in a rectilinear grid. *Combinatorica 2* (1982) 361-371
6. M. C. Golumbic. Algorithmic Graph Theory and Perfect Graphs. *Academic Press* (1980)
7. T. Gonzales, S. Zheng. Simple Three-layer channel routing algorithms. *Proc. of Aegean Workshop on Computing LNCS 319 (ed. J. H. Reif)* (1988) 237-246
8. T. Gonzales, S. Zheng. On ensuring three-layer wirability by stretching planar layouts. *INTEGRATION: The VLSI Journal 8* (1989) 111-141
9. M. Kaufmann, P. Molitor. Minimal stretching of a layout to ensure 2-layer wirability. *to appear in INTEGRATION: The VLSI Journal*
10. R. Kuchem, D. Wagner, F. Wagner. Area-optimal three-layer channel routing. *Proc. of the 30th Ann. Symposium on Foundations of Computer Science* (1989) 506-511
11. W. Lipski, Jr. On the structure of three-layer wirable layouts. In: *Advances in Computing Research 2 (VLSI Theory) (ed. F.P. Preparata) JAI Press* (1984) 231-243
12. W. Lipski, Jr., F. P. Preparata. A unified approach to layout wirability. *Mathematical Systems Theory 19* (1987) 189-203
13. K. Mehlhorn, F .P. Preparata. Routing through a rectangle. *J. ACM 33* (1986) 60-85
14. K. Mehlhorn, F. P. Preparata, M. Sarrafzadeh. Channel routing in knock-knee mode: Simplified algorithms and proofs. *Algorithmica 1* (1986) 213-221
15. F. P. Preparata, W. Lipski, Jr. Optimal three-layer channel routing. *IEEE Trans. on Computers 33* (1984) 427-437
16. I. G. Tollis. A new algorithm for wiring layouts. *Proc. of the Aegean Workshop on Computing LNCS 319 (ed. J.H. Reif)* (1988) 257-267
17. D. Wagner. A new approach to knock-knee channel routing. *Proc. of the International Symposium on Algorithms LNCS 557 (eds. W. L. Hsu, R. C. T. Lee)* (1991) 83-93
18. C. Wieners-Lummer. Three-layer channel routing in knock-knee mode. Preprint Universität Paderborn

Appendix We have implemented the methods described in Section 3. Applying this approach, examples of layouts constructed by the algorithms from [15] and [10] have been found where layout modifications are really necessary to obtain a legal V-partition. There also exist layouts containing only two knock-knees per column, e.g. computed by the algorithm from [17], that admit no such legal partition. But these examples are fairly large.

Experiments show that in most cases three-layer wirings exist for layouts containing only a small number of knock-knees, as e.g. those guaranteed by the algorithms from [17] and [13]. Even for layouts from [5], [4], where the total number of knock-knees may be linear in the number of grid points, only a small number of stretchings is necessary to derive three-layer wirable stretched layouts.

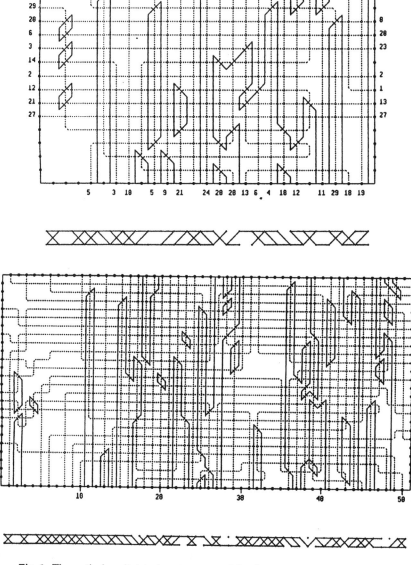

Fig. 6. The method applied to layouts computed by the rectangle routing algorithm of Mehlhorn and Preparata

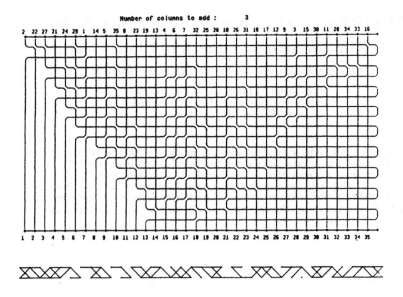

Fig. 7. The method applied to a layout computed by the algorithm of Frank.

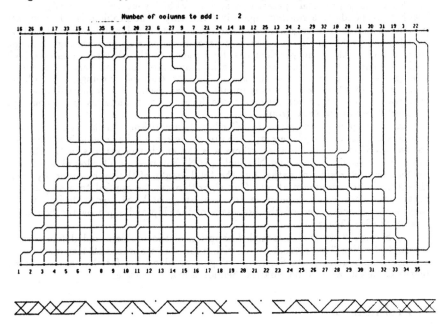

Fig. 8. The method applied to a layout computed by the wire length optimal algorithm of Formann, Wagner and Wagner.

Algorithms for Finding Non-Crossing Paths with Minimum Total Length in Plane Graphs

Jun-ya Takahashi, Hitoshi Suzuki and Takao Nishizeki

Department of Information Engineering
Faculty of Engineering, Tohoku University
Sendai 980, Japan

Abstract. Let G be an undirected plane graph with non-negative edge length, and let k terminal pairs lie on two specified face boundaries. This paper presents an algorithm for finding k "non-crossing paths" in G, each connecting a terminal pair, whose total length is minimum. Here "non-crossing paths" may share common vertices or edges but do not cross each other in the plane. The algorithm runs in time $O(n \log n)$ where n is the number of vertices in G.

1. Introduction

The shortest disjoint path problem, that is, to find k vertex-disjoint paths with minimum total length, each connecting a specified terminal pair, in a plane graph has many practical applications such as VLSI layout design. An edge in the graph corresponds to a routing region in VLSI chip. The problem is NP-complete [Lyn,KL], and so it is very unlikely that there exists a polynomial-time algorithm to solve the problem. If, however, two or more wires may pass through a single routing region, then the problem can be reduced to the shortest "non-crossing" path problem, where "non-crossing" paths may share common vertices or edges but do not cross each other in the plane. The shortest non-crossing path problem is expected to be solvable in polynomial time at least for a restricted case, for example, a case when all terminals are located on boundaries of a constant number of faces in a plane graph.

In this paper we give an $O(n \log n)$ algorithm which finds shortest non-crossing paths in a plane graph for the case when all the terminals are located on two specified face boundaries, where n is the number of vertices in G. For the same case Suzuki, Akama and Nishizeki [SAN] obtained an $O(n \log n)$ algorithm for finding *vertex-disjoint* paths, but the total length of the paths found by their algorithm is not minimum at all. Our algorithm can be applied to a single-layer routing problem which appears in the final stage of VLSI layout design, where each wire connects a pad on the boundary of the chip and a pin on the boundary of a block (See Figure 1).

In Section 2 we give a formal description of the problem and define several terms. In Section 3 we present an algorithm for the case where all terminals lie on a single face boundary. In Section 4 we give an algorithm for the case where all terminals lie on two face boundaries. Section 5 is a conclusion.

2. Preliminaries

In this section we give a formal description of the non-crossing path problem and define several terms. We denote by $G = (V, E)$ the graph consisting of a vertex set V and an edge set E. We sometimes denote by $V(G)$ and $E(G)$ the

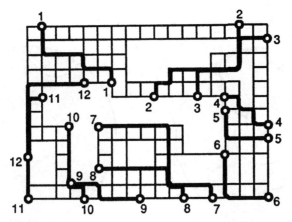

Fig.1. Shortest non-crossing paths in a grid graph.

vertex and edge sets of G, respectively. We assume that G is a 2-connected undirected plane graph and every edge in G has a non-negative edge length. Furthermore we assume that G is embedded in the plane $\mathbb{R}^2$. The image of G in $\mathbb{R}^2$ is denoted by $Image(G) \subset \mathbb{R}^2$. A *face* of G is a connected component of $\mathbb{R}^2 - Image(G)$. The *boundary* of a face is the maximal subgraph of G whose image is included in the closure of the face. A pair of vertices s_i and t_i which we wish to connect by a path is called a *terminal pair* (s_i, t_i). Let k be the number of terminal pairs. In this paper we do not assume that k is a constant. Suppose that all the terminals are located on boundaries B_1 and B_2 of two specified faces f_1 and f_2. Assume for simplicity that $V(B_1) \cap V(B_2) = \phi$ and all terminals are distinct from each other.

Let $P_1, P_2, \cdots, P_k$ be paths connecting the k terminal pairs. Let G^+ be a plane graph obtained from G as follows: add a new vertex v_{f_1} in face f_1, and join v_{f_1} to each terminal on B_1; similarly, add a new vertex v_{f_2} in face f_2, and join v_{f_2} to each terminal on B_2; the resulting graph is G^+. Let $P'_i, 1 \leq i \leq k$, be a path (or a cycle) in G^+ obtained from P_i by adding two new edges: one joining s_i to v_{f_1} or v_{f_2}, and the other joining t_i to v_{f_1} or v_{f_2}. We define paths $P_1, P_2, \cdots, P_k$ in a plane graph G to be *non-crossing* (for faces f_1 and f_2) if $Image(P'_i), 1 \leq i \leq k$, do not cross each other in the plane. Figure 2(a) depicts non-crossing paths P_1, P_2, P_3 and P_4. Non-crossing paths $P_1, P_2, \cdots, P_k$ are *shortest* if the sum of the lengths of $P_1, P_2, \cdots, P_k$ is minimum.

This paper presents an algorithm to solve the following *non-crossing path problem*.

Non-crossing path problem: Find shortest non-crossing paths, each connecting a terminal pair on two specified face boundaries in a plane graph G.

Figure 1 depicts shortest non-crossing paths in a grid graph where each edge has length 1.

Suppose that path P_1 connecting s_1 and t_1 has been decided. Then paths $P_1, P_2, \cdots, P_k$ are non-crossing (for faces f_1 and f_2) if paths $P_2, P_3, \cdots, P_k$ are non-crossing in a slit graph of G for P_1 defined as follows. A *slit graph* $G(P_1)$ *of* G *for path* P_1 is generated from G by slitting apart path P_1 into two paths P'_1 and P''_1, duplicating the vertices and edges of P_1 as follows (See Figure 2). Each vertex

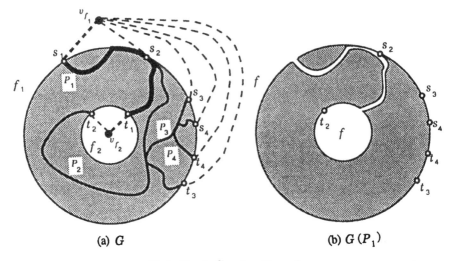

Fig.2. Graph G and a slit graph.

v in P_1 is replaced by new vertices v' and v''. Each edge (v_j, v_{j+1}) in P_1 is replaced by two parallel edges (v'_j, v'_{j+1}) and (v''_j, v''_{j+1}). Any edge (v, w) that is not in P_1 but is incident with a vertex v in P_1 is replaced by (v', w) if (v, w) is to the right of a path P_1 going from s_1 to t_1 through $image(P_1)$, and by (v'', w) if (v, w) is to the left of the path. The operation above is called *slitting G along P_1*. If a vertex $v \in V(B_i)$, $i = 1$ or 2, in P_1 is designated as a terminal in G, either v' or v'', that is incident with v_{f_i} in G^+, is designated as a terminal in the slit graph $G(P_1)$.

3. The Case when All the Terminals Lie on a Single Face Boundary

In this section we present an algorithm to solve the non-crossing path problem for the case when all the terminals are located on the boundary B of a single face f. We assume w.l.o.g. that f is the outer face of G. Let S be the set of terminal pairs. We separate this case into the following two cases:

CASE 1: the terminals $s_1, t_1, s_2, t_2, \cdots, s_k, t_k$ appear on B clockwise in this order when we interchange starting terminals s_i and ending terminals t_i and/or indices of terminal pairs if necessary.

CASE 2: otherwise.

We first present an algorithm PATH1(G, S) for Case 1 and then an algorithm PATH2(G,S) for Case 2. PATH1 first decomposes the graph G into k subgraphs $G_1, G_2, \cdots, G_k$ so that each subgraph G_i contains terminals s_i and t_i. It then finds a shortest path P_i between s_i and t_i in each graph G_i, and finally outputs the k shortest non-crossing paths $P_1, P_2, \cdots, P_k$. Denote by $P[v, w]$ the path connecting vertices v and w in a path or tree P.

procedure PATH1(G, S);
    begin
1.   let T be a shortest path tree containing shortest paths from s_1 to all s_i, $2 \leq i \leq k$;
2.   **for** $i := 1$ **to** k **do**
            begin
3.   let G_i be the maximal subgraph of G whose image is in the cycle consisting of two paths, the path $T[s_i, s_{i+1}]$ from s_i to s_{i+1} on tree T and the path on B counterclockwise going from s_{i+1} to s_i; $\{s_{k+1} = s_1\}$

4.   find a shortest path P_i between s_i and t_i in G_i;
            end;
5.   output $\{P_i | 1 \leq i \leq k\}$ {the shortest non-crossing paths}
    end;

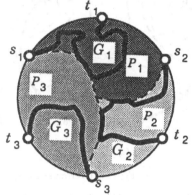

    In Figure 3 tree T is drawn in dotted lines and paths P_i in thick lines, and subgraphs G_1, G_2 and G_3 are colored in different gray tones. The following lemma guarantees the correctness of procedure PATH1.

Fig.3. Illustration for PATH1.

LEMMA 1. Let G_i, $1 \leq i \leq k$, be the subgraphs found in the procedure PATH1. Then graph G_i contains at least one of the shortest paths in G between terminals s_i and t_i.

    We now consider the execution time of PATH1. All the steps except lines 1 and 4 can be done in time $O(n)$. Line 1 which finds shortest paths from s_1 to all other vertices can be done in time $O(T(n))$, where $T(n)$ is the time required for finding shortest paths from a single vertex to all other vertices in a plane graph of n vertices. We claim that line 4 can be executed in time $O(T(n))$ in total. At line 4 each of the k shortest paths is found in a region of G bounded by B and tree T. Therefore every edge on T appears in at most two of the subgraphs $G_1, G_2, \cdots, G_k$, and any other edge of G appears in exactly one of them. Thus line 4 can be done in time $O(T(n))$ in total. Therefore the total running time of procedure PATH1 is $O(T(n))$.

    We next present an algorithm PATH2 for Case 2 using PATH1. Let $v_1, v_2, \cdots, v_b$ be the vertices on B, and assume that they appear on B clockwise in this order. We may assume w.l.o.g. that $s_1 = v_1$ and no terminals appear in the subpath of B counterclockwise going from $s_1(= v_1)$ to t_1. We may assume that, for each terminal pair (s_i, t_i), v_1, s_i and t_i appear on B clockwise in this order and that $s_1, s_2, \cdots, s_k$ appear on B clockwise in this order (see Figure 4(a)). For each vertex $v \in V(B)$, $index(v)$ denotes the index of v, that is, $index(v) = i$ if $v = v_i$. If $index(s_i) < index(s_j) < index(t_j) < index(t_i)$, then (s_i, t_i) is an *ancestor* of (s_j, t_j) and (s_j, t_j) is a *descendant* of (s_i, t_i). Note that non-crossing paths do not exist if $index(s_i) < index(s_j) < index(t_i) < index(t_j)$. Let (s_l, t_l) be the ancestor of (s_i, t_i) having the maximum index. Then (s_l, t_l) is the *parent* of (s_i, t_i), and (s_i, t_i) is a *child* of (s_l, t_l). Let T_g be the (genealogy) tree whose nodes correspond to terminal pairs and edges correspond to the relation of parent and child. Thus, if the terminal pair corresponding to a node p in T_g has a child, then an edge in T_g joins p to the node corresponding to the child. The terminal pair (s_1, t_1) does not have a parent, and is called the *root* of T_g. The *generation* of terminal pair (s_i, t_i) is the depth of the node p_i in T_g corresponding to (s_i, t_i) plus 1. See Figure 4(b).

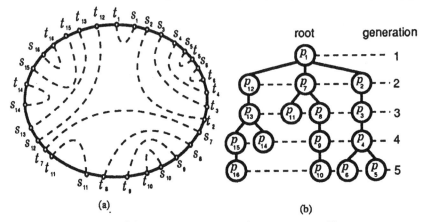

(a)   (b)

Fig.4. (a) Terminal pairs, and (b) genealogy tree T_g.

There are two ideas in the algorithm PATH2 for Case 2. The first idea is to find non-crossing paths for the terminal pairs of the same generation by using PATH1. Note that such terminal pairs satisfy the requirement for Case 1. We divide G into several components by slitting G along the found paths. For each terminal pair in a component, at least one of the shortest paths connecting the terminal pair in G is contained in the component. Thus we can find shortest non-crossing paths by applying PATH1 to each generation one by one from the first generation to the last. However such a naive implementation of the algorithm above spends time $O(kT(n))$. The second idea is to use the divide-and-conquer method. Our algorithm first finds non-crossing paths for the middle generation, slits the graph along the found paths, and recursively finds non-crossing paths in each connected component.

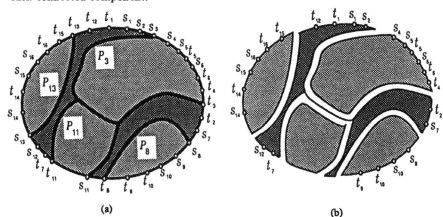

(a)   (b)

Fig.5. Illustration for PATH2.

Figure 5 illustrates the idea; Figure 5(a) depicts non-crossing paths for the third generation, that is, the middle generation in thick lines; and Figure 5(b) depicts a graph obtained by slitting G along the found paths, where all the terminal pairs of older generations are contained in the dark region and the younger in the light region. In such a way we can obtain an algorithm whose running time is $O(T(n)\log k)$, but need more definitions to present a formal description of the algorithm.

The *inside* of path P_i connecting terminal pair (s_i, t_i) is the inside of the cycle consisting of P_i and the subpath of B counterclockwise going from t_i to s_i, and is denoted by $in(P_i)$. The *outside* of P_i is the inside of the cycle consisting of P_i and the subpath of B clockwise going from t_i to s_i, and is denoted by $out(P_i)$. The *inside* of a set $\mathcal{P}$ of paths connecting terminal pairs is the union of the insides of paths in $\mathcal{P}$, and is denoted by $in(\mathcal{P})$. The *outside* of $\mathcal{P}$ is the intersection of the outsides of paths in $\mathcal{P}$, and is denoted by $out(\mathcal{P})$.

The output of our algorithms is not a set of k paths but is a set $\mathcal{F}$ of trees which contain the k terminal pairs. The set of paths connecting s_i and $t_i, 1 \le i \le k$, on trees in $\mathcal{F}$ is a solution of the non-crossing path problem. Since the total number of vertices of trees in $\mathcal{F}$ is $O(n)$, we can compute the total length of the k paths by solving the nearest common ancestor problem [GT] for trees in $\mathcal{F}$ total in time $O(n)$ [SAN].

procedure PATH2(G, S);
  begin
1.   let g be the maximum generation of terminal pairs;
2. $\mathcal{F} := \phi$;
3.   REDUCE$(G, [1, g], \mathcal{F})$
  end;

procedure REDUCE$(G, [l, h], \mathcal{F})$;
  begin
1.   **if** $l = h$ **then** {there is only one generation}
          begin
2.   let S^l be the set of terminal pairs of generation l;
3.   execute PATH1(G, S^l) and let $\mathcal{P}_l$ be the set of found paths;
4. $\mathcal{F} := \mathcal{F} \cup \mathcal{P}_l$
          end
5.   **else**
          begin
6. $m := \lfloor (l + h)/2 \rfloor$;
7.   let S^m be the set of terminal pairs of generation m;
8.   execute PATH1(G, S^m), and let $\mathcal{P}_m$ be the set of found paths;
9. $\mathcal{F} := \mathcal{F} \cup \mathcal{P}_m$;
10.   let G_{in} and G_{out} be the maximal subgraphs of G which are in $in(\mathcal{P}_m)$ and in $out(\mathcal{P}_m)$, respectively;
11.   REDUCE$(G_{in}, [m + 1, h], \mathcal{F})$;
12.   REDUCE$(G_{out}, [l, m - 1], \mathcal{F})$
          end
  end;

The running time of PATH2 is dominated by that of REDUCE. REDUCE uses a divide-and-conquer method on generations of terminal pairs. REDUCE first finds non-crossing paths connecting the terminal pairs of the middle generation by using PATH1 in time $O(T(n))$. Then REDUCE slits G along the found paths, divides the problem into two, one for older generations and the other for younger generations, and recursively solves the problem by calling REDUCE itself. Clearly the depth of the recursive calls is at most $\log k$. Procedure PATH1 executed for all subgraphs in the recursive calls of the same depth can be done in time $O(T(n))$ in total, because one can implement REDUCE so that every edge in G appears at most three of the subgraphs. Therefore the total execution time of REDUCE is $O(T(n) \log k)$. Thus we can conclude that the shortest non-crossing paths $P_1, P_2, \cdots, P_k$ can be found in $O(T(n) \log k)$ time for Case 2. Note that each path P_i is a shortest path connecting s_i and t_i in G.

4. The Case where All the Terminals Lie on Two Face Boundaries

In this section we present an algorithm for the case when all the terminals lie on two face boundaries B_1 and B_2. For each terminal pair (s_i, t_i) of one on B_1 and the other on B_2, one may assume w.l.o.g. that $s_i \in V(B_1)$ and $t_i \in V(B_2)$. Let

$$S_{12} = \{(s_i, t_i)|s_i \in V(B_1) \text{ and } t_i \in V(B_2)\},$$
$$S_1 = \{(s_i, t_i)|s_i, t_i \in V(B_1)\} \text{ and}$$
$$S_2 = \{(s_i, t_i)|s_i, t_i \in V(B_2)\}.$$

One may assume that $S_{12} \neq \emptyset$: otherwise, the non-crossing path problem can be easily solved by executing PATH2 twice: once for G to find paths for S_1, and then once for the graph obtained by slitting G along the found paths to find paths for S_2.

Assume that $(s_i, t_i) \in S_{12}$ if $1 \leq i \leq l$, and $(s_i, t_i) \in S_1 \cup S_2$ otherwise. One may assume w.l.o.g. that terminals $s_1, s_2, \cdots, s_l$ appears on B_1 counterclockwise in this order and $t_1, t_2, \cdots, t_l$ appears on B_2 counterclockwise in this order. For the sake of simplicity, we assume that graph G is embedded in the plane region Σ surrounded by two circles C_1 with radius 1 and C_2 with radius $1/2$ both having the center at the origin O of the xy-plane. We may assume w.l.o.g. that, for each terminal pair (s_i, t_i), $Image(s_i) = (\cos(\frac{2\pi}{l}i), \sin(\frac{2\pi}{l}i))$ and $Image(t_i) = (\frac{1}{2}\cos(\frac{2\pi}{l}i), \frac{1}{2}\sin(\frac{2\pi}{l}i))$, and that $Image(G) \cap (C_1 \cup C_2) = \{Image(s_i), Image(t_i)|1 \leq i \leq l\}$.

Let P be a path going from point a to point b in Σ. Let θ be the total angle turned through (measured counterclockwise) by the line OX when point X moves on P from a to b. Possibly $|\theta| > 2\pi$. We define the (normalized) angle $\theta(P)$ of path P by $\theta(P) = \frac{l}{2\pi}\theta$. If $P_1, P_2, \cdots, P_k$ are non-crossing paths in G, then clearly $\theta(P_1), \theta(P_2), \cdots, \theta(P_l)$ are all equal to the same integral multiple of l.

The following lemma holds. We denote by $length(P)$ the length of path P.

LEMMA 2. Let P_1^* be a shortest path connecting s_1 and t_1, and let P_i, $1 \leq i \leq l$, be an arbitrary path connecting s_i and t_i. Then there exists a path P_i' connecting s_i and t_i such that $length(P_i') \leq length(P_i)$ and

$$\theta(P_i') - \theta(P_1^*) = \begin{cases} l & \text{if } \theta(P_i) - \theta(P_1^*) \geq 2l \\ -l & \text{if } \theta(P_i) - \theta(P_1^*) \leq -2l. \end{cases}$$

Proof: Assume that $\theta(P_i) - \theta(P_1^*) = rl$ for an integer $r \geq 2$: a proof for the other case is similar to one for this case. Every intersecting vertex v of paths P_i and P_1^* satisfies

$$\theta(P_i[s_i, v]) - \theta(P_1^*[s_1, v]) = l - (i-1) \bmod l$$

and

$$\theta(P_i[v, t_i]) - \theta(P_1^*[v, t_1]) = i - 1 \bmod l.$$

Assume that the intersecting vertices $v_1, v_2, \cdots, v_q$ of P_i and P_1^* appear in this order on P_i going from s_i to t_i. Define r_x, $1 \le x \le q$: $r_x = \frac{1}{l}\{\theta(P_i[s_i, v_x]) - \theta(P_1^*[s_1, v_x]) + i - 1\}$. Then the sequence of integers $r_1, r_2, \cdots, r_q$ satisfies

$$\begin{cases} r_1 = 0, 1; \\ r_q = r, r + 1; \text{ and} \\ r_x - r_{x+1} = 0, \pm 1, \text{ for every } x, \ 1 \le x \le q - 1. \end{cases}$$

We prove this lemma separating into the following two cases.
Case 1: $r_q = r$.

In this case, $\theta(P_i[v_q, t_i]) - \theta(P_1^*[v_q, t_1]) = i - 1$. There exists an intersecting vertex v_a of P_i and P_1^* such that

i) $r_a = 1$; and

ii) $r_x \ge 1$ for every x, $a \le x \le q$.

Especially, let v_a be the vertex closest to v_q on $P_1^*[s_1, v_q]$ among such vertices. Then we claim that v_a satisfies

iii) $P_i[s_i, v_a]$ and $P_1^*([v_a, v_q])$ intersect only at v_a.

Assume for a contradiction that $P_i[s_i, v_a]$ and $P_1^*[v_a, v_q]$ would intersect at a vertex $v_b \ne v_a$. Let v_b be the intersecting vertex that appears first on P_i going from v_a to s_i (See Figure 6). Let C be the cycle consisting of $P_1^*[v_a, v_b]$ and $P_i[v_b, v_a]$. (In Figure 6, C is drawn in thick line.) Since v_q lies in C, $P_i[v_a, v_q]$ intersects C at a vertex $v_c \ne v_a$. Of course, v_c is on $P_1^*[v_a, v_b]$ and $1 \le c \le q$. Choose as v_c the intersecting vertex that appears first on $P_i[v_a, v_q]$ going from v_a to v_q. Then $r_c = 1$ because $P_i[v_a, v_c] + P_1^*[v_a, v_c]$ is a simple cycle and $\theta(P_i[v_a, v_c]) = \theta(P_1^*[v_a, v_c])$. Of

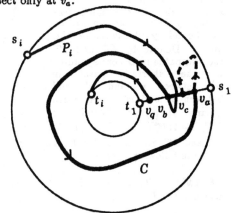

Fig.6. Illustration for the proof of Lemma 2.

course, $r_x \ge 1$ for every x, $c \le x \le q$. However, v_c is closer to v_q than v_a on $P_1^*[s_1, v_q]$, contradicting the selection of v_a.

Let $P_i' = P_i[s_i, v_a] + P_1^*[v_a, v_q] + P_i[v_q, t_i]$, then P_i' is a simple path connecting s_i and t_i and $\theta(P_i') - \theta(P_1^*) = l$. Since $length(P_1^*[v_a, v_q]) \le length(P_i[v_a, v_q])$, we have $length(P_i') \le length(P_i)$.

Case 2: $r_q = r + 1$.

In this case $\theta(P_i[v_q, t_i]) - \theta(P_1^*[v_q, t_1]) = (i - 1) - l$. We can prove that there exists an intersecting vertex v_a of P_i and P_1^* such that

i) $r_a = 2$;

ii) $r_x \ge 2$ for every x, $a \le x \le q$; and

iii) $P_i[s_i, v_a]$ and $P_1^*[v_a, v_q]$ intersects only at v_a.

Then $P_i' = P_i[s_i, v_a] + P_1^*[v_a, v_q] + P_i[v_q, t_i]$ is a simple path connecting s_i and t_i such that $\theta(P_i') - \theta(P_1^*) = l$ and $length(P_i') \le length(P_i)$. Q.E.D.

LEMMA 3. Let $\theta = 0, \pm l, \pm 2l, \cdots$, and let P_1 be a shortest path among ones which connects s_1 and t_1 and have angle θ. Then G contains paths $P_2, P_3, \cdots, P_l$ such that
(a) $\theta(P_2) = \theta(P_3) = \cdots = \theta(P_l) = \theta$, and $P_1, P_2, \cdots, P_l$ are non-crossing; and
(b) each P_i, $2 \leq i \leq l$, is a shortest path among ones which connect s_i and t_i and have angle θ.

From Lemmas 2 and 3 we have the following lemma.

LEMMA 4. Let P_1^* be an arbitrary shortest path connecting s_1 and t_1 in G. Then G contains shortest non-crossing paths $P_1, P_2, \cdots, P_k$ such that $\theta(P_1) - \theta(P_1^*)$ is either 0, l, or $-l$.

Let P_1^* be a shortest path between s_1 and t_1 in G, and let G_0' be the slit graph of G for P_1^*. G_0' has two vertices v' and v'' corresponding to s_1 and two vertices w' and w'' corresponding to t_1. Vertices v', w', w'' and v'' lie on the same face boundary in G_0' and appear on the boundary clockwise in this order. Denote by P_1^+ and P_1^- the two paths in G corresponding to the shortest paths in G_0' between v' and w'' and between v'' and w', respectively. Clearly $\theta(P_1^+) - \theta(P_1^*) = l$ and $\theta(P_1^-) - \theta(P_1^*) = -l$. In Figure 7

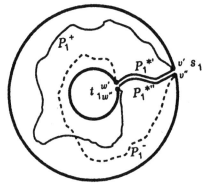

Fig.7. Slit graph G_0'.

P_1^+ and P_1^- are drawn in solid and dotted lines, respectively. From Lemma 4 we have the following lemma.

LEMMA 5. There exist shortest non-crossing paths $P_1, P_2, \cdots, P_k$ such that P_1 is either P_1^*, P_1^+ or P_1^-.

The discussion above leads to the following algorithm.

procedure PATH(G);
    begin
1.   find a shortest path P_1^* between s_1 and t_1 in G;
2.   construct the slit graph $G_0' = G(P_1^*)$, and find paths P_1^+ and P_1^-;
3. $\mathcal{P}_0 := \{P_1^*\}, \mathcal{P}_1 := \{P_1^+\}, \mathcal{P}_2 := \{P_1^-\}$;
        {each $\mathcal{P}_i, 0 \leq i \leq 2$, becomes a set of non-crossing paths}
4.   construct the slit graph $G_1' = G(P_1^+)$ and the slit graph $G_2' = G(P_1^-)$;
5.   **for** $i := 0$ **to** 2 **do**
            begin
6.   PATH2(G_i', $S - (s_1, t_1)$);
7.   add the found paths $P_2, P_3, \cdots, P_k$ to $\mathcal{P}_i$;
            end;
8.   output as a solution one of the sets $\mathcal{P}_0, \mathcal{P}_1$ and $\mathcal{P}_2$ whose total length is minimum
    end;

The dominating part of the execution time of algorithm PATH is one for executing PATH2 three times. Therefore the running time of PATH is $O(T(n)\log k)$. Thus we have the following theorem.

THEOREM 1. Given a plane graph G with k terminal pairs on its two face boundaries, one can find shortest non-crossing paths in time $O(T(n)\log k)$, where $T(n)$ is the time required for finding shortest paths from a single vertex to all other vertices in a plane graph of n vertices.

A usual shortest path algorithm, that is, Dijkstra's method with a heap, spends time $T(n) = O(n\log n)$ for a plane graph [AHU,Tar]. On the other hand Frederickson's method spends time $T(n) = O(n)$ with preprocessing time $O(n\sqrt{\log n})$ [Fre]. Therefore our algorithm can be done in time $O(n(\sqrt{\log n} + \log k)) = O(n\log n)$.

5. Conclusion

In this paper, we presented an efficient algorithm for finding shortest non-crossing paths for the case when k terminal pairs are located on two specified face boundaries of a plane graph, and proved that the running time is $O(n\log n)$. Our algorithm works well even for the case where edge length may be negative if there are no negative cycles. We are now extending the algorithm to a more general case where terminals lie on three or more face boundaries.

Acknowledgement. This research is partly supported by Grant in Aid for Scientific Research of the Ministry of Education, Science, and Culture of Japan under a grant number: General Research (C) 04650300.

References

[AHU]   A. V. Aho, J. E. Hopcroft and J. D. Ullman, The Design and Analysis of Computer Algorithms, Addison-Wesley, Reading, MA (1974).

[Fre]   G. N. Frederickson, Fast algorithms for shortest path in planar graphs, with applications, SIAM J. Comput., 16, pp. 1004-1022 (1987).

[GT]   H. N. Gabow and R. E. Tarjan, A linear-time algorithm for a special case of disjoint set union, Journal of Computer and System Sciences, 30, pp. 209-221 (1985).

[KL]   M. R. Kramer and J. van Leewen, Wire-routing is NP-complete, Report No. RUU-CS-82-4, Department of Computer Science, University of Utrecht, Utrecht, the Netherlands (1982).

[Lyn]   J. F. Lynch, The equivalence of theorem proving and the interconnection problem, ACM SIGDA, the Netherlands (1982).

[SAN]   H. Suzuki, T. Akama and T. Nishizeki, Finding Steiner forests in planar graphs, Proc. of First SIAM-ACM SODA, pp. 444-453 (1990).

[Tar]   R. E. Tarjan, Data Structures and Network Algorithms, SIAM,Philadelphia, PA (1983).

On Symmetry of Information and Polynomial Time Invertibility*

Luc Longpré[1] and Osamu Watanabe[2]

[1] Dept. of Computer Science, Northeastern University
Boston MA 02115, USA
(e-mail: luc@corwin.ccs.northeastern.edu)

[2] Dept. of Computer Science, Tokyo Institute of Technology
Meguro-ku Ookayama, Tokyo 152, Japan
(e-mail: watanabe@cs.titech.ac.jp)

Abstract. Symmetry of information states that for two strings x and y, $\mathrm{K}(xy) = \mathrm{K}(x) + \mathrm{K}(y \mid x) \pm O(\log|xy|)$. We consider the statement of whether symmetry of information holds in a polynomial time bounded environment. Intuitively, this problem is related the complexity of inverting a polynomial time computable function. We give some evidence supporting this intuition, by proving the following relations:

1. If the polynomial time symmetry of information holds, then there is a polynomial time algorithm that computes the shortest description of a string for "almost all" strings.
2. If the polynomial time symmetry of information holds, then every polynomial time computable function is probabilistic polynomial time invertible for "almost all" strings in its domain.
3. If P = NP (i.e., every polynomial time computable function is polynomial time invertible), then the polynomial time symmetry of information holds.

1. Introduction

The amount of information contained in a string x about a string y — $\mathrm{I}(x, y)$ — can be defined as the difference between the quantity of information in x when x has to be printed from nothing and when x has to be printed using the string y:

$$\mathrm{I}(x, y) \overset{\text{def}}{=} \mathrm{K}(x) - \mathrm{K}(x \mid y).$$

A crucial observation is that this quantity is symmetric, within small quantities:

$$\mathrm{I}(x, y) = \mathrm{I}(y, x) \pm O(\log|xy|). \tag{1}$$
$$\text{(Or equivalently, } \mathrm{K}(xy) = \mathrm{K}(x) + \mathrm{K}(y \mid x) \pm O(\log|xy|).)$$

This symmetry is called *symmetry of information for the resource unbounded case.* and for this reason, $\mathrm{I}(x, y)$ is called the *mutual information*. The symmetry of information has been discovered independently by Levin and Kolmogorov (see [ZL70]) in

---

* Part of this research was done while the first author was visiting Tokyo Institute of Technology. The second author was supported in part by Takayanagi-Kinen Zaidan (1992).

the late 60's. When the appropriate variant of Kolmogorov complexity is used, the difference of $O(\log |xy|)$ can be accounted for [Gác74].

It can be easily seen that symmetry of information still holds in space bounded environments (see [LM91]). But it is not believed to hold in polynomial time bounded environments. The question of whether the polynomial time symmetry of information holds seems to have close connection to several complexity theoretic questions. For example, it seems to be a nice characterization of the search versus recognition problem, it may be the precursor idea of the P = NP or P = P$^{\#P}$ problems, etc. Thus, we believe that the question is an interesting subject in computational complexity theory. (Indeed, Kolmogorov stated in various occasions that it would be very interesting to solve the question.) However, it is surprising that to this day, very few extensions or references have been made to this essential question. This paper studies the relation between the polynomial time symmetry of information and the complexity of computing the inverse of polynomial time computable functions. (From now on throughout this abstract, by "the symmetry of information", we mean the polynomial time symmetry of information.)

Equation (1) can be restated as

$$K(x) + K(y\,|\,x) = K(y) + K(x\,|\,y) \pm O(\log |xy|).$$

This intuitively means that, when computing x and y, there is not so much difference between computing x first then computing y from x, and computing y first then computing x from y. For polynomial time bounded environments, however, it seems that computation of x and y is not always symmetric, and hence we conjecture that the symmetry of information does not hold.

We may naturally consider that this nonsymmetry has something to do with the existence of polynomial time one-way functions. Roughly speaking, a *polynomial time one-way function* is a function f such that f itself is polynomial time computable, but its inverse f^{-1} is much "harder" than f, e.g., not polynomial time computable. Intuitively, the above nonsymmetry may characterize the one-wayness of certain polynomial time computable functions. In this paper, we give some evidence supporting this intuition.

The relation between the symmetry of information and the existence of one-way functions was first studied in [LM91]. They proved that the symmetry of information implies that every polynomial time computable *one-to-one* function is polynomial time invertible on a "large" portion of its domain[3]. In this paper, we can prove that the symmetry of information implies much stronger types of invertibilities:

(1) If the symmetry of information holds, then there is a polynomial time algorithm that computes the shortest description of a string from a "large" portion of the strings. (That is, the algorithm computes the particular inverse image of the function computed by the universal Turing machine.)

(2) If the symmetry of information holds, then every polynomial time computable function is probabilistic polynomial time invertible on a "large" portion of its domain. (In a sense, we can remove the one-to-oneness condition required in the result by Longpré-Mocas.)

---

[3] More precisely, a "large" means more than $(1 - 1/p(n))$ fraction for any polynomial p.

Concerning the converse relation, the only known condition that implies polynomial time symmetry is P=PP (see [LM91]). In this paper, we can show the following sufficient condition, which again has some connection with polynomial time invertibility.

(3) If P = NP, then the polynomial time symmetry of information holds.

Notice that P = NP if and only if every honest polynomial time computable function is polynomial time invertible. Thus, our result states that the symmetry of information is implied from this general type of polynomial time invertibility.

In order to simplify our discussion, we consider only total and *length-preserving* functions, i.e., total functions that maps each string to some string of the same length. However, similar results hold for any reasonable (i.e., honest) functions.

2. Preliminaries

In this paper we follow standard definitions and notations in computational complexity theory; the reader will find them in standard textbooks such as [BDG88].

We use $\{0,1\}$ for our alphabet, and by *string* we mean an element of $\{0,1\}^*$. The length of a string x is denoted by $|x|$. The cardinality of a finite set A is written as $\|A\|$. Let $\langle \cdot, \cdot \rangle$ denote a standard pairing function on $\{0,1\}^*$, i.e., a one-to-one function from $\{0,1\}^* \times \{0,1\}^*$ to $\{0,1\}^*$ that is polynomial time computable and polynomial time invertible.

In order to define "Kolmogorov complexity", we consider any reasonable universal Turing machine, and let it be fixed throughout this paper. Let U denote this machine, and for any string x, let $U(x)$ denote the output of U on input x. We consider the following Kolmogorov complexity measures (in the following, let u and v be any strings, and let t be any nonnegative integer):

$K(u) \overset{\text{def}}{=}$ the length of the shortest w such that $U(w) = u$.

$K(u|v) \overset{\text{def}}{=}$ the length of the shortest w such that $U(\langle w, v \rangle) = u$.

$KT(u;t) \overset{\text{def}}{=}$ the length of the shortest w such that $U(w) = u$ in t steps.

$KT(u|v;t) \overset{\text{def}}{=}$ the length of the shortest w such that $U(\langle w, v \rangle) = u$ in t steps

The quantity $K(u|v)$ is called the *conditional* Kolmogorov complexity. Intuitively, this is the amount of information necessary to describe u when v is given. We can naturally extend this to, e.g., $K(u|v_1, v_2)$.

In the following discussion, we usually assume that the length of a target string is given. That is, we consider, e.g., the Kolmogorov complexity $K(u | |u|)$. Since the difference between $K(u)$ and $K(u | |u|)$ is at most $\log |u| + 1$[4], and most of our equations are independant of log additive factors, we do not lose generality by this assumption. But we do not want to lose correctness, so we denote $K(u | |u|)$ as $K_\ell(u)$. $K_\ell(u|v)$, $KT_\ell(u;t)$ and $KT_\ell(u|v;t)$ are defined similarly. For more detailed definitions and discussion on Kolmogorov complexity, see [LV88].

In Kolmogorov complexity theory, small additive factors are often ignored. (Otherwise, the discussion becomes unnecessarily complicated, and we lose the focus on

---

[4] The base of log is 2 in this paper. In order to simplify statements, by $\log n$, we mean $\lceil \log n \rceil$, and we consider $\log 0 = 1$.

important points.) We use some notations that allow us to ignore small difference of certain types (in the following, let α and β be any functions from $\{0,1\}^*$ to the set of natural numbers, $\mathbf{N}$, and let X and Y be any sets of strings):

- $\forall x \in X, \forall y \in Y, [\, \alpha(x) \overset{\text{const}}{\lesssim} \beta(y)\,] \overset{\text{def}}{\leftrightarrow} \exists c \in \mathbf{N}, \forall x \in X, \forall y \in Y\, [\, \alpha(x) \leq \beta(y) + c\,]$.

- $\forall x \in X, \forall y \in Y, [\, \alpha(x) \overset{\text{log}}{\lesssim} \beta(y)\,]$
  $\overset{\text{def}}{\leftrightarrow} \exists d \in \mathbf{N}, \forall x \in X, \forall y \in Y\, [\, \alpha(x) \leq \beta(y) + d \log \max\{\alpha(x), \beta(y)\}\,]$.

We write $\alpha(x) \overset{\text{const}}{\approx} \beta(y)$ if $\alpha(x) \overset{\text{const}}{\lesssim} \beta(y)$ and $\beta(x) \overset{\text{const}}{\lesssim} \alpha(y)$. The relation $\overset{\text{log}}{\approx}$ is defined similarly.

Now "symmetry of information" is defined as follows for the resource unbounded case and for the polynomial time bounded case.

Definition 2.1.

(1) *Symmetry of information for the resource unbounded case* is the following relation:
$$\forall n, \forall x, y \in \{0,1\}^n\, [\, K_\ell(xy) \overset{\text{log}}{\approx} K_\ell(x) + K_\ell(y\,|\,x)\,].$$

(2) *Symmetry of information for the polynomial time bounded case* is the following pair of relations:

$\forall p$: polynomial, $\exists q$: polynomial,
$$\forall n, \forall x, y \in \{0,1\}^n\, [\, KT_\ell(xy\,;\,p(n)) \overset{\text{log}}{\lesssim} KT_\ell(x\,;\,q(n)) + KT_\ell(y\,|\,x\,;\,q(n))\,]$$

and

$\forall p$: polynomial, $\exists q$: polynomial,
$$\forall n, \forall x, y \in \{0,1\}^n\, [\, KT_\ell(xy\,;\,q(n)) \overset{\text{log}}{\gtrsim} KT_\ell(x\,;\,p(n)) + KT_\ell(y\,|\,x\,;\,p(n))\,].$$

What follows is some of the basic facts on Kolmogorov complexity that will be used later.

Proposition 2.2.
(1) $\exists c, \forall x\, [\, K(x) \leq |x| + c\,]$.
(2) $\forall x, y\, [\, K(y) \overset{\text{const}}{\lesssim} K(x) + K(y\,|\,x)\,]$.
(3) $\forall x, y\, [\, K(xy) \overset{\text{const}}{\lesssim} K(x) + K(y\,|\,x)\,]$.
(4) $\exists d_{\text{len}}, \forall x\, [\, K_\ell(x) \leq K(x) \leq K_\ell(x) + d \log |x|\,]$.
Remark. Corresponding relations hold for the polynomial time bounded case.

Proposition 2.3. [ZL70] Symmetry of information for the resource unbounded case holds.
Remark. One direction is part (3) of Proposition 2.2. The other direction of the proof is given as part of the proof of Theorem 4.1

3. What if Polynomial Time Symmetry of Information Holds?

First we discuss the polynomial time invertibility of the function computed by our universal Turing transducer U. For any time bound t, we can consider that U computes the following partial function in time bound t:

$$f_{U,t}(x) \overset{\text{def}}{=} \begin{cases} y & \text{if } U(x) \text{ outputs } y \text{ in } t(|y|) \text{ steps} \\ \bot & \text{otherwise} \end{cases}$$

Here we discuss, for a given polynomial p, the polynomial time invertibility of $f_{U,p}$.

Notice that for some string a, $U(ay)$ outputs y in linear time. (That is, a is the code of a program computing the identity function.) Hence, for a sufficiently large polynomial p and every $y \in \{0,1\}^*$, there exists a trivial inverse of y in $f_{U,p}^{-1}(y)$, namely, ay. Thus, the question of interest is to compute the shortest string in $f_{U,p}^{-1}(y)$ for a given y, i.e., some particular elements of $f_{U,p}^{-1}(y)$. We assume that polynomials p are sufficiently large so that every $f_{U,p}^{-1}(y)$ has a trivial element.

For any string y and any time bound t, we denote the shortest element of $f_{U,t}^{-1}(y)$ as $x_{y,t}^*$. If $f_{U,t}^{-1}(y)$ is empty, then $x_{y,t}^*$ is undefined. If $f_{U,t}^{-1}(y)$ contains many values of minimal length, then $x_{y,t}^*$ is the lexicographically smallest one. Note that $|x_{y,t}^*| = $ KT$(y; t(|y|))$.

The purpose of the following theorem is to show how to find the shortest description of a string, assuming polynomial time symmetry. Because of some technicality, we also have to assume that the Kolmogorov complexity of strings does not change too much when we vary the polynomial time bound. We don't know if we can get away without this technicality.

Theorem 3.1. Suppose that polynomial time symmetry of information holds. Then for every polynomial p, there exist some polynomial q and some polynomial time deterministic algorithm A such that for all $y \in \{0,1\}^*$,

$$\text{KT}_\ell(y; p(|y|)) \overset{\log}{\lesssim} \text{KT}_\ell(y; q(|y|)) \;\to\; A(y) = x_{y,p}^*.$$

Remark. (Precise statement)

$$\forall p : \text{polynomial}, \exists q : \text{polynomial}, \forall d, \exists A : \text{poly. time algorithm}, \forall y \in \{0,1\}^*$$
$$\left[\begin{aligned} &(*) \; \text{KT}_\ell(y; p(|y|)) \leq \text{KT}_\ell(y; q(|y|)) \\ &\qquad\qquad\qquad + d\log\max\{\text{KT}_\ell(y; p(|y|)), \text{KT}_\ell(y; q(|y|))\} \\ &\to A(y) = x_{y,p}^* \end{aligned} \right].$$

Proof. Let p be any polynomial, and let y be any string of length n. First consider the case that $|x_{y,p}^*| < |y|$. Let x be the string of length n that is obtained by padding $x_{y,p}^*$. (I.e, $x = x_{y,p}^* 10^m$ for some $m \geq 0$.) By definition of $x_{y,p}^*$ $U(x_{y,p}^*) = y$ in $p(n)$ steps; hence, there exist some fixed string a_1 and polynomial b_1 such that $U(a_1 x_{y,p}^*)$ outputs yx in $r(n) \overset{\text{def}}{=} b_1(p(n) + n)$ steps. That is, for some e_1,

$$\text{KT}_\ell(yx; r(n)) \leq |x_{y,p}^*| + e_1 = \text{KT}_\ell(y; p(n)) + e_1. \tag{2}$$

By the symmetry of information, for some polynomial q and constant e_2, and for every n and every $u, v \in \{0,1\}^n$, we have $\mathrm{KT}_\ell(v \mid u; q(n)) \leq \mathrm{KT}_\ell(uv; r(n)) - \mathrm{KT}_\ell(u; q(n)) + e_2 \log n$. Hence,

$$\mathrm{KT}_\ell(x \mid y; q(n)) \leq \mathrm{KT}_\ell(yx; r(n)) - \mathrm{KT}_\ell(y; q(n)) + e_2 \log n. \tag{3}$$

Now from (2) and (3), we have

$$\mathrm{KT}_\ell(x \mid y; q(n)) \leq \mathrm{KT}_\ell(y; p(n)) - \mathrm{KT}_\ell(y; q(n)) + e_2 \log n + e_1.$$

Hence, for any constant $d > 0$, if $\mathrm{KT}_\ell(y; p(n)) \leq \mathrm{KT}_\ell(y; q(n)) + d \log n$, then $\mathrm{KT}_\ell(x \mid y; q(n)) \leq (d + e_2) \log n + e_1$. (Note that $(*)$ implies $\mathrm{KT}_\ell(y; p(n)) \leq \mathrm{KT}_\ell(y; q(n)) + d \log n$.) Furthermore, since for some polynomial p', $\mathrm{KT}(x \mid y; p'(n)) \leq \mathrm{KT}_\ell(x \mid y; p(n)) + d_{\text{len}} \log n$, we have, for some polynomial p'', $\mathrm{KT}(x \mid y; p''(n)) \leq (d + d_{\text{len}} + e_2) \log n + e_1$. This means that there exists some string w of length $\leq (d + d_{\text{len}} + e_2) \log n + e_1$ such that $\mathrm{U}(\langle w, y \rangle) = x_{y,p}^* 10^m$. Thus, by trying every string w of length $\leq (d + d_{\text{len}} + e_2) \log n + e_1$, one can find $x_{y,p}^*$. (Note that $x_{y,p}^*$ is (one of) the shortest x' such that $\mathrm{U}(\langle w, y \rangle) = x' 10^m$ and $\mathrm{U}(x')$ outputs y in $p(|y|)$ steps.) Therefore, some polynomial time algorithm A_1 can compute $x_{y,p}^*$ from y (if $|x_{y,p}^*| < |y|$ and $(*)$ holds).

By similar argument, we can show a polynomial time algorithms A_2 and A_3 that obtains $x_{y,p}^*$ respectively in the case that $|x_{y,p}^*| = |y|$ and $|x_{y,p}^*| > |y|$. Therefore, using A_1, A_2, and A_3, $x_{y,p}^*$ is polynomial time computable from a given input y. $\quad\square$

For any $d > 0$, define $\mathrm{RAND}[d \log n]$ to be the set of strings x such that $\mathrm{K}(x) > |x| - d \log |x|$. That is, $\mathrm{RAND}[d \log n]$ is the set of strings of high Kolmogorov complexity. Consider any $y \in \mathrm{RAND}[d \log n]$ of length n. Then for any time bounds t_1 and t_2, $n - (d + d_{\text{len}}) \log n < \mathrm{KT}_\ell(y; t_1)$, $\mathrm{KT}_\ell(y; t_2) < n$. Thus, clearly $\mathrm{KT}_\ell(y; t_1) \overset{\log}{\approx} \mathrm{KT}_\ell(y; t_2)$. Hence, we have the following corollary.

Corollary 3.2. Assuming polynomial time symmetry, for any polynomial p and any constant $d > 0$, there exists a polynomial time deterministic algorithm A such that for every $y \in \mathrm{RAND}[d \log n]$, $A(y)$ outputs $x_{y,p}^*$.

Notice that for any polynomial s, there exists some $d > 0$ such that

$$\| \mathrm{RAND}[d \log n] \cap \{0,1\}^n \| / \| \{0,1\}^n \| \geq 1 - 1/s(n).$$

Then the following corollary is immediate.

Corollary 3.3. Assuming polynomial time symmetry, for any polynomial p and any polynomial s, there exists a polynomial time deterministic algorithm A such that for every $n \geq 0$,

$$\frac{\| \{ y \in \{0,1\}^n : A(y) = x_{y,p}^* \} \|}{\| \{0,1\}^n \|} \geq 1 - \frac{1}{s(n)}.$$

Next we consider the polynomial time invertibility of a given polynomial time computable function. In this case also, we can show a result with similar flavor as Corollary 3.2.

Theorem 3.4. Assuming polynomial time symmetry, let f be any polynomial time computable length-preserving function. For any constant $d > 0$ and any $\varepsilon > 0$, there exists a polynomial time randomized algorithm B such that for every $x \in$ RAND$[d \log n]$ (letting $y = f(x)$),

$$\Pr\{\ B(y) \text{ outputs some } x' \text{ in } f^{-1}(y)\ \} \ \geq\ 1 - \varepsilon.$$

Let f be any polynomial time computable length-preserving function, and let $d > 0$ be any constant. The theorem follows from the following lemmas. In the following, we consider any x of length n in RAND$[d \log n]$, and let it be fixed. Let $y = f(x)$ and $m = \|f^{-1}(y)\|$. Constants d_1, d_2, d_3 and polynomial q_1 appearing in the following lemmas depend on only f and d. (The proofs of Lemmas 3.5, 3.6, 3.7 are standard counting and enumeration arguments and/or some application of the symmetry hypothesis. They are omitted in this version of the paper. Contact the authors for a more complete version.)

Lemma 3.5. $n - \log m - d_1 \log n \ \leq\ \mathrm{K}(y) \ \leq\ n - \log m + d_1 \log n$.

Lemma 3.6. For at least half of x' in $f^{-1}(y)$, $\mathrm{K}(x') \geq n - d_2 \log n$.

Lemma 3.7. Assuming polynomial time symmetry, let x' be any string of length n such that $f(x') = y$ and $\mathrm{K}(x') \geq n - d_2 \log n$. Then there exists some string w of length $\leq \log m + d_3 \log n$ such that $M_U(\langle w, y \rangle)$ outputs x' in $q_1(n)$ steps.

Lemma 3.8. Assuming polynomial time symmetry, let

$$W_y = \{\ w \in \{0,1\}^{\leq \log m + d_3 \log n} : \mathrm{U}(\langle w, y \rangle) \text{ outputs} \\ \text{some } x' \in f^{-1}(y) \text{ in } q_1(n) \text{ steps }\}.$$

Then, for some polynomial s that depends only on f and d,

$$\frac{\|W_y\|}{\|\ \{0,1\}^{\leq \log m + d_3}\ \|} \ \geq\ \frac{1}{s(n)}.$$

Proof. From Lemma 3.7, every x' in $f^{-1}(y)$ with $\mathrm{K}(x') \geq n - d_2 \log n$ has some $w_{x'}$ such that $|w_{x'}| \leq \log m + d_3 \log n$ and $\mathrm{U}(\langle w_{x'}, y \rangle)$ outputs x' in $q_1(n)$ steps. Since each $w_{x'}$ is distinct, and at least half (i.e., 2^{m-1}) of x' in $f^{-1}(y)$ satisfies $\mathrm{K}(x') \geq n - d_2 \log n$ (Lemma 3.6), we can conclude that there are more than 2^{m-1} w of length $\leq \log m + d_3 \log n$ such that $\mathrm{U}(\langle w, y \rangle)$ outputs some element of $f^{-1}(y)$ in $q_1(n)$ steps. Thus, the ratio of such w in $\{0,1\}^{\leq \log m + d_3 \log n}$ is larger than $2^{m-1}/2^{\log m + d_3 \log n + 1} \geq 1/4n^{d_3}$. $\square$

Proof of Theorem 3.4. Let y be a given input, and let $n = |y|$. The basic process of our randomized algorithm B is as follows:
(1) Randomly guess $l = \log \|f^{-1}(y)\|$, and guess string w of length $\leq l + d_3 \log n$.
(2) Compute $\mathrm{U}(\langle w, y \rangle)$ in $q_1(n)$ steps. If U outputs some x such that $f(x) = y$, then output x.

B iterates this process for $r(n)$ times for some polynomial r. Now, from Lemma 3.8, it is easy to prove that B works correctly. $\square$

Corollary 3.9. Let f be any polynomial time computable length-preserving function. For any polynomial s, there exists a polynomial time randomized algorithm B such that for every $n \geq 0$ (letting $y = f(x)$ for each x),

$$\frac{\| \{ x \in \{0,1\}^n : \Pr\{B(y) \in f^{-1}(y)\} \geq 1 - \epsilon \} \|}{\|\{0,1\}^n\|} \geq 1 - \frac{1}{s(n)},$$

Although Corollaries 3.3 and 3.9 have some similarity, there is an important difference between them. Corollary 3.3 shows that some algorithm A computes a (certain) inverse of y for almost all y in $\{0,1\}^n$. On the other hand, the algorithm B stated in Corollary 3.9 computes an inverse of $y = f(x)$ for almost all x in $\{0,1\}^n$. That is, in Corollary 3.3, "almost all" is considered in the *range* (of $f_{U,p}$), while "almost all" in Corollary 3.9 is about the proportion in the *domain* of f. Notice that Corollary 3.9 implies "almost all invertibility in the range" for any length-preserving one-to-one function f. (Indeed, in this case, we have a much simpler proof of Corollary 3.9 [LM91].) However, in the general case, "almost all invertibility in the range" is an interesting open question.

4. When can the Symmetry of Information Hold?

Theorem 4.1. If P = NP, then polynomial time symmetry of information holds.

Outline of the Proof. Here we reproduce a proof of symmetry of information for the resource unbounded case. Then we look at how it could be modified (or, could not be modified) for the polynomial time bounded case. It will be shown that the problems in the modification are solvable if we have access to a Σ_2^P oracle. Thus, if P = NP, then the oracle can be simulated in polynomial time, and we have the theorem.

Notice that from Proposition 2.2 (3), we already have $K_\ell(xy) \overset{\log}{\lesssim} K_\ell(x) + K_\ell(y \mid x)$. (In fact, the proof even for the polynomial time bounded case is relatively straightforward.) Thus, in the following, we discuss the proof of $K_\ell(y \mid x) \overset{\log}{\lesssim} K_\ell(xy) - K_\ell(x)$ (which is equivalent to $K_\ell(x) + K_\ell(y \mid x) \overset{\log}{\lesssim} K_\ell(xy)$).

In the following, we fix any $n \in \mathbf{N}$, and consider only strings in $\{0,1\}^n$. Let x and y be any strings in $\{0,1\}^n$, for which we would like to show $K_\ell(y \mid x) \leq K_\ell(xy) - K_\ell(x) + d \log n$. We use m to denote $K_\ell(xy)$, and $e_1, e_2, ...$ to denote constants that are independent of choice of n, x, or y.

Consider set $S_{x,m} \overset{\text{def}}{=} \{y' : K_\ell(xy') \leq m\}$. This set can be determined by specifying parameters x and m. Also, $y \in S_{x,m}$. This means that y can be identified by specifying x, m, and the index of y in the enumeration of $S_{x,m}$. In order words, $K_\ell(y \mid x) \leq \log n + \log \|S_{x,m}\| + e_1$ because m can be coded in $\leq \log n + e_2$ bits, and the index of y in $S_{x,m}$ is coded in $\log \|S_{x,m}\|$ bits.

Suppose $2^k \leq \|S_{x,m}\| < 2^{k+1}$. Let $V_{n,m,k}$ be a set of strings such that for every x', $\|S_{x',m}\| \geq 2^k \to x' \in V_{n,m,k}$, and $\|S_{x',m}\| < 2^{k-2} \to x' \notin V_{n,m,k}$. (The definition of $V_{n,m,k}$ is given later.) From this property, we have $x \in V_{n,m,k}$. Hence, x is can be identified by specifying n, m, k, and the index of x in $V_{n,m,k}$. In other words, $K_\ell(x) \leq \log \|V_{n,m,k}\| + 3 \log n + e_3$. On the other hand, consider the set of all strings of length $2n$ that can be printed by U from an input of length m. This set has $\leq 2^m$ strings. Each of these strings contributes exactly one string to some $S_{x',m}$. Thus, $\| \cup_{x' \in \{0,1\}^n} S_{x',m} \| \leq 2^m$, and clearly, $\| \cup_{x' \in V_{n,m,k}} S_{x',m} \| \leq 2^m$. Hence, $\|V_{n,m,k}\| \leq 2^m/2^{k-2}$ since $\|S_{x,m}\| \geq 2^{k-2}$. From this and the above estimation of $K_{\ell}(x)$, we have $K_{\ell}(x) \leq \log \|V_{n,m,k}\| + 3 \log n + e_3 \leq m - k + 3 \log n + e_4$; that is, $k \leq m - K_{\ell}(x) + 3 \log n + e_4$, and hence, $\log \|S_{x,m}\| \leq m - K_{\ell}(x) + 3 \log n + e_5$. Therefore, $K_{\ell}(y \mid x) \leq m - K_{\ell}(x) + 4 \log n + e_1 + e_5$; thus, we get the desired result.

The problem with using this argument for the polynomial time bounded case is that even though a string $y \in S_{x,m}$ can be identified by specifying x, m, and the index w of y in $S_{x,m}$, it may take more than polynomial time to produce y from x, m, and w. (The same problem occurs for producing x in $V_{n,m,k}$.) However, we will show in the following lemmas that this type of problem is solvable with a help of some Σ_2^P oracle. $\square$

We use an encoding technique introduced by Sipser [Sip83]. Let A be any "easy" set that is specified by some parameters. Roughly speaking, we can give a code of length $\log \|A\| + \log \log \|A\| + O(1)$ to any string x in A, from which code (together with the parameters specifying A), one can produce x in polynomial time by using some Σ_2^P oracle. Notice that this code length is quite close to the optimal, $\log \|A\|$.

Consider any family $\mathcal{L} = \{L_\alpha\}$ of sets, where each set in $\mathcal{L}$ is determined by specifying a tuple α of parameters. (Here we are considering $\{S_{x,m}\}$ and $\{V_{n,m,k}\}$ in our minds.) We say that $\mathcal{L}$ has a *polynomial time scheme* if there is a polynomial time Turing machine M such that for every tuple α of parameters, and for all $x \in \{0,1\}^*$, $x \in L_\alpha \leftrightarrow M$ accepts $\langle \alpha, x \rangle$. (That is, M can recognize each L_α when α is given.)

Now the lemma we need is formally stated as follows. (The proof is essentially given in [Sip83].)

Lemma 4.2. Let $\mathcal{L}$ be a family of sets with some polynomial time scheme. Then there exist some polynomial time oracle Turing machine transducer M, some oracle set X in Σ_2^P, and some constant c with the following property: for any set L_α in $\mathcal{L}$, and any string x in L_α, there exists a string w of length $\leq \log \|A\| + \log \log \|A\| + c$ such that $M^X(\langle \alpha, w \rangle)$ yields x.

To complete the proof of Theorem 4.1, we need to define $V_{n,m,k}$, and to show that both $\{S_{x,m}\}$ and $\{V_{n,m,k}\}$ have polynomial time schemes. (Here we can use the assumption P = NP.)

First define $V_{n,m,k}$. Intuitively, we want to define $V_{n,m,k}$ as $\{x \in \{0,1\}^n : \|S_{x,m}\| \geq 2^k\}$. However, this set, which is in PP, may not belong to P even if we assume P = NP. On the other hand, by using the hashing technique [Sip83], we can define a set in PH with the desired property.

A (n,p)-*hash family of size* s is a collection of s linear transformations from $\{0,1\}^n$ to $\{0,1\}^p$. We say that a (n,p)-hash family $H = \{h_1, ..., h_s\}$ *hashes* a set $A \subseteq \{0,1\}^n$ if for every $u \in A$, there is an $h_i \in H$ such that $h_i(u) \neq h_i(u')$ for every $u' \in A$. Now, $V_{n,m,k} \subseteq \{0,1\}^n$ is defined as follows:

$$V_{n,m,k} = \{ x : \text{no } (n, k-1)\text{-family of hash functions of size } k-1 \text{ hashes } S_{x,m} \}.$$

Then from the lemmas in [Sip83], we can show that $\|S_{x,m}\| \geq 2^k \to x \in V_{n,m,k}$, and $\|S_{x,m}\| < 2^{k-2} \to x \notin V_{n,m,k}$.

Finally, we show that both $\{S_{x,m}\}$ and $\{V_{n,m,k}\}$ have polynomial time scheme. (The proof is omitted in this abstract.)

Lemma 4.3. Suppose that P = NP. Then both $\{S_{x,m}\}$ and $\{V_{n,m,k}\}$ have polynomial time schemes.

Since it is our ultimate goal to prove some if and only if statement regarding the symmetry of information, and we have a weak if and only if, it would be nice to investigate which direction of our proofs can be improved. A step in that direction is to look at relativized computations. In the following theorem, we show that the symmetry of information cannot be made equivalent to the P = NP question under every oracles.

Theorem 4.4. There is an oracle X such that the symmetry of information holds relative to X but $P^X \neq UP^X$ (hence $P^X \neq NP^X$).

Proof. Let $X = \Sigma_2^P \oplus B$, where B contains one Kolmogorov random string for each size n such that n is a power of 2.

Consider set $A = \{xy : y \in 10^*, |xy| = 2^m,$ and x is a prefix of a string in B of size $2^m\}$. Clearly, $A \in UP^X$. Also, if $A \in P$, then it can be shown that strings in B could be printed by short programs, contradiction our choice of Kolmogorov random strings. $\square$

References

[BDG88] J. Balcázar, J. Díaz, and J. Gabarró. *Structural Complexity I.* Springer-Verlag, 1988.

[Gác74] P. Gács. On the symmetry of algorithmic information. *Soviet Math. Dokl.*, 15:1477, 1974.

[LM91] L. Longpré and S. Mocas. Symmetry of information and one-way functions. In *ISA '91 Algorithms, Lecture notes in Computer Science*, pages 308–315. Springer-Verlag, 557, 1991.

[LV88] M. Li and P.M.B. Vitányi. Two decades of applied kolmogorov complexity. In *Proc. Structure in Complexity Theory third annual conference*, pages 80–101, 1988.

[Sip83] M. Sipser. A complexity theoretic approach to randomness. In *Proc. 15th ACM Symposium on Theory of Computing*, pages 330–335, 1983.

[ZL70] A.K. Zvonkin and L.A. Levin. The complexity of finite objects and the development of the concepts of information and randomness by means of the theory of algorithms. *Russ. Math. Surv.*, 25:83–124, 1970.

On Probabilistic ACC Circuits
with an Exact-Threshold Output Gate

Richard Beigel[*]

Department of Computer Science, Yale University
New Haven, CT 06520-2158, USA. beigel-richard@cs.yale.edu

Jun Tarui[†]

Department of Computer Science, University of Warwick
Coventry, CV4 7AL, United Kingdom. jun@dcs.warwick.ac.uk

Seinosuke Toda

Department of Computer Science and Information Mathematics
University of Electro-Communications
Chofu-shi, Tokyo, 182, Japan. toda@pspace.cs.uec.ac.jp

Abstract

Let SYM^+ denote the class of Boolean functions computable by depth-two size-$n^{\log^{O(1)} n}$ circuits with a symmetric-function gate at the root and AND gates of fan-in $\log^{O(1)} n$ at the next level, or equivalently, the class of Boolean functions f such that $f(x_1, \ldots, x_n)$ can be expressed as $f(x_1, \ldots, x_n) = h_n(p_n(x_1, \ldots, x_n))$ for some polynomial p_n over Z of degree $\log^{O(1)} n$ and norm (the sum of the absolute values of its coefficients) $n^{\log^{O(1)} n}$ and some function $h_n : Z \to \{0, 1\}$. Building on work of Yao [Yao90], Beigel and Tarui [BT91] showed that $ACC \subseteq SYM^+$, where ACC is the class of Boolean functions computable by constant-depth polynomial-size circuits with NOT, AND, OR, and MOD_m gates for some fixed m.

In this paper, we consider augmenting the power of ACC circuits by allowing randomness and allowing an exact-threshold gate as the output gate (an exact-threshold gate outputs 1 if exactly k of its inputs are 1, where k is a parameter; it outputs 0 otherwise), and show that every Boolean function computed by this kind of augmented ACC circuits is still in SYM^+.

Showing that some "natural" function f does not belong to the class ACC remains an open problem in circuit complexity, and the result that

[*]Supported in part by NSF grant CCR-8958528

[†]Supported in part by the ESPRIT II BRA Programme of the EC under contract # 7141 (ALCOM II). Part of the work was done while the author was a student at University of Rochester, and was supported in part by NSF grant CDA-8822724.

ACC $\subseteq$ SYM$^+$ has raised the hope that we may be able to solve this problem by exploiting the characterization of SYM$^+$ in terms of polynomials, which are perhaps easier to analyze than circuits, and showing that $f \notin$ SYM$^+$. Our new result and proof techniques suggest that the possibility that SYM$^+$ contains even more Boolean functions than we currently know should also be kept in mind and explored.

By a well-known connection [FSS84], we also obtain new results about some classes related to the polynomial-time hierarchy.

1 Introduction

Consider a sequence $\langle C_n : n = 1, \ldots \rangle$ of circuits computing a Boolean function $f : \{0,1\}^* \rightarrow \{0,1\}$, where C_n is a circuit for n Boolean variables computing f_n, the restriction of f to $\{0,1\}^n$. In *constant-depth* circuits, the depth of C_n is *fixed* as $n \rightarrow \infty$, and gates of *unbounded fan-in* are allowed.

In the last decade, strong lower bounds were established for the size of constant-depth circuits that compute explicit Boolean functions, in the case where the allowable gates are NOT, AND, OR, and MOD$_q$ for some fixed prime power q. (A MOD$_m$ gate on inputs $y_1, \ldots, y_\nu$ outputs 1 if $\sum y_i \equiv 0 \pmod{m}$; 0 otherwise.)

Furst, Saxe, and Sipser [FSS84] and, independently, Ajtai [Ajt83] first showed that constant-depth circuits with NOT, OR, and AND gates require superpolynomial size to compute PARITY. Yao [Yao85] improved this and gave an exponential bound, and later Hastad [Has87] simplified a proof, and gave a further improved near-optimal bound.

Razborov [Raz87] showed that to compute MAJORITY, constant-depth circuits with NOT, AND, OR, *and* PARITY gates require exponential size. Smolensky later extended this and showed that to compute the MOD$_q$ function, constant-depth circuits with NOT, AND, OR, and MOD$_{q'}$ gates require exponential size if q and q' are powers of distinct primes. (For more information about these results and general discussions on this line of research including history, motivations, and applications, see [Sip92] and [BS90].)

It remains an open problem, however, to show a limitation for constant-depth circuits with MOD$_m$ gates for some fixed composite m: The class ACC, first considered by Barrington [Bar89], consists of Boolean functions computable by constant-depth polynomial-size circuits with NOT, AND, OR, and MOD$_m$ gates for some fixed m. At present, we cannot prove that some explicit function, e.g., a function in NP, does not belong to ACC.

The first nontrivial *upper* bound on the computing power of ACC circuits was shown by Yao [Yao90], and it was later improved by Beigel and Tarui [BT91]. We first explain some conventions and definitions.

For a multivariate polynomial over $\mathbf{Z}$, define its *norm* to be the sum of the absolute values of its coefficients. A function $g(y_1, \ldots, y_\nu)$ is *symmetric* if g is fixed under any permutation of $\{1, \ldots, \nu\}$, which is equivalent to saying that $g(y_1, \ldots, y_\nu)$

only depends on $\sum y_i$ in the case y_i's are Boolean. As usual, we assume that NOT gates appear in a circuit only in the form of a negated input literal $\overline{x_i}$.

Let SYM$^+$ denote the class of Boolean functions computable by depth-two size-$n^{\log^{O(1)} n}$ circuits with an output gate labelled by a Boolean-valued symmetric function and AND gates of fan-in $\log^{O(1)} n$ at the next level. Similarly define the class FSYM$^+$ of integer-valued functions on $\{0,1\}^*$ by allowing an output gate labelled by an integer-valued symmetric function. (In [BT91], the class SYM$^+$ was called SYMMC. For general discussions of *quasipolynomial*-size (size $n^{\log^{O(1)} n}$) circuit classes, see the recent survey by Barrington [Bar92], which also includes the suggestion to use the name SYM$^+$.)

The following are immediate: A Boolean function f is in SYM$^+$ if and only if f can be expressed as $f(x_1, \ldots, x_n) = h_n(p_n(x_1, \ldots, x_n))$ for some polynomial p_n over $\mathbf{Z}$ of degree $\log^{O(1)} n$ and norm $n^{\log^{O(1)} n}$ and some function $h_n : \mathbf{Z} \to \{0,1\}$. By allowing $h_n : \mathbf{Z} \to \mathbf{Z}$, a similar characterization of FSYM$^+$ is obtained.

Improving the earlier result by Yao [Yao90], Beigel and Tarui [BT91] showed that ACC $\subseteq$ SYM$^+$. This result says that low-degree small-norm polynomials can represent a surprising rich class of Boolean functions. At the same time, it has raised the hope that we may be able to solve the ACC problem by exploiting the characterization of SYM$^+$ in terms of polynomials, which are perhaps easier to analyze than circuits, and showing that some explicit function does not belong to SYM$^+$.

In this paper, we show some additional representational power of low-degree small-norm polynomials in two settings. Our new results and the techniques we develop suggest that the possibility that SYM$^+$ contains even more Boolean functions than we currently know should also be kept in mind and explored.

In Section 2, we state our two results and the key lemma we use in proving them. In Section 3, we prove the two results assuming the key lemma, and we prove the key lemma in Section 4.

2 Results

For Boolean inputs $y_1, \ldots, y_\nu$, an *Exact$_k$* gate outputs 1 if $\sum y_i = k$, and 0 otherwise; and a *Count* gate outputs the nonnegative integer $\sum y_i$. When we consider an Exact$_k$ gate, the parameter k may grow as the size of input approaches infinity. A circuit with a Count gate as its single output gate is a natural circuit analogue of a counting Turing machine, the notion introduced by Valiant [Val79] to define the class #P; and such a circuit computes a function from $\{0,1\}^n$ to $\mathbf{Z}^{\geq 0}$, the set of nonnegative integers. *ACC circuits* consists of AND, OR, and MOD$_m$ gates for some fixed m. We consider augmenting the power of ACC circuits by allowing randomness and allowing an Exact gate or a Count gate as the output gate.

Theorem 1. *Suppose that a Boolean function $f : \{0,1\}^* \to \{0,1\}$ is computed with error probability bounded away from $1/2$ by some constant-depth polynomial-size probabilistic ACC circuits with an Exact gate as the output gate. Then $f \in SYM^+$.*

It is known [ABO84] that allowing randomness alone does not enable constant-depth polynomial-size ACC circuits to compute a function outside of ACC; and in [BT91], it was shown that allowing a symmetric-function output gate does not enable constant-depth polynomial-size deterministic ACC circuits to compute a function outside SYM^+. But to prove Theorem 1, which concerns augmented ACC circuits in which randomness *and* an Exact output gate are allowed, we need to develop new techniques.

Theorem 2. *Suppose that a function $f : \{0,1\}^* \to Z^{\geq 0}$ is computed with error probability bounded away from $1/2$ by some constant-depth polynomial-size probabilistic ACC circuits with a Count gate as the output gate. Then $f \in FSYM^+$.*

The following lemma will be the key to prove the two theorems above. Let p_j denote the j-th smallest prime. For a finite multiset S of integers, a prime p_j, and a residue $r \in \{0,\ldots,p_j - 1\}$, let $M_{j,r}(S)$ denote the number of elements in S that are congruent to $r \bmod p_j$.

Lemma 3. *Let a sequence $\langle l_n : n = 1,\ldots\rangle$ of positive integers of order $\log^{O(1)} n$ be given. Then, there is a sequence $\langle q_n(x_1,\ldots,x_n) : n = 1,\ldots\rangle$ of polynomials over Z of degree $\log^{O(1)} n$ and norm $n^{\log^{O(1)} n}$ such that the following holds for each n: For any integers $a_1,\ldots,a_n$, every prime p_j with $j \leq l_n$, and every residue $r \in \{0,\ldots,p_j-1\}$, $M_{j,r}(a_1,\ldots,a_n)$ can be determined from $q_n(a_1,\ldots,a_n)$.*

There is a well-known connection [FSS84] between constant-depth circuit classes and some complexity classes between P and PSPACE. In particular, the class of Boolean functions computable by constant-depth quasipolynomial-size AND/OR circuits corresponds to the class PH, the polynomial-time hierarchy (for a definition, see, e.g., [Joh90]). It will be clear from the proofs that the conclusions of Theorems 1 and 2 in fact hold for augmented ACC circuits of constant depth and *quasipolynomial* size; and we can obtain new results about some classes related to PH. The classes SYM^+ and $FSYM^+$ correspond to $P^{\#P[1]}$ and $FP^{\#P[1]}$, respectively. Below we state two new results corresponding to Theorems 1 and 2. (More precisely, they correspond to the versions of Theorems 1 and 2 stated in terms of constant-depth quasipolynomial-size AND/OR circuits.) Further discussions of these results are omitted from this paper.

Proposition 4.

$$BP \cdot C_= \cdot PH \subseteq P^{\#P[1]}.$$

Proposition 5.

$$BP \cdot \# \cdot PH \subseteq FP^{\#P[1]}.$$

3   Proofs of Theorems from the Key Lemma

For a circuit C for n variables $x_1, \ldots, x_n$, we let $C(x_1, \ldots, x_n)$ denote the function that C computes. First we explain the following nonconstructive argument:

Proposition 6. *Let D_n be a probabilistic circuit for n Boolean variables that computes $f_n : \{0,1\}^n \to \{0,1\}$ with error probability at most ε_n. Then, for any constant $\delta > 0$, there is a multiset of $N = O(1/\varepsilon_n \cdot n)$ deterministic circuits $C_1, \ldots, C_N$, each obtained by fixing a setting of random bits in D_n, such that for every $x \in \{0,1\}^n$, $C_i(x) = f_n(x)$ except for at most $(1 + \delta)\varepsilon_n$ fraction of i's $(1 \le i \le N)$.*

Proof. Fix $x \in \{0,1\}^n$ and consider $\overline{D}_1(x), \ldots, \overline{D}_N(x)$, where $\overline{D}_i$'s are N independent copies of D_n. Let E be the number of errors among $\overline{D}_i$'s, i.e., let $E = |\{i : 1 \le i \le N : \overline{D}_i(x) \ne f_n(x)\}|$. By the Chernoff bound [Che52] on the tails of Bernoulli trials, we can take $N = O(1/\varepsilon_n \cdot n)$ and have

$$\text{Prob}(E > (1 + \delta)\varepsilon_n N) < 2^{-n}.$$

¿From this, the proposition readily follows. ∎

Proof of Theorem 1. Let D_n be a constant-depth polynomial-size probabilistic ACC circuit with an Exact_{b_n} output gate of fan-in K and AND/OR/MOD$_m$ gates elsewhere that computes $f_n : \{0,1\}^n \to \{0,1\}$ with error probability at most $1/4$. (It will be clear that the proof works for any error probability bounded away from $1/2$.) By using a standard probabilistic simulation of AND/OR by MOD$_m$ gates (we can either use Valiant and Vazirani's lemma [VV86] as in [AH90] or the Razborov-Smolensky simulation [Raz87, Smo87]) and by the argument explained above, obtain $N = O(n)$ deterministic ACC circuits $C_1, \ldots, C_N$ consisting of MOD$_m$ gates and fan-in-$\log^{O(1)} n$ AND gates such that for every $x \in \{0,1\}^n$,

$$f_n(x) = 0 \implies |\{i : 1 \le i \le N : C_i(x) = 1\}| \le \frac{1}{3}N;$$
$$f_n(x) = 1 \implies |\{i : 1 \le i \le N : C_i(x) = 1\}| \ge \frac{2}{3}N.$$

Thus, if we let $\{g_i^{(1)}, \ldots, g_i^{(K)}\}$ be the set of MOD$_m$ gates that are connected to the output gate of C_i $(1 \le i \le N)$, one of the following two cases holds for each input x, where $g_i^{(j)}$ below denotes the $\{0,1\}$-value of that gate on input x.

$$\text{Case 1:} \quad |\{i : 1 \le i \le N, \sum_{j=1}^{K} g_i^{(j)} = b_n\}| \le \frac{1}{3}N.$$

$$\text{Case 2:} \quad |\{i : 1 \le i \le N, \sum_{j=1}^{K} g_i^{(j)} = b_n\}| \ge \frac{2}{3}N.$$

If there is a degree-$\log^{O(1)} n$ norm-$n^{\log^{O(1)} n}$ polynomial q in KN variables such that which of Cases (1) and (2) holds can be determined from the value of $q(g_1^{(1)}, \ldots, g_1^{(K)}, \ldots, g_N^{(1)}, \ldots, g_N^{(K)})$, then by the same argument as in [BT91, Proof of Lemma 5], we can conclude that $\{f_n\}_{n=0}^{\infty} \in \mathrm{SYM}^+$.

The following lemma says that such a polynomial indeed exists (substitute $\sum_{j=1}^{K} g_i^{(j)}$ for each variable x_i in a polynomial given in the lemma). ∎

Lemma 7. *Let $k \geq 1$ and let $\langle b_n : n = 1, \ldots \rangle$ be a sequence of integers with $|b_n| \leq n^k$. Then, there is a sequence $\langle q_n(x_1, \ldots, x_n) : n = 1, \ldots \rangle$ of polynomials of degree $\log^{O(1)} n$ and norm $n^{\log^{O(1)} n}$ and a sequence $\langle h_n : n = 1, \ldots \rangle$ of functions from $\mathbf{Z}$ to $\{0, 1\}$ with the following property: For any integers $a_1, \ldots, a_n$ with $|a_i| \leq n^k$,*

$$|\{i : a_i = b_n\}| \leq \frac{1}{3} n \implies h_n(q_n(a_1, \ldots, a_n)) = 0;$$

$$|\{i : a_i = b_n\}| \geq \frac{2}{3} n \implies h_n(q_n(a_1, \ldots, a_n)) = 1.$$

Proof. Let b_n and $a_1, \ldots, a_n$ be integers whose absolute values are at most n^k. Let S be the set of a_i's such that $a_i \neq b_n$. Take $l = 4k \log_2 n + O(1)$ so that the following holds: For each $a \in S$, if j is picked at random from $\{1, \ldots, l\}$, $a \equiv b_n$ (mod p_j) with probability at most $1/4$. Then, there exists $j \in \{1, \ldots, l\}$ such that

$$|\{a \in S : a \equiv b_n \pmod{p_j}\}| \leq \frac{1}{4} |S|.$$

Now assume that either (1) $|S| \geq \frac{2}{3} n$ or (2) $|S| \leq \frac{1}{3} n$ holds. The argument above shows that Case (2) holds if and only if $M_{j,(b_n \bmod p_j)}(a_1, \ldots, a_n) \geq \frac{2}{3} n$ for all $j = 1, \ldots, l$. Using Lemma 3, we can finish the proof. ∎

Proof of Theorem 2. Similar reasoning as in the proof of Theorem 1 yield the following. Below we use the same notation as in the proof of Theorem 1. For each input x, there exists c_n such that

$$|\{i : 1 \leq i \leq N, \sum_{j=1}^{K} g_i^{(j)} = c_n\}| \geq \frac{2}{3} N.$$

The following lemma says that there is a degree-$\log^{O(1)} n$ norm-$n^{\log^{O(1)} n}$ polynomial q in KN variables such that c_n can be determined from the value of $q(g_1^{(1)}, \ldots, g_1^{(K)}, \ldots, g_N^{(1)}, \ldots, g_N^{(K)})$. ∎

Lemma 8. *Let $k \geq 1$ and let $\langle b_n : n = 1, \ldots \rangle$ be a sequence of integers with $|b_n| \leq n^k$. Then, there is a sequence $\langle q_n(x_1, \ldots, x_n) : n = 1, \ldots \rangle$ of degree-$\log^{O(1)} n$ norm-$n^{\log^{O(1)} n}$ polynomials and a sequence $\langle h_n : n = 1, \ldots \rangle$ of functions from $\mathbf{Z}$ to $\mathbf{Z}$ with the following property: For any integers $a_1, \ldots, a_n$ with $|a_i| \leq n^k$,*

$$(\exists c) \, |\{i : a_i = c\}| \geq \frac{2}{3} n \implies h_n(q_n(a_1, \ldots, a_n)) = c.$$

Proof. Take $l = k \log_2 n + O(1)$ so that we have $P = \prod_{j=1}^{l} p_j > 2n^k + 1 = |\{-n^k, \ldots, n^k\}|$. By determining $M_{j,r}(a_1, \ldots, a_n)$ for each prime p_j $(1 \leq j \leq l)$, we can determine such c as above if it exists. Using Lemma 3, we can finish the proof. ∎

4 Proof of the Key Lemma

Say that a singe-variate polynomial $A(x)$ over $\mathbb{Z}$ is *degree-d modulus-amplifying* if for arbitrary integers N and m,

$$N \equiv 0 \pmod{m} \implies A(N) \equiv 0 \pmod{m^d};$$
$$N \equiv 1 \pmod{m} \implies A(N) \equiv 1 \pmod{m^d}.$$

The following lemma is essentially due to Toda [Tod91], and was proved in the form below with simplified analysis and optimal degree in [BT91].

Lemma 9. *For each $d \geq 1$, there is a unique degree-$(2d - 1)$ polynomial that is degree-d modulus-amplifying.*

In what follows, we let $A_d(x)$ denote the unique degree-d modulus-amplifying polynomial given in Lemma 9. When d is of order $\log^{O(1)} n$, the norm of A_d is of order $n^{\log^{O(1)} n}$.

For convenience, we restate Lemma 3.

Lemma 3. *Let a sequence $\langle l_n : n = 1, \ldots \rangle$ of positive integers of order $\log^{O(1)} n$ be given. Then, there is a sequence $\langle q_n(x_1, \ldots, x_n) : n = 1, \ldots \rangle$ of polynomials over $\mathbb{Z}$ of degree $\log^{O(1)} n$ and norm $n^{\log^{O(1)} n}$ such that the following holds for each n: For any integers $a_1, \ldots, a_n$, every prime p_j with $j \leq l_n$, and every residue $r \in \{0, \ldots, p_j - 1\}$, $M_{j,r}(a_1, \ldots, a_n)$ can be determined from $q_n(a_1, \ldots, a_n)$.*

Proof. We will gradually explain a construction of polynomials satisfying the conclusion of the lemma. Below we use "mod" as in "$a \equiv 0 \pmod{m}$" and also as a binary operator as in "$a \bmod m \in \{0, \ldots, m-1\}$".

Let $R_p(x, r)$ be the polynomial in x and r defined by $R_p(x, r) = 1 - (x - r)^{p-1}$. Then, by Fermat's little theorem, for an integer a, a prime p, and a residue $r \in \{0, \ldots, p-1\}$,

$$R_p(a, r) \bmod p = (1 - (a - r)^{p-1}) \bmod p = \begin{cases} 1 & \text{if } a \equiv r \pmod{p}; \\ 0 & \text{otherwise.} \end{cases} \tag{1}$$

First we explain a construction for *one* prime p_j; write p for p_j and $M_r(a_1, \ldots, a_n)$ for $M_{p_j,r}(a_1, \ldots, a_n)$. Define the polynomial $Q_{p,d}(x_1, \ldots, x_n)$ by

$$Q_{p,d}(x_1, \ldots, x_n) = \sum_{i=1}^{n} \sum_{r=0}^{p-1} A_d(R_p(x_i, r))(n + 1)^r.$$

Assume that $p^d > (n+1)^p$. Let $a_1, \ldots, a_n$ be arbitrary integers. We claim that $M_r(a_1, \ldots, a_n)$ can be determined from $Q_{p,d}(a_1, \ldots, a_n)$.

Fix $i \in \{1, \ldots, n\}$ and consider the value $\sigma(a_i)$ of the inner sum. By (1) and degree-d modulus amplification of A_d,

$$\sigma(a_i) \bmod p^d = (n+1)^{(a_i \bmod p)} \tag{2}$$

Since $(n+1)^{(a_i \bmod p)} \leq (n+1)^{p-1}$ and we have assumed that $(n+1)^p < p^d$,

$$\sum_{i=1}^{n} (\sigma(a_i) \bmod p^d) \leq n(n+1)^{p-1} < p^d,$$

and thus

$$\left(\sum_{i=1}^{n} \sigma(a_i) \right) \bmod p^d = \sum_{i=1}^{n} \left(\sigma(a_i) \bmod p^d \right). \tag{3}$$

Let $N = n+1$. Then, using (3) and (2) to get, respectively, the second and the third equalities below, we have

$$
\begin{aligned}
Q_{p,d}(a_1, \ldots, a_n) \bmod p^d &= \left(\sum_{i=1}^{n} \sigma(a_i) \right) \bmod p^d \\
&= \sum_{i=1}^{n} (\sigma(a_i) \bmod p^d) \\
&= \sum_{i=1}^{n} N^{(a_i \bmod p)} \\
&= \sum_{r=1}^{p-1} M_r(a_1, \ldots, a_n) N^r.
\end{aligned}
$$

Since $0 \leq M_r(a_1, \ldots, a_n) \leq n < N = n+1$, we can write $(Q_{p,d}(a_1, \ldots, a_n) \bmod p^d)$ in base N and obtain $M_r(a_1, \ldots, a_n)$ as the $(r+1)$-th least significant digit for each $r = 0, \ldots, p-1$.

Using the Chinese remainder theorem, we can construct a polynomial from which $M_{j,r}$ can be determined for *all* the primes $p_1, \ldots, p_{l_n}$.

Now we describe the full construction using the notation above.

Let $\langle l_n : n = 1, \ldots \rangle$ be as in the lemma. Consider n fixed and write l for l_n. Let $N = n+1$ as above, and take $d = p_l \cdot \log N$ $(= \log^{O(1)} n)$ so that for each prime p_j $(1 \leq j \leq l)$, we have $(p_j)^d > N^{p_j}$.

Let $P = \prod_{j=1}^{l} (p_j)^d$ $(= n^{\log^{O(1)} n})$. By the Chinese remainder theorem, for each prime p_j $(1 \leq j \leq l)$, find a unique $\overline{p_j} \in \{0, \ldots, P-1\}$ such that

$$\overline{p_j} \equiv \begin{cases} 1 & (\bmod (p_j)^d); \\ 0 & (\bmod (p_{j'})^d) \quad \text{for each } j' \neq j, \ 1 \leq j' \leq l. \end{cases} \tag{4}$$

Put

$$q_n(x_1,\ldots,x_n) = \sum_{j=1}^{l} \overline{p_j}\, Q_{p_j,d}(x_1,\ldots,x_n)$$

$$= \sum_{j=1}^{l} \sum_{i=1}^{n} \sum_{r=0}^{p_j-1} \overline{p_j}\, A_d\big(R_{p_j}(x_i,r)\big)\, N^r.$$

By (4), for any integers $a_1,\ldots,a_n$ and for each $j \in \{1,\ldots,l\}$,

$$q_n(a_1,\ldots,a_n) \bmod (p_j)^d = Q_{p_j,d}(a_1,\ldots,a_n) \bmod (p_j)^d,$$

and thus $M_{j,r}(a_1,\ldots,a_n)$ can be determined as explained above.

Finally, it is easy to check that q_n has degree $\log^{O(1)} n$ and norm $n^{\log^{O(1)} n}$. ∎

References

[ABO84] M. Ajtai and M. Ben-Or. A theorem on probabilistic constant depth circuits. In *Proceedings of the 16th Annual ACM Symposium on Theory of Computing*, pages 471–474. ACM Press, 1984.

[AH90] E. Allender and U. Hertrampf. On the power of uniform families of constant depth threshold circuits. In *Proceedings of the 15th International Symposium on Mathematical Foundations of Computer Science*, pages 158–164. Springer-Verlag, 1990. Lecture Notes in Computer Science, vol. 452.

[Ajt83] M. Ajtai. Σ_1^1 formulae on finite structures. *Ann. Pure Appl. Logic*, 24:1–48, 1983.

[Bar89] D. Barrington. Bounded-width polynomial-size branching programs recognize exactly those languages in NC^1. *J. Comput. System Sci.*, 38(1):150–164, February 1989.

[Bar92] D. Barrington. Quasipolynomial size circuit classes. In *Proceedings of the 7th Annual Conference on Structure in Complexity Theory*, pages 86–93. IEEE Computer Society Press, 1992.

[BS90] R. Boppana and M. Sipser. The complexity of finite functions. In J. van Leeuwen, editor, *Handbook of Theoretical Computer Science: Vol. A*, pages 757–804. North-Holland, Amsterdam, 1990.

[BT91] R. Beigel and J. Tarui. On ACC. In *Proceedings of the 32nd IEEE Symposium on Foundations of Computer Science*, pages 783–792. IEEE Computer Society Press, 1991.

[Che52]   H. Chernoff. A measure of asymptotic efficiency for tests of a hypothesis based on the sum of observations. *Annals of Mathematical Statistics*, 23:493–507, 1952.

[FSS84]   M. Furst, J. Saxe, and M. Sipser. Parity, circuits and the polynomial-time hierarchy. *Math. Systems Theory*, 17:13–27, 1984.

[Has87]   J. Hastad. *Computational Limitations of Small-Depth Circuits*. MIT Press, 1987.

[Joh90]   D. Johnson. A catalog of complexity classes. In J. van Leeuwen, editor, *Handbook of Theoretical Computer Science: Vol. A*, pages 67–161. North-Holland, Amsterdam, 1990.

[Raz87]   A. Razborov. Lower bounds for the size of circuits of bounded depth with basis $\{\wedge, \oplus\}$. *Mathematical Notes of the Academy of Sciences of the USSR*, 41(4):333–338, September 1987.

[Sip92]   M. Sipser. The history and status of the P versus NP question. In *Proceedings of the 24th Annual ACM Symposium on Theory of Computing*, pages 603–618. ACM Press, 1992.

[Smo87]   R. Smolensky. Algebraic methods in the theory of lower bounds for Boolean circuit complexity. In *Proceedings of the 19th Annual ACM Symposium on Theory of Computing*, pages 77–82. ACM Press, 1987.

[Tod89]   S. Toda. On the computational power of PP and $\oplus$P. In *Proceedings of the 30th IEEE Symposium on Foundations of Computer Science*, pages 514–519. IEEE Computer Society Press, 1989.

[Tod91]   S. Toda. PP is as hard as the polynomial-time hierarchy. *SIAM J. Comput.*, 20(5):865–877, 1991. Earlier version appeared as [Tod89].

[Val79]   L. Valiant. The complexity of computing the permanent. *Theoret. Comput. Sci.*, 7:189–201, 1979.

[VV86]   L. Valiant and V. Vazirani. NP is as easy as detecting unique solutions. *Theoret. Comput. Sci.*, 47:85–93, 1986.

[Yao85]   A. Yao. Separating the polynomial-time hierarchy by oracles. In *Proceedings of the 26th IEEE Symposium on Foundations of Computer Science*, pages 1–10. IEEE Computer Society Press, 1985.

[Yao90]   A. Yao. On ACC and threshold circuits. In *Proceedings of the 31st IEEE Symposium on Foundations of Computer Science*, pages 619–627. IEEE Computer Society Press, 1990.

Computational and Statistical Indistinguishabilities *

Kaoru Kurosawa[1] and Osamu Watanabe[2]

[1] Dept. of Electrical & Electronic Eng., Tokyo Institute of Technology
Meguro-ku Ookayama, Tokyo 152, Japan
(e-mail: kkurosaw@ss.titech.ac.jp)

[2] Dept. of Computer Science, Tokyo Institute of Technology
Meguro-ku Ookayama, Tokyo 152, Japan
(e-mail: watanabe@cs.titech.ac.jp)

Abstract. We prove that a pair of polynomially samplable distributions are statistically indistinguishable if and only if no polynomial size circuits relative to NP sets (P_{nu}^{NP}-distinguishers) can tell them apart. As one application of this observation, we classify "zero-knowledge" notions that are used for interactive systems.

1 Introduction

For any pair of probability distributions, we say that they are *computationally indistinguishable* if no polynomial size circuits (which is called P_{nu}-*distinguishers*) can tell them apart, and we say that they are *statistically indistinguishable* if no distinguishers (that could be infinitely powerful) can tell them apart. (See Section 2.1 for the precise definition.) Intuitively, a pair of statistically indistinguishable distributions are "statistically" so close each other that no one can find their difference, while a pair of computationally indistinguishable distributions may be statistically different, but such difference cannot be detected in reasonable amount of time. These indistinguishability notions are fundamental in complexity theory, for example, in modern cryptology. In particular, the problem of detecting the difference (if it exists) between two "polynomial time samplable" distributions is essential in several situations. (A distribution is called *polynomial time samplable* if it is defined as output of some polynomial time randomized algorithm.)

Here we may naturally think of the following two questions: (1) Is the statistical indistinguishability really different from the computational indistinguishability? (2) If so, how much computational power is necessary and/or sufficient to find statistical difference. For the first question, by using the technique and results established in [ILL89], Goldreich [Gol90] essentially proved that the computational and statistical indistinguishability are equivalent if and only if no cryptographic *i.o.* one-way function exists. (Precisely speaking, the statistical indistinguishability considered

* The second author was supported in part by Grant in Aid for Scientific Research of the Ministry of Education, Science and Culture of Japan under Grant-in-Aid for Research (A) 04780027 (1992).

in [Gol90] is *i.o.* type whereas ours is *a.e.* type. Nevertheless, his argument and
the proof technique adapted from [ILL89] work even in our context.) On the other
hand, their proof technique do not seem to answer to the second question. This
paper presents one explicit algorithm — distinguisher — to detect the difference
between two statistically different distributions, thereby showing one upper bound
for the second problem.

We introduce a new class of distinguishers that are (supposed to be) more powerful than polynomial size circuits. A P_{nu}^{NP}-*distinguisher* is a family of polynomial size
circuits that can ask queries to some fixed NP oracle set. Our main result shows that
P_{nu}^{NP}-distinguishers are powerful enough to detect the statistical difference between
the outputs of two polynomial time randomized algorithms. That is, we prove that
two polynomially samplable distributions are statistically indistinguishable if and
only if no P_{nu}^{NP}-distinguishers can tell them different. (Clearly, it follows from this
observation that if P = NP, then the computational and statistical indistinguishabilities coincide for polynomial samplable distributions. However, this consequence
is weaker than the above mentioned observation on Goldreich's result.)

We also consider an application of the above result to the study of interactive proof systems [GMR89]. The security of the communication by an interactive
proof system is discussed by its zero-knowledgeness. Roughly speaking, the zero-
knowledgeness of an interactive system is defined by the "similarity" of real communications and pseudo communications generated by some polynomial time simulator. In the literature, three zero-knowledge notions have been considered:- computational, statistical, and perfect zero-knowledge. Among them, computational and statistical zero-knowledge notions are respectively defined by using computational and
statistical indistinguishabilities for the formal definition of "similarity". Here we propose to investigate computational classification of the statistical zero-knowledgeness
by using various types of indistinguishabilities. Then from our main result and the
recent result by [BP92], we can show that the statistical zero knowledgeness and the
$P_{nu}^{\Sigma_j^P}$-zero knowledgeness coincide. As a corollary, we obtain that if P = NP, then
the statistical zero knowledge is equivalent to the computational zero knowledge. A
similar result is obtained for interactive proof systems of knowledge (where a prover
is given some secret and runs in probabilistic polynomial time. [FFS87, TW87]). For
such proof systems, the statistical zero knowledgeness and the P_{nu}^{NP}-zero knowledgeness coincide.

2 Preliminaries

In this paper, we assume that every "object" is encoded in $\{0,1\}^*$. Thus, by a
language we mean a subset of $\{0,1\}^*$, and the value of a random variable is in
$\{0,1\}^*$. For any predicate $\Phi(X)$ on random variable X, we write $\Pr_X\{\Phi(X)\}$ or
$\Pr_X\{\Phi(X)\}$ to denote the probability that $\Phi(X)$ holds.

We define the "indistinguishability" notions of two different types: namely, statistical and computational indistinguishabilities.

In the following, we use the notation $\nu(n)$ to denote any function vanishing faster
than the inverse of any polynomial. Formally, μ is any function such that $\forall k, \exists N,$
$\forall n > N \ [\ \nu(n) < n^{-k}\ ]$. For any string x, let $C(x)$ and $E(x)$ denote any random

variables that are parameterized by x. We assume that both $C(x)$ and $E(x)$ take values in $\{0,1\}^{f(|x|)}$, for some function f. In other words, the length of both $C(x)$ and $E(x)$ is determined by that of x.

Definition 1. Two families of random variables $\{C(x)\}_{x \in \{0,1\}^*}$ and $\{E(x)\}_{x \in \{0,1\}^*}$ are *statistically indistinguishable on L* if for any sufficiently large $x \in L$, we have

$$\sum_{y \in \{0,1\}^*} | \Pr_{C(x)} \{C(x) = y\} - \Pr_{E(x)} \{E(x) = y\}| < \nu(|x|).$$

Remark. Because of "for *any* sufficiently large $x \in L$", this indistinguishability notion is regarded as *a.e.* type. On the other hand, the one considered in [Gol90] uses "for *infinitely many* $x \in L$", and thus is regarded as *i.o.* type.

The computational indistinguishability was introduced in [GMR89]. Here we extend this notion by considering more powerful distinguishers than polynomial size circuits. For this purpose, we consider *query circuits*, circuits that can access to some oracle set. For any set of strings Q, let D_n^Q denote a query circuit with n input gates that queries some strings to Q. A family $\{D_n^Q\}_{n \geq 0}$ of polynomial size circuits is called *polynomial size circuits relative to Q*. Now we define two classes of predicates on $\{0,1\}^*$. Define P_{nu} to be the class of predicates computable by some polynomial size circuits, and define P_{nu}^{NP} to be the class of predicates computable by some polynomial size circuits relative to some set in NP. Clearly, P_{nu} and P_{nu}^{NP} are nonuniform versions of classes P and P^{NP}.

Definition 2. Let $\mathcal{C}$ be one of the above predicate classes. Two families of random variables $\{C(x)\}_{x \in \{0,1\}^*}$ and $\{E(x)\}_{x \in \{0,1\}^*}$ are $\mathcal{C}$-indistinguishable on L if for every predicate $D = \{D_n\}_{n \geq 0}$ in $\mathcal{C}$, and for all sufficiently large $x \in L$, we have

$$\Pr_{C(x)} \{ D_{|x|}(C(x)) \} - \Pr_{E(x)} \{ D_{|x|}(E(x)) \}| < \nu(|x|).$$

A predicate like D is called a $\mathcal{C}$-*distinguisher*.

Remark. P_{nu}-indistinguishability corresponds to the "computational indistinguishability" [GMR89] under the nonuniform computation model. Notice that distinguishers considered in [GMR89] are defined as $D = \{D_x\}_{x \in \{0,1\}^*}$; that is, they can change a predicate depending on x. Thus, in this sense, distinguishers in [GMR89] are more powerful than P_{nu}-distinguishers.

3 P_{nu}^{NP}-distinguisher

In this section, we show that in some context, the P_{nu}^{NP}-distinguishability and statistical distinguishability coincide. More specifically, we prove that P_{nu}^{NP}-distinguishers are powerful enough to find the statistical difference between the outputs of two polynomial time randomized algorithms.

Let R be any randomized algorithm on $\{0,1\}^*$. For any input x, we can regard the outputs $R(x)$ as a parameterized random variables; thus, R defines a family $\{R(x)\}_{x \in \{0,1\}^*}$ of random variables. Hence, by using the same definition, we can discuss the "indistinguishability" of a pair of randomized algorithms.

In this paper, a pair of (randomized) algorithms A and B are called *length-regular* if for any input x, both A and B output strings of the same length, which is uniquely determined by $|x|$.

Theorem 3. Let A and B be any length-regular pair of polynomial time randomized algorithms. $\{A(x)\}$ and $\{B(x)\}$ are statistically indistinguishable on L if and only if they are P_{nu}^{NP}-indistinguishable on L.

The only if part is obvious. Thus, in the following, we show the if part; that is, if $\{A(x)\}$ and $\{B(x)\}$ are statistically distinguishable, then they are indeed P_{nu}^{NP}-distinguishable. The following lemma plays a key role here.

Lemma 4. Let X and Y be any two random variables whose range is $\{0,1\}^m$ for some $m \geq 0$. Consider any δ, $0 < \delta < 1$, and define sets S_{++}, S_+, S_-, S_{--} as follows:

$$S_{++} = \{y \in \{0,1\}^m : \Pr_X\{X = y\} > (1 + \delta/8)\Pr_Y\{Y = y\}\}\},$$
$$S_+ = \{y \in \{0,1\}^m : (1 + \delta/8)\Pr_Y\{Y = y\} \geq \Pr_X\{X = y\} > \Pr_Y\{Y = y\}\}\},$$
$$S_- = \{y \in \{0,1\}^m : (1 + \delta/8)\Pr_X\{X = y\} \geq \Pr_Y\{Y = y\} \geq \Pr_X\{X = y\}\}\},$$
$$S_{--} = \{y \in \{0,1\}^m : \Pr_Y\{Y = y\} > (1 + \delta/8)\Pr_X\{X = y\}\}\}.$$

Then

$$\sum_{y \in \{0,1\}^m} |\Pr_X\{X = y\} - \Pr_Y\{Y = y\}| > \delta$$

implies

$$\sum_{y \in S_{++} \cup S_-} \Pr_X\{X = y\} - \sum_{y \in S_{++} \cup S_-} \Pr_Y\{Y = y\} > \delta/4.$$

Proof. Define $p_{++}, p_+, p_-, p_{--}, p_0, p_1$ as follows:

$$p_{++} = \sum_{y \in S_{++}} \Pr_X\{X = y\} - \Pr_Y\{Y = y\}, \quad p_+ = \sum_{y \in S_+} \Pr_X\{X = y\} - \Pr_Y\{Y = y\},$$
$$p_- = \sum_{y \in S_-} \Pr_Y\{Y = y\} - \Pr_X\{X = y\}, \quad p_{--} = \sum_{y \in S_{--}} \Pr_Y\{Y = y\} - \Pr_X\{X = y\},$$
$$p_0 = p_{++} + p_+, \quad \text{and} \quad p_1 = p_- + p_{--}.$$

Then we have

$$p_0 + p_1 = p_{++} + p_+ + p_- + p_{--} = \sum_{y \in \{0,1\}^m} |\Pr_X\{X = y\} - \Pr_Y\{Y = y\}| > \delta.$$

On the other hand, $p_0 - p_1 = \sum_{y \in \{0,1\}^m} \Pr_X\{X = y\} - \sum_{y \in \{0,1\}^m} \Pr_Y\{Y = y\} = 0$. Hence, $p_0 = p_1 > \delta/2$.

Here notice that

$$p_+ = \sum_{y \in S_+} \Pr_X\{X = y\} - \Pr_Y\{Y = y\} \leq \delta \frac{\sum_{y \in S_+} \Pr_Y\{Y = y\}}{8} \leq \frac{\delta}{8}, \quad \text{and}$$
$$p_- = \sum_{y \in S_-} \Pr_Y\{Y = y\} - \Pr_X\{X = y\} \leq \delta \frac{\sum_{y \in S_-} \Pr_X\{X = y\}}{8} \leq \frac{\delta}{8}.$$

Thus,

$$\sum_{y \in S_{++} \cup S_-} \Pr_X\{X = y\} - \sum_{y \in S_{++} \cup S_-} \Pr_Y\{Y = y\}$$
$$= p_{++} - p_- = (p_0 - p_+) - p_- > \frac{\delta}{2} - (p_+ + p_-) \geq \frac{\delta}{2} - \frac{\delta}{4} = \frac{\delta}{4}. \quad \Box$$

Now by using this lemma, we can state the outline of our proof of Theorem 3.

Let A and B are any length-regular pair of polynomial time randomized algorithms. Suppose that $\{A(x)\}$ and $\{B(x)\}$ are not statistically indistinguishable on L. Then by definition, there exists some $k \geq 0$ and infinitely many x in L such that (letting $n = |x|$, and letting m be the length of the output of A and B on input of length n)

$$\sum_{y \in \{0,1\}^m} |\Pr_A\{A(x) = y\} - \Pr_B\{B(x) = y\}| \geq n^{-k}.$$

Consider any x that satisfies the above inequality, and let n and m be as above. Then we define sets S_{++}, S_+, S_-, S_{--} in the same way as Lemma 4 by using n^{-k} for δ. For example, the set S_{++} is defined to be $\{y \in \{0,1\}^m : \Pr_A\{A(x) = y\} > (1 + n^{-k}/8)\Pr_B\{B(x) = y\}\}$. In what follows, we show a predicate $D = \{D_n\}_{n \geq 0}$ in P_{nu}^{NP} such that D_m accepts every $y \in S_{++}$ and rejects every $y \in S_{--}$. Then

$$\Pr\{D_m(A(x))\} - \Pr\{D_m(B(x))\} \geq \sum_{y \in S_{++} \cup S_-} \Pr_A\{A(x) = y\} - \sum_{y \in S_{++} \cup S_-} \Pr_B\{B(x) = y\},$$

which is, from Lemma 4, larger than $n^{-k}/8$. Thus, D_m distinguishes the outputs from A and B on input x. This proves that $\{A(x)\}$ and $\{B(x)\}$ are P_{nu}^{NP}-distinguishable.

The construction of this P_{nu}^{NP}-distinguisher D is based on the hashing technique [Sip83]. Notice that every y in S_{++} satisfies $\Pr_A\{A(x) = y\} > (1+n^{-k}/8)\Pr_B\{B(x) = y\}$ and that every y in S_{--} satisfies $\Pr_B\{B(x) = y\} > (1 + n^{-k}/8)\Pr_A\{A(x) = y\}$. Our P_{nu}^{NP} predicate D uses the hashing technique to detect this difference, and accepts every $y \in S_{++}$ and rejects every $y \in S_{--}$.

In what follows, we show how to use the hashing technique for our purpose. The key idea is stated in the following lemma. (Since the construction of D is clear from the proof of this lemma, we omit describing our D in detail.)

Lemma 5. Let A and B be any length-regular pair of polynomial time randomized algorithms. Then there exist some polynomial p and some set $Q \in NP$ with the following properties: For every ϵ, $0 < \epsilon < 1$, there is a query circuit E such that (letting $n = |x|$, and letting m be the length of the outputs of A and B on x),
(a) E is of size $\leq p(n + m + 1/\epsilon)$,
(b) E^Q accepts every y in $S_A = \{y \in \{0,1\}^m : \Pr_A\{A(x) = y\} > (1+\epsilon)\Pr_B\{B(x) = y\}\}$, and
(c) E^Q rejects every y in $S_B = \{y \in \{0,1\}^m : \Pr_B\{B(x) = y\} > (1+\epsilon)\Pr_A\{A(x) = y\}\}$.

Before stating the proof, we recall some notions on hashing, and show necessary properties for our proof.

Definition 6. [Sip83]

(1) For any $u, v \geq 1$, a (u, v)-*linear transformation* is a mapping $h : \{0,1\}^u \to \{0,1\}^v$ that is defined by some $u \times v$ binary matrix M as $h(x) = Mx$.

(2) A (u, v)-*hash family of size* s is a collection of s (u, v)-linear transformations.

(3) A (u, v)-hash family $H = \{h_1, ..., h_s\}$ *hashes* a set $Z \subseteq \{0,1\}^u$ if for every $z \in Z$, there is an $h_i \in H$ such that $h_i(z) \neq h_i(z')$ for every $z' \in Z$.

The following two facts are from [Sip83]. (For Proposition 8, a stronger version is proved in [Gav92].)

Proposition 7. For any $u, v \geq 1$ and any set $Z \subseteq \{0,1\}^u$, if some (u, v)-hash family of size s hashes Z, then $\|Z\| \leq s2^v$.

Proposition 8. Let $\mathcal{Z}$ be any collection of 2^t sets $Z_1, ..., Z_{2^t} \subseteq \{0,1\}^u$. Then for any w, there exits a $(u, w + 1)$-hash family H of size $t(w + 1)$ such that $\forall Z_i \in \mathcal{Z}$ $[\, \|Z_i\| \leq 2^w \;\to\; H$ hashes $Z_i \,]$.

Proof of Lemma 5. Let A and B be any length-regular pair of polynomial time randomized algorithms. Let r be some polynomial such that neither A nor B make more than $r(n)$ coin tosses on input of length n. Consider any input string x, and any ϵ, $0 < \epsilon < 1$, and let them fixed in the rest of the proof. (Since we fix the input string, we often omit indicating x in the following discussion.) We use n and m to denote $|x|$ and the length of the outputs of A and B on input x respectively; in the following, y always denotes a string in $\{0,1\}^m$. Finally, let S_A and S_B be the sets defined in the lemma.

In order to discuss randomized computation of A and B, the following notations are used. Since there would be no confusion, we use r to denote $r(n)$. A sequence of random coin tosses of A or B is regarded as a string in $\{0,1\}^r$, which is denoted by τ. ($\tau, \tau_1, ...$ always denote some strings in $\{0,1\}^r$.) For any sequence τ, we write $A(x; \tau)$ (resp., $B(x; \tau)$) to denote the output or execution of A (resp., B) following τ. For any y, let $R_A(y) = \{\tau : A(x; \tau) = y\}$, and $R_B(y) = \{\tau : B(x; \tau) = y\}$. Note that, for any y,

$$y \in S_A \;\to\; \|R_A(y)\| > (1 + \epsilon)\|R_B(y)\|, \quad \text{and}$$
$$y \in S_B \;\to\; \|R_B(y)\| > (1 + \epsilon)\|R_A(y)\|.$$

We can amplify the above difference by considering series of random sequences. More precisely, we consider the following sets: $R_{A^*}(y) = \{\tau_1 \cdots \tau_k : \bigwedge_{1 \leq j \leq k} A(x; \tau_j) = y\}$, and $R_{B^*}(y) = \{\tau \cdots \tau_k : \bigwedge_{1 \leq j \leq k} B(x; \tau_j) = y\}$. Here for our purpose, it suffices to assume that $k > 16r(m + 1)/\epsilon^2$ and $k > 3$. Then we have

$$\|R_A(y)\| > (1 + \epsilon)\|R_B(y)\| \;\to\; \|R_{A^*}(y)\| > 2^2(m + 1)(rk + 1)\|R_{B^*}(y)\|, \quad \text{and}$$
$$\|R_B(y)\| > (1 + \epsilon)\|R_A(y)\| \;\to\; \|R_{B^*}(y)\| > 2^2(m + 1)(rk + 1)\|R_{A^*}(y)\|.$$

Consider a collection $\mathcal{Z}$ of sets $R_{A^*}(y)$ and $R_{B^*}(y)$; i.e., $\mathcal{Z} = \{R_{A^*}(y) : y \in \{0,1\}^m\} \cup \{R_{B^*}(y) : y \in \{0,1\}^m\}$. Notice that there are 2^{m+1} sets in $\mathcal{Z}$, and that every set in $\mathcal{Z}$ is a subset of $\{0,1\}^{rk}$. Then from Proposition 8, for each w, $1 \leq w \leq rk$, there exists $(rk, w + 1)$-hash family H_w of size $(m + 1)(w + 1)$ that hashes every $Z \in C$ of size $\|Z\| \leq 2^w$. We use these hash families $H_1, ..., H_{rk}$.

Now we are ready to describe our desired circuit E and oracle set Q. The circuit E is designed to accept or reject a given input y; the execution of E on y is stated as follows:

(a) **if** $R_{A^*}(y) \neq \phi \wedge R_{B^*}(y) = \phi$ **then** accept y **end-if;**
   for $l := 1$ **to** rk **do** {
      (b) **check if** H_l hashes $R_{A^*}(y)$;   (c) **check if** H_l hashes $R_{B^*}(y)$;
      if $\neg$ (b) $\wedge$ (c) **then** accept y **end-if** };
   reject y

For checking (a), (b), and (c), E uses the oracle Q. Roughly speaking, Q is an oracle that tells whether $R_{A^*}(y)$ (resp., $R_{B^*}(y)$) is empty for a given k, x, y, and indication of A or B, and whether H_l hashes $R_{A^*}(y)$ (resp., $R_{B^*}(y)$) for given H_l, k, x, y, and indication of A or B. It is not hard to show that such an oracle set is constructed in NP. Note that the hash families $H_1, ..., H_{rk}$ are encoded in E, which does not need so many gates. Thus, it is also easy to show that the size of E is bounded by $p(n + m + 1/\epsilon)$ for some fixed polynomial p.

Finally, we show that every $y \in S_A$ is accepted and every $y \in S_B$ is rejected by the above execution. Let y be any string in S_A, and w be an integer, $1 \leq w \leq rk$, such that $2^{w-1} < \|R_{B^*}(y)\| \leq 2^w$. (We assume that $R_{B^*}(y)$ is not empty, which case is handled correctly at (a).) Consider the wth loop (i.e., $l = w$) during the execution of E on y. ¿From our assumption on H_w, H_w hashes $R_{B^*}(y)$; that is, (c) holds. On the other hand, we show in the following that H_w does not has $R_{A^*}(y)$. Hence, y is accepted. Recall that $y \in S_A$ implies $(m + 1)(rk + 1)2^2 \|R_{B^*}(y)\| < \|R_{A^*}(y)\|$. Thus, from $2^{w-1} < \|R_{B^*}(y)\|$, we have $(m + 1)(w + 1)2^{w+1} < \|R_{A^*}(y)\|$. Therefore, it follows from Proposition 7 that H_w cannot hash $R_{A^*}(y)$. Similarly, we can show that every $y \in S_B$ is rejected. □

Notice that Theorem 3 clearly holds when we consider distinguishers of the form $\{D_x\}_{x \in \{0,1\}^*}$. (See Remark of Definition 1).

4 Definition of "Zero-Knowledge"

In this section, we discuss the definition of "zero-knowledge" notions for interactive proof systems.

First we recall basic definitions [GMR89] concerning interactive proof systems. In the following definitions, let A be a probabilistic Turing machine (*prover*), and B be a probabilistic polynomial time Turing machine (*verifier*).

Definition 9. A pair (A, B) is an *interactive proof system* for L if it satisfies the following two conditions:

- *Completeness*: For every $x \in L$, (A, B) halts and accepts x with probability at least $1 - \nu(n)$, where the probability is taken over the coin tosses of A and B.
- *Soundness*: For every $x \notin L$ and for any A^*, (A^*, B) halts and accepts x with probability at most $\nu(n)$, where the probability is taken over the coin tosses of A^* and B.

Let $(A, B)(x)$ denote a random variable whose value is $(r_B, b_1, a_1, b_2, a_2, \ldots,)$, where r_B is the string contained in the random tape of B and b_i (resp., a_i) is the ith message of B (resp., A). Now, two zero-knowledge notions are defined as follows.

Definition 10. For any set L of strings, a pair (A, B) is *computational* (resp., *statistical*) *zero-knowledge* (in short, *computational-ZK* and *statistical-ZK* respectively) on L if it satisfies the following condition:

- *Zero-Knowledgeness*: For every probabilistic polynomial time Turing machine B^*, there exists an expected polynomial time M_{B^*} such that $\{(A, B^*)(x)\}$ and $\{M_{B^*}(x)\}$ are computationally (resp., statistically) indistinguishable on L.

Machine M_{B^*} is called a *simulator*.

Remark. Here indistinguishability is defined by considering distinguishers of the form $\{D_x\}_{x \in \{0,1\}^*}$.

An interactive proof system (A, B) for L is called *computational* (resp., *statistical*) *zero-knowledge interactive proof system* for L if it is computational- (resp., statistical-) ZK on L.

The above two zero-knowledge notions have been used in the literature. However, by using our extension of indistinguishability, we can define finer classification of zero-knowledgeness, and thereby discussing degree of zero-knowledgeness. Thus, we propose the following extension: For any predicate class C, an interactive proof system (A, B) for a language L is *C-zero-knowledge* (in short, *C-ZK*) if for every probabilistic polynomial time Turing machine B^*, there exists an expected polynomial time M_{B^*} such that (A, B^*) and M_{B^*} are C-indistinguishable on L.

By considering some hierarchy of predicate classes such as P_{nu}, P_{nu}^{NP}, $P_{nu}^{NP^{NP}}$, ..., we can discuss degree of zero-knowledgeness. In general, the statistical-ZK may locate in very high level in such a hierarchy. However, by using essentially the same proof technique explained in Section 3, we can show that the statistical-ZK is equivalent to the zero-knowledgeness of fairly low level.

Theorem 11. The $P_{nu}^{\Sigma_2^P}$-ZK is equivalent to the statistical-ZK.

The proof of this theorem makes use of the following recent result of [BP92].

Proposition 12. Suppose that L has a statistical-ZK proof system (A, B). Then there is a probabilistic polynomial time oracle machine A_{eff} such that (A_{eff}^{NP}, B) is a statistical-ZK interactive proof system for L.

Remark Indeed, they showed that for every probabilistic polynomial time Turing machine B^*, $\{(A, B^*)(x)\}$ and $\{(A_{\text{eff}}^{NP}, B^*)(x)\}$ are statistically indistinguishable.

Proof Sketch of Theorem 11. It suffices to show that any interactive proof system that is not statistical-ZK is indeed not $P_{nu}^{\Sigma_2^P}$-ZK either. Thus, for given interactive proof system (A, B^*) for L and simulator M that are statistically different, we need to construct a $P_{nu}^{\Sigma_2^P}$-distinguisher that detects the difference between $\{(A, B^*)(x)\}$ and $\{M(x)\}$ on L. From the proposition above, we know some A_{eff} exists such that $\{(A, B^*)(x)\}$ and $\{(A_{\text{eff}}^{NP}, B^*)(x)\}$ are statistically indistinguishable. Thus, a distinguisher of $\{(A_{\text{eff}}^{NP}, B^*)(x)\}$ and $\{M(x)\}$ satisfies our purpose.

Recall that the distinguisher constructed in Section 3 is for a pair of distributions generated by polynomial time randomized algorithms. However, one of the distributions, i.e., $\{(A_{\text{eff}}^{\text{NP}}, B^*)(x)\}$, is generated by a polynomial time randomized algorithm with the help of some NP oracle. However, it is easy to modify our argument in Section 3 to construct a distinguisher for $\{(A_{\text{eff}}^{\text{NP}}, B^*)(x)\}$ and $\{M(x)\}$, by using $\text{NP}^{\text{P}^{\text{NP}}}$ oracle, which is clearly in NP^{NP} $(= \Sigma_2^{\text{P}})$. The detail is left to the reader. $\square$

As an immediate corollary of Theorem 11, it is shown that the computational-ZK is equivalent to the statistical-ZK if P = NP.

For "interactive proof systems of knowledge", we also have a similar result.

Interactive systems of Definition 9 often need computationally powerful provers. On the other hand, one variant of interactive proof system — interactive proof system of knowledge — considered in [FFS87, TW87] does not require such a powerful prover. Intuitively, provers are just polynomial time algorithm that happen to know some secret. More precisely, for any binary relation R, an *interactive proof system of knowledge* for R is a pair of prover and verifier, a prover is defined as a probabilistic polynomial time Turing machine that has one auxiliary tape, on which some secret about an input string is written. For given input x and secret s, the goal of the prover is to convince the verifier that there exists some y such that $R(x, y)$ holds. (See [FFS87, TW87] for the formal definition.)

Theorem 13. In the context of interactive proof system of knowledge, the $P_{\text{nu}}^{\text{NP}}$-ZK is equivalent to the statistical-ZK.

Acknowledgments
The authors would like to thank Mr. Kouichi Sakurai for drawing their attention to [Gol90], and Mr. Hiroyuki Sato for pointing out some errors.

References

[BP92] M. Bellare and E. Petrank, Making zero-knowledge provers efficient, in *Proc. 24th ACM Sympos. on Theory of Computing*, ACM (1992), 711-722.

[FFS87] U. Feige, A. Fiat, and A. Shamir, Zero-knowledge proofs of identity, in *Proc. 19th ACM Sympos. on Theory of Computing*, ACM (1987), 210-217.

[Gav92] R. Gavaldà, Bounding the complexity of advice functions, in *Proc. 7th Structure in Complexity Theory Conference*, IEEE (1992), 249-254.

[Gol90] O. Goldreich, A note on computational indistinguishability, *Inform. Process. Lett.*, 34 (1990), 277-281.

[GMR89] S. Goldwasser, S. Micali, and C. Rackoff, The knowledge complexity of interactive proof systems, *SIAM Journal of Computing*, 18 (6) (1989), 186-208.

[ILL89] R. Impagliazzo, L. Levin, and M. Luby, Pseudorandom generation from one-way functions, in *Proc. 21st ACM Sympos. on Theory of Computing*, ACM (1989), 12-24.

[Sip83] M. Sipser, A complexity theoretic approach to randomness, in *Proc. 15th ACM Sympos. on Theory of Computing*, ACM (1983), 330-335.

[TW87] M. Tompa and H. Woll, Random self-reducibility and zero-knowledge interactive proofs of possession of information, in *Proc. 28th IEEE Sympos. on Foundations of Computer Science*, IEEE (1987), 472-482.

On Symmetric Differences of NP-hard Sets with Weakly-P-Selective Sets†

Bin Fu*   Hong-zhou Li**

Abstract. The symmetric differences of $NP-hard$ sets with *weakly $-P-selective$* sets are investigated in this paper. We show that if there exist a weakly-P-selective set A and a $NP- \leq^P_m -hard$ set H such that $H - A \in P_{btt}(Sparse)$ and $A - H \in P_m(Sparse)$ then $P = NP$ So no $NP- \leq^P_m -hard$ set has sparse symmetric difference with any *weakly $-P-selective$* set unless $P = NP$. In addition we show there exists a P-selective set which has exponentially dense symmetric difference with every set in $P_{btt}(Sparse)$.

1.Introduction

The low sets (such as sparse sets, P-selective sets , cheatable sets, etc.) play very important roles in the study of the structures of hard sets in complexity theory. Using low sets to study hard sets one can obtain structure properties of intractable sets and relationships among complexity classes. The study that hard sets can be reducible to sparse sets has long and rich history. The most notable early result is due to Mahaney[M82], who proved that if $NP \subseteq P_m(Sparse)$ then $P = NP$. Ogiwara and Watanabe[OW90] improved Mahaney's result by showing that if $NP \subseteq P_{btt}(Sparse)$ then $P = NP$. Similar results on relationships between $NP-hard$ sets and other low sets have been found. $P-selective$ sets introduced by Selman[Se79, Se82] as polynomial-time analogue of the semirecursive sets. He used them to distinguish polynomial-time m-reducibility from Turing-reducibility in NP under the assumption $E \neq NE$, and he proved that if every set in NP is $\leq^p_{ptt}$- reducible to $P-selective$ set, then $P = NP$. $Weakly-P-selective$ sets were introduced by Ko[K83] as a generalization of both the $P-selective$ structure and the real number structure based on the charaterization of $P-selective$ sets. He showed that $(weakly-)P-selective$ sets cannot distinguish polynomial- time m-completeness from T-completeness in NP unless the polynomial time hierarchy collapses to Σ^P_2, and if every set in NP is $\leq^P_{dtt}$- reducible to a $weakly-P-selective$ set then $P = NP$. Further results about P-selective sets can be found in [T91].

In this paper, we investigate the structural properties of $NP- \leq^P_m -hard$ sets by combining (weakly) P-selective sets with sparse sets.

Yesha[Y83] first considered the symmetric difference between $NP- \leq^P_m -hard$ sets and the sets in P. He proved that no $NP- \leq^P_m -hard$ sets has $O(\log \log n)$

†This research is supported in part by HTP863.

*Department of Computer Science, Beijing Computer Institute, Beijing, 100044, P.R.China & Beijing Laboratory of Cognitive Science, The University of Science and Technology of China.

**This research was performed while this author was visiting Beijing Computer Institute. Department of Mathematics, Yunnan Education College, Kunming, 650031, P.R.China.

dense symmetric difference with any set in P unless $P = NP$. Schöning showed that no paddable $NP- \leq_m^P -hard$ sets has sparse symmetric difference with any set in P unless $P = NP$. From Ogiwara and Watanabe's[OW90] theorem that $NP \subseteq P_{btt}(A) \implies P = NP$, we know no $NP - hard$ set has sparse symmetric difference with any set in P unless $P = NP$. Fu[F91] investigated lower bounds of closeness between many complexity classes. He showed that if an $NP- \leq_m^P -hard$ set is the union of a set in $P_{btt}(Sparse)$ with set A then $NP \subseteq P_{dtt}(A)$. Thus no $NP - hard$ set can be the union of a sparse set and a set in $co - NP$ $(FewP)$ unless $NP = co - NP$ $(NP = FewP)$. Recently Fu and Li[FL92] showed that if a $NP - hard$ set has sparse symmetric difference with the set A then $NP \subseteq P_{ptt}(A)$. Since both $co-NP$ and R are closed under positive truth-table reductions, thus no $NP- \leq_m^P -hard$ set has sparse symmetric difference with any set in $co - NP$ (R) unless $NP = co - NP$ $(NP = R)$. The symmetric difference between $E - hard$ sets and subexponential time computable sets were studied by Tang, Fu and Liu[TFL91]. They proved that the symmetric difference of a $E- \leq_m^P$ -hard set and a subexponential time computable set is still $E- \leq_m^P -hard$.

The symmetric diffences of $NP- \leq_m^P -hard$ sets with weakly-P-selective sets are investigated in this paper. We show that if there exist a $NP- \leq_m^P -hard$ set H and a weakly-P-selective set A such that $H - A \in P_{btt}(Sparse)$ and $A - H \in P_m(Sparse)$ then $P = NP$. In the last section, we separate P-selective sets from $P_{btt}(Sparse)$ by constructing a $P - selective$ set that has exponentially dense symmetric difference with any set in $P_{btt}(Sparse)$.

2.Preliminaries

We fix $\Sigma = \{0,1\}$ as our alphabet. By a "string" we mean an element of Σ^*. For a string x in Σ^*, $|x|$ denotes the length of x. Let $S \subseteq \Sigma^*$, the cardinality of S is denoted by $\| S \|$, set $S^{=n}(S^{\leq n})$ consists of all words of length $= n(\leq n)$ in S. In particular, let $\Sigma^n = \{x | x \in \Sigma^* \text{ and } |x| = n\}$ and $\Sigma^{\leq n} = \{x | x \in \Sigma^* \text{ and } |x| \leq n\}$. For any $u \in \Sigma^*$ and $A \subseteq \Sigma^*$, $uA = \{ux | x \in A\}$. We use λ to denote the null string. N reprents the set $\{0, 1, \cdots\}$.

We use the pairing function $< ., . >: \Sigma^* \times \Sigma^* \longrightarrow \Sigma^*$. It is convenient to assume for any x, y in Σ^*, $| < x, y > | \leq 2(|x| + |y|)$.

Our computation model is the Turing machine. $P(resp. NP)$ denotes the class of languages accepted by deterministic (resp. nondeterministic) Turing machines in polynomial time. PF denotes the class of polynomial time computable functions. We now define some notions of polynomial-time reducibilities.

A $\leq_m^P -reduction$ of A to B is a polynomial time computable function f such that for each $x \in \Sigma^*$, $x \in A \iff f(x) \in B$.

For function $e : N \longrightarrow N$, A $\leq_{e(n)-tt}^P$-reduction of A to B is a pair $< f, g >$ of polynomial time computable functions such that the the following hold for each $x \in \Sigma^*$, (1) $f(x) =< x_1, \cdots, x_k >$ is an ordered k-tuple of strings and $k \leq e(|x|)$, (2)$g(x)$ is a k-argument truth-table: $\{0,1\}^k \longrightarrow \{0,1\}$, (3)$x \in A \iff g(x)(\chi_B(x_1), \cdots, \chi_B(x_k)) = 1$.

A $\leq_{dtt}^P -reduction$ of A to B is polynomial time computable function f such that for each $x \in \Sigma^*$, $f(x) =< x_1, \cdots, x_k >$ and $x \in A \iff x_i \in B$ for some $i \leq k$.

Let $\leq_r^P$ be one of kinds of reductions defined above and C be a class of languages.

We say language H is $C- \leq_r^P -hard$ if each language in C is $\leq_r^P$-reducible to H.

Let $A, B \subseteq \Sigma^*$, we say A is spares if there exists a polynomial p such that $\| A^{\leq n} \| \leq p(n)$ for all $n \in N$. We define $A \triangle B = (A - B) \cup (B - A)$. The function $dist_{A,B} : N \longrightarrow N$ is called the distance function of A and B, where $dist_{A,B}(n) = \| (A \triangle B)^{\leq n} \|$.

A set A is $P - selective$ if there is a polynomial time computable function $f : \Sigma^* \times \Sigma^* \longrightarrow \Sigma^*$ such that (1)for all $x, y \in \Sigma^*, f(x, y) \in \{x, y\}$; (2)If $x \in A$ or $y \in A$ then $f(x, y) \in A$. f is called the selector function for A.

A set A is $weakly - P - selective$ if there is a polynomial time computable function $f : \Sigma^* \times \Sigma^* \longrightarrow \Sigma^* \cup \{\#\}$ and a polynomial $q(n)$ such that for each integer n, (1) $\Sigma^{\leq n}$ is the disjoint union of $m_n < q(n)$ sets :$\Sigma^{\leq n} = C_{n,1} \cup C_{n,2}... \cup C_{n,m_n}$; (2)if x, y are in $C_{n,i}$,then $f(x, y) \in \{x, y\}$; (3)if $x \in C_{n,i}$ and $y \in C_{n,j}, i \neq j$, then $f(x, y) = \#$; (4) if $x, y \in C_{n,i}$ and $(x \in A$ or $y \in A)$ then $f(x, y) \in A$. f is called the selector function for A and $C_{n,i}$ is called a chain obtained from f in $\Sigma^{\leq n}$.

Ko[K83] involved "polynomial time computable linear order" and "partially polynomial time computable partial order" to characterize $P - selective$ sets and $weakly - P - selective$ sets respectively.

If f is the selector function for $P - selective$ set A, a linear order $\leq_f$ can be induced from f: for each $x, y \in \Sigma^*, x \leq_f y$ if and only if there exists a finite sequence $z_0 = x, z_1, ..., z_n, z_{n+1} = y$ of strings in Σ^* such that $f(z_i, z_{i+1}) = z_i$ for all $i \leq n$. $\leq_f$ is a linear order in Σ^*. A is an initial segement of Σ^* using order $\leq_f$ [K83]. Let $x_1, ..., x_t \subseteq \Sigma^*$. Since $f(x, y) = x$ implies $x \leq_f y$, we only need to calculate function f $t - 1$ times to obtain an element x in $\{x_1, ..., x_t\}$ such that $x \leq_f x_i$ for all $i \leq t$.

If f is the selector function for $weakly - P - selective$ set A, we define $\leq_f$ in the same way as above. $\leq_f$ is a partial order in Σ^*. $A^{\leq n}$ is the union of initial segement of at most polynomial chains in $\Sigma^{\leq n}$[Ko83]. Clearly two strings u, v are in the same chain if and only if $f(u, v) \in \{u, v\}$. Let all elements of $\{x_1, ..., x_t\}$ are in the same chain. Since $f(x, y) = x$ implies $x \leq_f y$, we only need to calculate function f $t - 1$ times to obtain an element x in $\{x_1, ..., x_t\}$ such that $x \leq_f x_i$ for all $i \leq t$.

We assume all polynomials involved in this paper are with positive coefficients.

3.Symmetric Difference between NP-hard Sets and Weakly-P- selective Sets

In this section we study the symmetric difference between a $NP- \leq_m^P -hard$ set and a $weakly - P - selective$ set . The main result (Theorem 3.3) is obtained from a result in [F91] and the following lemma.

Lemma 3.1 Let H be $NP- \leq_{dtt}^P -hard$. If there exist a $weakly - P - selective$ set A and a set $B \in P_m(Sparse)$ such that $H = A - B$, then $P = NP$.

Proof. Let H be $NP- \leq_{dtt}^P -hard$, A be $weakly - P - selective$, and $g \in PF$ be the selector function for A. Let $B \leq_m^P S$ via $h_1 \in PF$, where S is a sparse set. Since H is $NP- \leq_{dtt}^P -hard$, let $SAT \leq_{dtt}^P H$ via $h_2 \in PF$. For each $x \in \Sigma^*, h_2(x)$ is considered as a subset of Σ^*.

Let $p(n)$ be a polynomial such that $\| S^{\leq n} \| \leq p(n)$, $g, h_1, h_2 \in DTIME(p(n))$ and $\Sigma^{\leq n}$ can be divided into at most $p(n)$ chains by the selector function g.

Algorithm

Input formula f with length n.

Let $F_0 = \{f\}$ and $i = 0$.

Repeat

Let $C_i = \bigcup_{x \in F_i} h_2(x)$.

Divide C_i into following divisions: $L_{i,1}, L_{i,2}, \ldots$

For $u, v \in C_i$, u, v are in the same division if and only if u, v are in the same chain $(g(u, v) \in \{u, v\})$.

Let n_i be the number of divisions divided from C_i.

For each $L_{i,j}, j \leq n_i$, $L_{i,j}$ is divided into the following divisions:

$M_{i,j,1}, M_{i,j,2}, \cdots$

which satisfy the three condition as follows.

1) $L_{i,j} = M_{i,j,1} \cup M_{i,j,2} \cup \ldots$

2) For every $x, y \in L_{i,j}$, x, y are in the same division if and only if $h_1(x) = h_1(y)$.

3) If $s < t$, then $m_{i,j,s} \leq_g m_{i,j,t}$, where $m_{i,j,s}, m_{i,j,t}$ are the least elements (under the order $\leq_g$) of $M_{i,j,s}, M_{i,j,t}$ respectively.(Note: if there are more than one least elements in a division we choose one arbitrarly).

Let $n_{i,j}$ be the number of divisions divided from $L_{i,j}$.

Let $f_{i,j,s}$ be a formula in F_i such that $m_{i,j,s} \in h_2(f_{i,j,s})$.

$G_{i,j} = \{f_{i,j,s} | s \leq p(p(p(n))) + 1\}$.

$G_i = \bigcup_{j=1}^{n_i} G_{i,j}$.

$F_{i+1} = \{g_1, g_2 | g_1, g_2$ are obtained by fixing one variable of $g \in G_i$ to 0,1 respectively $\}$.

$i = i + 1$.

Until all formulas in F_i are of length ≤ 5.

Let i_0 be the value i after executing the above cycle.

Accept x if and only if one of the formulas in F_{i_0} is satisfiable.

End of the algorithm.

Claim 3.1 1) For each $i \leq i_0$ $n_i \leq p(p(n))$. 2) For each $i < i_0$, $\| F_{i+1} \| \leq 2 \| G_i \| \leq 2(p(p(p(n))) + 1)^2$.

Proof. Since $h_2 \in DTIME(p(n))$, so $C_i \subseteq \Sigma^{\leq p(n)}$. $\Sigma^{\leq p(n)}$ can be divided into at most $p(p(n))$ chains by the selector function g, so $n_i \leq p(p(n))$. Since $\| G_{i,j} \| \leq p(p(p(n))) + 1$, therefore $\| F_{i+1} \| \leq 2 \| G_i \| \leq 2n_i \cdot (p(p(p(n))) + 1) \leq 2(p(p(p(n))) + 1)^2$. ∎

Claim 3.2 For each $i \leq i_0$, there exists a satisfiable formula in F_i $\Longleftrightarrow$ there exists a satisfiable formula in G_i.

Proof. We assume a formula x in F_i is satisfiable. Thus there exists an element y in $h_2(x)$ having $y \in H$(for $SAT \leq_{dtt}^P H$ via h_2). Let $y \in L_{i,j}$ and $y \in M_{i,j,s}$, where $j \leq n_i$ and $s \leq n_{i,j}$.

Case 1 $s \leq p(p(p(n))) + 1$.

In this case $f_{i,j,s} \in G_{i,j} \subseteq G_i$, $m_{i,j,s} \leq_g y$ and $m_{i,j,s} \in h_2(f_{i,j,s})$.

On the other hand, since $y \in H$ and $H = A - B$. So $y \in A$ implies $m_{i,j,s} \in A$ (For A is $weakly - P - selective$ and $y, m_{i,j,s}$ are in the same chain). Because both $m_{i,j,s}$ and y are in $M_{i,j,s}$, $h_2(m_{i,j,s}) = h_1(y) \notin S$. Thus $m_{i,j,s}$ is in $A - B$. Therefore $f_{i,j,s}$ is satisfiable. So there exists a satisfiable formula in G_i.

Case 2 $s > p(p(p(n))) + 1$.

In this case for each $t \leq p(p(p(n))) + 1, m_{i,j,t} \leq_g m_{i,j,s} \leq_g y$. Since A is $weakly - P - selective$ and $y, m_{i,j,s}$ are in the same chain, thus $m_{i,j,t} \in A$. On the other hand if $t \neq t'$ then $h_1(m_{i,j,t}) \neq h_1(m_{i,j,t'})$. We know the formulas in F_i are of length $\leq n$, thus the strings in C_i are of length $\leq p(n)$. Since $\| S^{\leq p(p(n))} \| \leq p(p(p(n)))$. Therefore there exists a $t_0 \leq p(p(p(n))) + 1$ such that $h_1(m_{i,j,t_0}) \notin S$. So $m_{i,j,t_0} \in A - B = H$. thus f_{i,j,t_0} is satisfiable. There exists a satisfiable formuila in G_i. ∎

Since initially $F_0 = \{f\}$, by the above claim f is satisfiable if and only if there exists a satisfiable formula in F_{i_0}. It is easy to see that the algorithm will stop in polynomial times. We have $SAT \in P$, hence $P = NP$ ∎

Lemma 3.2[F91] Let H be $NP- \leq_m^P$ $-hard$ set and $A \subseteq \Sigma^*$. If there exists $B \in P_{btt}(Sparse)$ such that $H = A \cup B$, then $NP \subseteq P_{dtt}(A)$. ∎

Theorem 3.3 If there exist an $NP- \leq_m^P$ $-hard$ set H and a $weakly - P - selective$ set A such that $H - A \in P_{btt}(Sparse)$ and $A - H \in P_m(Sparse)$ then $P = NP$.

Proof. Let H and A satisfy the condition of the theorem. Let $H' = H \cap A$. Thus $H' \cup (H - A) = H$ is $NP- \leq_m^P$ $-hard$. By Lemma 3.2 we have $NP \subseteq P_{dtt}(H')$. So H' is $NP- \leq_{dtt}^P$ $-hard$. Since A is $weakly - P - selective$ and $A - H = A - H' \in P_m(Sparse)$, so we have $P = NP$ by Lemma 3.1. ∎

Corollary 3.4 Let H be $NP- \leq_m^P$ $-hard$. If there exists a $weakly - P - selective$ set A such that $A \triangle H$ is sparse, then $P = NP$. ∎

4. Separating P-selective sets from $P_{btt}(Sparse)$

Ko[K83] showed that each $weakly - P - selective$ set belongs to $P_{tt}(Sparse)$, the following theorem give a lower bound for the number of queries to the sparse sets that may be required by such a reduction. In particular we show that there are weakly-P-selective sets A such that $A \notin P_{btt}(Sparse)$. This implies that Theorem 3.3 is not a trivial consequence of Ogiwara and Watanabe's theorem. The techniques used here are from [BK87] and [F91].

Let $< i, j >, < i', j' > \in N \times N$, we say $< i, j > \ll < i', j' >$ if $(i < i')$ or $(i = i'$ and $j \leq j')$; if $< i, j > \ll < i', j' >$ and $< i, j > \neq < i', j' >$, then we say $< i, j > \ll < i', j' >$. Let strings $s_1, s_2 \in \Sigma^*$, we say $s_1 \subset s_2$ if $s_2 = s_1 x$ for some string x with $|x| > 0$. The proof of Theorem 4.1 will employ ordinary dictionary ordering $\leq_d$ of binary strings with $0 \leq_d 1$. Let x and y belong to $\{0, 1\}^*$, $x = x_1...x_m$ and $y = y_1...y_n$, $x \leq_d y$ if and only if

(1) $m = n$ and $\exists i \leq m \forall j < i[x_j = y_j$ and $x_i = 0$ and $y_i = 1]$ or

(2) $m < n$ and $x \leq_d y_1...y_m$, or

(3) $m > n$ and $(x_1...x_n \leq_d y \wedge x_1 \cdots x_n \neq y)$.

For strings $x, y \in \Sigma^*$, we say $x <_d y$ if $(x \leq_d y)$ and $(x \neq y)$.

Let $f : N \to N$. $S(f)$ is the class of languages A that $\| A^{\leq n} \| \leq f(n)$ for all large n.

Theorem 4.1 There exists a $P-selective$ set A such that for any $B \in P_{loglog-tt}$ (*Sparse*), $dist_{A,B}(n) > 2^{n^{1/8}}$ for all large n.

Proof. Let $< f_1, g_1 >, < f_2, g_2 >, \cdots$ be an effective enumeration of all $\leq_{loglog-tt}^P$ $-reductions$. For each reduction $< f_i, g_i >$, f_i, g_i are computable functions under the time bound $n^{logi} + logi$ and for each $x \in \Sigma^*$, $f_i(x) =< x_1, ..., x_k >$ with $k \leq \log \log |x|$, $g(x)$ is a k-arguement truth-table.

In the following construction, we will construct a set A which is a initial segement of Σ^* under the order $\leq_d$ and make $dist_{A,B}(n)$ have exponential lower bound for every B which is $loglog - tt$-reducible to a sparse set via reduction $< f_j, g_j >$. The construction of stage $< n, i >$ will guarantee that there are at least $2^{n^2/2}$ elements in $(A \triangle B) \cap D_{n,i}$ for all sufficiently large n, where $D_{n,i} \subseteq \Sigma^{n^4 + in^2}$ and will be defined in the construction.

Let function $d(n) = n^{logn}$ which is subexponential and dominates all polynomials. Every sparse set belongs to $S(d)$.

Construction

Let n_0 be the least number such that for all $n > n_0$: $(\log \log n^5) \cdot n^5 < 2^{n^{1/2}}$, $\log \log n^5 \cdot d(((n^5)^{\log n} + \log n) < 2^{n^{1/2}}$ and $3(3 \log \log n^5 + 1) < n^{1/2}$.

$d_{n_0,n_0} = 0^{n_0^4}$.

For an integer pair $< n, i >$, we say $< n, i >$ is active if $(n_0 < n)$ and $(1 \leq i \leq n)$. Stage $< n, i >$ will be processed in the construction if and only if $< n, i >$ is active. For two active pairs $< n, i >, < n', i' >$, stage $< n, i >$ will be processed before stage $< n', i' >$ if and only if $< n, i > << n', i' >$.

Stage $< n, i >$

Let $\sigma = \frac{1}{3(3 \log \log n^5 + 1)}$.

If $i = 1$ then let $d_{n,0} = d_{n-1,n-1} 0^{n^4 - |d_{n-1,n-1}|}$.

For each $x \in \Sigma^*$, let x_r be the r-th element of $f_i(x)$, where $f_i(x) =< x_1, \cdots, x_k >$.

Let $D_{n,i} = d_{n,i-1} \Sigma^{n^2}$.

Let t_0 be one of the truth-tables t such that $\| \{x | x \in D_{n,i} \wedge g_i(x) = t\} \|$ is the largest.

Let r_0 be the dimension of t_0.

Define two sets $G_0^{(0)}, G_0^{(1)}$: $G_0^{(0)} = G_0^{(1)} = \{x | x \in D_{n,i} \wedge g_i(x) = t_0\}$.

$e = 1, J_0 = \phi$

Substage e:

    Case 1 There exist $v \in \Sigma^*, r \in \{1, \cdots, r_0\} - J_{e-1}$ and $b \in \{0, 1\}$ such that $\| \{x | x \in G_{e-1}^{(b)} \wedge x_r = v\} \| \geq 2^{(n-(e-1)-3e\sigma)n} \cdots \cdots (1)$

Fix such v, r and b, let $H_e = \{x | x \in G_{e-1}^{(b)} \wedge x_r = v\}$. If $e > 1$ then let

$$u_{e-1} = \begin{cases} y_{e-1} & if \ b = 0 \\ z_{e-1} & if \ b = 1. \end{cases}$$

$J_e = J_{e-1} \cup \{r\}$.

Let y_e be a $y \in \Sigma^n$ such that $\parallel H_e \cap d_{n,i-1}u_1 \cdots u_{e-1}y\Sigma^{(n-e)n} \parallel$ is the largest and

z_e be a $z \in \Sigma^n - \{y_e\}$ such that $\parallel H_e \cap d_{n,n-1}u_1 \cdots u_{e-1}z\Sigma^{(n-e)n} \parallel$ is the largest.

Let

$$G_e^{(0)} = H_e \cap d_{n,i-1}u_1 \cdots u_{e-1}y_e\Sigma^{(n-e)n}$$

$$G_e^{(1)} = H_e \cap d_{n,i-1}u_1 \cdots u_{e-1}z_e\Sigma^{(n-e)n}$$

Subcases 1.1 $e < r_0$

$e=e+1$ and enters the next substage.

Subcase 1.2 $e = r_0$

$$d_{n,i} = \begin{cases} d_{n,i-1}u_1 \cdots u_{e-1}y_e 1^{(n-e)n} & if \ y_e <_d z_e \\ d_{n,i-1}u_1 \cdots u_{e-1}z_e 1^{(n-e)n} & otherwise \end{cases}$$

Case 2 There exist no v,r and b to satisfy the inequality (1),

Subcase 2.1 $e = 1$

$$d_{n,i} = \begin{cases} d_{n,i-1}0^{n^2} & if \ t_0(0,\cdots,0) = 1 \\ d_{n,i-1}1^{n^2} & if \ t_0(0,\cdots,0) = 0 \end{cases}$$

Subcases 2.2 $e > 1$

$$d_{n,i} = \begin{cases} d_{n,i-1}u_1 \cdots u_{e-2}y_{e-1}1^{(n-(e-1))n} & if \ y_{e-1} <_d z_{e-1} \\ d_{n,i-1}u_1 \cdots u_{e-2}z_{e-1}1^{(n-(e-1))n}, & otherwise \end{cases}$$

End of stage $< n,i >$.

We define the set A as follows.

$x \in A \Longleftrightarrow x \leq_d d_{n,i}$ for some active integer pair $< n,i >$.

We shall show that for every $B \leq_{\log\log -tt}^{P} S$ via $< f_i, g_i >$, where $S \in S(d)$, $\parallel (A \bigtriangleup B) \cap D_{n,i} \parallel > 2^{n^2/2}$ for all large n and A is $P - selective$.

We only consider the stage $< n,i >$ in the construction. We assume n is sufficiently large and $< n,i >$ is active. Claim 4.1-4.2 can be verified easily from the construction. The detailed proofs are omitted here.

Claim 4.1 If stage $< n,i >$ ends at substage e_0, then for each $e < e_0$ we have the following facts:

1) $H_e \subseteq d_{n,i-1}u_1 \cdots u_{e-1}\Sigma^{(n-(e-1))n}$.

2) $G_e^{(0)} \subseteq d_{n,i-1}u_1 \cdots u_{e-1}y_e\Sigma^{(n-e)n}$.

3) $G_e^{(1)} \subseteq d_{n,i-1}u_1 \cdots u_{e-1}z_e\Sigma^{(n-e)n}$.

4) For any $x, x' \in H_e$, $x_r = x'_r$ for all $r \in J_e$.

5) $G_e^{(0)}, G_e^{(1)} \subseteq H_e$.

6) $J_e \subseteq \{1, 2, \cdots, r_0\}$ and $\parallel J_e \parallel = e$.

7) $H_e, G_e^{(0)}, G_e^{(1)} \subseteq D_{n,i}$.

8) If the condition of case 1 is true at substage e_0, then 1)–7) still hold when e_0 replaces e. ∎

Claim 4.2 1) For two active integer pairs $< n, i >, < n', i' >$, if $< n, i > << n', i' >$, then $d_{n,i} \subset d_{n',i'}$.

2) For each $x \in D_{n,i}, x \in A \iff x \leq_d d_{n,i}$.

3) $d_{n,i} \in D_{n,i} \subseteq \Sigma^{n^4 + in^2} \subseteq \Sigma^{\leq n^5}$. ∎

Claim 4.3 At stage $< n, i >$, 1) If the condtion of Case 1 is satisfied at substage e, then $\| G_e^{(b)} \| \geq 2^{(n-e-(3e+1)\sigma)n}$ for each $b \in \{0, 1\} \cdots \cdots (2)$

2) (2) is also true for $e = 0$.

Proof of Claim 4.3. At stage $< n, i >$, (1) $D_{n,i} \subseteq \Sigma^{\leq n^5}$ (by Claim 4.2 3)) and (2) initially t_0 is one of the truth-tables t with dimension $\leq \log \log n^5$ such that $\| \{x | x \in D_{n,i} \wedge g_i(x) = t\} \|$ is the largest. Hence $G_0^{(0)} = G_0^{(1)} = \{x | x \in D_{n,i} \wedge g_i(x) = t_0\}$.

The number of truth-tables with dimension $\leq \log \log n^5$ is not more than $(\log \log n^5) \cdot 2^{2^{\log \log n^5}} = (\log \log n^5) \cdot n^5 \leq 2^{n\sigma}$. So $\| G_0^{(0)} \| = \| G_0^{(1)} \| \geq \frac{\|D_{n,i}\|}{2^{n\sigma}} = \frac{2^{n^3}}{2^{n\sigma}} = 2^{(n-\sigma)n}$. Thus (2) holds for $e = 0$.

We consider $e = m > 0$. At substage m there exist v, r and b such that : $\| H_m \| \geq 2^{(n-(m-1)-3m\sigma)n}$, where $H_m = \{x | x \in G_{m-1}^{(b)}$ and $x_r = v\}$. Since $G_m^{(0)} = H_m \cap d_{n,i-1} u_1 \cdots u_{m-1} y_m \Sigma^{(n-m)n}$, $G_m^{(1)} = H_m \cap d_{n,i-1} u_1 \cdots u_{m-1} z_m \Sigma^{(n-m)n}$, where y_m is such a $y \in \Sigma^n$ that $\| H_m \cap d_{n,i-1} u_1 \cdots u_{m-1} y \Sigma^{(n-m)n} \|$ is the largest. and z_m is such a $z \in \Sigma^n - \{y_m\}$ that $\| H_m \cap d_{n,i-1} u_1 \cdots u_{m-1} z \Sigma^{(n-m)n} \|$ is the largest.

Because $H_m \subseteq d_{n,i-1} u_1 \cdots u_{m-1} \Sigma^{(n-(m-1))n}$ (By Claim 4.1 1)). Thus
$$\| G_m^{(0)} \| = \| H_m \cap d_{n,i-1} u_1 \cdots u_{m-1} y_m \Sigma^{(n-m)n} \| \geq \frac{\|H_m\|}{\|\Sigma^n\|} \geq \frac{2^{(n-(m-1)-3m\sigma)n}}{2^n} = 2^{(n-m-3m\sigma)n} > 2^{(n-m-(3m+1)\sigma)n}.$$
$$\| G_m^{(1)} \| = \| H_m \cap d_{n,i-1} u_1 \cdots u_{m-1} z_m \Sigma^{(n-m)n} \| \geq \frac{\|H_m - d_{n,i-1} u_1 \cdots u_{m-1} y_m \Sigma^{(n-m)n}\|}{\|\Sigma^n\|} \geq \frac{2^{(n-(m-1)-3m\sigma)n} - 2^{(n-m)n}}{2^n}$$
$$> 2^{(n-m-(3m+1)\sigma)n}. \quad ∎$$

Claim 4.4 If stage $< n, i >$ ends at subcase 1.2 of substage e, then $\| (A \triangle B) \cap D_{n,i} \| > 2^{n^3}/2$.

Proof of Claim 4.4. We know for each $x \in D_{n,i}, x \in A \iff x \leq_d d_{n,i}$ (By Claim 4.2 2)).

$$G_{r_0}^{(0)} \subseteq d_{n,i-1} u_1 \cdots u_{r_0-1} y_{r_0} \Sigma^{(n-r_0)n}, \quad G_{r_0}^{(1)} \subseteq d_{n,i-1} u_1 \cdots u_{r_0-1} z_{r_0} \Sigma^{(n-r_0)n}$$

, (By Claim 4.1 2), 3)). By the definition of $d_{n,i}$ at substage 1.2 all of the strings in $G_{r_0}^{(b)}$ belong to A and none of the strings of $G_{r_0}^{(1-b)}$ is in A, where

$$b = \begin{cases} 0 & if \ y_{r_0} <_d z_{r_0} \\ 1 & otherwise \end{cases}$$

On the other hand, for any x, y in $G_{r_0}^{(0)} \cup G_{r_0}^{(1)}$, $< f_i(x), g_i(x) > = < f_i(y), g_i(y) >$ By Claim 4.1 4),5),6)). Hence, either all of the strings in $G_{r_0}^{(0)} \cup G_{r_0}^{(1)}$ are in B, or none of them is in B. Therefore $\| (A \triangle B) \cap D_{n,i} \| \geq \text{Min}(\| G_{r_0}^{(0)} \|, \| G_{r_0}^{(1)} \|)$ $\geq 2^{(n-r_0-(3r_0+1)\sigma)n}$ (by Claim 4.3) $\geq 2^{n^2/2}$. ∎

Claim 4.5 If stage $< n, i >$ ends at subcase 2.1 or 2.2 for some e, then $\| (A \triangle B) \cap D_{n,i} \| \geq 2^{n^2/2}$.

Proof of Claim 4.5. At the substage e of stage $< n, i >$, for every $v \in \Sigma^*, r \in \{1, \cdots, r_0\} - J_{e-1}$ and $b \in \{0, 1\}$ $\| \{x | x \in G_{e-1}^{(b)}$ and $x_r = v\} \| < 2^{(n-(e-1)-3e\sigma)n}$.

Because $S \in S(d), f_i, g_i \in DTIME(n^{\log i} + \log i)$ and $G_{e-1}^{(b)} \subseteq \Sigma^{\leq n^s}$ there are at most $r_0 \cdot d((n^5)^{\log i} + \log i) \cdot 2^{(n-(e-1)-3e\sigma)n} < 2^{\sigma n} \cdot 2^{(n-(e-1)-3e\sigma)n} = 2^{(n-(e-1)-(3e-1)\sigma)n}$ strings x in $G_{e-1}^{(b)}(b \in \{0, 1\})$ having $x_{r_0} \in S$ for some $r \in \{1, \cdots, r_0\} - J_{e-1}$, where $f_i(x) = < x_1, \cdots, x_{r_0} >$. By Claim 4.3, $\| G_{e-1}^{(b)} \| \geq 2^{(n-(e-1)-(3(e-1)+1)\sigma)n}$ for each $b \in \{0, 1\}$, hence in $G_{e-1}^{(b)}(b \in \{0, 1\})$ there are at least

$2^{(n-(e-1)-(3(e-1)+1)\sigma)n} - 2^{(n-(e-1)-(3e-1)\sigma)n} \geq$
$2^{(n-(e-1)-((3e-1)+2)\sigma)n} = 2^{(n-(e-1)-(3e-1)\sigma)n}$ strings x having $x_r \notin S$ for every $r \in \{1, \cdots, r_0\} - J_{e-1}$.

Let $F_{e-1}^{(b)} = \{x | x \in G_{e-1}^{(b)}$ and $x_r \notin S$ for all $r \in \{1, \cdots, r_0\} - J_{e-1}\}$. So $\| F_{e-1}^{(b)} \| \geq 2^{(n-(e-1)-(3e-1)\sigma)n}$.

It is easy to see that for any $x, y \in F_{e-1}^{(0)} \cup F_{e-1}^{(1)}$
$g_i(x) = g_i(y) = t_0$ and $< \chi_S(x_1), \cdots, \chi_S(x_{r_0}) > = < \chi_S(y_1), \cdots, \chi_S(y_{r_0}) >$ (By Claim4.1 4),5), 6) and the above discussion).

Let $c = \begin{cases} 0 & if \ y_{e-1} <_d z_{e-1} \\ 1 & otherwise \end{cases}$

If the stage $< n, i >$ ends at subcase 2.1 $t_0(0, \cdots, 0) = 0 \implies$ All of the strings in $F_0^{(c)}$ belong to A and none of them is in B. $t_0(0, \cdots, 0) = 1 \implies$ All of the strings in $F_0^{(c)} - \{d_{n,i}0^{n^2}\}$ belong to B and none of them is in A. So $\| (A \triangle B) \cap D_{n,i} \| \geq \| F_0^{(c)} \| - 1 \geq 2^{n^2/2}$.

If the stage $< n, i >$ ends at subcase 2.2 of substage e. It is easy to see that all of the strings in $F_{e-1}^{(c)}$ are in A and none of the strings in $F_{e-1}^{(1-c)}$ is in A. On the other hand, either all of the strings in $F_{e-1}^{(0)} \cup F_{e-1}^{(1)}$ are in B or none of them is in B. $\| (A \triangle B) \cap D_{n,i} \| \geq \text{Min}(\| F_{e-1}^{(0)} \|, \| F_{e-1}^{(1)} \|) \geq 2^{n^2/2}$. ∎

Claim 4.6 A is $P - selective$.

Proof of Claim 4.6 By Claim 4.2 1) and the definition of A that $x \in A \iff x \leq_d d_{n,i}$ for some active integer pair $< n, i >$, it is easy to verify this claim. ∎

By Claim 4.4 and Claim 4.5, it is easy to see that for all large n. $dist_{A,B}(n^5) \geq dist_{A,B}(n^4 + in^2) \geq \| (A \triangle B) \cap D_{n,i} \| \geq 2^{n^2/2}$. Hence, for almost every $n, dist_{A,B}(n) \geq 2^{n^{1/5}}$. ∎

Corollary 4.2 There exists a $P - selecrive$ set A such that for any $B \in P_{btt}(Sparse)$, $dist_{A,B}(n) > 2^{n^{1/5}}$ for all large n. ∎

Acknowledgement

We sincerely thank Prof.Shouwen Tang for his encouragement in this research. The first would also like to thank Prof.Qiongzhang Li for his encouragement.

Reference

[BGS88] J.Balcarzar, J.Diaz, and J.Gabarro: Structural Complexity I,II. Springer- Verlag, 1988 and 1990.

[BK87] R.Book and K.Ko: On Sets Truth-table Reducible to Sparse Sets. SIAM J.Computing 17(1988),903-919.

[F91] B.Fu: On Lower Bounds of the Closeness between Complexity Classes. To appear in Math. Sys.Theory.

[FL92] B.Fu and H.Li: On Closeness of $NP - Hard$ sets to other Complexity Classes. Proc.7th Annual Conference in structural complexity theory, IEEE, 1992, 243-248.

[K83] K.Ko: On Self-reducibility and Weakly-P-selectivity. J.of Comp. and Sys.Sci. 26(1983), 209-211.

[K88] K.Ko: Distinguishing Bounded Reducibility by Sparse Sets. Proc.3th Conference on Structural Complexity Theory, IEEE, 1988, 181-191.

[Ma82] S.Mahaney: Sparse Complete Sets for NP:Solution of a Conjecture of Berman and Hartmanis. J.Comput.Sys.Sci. 25(1982), 130-143.

[OW90] M.Ogiwara and O.Watanabe: On Polynomial-Time Bounded Truth-table Reducibility of NP sets to Sparse Sets. Proc. of the 22th ACM Symposium on Theory of Computing, 1990, 457-467.

[Sc86] U.Schöning: Complete Sets and Closeness to Complexity Classes. Math.Sys. Theory, 13(1986), 55-65.

[Se79] A.L.Selman: $P-selective$ Sets, Tally Languages and the Behavior of Polynomial -Time Reducibility on NP. Math.Sys.Theory 13(1981), 326-332.

[Se81] A.L.Selman: Some Observation on NP Real Numbers and $P-Selective$ Sets. J.Comput. Sys. Sci. 19(1981), 326-332.

[Se82] A.L.Selman:Reductions on NP and P-selective Sets. Theoret.Comput.Sci. 19 (1982), 287-304.

[TFL91] S.Tang, B.Fu and T.Liu: Exponential Time and Subexponential Time Sets. Proc.6th Annual Conference in structural complexity theory, IEEE, 1991, 230-237.

[T91] S.Toda: On Polynomial-Time Truth Table Reducibility of Intractable Sets to P-Selective Sets. Math.Sys.Theory 24(1991), 69-82.

[Y83] Y.Yesha: On Certain Polynomial-Time Truth-table Reducibilities of Complete Sets to Sparse Sets. SIAM J.Comput. 12(1983), 411-425.

RESTRICTED TRACK ASSIGNMENT WITH APPLICATIONS[*]

Majid Sarrafzadeh and D. T. Lee

Department of Electrical Engineering and Computer Science
Northwestern University, Evanston, IL 60208

Abstract. Consider a set of intervals $S = \{I_1, I_2, \ldots, I_n\}$, where $I_i = (l_i, r_i), l_i$, and r_i are real numbers, and $l_i < r_i$. We study a restricted track assignment problem (RTAP): if an interval I_a contains another interval I_b then I_a must be assigned to a higher track than I_b, and the goal is to minimize the number of tracks used. The problem RTAP is shown to be NP-hard. An approximation algorithm that produces a solution within twice of the optimal is also presented and the bound is shown to be tight. The algorithm, uses a segment tree as the basic structure, runs in $O(n \log n)$ time and requires linear space. The proposed approximation algorithm is employed to solve the problem of finding a maximum-weighted independent set in a circle graph, and related problems.

1 Introduction

Consider a set $\mathcal{I} = \{I_1, \ldots, I_n\}$ of intervals, where $I_i = (l_i, r_i)$ is specified by its two end-points. The *left point* l_i and the *right point* $r_i, l_i < r_i$. We assume all end-points are distinct real numbers (the assumption simplifies our analysis and can be trivially removed). Two intervals $I_i = (l_i, r_i)$ and $I_j = (l_j, r_j)$ are *independent* if $r_i < l_j$ or $r_j < l_i$; otherwise, they are *dependent*. Two dependent intervals are *crossing* if $l_i < l_j < r_i < r_j$ or $l_j < l_i < r_j < r_i$. If $l_i < l_j < r_j < r_i$ then we say I_i *contains* I_j.

The track assignment problem (TAP) is to assign the intervals into tracks 1 to t such that in each track, intervals are pairwise independent. The goal is to minimize the number of tracks t. The *density* of a column c, denoted d_c, is the number of intervals in $\mathcal{I}$ that contain c. The density of the set $\mathcal{I}$ is $d = max_c d_c$. Certainly, d is a lower-bound on the number of tracks for TAP; that is, $t \geq d$. It was shown that d is also an upper-bound on t [9], where a $\Theta(n \log n)$ time track assignment algorithm was proposed. The track assignment problem finds applications, for example in, job scheduling and VLSI layout.

Consider a set of points $\mathcal{C} = \{c_1, \ldots, c_m\}$, called a *restriction set*. The maximal subset of $\mathcal{I}$ that contains a point of $\mathcal{C}$ (an interval $I_i = (l_i, r_i)$ contains a point c_a if $l_i \leq c_a \leq r_i$) is denoted by $\mathcal{I}_c$; intervals in $\mathcal{I}_c$ are called *restricted*

---

[*] Supported in part by the National Science Foundation under Grants MIP-8921540 and CCR-8901815.

intervals. Assume track $i + 1$ is above track i, $1 \leq i \leq t - 1$. Given $\mathcal{I}$ and $\mathcal{C}$, the restricted track assignment problem (RTAP) is the problem of assigning intervals into tracks 1 to t such that: **p1)** In each track, intervals are pairwise independent. **p2)** If a restricted interval I_i contains another restricted interval I_j then $\tau_i > \tau_j$, where τ_a is the track to which I_a is assigned.

The goal is, as before, to minimize the number of tracks t. Figure 1 shows an example. Note that when the cardinality of the restriction set is one (i.e., $\mathcal{C} = \{c_1\}$) then RTAP is equivalent to TAP: An assignment of $\mathcal{I}$, employing the algorithm in [9], is obtained. Then, the tracks are permuted to satisfy the restriction imposed at column c_1. Here, we will show RTAP is NP–hard for $|\mathcal{C}| = 2$ (and $|\mathcal{C}| > 2$). Then, we propose an approximation algorithm for solving an arbitrary instance $(\mathcal{I}, \mathcal{C})$ of RTAP. The proposed algorithm will be used to efficiently solve the maximum-weighted independent set problem in a circle graph and related problems.

$$I_1 = (1, 8) \qquad I_3 = (3, 7) \qquad I_5 = (6, 11)$$
$$I_2 = (2, 12) \qquad I_4 = (4, 5) \qquad I_6 = (9, 10)$$

a) Input

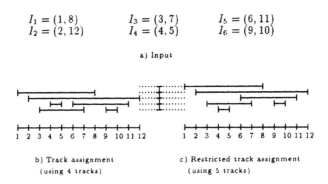

b) Track assignment   c) Restricted track assignment

(using 4 tracks)   (using 5 tracks)

Fig. 1. Track assignment of a set of intervals

This paper is organized as follows. In Section 2, we will show RTAP is NP-hard even for $|\mathcal{C}| = 2$. In Section 3, we propose a 2-approximation algorithm for an arbitrary instance $(\mathcal{I}, \mathcal{C})$ of RTAP and show this bound is existentially tight. We employ, in Section 4, the proposed approximation algorithm to effectively solve the weighted independent set problem in a circle graph and related problems.

2 NP-hardness of RTAP

Consider an arbitrary instance $(\mathcal{I}, \mathcal{C})$ of RTAP, where $\mathcal{I} = \{I_1, \ldots, I_n\}$ and $\mathcal{C} = \{c_1, \ldots, c_m\}$. Certainly, the number of tracks is lower-bounded by the density d. We will show that it is NP-complete to decide if d tracks are sufficient for assignment of intervals, when $m = 2$. We shall transform the problem of coloring a circular-arc graph to RTAP.

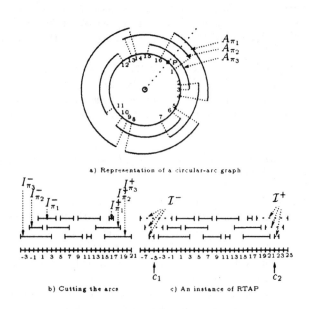

a) Representation of a circular-arc graph

b) Cutting the arcs   c) An instance of RTAP

Fig. 2. Opening a circular-arc representation

A circular-arc graph is defined on a circle with $2s$ points. Points are numbered with respect to an origin in a clockwise manner. Consider a set $\mathcal{A} = \{A_1, \ldots, A_s\}$ of arcs, where each arc $A_i = (a_i, b_i)$ is specified by its *first point* a_i and its *second point* b_i, $1 \leq a_i, b_i \leq 2s$. A_i is realized by connecting its first point to its second point in a clockwise manner. See Figure 2a. The intersection graph of $\mathcal{A}$ is called a *circular-arc graph* (see [8] for more definitions; two arcs intersect if and only if they share at least one point on the circle). Consider a point P on the circle. The set of arcs cut by a line segment starting at the center of the circle and passing through P is denoted by $\mathcal{A}_P$. The arcs in $\mathcal{A}_P$ are said to be *P-equivalent*.

The circular-arc coloring problem (CACP) is to assign the minimum number of colors to arcs of $\mathcal{A}$ such that every two P-equivalent arcs, for all P, are assigned distinct colors. The maximum of $|\mathcal{A}_P|$, over all P, is called the *density of $\mathcal{A}$* and is denoted by δ. It is NP-complete to decide if δ colors are sufficient to color an arbitrary instance $\mathcal{A}$ of CACP [6].

We will show CACP is polynomial-time transformable to RTAP. A similar reduction was used in [10] to prove the NP-hardness of a channel routing problem. Consider an arbitrary instance $\mathcal{A}$ of CACP and let P be a point with $\delta = |\mathcal{A}_P|$, where $\mathcal{A}_P = \{A_{\pi_1}, \ldots, A_{\pi_\delta}\}$. The point P is selected such that it is not an end-point of any arcs. Assume P is at the origin; otherwise, the origin can be re-defined and the input is accordingly modified. The arcs of $\mathcal{A}_P$ are "cut" at point P to obtain a set of intervals $\mathcal{I}_P = (\mathcal{A} - \mathcal{A}_P) \cup \mathcal{A}_P^*$. The set $\mathcal{A}_P^*$ is obtained by partitioning each arc $A_{\pi_i} = (a_{\pi_i}, b_{\pi_i})$ in $\mathcal{A}_P$ into

Consider a track assignment in $\mathcal{I}_P$. Such an assignment would correspond to a coloring in $\mathcal{A}$ (i.e., if I_a is assigned to track b then $\mathcal{A}_a$ would be assigned color number b) provided that $I_{\pi_i}^-$ and $I_{\pi_i}^+$ are assigned to the same track, for $1 \le i \le \delta$.

We construct the following instance of RTAP. $\mathcal{I}_{\mathcal{A}} = \mathcal{I}^- \cup \mathcal{I}^+ \cup \mathcal{I}_P$, where $\mathcal{I}^- = \{(i - 3\delta, 1 - 2i)|1 \le i \le \delta\}$ and $\mathcal{I}^+ = \{(2s + 2i, 2s + 3\delta + 1 - 2i)|1 \le i \le \delta\}$. See Figure 2c. $\mathcal{I}^-$ and $\mathcal{I}^+$ are called *enforcing intervals* (shown by dashed lines in Figure 2c). The restriction set is $\mathcal{C}_{\mathcal{A}} = (c^-, c^+)$, where $c^- = 1 - 2\delta$ and $c^+ = 2s + 2\delta$.

Lemma 1. *An arbitrary instance $\mathcal{A}$ of CACP is δ-colorable if and only if $(\mathcal{I}_{\mathcal{A}}, \mathcal{C}_{\mathcal{A}})$ can be assigned into δ tracks.*

Proof: *(only if)* If all the arcs can be colored with δ colors, the intervals are assigned as follows. Assume A_{π_i} has been assigned color number i, for $1 \le i \le \delta$; otherwise, the colors are renamed. The intervals $I_{\pi_i}^- = (2 - 2i, b_{\pi_i})$, and $I_{\pi_i}^+ = (a_{\pi_i}, 2s + 2i - 1)$, and the i-th enforcer interval in each side are assigned to track i. The rest of intervals will be assigned the same color as the corresponding arcs. A restricted track assignment of $\mathcal{I}_{\mathcal{A}}$ is thus obtained.

(if) Consider a legal assignment of $(\mathcal{I}_{\mathcal{A}}, \mathcal{C}_{\mathcal{A}})$. If an interval is assigned to track j, the corresponding arc is assigned color number j. Also, $A_{\pi_i} = (a_{\pi_i}, b_{\pi_i})$ is assigned color number j if $I_{\pi_i}^- = (2 - 2i, b_{\pi_i})$ (and thus $I_{\pi_i}^+ = (a_{\pi_i}, 2s + 2i - 1)$) is assigned to track j. A legal coloring is obtained. $\square$

RTAP is in class NP, for we can non-deterministically assign the intervals into tracks and check to see if it is legal. Based on Lemma 1, we conclude:

Theorem 1 *It is NP-complete to decide if an arbitrary instance $(\mathcal{I}, \mathcal{C})$ of RTAP can be assigned into δ tracks, even for $|\mathcal{C}| = 2$.*

3 An Approximation Algorithm

In this section, first we show there are instances of RTAP with $t \ge 2d - 1$, where d is the problem density. Then, we propose an approximation algorithm for an arbitrary instance of RTAP achieving $t \le 2d - 1$.

We define an instance $(\mathcal{I}, \mathcal{C})$ of RTAP that requires at least $2d - 1$ tracks in a recursive manner. It consists of a collection of blocks. Block 1, denoted by B_1, has a single unit-length interval. We obtain B_i, $i > 1$, by combining two copies of B_{i-1} and two intervals, called *top-intervals*, as shown in Figure 3a. Note that the density of B_i is equal to the density of B_{i-1} plus one, that is, B_i has density i, for all i. For consistency, the interval in B_1 is also called a top-interval. (Note that top-intervals are defined in conjunction with a block.) The instance B_d with

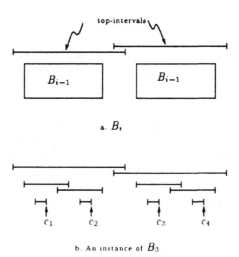

a. B_i

b. An instance of B_3

Fig. 3. A "difficult" instance

C being the set of right points of all unit intervals is called a "difficult" example. An example of B_3, with $|C| = 4$, is shown in Figure 3b.

Lemma 2. *In an arbitrary instance of difficult example B_d, $t \geq 2d - 1$.*

Proof: We show, by induction on i, that B_i requires $2i - 1$ tracks. In our discussion below, by a legal (track) assignment we mean one that does not contain any empty tracks. The two top-intervals of B_i are called the left top-interval and the right top-interval and the two copies of B_{i-1} used in constructing B_i are called the left B_{i-1} and the right B_{i-1} with obvious meanings. Note that in any legal assignment of B_i there must be a top-interval on the top-most track; or else, there is at least one empty track. We establish the following invariants on B_i: *A legal assignment of B_i requires at least $2i - 1$ tracks.*

Basis: The basis is trivial, as the interval in B_1 requires at least one track. *Inductive step:* Assume the invariant is maintained for $B_1, \ldots, B_{i-1}$. Now consider B_i. Note that the two copies of B_{i-1} used in constructing B_i are (legally) assigned to tracks independently, as they are disjoint. Furthermore, by induction, each copy requires at least $2i - 3$ tracks (and in each copy there must be a top-interval on the top-most track).

Case 1. *At least one copy of B_{i-1} occupies more than $2i - 3$ tracks.*
Without loss of generality assume the left B_{i-1} has more than $2i - 3$ tracks, that is, $2i - 2$ tracks or more. Thus, the left top-interval of B_i occupies track $2i - 1$ or a higher track. We conclude, B_i requires at least $2i - 1$ tracks.

Case 2. *Each copy B_{i-1} of B_i occupies $2i - 3$ tracks.*
Since each copy has a top-interval in track $2i - 3$ then none of the top-intervals of B_i can be assigned to track $2i - 3$. They must be assigned to tracks $2i - 2$ and $2i - 1$ (or to higher tracks). Therefore, a total of $2i - 1$ tracks are used.

□

Next, we propose a restricted track assignment algorithm. Consider an arbitrary instance $(\mathcal{I}, \mathcal{C})$ of the problem with density d. Let $\{k_1, \ldots, k_s\}$ denote the set of max-density columns, that is, columns where the local density is d, assuming $k_1 < k_2 < \ldots < k_s$. Among all intervals containing k_1 we select the one with the rightmost right point. This interval is denoted by I_1^1. Column k_1 is the *key* column associated with I_1^1. Assume I_1^1 contains columns $k_1 \ldots k_j$ and does not contain column k_{j+1}, unless k_j is the last column. Then among all intervals containing k_{j+1} we select the one with the rightmost right point. This interval is denoted by I_1^2. Column k_{j+1} is the *key* column associated with I_1^2. This task is repeated until all columns $\{k_1, \ldots, k_s\}$ are processed (see Figure 4). The set of intervals selected is denoted by $\mathcal{I}_1 = \{I_1^1, I_1^2, \ldots\}$ and the associated set of *key* columns denoted by $\mathcal{K}_1 = \{k_1, k_{j+1}, \ldots\}$. The set $\mathcal{I}_1$ is removed and the process is repeated for the rest of intervals – we obtain $\mathcal{I}_1, \mathcal{I}_2, \ldots$ and the associated key column sets $\mathcal{K}_1, \mathcal{K}_2, \ldots$ respectively until no interval is remained. We establish three properties of the just described selection process.

- **Prop 1.** *Desnity of each $\mathcal{I}_i$ is at most two:* Consider the set $\mathcal{K}_i = \{k_{i_1}, k_{i_2}, \ldots, k_{i_t}\}$ of key columns associated with $\mathcal{I}_i = \{I_i^1, I_i^2, \ldots, I_i^t\}$. It is obvious that key column k_{i_j} is contained in exactly one interval I_i^j in $\mathcal{I}$ for $1 \leq j \leq t$. Thus, $I_i^{j+2} \cap I_i^j = \emptyset$, for $1 \leq j \leq t - 2$, (see Figure 4), that is, $\mathcal{I}_i$ has density at most two.
- **Prop 2.** *After removing each $\mathcal{I}_i$ the density of $\mathcal{I}$ is reduced by at least one:* Since from each max-density column we remove at least one interval, then after removing $\mathcal{I}_i$ the density is reduced by at least one.
- **Prop 3.** *Consider any interval $I_i^a \in \mathcal{I}_i$ and any interval $I_j^b \in \mathcal{I}_j$, for $i < j$. Then I_i^a is not contained in I_j^b:* Let k_i^a be the key column of I_i^a. If I_j^b contains I_i^a, then at column k_i^a when we compute $\mathcal{I}_i$, we would have selected I_j^b as the interval with the rightmost right point. That is a contradiction.

The proposed restricted track assignment algorithm, called Density-Reduction-Algorithm, works as follows. We assign $\mathcal{I}_i$, $1 \leq i \leq d - 1$ to tracks $2d - 2i + 1$ and $2d - 2i$. $\mathcal{I}_d$ has density 1, and thus we assign it to track 1. A total of $2d - 1$ tracks are used. Based on Property 3 the just described track assignment is a legal one.

We now give an algorithm that performs the *label* assignment in $O(n \log n)$ time and space using segment trees [12]. Given a set $\mathcal{I}$ of n intervals $(l_i, r_i), i = 1, 2, \ldots, n$, the end point values define $2n - 1$ *basic* intervals. A segment tree defined on this set of basic intervals has $2n - 1$ leaves, each corresponding to a basic interval, and each internal node v represents a *standard* interval, which

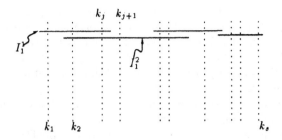

Fig. 4. Assignment of intervals by Density-Reduction-Algorithm

is the *union* of the basic intervals, corresponding to the leaves of the subtree rooted at v. Let $INT(v)$ denote the basic or standard interval associated with a leaf or internal node v respectively. It is well known that any interval $I_i = (l_i, r_i)$ can be decomposed into at most $O(\log n)$ maximal standard intervals. That is, $I_i = \cup_{j=1}^{k} INT(v_j)$, for some $k = O(\log n)$, such that $INT(v_j)$'s are pairwise disjoint, except possibly at a common endpoint. The set of nodes, v is collectively referred to as the *canonical covering* of I_i. Associated with each node v there are four fields, $LIST(v)$, $NUM(v)$, $MAX(v)$ and $flag(v)$. $LIST(v)$ is a list of intervals in $\mathcal{I}$, whose canonical covering contains node v, and $NUM(v)$ denotes the cardinality of $LIST(v)$. $MAX(v)$ denotes the maximum number of intervals whose common intersection is nonempty in $INT(v)$ (excluding those intervals that contain $INT(v)$). We shall maintain the invariant condition that $MAX(v) = \max\{(MAX(v_l) + NUM(v_l)), (MAX(v_r) + NUM(v_r))\}$, where v_l and v_r are the left and right sons of v respectively, and $MAX(v) = 0$ if v is a leaf. $flag(v) = left$ if $MAX(v_l) + NUM(v_l) \geq MAX(v_r) + NUM(v_r)$, and is *right* otherwise. Thus, $MAX(r)$ of the root node r of T gives the maximum number of intervals of $\mathcal{I}$ with a nonempty common intersection, *i.e.*, the density of $\mathcal{I}$. The *flag* field of the root and that of subsequent nodes will provide information leading to the position of the *leftmost* key column.

Algorithm Label

1. Sort the intervals in $\mathcal{I}$ into a list L in ascending order according to their right endpoint.
2. Take the next interval I_i from L and insert into the segment tree T by doing the following.
   - **2.1** For each node v in the canonical covering of I_i insert I_i at front of the list $LIST(v)$, and increase $NUM(v)$ by 1.
   - **2.2** Update the $MAX(u)$ of all the nodes u on the path from node v in the canonical covering to the root of T (a node is updated after its children have been updated or if its children do not need to be updated.) Let uf and ub denote the father and brother of node u respectively. If $MAX(u)$

+ NUM(u) $\geq$ MAX(ub) + NUM(ub) and u is the *left* son of uf, then MAX(uf) is incremented by 1 and *flag*(uf) is set to *left*. If MAX(u) + NUM(u) > MAX(ub) + NUM(ub) and u is the *right* son of uf, then MAX(uf) is incremented by 1 and *flag*(uf) is set to *right*. Otherwise, do nothing. Note that the invariant condition is maintained.

3. Repeat the following until T is empty.

   3.1 Beginning with the root node r, traverse a path according to *flag*(v) of each node visited and do the following.

   3.2 For each node v, put the interval at the front of LIST(v) (i.e., the interval with the rightmost right point) into a list Q.

   3.3 Let I_i denote the interval with the rightmost right point among those intervals in Q. Assign label d_j to I_i, where d_j is the density (d_j = MAX(r).)

   3.3a For each node w in the canonical covering of I_i delete I_i from LIST(w), and decrease NUM(w) by 1.

   3.3b Update the MAX(u) of all the nodes u on the path from node w in the canonical covering to the root of T (again, a node is updated after its children are or if its children does not have to be updated). As before, uf and ub denote the father and brother of node u respectively. If u is the *left* son and *flag*(uf)= *left* or if u is the *right* son and *flag*(uf)= *right*, then MAX(uf) is decreased by 1. There is one exception to this case: if MAX(u) + NUM(u) = MAX(ub) + NUM(ub) -1 and u is the *left* son, and thus *flag*(uf)= *left*, MAX(uf) remains unchanged. If MAX(u) + NUM(u) $\geq$ MAX(ub) + NUM(ub), and u is the *left son*, then *flag*(uf) is set to *left*, otherwise, *flag*(uf) is set to *right*. Note that the invariant condition is maintained.

4. For those intervals with label $d_j \neq 1$, assign to tracks $2d_j - 2$ and $2d_j - 1$, and for those intervals with label $d_j = 1$, assign them to track 1. Note that some tracks may be empty as density of a set $\mathcal{I}_i$ may be one. Such tracks are simply removed. As mentioned before, the just described track assignment is called Density-Reduction-Algorithm.

Lemma 3. *Density-Reduction-Algorithm assigns the intervals into t tracks, $t \leq 2d - 1$, and runs in $O(n \log n)$ time and requires $O(n)$ space, where d is the density of the problem and n is the number of intervals.*

Proof: Note that the *leftmost* key column is always maintained by the *flag* field of the root and that of subsequent node. That is, let $(v_0, v_1, \ldots, v_k)$ be a sequence of nodes such that v_0 is the root, and v_i is the *left* or *right* son of v_{i-1} depending on the value *flag*(v_{i-1}) respectively, for $i = 1, 2, \ldots, k$. Then v_k is the leftmost key column that gives the maximum density d, which is stored at MAX(v_0). The correctness of the algorithm was established by Properties 1-3. Each interval is partitioned into $O(\log n)$ standard intervals and "affects" $O(\log n)$ nodes of the tree T (see [12]). Step 1 and Step 2.1 take $O(\log n)$ time per interval. In Step 2.2, $O(\log n)$ nodes are affected. Updating each requires constant time – a simple comparison. In Step 3, each element of Q can be obtained in constant time, size of Q is $O(\log n)$, and thus, the interval with the rightmost right point can be found in $O(\log n)$ time. Again, deleting I_i (Step 3.3a) and updating all intervals affected by it (Step 3.3b) requires $O(\log n)$ time.

The construction requires $O(n \log n)$ space. However, instead of storing intervals in $LIST(v)$ we can employ a binary search tree to find the rightmost interval that contains the desired key point. The space can be reduced to $O(n)$ [4].

□

4 Applications

In this section, we briefly discuss several applications of the proposed approximation algorithm. For more details, see the full manuscript. We study the maximum-weighted independent set problem for circle graphs (or equivalently, overlap graphs), proper circular-arc graphs, and track assignment with limited capacity. Also, an application in VLSI layout is discussed.

In a circle, consider a set $K = \{k_1, \ldots, k_n\}$ of chords, where $k_i = (f_i, s_i)$. Each chord has a positive weight. Intersection representation of chords of K is called a circle graph. The circle can be "opened" at any point to obtain a set of intervals $\mathcal{I}$ (see [7] for a formal definition). $\mathcal{I}$ is called an *overlap representation* of K. In the overlap graph of $\mathcal{I}$, there is an edge between two intervals if and only if they properly intersect. That is, no edge is created if one interval contains the other. Certainly, any independent set in K is an independent set in $\mathcal{I}$ and vice versa. Thus, a maximum weighted independent set (MWIS) of K is also an MWIS of $\mathcal{I}$. An MWIS of K is a maximum weighted subset of pairwise non-intersecting chords.

A given family K of chords has a number of distinct overlap representations, in general, $2n$ such representations. An overlap representation of K with minimum density is called a *min-density representation* of K. The following has been proved (see the complete manuscript).

Lemma 4. *A min-density representation of K can be obtained in $O(n \log n)$ time.*

Given a min-density representation $\mathcal{I}^*$ of K, we want to find a maximum weighted independent set (MWIS) of $\mathcal{I}^*$. We have shown the following lemmas:

Lemma 5. *A maximum-weighted independent set of $\mathcal{I}^*$ (and K) can be obtained in $O(d^*n + n \log n)$, where d^* is the density of $\mathcal{I}^*$ and $n = |\mathcal{I}^*|$.*

Lemma 6. *A maximum-weighted independent set of a family of proper circular arcs can be obtained in $O(\delta n + n \log n)$, where δ is the density of the arcs and n is the number of arcs. The expected value of δ is $\Theta(\sqrt{n})$.*

Lemma 7. *An arbitrary instance $\mathcal{I}$ of MIS-K (i.e., maximum independent set with density at most K) for overlap graphs can be solved in $O(dn + n \log n)$ time, where d is the density of $\mathcal{I}$ and $n = |\mathcal{I}|$. The expected value of d is $\Theta(\sqrt{n})$.*

References

1. Apostolico, A., Atallah, M.J., and Hambrusch, S.E., "New Clique and Independent Set Algorithms for Circle Graphs," to appear in Discrete Applied Mathematics.
2. Asano, T, Asano, T., and Imai, H., "Partitioning a Polygonal Region into Trapezoids," *Journal of the ACM*, Vol. 33, No. 2, April 1986, pp. 290-312.
3. Buckingham, M., *Circle Graphs*, Ph. D. Thesis (Report No. NSO-21), Courant Institute of Mathematical Sciences, Computer Science Department, October 1980.
4. Danny Z. Chen, Private communication.
5. Cong, J., *Routing Algorithms in Physical Design of VLSI Circuits*, Ph. D. Thesis, University of Illinois at Urbana-Champaign, Department of Computer Science, August 1990.
6. Garey, M.R., Johnson, D.S, Miller, G.L., and Papadimitriou, C.H., "The Complexity of Coloring Circular Arcs and Chords," *SIAM Journal on Algebraic Discrete Methods*, Vol. 1, No. 20, June 1980, pp. 216-227.
7. Gavril, F. "Algorithms for a Maximum Clique and a Maximum Independent Set of a Circle Graph," *Networks*, Vol. 3, 1973, pp. 261-273.
8. Golumbic, M.C., *Algorithmic Graph Theory and Perfect Graphs*, Academic Press, 1980.
9. Gupta, U., Lee, D.T., Leung, J., "An Optimal Solution for the Channel Assignment Problem," *IEEE Transactions on Computers*, November 1979, pp. 807-810.
10. LaPaugh, A. S., *Algorithms for Integrated Circuit Layout: An Analytic Approach*, Department of Electrical Engineering and Computer Science, M.I.T. 1980.
11. G. Manacher, personal communication.
12. Preparata, F. and Shamos, M. I., *Computational Geometry: an Introduction*, Springer-Verlag, 1986.

A Simple Test for the Consecutive Ones Property

Wen-Lian Hsu
Institute of Information Science
Academia Sinica
Nankang, Taipei, Taiwan, Republic of China

ABSTRACT

A (0,1)-matrix satisfies the consecutive ones property if there exists a column permutation such that the one's in each row of the resulting matrix are consecutive. Booth and Lueker [1976] designed a linear time testing algorithm for this property based on a data structure called "PQ-trees". This procedure is very complicated and the linear time amortized analysis is also rather involved. We develop an off-line testing algorithm for the consecutive ones property without using PQ-trees. Our approach is based on a decomposition technique which separates the rows into "prime" subsets which admit unique column orderings that realize the consecutive ones property. The success of this approach is based on finding a good "row ordering" to be tested iteratively.

1. Introduction

A (0,1)-matrix satisfies the *consecutive ones property* for the rows if there exists a column permutation such that the ones in each row of the resulting matrix are consecutive. We interchange row and column in the conventional definition for notational convenience. A consecutive ones test can be used to recognize interval graphs [1,2] and planar graphs [1]. Booth and Lueker [1] invented a data structure called PQ-trees to test the consecutive ones property of (0,1)-matrices in linear time. They associate a PQ-tree to each (0,1)-matrix such that each column permutation generated by the PQ-tree corresponds to a valid ordering satisfying the consecutive ones property for the rows. This is implemented in an on-line fashion by adding one row at a time and adjusting the PQ-trees accordingly. At each iteration, the algorithm checks eleven templates to see if there is a match with part of the current PQ-tree. This procedure is very complicated since large part of the tree may be affected. The linear time amortized analysis is also rather involved. Let K be an $m \times n$ (0,1)-matrix whose total number of 1's is r. We develop a simple $O(m+n+r)$ time consecutive ones testing algorithm which adopts a set partitioning strategy. The key idea is to determine a "good" ordering of the rows to be tested.

Denote by R the set of rows of K and CL the set of columns. Then $|R| = m$ and $|CL| = n$. We use u to denote a general column and v to denote a general row. Label the columns in K arbitrarily from 1 to n. For convenience, we shall not distinguish between the two terms, "row" (respectively, "column") and its corresponding "row index" (respectively, "column index"). For each row v in R, let $CL(v)$ be the set of columns that contain a non-zero in this row. Denote the *size* of a row v by $|CL(v)|$. Label the rows according to the ascending order of their sizes. For any subset R' of R define $CL(R')$ to be the set of columns that contain a 1 in a row of R'. For each column u in K, let $R(u)$ be the set of all rows that contain a "1" in this column. A row v is said to be *singular* if $|CL(v)| = 1$.

Each row v (respectively, column u) determines a partition of CL (respectively, R) into $CL(v)$ and CL-$CL(v)$ (respectively, $R(u)$ and R-$R(u)$). Suppose we refine such a partition based on all the rows one by one (an efficient way is described in Section 2), then two columns of K are identical iff they are contained in the same set of the final partition. Identical rows of K can be determined similarly through a row partition. We shall assume the given matrix K does not contain identical rows, columns or singular rows. Such a matrix is said to be *normalized*. Two sets of columns CL_1 and CL_2 are said to *overlap* if they have a column in common. They are said to *overlap strictly* if they overlap but none is contained in the other (or each contains a column whose element is greater that of the other in the more general case when there are (1/2)"s in the entries). Two rows x and y are said to *overlap* (resp., *overlap strictly*) if CL(x) and CL(y) overlap (resp., overlap strictly). The strictly-overlapping relationships on the columns play an important role in the decomposition.

Associate with matrix K the graph *G(K)*, in which each vertex denotes a row in R and each edge connects two strictly overlapping pairs of rows in R. Denote the connected components of G(K) by $G(K_1)$, ..., $G(K_p)$, where K_1, ..., K_p are the submatrices induced by the corresponding row sets of K. For any subset R' of R, denote by *G(R')* the subgraph of G(K) induced on R'. A set R' of rows is said to be *connected* if its corresponding induced subgraph G(R') is connected. A matrix K is said to be *prime* if its associated G(K) is connected. Note that a prime matrix is not necessarily normalized.

The consecutive ones test for prime matrices is straightforward and will be the subject of the next section. The test problem for a general nomalized matrix can be reduced to those of its prime submatrices. We shall illustrate how to perform such a decomposition efficiently in Section 3.t

2. Testing the Consecutive Ones Property for Prime Matrices

Let M be a prime matrix satisfying the consecutive ones property. Let G(M) be its associated connected graph. Let T be a spanning tree of G(M). Apply a preorder traversal on T to obtain the ordering $\pi' = (v_1', ..., v_n')$ for rows of M. The ordering π' satisfies that for each i, v_i' strictly overlaps with some row in $\{v_1',...,v_{i-1}'\}$.

The consecutive ones test can be easily carried out as follows. First, by considering v_1' and v_2', we obtain the following unique partition of $CL(v_1') \cup CL(v_2')$ as $CL(v_1')$-$CL(v_2')$, $CL(v_1') \cap CL(v_2')$ and $CL(v_2')$-$CL(v_1')$, ordered consecutively from left to right. Such a partition dictates the left-right relationships of those three sets. However, the columns within each set could still be permuted freely. In general, the rows of $\{v_1',...,v_{i-1}'\}$ determine a **unique partition** of the columns in $CL(v_1') \cup \cdots \cup CL(v_{i-1}')$ with fixed left-right relations among the sets. When v_i' is considered, the consecutive requirement of columns in $CL(v_i')$ further refines the above partition (which can be easily carried out by marking) affecting at most two sets. At the end, after v_n' has been considered, we obtain a unique partition of all columns of M that will satisfy the consecutive ones requirement of the rows. This process can be easily seen to take $O(r)$ time.

Note that in the final unique partition a set could still contain more than one columns. However, since these columns must appear in exactly the same set of rows, they must be identical.

3. Decomposition into Prime Submatrices

Consider a normalized matrix K. We show that the consecutive ones testing problem for K can be reduced to that of each component submatrices in $O(r)$ time.

Theorem 3.1. *Matrix K satisfies the consecutive ones property if and only if each of its component submatrices K_i, i = 1, ..., p satisfies that property.*

Proof. The "only if" part is trivial. Hence, assume each K_i satisfies the consecutive ones property. Define a relation "»" on the component submatrices as follows: $K_i » K_j$ if $CL(v) \supseteq CL(K_j)$ for every v in K_i such that $CL(v) \cap CL(K_j) \neq \emptyset$. It is easy to see that this is a partial order.

We claim that if $CL(K_i) \cap CL(K_j) \neq \emptyset$ for some i and j, then either $K_i » K_j$ or $K_j » K_i$. Let v_i, v_j be two overlapping rows in K_i, K_j, respectively. Since v_j does not strictly overlap with v_i, one of them must contain the other. Suppose we have v_j contains v_i (the converse would lead to the conclusion that $K_i » K_j$). Then all rows of K_i that strictly overlaps with v_i must also overlap with v_j and hence, be contained in v_j. Thus, one can argue, by breadth-first-search, that all rows in K_i must be contained in v_j. On the other hand, no row in K_j overlapping with v_i can be contained in v_i; for otherwise, let P be a path in K_j from v_j to such a row, then v_i must strictly overlap with one row in P. Therefore, we have $CL(v) \supseteq CL(K_i)$ for every v in K_j such that $CL(v) \cap CL(K_i) \neq \emptyset$ and $K_j » K_i$.

Since each K_i satisfies the consecutive ones property, there exists an ordering of the columns that realizes it. If $K_i » K_j$, then an ordering π_i for K_i is said to *conform* to an ordering π_j for K_j if the column subset $CL(K_j)$ is ordered consecutively according to π_j in the ordering π_i for K_i. Thus, in a bottom-up fashion, one can obtain an ordering π_i^* for each K_i that conforms to all ordering π_j^* such that $K_i » K_j$. Then the ordering π_p^* for K_p is the desired ordering for K. $\square$

Although Theorem 3.1 provides a useful way for finding prime submatrices, we cannot afford to compute the entire graph $G(K)$ since its number of edges could be $O(r^2)$. Instead, we shall identify each submatrix K_i by constructing a spanning forest of $G(K)$. We shall use G' to denote the subgraph constructed at the end of the algorithm.

We process the rows according to their ascending sizes. Thus, a row properly contained in another row must be processed before that row. When a row v is processed, the algorithm scans every nonzero in the columns of $CL(v)$. At the end, all columns of $CL(v)$ will be deleted and replaced by a single representative column. During the scanning process, we keep enough information about those crucial adjacency relationships that could be missing after $CL(v)$ is deleted.

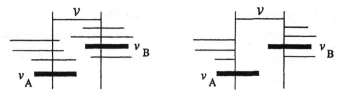

Figure 1. Truncation of intervals

To be more specific, consider a column ordering for K that realizes the consecutive ones property for the rows. Let $CL(v)$ divides the current column set of K into three parts, namely, $CL(v)$, those columns on the left of $CL(v)$ (denoted by $LF(v)$), and those on the right (denoted by $RT(v)$). Let $CL'(v)$ be the set of all columns in $CL(v)$ that contain a 1 in row v. Consider the following sets of rows:

$A(v) = \{ v' \mid CL(v') \cap LF(v) \neq \emptyset, 0 < |CL(v') \cap CL(v)| < |CL(v)| \}$

$B(v) = \{ v' \mid CL(v') \cap RT(v) \neq \emptyset, 0 < |CL(v') \cap CL(v)| < |CL(v)| \}$

$C(v) = \{ v' \mid CL(v') \subseteq CL'(v) \}$

$D(v) = \{ v' \mid \text{both } v \text{ and } v' \text{ contain } 1/2 \text{ in some column} \}$

Denote the row in A (respectively, B) with the smallest π-index by v_A (respectively, v_B). These two rows will be the representatives for A, B, respectively.

Make v adjacent to every row in $[A(v) \cup B(v) \cup C(v)] - FB(v)$, where $FB(v)$ is a list of forbidden vertices to be discussed later. Eliminate all columns in $CL(v)$ and add a **representative column [v]**, which contains 1's in those rows of $CL(v) - [A(v) \cup B(v) \cup C(v) \cup D(v)]$ and (1/2)'s in rows v_A and v_B.

To avoid the creation of adjacency relationship that does not appear in G_K (for example, among some rows in $A(v)$), we need to list some forbidden pairs. Let $FB(v)$ denote a collection of rows that are forbidden to be made adjacent to v. Concatenate $A(v) \cup B(v)$ to $FB(v_A)$ and $FB(v_B)$ respectively, to form the new $FB(v_A)$ and $FB(v_B)$.

Processing a row v_l (initialize all $FB(v_i)$ to be the empty set)

1. If $|CL(v_i)| > 1$, calculate $|CL(v_i) \cap CL(v')|$ for each row v' that overlaps with v_i and ordered before v_i by scanning each column u of $CL(v_i)$ and adding 1 to the counter of each row that has a non-zero in this column.

2. Let $C(v_l) = \{ v' \mid CL(v') \subseteq CL'(v_i) \}$ and $D(v_l) = \{ v' \mid \text{both } v_i \text{ and } v' \text{ contain } 1/2 \text{ in some column} \}$. Let $P(v_l) = \{ v' \mid v' \notin D(v_i), 1 < |CL(v_i) \cap CL(v')| < |CL(v_i)|, CL(v')-CL(v_i) \neq \emptyset \}$. Make v_i adjacent to all rows in $[C(v_i) \cup P(v_i)]$-$FB(v_i)$. Let $(v_l)_A$ be the lowest indexed row in $P(v_i)$, if it is nonempty. Concatenate $P(v_i)$ with $FB((v_i)_A)$ to form the new $FB((v_i)_A)$.

3. Let $A(v_l) = \{ v' \mid v' \in P(v_i)$, either $CL(v') \cap CL(v_i) \subseteq CL((v_i)_A) \cap CL(v_i)$ or $CL((v_i)_A) \cap CL(v_i) \subseteq CL(v') \cap CL(v_i) \}$ (this can be obtained by marking all columns in $CL((v_i)_A)$). Let $B(v_l) = P(v_i)-A(v_i)$. If $B(v_i)$ is nonempty, let $(v_l)_B$ be its lowest indexed row. Concatenate $P(v_i)$ with $FB((v_i)_B)$ to form the new $FB((v_i)_B)$.

4. Let $F(v_l) = CL(v_i) - [A(v_i) \cup B(v_i) \cup C(v_i) \cup D(v_i)]$. If $F(v_i) \neq \emptyset$, construct a **representative column [v_l]** with 1's in the rows of $F(v_i)$ and (1/2)'s in the rows $(v_i)_A$ and $(v_i)_B$. Delete row v_i and the current columns in $CL(v_i)$.

One can easily derive the following observation:

(3.2) *If $CL(v_i) \subseteq CL(v_j)$ originally but at iteration i, $CL(v_i)$ strictly overlaps with $CL(v_j)$, then there must exist a row v_k with $k < i$, such that both $CL(v_i)$ and $CL(v_j)$ strictly overlap with $CL(v_k)$ at iteration k and row v_i has 1/2 in the representative column $[v_k]$. Thus, v_i is made connected to v_j through v_k by the algorithm.*

On the other hand, it is also possible that $CL(v_i)$ strictly overlaps with $CL(v_j)$ originally but at iteration i $CL(v_i) \subseteq CL(v_j)$ as shown in Figure 2.

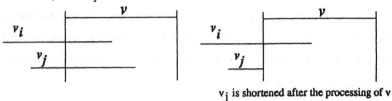

v_j is shortened after the processing of v

Figure 2. Change of overlapping relationships

A row v_i is said to be *made connected* (respectively, *adjacent*) to another row v_j in G' if there exists a path (respectively, edge) in G' from v_i to v_j.

Theorem 3.3. *The graph G' constructed is a subgraph of G(K) and the set of connected components in G' is identical to that of G(K).*

Proof. We prove this theorem by showing the following two claims:

Claim 1. *If $(v_i, v_j) \notin E(G(K))$ $(i < j)$, then $(v_i, v_j) \notin E(G')$.*

Proof of Claim 1. Since edge connection is always made from a low indexed vertex to a higher indexed one, it suffices for us to show that v_i is not made adjacent to v_j at iteration i. If v_i does not overlap with v_j originally, then the algorithm will never make them overlap. Hence, we must have that $CL(v_i) \subseteq CL(v_j)$ originally. If, at iteration i, $CL(v_i)$ strictly overlaps with $CL(v_j)$, then by (3.2), there must exist a row v_k with $k < i$, such that both $CL(v_i)$ and $CL(v_j)$ strictly overlap with

$CL(v_k)$ at iteration k and row v_i has 1/2 in the representative column $[v_k]$. However, in this case, v_j would belong to $FB(v_i)$. Hence, v_i will not be made adjacent to v_j at iteration i.

Claim 2. *If $(v_i, v_j) \in E(G(K))$, then they belong to the same connected component in G'.*

Proof of Claim 2. Suppose the Claim is false. Let v_i, v_j be a strictly overlapping pair in $G(K)$ violating this lemma such that $i < j$ and i is the smallest possible and j is the smallest with respect to i. Since the processing of the first row v_1 will find all rows strictly overlapping with v_1, we must have $i > 1$. Hence, at the i-th iteration, v_i is not made adjacent to v_j. We shall show that they are made connected through other vertices. Consider the first iteration, say the k-th, that their overlapping relationship change at the end of that iteration. Then k must be less than i. We have the following cases at the end of iteration k:

(a) $CL(v_i) \cap CL(v_j) = \emptyset$. This must happen when, at the beginning of this iteration, one of $CL(v_i)$ and $CL(v_j)$ contains $CL(v_k)$ but the other one strictly overlaps with $CL(v_k)$ (and is reduced at the end of the iteration). Suppose $CL(v_i)$ contains $CL(v_k)$ and $CL(v_j)$ strictly overlaps with $CL(v_k)$ (the other case can be argued similarly). Without loss of generality, assume $v_j \in A(v_k)$. Then we have $CL((v_k)_A)$ strictly overlaps with $CL(v_i)$ at this iteration. If $(v_k)_A$ strictly overlaps with v_i originally, then by the definition of indices i and j, v_i must be made connected to $(v_k)_A$. Otherwise, $(v_k)_A$ is contained in v_i originally. By (3.2), v_i must be made connected to $(v_k)_A$. Hence, v_i is connected to v_j through $(v_k)_A$ and v_k.

(b) One of $CL(v_i)$ or $CL(v_j)$ is contained in the other at the end of the k-th iteration. This happens when one of the following happens at the beginning of the k-th iteration:

(i) Both $CL(v_i)$ and $CL(v_j)$ strictly overlap with $CL(v_k)$. But then, v_i and v_j are both made adjacent to v_k.

(ii) $CL(v_i)$ strictly overlaps with $CL(v_k)$ but $CL(v_i)-CL(v_j) \subseteq CL(v_k)$ and $CL(v_j) \cap CL(v_k) = \emptyset$. Without loss of generality, assume $v_i \in A(v_k)$. Then $(v_k)_A$ has an index smaller than i and strictly overlaps with $CL(v_j)$. Hence, v_i is connected to v_j through $(v_k)_A$ and v_k.

(iii) $CL(v_j)$ strictly overlaps with $CL(v_k)$ but $CL(v_j)-CL(v_i) \subseteq CL(v_k)$ and $CL(v_i) \cap CL(v_k) = \emptyset$. So, $CL(v_j) \subseteq CL(v_i)$ at the end of this iteration. Unless both of them belong to some $C(v_m)$ (in which case they are adjacent through v_m), this relation must remain so through the i-th iteration. Hence v_j will belong to $C(v_i)$ and made adjacent to v_i. □

The corresponding PQ-tree for K can also be obtained from this decomposition. Section 2 shows that the PQ-tree for a normal prime matrix is a Q-node with all columns as leaves. In general, it suffices to note that after each K_i is processed its corresponding column set is reduced to a single representative column.

The total number of operations in our decomposition is proportional to the number of nonzero entries in K and the number of new 1's and (1/2)'s generated. Since there can be at most two (1/2)'s generated for a row, the total number of (1/2)'s generated is at most 2m. Note also that the total size of all forbidden pairs is no more than the total number of 1's and (1/2)'s.

Lemma 3.4. *When a row v_i is processed, $|CL(v_i)|$ is either 0 or ≥ 2.*

Proof. Suppose $|CL(v_i)|$ is > 0. Then at any iteration $s < i$, the column set of v_i cannot be contained in that of v_s (otherwise, $CL(v_i)$ would be empty). We show that it is impossible for $|CL(v_i)|$ to be 1. Suppose otherwise. Let v_j be the first row such that $v_i \in P(v_j)$ and the column set for v_i is reduced to 1 after v_j is processed. Denote by $CL^j(v)$ the column set of a row at the beginning of iteration j. Without loss of generality, assume $v_i \in A(v_j)$. Since the column set of v_i is reduced to 1 at the the end of the j-th iteration (whereas $CL((v_j)_A)$ is at least 2), $v_i \neq (v_j)_A$ (namely, $(v_j)_A = v_k$ with $k < i$). Since $|CL^j((v_j)_A) - CL^j(v_j)| > 0$, we must have that at the end of the j-th iteration, $CL(v_i) = CL^j(v_i) - CL^j(v_j) \subseteq CL^j((v_j)_A) - CL^j(v_j)$. Therefore, either $CL(v_i)$ was already reduced to 0 before the k-th iteration or $v_i \in C((v_j)_A) \cup D((v_j)_A)$ at the k-th iteration. Hence, $CL(v_i)$ should have been reduced to 0 after the k-th iteration. □

Lemma 3.5. *The total number of new 1's generated is $O(r)$.*

Proof. Consider the creation of a representative column $[v]$. Let n_1 be the number of 1's in $[v]$. By Lemma 3.4, for each entry of 1 generated in $[v]$, there are at least two 1's deleted in $CL(v)$.

Hence, the net decrease of 1's is at least n_1. Therefore, the total number of new 1's generated can never be greater than the total number of original 1's in K, which is $O(r)$. ▢

By the above discussion, the complexity of our algorithm is $O(m+n+r)$.

References

1. K. S. Booth and G. S. Lueker, *Testing of the consecutive ones property, interval graphs, and graph planarity using PQ-tree algorithms*, **J. Comptr. Syst. Sci.** 13, 3 (1976), 335-379.

2. N. Korte and R. H. Möhring, *An Incremental Linear-Time Algorithm for Recognizing Interval Graphs*, **SIAM J. Comput.** 18, (1989), 68-81.

The Longest Common Subsequence Problem for Small Alphabet Size Between Many Strings

Koji Hakata   Hiroshi Imai

Department of Information Science, University of Tokyo
Tokyo 113, Japan
E-mail: hakata@is.s.u-tokyo.ac.jp

Abstract. Given two or more strings (for example, DNA and amino acid sequences), the longest common subsequence (LCS) problem is to determine the longest common subsequence obtained by deleting zero or more symbols from each string. The algorithms for computing an LCS between two strings were given by many papers, but there is no efficient algorithm for computing an LCS between more than two strings. This paper proposes a method for computing efficiently the LCS between three or more strings of small alphabet size. Specifically, our algorithm computes the LCS of $d(\geq 3)$ strings of length n on alphabet of size s in $O(nsd + Dsd(\log^{d-3} n + \log^{d-2} s))$ time, where D is the number of dominant matches and is much smaller than n^d. Through computational experiments, we demonstrate the effectiveness of our algorithm.

1   Introduction

The longest common subsequence problem is stated as follows: Let $A_1, A_2, \ldots, A_d$ be d strings of length $n_1, n_2, \ldots, n_d$ on an alphabet $\Sigma = \{\sigma_1, \sigma_2, \ldots, \sigma_s\}$ of size s. A *subsequence* of A_i can be obtained by deleting zero or more (not necessarily consecutive) symbols from A_i. String B is a *common subsequence* of $A_1, A_2, \ldots, A_d$ iff B is a subsequence of each $A_i, i = 1, 2, \ldots, d$. The *longest common subsequence* (LCS in short) problem is to find a common subsequence of $A_1, A_2, \ldots, A_d$ of maximal length.

The LCS problem is a common task in DNA sequence analysis, and has applications to genetics and molecular biology. In such applications, the number s of alphabets is constant independent of the length n_i of each string. In fact, $s = 4$ and 20 for DNA sequences and amino acid sequences, respectively. Also, in such applications, the LCS problem for many strings, i.e. for large d, arises. This paper presents an efficient algorithm for the LCS problem when the number s of alphabets is constant as in the above cases and the number d of strings is more than two.

Before stating our results, we first explain the dynamic programming approach to LCS, which is a basis of most algorithms for the problem. Throughout this paper, l is the length of an LCS. We assume $n_1 = n_2 = \cdots = n_d = n$ only for convenience. For string $A = a_1 a_2 \cdots a_n$, $A[p \ldots q]$ is $a_p a_{p+1} \cdots a_q$.

Define an L-matrix for the d strings $A_1, A_2, \ldots, A_d$ as an integer array $L[0 \ldots n_1, \cdots, 0 \ldots n_d]$ such that $L[p_1, p_2, \ldots, p_d]$ is the length of an LCS for

$A_i[1 \ldots p_i], i = 1, 2, \ldots, d$. For $d = 2$, $L[i, 0] = L[0, j] = L[0, 0] = 0$ and for $1 \leq i, j \leq n$

$$L[i, j] = \begin{cases} L[i-1, j-1] + 1 & \text{if } A_1[i] = A_2[j] \\ \max\{L[i-1, j], L[i, j-1]\} & \text{otherwise} \end{cases}$$

Since L is nondecreasing in every argument, we can draw contours on L to separate regions of different values. The entire matrix is specified by its contours (there are l contours), and the contours can be completely specified by their corner points, which are called *dominants*. An LCS can be obtained by finding the set of dominants, instead of filling all the $(n+1)^d$ entries in L.

For $d = 2$, various improvements to the simple $O(n^2)$ algorithm were made and they concentrate on efficient generation of the dominants. Their complexity is shown in Table 1.

Table 1. The algorithms of the LCS problem for two strings. R is the total number of matches between the two strings. $D(\leq R)$ is the total number of dominants between the two strings. l is the length of an LCS.

(Algorithm)	(Time)	(Year)
Wagner and Fischer [9]	$O(n^2)$	1974
Hirschberg[3]	$O(n^2)$	1975
Hirschberg [4]	$O(n \log s + ln)$	1977
Hunt and Szymanski [5]	$O(n \log s + R \log n)$	1977
Masek and Paterson [7]	$O(n^2 / \log n)$	1980
Apostolico and Guerra [1]	$O(n \log n + D \log(n^2/D)))$	1987
Chin and Poon [2]	$O(ns + \min(Ds, ln))$	1990

Hirschberg's algorithm [4] generates dominants of each contour after each scan of A_1. Hunt and Szymanski's algorithm [5] finds dominants by binary search. Chin and Poon [2] propose to compute dominants in each contour from dominants with a lower value, instead of scanning A_1. Since each dominant can generate at most s dominants of the next contour, no more than $O(Ds)$ time will be required. Their algorithm takes $O(ns + \min(Ds, ln))$ time and $O(ns + D)$ space. Thus for two strings many algorithms have been proposed.

For three strings, the dynamic programming strategy yields an $O(n^3)$ algorithm which computes L row by row. Some of the algorithms for two strings may be extended to this case, but the time complexity becomes higher than in the two-dimensional case. There has been no efficient algorithm for three or more strings which outperforms the naïve dynamic programming algorithm.

In this paper, we present an efficient algorithm that, for three strings and small alphabet size, runs in time $O(ns + Ds \log s)$, where D is again the number of dominants between the three strings and $O(ns)$ is the preprocessing time. This algorithm is more efficient when the number s of alphabets is small, which is the case in the above-mentioned applications. Furthermore, our algorithm can be generalized to the case for d strings ($d \geq 3$), and then the time complexity becomes $O(nsd + Dsd((\log^{d-3} n) + \log^{d-2} s))$. The running time depends on the number of dominants D, and, since roughly speaking D becomes relatively small to n^d when d becomes large, our algorithm remains efficient for moderate size of d. The basic idea of our algorithm is based on the algorithm of Chin and Poon, but more properties of dominant matches are utilized to enhance the efficiency of the algorithm. Our algorithm expands the candidates of next dominants which is few when the alphabet size is small, and then eliminates redundant matches from the candidates. The analysis and experiments show that the algorithm performs better than the simple dynamic programming.

The paper proceeds as follows. Section 2 gives the definitions and the theorems for the algorithm. Section 3 gives the algorithm, and section 4 gives the analysis of the algorithm. Section 5 presents the result of computational experiments. Section 6 mentions the higher dimensional case. Section 7 concludes the paper.

2 Preliminaries

Denote the i-th coordinate of the position p in L by p_i. p is a *match* iff $A_1[p_1] = A_2[p_2] = \cdots = A_d[p_d]$. Point p *dominates* point q if $p_i \le q_i$ for $i = 1, 2, \ldots, d$ (denoted by $p \preceq q$). $\preceq$ is a partial order for $d > 1$. For a point set S, by $p \preceq S$ we mean that there is a point $q \in S$ such that $p \preceq q$. If $p_i < q_i$ for $i = 1, 2, \ldots, d$, we write $p \prec q$. A match p is a *k-dominant* iff $L[p] = L[p_1, p_2, \ldots, p_d] = k$ and $L[q] < k$ for $\{q \mid [0, 0, \ldots, 0] \prec q \preceq p, q \ne p\}$. Points p and q are *independent* iff $p \not\preceq q$ and $p \not\preceq q$. If p and q are k-dominants, then p and q are independent of each other.

Lemma 1. *A match q is a $(k + 1)$-dominant iff there is a k-dominant p with $p \prec q$, and there is no match r such that $p \prec r \preceq q, r \ne q$ for any k-dominant p with $p \prec q$.*

Proof. This lemma is evident from the definition of dominants. □

A point p in a set S is a *maximal* element of S, if $p \not\preceq S$. Given a set S of n points, the *maxima* problem consists of finding all the maximal elements of S. A *minimal* element and the *minima* problem are also defined similarly. The following algorithm M, given in [6,8], for solving the maxima problem is well-known, but we here explain it briefly since this is helpful to understand our algorithm. Here, maxima($A^* \cup \{p^*\}$) is a set of maximal elements of $A^* \cup \{p^*\}$.

ALG M(S, d)
1. $A := \phi$ (A is the set of maxima of dimension d)
   $A^* := \phi$ (A^* is the set of maxima of dimension $d - 1$)
   sort S into a list in the decreasing order of the coordinate x_d (if necessary)
2. **while** $S \ne \phi$ **do**
2.1   pick a point p from S
2.2 $p^* :=$ project p on $(x_1, \ldots, x_{d-1})$

2.3   if $p^* \npreceq A^*$ then $A^* :=$ maxima$(A^* \cup \{p^*\})$; $A := A \cup \{p\}$

2.4 **end-while**

2.5 **return** A

M sweeps the point set S in the decreasing order of x_d, and, for each point p, it determines whether or not it is dominated with respect to the coordinates $x_1, x_2, \ldots, x_{d-1}$ by any of the previously scanned points.

Theorem 1. [6,8] *For $d = 2$, the maxima problem requires time $\Omega(n \log n)$. However, if S is already sorted, it can be solved in $O(n)$ time.*

Proof. We show that the step 2 of M$(S, 2)$ takes $O(n)$ time. Step 2.1 and 2.2 can be done in $O(1)$ time. For $d = 2$, the set A^* is one-dimensional and $p^* \preceq A^* \Leftrightarrow p_1 \leq \max\{q_1 \mid q \in A\}$. This test (step 2.3) can be carried out in $O(1)$ time. M loops $|S| = n$ times, then M takes $O(n)$ time in total.   □

In the case of $d = 3$, using a proper data structure, we have the following.

Theorem 2. [6,8] *For $d = 3$, the maxima problem can be solved in $O(n \log n)$ time.*

Define D^k as the set of k-dominants, that is, all the dominants on the contour with value k. The dominants is composed of l disjoint subsets $D^1, D^2, \ldots, D^l$. Let $p_i(\sigma)$ be the position of the first σ in $A_i[p_{i+1} \ldots n]$, and for $p = [p_1, \ldots, p_d]$, let $p(\sigma)$ be $[p_1(\sigma), p_2(\sigma), \ldots, p_d(\sigma)]$ and let $p(\Sigma)$ be $\{p(\sigma) \mid \sigma \in \Sigma\}$. Define $D^k(\sigma)$ as $\{p(\sigma) \mid p \in D^k\}$ and $D^k(\Sigma)$ as $\{p(\Sigma) \mid p \in D^k, \sigma \in \Sigma\}$. $D^k(\Sigma)$ is the set of candidates of $(k + 1)$-dominants.

We now provide nice properties of dominants, which will be used in developing our efficient algorithm to LCS.

Lemma 2. *For any match p, $p \prec p(\sigma)$, and there is no match r of σ such that $p \prec r \prec p(\sigma)$.*

Proof. This lemma is evident from the definition of $p(\sigma)$.   □

Lemma 3. *For any i, if $p_i < q_i$ then $p_i(\sigma) \leq q_i(\sigma)$.*

Proof. Since $p_i < q_i$ and there is no match of σ between p_i and $p_i(\sigma)$, if $q_i < p_i(\sigma)$, there is also no match of σ between q_i and $p_i(\sigma)$, and hence $p_i(\sigma) \leq q_i(\sigma)$. Otherwise, $p_i(\sigma) \leq q_i \leq q_i(\sigma)$.   □

Theorem 3. *If D^k is sorted in respect to a coordinate, for any $\sigma \in \Sigma$, $p(\sigma)$'s for $p \in D^k$ in the sorted order are again sorted in respect to the same coordinate.*

Proof. This theorem is evident from Lemma 3.   □

Theorem 4. *For a pair of k-dominants p and q, if $p(\sigma)$ and $q(\sigma)$ are not independent, at least one of $p_i(\sigma) = q_i(\sigma)$ $(i = 1, 2, \ldots, d)$ holds.*

Proof. If $p_1 = q_1$, trivially $p_1(\sigma) = q_1(\sigma)$. Otherwise, without loss of generality, we may assume $p_1 < q_1$. Then, since $p_1 < q_1$ and the k-dominants are independent, for some $i \in \{2, \ldots, d\}$, $p_i > q_i$ holds. For this i, from Lemma 3, $p_i(\sigma) \geq q_i(\sigma)$. If $p(\sigma) \preceq q(\sigma)$, $p_i(\sigma) \leq q_i(\sigma)$ and hence $p_i(\sigma) = q_i(\sigma)$. If $p(\sigma) \succeq q(\sigma)$ then $p_1(\sigma) = q_1(\sigma)$, since $p_1(\sigma) \leq q_1(\sigma)$ from $p_1 < q_1$. $\square$

Theorem 5. *Assume p, q are k-dominants. If $p(\sigma_i) \prec q(\sigma_j)$ for $i \neq j$, then $p(\sigma_j) \preceq q(\sigma_j)$.*

Proof. Since $p \prec p(\sigma_i) \prec q(\sigma_j)$, $q(\sigma_j)$ is a match of σ_j in $S = \{r \mid p \prec r \preceq [n_1, n_2, \ldots, n_d]\}$. From Lemma 2, $p(\sigma_j)$ is the first match of σ_j in S, and hence $p(\sigma_j) \preceq q(\sigma_j)$. $\square$

Theorem 6. D^{k+1} *is the minima of* $D^k(\Sigma)$.

Proof. First we prove that D^{k+1} is a subset of $D^k(\Sigma)$. Assume that a match q is not in $D^k(\Sigma)$. If q is a $(k+1)$-dominant, from Lemma 1, then there must exist $p \in D^k$ such that $p \prec q$. Then, for this p, from Lemma 3, $p(A_1[q_1]) \preceq q$ holds. Since $p(A_1[q_1]) \in D^k(\Sigma)$ and $q \notin D^k(\Sigma)$, $p(A_1[q_1]) \neq q$. Then, again from Lemma 1, q can not be a $(k+1)$-dominant. Thus $D^{k+1} \subset D^k(\Sigma)$.

If $q \in D^k(\Sigma)$ is not the minima of $D^k(\Sigma)$, there is $r \in D^k(\Sigma)$ such that $p \prec r \preceq q, r \neq q$ for a point $p \in D^k$. From Lemma 1, q can not be a $(k+1)$-dominant. $\square$

3 Algorithm

We present algorithm A which obtains an LCS of input strings A_1, A_2 and A_3 in time $O(ns + Ds \log s)$ and space $O(ns + D)$. The principle of the algorithm is to compute repeatedly the minima of D^k, as proved in Theorem 6. If we use simply Theorem 2 to obtain the minima of $D^k(\Sigma)$, $O(|D^k(\Sigma)| \log |D^k(\Sigma)|)$ time is needed and the linearity of D in the time complexity can not be accomplished. To make the time complexity linear in D, we have to make full use of properties derived in the previous section.

The algorithm is based on three ideas. The first idea is due to Theorem 5. In Theorem 5, if $p(\sigma_i) \prec q(\sigma_j)$, $q(\sigma_j)$ cannot be a k-dominant. By the theorem, in this case, $p(\sigma_j) \preceq q(\sigma_j)$ and hence $q(\sigma_j)$ can be excluded from the candidates $D^k(\Sigma)$ of $(k+1)$-dominants simply by checking the dominance relations among $D^k(\sigma_j)$. That is, after we examined the dominance relations between $p(\Sigma)$ for $p \in D^k$, we have only to compute the minima of each $D^k(\sigma_j)$, not the whole $D^k(\Sigma)$.

The second idea is the use of Theorem 3. Every $D^k(\sigma)$ is already sorted in the same order as D^k in some coordinate, and then by Theorem 1 the two dimensional maxima problem concerning $D^k(\sigma)$ can be solved in linear time $O(|D^k(\sigma)|)$.

The third idea is as follows. For a point p in the d-dimensional space, define a point $p[i]$ in the $(d-1)$-dimensional space to be $(p_1, \ldots, p_{i-1}, p_{i+1}, \ldots, p_d)$. Also, define $D^k(\sigma)[i,j]$ to be $\{p[i] \mid p \in D^k(\sigma), p_i = j\}$. From Theorem 4, we have only to compute the two-dimensional maxima of $D^k(\sigma)[i,j]$ for $\sigma \in \Sigma$, $1 \le i \le 3$ and $1 \le j \le n$, instead of the three-dimensional maxima. Since these sets can be sorted in linear time according to the coordinate x_1 or x_2 by virtue of the second idea, the computation for the two-dimensional maxima can be performed in $O(|D^k(\sigma)[i,j]|)$ by Theorem 1.

We now describe the algorithm. In the preprocessing stage, we construct a data structure $T[\sigma_1 \ldots \sigma_s, 0 \ldots n, 1 \ldots d]$ to enumerate $p(\Sigma)$ efficiently. $T[\sigma, i, j]$ specifies the position of the first σ in $A_j[i+1 \ldots n]$. If such σ does not exist, $T[\sigma, i, j] = n+1$. Therefore $T[\sigma, p_i, i]$ stores $p_i(\sigma)$. The procedure which computes the T table is as follows.

1. **for** $1 \le i \le d$ **do**
   for $\sigma \in \Sigma$ **do** $T[\sigma, n, i] := n+1$
   for $j := n-1$ **to** 0 **do**
      for $\sigma \in \Sigma$ **do** $T[\sigma, j, i] := T[\sigma, j+1, i]$
      $T[A_i[j+1], j, i] := j+1$

The algorithm A needs $O(ns)$ space for T and a working area storing $D^k(\Sigma)$ and $O(D)$ space for storing the position of dominants, and the parent pointer and the ordering number according to the coordinate x_2. The algorithm A is as follows.

ALG $A(n, s, \{A_1, A_2, A_3\})$
1. compute the T table; $D^0 := \{[0,0,0]\}$; $k := 0$
2. **while** D^k not empty **do**
   $A := B := \phi$
2.1   **for** $p \in D^k$ **do**
   the parents of $p(\Sigma)$ are set to p
   $A := A \cup M(p(\Sigma), 3)$
2.2   **for** $\sigma \in \Sigma$ **do**
   sort $D^k(\sigma)$ concerning the coordinate x_1
   sort $D^k(\sigma)$ concerning the coordinate x_2
   for $1 \le i \le 3$ **do**
      sort $D^k(\sigma)[i,j]$ for all $j \in \{q_i \mid q \in D^k(\sigma)\}$
         concerning the coordinate x_1 when $i = 2, 3$ and x_2 when $i = 1$
      for $j \in \{q_i \mid q \in D^k(\sigma)\}$ **do** $B := B \cup M(D^k(\sigma)[i,j], 2)$
2.3   merge sort $\{D^k(\sigma) \mid \sigma \in \Sigma\}$ into $D^k(\Sigma) = \{D^k(\sigma) \mid \sigma \in \Sigma\}$ concerning x_1
   merge sort $\{D^k(\sigma) \mid \sigma \in \Sigma\}$ concerning x_2
2.4 $D^{k+1} :=$ extract $A \cap B$ from $D^k(\Sigma)$ keeping these orderings
2.5 $k := k+1$
   end-while
3. pick a point p in D^{k-1}
   while $k-1 > 0$ **do**
      output p and set p to the parent of p by the parent pointer

$$k := k - 1$$
end-while

4 Analysis of Algorithm

We now prove the validity of the algorithm and analyze its time complexity.

Theorem 7. *The algorithm A correctly computes the LCS of strings A_1, A_2 and A_3.*

Proof. By Theorem 6, D^{k+1} is a subset of $D^k(\Sigma)$, which is computed in steps 2.2 and 2.3. The remaining task is to exclude points from $D^k(\Sigma)$ which are dominated by other points in $D^k(\Sigma)$.

Dominance relations among points $p(\sigma_i)$ $(1 \le i \le s)$ in $D^k(\Sigma)$ for each $p \in D^{k-1}$ are all checked at step 2.1. Then, by Theorem 5, we do not have to check dominance relations between $p(\sigma_i)$ and $q(\sigma_j)$ for $p \ne q \in D^{k-1}$ and $i \ne j$ in order to exclude dominated points from $D^k(\Sigma)$, because, even if $p(\sigma_i) \prec q(\sigma_j)$, $p(\sigma_j) \preceq q(\sigma_j)$ and we have already checked the relation between $p(\sigma_i)$ and $p(\sigma_j)$ in step 2.1.

Thus, we have only to check dominance relations among $p(\sigma)$ $(p \in D^{k-1})$ for each σ. By Theorem 4, if $p(\sigma)$ and $q(\sigma)$ are not indepedent, there is i such that $p_i(\sigma) = q_i(\sigma)$. Then $p[i], q[i] \in D^k(\sigma)[i, p_i(\sigma)]$, and dominance relations between $p[i]$ and $q[i]$ (and accordingly $p(\sigma)$ and $q(\sigma)$) can be checked in $M(D^k(\sigma)[i, p_i(\sigma)], 2)$. The theorem follows. $\square$

Lemma 4. *If sorted orders of D^k with respect to both x_1 and x_2 are at hand, we can sort $D^k(\sigma)$ in both x_1 and x_2 in $O(|D^k|)$ time for each σ, and sort D^{k+1} in both x_1 and x_2 in time $O(|D^k|s \log s)$.*

Proof. For the former, simply use Theorem 3. For the latter, we sort D^{k+1} with respect to x_1 and x_2 by merging s sorted lists of $D^k(\sigma)$ $(\sigma \in \Sigma)$. Each level of merges requires $O(|D^k|s)$ time, and the level of merges is $\lceil \log s \rceil$. Then the whole time needed to sort D^{k+1} is $O(|D^k|s \log s)$. $\square$

Lemma 5. *If the sorted orders of $D^k(\sigma)$ in both x_1 and x_2 are given, all $D^k(\sigma)[i, j]$ for $j \in \{q_i \mid q \in D^k(\sigma)\}$ can be computed in the sorted order of x_1 for $i = 2, 3$ and in the sorted order of x_2 for $i = 1$ in $O(|D^k(\sigma)|)$ time. Further, $M(D^k(\sigma)[i, j], 2)$ can then be solved in $O(|D^k(\sigma)[i, j]|)$ time.*

Proof. We adopt the following simple algorithm for each i to compute the sorted list of $D^k(\sigma)[i, j]$: Initially, an empty list is made to correspond to each $D^k(\sigma)[i, j]$ for each $j \in \{q_i \mid q \in D^k(\sigma)\}$; we then enumerate points p of $D^k(\sigma)$ in the sorted order of x_1 when $i = 2, 3$ and x_2 when $i = 1$ with adding $p[i]$ to the list of $D^k(\sigma)[i, p_i]$ in the same order. This obviously produces sorted lists for all $D^k(\sigma)[i, j]$ and takes $O(D^k(\sigma))$ time in total.

Then, in solving the two-dimensional maxima problem $M(D^k(\sigma)[i,j], 2)$, points in $D^k(\sigma)[i,j]$ are sorted in either x_1 or x_2, and hence applying Theorem 1 we obtain the theorem. $\qquad\square$

Theorem 8. *The algorithm A requires time of $O(ns + Ds\log s)$, where n is the length of strings A_1, A_2 and A_3 and s is the number of different symbols that appear in strings A_1, A_2 and A_3 and D is the number of dominant matches, assuming that symbols can be compared in one time unit.*

Proof. Step 1 is the preprocessing step that builds the T table, which takes time $O(ns)$. In Step 2, the outer loop is repeated l times. Step 2.1 loops $|D^k|$ times. Therefore, there should be $|D^1| + |D^2| + \cdots + |D^l| = D$ executions to compute $A := A \cup M(p(\Sigma), 3)$. By Theorem 2 and $|p(\Sigma)| = s$, step 2.1 requires $O(Ds\log s)$ time. By Lemma 9, the innermost loop of step 2.2 takes time proportional to $\sum |D^k(\sigma)[i,j]| = |D^k(\sigma)| = |D^k|$, and combining this with Lemma 8 step 2.2 takes $O(|D^k|s)$ time for each iteration and the whole work is $O(Ds)$ time. By Lemma 8, step 2.3 takes $O(Ds\log s)$ time over the whole algorithm. Step 2.4 requires $O(|D^k|s)$ time. After all, step 2 takes $O(Ds\log s)$ time. Step 3 takes $O(l)$ time, which is less than $O(ns)$. $\qquad\square$

5 Experimental Results

We have run experiments for the algorithm in the paper on three strings over alphabet of size $s=4$. Strings are randomly and independently generated from the alphabet. Programs are written in C, and tests are run on a Sun SPARC station ELC using the time command. Table 2 shows the running time of two LCS algorithms, the naive dynamic programming algorithm (DP) and our algorithm (A).

As is seen from the table, our algorithm runs more than 10 times faster than the DP algorithm. This is mainly because the number D of dominants is much smaller than the size n^3 of the matrix. We also confirmed this fact for the LCS problem of more than three strings by computational experiments. This indicates the usefulness of devising algorithms whose complexity depends mainly on D if there are many strings.

Table 2. The running time (s) of DP, and the maximum A_{max}, average A_{ave} and minimum A_{min} running time of our algorithm A for 20 different test sets of three strings of length n.

n	100	200	300	400	500	600	700
DP	6	54	189	423	826	1516	2423
A_{max}	1	6	17	45	79	142	241
A_{ave}	0.1	4.9	16.2	42.8	75.4	135.2	222.3
A_{min}	0	4	15	39	71	123	213

6 Extensions

Our algorithm for three strings can be generalized to the problem for d strings ($d > 3$) by using an algorithm for the higher-dimensional maxima problem. Concerning the maxima problem in general dimension d, the following theorem is well known.

Theorem 9. [6,8] *For $d \geq 3$, the maxima of n points in the d-dimensional space can be computed in $O(dn \log^{d-2} n)$ time by a divide-and-conque algorithm.*

This theorem is too general for our purpose, because in our problem points are not general points in the d-dimensional space but are grid points in the

$$\overbrace{n \times n \times \cdots \times n}^{d \text{ times}} = n^d$$

grid. For this case, we have the following theorem.

Theorem 10. *For $d \geq 3$, the maxima of n grid points in the d-dimensional N^d grid can be computed in $O(dn \log^{d-2}(\min\{n, N\}))$ time by using a modified divide-and-conquer algorithm.*

Proof. Due to the space limitation, we here just give a main idea. At the topmost recursion of the original divide-and-conquer algorithm in [6], the point set is equipartitioned with respect to a coordinate and the maxima of each partitioned subset is recursively computed. Here, after $O(\log N)$ recursions, points in subproblems come to have the same coordinate, since they are grid points in the N^d grid. Therefore, the maxima problem for this point subset is really a $(d-1)$-dimensional problem. At this point, instead of dividing this point subset further, we call the $(d-1)$-dimensional maxima algorithm for it.

Since there are several types of recursions in the original algorithm, we have to modify it to this grid case in a more elaborate manner, but through careful analysis the theorem can be shown. □

Using the algorithm in this theorem, we can develop an efficient LCS algorithm A' for d strings ($d > 3$) by generalizing the three-dimensional algorithm A into A', where in A' no sorting is done to maintain D^k and the maxima problem is solved from scratch by the specialized maxima algorithm.

For this algorithm A', we have the following theorem.

Theorem 11. *The LCS problem for $d(\geq 3)$ strings of length n can be solved in time $O(nsd + Dsd(\log^{d-3}(\min[(\max |D^k(\sigma)[i,j]|), n]) + \log^{d-2} s)) = O(nsd + Dsd(\log^{d-3} n + \log^{d-2} s))$.*

Proof. Omitted in this version. □

Note that if the original maxima algorithm in [6] is used in A', the dominant term in the time complexity becomes $O(Dsd \log^{d-3}(\max |D^k(\sigma)[i,j]|))$, which becomes much worse than the bound in Theorem 11 if we only use a trivial bound of $|D^k(\sigma)[i,j]| \leq n^{d-1}$. Although we have not implemented the algorithm A', A' is likely to be fairly faster than the dynamic programming algorithm, since the number D of dominants becomes much smaller to n^d as d becomes larger.

7 Conclusion

The LCS problem has been studied by a number of researchers and its complexity has been improved in different respects. We have presented a better solution when the alphabet size is small and the number of strings are more than two.

Computational experiments have been done for the case of $s = 4$, corresponding to the DNA sequence, and the results show that our algorithm is quite promising. This is also observed by preliminary experiments for the case of $s = 20$, that is, the case where the protein consisting of 20 kinds of amino acids.

Regarding the case of $d \geq 4$, the time complexity of our algorithm is not linear in D but proportional to $D \log^{d-3} n$. It is left as a future problem to make it linear by making use of more properties of the LCS problem.

Acknowledgment

This work is supported in part by the Grand-in-Aid for Scientific Research on Priority Areas, "Genome Informatics", of the Ministry of Education, Science and Culture of Japan.

References

[1] Apostolico, A. and C. Guerra, The longest common subsequence problem revisited, *Algorithmica*, Vol.2, 1987, pp.315-336.

[2] Chin, F. Y. L. and C. K. Poon, A fast algorithm for computing longest common subsequences of small alphabet size, *J. of Info. Proc.*, Vol.13, No.4, 1990, pp.463-469.

[3] Hirschberg, D. S., A linear space algorithm for computing maximal common subsequences, *Comm. ACM*, Vol.18, 1975, pp.341-343.

[4] Hirschberg, D. S., Algorithms for the longest common subsequence problem, *J. ACM*, Vol.24, 1977, pp.664-675.

[5] Hunt, J. W. and T. G. A. Szymanski, A fast algorithm for computing longest common subsequences, *Comm. ACM*, Vol.20, 1977, pp.350-353.

[6] Kung, H. T., F. Luccio, and F. P. Preparata, On finding the maxima of a set of vectors, *J. ACM*, Vol.22, No.4, 1975, pp. 469-476.

[7] Masek, W. J., and M. S. Paterson, A faster algorithm computing string edit distances, *JCSS*, 1980, pp.18-31.

[8] Preparata, F. P., and M. Shamos, *Computational Geometry*, Springer-Verlag, 1985.

[9] Wagner, R. A., and M. J. Fischer, The string-to-string correction problem, *J. ACM*, Vol.21, No.1, 1974, pp.168-173.

The Implicit Dictionary Problem Revisited

Tak Wah Lam   Ka Hing Lee

Department of Computer Science
The University of Hong Kong
Pokfulam, Hong Kong

Abstract. In this paper we give an *implicit* data structure for storing a set of 2-key records; it can support the operations insert, delete, and search under either key in $O((\lg^3 n(\lg \lg n)^2)$ time, where n is number of records currently in the data structure. The data structure can be generalized to handle k-key records for any $k > 2$ and the time complexity of each operation becomes $O(\lg^{k+1} n(\lg \lg n)^2)$. Prior to our work, finding an implicit data structure to support such operations in polylogarithmic time was an open problem [8, 4].

1. Introduction

Implicit data structures refer to those data structures that use no extra storage beyond the space needed for the data and a constant number of parameters. That is, no explicit pointers are allowed to represent the relationship among the data. A classical example is the *heap* which gives a very efficient implementation of a priority queue on an array without pointers. It is, of course, not obvious whether explicit pointers are dispensable for more complicated problems [9, 3, 6]. In particular, the dictionary problem, which asks for a data structure to keep track of a set of records such that records can be inserted, deleted, and searched efficiently, is the most challenging to the research on implicit data structures.

Let us first review implicit data structures for the *static* dictionary problem which assumes records are fixed in advance and asks for the search operation only. If every record has only one search key, we simply maintain the records as an array sorted in ascending order with respect to the only key. To search for any key, we perform a binary search in $O(\lg n)$ time. When each record has two or more search keys, a search may be requested under any key and it is no longer trivial how records can be arranged to support a search operation in polylogarithmic time. The multi-key static dictionary problem has been studied by a number of researchers [9, 1, 8, 4]. Munro [8] was the first to give a method for arranging an array of 2-key records such that a search can be performed under either key in polylogarithmic time. Recently, Fiat *et al.* [4] constructed a more general structure for records with $k \geq 2$ keys. Their scheme supports a search under any one of the k keys in $O(\lg n)$ time.

Previous works on implicit data structures for the general dictionary problem were limited to single-key records only. Munro and Suwanda [10] were the first to study this problem. They devised a scheme of organizing an array of n single-key records to support the operations insert, delete, and search in $O(n^{1/3} \lg n)$ time. Their work was later improved by Frederickson [5]. Now the best known result is due to Munro [7] who has found an implicit data structure supporting each operation in $O(\lg^2 n)$ time. Notice that if explicit pointers are permitted, organizing the records as an AVL tree or any other balanced tree would support each operation in $O(\lg n)$ time. On the other hand, Borodin *et al.* [2] studied the tradeoff between search and update time required by any implicit data structure for the single-key dictionary problem. They showed that if the update time is constant, then the search time is $\Omega(n^\epsilon)$ for some $\epsilon > 0$. Other than the foregoing results, little was previously known about the implicit dictionary problem for records having $k \geq 2$ keys. As in the static case, the problem becomes very complicated when each record has two or more keys, and there is no simple way to generalize the techniques for single-key records. Without explicit pointers, it seems to be very difficult to maintain the *dynamic* relationship among the records under two or more keys. In fact, even assuming the number of records is fixed (i.e., no insertion and deletion), finding an implicit data structure for a set of 2-key records to support the operations *change* of key value and search under either key in polylogarithmic time was an open problem [8, 4].

The difficulty in extending the static solutions [8, 4] to handle a change of key value is that their ways of arranging records are very "fragile"—if we change a key value of a record, we might need to rearrange a large number of records in order to restore the structure. For instance, Fiat *et al.* [4], using a complicated encoding scheme, were able to store a large number of bits of extra information which are essential to their efficient searching algorithm, yet it looks almost impossible to change even one of such bits in polylogarithmic time. In deriving a solution for the dynamic case, we cannot rely on such a powerful encoding scheme and we are unable to take advantage of storing a huge amount of extra information like pointers. Moreover, we must ensure that changing a key value would only affect the arrangement of a small number of records.

In this paper we give the first implicit data structure for 2-key records to support change of key value and search under either key in polylogarithmic time. More precisely, each operation requires $O(\lg^3 n(\lg \lg n)^2)$ time. In the full paper we will give the details of supporting the operations insert and delete in $O(\lg^3 n(\lg \lg n)^2)$ time, and we will show how to extend the data structure to handle k-key records in $O(\lg^{k+1} n(\lg \lg n)^2)$ time.

Unlike the previous works, we conceptually do not attempt to reorganize all records globally after changing the key value of a record (or inserting/deleting a record). We use an ad hoc method to tolerate a change, yet we guarantee that all records can still be searched efficiently. Using an ad hoc approach, we of course cannot tolerate an infinite number of changes. The major innovation, on which the time bound stated above depends, is a robust implicit data structure that can tolerate a large number (i.e., a constant fraction of $n/\lg n$) of changes without worsening the search time. For this implicit data structure, polylogarithmic time suffices to tolerate a change in an ad hoc manner, and a global reorganization is needed only after $\Theta(n/\lg n)$ changes. It can be shown that reorganization can be done in $O(n \lg n)$ time. A change can thus be handled in polylogarithmic *amortized* time, though this is not our aim. To achieve polylogarithmic *worst-case* time, we observe that our structure permits the work of reorganization to spread over the following changes. That is, we are going to reorganize the records while we process the coming searches and changes. We ensure that the reorganization will be completed before we have accumulated too many changes and need another reorganization.

The rest of the paper is organized as follows. Section 2 describes a simple implicit data structure which supports searching under either key. It is based on the works of Munro [7, 8] and Fiat *et al.* [4]. Section 3 extends this simple structure to tolerate a large number of changes of key values, and Section 4 and 5 give the details of supporting the operations change and search in $O((\lg n \lg \lg n)^2)$ time. Section 6 shows how we can reorganize the records while we process a change or a search in $O(\lg^3 n (\lg \lg n)^2)$ time.

Definitions and Notations: As with most prior work on this topic, we confine our attention to comparison-based data structures, which use comparison between keys of records to gain information. In other words, we make no assumption on the representation of keys.

Let $A[0..n-1]$ be an array of n records. Each record has two keys. All key values are assumed to be distinct, and each record can be identified by either key. We make no assumption on the size of the keys, except that we can compare any two keys in one unit of time. For any $l \in \{1, 2\}$ and any $0 \le i, j < n$, $A[i] <_l A[j]$ if the lth key of $A[i]$ is less than that of $A[j]$.

Several AVL trees will be used in the data structure described below. Each tree is encoded in a bit format, and the content of a node or a pointer is retrieved bit by bit. Each node in an AVL tree stores an index (not a key value) of some record in A. The ordering maintained in a tree is according to either the indices stored in the tree or the key values of the records referred by these indices.

2. A Simple Implicit data structure

In this section we describe a data structure for 2-key records which supports searching under either key in $O((\lg n \lg \lg n)^2)$ time and requires less than $0.3n$ bits of extra storage. To give a fully implicit structure, we show a simple scheme to encode the extra storage in A itself. In the following, let α denote some sufficiently large constant.

Organizing A itself: Firstly, sort the array A under the 1st key, and partition the n locations in A logically into p chains $Y_0, Y_1, \cdots, Y_{p-1}$, where $p = \lceil \alpha \lg n \rceil$. For $0 \le j < p$, Y_j is comprised of all locations i such that $i \bmod p = j$. The locations in Y_0 are also called guide-positions, as records in these locations are to be fixed in their current positions. Other locations are called non-guide-positions.

Secondly, each chain except Y_0 is further sorted under the 2nd key separately.

Extra data structures: To keep track of the order of records in guide-positions under the 2nd key, we create an AVL tree T_{g_2} which stores the indices of all guide-positions ordered according to their 2nd keys. T_{g_2} requires $\lceil n/p \rceil (3 \lg n + 2)$ bits of storage.

Next, we show how to recover the order of records in non-guide-positions under the 1st key. Let $\mathcal{X}$ denote the set of non-guide-positions. Following Fiat *et al.*, we define $\pi : \{0, \cdots, n-1\} \to \{0, \cdots, n-1\}$ be the permutation such that the record in $A[i]$ was at location $\pi(i)$ when all records were sorted under the 1st key. Obviously, for any guide-position i, $\pi(i) = i$. Note that records have been "permuted" only within each chain; for any $i \in \mathcal{X}$, i and $\pi(i)$ belong to the same chain. Records in guide-positions are still in order under the 1st key. For any $i \in \mathcal{X}$, $\pi(i)$ can be determined in $O(\lg n)$ time by performing a binary search for $A[i]$'s current 1st key on guide-positions.

In order to search for the 1st key efficiently, we will need to compute the inverse of π (denoted π^{-1} below) for some non-guide-positions. Intuitively, $\pi^{-1}(i)$ tells us which record would be at location i when A was sorted under the 1st key, and $A[\pi^{-1}(0)] <_1 A[\pi^{-1}(1)] <_1 \cdots <_1 A[\pi^{-1}(n-1)]$. Computing $\pi^{-1}(i)$ requires following the permutation cycle starting at i until i is reached again. As a cycle can be as long as n/p, following the cycle step by step would be too slow. A natural way to speed up this process is adding shortcuts [4] into the cycles. Starting at an arbitrary point on a cycle of length greater than $\lceil \alpha \lg n \rceil$, we give every $\lceil \alpha \lg n \rceil^{th}$ element i a shortcut to the element j that would be reached in moving from i backward along the cycle for $\lceil \alpha \lg n \rceil$ steps. In other words, $\pi^{\lceil \alpha \lg n \rceil}(j) = i$. The length of a cycle may not be divisible by $\lceil \alpha \lg n \rceil$, and the shortcut of the starting point may span less than $\lceil \alpha \lg n \rceil$ elements. Nevertheless, the total number of shortcuts over all permutation cycles is at most $\lceil 2n/\alpha \lg n \rceil$.

The data structures used by Fiat *et al.* [4] to store shortcuts allow very efficient retrieval of shortcuts, yet the storage required exceeds the capacity of a simple encoding scheme. We store shortcuts in a more space-efficient way which involves $\lceil n/p \rceil$ AVL trees $C_0, C_1, \cdots, C_{\lceil n/p \rceil - 1}$, each of size less than p. If there is a shortcut associated with a location i, we insert a node storing i as well as its shortcut in $C_{\lfloor i/p \rfloor}$ with respect to the value of i. There are only $\lceil 2n/\alpha \lg n \rceil$ shortcuts; the storage required by these trees is $\lceil 2n/\alpha \lg n \rceil (4 \lg n + 2)$ bits. We also need an array of pointers $X_C[0 \cdots \lceil n/p \rceil - 1]$ to keep track of the roots of the AVL trees. X_C requires $\lceil n/p \rceil \lg n \leq \frac{n}{\alpha}$ bits. Given any location i, we can detect and retrieve its shortcut by searching for i in $C_{\lfloor i/p \rfloor}$. This requires $O(\lg n \lg \lg n)$ time since the height of $C_{\lfloor i/p \rfloor}$ is $O(\lg \lg n)$.

To move swiftly along a permutation cycle, we take advantage of the shortcut whenever it is available. At a point, say i, where a shortcut is not found, we move one step forward by computing $\pi(i)$ through a binary search on guide-positions. Note that shortcuts reduce the effective cycle length to at most $\lceil \alpha \lg n \rceil + 1$. Thus, computing π^{-1} requires $O(\lg^2 n \lg \lg n)$ time.

Searching under the 1st key: We perform a binary search on the 1st keys in the guide-positions. If there is no match, the search value lies between two consecutive guide-positions $A[l]$ and $A[l + p]$. When A was sorted under the 1st key, the search value could only appear at locations $l + 1, \cdots, l + p - 1$. Therefore, the search value can now be found at locations $\pi^{-1}(l+1), \cdots, \pi^{-1}(l+p-1)$. Note that $A[\pi^{-1}(l+1)] <_1 A[\pi^{-1}(l+2)] <_1 \cdots <_1 A[\pi^{-1}(l+p-1)]$. We can perform another binary search on these locations, and hence, π^{-1} is computed for at most $\lceil \lg p \rceil$ locations. The time required is $O((\lg n \lg \lg n)^2)$.

Searching under the 2nd key: Searching under the 2nd key is straightforward. Each chain is searched separately. All chains except Y_0 are sorted under the 2nd key. We perform a binary search on each of them. They altogether require $(p-1) \left\lceil \lg \frac{n}{p} \right\rceil =$

$O(\lg^2 n)$ time. For Y_0, we search the tree T_{g_2}. It again takes $O(\lg^2 n)$ time.

Encoding scheme: We encode extra data structures by the relative ordering of records in non-guide-positions. If we choose a suitable α, the storage required for $T_{g_2}, X_C, C_0, \cdots, C_{\lfloor n/p \rfloor}$ is less than $0.3n$ for any sufficiently large n.

Let $Y = \{i_1, i_2, \cdots, i_{2x}\}$ be a sequence of locations in A. Suppose that records in Y are *odd-even* sorted [7, 8] under the lth key, that is, for any $1 \leq j \leq x$, $A[i_{2j-1}] \leq_l A[i_{2j}]$ (and i_{2j-1} is called the mate of i_{2j} and *vice versa*). Then Y can be used to encode x bits in such a way that the current ordering of records in Y can be recovered easily. For any $1 \leq j \leq x$, records in $A[i_{2j-1}]$ and $A[i_{2j}]$ are kept in ascending order to denote a "0" and swapped to denote a "1". We can retrieve a bit represented in any pair of locations by one single comparison. More importantly, it needs one comparison to find out the record that was in any location i of Y before the encoding process, and we can indeed treat records in Y as if no encoding is done.

Any chain Y_j except Y_0 in A are sorted (and hence odd-even sorted) under the 2nd key and can be used to encode extra data structures. If we choose a suitable α, we can encode at least $0.4n$ bits in the $p - 1$ chains for any sufficiently large n.

3. A more robust structure

The initial set-up: Initially, the array A is organized as described in the previous section. That is, A is first sorted under the 1st key; each chain other than Y_0 is further sorted under the 2nd key separately; the AVL tree T_{g_2} is built to store the order of guide-positions under the 2nd key and finally shortcuts of permutation cycles are computed according to the permutation function π.

Notice that changing even one key value in A might have disrupted the initial set-up. If we reorganize A globally to restore the initial structure after a change of key values, it would take a lot of time. In the following, we describe less stringent requirements on the structure of A which can be maintained easily after every change of key value. These requirements together with some extra data structures would still guarantee the same search time as before. Since some of the extra data structures will become too big as more changes come in, A can only tolerate up to $n/\alpha \lg n$ changes for some sufficiently large constant α.

Before stating the new requirements on A, we need to clarify the notion of *updated* locations. To serve a request for changing the key value of a record, we find the location (denoted x below) containing the record and modify the key value in the record accordingly. The modified record may then be stored in a location other than x. In other words, x and possibly a few other locations will get *updated* in serving the request. Formally speaking, a location i in A (or an array entry $A[i]$) has been *updated* if we have ever modified or replaced the record stored in $A[i]$ after the initial set-up. Note that the replacement may not be a record with keys modified. As mentioned in the previous section, the encoding of extra data structures in A is "transparent" to the structure of A, and record movements due to altering the content of the extra data structures are not considered to cause any locations to be *updated*.

A. Records in un-updated locations are always in proper order: Once the permutation function π and the shortcuts are computed during the initial set-up, we will not recompute them even though some records in A are changed later. Recall that just after the initial set-up, the order of records in A under the 1st key is according to the sequence $Z = (\pi^{-1}(0), \pi^{-1}(1), \cdots, \pi^{-1}(n-1))$. At any time later, records in un-updated locations of A, in respect of their 1st keys, will still be kept in the order as they appear in Z. That is, for any un-updated locations i_1, i_2, if $\pi(i_1) < \pi(i_2)$ then $A[i_1] <_1 A[i_2]$. On the other hand, each chain Y_j (except Y_0) was sorted under the 2nd key initially. At any time later, records in un-updated locations of Y_j will still be sorted under the 2nd key.

B. Keeping track of records at updated locations: When a location i on A is updated, we would like to conceptually "mark" $A[i]$ being updated and use a different method to search for the record in $A[i]$. To be more specific, we insert i into two AVL trees T_{u_1} and T_{u_2} which are ordered with respect to $A[i]$'s 1st key and 2nd key respectively. To search for the 1st key (or the 2nd key) in any updated locations, we simply search T_{u_1} (T_{u_2}).

C. Computing $\pi(i)$ for any non-guide-position i in $O(\lg n \lg \lg n)$ time: Recall that to search for the 1st key in any un-updated non-guide-position, we rely on the efficient computation of π^{-1}. The effective length of any permutation cycle augmented with shortcuts is $O(\lg n)$. We must be able to compute π for any location in $O(\lg n \lg \lg n)$ time so that π^{-1} can be computed in $O(\lg^2 n \lg \lg n)$ time.

C-1. *A requirement on guide-positions for computing π on un-updated locations:* If some guide-positions are updated arbitrarily, we will not be able to compute $\pi(i)$ by performing a binary search on guide-positions. We require records in guide-positions to satisfy the following invariant after every change of key values is processed. This would enable us to compute $\pi(i)$ for any location i that has not been updated.

Invariant 1: Let l be any guide-position. Then for any non-guide-position i which has not been updated, $A[i] <_1 A[l]$ if $\pi(i) < l$, and $A[i] >_1 A[l]$, otherwise.

Immediately after the initial set-up, the invariant is satisfied. The invariant does not guarantee records in guide-positions are sorted themselves, yet for any location i that has not been updated, $A[i] >_1 A[l]$ for all guide-positions $l < \pi(i)$, and $A[i] <_1 A[l]$ for all guide-positions $l > \pi(i)$. To determine $\pi(i)$, we can perform a *"binary search"* for $A[i]$'s 1st key on guide-positions as if records in guide-positions are sorted. This takes $O(\lg n)$ time.

C-2. *Additional data structures for computing π on updated location:* For an updated location i, the key value initially in $A[i]$ was lost and we can no longer compute $\pi(i)$ by performing a binary search on guide-positions. Thus, the value of $\pi(i)$ must be stored up when $A[i]$ is first updated. Again, we keep $\lceil n/p \rceil$ AVL trees $D_0, D_1, \cdots, D_{\lceil n/p \rceil - 1}$; i and $\pi(i)$ are stored together in a node of $D_{\lfloor i/p \rfloor}$. Each tree is of size less than p, and it takes $O(\lg n \lg \lg n)$ time to find a node.

D. Structure of any chain Y_j, for $0 < j < p$: When some locations on Y_j are updated, we will not insist to have all records in Y_j sorted under the 2nd key. To search Y_j under the 2nd key, it is sufficient to maintain records in a small fraction of locations known as *key-positions* to be arranged properly.

Consider Y_j to be composed of many consecutive regions. Each region except possibly the last one consists of q consecutive locations on the chain, where $q =$

$\lceil \alpha \lg n \rceil$. For example, Y_j's first region consists of locations $j, p+j, 2p+j, \cdots, (q-1)p+j$. Furthermore, the first location of each such region is called a key-position; other locations are called non-key-positions. After processing a request for changing key values, we require records in the key-positions to satisfy the following invariant.

Invariant 2: Let l be any key-position on Y_j. Then for any non-key-position i on Y_j which has not been updated, $A[i] <_2 A[l]$ if $i < l$, and $A[i] >_2 A[l]$, otherwise.

Y_j initially has its records sorted under the 2nd key, and the invariant is satisfied. The invariant does not require records in the key-positions to be sorted under the 2nd key. Nevertheless, for any un-updated non-key-position i, $A[i] >_2 A[l]$ for all key-positions $l < i$, and $A[i] <_2 A[l]$ for all key-positions $l > i$. Thus, we can search in $O(\lg n)$ time for the 2nd key in any un-updated non-key-position of Y_j as follows: Perform a *"binary search"* on key-positions of Y_j to narrow the search down to a region of $q-1$ consecutive locations, then examine records in the region one by one.

E. Keeping track of records at guide-positions and key-positions: Recall that records in guide-positions are not required to be sorted under the 1st key. To search for a 1st key in a guide position, we cannot use a binary search on guide-positions themselves. We store all guide-positions in an AVL tree T_{g_1}, ordered according to the 1st keys in them. T_{g_1} has $\lceil n/p \rceil$ nodes. When a guide-position is updated, T_{g_1} is modified accordingly. At any time, we can search T_{g_1} for a 1st key in any guide-position in $O(\lg^2 n)$ time.

Records in key-positions of any chain Y_j may also not be in order under the 2nd key. To search for a 2nd key in any key-position of any chain, we build another AVL tree T_k to store key-positions of *all* chains except Y_0. Nodes in T_k are ordered according to their corresponding 2nd keys. It takes $O(\lg^2 n)$ time to search T_k.

F. Requirements due to encoding: To encode extra data structures, we require that on every chain except Y_0, each region excluding the first location is odd-even sorted under the 2nd key. The requirement is enforced no matter the locations are updated or not.

Remarks: The data structures described so far fall into two categories. (a) $T_{g_1}, T_{g_2}, T_k, T_{u_1}, T_{u_2}$ are search trees ordered in respect of key values in some locations of A. They are used to search the records in these locations. (b) Other data structures including C_i and D_i are not related to key values. Instead, they provide information like the shortcut or the value of π for some locations.

4. Algorithms for searching under either key

The 1st key: Suppose we are looking for a 1st key v. We search AVL trees T_{u_1} and T_{g_1}. If v appears in an updated location or a guide-position, we would report a hit here. Otherwise, we perform a binary search on guide-positions to determine two consecutive guide-positions $A[l]$ and $A[l+p]$ such that v is between their 1st keys. If v is indeed in a non-guide-position i that is not updated, then by Invariant 1, $l < \pi(i) < l + p$. Thus, we only need to examine un-updated locations among the locations $\pi^{-1}(l+1), \pi^{-1}(l+2), \cdots, \pi^{-1}(l+p-1)$. As discussed in Section 3, we can compute $\pi^{-1}(j)$ for any location j in $O(\lg^2 n \lg \lg n)$ time. However, we do not want to compute π^{-1} for all locations $l+1, \cdots, l+p-1$ because it would require $O(\lg^3 n \lg \lg n)$ time. Let Z' denote the sequence $(\pi^{-1}(l+1), \pi^{-1}(l+2), \cdots, \pi^{-1}(l+p-1))$. Since records in all un-updated locations in Z' are still sorted under the 1st key in the order as they appear in Z', we would like to perform a binary search on them. There is, however, no efficient way to retrieve the middle or any k^{th} un-updated location in Z'.

In order to determine any k^{th} un-updated location in Z' efficiently, we use additional $\lceil n/p \rceil$ *sorted* linked lists $L_0, \cdots, L_{\lceil n/p \rceil - 1}$. When a non-guide-position h is first updated, we compute $\pi(h)$ and insert $\pi(h)$ into $L_{\lfloor \pi(h)/p \rfloor}$. Therefore, for any guide-position l, $L_{l/p}$ stores all the values j such that $l < j < l + p$ and the location $\pi^{-1}(j)$ has been updated. $L_{l/p}$ is of length less than p and $O(\lg^2 n)$ time suffices to execute operations like traversing $L_{l/p}$, counting the number of un-updated locations in Z', and for any k, finding the value of j such that $\pi^{-1}(j)$ is the kth un-updated locations in Z'. With $L_{l/p}$, we can perform a binary search for v on the un-updated locations in Z'. The search requires computing π^{-1} for at most $\lceil \lg p \rceil$ locations and takes $O((\lg n \lg \lg n)^2)$ time.

The 2nd key: To search a 2nd key v, we first search the AVL trees T_{u_2}, T_{g_2}, and T_k to check if v appears in an updated location, or in an guide-position, or in an key-position of a chain other than Y_0. If all searches fail, v can only appear in an un-updated non-key-position of some chain Y_j. By Invariant 2, we can find such a key by a separate binary search on key-positions of each chain. That is, on each Y_j, we determine a region in which v may appear, and then examine the 2nd keys in the region one by one. The whole process requires $O(\lg^2 n)$ time.

5. Algorithms for changing key values

Suppose the record currently in $A[i]$ has key values (α_1, α_2). We want to change the key values of this record to (β_1, β_2).

First of all, if i is not a guide-position, we search $D_{\lfloor i/p \rfloor}$ for the appearance of i. If i a non-guide-position to be first updated, we need to store $\pi(i)$ into the data structures $D_{\lfloor i/p \rfloor}$ and $L_{\lfloor \pi(i)/p \rfloor}$. Invariant 1 enables us to compute $\pi(i)$ in $O(\lg n)$ time by a binary search for $A[i]$'s 1st key (i.e., α_1) on the guide-positions.

Next, we replace the key values of the record in $A[i]$ by (β_1, β_2) and modify T_{u_1} and T_{u_2} to reflect the update of $A[i]$. As the structure of A might have been disrupted, we reorganize A as follows. Note that the trees $T_{g_1}, T_{g_2}, T_k, T_{u_1}, T_{u_2}$ are not useful to the algorithm below, yet we need to update them for the searching algorithms.

I. i is not a guide-position nor a key-position: Recall that for such location, we only require that i must be odd-even sorted under the 2nd key with respect to its mate i'. Let (γ_1, γ_2) be the current key values in $A[i']$. If changing $A[i]$'s 2nd key to β_2 can still keep $A[i]$ and $A[i']$ odd-even sorted, we are done. Otherwise, we swap the records in $A[i]$ and $A[i']$. Again, if this is the first time to update the location i', we store $\pi(i')$ into $D_{\lfloor i'/p \rfloor}$ and $L_{\lfloor \pi(i')/p \rfloor}$ before the swap. T_{u_1} and T_{u_2} must also be modified to capture the update of both locations i and i'. The time required for all the updates is $O(\lg^2 n)$.

II. i **is a key-position on a chain** Y_j **for some** $j > 0$: To maintain Invariant 2, we cannot allow the record in $A[i]$ to have an arbitrary 2nd key. We find a non-key-position k which contains a record satisfying the requirement to be in the key-position i, and swap the records in $A[i]$ and $A[k]$. Since k is a non-key-position on Y_j, we can use Step I to accommodate the new record in $A[k]$.

Let k be the first non-key-position after i on Y_j which has not been updated. We can prove that if we swap the records in $A[i]$ and $A[k]$ and get $A[k]$ updated, Invariant 2 would be satisfied. The location k can be found in $O(\lg^2 n)$ time with some extra bookkeeping of un-updated locations. Details are omitted here. We must also update T_k, T_{u_1}, T_{u_2}, and if necessary, $D_{\lfloor k/p \rfloor}$ and $L_{\lfloor \pi(k)/p \rfloor}$ to reflect the change of $A[i]$ and $A[k]$ due to the swap. These updates altogether require $O(\lg^2 n)$ time. If all non-key-positions after i have been updated, we instead find the last un-updated non-key-position k' before i. It is easy to see that swapping $A[i]$ with $A[k']$ has the same effect as with $A[k]$. If there is no un-updated non-key-positions on Y_j, Invariant 2 is already satisfied and no swapping is required.

III. i **is a guide-position:** The situation is similar to the previous case. Changing the 1st key in $A[i]$ to β_1 might have violated Invariant 1. We swap the record in $A[i]$ with another record in some suitable non-guide-position k, and rely on Step I or Step II to handle the new record in $A[k]$.

Let l be the first location after i such that $k = \pi^{-1}(l)$ is an un-updated not-guide-position. Then for any other un-updated non-guide-position j not equal to k, if $\pi(j) < i$, $A[j] <_1 A[k]$; if $\pi(j) > i$, $A[j] >_1 A[k]$. If we swap the records in $A[i]$ and $A[k]$, Invariant 1 would be satisfied. The details of finding k in $O(\lg^2 n)$ time are omitted. If all non-guide-positions after i have been updated, we use the last un-updated non-guide-position before i.

6. Allowing any number of changes

The structure described in previous sections supports a search in $O((\lg n \lg \lg n)^2)$ time and a change in $O(\lg^2 n)$ time, but it can tolerate $O(n/\lg n)$ changes. In this section, we generalize this structure to allow any number of changes.

The array A is partitioned into $r = \lceil \beta \lg n \rceil$ subarrays for some sufficiently large constant β. Each subarray has $m = \lfloor n/r \rfloor$ elements and is organized as an individual structure according to Section 3. To search for any key value, we search the subarrays separately; the time required is $O(r(\lg m \lg \lg m)^2) = O(\lg^3 n(\lg \lg n)^2)$.

Note that each subarray can tolerate at most $m/\alpha \lg m$ changes of its key values for some sufficiently large α. If we keep on accepting changes on A, some subarrays would eventually accumulate too many changes. Thus, we need regular clean-ups to ensure that no more than $m/\alpha \lg m$ changes are accumulated in any subarray. Let $\rho = \frac{m}{4\alpha \lg m \lg r}$. After processing every ρ changes on A, we take the subarray A' with the largest number of accumulated changes and reorganize A' globally to clean up its accumulated changes. The work of reorganizing A' will indeed spread over the processing of next ρ changes on A. Note that while A' is reorganized, some of the coming changes (say, s of them) may occur on A'. Actually, when the reorganization of A' completes, A' would look like a subarray that has accepted s (or less) changes after the initial set-up. In the full paper we will show a pebble game modelling the periodic clean-ups of A, and we can prove that at any time the number of accumulated changes in any subarray is less than $m/\alpha \lg m$.

In the course of reorganizing A', a change on a subarray other than A' can be handled in $O((\lg m \lg \lg m)^2)$ time by the algorithms in Section 5, and a change on A' will be processed differently in $O(\lg^2 m \lg n)$ time. After serving a change in any case, we spend $O(\lg^3 n \lg \lg n)$ time to perform the extra work for reorganizing A'. The details of reorganizing A' will be given in the full paper.

The motivation of dividing A into subarrays is as follows. When we reorganize a subarray A', other subarrays can be used to encode extra data structures for A'. In fact, if we choose sufficiently large α and β, the subarrays can provide $O(m \lg n)$ bits of *external* storage for A'. Each record in A' can be conceptually given several extra fields, each of size $\lceil \lg n \rceil$ bits. However, we cannot assume key values can be encoded in external storage.

The major difficulty of reorganizing A' is due to the presence of changes on A'. Notice that requests for search or change on A' must still be processed properly while we are reorganizing A'. Worst of all, unpredictable changes on A' might disrupt the part of A' that has been reorganized. In the full paper we will show how these difficulties can be resolved.

References

[1] H. Alt, K. Mehlhorn, and J.I. Munro, Partial Match Retrieval in Implicit Data Structures, *Information Processing Letters*, 19, 1984, 61-65.

[2] A. Borodin, F.E. Fich, F. Meyer Auf Der Heidej, E. Upfal, and A. Wigderson, Tradeoff Between Search and Update Time for the Implicit Dictionary Problem, *Theoretical Computer Science*, 58, 1988, 57-68.

[3] S. Carlsson, J.I. Munro, and P.V. Poblete, An Implicit Binomial Queue with Constant Insertion time, *Proceedings of the Scandinavian Workshop on Algorithm Theory*, 1988, 1-13.

[4] A. Fiat, M. Naor, A.A. Schäffer, J.P. Schmidt, and A. Siegel, Storing and Searching a Multikey Table, *Proceedings of the Twentieth Annual ACM Symposium on Theory of Computing*, 1988, 344-353.

[5] G.N. Frederickson, Implicit Data Structures for Dictionary Problem, *Journal of the ACM*, 30, 1983, 80-94.

[6] G. Gambosi, E. Nardelli, and M. Talamo, A Pointer-free Data Structure for Merging Heaps and Min-max Heaps, *Theoretical Computer Science*, 84, 1991, 107-126.

[7] J.I. Munro, An Implicit Data Structure Supporting Insertion, Deletion, and Search in $O(\log^2 n)$ Time, *Journal of Computer and System Sciences*, 33, 1986, 66-74.

[8] J.I. Munro, Searching a Two Key Table Under a Single Key, *Proceedings of the Nineteenth Annual ACM Symposium on Theory of Computing*, 1987, 383-387.

[9] J.I. Munro, A Multikey Search Problem, *Proceedings of the Allerton Conference on Communication, Control, and Computing*, 1979, 241-244.

[10] J.I. Munro and H. Suwanda, Implicit Data Structures for Fast Search and Update, *Journal of Computer and System Sciences*, 21, 1980, 236-250.

Sorting *In-Place* with a *Worst Case* Complexity of $n \log n - 1.3n + O(\log n)$ Comparisons and $\varepsilon\, n \log n + O(1)$ Transports *

Klaus Reinhardt

Institut für Informatik, Universität Stuttgart
Breitwiesenstr.22, D-7000 Stuttgart-80, Germany
e-mail: reinhard@informatik.uni-stuttgart.de

Abstract. First we present a new variant of Merge-sort, which needs only 1.25n space, because it uses space again, which becomes available within the current stage. It does not need more comparisons than classical Merge-sort.

The main result is an easy to implement method of iterating the procedure in-place starting to sort 4/5 of the elements. Hereby we can keep the additional transport costs linear and only very few comparisons get lost, so that $n \log n - 0.8n$ comparisons are needed.

We show that we can improve the number of comparisons if we sort blocks of constant length with Merge-Insertion, before starting the algorithm. Another improvement is to start the iteration with a better version, which needs only $(1 + \varepsilon)n$ space and again additional $O(n)$ transports. The result is, that we can improve this theoretically up to $n \log n - 1.3289n$ comparisons in the worst case. This is close to the theoretical lower bound of $n \log n - 1.443n$.

The total number of transports in all these versions can be reduced to $\varepsilon\, n \log n + O(1)$ for any $\varepsilon > 0$.

1 Introduction

In regard to well known sorting algorithms, there appears to be a kind of trade-off between space and time complexity: Methods for sorting like Merge-sort, Merge-Insertion or Insertion-sort, which have a small number of comparisons, either need $O(n^2)$ transports or work with a data structure, which needs $2n$ places (see [Kn72] [Me84]). Although the price for storage is decreasing, it is a desirable property for an algorithm to be *in-place*, that means to use only the storage needed for the input (except for a constant amount).

In [HL88] Huang and Langston gave an upper bound of $3n \log n$ for the number of comparisons of their *in-place* variant of merge-sort. Heap-sort needs $2n \log n$ comparisons and the upper bound for the comparisons in Bottom-up-Heapsort of $1.5n \log n$ [We90] is tight [Fl91]. Carlsson's variant of heap-sort

---

* this research has been partially supported by the EBRA working group No. 3166 ASMICS.

[Ca87] needs $n \log n + \Theta(n \log \log n)$ comparisons. The first algorithm, which is nearly in-place and has $n \log n + O(n)$ as the number of comparisons, is Mc Diarmid and Reed's variant of Bottom-up-Heap-sort. Wegener showed in [We91], that it needs $n \log n + 1.1n$ comparisons, but the algorithm is not in-place, since n additional bits are used.

For in-place sorting algorithms, we can find a trade-off between the number of comparisons and the number of transports: In [MR91] Munro and Raman describe an algorithm, which sorts in-place with only $O(n)$ transports but needs $O(n^{1+\epsilon})$ comparisons.

The way the time complexity is composed of transports and essential[2] comparisons depends on the data structure of the elements. If an element is a block of fixed size and the key field is a small part of it, then a transport can be more expensive than a comparison. On the other hand, if an element is a small set of pointers to an (eventually external) database, then a transport is much cheaper than a comparison. But in any way an $O(n \log n)$ time algorithm is asymptotically faster than the algorithm in [MR91].

In this paper we describe the first sorting algorithm, which fulfills the following properties:

- It is **general**. (Any kind of elements of an ordered set can be sorted.)
- It is **in-place**. (The used storage except the input is constant.)
- It has the **worst-case** time complexity $O(n \log n)$.
- It has $n \log n + O(n)$ as the number of (essential) comparisons in the worst case (the constant factor 1 for the term $n \log n$).

The negative linear constant in the number of comparisons and the possibility of reducing the number of transports to $\epsilon n \log n + O(1)$ for every fixed $\epsilon > 0$ are further advantages of our algorithm.

In our description some parameters of the algorithm are left open. One of them depends on the given ϵ. Other parameters influence the linear constant in the amount of comparisons. Regarding these parameters, we can again find a trade-off between the linear component in the number of comparisons and a linear amount of transports. Although a linear amount of transports does not influence the asymptotic transport cost, it is surely important for a practical application. We will see that by choosing appropriate parameters, we obtain a negative linear component in the number of comparisons as close as we want to 1.328966, but for a practical application, we should be satisfied with at most 1. The main issue of the paper is theoretical, but we expect that a good choice of the parameters leads to an efficient algorithm for practical application.

2 A Variant of Merge-sort in $1.25n$ Places

The usual way of merging is to move the elements from one array of length n to another one. We use only one array, but we add a gap of length $\frac{n}{4}$ and move

---

[2] Comparisons of pointers are not counted

the elements from one side of the gap to the other. In each step pairs of lists are merged together by repeatedly moving the bigger head element to the tail of the new list. Hereby the space of former pairs of lists in the same stage can be used for the resulting lists. As long as the lists are short, there is enough space to do this (see Figure 1). The horizontal direction in a figure shows the indices of the

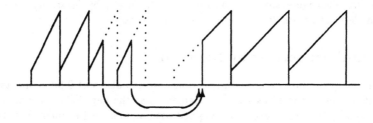

Fig. 1. Merging two short lists on the left side to one list on the right side

array and the vertical direction expresses the size of the contained elements.

In the last step it can happen, that in the worst case (for the place) there are two lists of length $\frac{n}{2}$. At some time during the last step the tail of the new list will hit the head of the second list. In this case half of the first list (of length $\frac{n}{4}$) is already merged and the other half is not yet moved. This means, that the gap begins at the point, where the left end of the resulting list will be. We can then start to merge from the other side (see Figure 2). This run ends exactly when all elements of the first list are moved into the new list.

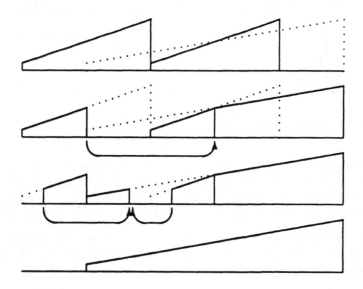

Fig. 2. The last step, merging two lists of length $\frac{n}{2}$ using a gap of length $\frac{n}{4}$

Remark. In general, it suffices to have a gap of half of the size of the shorter list (If the number of elements is not a power of 2, it may happen that the other list has up to twice this length.) or $\frac{1}{2+r}$, if we don't mind the additional transports for shifting the second list r times. The longer list has to be between the shorter list and the gap.

Remark. This can easily be performed in a stable way by regarding the element in the left list as smaller, if two elements are equal.

3 In-Place sorting by Iteration

Using the procedure described in Section 2 we can also sort $0.8n$ elements of the input in-place by treating the $0.2n$ elements like the gap. Whenever an element is moved by the procedure, it is swapped with an unsorted element from the gap similar to the method in [HL88].

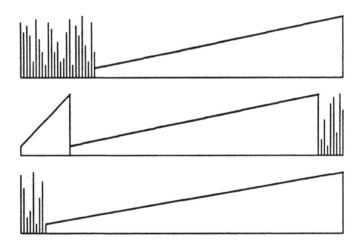

Fig. 3. Reducing the number of unsorted element to $\frac{1}{3}$

In one iteration we sort $\frac{2}{3}$ of the unsorted elements with the procedure and merge them into the sorted list, where we again use $\frac{1}{3}$ of the unsorted part of the array, which still remains unsorted, as gap (see Figure 3). This means that we merge $\frac{2}{15}n$ elements to $\frac{4}{5}n$ elements, $\frac{2}{45}n$ to $\frac{14}{15}n$, $\frac{2}{135}n$ to $\frac{44}{45}n$ and so on.

Remark. In one iteration step we must either shift the long list once, which costs additional transports, or address the storage in a cyclic way in the entire algorithm.

Remark. This method mixes the unsorted elements completely, which makes it impossible to perform the algorithm in a stable way.

3.1 Asymmetric Merging

If we want to merge a long of length l and a short list of length h together, we can compare the first element of the short list with the 2^k'th element of the long list and then we can either move 2^k elements of the long list or we need another k comparisons to insert and move the element of the short list. Hence, in order to merge the two lists one needs in the worst case $\lfloor \frac{l}{2^k} \rfloor + (k+1)h - 1$ comparisons. This was first described by [HL72].

3.2 The Number of Additional Comparisons

Let us assume, that we can sort n elements with $n \log n - dn + O(1)$ comparisons, then we get the following: Sorting $\frac{4}{5}n$ elements needs $\frac{4}{5}n(\log \frac{4}{5}n - d) + O(1)$ comparisons. Sorting $\frac{2}{5 \cdot 3^k}n$ elements for all k needs

$$\sum_{k=1}^{\log n} \left(\frac{2}{5 \cdot 3^k}n(\log(\frac{2}{5 \cdot 3^k}n) - d) + O(1) \right)$$

$$= \frac{2}{5}n \left(\sum_{k=1}^{\log n} \frac{1}{3^k}(\log \frac{2}{5}n - d) - \log 3 \sum_{k=1}^{\log n} \frac{k}{3^k} \right) + O(\log n)$$

$$= \frac{2}{5}n \left(\frac{1}{2}(\log \frac{2}{5}n - d) - \log 3\frac{3}{4} \right) + O(\log n)$$

$$= \frac{1}{5}n \left(\log n - d + \log \frac{2}{5} - \log 3\frac{3}{2} \right) + O(\log n)$$

comparisons. Together this yields $n(\log n - d + \frac{4}{5} \log \frac{4}{5} + \frac{1}{5} \log \frac{2}{5} - \frac{3}{10} \log 3) + O(\log n) = n(\log n - d - 0.9974) + O(\log n)$ comparisons. Merging $\frac{2}{5}n$ elements to $\frac{4}{5}n$ elements, $\frac{2}{5 \cdot 3^m}n$ to $(1 - \frac{1}{5 \cdot 3^{m-1}})n$ with $k = 3m + 1$ and $\frac{2}{5 \cdot 3^{m+1}}n$ to $(1 - \frac{1}{5 \cdot 3^m})n$ with $k = 3m + 3$ needs

$$n\frac{1}{5} + n\frac{3 \cdot 2}{15} +$$

$$n \sum_{m=1}^{\log n} \left(\frac{1 - \frac{1}{5 \cdot 3^{m-1}}}{2^{3m+1}} + \frac{(3m+2)2}{5 \cdot 3^{2m}} + \frac{1 - \frac{1}{5 \cdot 3^m}}{2^{3m+3}} + \frac{(3m+4)2}{5 \cdot 3^{2m+1}} \right) + O(\log n)$$

$$= n \left(\frac{3}{5} + \frac{5}{8} \sum_{m=1}^{\log n} \frac{1}{8^m} - \frac{13}{40} \sum_{m=1}^{\log n} \frac{1}{72^m} + \frac{8}{5} \sum_{m=1}^{\log n} \frac{m}{9^m} + \frac{4}{3} \sum_{m=1}^{\log n} \frac{1}{9^m} \right) + O(\log n)$$

$$= n \left(\frac{3}{5} + \frac{5}{8} \cdot \frac{1}{7} - \frac{13}{40} \cdot \frac{1}{71} + \frac{8}{5} \cdot \frac{9}{64} + \frac{4}{3} \cdot \frac{1}{8} \right) + O(\log n)$$

$$= 1.076n + O(\log n)$$

comparisons. Thus we need only $n \log n - d_{ip}n + O(\log n) = n \log n - (d - 0.0785)n + O(\log n)$ comparisons for the in-place algorithm. For the algorithm described so far $d_{ip} = 0.8349$; the following section will explain this and show how this can be improved.

4 Methods for Reducing the Number of Comparisons

The idea is, that we can sort blocks of chosen length b with c comparisons by Merge-Insertion. If the time costs for one block are $O(b^2)$, then the time we need is $O(bn) = O(n)$ since b is fixed.

4.1 Good Applications of Merge-Insertion

Merge-Insertion needs $\sum_{k=1}^{b} \lceil \log \frac{3}{4} k \rceil$ comparisons to sort b elements [Kn72]. So it works well for $b_i := \frac{4^i - 1}{3}$, where it needs $c_i := \frac{2i4^i + i}{3} - 4^i + 1$ comparisons. We prove this in the following by induction: Clearly $b_1 = 1$ element is sorted with $c_1 = 0$ comparisons. To sort $b_{i+1} = 4b_i + 1$ we need

$$
\begin{aligned}
c_{i+1} &= c_i + b_i \lceil \log \frac{3}{2} b_i \rceil + (2b_i + 1) \lceil \log \frac{3}{4} b_{i+1} \rceil \\
&= c_i + b_i \lceil \log \frac{4^i - 1}{2} \rceil + (2b_i + 1) \lceil \log \frac{4^{i+1} - 1}{4} \rceil \\
&= \frac{2i4^i + i}{3} - 4^i + 1 + \frac{4^i - 1}{3}(2i - 1) + \frac{2 \cdot 4^i + 1}{3} 2i \\
&= \frac{8i4^i + i - 4 \cdot 4^i + 4}{3} \\
&= \frac{2(i+1)4^{i+1} + i + 1}{3} - 4^{i+1} + 1.
\end{aligned}
$$

Table 1 shows some instances for b_i and c_i.

i	b_i	c_i	$d_{best,i}$	d_i	$d_{ip,i}$
1	1	0	1	0.9139	0.8349
2	5	7	1.1219	1.0358	0.9768
3	21	66	1.2971	1.2110	1.1320
4	85	429	1.3741	1.2880	1.2090
5	341	2392	1.4019	1.3158	1.2368
6	1365	12290	1.4118	1.3257	1.2467
$\vdots$	$\vdots$	$\vdots$	$\vdots$	$\vdots$	$\vdots$
∞	∞	∞	1.415037	1.328966	1.250008

Table 1. The negative linear constant depending on the block size.

4.2 The Complete Number of Comparisons

In order to merge two lists of length l and h one needs $l + h - 1$ comparisons. In the best case where $n = b2^k$ we would need $2^k c$ comparisons to sort the blocks first and then the comparisons to merge 2^{k-i} pairs of lists of the length $2^{i-1}b$ in the i'th of the k steps. These are

$$2^k c + \sum_{i=1}^{k} 2^{k-i}(2^{i-1}b + 2^{i-1}b - 1) = 2^k c + kn - 2^k + 1$$

$$= n \log n - n \underbrace{\left(\log b + \frac{1-c}{b}\right)}_{d_{best}:=} + 1$$

$$= n \log n - n d_{best} + 1$$

comparisons. In the general case the negative linear factor d is not so good as d_{best}. Let the number of elements be $n = (2^k + m)b - j$ for $j < b$ and $m < 2^k$, then we need $(2^k + m)c$ comparisons to sort the blocks, $m(2b-1) - j$ comparisons to merge m pairs of blocks together and $\sum_{i=1}^{k}(n - 2^{k-i}) = kn - 2^k + 1$ comparisons to merge everything together in k steps. The total number of comparisons is

$$c_b(n) := (2^k + m)c + m(2b - 1) - j + kn - 2^k + 1$$

$$= n \log n - n \log \frac{n}{2^k} + (2^k + m)(c + 2b - 1) - 2b2^k - j + 1$$

$$= n \log n - n \left(\log \frac{n}{2^k} + 2b\frac{2^k}{n} - \frac{c + 2b - 1}{b}\right) + \frac{c + 2b - 1}{b}j - j + 1$$

$$\leq n \log n - n \underbrace{\left(\log(2b \ln 2) + \frac{1}{\ln 2} - \frac{c + 2b - 1}{b}\right)}_{d:=} + \frac{c + b - 1}{b}j + 1$$

$$= n \log n - nd + \frac{c + b - 1}{b}j + 1.$$

The inequality follows from the fact that the expression $\log x + \frac{2b}{x}$ has its minimum for $x = 2b \ln 2$, since $(\log x + \frac{2b}{x})' = \frac{1}{x \ln 2} - \frac{2b}{x^2} = 0$. We have used $x := \frac{n}{2^k}$. Hence we loose at most $(\log(2 \ln 2) - 2 + \frac{1}{\ln 2})n = -0.086071n = (d - d_{best})n$ comparisons in contrast to the case of an ideal value of n.

Table 1 shows the influence of the block size b_i and the number of comparisons c_i to sort it by Merge-Insertion on the negative linear constant d_i for the comparisons of the algorithm in Section 2 and $d_{ip,i}$ for the algorithm in Section 3. It shows that d_{ip} can be improved as close as we want to 1.250008. As one can see looking at Table 1, most of the possible improvement is already reached with relatively small blocks.

Another idea is that we can reduce the additional comparisons in Section 3 to a very small amount, if we start using the algorithm in Section 4.3. This allows us to improve the linear constant for the in-place algorithm as close as we want to d_i and in combination with the block size we can improve it as close as we want to 1.328966.

4.3 A variant of Merge-sort in $(1+\varepsilon)n$ places

We can change the algorithm of Section 2 in the following way: For a given $\varepsilon > 0$ we have to choose appropriate r's for the last $\lceil \log \frac{1}{\varepsilon} \rceil - 2$ steps according to the first remark in that section. Because the number of the last steps and the r's are constants determined by ε, the additional transports are $O(n)$ (of course the constant becomes large for small ε's).

5 The Reduction of the Number of Transports

The algorithms in Section 2 and Section 4.3 perform $n \log n + O(n)$ transports. We can improve this to $\varepsilon n \log n + O(1)$ for all constants $\varepsilon > 0$, if we combine $\lceil \frac{1}{\varepsilon} + 1 \rceil$ steps to 1 by merging $2^{\lceil \frac{1}{\varepsilon} + 1 \rceil}$ (constantly many) lists in each step, as long as the lists are short enough. Hereby we keep the number of comparisons exactly the same using for example the following technique: We use a binary tree of that (constant) size, which contains on every leaf node the pointer to the head of a list ('nil' if it is empty, and additionally a pointer to the tail) and on every other node that pointer of a son node, which points to the bigger element. After moving one element we need $\lceil \frac{1}{\varepsilon} + 1 \rceil$ comparisons to repair the tree as shown in this example:

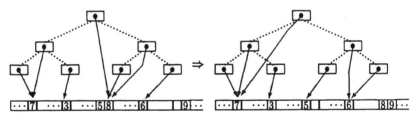

Note that if we want to reduce the number of transports to $o(n \log n)$, then the size of the tree must increase depending on n, but then the algorithm would not be in-place.

Remark. This method can also be used to reduce the total number of I/Os (for transports and comparisons) to a slow storage to $\varepsilon n \log n + O(1)$, if we can keep the $2^{\lceil \frac{1}{\varepsilon} + 1 \rceil}$ elements in a faster storage.

5.1 Transport Costs in the In-Place Procedure

Choosing ε of the half size eliminates the factor 2 for swapping instead of moving. But there are still the additional transports in the iteration during the asymmetric merging in Section 3:

Since $\approx \log_3 n$ iterations have to be performed, we would get $O(n \log n)$ additional transports, if we perform only iterations of the described kind, which need $O(n)$ transports each time. But we can reduce these additional transports to $O(n)$, if we perform a second kind of iteration after each i iterations for any chosen i. Each time it reduces the number of elements to the half, which have to be moved in later iterations. This second kind of iteration works as follows (see Figure 4):

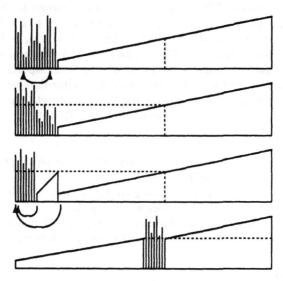

Fig. 4. Moving half of the elements to their final position.

One Quick-Sort iteration step is performed on the unsorted elements, where the middle element of the sorted list is chosen as reference. Then one of the lists is sorted in the usual way (the other unsorted list is used as a gap) and merged with the corresponding half of the sorted list. These elements will never have to be moved again.

Remark. It is easy to see that we can reduce the number of additional comparisons to εn, if we choose i big enough. Although a formal estimation is quite hard, we conjecture, that we need only a few comparisons more, if we mainly apply iterations of the second kind, even in the worst case, where all unsorted elements are on one side. Clearly this iteration is good in the average.

6 Possible Further Improvements

The algorithm in Section 2 works optimal for the case that $n = b2^m$ and has its worst behavior (loosing $-0.086n$) in the case that $n = 2 \ln 2 b2^m = 1.386b2^m$. But we could avoid this for the start of the iteration, if we are not fixed to sort 0.8

of the elements in the first step but exactly $b2^m$ elements for $\frac{2}{5} < b2^m \leq \frac{4}{5}$. A simpler method would be to sort always the biggest possible $b2^m$ in each iteration step. In both cases the k's for the asymmetric merging in the further iterations have to be found dynamically, which makes a proper formal estimation difficult.

For practical applications it is worth to note, that the average number of comparisons for asymmetric merging is better than in the worst case, because for each element of the smaller list, which is inserted in $2^k - 1$ elements, on the average some elements of the longer list also can be moved to the new list, so that the average number of comparisons is less than $\lfloor \frac{1}{2^k} \rfloor + (k+1)h - 1$. So we conjecture, that an optimized in-place algorithm can have $n \log n - 1.4n + O(\log n)$ comparisons in the average and $n \log n - 1.3n + O(\log n)$ comparisons in the worst case avoiding large constants for the linear amount of transports.

If there exists a constant $c > 0$ and an in-place algorithm, which sorts cn of the input with $O(n \log n)$ comparisons and $O(n)$ transports, then the iteration method could be used to solve the open problem in [MR91].

Open Problems:

- Does there exists an in-place sorting algorithm with $O(n \log n)$ comparisons and $O(n)$ transports [MR91]?
- Does there exists an in-place sorting algorithm, which needs only $n \log n + O(n)$ comparisons and only $o(n \log n)$ transports?
- Does there exists a *stable* in-place sorting algorithm, which needs only $n \log n + O(n)$ comparisons and only $O(n \log n)$ transports?

References

[Ca87] S. Carlsson: A variant of HEAPSORT with almost optimal number of comparisons. Information Processing Letters 24:247-250,1987.

[Fl91] R. Fleischer: A tight lower bound for the worst case of bottom-up-heapsort. Technical Report MPI-I-91-104, Max-Planck-Institut für Informatik, D-W-6600 Saarbrücken, Germany April 1991.

[HL88] B-C. Huang and M.A. Langston: Practical in-place merging. CACM, 31:248-352,1988.

[HL72] F. K. Hwang and S. Lin: A simple algorithm for merging two disjoint linearly ordered sets. SIAM J. Computing 1:31-39, 1972.

[Kn72] D. E. Knuth: The Art of Computer Programming Volume 3 / Sorting and Searching. Addison-Wesley 1972.

[Me84] K. Mehlhorn: Data Structures and Algorithms, Vol 1: Sorting and Searching. Springer-Verlag, Berlin/Heidelberg, 1984.

[MR91] J. I. Munro and V. Raman: Fast Stable In-Place Sorting with $O(N)$ Data Moves. Proceedings of the FST&TCS, LNCS 560:266-277, 1991.

[We90] I. Wegener: BOTTOM-UP-HEAPSORT, a new variant of HEAPSORT beating on average QUICKSORT (if n is not very small). Proceedings of the MFCS90, LNCS 452, 516-522, 1990.

[We91] I. Wegener: The worst case complexity of Mc Diarmid and Reed's variant of BOTTOM-UP-HEAP SORT is less than $n \log n + 1.1n$. Proceedings of the STACS91, LNCS 480:137-147, 1991.

Sorting and/by Merging Finger Trees

Alistair Moffat[1], Ola Petersson[2], and Nicholas C. Wormald[3]

[1] Department of Computer Science,
The University of Melbourne, Australia 3052.
alistair@cs.mu.oz.au
[2] Department of Computer Science,
Lund University, Box 118, Lund S-221 00, Sweden.
ola@dna.lth.se
[3] Department of Mathematics,
The University of Melbourne, Australia 3052.
nick@maths.mu.oz.au

Abstract. We describe a sorting algorithm that is optimally adaptive with respect to several important measures of presortedness. In particular, the algorithm requires $O(n \log(k/n))$ time on sequences with k inversions; $O(n + k \log k)$ time on sequences X that have a longest ascending subsequence of length $n - k$ and for which $Rem(X) = k$; and $O(n \log k)$ time on sequences that can be decomposed into k monotone shuffles. The algorithm makes use of an adaptive merging operation implemented using finger search trees.

1 Introduction

An *adaptive* algorithm is one which requires fewer resources to solve 'easy' problem instances than it does to solve 'hard'. For sorting, an adaptive algorithm should run in $O(n)$ time if presented with a sorted n-sequence, and in $O(n \log n)$ time for all n-sequences, with the time for any particular sequence depending upon the 'nearness' of the sequence to being sorted. Mannila [8] established the notion of a *measure of presortedness* to quantify the disorder in any input sequence, and introduced the concept of *optimal adaptivity*. For example, if presortedness is measured by counting the number of pairwise inversions in the input n-sequence X, denoted $Inv(X)$, then an *Inv*-optimal algorithm must sort X in* $O(n \log(Inv(X)/n))$ time [8]. Several authors have introduced other measures of presortedness, including Altman and Igarashi [1]; Carlsson, Levcopoulos and Petersson [3]; Estivill-Castro and Wood [4]; Levcopoulos and Petersson [6, 7]; Petersson and Moffat [13]; and Skiena [15].

In this paper we consider an adaptive sorting algorithm first described by Moffat [11] and further analysed by Moffat, Petersson and Wormald [12]. The algorithm is a variant of the standard textbook Mergesort, but is implemented using finger search trees rather than arrays, and makes use of an adaptive merging operation on these trees. The new algorithm has already been shown to be optimally adaptive (in terms of comparisons performed) with respect to the measures *Runs*, the number of ascending runs in the input sequence; *Rem*, the minimum number of items that must be removed from X to leave a sorted subsequence; and *Inv*, the number of

---

* We assume that $\log x$ is defined to be $\log_2(\max\{2, x\})$.

inversions in the input sequence [11, 12]. Here we review the algorithm and those results, providing an analysis of the tree merging process that was sketched in [11], and showing that the running time of the algorithm is linear in the number of comparisons and thus that the algorithm is optimally adaptive for these measures in terms of execution time as well. We then provide an analysis to show that the algorithm is also optimal with respect to the measure *SMS*, the minimum value k such that the input sequence X can be composed as a shuffle of k monotone sequences. The algorithm is the first known algorithm that is optimally adaptive with respect to this combination of measures.

In Section 2 we consider the problem of merging two ordered lists, where each list is stored using a finger search tree. Section 3 describes the sorting algorithm that results when this data structure is used with standard Mergesort, and summarises the previous optimality results for the measures *Runs*, *Inv*, and *Rem*. Then in Section 4 we consider the measure *SMS* and prove the new optimality result mentioned above. Finally, Section 5 briefly discusses a number of related problems.

2   Merging Finger Trees

Mannila [8] and Mehlhorn [10] both used the locality properties of finger search trees to obtain adaptive insertion sorts. Here we consider other properties of finger search trees. For definiteness, our discussion will assume that level linked 2-3 trees are being used [2], but our results extend to other implementations of finger search trees. Indeed, our investigation was prompted by the merging algorithm given by Pugh [14] for Skip Lists, a probabilistic finger tree.

We suppose that X and Y are sorted sequences of n and m items respectively, stored as level linked 2-3 trees, where, without loss of generality, it is assumed that $m \leq n$. We wish to *meld* or *merge* these two trees, generating an output tree Z of $n + m$ items. The obvious technique of repeated insertion of items from Y (the smaller tree) into X (the larger) requires $\Theta(m \log n)$ time if each search starts at the root of X, or $O(m \log(n/m))$ time if the items are inserted in ascending or descending order and the search for each insertion point commences from the most recent insertion point (i.e., the current node is fingered) [2]. In a worst case sense this latter algorithm is asymptotically optimal, since $\lceil \log_2 \binom{n+m}{m} \rceil = \Omega(m \log(n/m))$.

However we can also consider instance easiness in the merging problem. For example, if X and Y are *disjoint*—the largest item in X is no greater than the smallest item in Y, denoted $X \leq Y$—then $O(\log n)$ time suffices to perform a simple concatenation and 'merge' the two trees. In the development below we first consider the complexity of concatenating a set of b disjoint 2-3 trees, and then consider the cost of partitioning two input trees X and Y into disjoint components that can be concatenated to form their merged output.

Theorem 1. *Suppose that Z_i, $1 \leq i \leq b$, is a forest of disjoint level-linked 2-3 trees, $Z_1 \leq Z_2 \leq \ldots \leq Z_b$. Then the b trees can be concatenated to make a single 2-3 tree Z in $O(\sum_{i=1}^{b} \log |Z_i|)$ time.*

Proof. (Sketch) The sequence of operations we consider is given by

$$Z := \emptyset$$
for $i := 1$ **to** b **do**
$\quad\quad Z := Join(Z, Z_i)$
endfor

where to *Join* two trees Z and Z_i we begin at the rightmost node of Z and the leftmost node of Z_i and 'zip' the two trees together, setting the level links, until we either come to the root of Z or the root of Z_i. In either case the time required during this 'zipping' stage is at most proportional to the depth of Z_i, which is $O(\log |Z_i|)$.

We then insert the root of Z_i as an additional rightmost child of the corresponding node on the right branch of Z, or, if it was the root of Z that was reached first, we insert Z as a leftmost child of the appropriate node on the left branch of Z_i. If this causes that node to have more than three children, we split the node and add a new child to the parent of that node, and so on up either the right branch of Z or the left branch of Z_i. If the root is split we add a new root, and Z grows by one level.

The cost of all of the 'zipping' operations is $O(\sum_{i=1}^{b} \log |Z_i|)$. We must also account for the node splittings that take place during the b *Joins*. The splitting of a node that became a 3-node as a result of a previous *Join* can be charged as an $O(1)$ overhead to that *Join*, since each *Join* creates at most one 3-node. Pre-existing 3-nodes that cannot be charged in this way might also be split, but any such nodes must lie on either a left branch or a right branch of one of the original input trees, and since the total number of such nodes is $O(\sum_{i=1}^{b} \log |Z_i|)$, we have, summed over the sequence of operations, that the cost of the node splittings does not dominate the cost of zipping the trees together. □

In the case when $b = 2$ this gives the familiar result that a *Join* of two 2-3 trees can be performed in $O(\log n + \log m) = O(\log n)$ time [10].

Splitting an input tree into b components is no more expensive:

Theorem 2. *Suppose that X is a 2-3 tree, and is to be partitioned into b disjoint 2-3 trees X_i, where sequence y_i determines the partition. That is, the items in X_i are to have values that are greater than y_{i-1} and less than or equal to y_i, with $y_0 = -\infty$ and $y_b = +\infty$. Then the partitioning can be performed in $O(\sum_{i=1}^{b} \log |X_i|)$ time.*

Proof. (Sketch) The sequence of operations we consider is:

for $i := 1$ **to** b **do**
$\quad\quad X_i := Split(X, y_i)$
endfor

where it is assumed that $Split(X, y_i)$ prunes from X all items less than or equal to y_i and returns them as a separate 2-3 tree, leaving X as a depleted 2-3 tree.

To locate the last node to be pruned we perform a finger search for y_i from the leftmost node of (what remains of) X. This will take $O(\log |X_i|)$ time. Then, to actually perform the *Split*, we start at the parent node of the rightmost item to be included in X_i and work our way up the tree, duplicating every ancestor until we reach either the left branch of X or the right branch of X. One of each of these pairs of nodes, together with all of the items to the left of the duplicated node, will eventually form the tree X_i. The other node of the pair, and items to the right of it, will remain in X. At each level the children of each duplicated node are partitioned between the original node and the duplicate node according to their range of key values, and the level links adjusted accordingly.

This will result in two trees, X_i and X, that are 2–3 trees *except* there may be violations down the right branch of X_i and down the left branch of X. For each of the at most $2\log_2 |X_i|$ nodes thus affected we perform a distribution step, starting from the parent nodes of the leaves of the tree:

- if the node (now) has zero children, or if it is the root node of either X_i or X and has only one child, it is deleted;
- if the node has one child, and the node immediately to the left (if we are in X_i) or right (if we are in X) has three children, a child is adopted from the sibling;
- if the node has one child, and the node immediately to the left (if we are in X_i) or right (in X) has only two children, the child is moved to the sibling, and the node is deleted;
- if the node has two or three children, no action is required.

This sequence of steps will take $O(1)$ time at each node along the cutting path from the leaf level up to either the left or right branch of X. The length of that cutting path is $O(\log|X_i|)$, and so the 'unzipping' phase will cost $O(\log|X_i|)$ time.

Then, at the node on the left or right branch that is at the top of the cutting path, a child must be removed to form the root of the new tree X_i (if we are on the left branch) or the root of the remainder of the tree X (if we are on the right branch). We call this operation a *branch deletion*.

The removal of this child during a branch deletion might cause a violation of the 'at least two children' rule, and a sequence of cascading node fusings that reaches right up to the root of the larger tree. The total number of node fusings over all *Split* operations is, however, bounded by the sum of the number of branch deletions plus the total length of the right and left branches of the trees in the final forest, and so the result follows. □

Figure 1 describes the new adaptive merging algorithm. It is similar to the simple linear merge algorithm, but in effect uses exponential and binary search rather than linear search to break the two input sequences into the minimum number of *blocks* that must be rearranged to produce the output sequence. In Figure 1, x_1 and y_1, the sequences of splitting elements, are always the smallest remaining items in the (shrinking) trees X and Y; and, for descriptive purposes, it is assumed that references to the first item of an empty tree will return the value $+\infty$.

If 2–3 trees are used as the finger search tree then *Joins* and *Splits* on the same tree cannot be interleaved without affecting the analysis, since the amortised bounds on the numbers of node splittings and fusings rely on there being no other intervening operations [10]. However the sequences of operations required by the algorithm described in Figure 1 meet this requirement. Theorems 1 and 2 then provide the following bound on the cost of merging two level linked 2–3 trees:

```
procedure Merge(X:list, Y:list):list
    Z := ∅; i := 0
    while X ≠ ∅ or Y ≠ ∅ do
        i := i + 1
        Xᵢ := Split(X, y₁)
        Z := Join(Z, Xᵢ)
        Yᵢ := Split(Y, x₁)
        Z := Join(Z, Yᵢ)
    endwhile
    return Z
end
```

Fig. 1. Tree-based merging algorithm

Theorem 3. *Suppose that two sequences X and Y of n and m items, $m \le n$, are represented as level-linked 2-3 trees. Suppose further that the result of merging X and Y is the tree Z,*

$$Z = X_1 Y_1 X_2 Y_2 \ldots X_{b-1} Y_{b-1} X_b Y_b,$$

where $X = X_1 X_2 \ldots X_b$, $Y = Y_1 Y_2 \ldots Y_b$, and X_1 and/or Y_b might be empty. Then tree Z can be constructed from X and Y in

$$O\left(\sum_{i=1}^{b} \log |X_i| + \sum_{i=1}^{b} \log |Y_i|\right) = O\left(b \log \frac{n}{b}\right) = O\left(m \log \frac{n}{m}\right)$$

time in the worst case.

The merge is thus worst-case optimal, but also better in some situations. For example, if $b = O(1)$ and the merge requires that only a constant number of blocks be formed, then the cost of the merge is $O(\log n)$. Counting the number of blocks formed—$Block(X, Y)$—is one possible *measure of premergedness* [3]. Another is to count the number of inversions between the two sequences, $Inv(X, Y) = \min\{Inv(XY), Inv(YX)\}$ [9]. The merging algorithm is adaptive with respect to this measure too:

Lemma 4. [12] *The merge of X and Y will consume $O(\log n + \sqrt{Inv(X, Y)})$ time.*

Another measure of premergedness is *Max* [3], the total number of items by which the two sequences overlap. It is not difficult to see that our merging algorithm runs in $O(\log n + Max(X, Y))$ time, and so is adaptive with respect to *Max* too.

Finally, it is worth noting that if we adopt the 'double-ended' searching technique employed by Carlsson, Levcopoulos, and Petersson [3] in their adaptive merging algorithm, the number of *comparisons* consumed can be reduced to

$$O\left(\min\left\{Block(X, Y) \log \frac{n}{Block(X, Y)}, \ \sqrt{Inv(X, Y)}, \ Max(X, Y)\right\}\right),$$

removing the $\log n$ time term caused by the two 'longest' blocks X_1 and Y_b.

3 Sorting by Merging Finger Trees

Our adaptive Mergesort is now obvious. The items in the input sequence X are inserted into n singleton 2-3 trees in $\Theta(n)$ time. The trees are then pairwise merged in the order they appeared in the input, building trees of size 2, then 4, then 8, and so on. After $\lceil \log_2 n \rceil$ passes we will have a single tree remaining. It then takes $\Theta(n)$ time to traverse that tree and copy the items to the desired output location. To distinguish between our adaptive Mergesort and the standard Mergesort (and other adaptive Mergesorts [3, 5]) we will refer to the new algorithm as *Margesort*.

In [11] it was shown that Margesort is optimally adaptive (in terms of comparisons) with respect to *Runs*, consuming $O(n \log k)$ comparisons to sort a sequence with $k = Runs(X)$ ascending runs**. The analysis was extended in [12], where we were able to show that Margesort is also optimal (again, counting comparisons only) with respect to the measures *Rem* and *Inv*. Those results, combined with Theorem 3, mean that we have now shown:

Theorem 5. *Margesort is Runs-, Rem-, and Inv-optimal. That is, the running time of Margesort on n-sequence X is given by*

$$O\left(\min\left\{\begin{array}{l} n \log Runs(X) \\ n + Rem(X) \log Rem(X) \\ n \log(Inv(X)/n) \end{array}\right\}\right).$$

Theorem 5, taken in conjunction with the results of Petersson and Moffat [13], is sufficient to prove that Margesort is also optimally adaptive with respect to several other commonly discussed measures of presortedness.

4 *SMS*-Optimality

We now consider the adaptivity of Margesort with respect to *SMS*:

$$SMS(X) = \min\{k \mid X \text{ can be partitioned into } k \text{ monotone sequences}\}.$$

Our main result in this section is that Margesort is *SMS*-optimal (and hence *Enc*-optimal and *SUS*-optimal [7, 13, 15]), running in $O(n \log k)$ time on a sequence that is the shuffle of k monotone sequences. The only other known *SMS*-optimal algorithm is the *Slabsort* of Levcopoulos and Petersson [7], and that algorithm is not optimal with respect to *Inv*. Margesort is thus optimally adaptive with respect to a unique combination of measures.

The analysis will be in an amortised sense [16], making use of a potential function Φ to measure the amount of 'indebtedness' currently in the data structure. Following Tarjan, we define the amortised time a_i of the i'th operation to be $a_i = t_i + \Phi_i - \Phi_{i-1}$ where t_i is the actual time of the i'th operation and Φ_i is the value of the potential

---

** In fact a very simple addition to the standard linear merge is sufficient to achieve this behaviour. All that is necessary is to compare the last item of X with the first of Y at each call to the merging procedure, and perform a (linear) merge only if the two sequences are non-disjoint [11].

function after the i'th operation has taken place. We let Φ_0 denote the potential before any of the operations have taken place.

The worst-case time T required by a sequence of n operations is then given by $T = \sum_{i=1}^{n} t_i = (\sum_{i=1}^{n} a_i) + \Phi_0 - \Phi_n$, that is, the sum of the amortised cost of the operations plus the net decrease in potential over the whole sequence.

Suppose that the decomposition of an input sequence X into k monotone shuffles is possible. Let $S_1, \ldots, S_k$ be any fixed decomposition of size k, such that item x_i is a member of shuffle S_{s_i}, $1 \leq s_i \leq k$.

We will say that x_i is *guarded* by shuffle S_j if there are elements x_l and x_g such that $s_l = s_g = j$; $x_l \leq x_i \leq x_g$; and x_l, x_i, and x_g are all currently contained in the same tree. That is, x_i is guarded by a shuffle when there are both smaller and larger items from that shuffle in the same tree as x_i. Initially, when x_i is the only item in its tree, the only shuffle that it is guarded by is its own, S_{s_i}.

Define the *guardedness* g_i of any item x_i to be the number of shuffles S_j that guard x_i. At the beginning of the sorting $g_i = 1$ for all i; during the mergings affecting x_i, g_i is non-decreasing; and at the end of the sorting $1 \leq g_i \leq k$ for all i. Finally, we take as our potential function

$$\Phi = \sum_i -c \log_e g_i,$$

where c is a constant that will become apparent below.

Consider one merge, in which two trees each of n items are merged to make a single tree of $2n$ items. For the purposes of the analysis we suppose that the least and greatest elements within each shuffle are identified for each input tree. That is, the (at most) $4k$ shuffle extrema are noted. Loosely, we define a *slab* to be the items in the input sequences whose values are between an adjacent pair of these shuffle extrema, including either or both of the extremal items if they belong to shuffles that guard the slab. More precisely, if x_a and x_b are shuffle extrema and there is no shuffle extremum x_c such that $x_a < x_c < x_b$, then the slab defined by (x_a, x_b) contains all items x_i from the two inputs to the merge such that $x_a < x_i < x_b$, plus x_a if it is in a shuffle guarding x_b, plus x_b if it is in a shuffle guarding x_a. Thus, if any item is the only item from its shuffle in the merge, it forms a slab all by itself.

Without explicitly identifying the slabs, the main merge can be thought of as first merging the two 'sides' in the first slab; then merging the two sides in the second slab; and so on until the left and right sides of the final slab have been merged.

Let us now consider one of these 'slab merges'. In Figure 2 we sketch a possible configuration, with items represented by circles, shuffles by looped lines, and a shuffle that continues from the left side to the right side shown by a dotted line.

We suppose that there are n' items in total within the slab, that r of them are on the right side, and, without loss of generality, that $r \leq n' - r$. By the definition of a slab, all items on the right side have the same guardedness, j say. Similarly, all items on the left side have the same guardedness which will be $k' - j$, where k' is the total number of shuffles guarding items in this slab merge. After the merge all n' items in the slab will have $g_i = k'$, since all items will be guarded by all shuffles guarding any items in the slab. No items can escape this remorseless increase in guardedness.

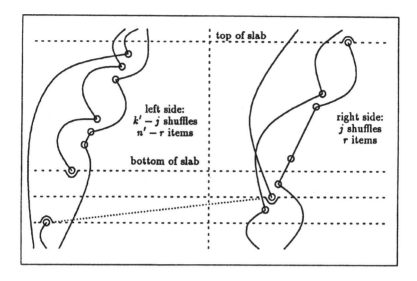

Fig. 2. Decomposition into slabs

By Theorem 3 the actual cost t_i of the slab merge is bounded by

$$t_i \leq c + c(r+1)\log_e \frac{n'}{r+1},$$

for some constant c. The amortised cost of a slab merge is then given by

$$a_i = t_i + \Phi_i - \Phi_{i-1}$$
$$\leq c + c(r+1)\log_e \frac{n'}{r+1} - cn'\log_e k' + c(n'-r)\log_e(k'-j) + cr\log_e j.$$

Considered as a function of j, the first derivative is

$$-c(n'-r)\frac{1}{k'-j} + cr\frac{1}{j},$$

which is zero when $j = rk'/n'$. Moreover, the second derivative is always negative, and so a global maximum has been identified. Substituting for j gives

$$\frac{a_i}{c} \leq 1 + (r+1)\log_e \frac{n'}{r+1} - n'\log_e k' + (n'-r)\log_e(k' - \frac{rk'}{n'}) + r\log_e \frac{rk'}{n'}$$
$$= 1 + (r+1)\log_e \frac{n'}{r+1} - n'\log_e k' + (n'-r)\log_e \frac{(n'-r)k'}{n'} + r\log_e \frac{rk'}{n'}$$
$$= 1 + r\log_e n' + \log_e n' - r\log_e(r+1) - \log_e(r+1)$$
$$\quad + (n'-r)\log_e \frac{n'-r}{n'} + r\log_e r - r\log_e n'$$
$$\leq 1 + \log_e n' - \log_e(r+1) + (n'-r)\log_e \frac{n'-r}{n'}$$
$$\leq 1 + \log_e n',$$

with the last inequality following from the observation that $(n'-r)/n' \leq 1$. That is, the amortised cost of each slab merge is logarithmic in the number of items contained in the slab.

Summed over the at most $4k$ slab merges that comprise an actual merge, the amortised cost of merging two n item lists is at most $4ck(1 + \log_e(n/4k))$, and the amortised time for the entire sort is then bounded by the recurrence

$$A(n) = \begin{cases} n \log k & n \leq 4k \\ 2 \cdot A(n/2) + 4ck(1 + \log_e(n/4k)) & n > 4k \end{cases}.$$

The actual time required by the sort (after $n-1$ merges) is given by

$$T(n) = A(n) + \Phi_0 - \Phi_{n-1},$$

and since $A(n) = O(n \log k)$, $\Phi_0 = 0$, and $\Phi_{n-1} \geq -cn \log_e k$, we have

$$T(n) = O(n \log k).$$

That is, we have proved:

Theorem 6. *Margesort is optimally adaptive with respect to SMS.*

5 Open Problems

In [12] we gave a sequence to show that Margesort is *not* optimally adaptive with respect to the measure *Block*, the minimum number of blocks of adjacent elements that the input sequence must be cut into to effect the sort [3]. (The adaptive mergesort of Carlsson, Levcopoulos and Petersson *is Block*-optimal [3]). The non-optimality of Margesort with respect to *Block* means that Margesort cannot be optimally adaptive with respect to any of the measures *Loc*, *Hist*, and *Reg* [13]. Thus the only previously defined measure of presortedness for which the optimality (or otherwise) of Margesort is yet to be decided is *Osc* [6].

Our second open problem concerns the bounds for adaptive merging. It is not difficult to show that there are at least $2^{\sqrt{I}}$ possible mergings of sequences X and Y for which $Inv(X,Y) \leq I$, and thus that, if not for the additive $\log n$ term, our merging algorithm would be 'optimally' adaptive. Our question then is this: is there a data structure for representing sorted lists that allows sequences X and Y to be merged in $O(\sqrt{Inv(X,Y)})$ time? Similarly for *Block*: we wonder if there is some data structure that allows merging in $O(b \log(n/b))$ time, where $b = Block(X,Y)$. For *Max* the answer to the corresponding question is 'yes'—doubly linked sorted lists with both head and tail pointers will allow merging in $O(Max(X,Y))$ time.

Finally, we ask if there is some other 'natural' measures of premergedness to which our merging algorithm fails to adapt. If there is, and there is some adaptive merging algorithm that exploits the measure, we wonder as to the extent of adaptivity of the Mergesort that results from the use of that method.

Acknowledgements

Thanks are due to Gary Eddy, who undertook an implementation and encouraged us with graphs of running times. This work was in part supported by the Australian Research Council.

References

1. T. Altman and Y. Igarashi. Roughly sorting: sequential and parallel approach. *Journal of Information Processing*, 12(2):154–158, 1989.
2. M.R. Brown and R.E. Tarjan. Design and analysis of a data structure for representing sorted lists. *SIAM Journal on Computing*, 9:594–614, 1980.
3. S. Carlsson, C. Levcopoulos, and O. Petersson. Sublinear merging and Natural Mergesort. *Algorithmica*. To appear. Prel. version in *Proc. Internat. Symp. on Algorithms*, pages 251–260, LNCS 450, Springer-Verlag, 1990.
4. V. Estivill-Castro and D. Wood. A new measure of presortedness. *Information and Computation*, 83(1):111–119, 1989.
5. V. Estivill-Castro and D. Wood. A generic adaptive sorting algorithm. Research Report CS-90-31, Department of Computer Science, University of Waterloo, Waterloo, Canada, August 1990.
6. C. Levcopoulos and O. Petersson. Adaptive Heapsort. *Journal of Algorithms*. To appear. Prel. version in *Proc. 1989 Workshop on Algorithms and Data Structures*, pages 499–509, LNCS 382, Springer-Verlag, 1989.
7. C. Levcopoulos and O. Petersson. Sorting shuffled monotone sequences. *Information and Computation*. To appear. Prel. version in *Proc. 2nd Scandinavian Workshop on Algorithm Theory*, pages 181–191, LNCS 447, Springer-Verlag, 1990.
8. H. Mannila. Measures of presortedness and optimal sorting algorithms. *IEEE Transactions on Computers*, C-34(4):318–325, 1985.
9. H. Mannila and E. Ukkonen. A simple linear-time algorithm for in situ merging. *Information Processing Letters*, 18(4):203–208, 1984.
10. K. Mehlhorn. *Data Structures and Algorithms, Vol. 1: Sorting and Searching*. Springer-Verlag, Berlin, Germany, 1984.
11. A. Moffat. Adaptive merging and a naturally Natural Mergesort. In *Proc. 14'th Australian Computer Science Conference*, pages 08.1–08.8. University of New South Wales, February 1991.
12. A. Moffat, O. Petersson, and N.C. Wormald. Further analysis of an adaptive sorting algorithm. In *Proc. 15'th Australian Computer Science Conference*, pages 603–613. University of Tasmania, January 1992.
13. O. Petersson and A. Moffat. A framework for adaptive sorting. In *Proc. 3'rd Scandinavian Workshop on Algorithm Theory*, pages 422–433. LNCS 621, Springer-Verlag, July 1992.
14. W. Pugh. Skip lists: a probabilistic alternative to balanced trees. *Communications of the ACM*, 33(6):668–676, 1990.
15. S.S. Skiena. Encroaching lists as a measure of presortedness. *BIT*, 28(4):775–784, 1988.
16. R.E. Tarjan. Amortised computational complexity. *SIAM Journal on Applied and Discrete Methods*, 6:306–318, 1985.

Author Index

Lecture Notes in Computer Science

For information about Vols. 1–570
please contact your bookseller or Springer-Verlag

Vol. 610: F. von Martial, Coordinating Plans of Autonomous Agents. XII, 246 pages. 1992. (Subseries LNAI).

Vol. 611: M. P. Papazoglou, J. Zeleznikow (Eds.), The Next Generation of Information Systems: From Data to Knowledge. VIII, 310 pages. 1992. (Subseries LNAI).

Vol. 612: M. Tokoro, O. Nierstrasz, P. Wegnèr (Eds.), Object-Based Concurrent Computing. Proceedings, 1991. X, 265 pages. 1992.

Vol. 613: J. P. Myers, Jr., M. J. O'Donnell (Eds.), Constructivity in Computer Science. Proceedings, 1991. X, 247 pages. 1992.

Vol. 614: R. G. Herrtwich (Ed.), Network and Operating System Support for Digital Audio and Video. Proceedings, 1991. XII, 403 pages. 1992.

Vol. 615: O. Lehrmann Madsen (Ed.), ECOOP '92. European Conference on Object Oriented Programming. Proceedings. X, 426 pages. 1992.

Vol. 616: K. Jensen (Ed.), Application and Theory of Petri Nets 1992. Proceedings, 1992. VIII, 398 pages. 1992.

Vol. 617: V. Mařík, O. Štěpánková, R. Trappl (Eds.), Advanced Topics in Artificial Intelligence. Proceedings, 1992. IX, 484 pages. 1992. (Subseries LNAI).

Vol. 618: P. M. D. Gray, R. J. Lucas (Eds.), Advanced Database Systems. Proceedings, 1992. X, 260 pages. 1992.

Vol. 619: D. Pearce, H. Wansing (Eds.), Nonclassical Logics and Information Proceedings. Proceedings, 1990. VII, 171 pages. 1992. (Subseries LNAI).

Vol. 620: A. Nerode, M. Taitslin (Eds.), Logical Foundations of Computer Science - Tver '92. Proceedings. IX, 514 pages. 1992.

Vol. 621: O. Nurmi, E. Ukkonen (Eds.), Algorithm Theory - SWAT '92. Proceedings. VIII, 434 pages. 1992.

Vol. 622: F. Schmalhofer, G. Strube, Th. Wetter (Eds.), Contemporary Knowledge Engineering and Cognition. Proceedings, 1991. XII, 258 pages. 1992. (Subseries LNAI).

Vol. 623: W. Kuich (Ed.), Automata, Languages and Programming. Proceedings, 1992. XII, 721 pages. 1992.

Vol. 624: A. Voronkov (Ed.), Logic Programming and Automated Reasoning. Proceedings, 1992. XIV, 509 pages. 1992. (Subseries LNAI).

Vol. 625: W. Vogler, Modular Construction and Partial Order Semantics of Petri Nets. IX, 252 pages. 1992.

Vol. 626: E. Börger, G. Jäger, H. Kleine Büning, M. M . Richter (Eds.), Computer Science Logic. Proceedings, 1991. VIII, 428 pages. 1992.

Vol. 628: G. Vosselman, Relational Matching. IX, 190 pages. 1992.

Vol. 629: I. M. Havel, V. Koubek (Eds.), Mathematical Foundations of Computer Science 1992. Proceedings. IX, 521 pages. 1992.

Vol. 630: W. R. Cleaveland (Ed.), CONCUR '92. Proceedings. X, 580 pages. 1992.

Vol. 631: M. Bruynooghe, M. Wirsing (Eds.), Programming Language Implementation and Logic Programming. Proceedings, 1992. XI, 492 pages. 1992.

Vol. 632: H. Kirchner, G. Levi (Eds.), Algebraic and Logic Programming. Proceedings, 1992. IX, 457 pages. 1992.

Vol. 633: D. Pearce, G. Wagner (Eds.), Logics in AI. Proceedings. VIII, 410 pages. 1992. (Subseries LNAI).

Vol. 634: L. Bougé, M. Cosnard, Y. Robert, D. Trystram (Eds.), Parallel Processing: CONPAR 92 - VAPP V. Proceedings. XVII, 853 pages. 1992.

Vol. 635: J. C. Derniame (Ed.), Software Process Technology. Proceedings, 1992. VIII, 253 pages. 1992.

Vol. 636: G. Comyn, N. E. Fuchs, M. J. Ratcliffe (Eds.), Logic Programming in Action. Proceedings, 1992. X, 324 pages. 1992. (Subseries LNAI).

Vol. 637: Y. Bekkers, J. Cohen (Eds.), Memory Management. Proceedings, 1992. XI, 525 pages. 1992.

Vol. 639: A. U. Frank, I. Campari, U. Formentini (Eds.), Theories and Methods of Spatio-Temporal Reasoning in Geographic Space. Proceedings, 1992. XI, 431 pages. 1992.

Vol. 640: C. Sledge (Ed.), Software Engineering Education. Proceedings, 1992. X, 451 pages. 1992.

Vol. 641: U. Kastens, P. Pfahler (Eds.), Compiler Construction. Proceedings, 1992. VIII, 320 pages. 1992.

Vol. 642: K. P. Jantke (Ed.), Analogical and Inductive Inference. Proceedings, 1992. VIII, 319 pages. 1992. (Subseries LNAI).

Vol. 643: A. Habel, Hyperedge Replacement: Grammars and Languages. X, 214 pages. 1992.

Vol. 644: A. Apostolico, M. Crochemore, Z. Galil, U. Manber (Eds.), Combinatorial Pattern Matching. Proceedings, 1992. X, 287 pages. 1992.

Vol. 645: G. Pernul, A M. Tjoa (Eds.), Entity-Relationship Approach - ER '92. Proceedings, 1992. XI, 439 pages, 1992.

Vol. 646: J. Biskup, R. Hull (Eds.), Database Theory - ICDT '92. Proceedings, 1992. IX, 449 pages. 1992.

Vol. 647: A. Segall, S. Zaks (Eds.), Distributed Algorithms. X, 380 pages. 1992.

Vol. 648: Y. Deswarte, G. Eizenberg, J.-J. Quisquater (Eds.), Computer Security - ESORICS 92. Proceedings. XI, 451 pages. 1992.

Vol. 650: T. Ibaraki, Y. Inagaki, K. Iwama, T. Nishizeki, M. Yamashita (Eds.), Algorithms and Computation. Proceedings, 1992. XI, 510 pages. 1992.